Boeotian Landscapes

A GIS-based study for the reconstruction and interpretation of the archaeological datasets of ancient Boeotia

Emeri Farinetti

BAR International Series 2195
2011

Published in 2016 by
BAR Publishing, Oxford

BAR International Series 2195

Boeotian Landscapes

ISBN 978 1 4073 0750 3

BAR Publishing is the trading name of British Archaeological Reports (Oxford) Ltd. British Archaeological Reports was first incorporated in 1974 to publish the BAR Series, International and British. In 1992 Hadrian Books Ltd became part of the BAR group. This volume was originally published by Archaeopress in conjunction with British Archaeological Reports (Oxford) Ltd / Hadrian Books Ltd, the Series principal publisher, in 2011. This present volume is published by BAR Publishing, 2016.

Printed in England

BAR
PUBLISHING

BAR titles are available from:

BAR Publishing
122 Banbury Rd, Oxford, OX2 7BP, UK
EMAIL info@barpublishing.com
PHONE +44 (0)1865 310431
FAX +44 (0)1865 316916
www.barpublishing.com

ad Eratò, dolce musa,
nata insieme a questo lavoro

CONTENTS

LIST of FIGURES

LIST of TABLES

PREFACE and ACKNOWLEDGEMENTS

My love for ancient Greece began early in my life on the occasion of a visit to ancient Olympia with my parents, but decisive for the directions that my studies took later was a university training trip in 1995. Then, the patient guide Professor Anna Sacconi and Professor Luis Godart along with the local expertise of Professor Vassileos Aravantinos, director of the local archaeological service (8th *ephoreia* of Classical Antiquities), introduced me to the intriguing and fascinating world of ancient Boeotia. My first meeting with Professor John Bintliff and his team, at that time surveying in the area of ancient Hyettos, widened my scientific horizon towards the potential of GIS and of the interdisciplinary and diachronical study of the landscape.

The random discoveries of traces of ancient activities which occurred in the wider Boeotia region, jointly with the several windows of the landscape hovered by intensive and systematic surveys, increased the urge to contextualise the intensive surveys carried out in various parts of Boeotia within the context of the wider region, organically collecting the available archaeological record coming from a variety of sources. In order to do this, a local level of analysis was to be combined with a broad regional perspective, focusing on changes in the settlement dynamics within the micro-regions of Boeotia. My personal involvement in the intensive survey project which was being conducted by the University of Leiden in the area of ancient Tanagra allowed me to personally experience the ground and the landscape of Boeotia, and gave me basic insights on the potential biases in the recovery of archaeological data in the region.

I would not have been able to pursue this challenge without the support of many people.

This work, presented as a PhD dissertation to the University of Leiden (the Netherlands), has benefited from discussions with Professor John Bintliff (Leiden University), academic supervisor of the work. He is gratefully thanked for introducing me to landscape archaeology and sharing with me his large knowledge of Boeotian landscape history and his thoughts on diverse approaches to landscape biographies.

I am most grateful to Dr. Hans Kamermans (Leiden University), co-supervisor of the work, who advised me on GIS techniques and literature and always encouraged me, and also to Professor Nicola Terrenato (University of Michigan), a later but precious and critical reader of this work, who helped me to improve and polish my text.

Special thanks should be expressed to Professor Vassileos Aravantinos, director of the local archaeological service for Classical antiquities, who first introduced me to Boeotian archaeology and gave me the opportunity to become involved in several excavations and research projects in the region and to experience life and study at the Thebes museum. For the latter I would like to thank also his wife, Professor Margherita Bonanno (Tor Vergata University – Rome), and all the archaeologists and conservatories of the 8th *ephoreia*. This research was planned since the beginning to also have as a final product a computerized interactive database, which could be made available to the local archaeological service as an easily searchable and updatable cultural heritage management tool. I hope to be able in this way to recompense them for their precious help.

My first Boeotian experience was led and had benefited from people who introduced me to historical thoughts: Professor Domenico Musti (La Sapienza University – Rome), precious scientific guide of my early research on Boeotia and always reference in the later development, Dr. Maria Rocchi (National Council of Research – Rome) and Professor Pietro Vannicelli (Urbino University). They all deserve my special thanks.

This work has also benefited from information kindly provided by archaeologists who have been working for years on niches of the Boeotian landscape, and in particular Professor Andreas Konečny (for the area of Plataia), Professor Mark Munn (for the Skourta plain), and Professor Jost Knauss (for the Copais), who provided me with valuable information on specific matters and areas.

I would like to warmly thank also Dr. Kalliopi Sarri, Professor Vladimir Stissi, Professor Jeroen Poblome, Professor Bozidar Slapšak, Dr. Branko Musič and other specialists from the Boeotia Survey Project, who shared with me their experience in the field and in the laboratory and gave me preliminary information on sites and material they were studying.

A special thank goes to Dr. Athanasios Vionis, who shared with me Ph.D. life in Leiden and fieldwork seasons in Boeotia, while both working on our PhD theses, and who never deprived me of his support. Many other colleagues and friends patiently supported me during the difficult writing-up stages of my thesis in Leiden, Athens, and Rome, and have been very encouraging in difficult times. In particular, special thanks should be expressed to Dr. Philip Bes, Drs. Hanna Stöger, Dr. Elephterios Sigalos, Dr. Thomas Tselios, Dr. Georgia Kordatsaki, Professor Mirella Mataranga, Dr. Francesca Boldrighini, and Drs. Elena Frisone.

A special thank is due also to Drs. Giacomo Pardini,

colleague and friend, who helped in the realization of the cover drawing, free rendering of an Orchomenian coin type from the 4th C BC showing the shield symbol of the Boeotian confederation with a grain ear along the vertical centreline, chosen as a symbol of Boeotia (the shield) and its human landscape (the grain ear).

I am grateful to Dr. Jonathan Tomlinson for checking the English language and proof-reading the text, and to Drs. Rinse Willet and Drs. Karin Schuitema for translating into Dutch the abstract of the thesis.

My research was mainly self-funded, but was financially supported partially in its early stage by the *Ernst-Kirsten-Gesellschaft* (Stuttgart, Germany), the *Ministry of Foreign Affairs* (Italy), and a *Faculty of Archaeology Fund* (Leiden University).

Finally, most of my gratitude goes to Walter and Nives, my dear parents, my brother Edgar, and my grandmother Lisetta, for their endless support in all of my choices and their continuous deep understanding. I thank them from the bottom of my heart, as well as my Greek family, Stamatios, Theodoros and Vassileios Sbonias, who always make me feel loved and as belonging to Greece.

Last but not least, my most special thank goes to my husband, Konstantinos Sbonias, whose precious help, moral and scientific support, and dedication through all the phases of the work goes behind any description. My work has profited greatly from our long discussions on diverse archaeological and landscape themes, and it would not have been possible without his love. I am very much indebted to him, as well as to our beloved daughter, Eratò - major motivation for my work.

INTRODUCTION

Aim of the present work is to illustrate a possible way of dealing with a regional landscape and its long-term settlement history based on the integration of archaeological data applying a GIS based approach to the social dimension of the landscape.
The large province area (ca 2,500 sqkm) of Boeotia (Central Greece) is examined by means of GIS (Geographical Information System), processing data from different archaeological, historical and environmental sources.

The methodology established, dealing jointly with material culture and the environment, follows a critical comparative regional approach and opts for both region and micro-regions as the analytical unit. It aims mainly to assess landscape characters and the interface between human and social actions and landscape by critically assessing, first of all, the available archaeological record constituted by diverse, variegate and often incoherent data sets. Main periods of interest are the historical periods from Archaic to Late Roman, while earlier (Neolithic to Geometric) and later periods are taken into account for the analysis and understanding of diachronical processes which took place at the micro-regional and regional levels.

In the first part, the theoretical background and the methodological framework on which the followed methodology of research is based are illustrated. Special focus is given first to the research framework, discussing the approach to the regional archaeological landscape and the use of GIS in regional studies. Secondly, the presentation of the methodology is followed in the collection, recording and management of physical and archaeological/cultural data sets in the GIS system created for this work. A relational database has been implemented (in a Microsoft Access 2004 environment) in order to record the archaeological information available, which has then been included into the GIS system, realized in an ArcGIS 8.1 environment.

The main part starts with the description of the Boeotian landscape and its geographical sub-regions followed by an examination of the state of archaeological research for the region. It follows the presentation of the archaeological record available for the individual *chorai*/micro-regions of Boeotia, after which aspects of the landscape and settlement of each *chora* are discussed in detail, and micro-region and micro-landscape trends are individuated. The marked territorial character of the ancient Greek *polis* requires a detailed examination of the different landscape issues each *chora* may present, and the separate treatment of each *chora* facilitates the analysis of the lower rank sites, mainly directly related to the city. Moreover, there are differences in intensity and quality of research in the various *chorai* affecting the available data and the knowledge of landscape and settlement history. A systematic analysis of the archaeological record is performed to produce aggregation into culturally meaningful entities, through info-critique processes and comparison of data sets, led by the challenge of dealing with a heterogeneous, poor, incoherent and spatio-temporally incomplete data body. From the examination of the landscape at the micro-regions and settlement chambers levels in the *chorai* chapters, I move in the final chapter to some landscape trajectories at the regional scale under thematic units by means of GIS analysis, concerning the ancient socio-political and cultural landscape within the Boeotia region.

Throughout the entire work, I compare and analyse similarities and differences among micro-regions, as well as diachronical regional trends, using data that include the diverse types of data sets involved, such as physical landscape features, data sets derived by computer modelling, archaeological information given by the topographical tradition, information from the intensive hovering of the landscape performed by systematic artefact surface survey, epigraphical evidence, and relevant historical sources. Analyses and comments give special focus on the association between landscape zones and activities, resulting in landscape character, on cultural landscape meanings, as well as on the settlement behaviours and community choices over the landscape.

The volume is completed by some appendices: the analytical description of the archaeological evidence known for the region structured according to the enucleated components of the archaeological landscape; a qualitative geographical description of Boeotian landscape sub-regions (available online); a list of units of archaeological evidence and their attribution to archaeological landscape components (available online).

I.1

Regional approaches to landscape studies

In undertaking the study of a regional landscape, one ought to explore the possible ways according to which the historical landscape of a region can be approached, and the landscape marks of past transformations can be read and interpreted. In this chapter I will examine the meaning of key concepts, such as *region* and *landscape* and related notions, and I will explore several approaches to regional landscape studies which concurred and contributed to the developing of the approach I followed in my examination of the historical landscape of ancient Boeotia.

THE GEOGRAPHICAL AND ARCHAEOLOGICAL APPROACHES TO *REGION* AND *LANDSCAPE*

REGION

As contemporary geographers state, the concept of *region* is relatively clear, compared with many troublesome social science concepts. Vance defines the region as "*a homogeneous area with physical and cultural characteristics distinct from those of neighbouring areas*" (Vance and Henderson 1968: 377), and his definition is close to that of Vitkovsky and Kolossov (1980: 539) as "*a complex spatial socioeconomic system, characterized by a stable combination of political forces and possessing a specific complex of features*". As the political geographers Janda and Gillies (1983) point out, perhaps the most critical issue in conceptualising a region revolves around the requirement of geographical contiguity.[1]

What is defined is the region in contemporary terms, primarily as an object of study for political and economic issues within human geography. In fact, the concept of region itself can be considered as having been born very much earlier within the discipline of geography.

Regional geography finds its origins in the 19th century, influenced by Darwin's work (*Origin of Species*, published in 1859), which broke the links between geography and natural theology and caused rejection of geographical work based on teleological approaches, laying the groundwork for new ways of conceptualising the relationship between people and their environment, and heightening interest in the region as a subject for geographical enquiry (Humboldt and Ritter[2]). Darwin's impact provided fresh stimulation for the doctrine of environmental determinism, according to which human activities are controlled by the environment. This would explain the pattern and processes of human habitation of the earth's surface, and what must be analysed are the effects of the natural environment on the human body and spirit of both individuals and entire social groups, as the work of Ratzel,[3] for instance, attests.

Afterwards, regional geography developed from 19th-century origins in different directions. In the United States, regional geography is seen as a descriptive rather than a systematic study, while in Europe, regional geography can combine both aspects. In France, in particular, regional geography came to be associated with the work of Vidal de la Blache, who approached the concept of region from its human and cultural dimensions,[4] seeing the region as the unique expression of the interaction between humanity and the physical environment (see Vallega 1999; Agnew et al. 1996). Regional approaches, initially very much influenced by environmental determinism, slowly moved within geographical studies in other directions, allowing for the study of the interaction between people and the environment without seeing causal links between the two, and offering human geographers a clear and well-defined object of study, having as key themes the concept of region and the interaction between peoples and

[1] Although many contemporary geographers do not require contiguity in defining a region (Russet 1967: 2-7; Vance and Henderson 1968: 378; Cox 1969: 71, 77), especially when considering the international and global sphere, in studying antiquity I believe that geographical contiguity has to be applied as a criterion in defining a region's boundaries, as scales are certainly smaller and interconnection processes much more localised.

[2] Alexander Von Humboldt (1769-1859; his work was *Kosmos*, 1845-1862, subtitled *Sketch of a physical description of the world*) and Carl Ritter (1779-1859; his unfinished work was *Erdkunde*, 1822-1859) developed regional and systemic approaches and laid the foundations for geography as a specific branch of knowledge based on scientific methods of enquiry, though still influenced by teleological approaches.

[3] Friedrich Ratzel (with his first work *Sein und Werden der organischen Welt*, 1869) can be considered the founder of environmental determinism in geography, and a pioneer in anthropogeography, which had three main aims: to describe the regions of the *ecumene* -habitable earth- and the distribution of mankind over it; to study human migratory movements of all types and their dependency on the land; to analyse the effects of the natural environment on the human body and spirit, both on individuals and entire social groups.

[4] Vidal de la Blache (1845-1918), one of the founders of modern geography, developed three key concepts for the understanding of regional geography: *milieu*, *genre de vie*, *circulation* (*Principles de géographie humaine*, 1922).

environments; in this way radically changing the discipline.[5]

LANDSCAPE

The word 'landscape' comes from the Dutch term *Landschap* and was borrowed by painters in the 16th century (Thomas 1984). Through the painted representations, *landscape* becomes "*something out there, to be appreciated in visual terms and constructed through the rational principles of perspective. Distanced geometrical images of landscape were reinforced by cartography* [...] *Landscape was not lived, but looked at, being seen as something external to people; nature versus their culture*" (Gosden 1999: 153). Initially, the landscape was therefore seen as a representation, either as a picture, or as a map. Still, the geographers Cosgrove and Daniels (1988: 1) define landscape as "*a cultural image, a pictorial way of representing, structuring or symbolising surroundings*".

Subsequently, focus was placed upon the binomial nature and culture within the landscape, and as such, the geographer Roberts (1987) defines the landscape as a set of real-world features, natural or cultural, which give character and diversity to the earth's surface. In more recent approaches, influenced by phenomenology, more and more focus has been given to the landscape as an holistic, all-embracing concept, definable as a dynamic and continuously transforming (in process) synergy of physical/natural and human/cultural factors mutually interrelating, within which, in the anthropologist Ingold's words (2000: 191), "*each component enfolds within its essence the totality of its relations with each and every other*".[6]

Also underlined within anthropological studies is the qualitative and holistic nature of the concept of landscape, not being an empty box prepared in advance for creatures to occupy and in which life and actions simply take place, but rather constituted by the unfolding of mutual interrelation between 'nature' and 'culture' (see among others Gosden 1999: 153ff; Ingold 2000: 199 and *passim*, within the phenomenological trend).[7]

In archaeological studies, landscape in archaeology was pioneered by Crawford and Fox in the 1920s-30s in Britain.[8] In America, in the 1960s, the archaeologists benefited from the emergence of ecology[9] and became involved in modelling subsistence, settlement, and culture change utilising an ecological perspective (Hassan 2004). That fitted with the objectives of the New Archaeology, within which, also in the 1960s, landscape entered the domain of archaeology in the form of spatial analysis, focusing on quantitative analysis of the distribution of sites and artefacts within space (Clarke 1977; Hodder and Orton 1976). More recently, approaches to landscape have represented a movement against spatial analysis, rejecting the risk of environmental determinism and stressing instead the mutual creation of people and landscape[10] (see below) and the notion of cultural experienced landscape, which was soon influenced by Heidegger and phenomenology, looking at the lived experience of landscapes (Ingold 1993 and then Ingold 2000, Gosden 1994, Tilley 1994).

Landscape archaeology at first looks mainly at the spatial relationships of artefacts and features in order to infer the past use of landscape, and later on delves further into the mutual interrelationships of natural and cultural characters and processes in order to infer meanings of ancient landscapes, considering the landscape not as a passive recipient of human activities, but as a dynamic and interactive element in the evolution of past societies (Roberts 1987; Fleming 1990; Kuna et al. 1993; Tilley 1994; Knapp and Ashmore 1999; Wilkinson 2004), and a spatial manifestation of the relations between humans and their environment. Landscape archaeologists deal mainly with the interpretation and the reconstruction of human behaviour over space through time.[11]

The two concepts illustrated above (*region* and *landscape*) come together whenever research is carried

[5] Agnew et al. 1996; Johnston 1991; Paasi 1991. Contemporary regional studies in geography involve: Historical and thematic approaches to regional structure; theory of the region; the origin and development of regional character; locality and place-making; nature and culture in regional settings; comparative study of regions.

[6] Contemporary landscape studies in geography involve: Landscape as an embodiment and shaper of social values and attitudes towards environment; theories of landscape structure and change; the historical development and regional construction of landscapes; thematic landscapes; landscape character; the role of institutions in environmental design and management; aesthetic landscape values; landscape and the sense of place; comparative landscape analysis.

[7] In anthropological studies, the notion of the cultural (or social) landscape goes back to Sauer (1963).

[8] O.G.S. Crawford represented the evolutionist-diffusionist school of thought in British archaeology, according to which the analogy and cause-and-effect relations in the interpretation of spatial arrangements was stressed out and discontinuity in the emergence of new cultural forms was often explained in terms of migration. His clearly positivist approach to archaeology regarded archaeological sources as objective, and lead him to form and formulate the principles of the application of aerial photography in archaeology. He was the pioneer of the use of aerial photography to enhance the archaeological understanding of the landscape. Sir Ciril Fox, inspired by Crawford, gave rise to a new trend called landscape archaeology, which chose aerial photographs to be its key source of information, also introducing a new perspective to the studies of past agriculture (based on the changeability of field systems) and showing how the pattern of settlement had changed in relation to natural vegetation in his archaeology of the Cambridge region (1923) (see on this Evans 1999 and 2003).

[9] Odum's *Fundamentals of Ecology* appeared in 1953.

[10] Bordieu is influential at this point, with landscapes and *habitus* being seen as mutually creative (see below in text). Bordieu's *habitus* is a term compounded from *habitude* (habit or custom) and *exis* (acquired ability or facility).

[11] 'How' and 'why' human behaviour changes over space and during time, the dynamics which occur... (Van Leusen 2002: 5.12). "*Space, time and function can be thought of as different axes along which the available data can be differentiated*" (Van Leusen 2002: 5.13).

out on the *landscape* of a *region* (either macro-region or micro-region). In the history of the archaeological discipline, along with geography (mainly; and anthropology later), one encounters several different landscape study traditions, which, to varying degrees, involve the focusing on regional patterns, trends and dynamics.

In archaeological research, interest in micro-regions, and consequently in micro-landscapes, seems to have been strong during the final decades of the 19th century (and in the first half of the 20th), within the framework of the German *Landeskunde* tradition (Jahnkuhn 1955). This probably occurred because past societies, the objects of archaeological studies, were based on rural economies as well as Southern European society in the 19th and the beginning of the 20th century. Rural communities still lived then according to micro-regional settlement organisations, and therefore archaeologists somehow naturally tended to focus their research on past societies around micro-landscapes. Within this framework and social context the concept of *Siedlungskammern*, settlement chambers, originated, which was then borrowed by geographical studies.

In modern times, economic geography, moving from the study of a principally rural economy to the study of much more differentiated economic processes (industrial-demographic planning, as well as the even wider service - tertiary - network), gave higher value to the wider regional trajectories, and this also led human geography to a different dimension in its approach to the landscape, moving from the micro-regions to wider regional and broader systems. Therefore, within human geography, focus was given mainly to the concept of macro-region (vs. micro-region), primarily because human behaviour and social and economic life had changed. Only physical geography still remained interested in micro-landscapes.

On the other hand, the management of heritage (cultural, historical, and archaeological) landscape has recently become a relevant issue as one of the European Union's strategic actions, and the latest trend is therefore to give emphasis to historical landscape preservation and management (Grenville 1999). It can be noted how, with the decline of farming, landscape is still identifiable with environmental interest (nature, leisure resource), but also with the need to document our farming past which is quickly disappearing.[12] In this perspective, the heritage landscape means something more than landscape of sites, and it is the region that has to be preserved rather than the individual historical-archaeological sites. The regions involved in this preservation, however, are mainly micro-regions, on which the rural landscape to be preserved was based. We therefore assist in an increased attention to landscape environmental and historical problems, and to the production of sub-regional historical maps, aiming to help the monitoring, and therefore the preservation, of historical landscapes (successful examples are, for instance, some areas in the United Kingdom: Landscape Character Assessment programme – LCA – Bell 1999, Hughes and Buchan 1999, Gustavsson 1999, Scottish Executive 2000; and in the United States[13]). Landscape heritage has thus become a new issue, and its introduction and developments have helped in some way to bring archaeology and geography together once again in their common renewed interest in small landscapes whose character ought to be preserved. In this heritage management perspective, the individuation and preservation of landscape character is enlightened[14] (Grenville 1999 among others), and archaeology and anthropology both argue on it. Tilley (1994: 25-26) refers to 'qualities' of landscape, that create part of their significance for those who use them and are constituted by 'locales'. The anthropologist Ingold, interested in archaeological issues also through his concern for the temporality of the landscape (Ingold 1993), in his main work gives a definition of 'landscape character' (landscape 'form' - Ingold 2000: 193, see below). The concept of 'landscape character' becomes a bridge between landscape and region (micro-region).

REGIONS AND MICRO-REGIONS IN THE READING OF THE GREEK LANDSCAPE

As noted by Bintliff (2000a: 148), the reading of Greek landscapes was achieved by the German *Landeskunde* tradition of historical geography between the final decades of the 19th century and the 1950s (Jahnkuhn 1955). A particularly relevant application is Lehmann's analysis of the long-term settlement history of the micro-regions of Crete (Lehmann 1939), where the concept of the *Siedlungskammer* was employed to identify in the landscape 'settlement chambers', as areas within which "*there is usually a single major settlement at any one period, but its precise location is a conjunction of natural geographic opportunities and the specific economic and political context of the culture concerned*" (Bintliff

[12] Terraces for agricultural purposes, for instance, are disappearing at a dramatic rate throughout the Mediterranean region, where they have been used since early times (often from the Bronze Age; see chapter I.2.1).

[13] See http://ceres.ca.gov/planning/genplan/sutter/landuse3.html on Sutter county (last visit in September 2008): "*[...] The physical environment is a key component in planning for future county growth since it contributes directly to the perceived desirability of the county as a place to live and work. [...] The concept of "physical form" encompasses the physical qualities of a place at different scales: regional, sub-regional, and local. In addition, individual perception of the county's form and character is influenced by context and intent. For example, individual sense of physical form can be influenced by the way a region appears on a map, by sub-regional geographic features observed during automobile travel, or by architectural qualities observed when walking a local neighbourhood street. All three experiences contribute to the sense of the County as a unit*".

[14] Interest is taken in the local landscape character (and documentation of landscape change) for management and planning aiming to preserve the special character and quality of the landscape and the features that contribute to this, to produce a historic landscape assessment and recognise the character of the local countryside and ensure any development would be appropriate to its context. These are, for instance, the main aims of the LCA programme – Bell 1999.

2000a: 148). In the earliest topographical accounts on Greek landscape,[15] therefore, one can appreciate a proximity between the two disciplines of geography and archaeology. Among them are exponents of historical geography, such as Philippson and Partsch (geographers, not by chance. Philippson –1892- and Partsch 1887 - 1889 – 1890), the aforementioned Lehmann (mainly Lehmann 1939), as well as Kirsten, who edited Philippson's work in 1951. In these works, the description of the past and present landscape focuses on settlement chambers or small landscapes, as far as physical characteristics and organisation of space are concerned.

A good description of both the physical and historical landscape of Greece was given by A. Philippson in his book '*Die Griechischen Landschaften. Eine Landeskunde*', published from 1950 to 1959 and written in elegantly old-fashioned German, with contributions by E. Kirsten for the descriptions of historical landscapes. Landscape archaeology concepts such as *settlement chambers* (*Siedlungskammern*), *cultural landscapes* and *Kunstlandschaften*, are recognisable in this important precursory work, combining the purely geographical perspectives and narratives with historic-archaeological ones.

In the meantime, the extensive/topographical survey tradition, originated with the travellers (15th/16th century onwards), focused on the discovery and knowledge of the main sites and was characterised by the search of visible remains of sites known from the Greco-Roman sources, without any real regional focus. By the 1960s, far more detailed regional surveys were carried out within a defined area. Their approach was much more diachronical and new locations were investigated together with every previously recorded find spots. The South Etruria survey in Italy (directed by J. Ward-Perkins in the 1950s to 1970s – Potter 1979) and the Messenia survey in Greece (McDonald and Rapp 1972) are well known examples of this kind of extensive regional survey in the Mediterranean area.[16]

Soon after, by the early 1970s, a new survey phase had been inaugurated with the intensive field-by-field fieldwalking surveys (the so-called 'New Wave Surveys' - Bintliff 1992c and Cherry 1994), approaching the landscape as a synergy between anthropic and environmental factors and as continuously changing/transforming products of the human-environment relationship in the long term. Research projects work mainly on micro-regions, initially fieldwalking the landscape in search of artefact concentrations (sites),[17] then focusing attention on the artefacts rather than sites, therefore performing an overall (or strategically sampled) covering of the surface with attention paid also to the meaning of off-site material.[18] Those projects gave an increasing importance to the absence of material,[19] as well as to the systematic analysis of on-site surface finds (date and functional analysis).[20]

The New Wave Surveys approach the region as the main context within which to carry out fieldwalking and to detect settlement patterns, and they therefore work on relatively small regions, diachronically examining landscape narratives of small landscapes that can be 'hoovered' intensively.

In more recent years there has been a build up in a trend towards comparative studies, and comparison seems to be a thrust in landscape studies based mainly on intensive field-by-field surveys (Cherry 2003; Osborne 2004). For Greece, in 1972 Renfrew had already tried to obtain results from a comparison of landscape and settlement data from extensive/topographical surveys (for the Prehistoric Aegean), while the first attempt at a comparative regional study using results mainly from intensive/New Wave surveys is Alcock's book (1993), which focuses mainly on the Roman period.[21] The challenge of comparing regions is that they are sometimes very different from each other in terms of history and environment, as well as regarding methodologies applied by the research carried out on them (on issues of integration and/or comparison of survey datasets see mainly Van Leusen 2002). A recent work in comparative regional studies is a book entitled *Side-by-side survey* (Alcock and Cherry 2004), which

[15] Mainly represented by the geographical-topographical accounts immediately following (and overlapping) the travellers tradition.

[16] Elsewhere, pioneer is the Willey's 1946 extensive regional archaeological survey project in Peru (Willey 1953), and later well known extensive regional surveys are Adams' research in ancient Mesopotamia (Adams 1958) and Sanders' research in the basin of Mexico (Sanders et al. 1979).

[17] Pioneer projects in this direction in Greece: Melos Survey (Renfrew and Wagstaff 1982); Agiofarango Survey - Crete (Blackman and Branigan 1977); SW Argolid project (Jameson et al.1994 and Van Andel - Runnels 1987).

[18] In the 1980s awareness grew of the limitations of site-oriented archaeology. Among others: Foley 1981; Dunnel-Dancey 1983; Gaffney-Tingle 1984; Rossignol-Wandsnider 1992. Pioneer projects in this direction in Greece: Kea project (Cherry et al. 1991); Boeotia Survey Project 1978-1989 (see Bintliff and Snodgrass publications).

[19] There are no empty or meaningless spaces between sites, and attention is paid to the archaeological record as having a continuous character within a dynamic geomorphological context. It is a notion of landscape as a continuously used space consisting of areas of variable functions and with variable density of archaeological remains.

[20] For an overview up to the early 1980s see Cherry 1984. The main survey projects carried out in mainland Greece to the present day are: McDonald - Rapp 1972 (Messenia); Wagstaff 1982 (Helos plain - Southern Greece); van Andel - Runnels 1987 and Jameson et al. 1994 (Southern Argolid); Bommeljé et al. 1987 (Aetolia); Wright et al.1990 (Nemea); Lohmann 1999 (Attica); Rizakis 1992 (Achaia); Cavanagh et al. 1996 (Laconia); Wells 1996 (Berbati-Limnes, Argolid); Stelios - Kotsakis 1994 (Macedonia); Cosmopoulos 2001 (Oropos); Forbes - Mee 1997 (Methana peninsula); Davis et al. 1997 and Davis 1998 (Pylos); for Boeotia see chapter II.2.

[21] For other smaller contributions to comparative survey studies in Greece, see Halstead 1994 and Bintliff 1997c.

critically and fruitfully compares survey projects taking place in the Mediterranean area from a methodological point of view rather than addressing the comparison to specific historical issues.[22]

New Wave Survey projects (field-by-field artefact surface surveys) were thus initially concerned with micro-regions (for both data collection and data analysis[23]), while more recently the tendency is to extract meaningful information, either from the comparison of the results coming from individual surveys still carried out at a micro-regional level or from the comparison of regional trends (see RPC project – Attema and Van Leusen 2004; Attema et al. 2002; Alcock and Cherry 2004; van Leusen 2002: chapter 2). Data collection is therefore focused on the micro-regional level, while landscape and location analysis work at both the micro and macro-regional levels.

In the meantime, the somewhat neglected settlement chamber approach has been developed further, in recent years, by contemporary Czech prehistorians as the 'Community Area' theory (Kuna 1991 and see below), though outside research on Greece. This approach always moves within the New Wave survey framework, emphasising the role of surface data intensively collected and the importance of off-site material. Bintliff applied the approach to the Greek mainland landscape, especially in the work on the Valley of the Muses (Bintliff 1996b).

THE *COMMUNITY AREA* THEORY

The concept of community area (and settlement area), as proposed by E.Neustupny (1991, and earlier in Bohemian/Czech journals[24]), initially created a theoretical framework for the study of Prehistoric communities, but also, in the development of the theory and methods, for the study of landscape regions in the long term.
A community area (or settlement area) is a space where the activities of one community took place (Neustupny 1991, Venclova 1995). Community areas, expressions of past living cultures, are only accessible to archaeologists through the study of the archaeological remains that represent them, which form what Neustupny calls the 'settlement area', the world of dead (archaeological) culture (Neustupny 1994: 248; Neustupny 1991: 327). Thus, the community area approach takes an interest in many areas of the landscape which were not considered to be 'sites', and therefore not of interest within the framework of traditional topographical archaeology, and illuminates the importance of off-site information (presence and absence of archaeological material) from intensive artefact surface surveys.
An archaeological map ceases to describe a landscape possibly filled with individual points representing 'sites', and rather becomes a richly structured space in which only parts are occupied by settlement habitation areas (Neustupny 1994; Kuna 1991). Various sub-areas, representing various activities, can be specified, forming components of the total settlement area: habitation, storage, burial, ritual, as well as production areas, fields, pastures, woodland, quarries, and mines (Venclova 1995; Kuna 2000). Communities performed many different activities which in turn structured the space in which the communities lived. Each of the activities could have occupied a different spatial unit, a different 'activity area' (Neustupny 1991). The various activities are considered within an "*explicit spatial model derived from the behavioural rules of living cultural systems*" (Kuna 2000: 31).
As Dreslerova (1995: 145) points out, the community area approach permits "*landscape to be understood as a series of particular localities (areas), presuming that any part of the landscape in the past had a certain specific function*". Landscape is divided into spatial segments, community areas, which were inhabited by basic economic and social units, ancient (prehistoric or historical) communities (Neustupny 1991 and 1994). Thus, the community area approach focuses on the individuation of settlements exploiting a particular micro-landscape, therefore giving attention to small landscapes within which the main settlement (or cores of habitation areas), and consequently other activities, might show either continuity or shift in location.
Attention is paid to the interrelation of individual components (especially to shifts of the dwelling component within the territory of the settlement area) as well as to their interrelations with the natural environment and the movements of the settlement area as a whole (Neustupny 1994). Groups of community areas were linked into so-called settlement zones, being connected as clusters of community areas within separate parts (referred to as micro-regions[25]) of larger territories, or wider regions (Kuna 1991; Neustupny 1994; Gojda 2004).[26] Within the community area framework, we assist in the transition from regions or micro-regions towards settlement areas, as a transition from natural/geographical units towards archaeological and historical units, with a considerable change in scale also (Neustupny 1994: 250).

While in the *Landeskunde* tradition mainly environmental factors (availability of fertile land, location of agricultural

[22] The five volumes of the Populus project on *The Archaeology of Mediterranean Landscapes* were already explicitly concerned with comparatively addressing methodological issues: Bintliff and Sbonias 1999; Leveau 1999; Gillings et al. 1999; Pasquinucci and Trément 2000; Francovich et al. 2000.
[23] From the micro-landscape level one would move to the regional level and the wider dynamics at that level, etc.
[24] Neustupny's pioneering article was the 1986 article, '*Sidelni arealy pravekych zemedelcu*' (Settlement areas of Prehistoric farmers), published in *Pamatky archaeologicke* 77: 226-34.

[25] Originally, the term 'micro-region' meant a small sample territory to be intensively studied, rather than a wider territory too large to be studied effectively. The concept was soon underpinned theoretically, receiving the meaning of a definable geographical and cultural unit where various aspects of past cultural systems can be studied (as summarised in Kuna 2000).
[26] See also Dewar 1986 and 1992 for the concept of dynamic settlement systems.

land and crops, presence of water sources) were used to individuate settlement chambers,[27] according to the Community Area approach, social and historical variables are also involved in the definition of community areas active in spatially more or less well-defined settlement chambers.
Therefore, the concept of *settlement chamber* does not correspond exactly to the concept of *community area*. The former can be considered more deterministic, while, according to the latter, 'behaviours' of community areas are also strongly influenced by historical/cultural/social factors. In Neustupny's words, the community areas "*cannot exist outside of what they are given by nature, but they are much more than a sum of environmental factors*" (Neustupny 1994: 251). Thus, community areas can also be detected by examining what still remained obscure, at divergences which still remained unexplained, having examined the physical/environmental factors, paying attention to possible social landscapes (for instance, the role and the influence of a *polis* in the area; small landscapes conditioned by the size of the settlement, etc.). As Neustupny (1994: 251) points out, the 'human dimension' of community areas can be studied at three levels: practical,[28] social[29] and symbolical[30] (most archaeological entities are characterised by all three dimensions).

Several scholars, even if outside or earlier than the main Czech 'Community Area' school, either explicitly or only in practice, followed and follow the same approach in building up their reasoning on the history of settlement in a particular area. For instance, Heidinga's work (1987) on the lower Rhine area in the medieval period can be considered one of first applications of the approach. Also, as pointed out earlier, Bintliff's article on the settlement chamber of the valley of the Muses (Bintliff 1996b) revitalises the *Landeskunde* tradition with approaches to the shifting of settlement closer to the Czech school.

THE 'SENSE' OF CONTINUITY

In Bintliff's words (2000b: 8), one could say that: "*The strong regularities exhibited in settlement networks of later prehistoric and ancient societies, may emanate as much from human ecological and sociobiological constraints as from the conscious planning programmes of ancient communities*". On the other hand, we must note how this approach mainly tends to follow the argument that the distribution of fertile agricultural land, ergonomic work constraints on territorial size, social factors affecting the dispersion of communal groups, and limited locational possibilities for settlement micro-location, might appear in the long-term more important than the conscious inheritance of traditional 'senses of place' or continuity of populations and cultures. This might sound "*as a warning against taking to extremes a currently-fashionable trend in landscape archaeology theory*" (from Bintliff 2000a: 148), which mainly emphasises the role of 'memory' in the interpretation of past settlement networks (cf. Barrett 1994; Tilley 1994; Bender et al. 1997), and explains the continuity of residential areas more through the potential continuity of social meanings ascribed to certain places by people, rather than through the formation and maintenance of cultural landscapes through time (Tilley 1994).

Sometimes the settlement chambers[31] are clearly geographically defined (either by physical boundaries – rivers, watersheds - or by the presence of outstanding environmental features – a lake, a marsh, a basin - or by man-made features in particular periods of history), but at other times it is harder to individuate them, and socio-political/historical variables play the most crucial role (see above). Within a community area, continuity of occupation of particular locations in the landscape does not automatically mean that the permanent settlements remain at the very same *locus* for centuries or millennia, but could correspond to a continuous oscillation of residential areas around some focal points in the landscape ('residential cores' - Kuna 2000: 41), which could explain striking long-term continuity. One could hypothesise that once the landscape had been structured by human behaviour, other activity areas (apart from the residential places) must have also retained some stability, as a result either of economic, or of symbolic and ideological reasons (Kuna 2000). The landscape of each period has to be considered as a social, cultural product (interconnected with a set of natural elements) of long-

[27] See, within this tradition, the creation of geographical models (von Thunen 1826 especially on the use of areas in agriculture, Weber 1909 for industrial location and patterns, Christaller 1933 for Central Place theory), then applied also in archaeology (many applications in the wave of New Archaeology). Site catchment theory was also generated within this framework, in archaeology, by Vita-Finzi and Higgs 1970: 1-37. For a review and elaboration of catchment analysis see Bintliff 1999c.

[28] With the 'practical' dimension, Neustupny (1004: 251-2) describes categories of practical activity, such as economic life and the use of facts as carriers of meaning and significance. At least some of the economic activities concentrated into particular areas, to be detected also by means of the non site-oriented archaeology. The practical activity of a community within its settlement area must frequently have gone beyond the self-regulation capacity of natural resources (see, for instance, the crisis at the end of Classical/Early Hellenistic period in Greece described in van Andel and Zangger 1990, Bintliff 2002 among others).

[29] "*The structure of each settlement area has a meaning which can be formulated in terms of the* social organisation *of the community which settled it, i.e., in terms of the interrelationships of individuals and their groups*" (Neustupny 1994: 252). Community areas represent the space within which social activities and relationships take place.

[30] "The structure of each settlement area may have some significance which can be formulated in terms of ideology adopted by the community that settled the region" (Neustupny 1994: 253). For instance, the location of dwelling areas had a certain function (linked to practical factors), a certain social meaning, and a certain significance (i.e. a set of ideological ideas related to that sphere of human activity).

[31] From now on in the volume the term 'settlement chamber' is used much more in the sense of the 'community area' of the Czech school, rather than recalling the 'settlement chamber' of the *Landeskunde* tradition.

term diachronic stability (Kuna 2000: 41) spaced out by radical/abrupt changes/shifts, due to some particular environmental, political, social, or economic factors that put an end to the 'continuity'.

To sum up, the application of the community area theory, which aims mainly to recognise possible shifting of the main nucleated settlement (towns and villages) within each 'chamber', implies a sort of 'continuity' of settlement at or beside the occupation of the previous phase. The examination of a small landscape, and especially of the continuity or shift of main settlement and other activities within it through time, would allow us to 'narrate' the story of the small 'chamber', without taking for granted stability of landscape conditions or of economic or social factors (obviously changing through time), and without 'idealising' continuity with the attempted emphasis on the role of memory, but rather focusing on recurrent natural and cultural factors and/or characters, which could be of a different nature.[32] As Bintliff (2000a) points out, a 'pragmatic' choice of place, (while avoiding being too deterministic), seems to appear more important than the conscious inheritance of traditional 'senses of place', or continuity of populations and cultures.[33]

As an alternative, Bintliff (2000a) would suggest recent developments in scientific thinking, in particular complexity theory and non linear system theory which, giving emphasis to the concept of 'strange attractors', considers not only small variations – due to cultural choices - that could lead to major changes, but especially the recurring of situations due to recurring of 'initial conditions' (i.e. small variations).[34] It recognises the importance of 'anomalies' and tries to deal with them instead of considering them as simple biases in the dataset (see Lewin 1993, Bintliff 1996a, as well as Spencer-Wood 2000, Stewart 1989: 268ff. For archaeological examples and models: Zubrow 1984, Spencer-Wood 1996, Reed - Harvey 1992; van der Leeuw - McGlade 1997, Bintliff 1996b, Uleberg 2004: 444, as well as McGlade 1995).

[32] For instance, a site in a particular period might sometimes be located exactly at the same spot where it was in previous times due to geographical locational advantages, but the fact could also be due to the presence of previous investment in land clearance, terracing, building stones and rubbish/manuring for vegetable gardens, etc.

[33] Sometimes, even if populations living in an area vary, the location of the main settlement or other activity areas continues to be the same (see Bintliff 2000a for the case of post-antique Boeotia).

[34] According to the complexity theory, "*potentially highly variable and historically very specific agglomerations of elements show recurrent tendencies towards systematic patterning of a complex form, as the result of the operation of factors which are 'enabling and constraining' (the 'strange attractors' which produce order out of chaos)*" – Bintliff 2000a: 148.

DEFINING MICRO-REGIONS AND SETTLEMENT AREAS

The aim of regional studies is to identify similarities and contrasts within a single landscape unit across time in order to analyse and interpret the history of the landscape.[35] The need of diachronic studies for the interpretation of landscape behaviours aiming at a landscape narrative is therefore evoked (see, among others, Neustupny 1994: 254). Thus, landscape archaeology cannot but adopt the time perspective of '*long durée*' (Braudel 1980), and any attempt to understand past societies must take into account the preceding and successive uses of the landscape within a landscape unit, occurring before and after the society under investigation.

Therefore, the main problem in regional studies becomes that of discovering the presumed unit of research in the archaeological record and defining its dimensions. As Kuna points out, "*the theoretical definition of the community area does not itself tell us anything about the expected size of the community, the number of its habitation areas, and the space belonging to them. It works, however, from the logic of the concept that a community area must be identified with such a minimal spatial segment of space (such a cluster of sites) within which the occupation can be supposed as continuous from the diachronic point of view and where the range of different activities is relatively complete (at least in the sense of activities that can be archaeologically traced)*" (Kuna 1991: 340-342). This works for areas that can be covered or sampled through intensive and systematic survey work. When these data are not available, and the archaeological record is constituted by palimpsests of data, based on information collected through time, within different research frameworks, and usually not systematically collected, as in the case of the present work, the application of the model is slightly different. It is focussed on the structure of settlement rather than on the mapping of every locality and its exact extent, and takes into consideration the poor information available and the characteristics of the landscape during time (physical and cultural), and later historical (Frankish-Ottoman) or modern landscape for which the record is less incomplete, or complete in the case of the modern village system, rather than the actual distribution of activities in the landscape for which we have only poor information (burial areas, sanctuaries, rural and production activities, etc.).

As seen earlier, sometimes the micro-landscapes are well defined by physical constraints, while in other cases borders are fading and only cultural elements or meanings would define and mark the border line between the territories of two communities. Working within the community area framework, archaeologists should develop the ability to detect the presence of settlement areas with sufficient accuracy, while keeping in mind that

[35] As seen above in the text, attention to the micro-region is, in the study of the Mediterranean landscape, the desire to enhance and examine small diversities among regularities (Cherry 2003; Alcock and Cherry 2004; Osborne 2004).

in addition to function (easier to define), meaning and significance, random decisions of individuals will also play a role. As stated by some proponents of the theory, the concept of community area and the presumed rules of distribution and co-occurrence of certain phenomena would allow us to make generalisations even from fragmentary data (Kuna et al. 1993), from an archaeological record that is naturally incomplete.

Despite the fact that, as Ingold points out (2000: 191), the concept of landscape is qualitative rather than quantitative,[36] and one can ask of a landscape what it is like, but not how much of it there is, boundaries of various kinds may be drawn in the landscape. They do not segment the landscape, for the features with which they are identified (either natural or anthropic, a river, an escarpment, a watershed, a defensive wall, etc.) are themselves, in theory, an integral part of it. They can also become a boundary, or indicators of a boundary, in relation to the activities of the people for whom they are recognised or experienced as such (Ingold 2000: 192-3).

In his contribution to the book (edited by Renfrew and Zubrow) on cognitive archaeology, Bradley (1994) criticises the studies of landscape history which are concerned "*with the organisation of fixed resources and presuppose the existence of a network of recognisable boundaries*".[37] He quotes Tuan 1977, who states: "*The recognition and differentiation of landscapes does not seem to be an old or common human trait*", pointing out how in pre- and non-literate peoples nature is recognised in local objects (plants, animals, rocky prominences) or in generalised phenomena (sky, moon, earth, water, etc.). On the other hand, among literate peoples, "*the reading of significance into arbitrarily selected spatial units of nature is remarkably rare*". In the case of the Greek *polis* territory, the relationship between city (and citizens) and land was so close that boundary definitions become desirable (boundary inscribed stones – *horoi* -, treaties, etc.). In fact, this can be due to the tremendous value given to land by farmer societies, which, from the beginning, define agricultural territories by enclosing them. On the other hand, the territorial behaviour is very complex, and can be seen as working even in earlier societies,[38] such as hunter-gatherer groups, who perceive their territories by monitoring the paths running between specific places, living in the landscape with its original objects and features.

THE CONCEPT OF *TASKSCAPE* FOR A RENEWED APPROACH TO THE LANDSCAPE

In the last decades, research in cultural anthropology has been emphasising material worlds and meanings worked out through material culture and landscape. Time and landscape, states the cultural anthropologist Ingold (2000: 189), are the essential points of contact between archaeology and anthropology.

While discussing the concept of landscape, Ingold (2000: 190-3) states how the forms of the landscape are not prepared in advance for creatures to occupy (and the landscape is not a neutral box in which life and actions take place), but both physical and human landscape forms are generated and sustained in and through the processual unfolding of relations and the continuous mutual transformations (see also Ingold 2000: 193 and Goodwin 1988). It is in the very process of 'dwelling' that these forms are constituted, as Ingold (2000: 199) points out, quoting Heidegger's remark "*to build is in itself already to dwell*" (Heidegger 1971: 146). The landscape therefore always has the nature of a work in progress.

Just as landscape is not a passive, neutral box in which human actions occur, space and time are not passive, abstract media in which action occurs, but are themselves created through social acts (Munn 1990; Gosden 1999: 158). According to Munn,[39] the regional world is not given but lived, and history is the temporal aspect of regionality. Regionality is therefore created in experience, it is a temporal as well as a spatial construct, and regions are the lived landscape, lived spaces whose character is given through action and material culture (Munn 1990).

Once again, evident here is the intimate correlation between the concepts of landscape and regions, space, time, temporality and history. These are all concepts that will help us in the definition of a particular approach to past landscapes.

We can move further, by referring to the phenomenological 'dwelling perspective',[40] according to which the landscape is constituted as "*an enduring record of – and testimony t o- the lives and works of past generations who have dwelt within it, and in so doing, have left there something of themselves*" (Ingold 2000: 189). Ingold's 'dwelling perspective' is given a spatial dimension through his image of taskscape, which is seen as an array of related activities spread across the physical landscape (Ingold 1993: 158). The idea of taskscape (formalised by Ingold 1993 and 2000) gives emphasis to the role of human actions within the landscape, but also helps to emphasise the role of time and history in landscape analysis. As noted by Gosden, "*activities which are separated in space must also be spread out in time and each taskscape has its own temporality, its rhythms of action and of rest*" (Gosden 1999: 128). An activity

[36] Unlike the concept of 'land', which is quantitative rather than qualitative.
[37] Bradley (1994) would consider as required a *more flexible archaeology of place*, as the landscape is a wider concept than that of land/territory.
[38] See the classic paper by Dyson-Hudson – Smith 1978.

[39] Nancy Munn, who works on communities on Gawa island (Papua New Guinea).
[40] The 'dwelling perspective' was formulated by Ingold, inspired by Heidegger, and then influenced by Bordieu (1977). The work of Bordieu has been crucial in inspiring much anthropological and archaeological research that stresses the importance of the mutual involvement of people, material culture and landscape. For details of the concept of 'dwelling perspective' (inspired by Heidegger), see Ingold 2000: 185-7.

carried out in one place "*refers explicitly or implicitly to a host of others carried out elsewhere*" (Gosden 1999: 128), and acts are therefore not isolated, but link into complex chains creating a sense of space and of time, which would constitute what we usually call 'landscape history'.

The temporality of the taskscape is essentially social, not because each task is carried out within a framework given by society, but rather because, in Ingold's (2000: 196) words, "*people, in the performance of their tasks, also attend to one another*" (by watching, listening, touching one another and the physical or 'built' landscape features), therefore acting socially. The taskscape exists not just as activity but as interactivity (Ingold 2000: 199). The passage of time is primarily the succession of tasks and their relations to one another. The taskscape must be populated with beings who are themselves agents, therefore giving emphasis to the role of human and social agency,[41] realised by humans making choices over the landscape (see chapter I.2.2 for an illustration of the concept in terms of human behaviour over the landscape).

According to Ingold, the forms of the landscape (see above – definition) carry signs of the tasks executed in particular areas, and therefore landscape becomes an 'embodied' form of the taskscape (Ingold 2000: 198).[42] In a less theoretical and less holistic view, one could say that different landscapes (i.e. different forms of landscape) carry signs of different taskscapes (different forms of taskscape) evolving in the temporal dimension. In Inglis' words, "*a landscape is the most solid appearance in which a history can declare itself*" (Inglis 1977: 489). In Ingold's view, the taskscape is the temporal dimension of the landscape (Ingold 2000: 200). By using the word in the plural ('taskscapes') one could define the different temporal forms of landscape, or landscape characters. This might correspond, in other words and within a different theoretical framework, to the issues expressed in the framework of the community area theory.

In addition, the relational view of the anthropologists Marilyn Strathern and Roy Wagner (Strathern 1995) stresses the concept of continuity, when some transformations (which occur as the relations between people and things shift) are regularly occurring ones, opposed to the concept of change, when some transformations are unexpected and bring about new sets of relations (Gosden 1999: 121). The consideration of continuity can be linked to the community area long-term analysis and to the diachronic perspective which inform regional archaeological landscape studies, as well as to the 'sense of continuity' discussed above. Landscape archaeologists try to identify a series of cycles, which build themselves into the forms of the landscape (see above).

[41] For a definition of agency, see Johnson 1999: 189 and Tilley-Shanks 1992: 122-6.

[42] Other anthropologists point out the necessity of isolation of potential landmarks associated with specific activities in the landscape (Stewart and Strathern 2003).

To summarise, historical geography, and the *Landeskunde* tradition, was concerned with settlement chambers due to their interest in the main settlements and their territories. Afterwards, the Community Area school also showed interest in settlement chambers, but this time due to interest in activities and the distribution of activities in the landscape. On the other hand, the Taskscape approach may also concern settlement chambers, as it considers the different areas of activities in the landscape, calling them all 'dwelling' (in the Heideggerian sense), and therefore also includes the actual activity of inhabiting, as well as all the other human activities.

Settlement chambers, community/settlement areas and taskscapes can all be considered concepts belonging to the same reasoning on landscape history and narratives, having as a basis the landscape as a social space, although expressed in different ways and with slightly different implications, and generated within different disciplines (archaeology with a strong influence from geography on one side, anthropology reasoning on archaeological and material culture issues - space and time - on the other). These concepts form the background on which my approach to the Boeotian landscape is based. It gives focus to the micro-region, enlightening the role of landscape character, as a result of the association, transforming through time, between landscape zones and activities within a wider region (see earlier, the revitalisation of landscape character in historical landscape management and Ingold's notion of landscape forms), and it re-proposes the settlement chamber approach born within the *Landeskunde* tradition, renewed and revisited with demanded modifications through the community area and taskscape approaches examined above.[43]

THE USE OF GIS IN REGIONAL LANDSCAPE STUDIES

The use of GIS by archaeologists for research on regional or 'wide-area' contexts, is strictly related to Landscape Archaeology,[44] mainly adding new 'technological ways' to satisfy the needs of the theoretical background belonging to the discipline, but pushing itself towards new approaches. The main GIS applications in regional archaeology include mapping, manipulating historical maps and remote sensing data, predictive modelling, processing of survey data and sampling techniques, analysis and simulation of physical and cultural landscapes, and landscape visualisation. These can differ in the type of data involved (collecting and recording phases of GIS), in use, framework and aims (management

[43] The theoretical approach illustrated in this chapter constitutes the theoretical basis of the whole thesis. In chapter I.2.2 the application of it to my work is discussed.

[44] This involves the study and reconstruction of complete past landscapes, and places the emphasis on the relationships between sites and their environment, rather than on individual sites as 'islands' within a landscape (Lock 1998).

and analysis phases of GIS), and in data presentation and visualisation (presentation phase of GIS).[45]

Regional archaeological GIS involve different types of data, such as vector and raster data (often integrated), remote sensing data, historical maps, survey data and interdisciplinary information (data coming from different disciplines). For the purpose of this work, the use of survey data and interdisciplinarity within GIS will be focussed upon.

Some GIS projects map and apply analyses of spatial and archaeological data from the bibliography and from local archaeological records (known sites, known environments, known materials). In most of these cases, GIS can help in integrating diverse datasets and make the use of existing non-systematically collected data more objective and monitored. In addition, GIS is strongly used in processing data collected by intensive artefact surface surveys in a systematic way. It provides an environment within which one is able to explore in its entirety the regional surface information and, therefore, to critically and reflexively evaluate issues such as complex relationships between ceramic density and identification of sites.[46]

A hallmark of fully-realised GIS is interdisciplinarity. Knowledge and information coming from related disciplines can be inserted into the system and analysed along with archaeological data. Thus, data produced by other scientists (especially geomorphological and environmental layers, as well as maps resulting from land evaluation) can also be profitably introduced, thus reflecting the complexity of the landscape. Hence, archaeological GIS often include analyses typical of earth-sciences, such as terrain modelling,[47] slope and aspect determination, hydrogeological modelling (see below), watersheds detection,[48] terrain dynamics, etc.

The GIS ability of integrating diverse landscape datasets led many researchers to employ it as an environment within which to explore and evaluate the correlation between archaeological parameters and the physical landscape. GIS applications have therefore been associated with a risk of 'ecological (or environmental) determinism' (Gaffney and Van Leusen 1995, Kvamme 1997, Wheatley 1998), as well as with the tendency to forget and/or exclude cultural aspects of the landscape as well as historical factors and social variables which would have had a significant effect on how the landscape was used and perceived. This soon resulted in an increasing involvement of 'cognitive'[49] variables in location analysis and predictive modelling (Kvamme 1999: 182), and European researchers began to focus on the incorporation of social variables into their predictive models as well as in spatial and locational analyses (Wheatley 1996; Stančič - Kvamme 1999; several papers in Lock 2000). In order to do this, they incorporate for instance qualitative evidence derived from oral history and ethnographic studies (Pilon et al. 1997), they infer some cultural parameters from historical sources and archival documents, which could be either included into the GIS as cognitive landscape parameters (mainly landscape visibility and accessibility involved in analyses such as visibility and cost surface analysis – Gillings and Wheatley 2002, Van Leusen 2002: chapter 6 – to explore past cognitive landscapes) or taken into account during the performing of the analysis (see below).

Archaeologists have been accused of using GIS 'as little more than a mapping system' (Goodchild 1995: 46). In fact, some work uses GIS simply as a method for computerising cartography, combining disparate datasets with speed and accuracy (thanks to the multiple layer logic of GIS). If often studies do not go really deeply into the analysis, it is also because there are a considerable number of applications[50] that emphasise the mapping capabilities of GIS rather than its analytical functionality (this has been pointed out by many authors; see, for example, Gillings and Wise 1999). Therefore, while any

[45] Since the late 1980's, GIS has gained popularity in Archaeology. The volume edited by Allen, Green and Zubrow (Allen et al. 1990) marked a crucial phase for the young discipline, opening its potential to the academic archaeological world. Since then there has been rapid growth in the application of GIS techniques, as a tool both for Cultural Resource Management (CRM) and for academic research. In 1995 Lock and Stančič published a European reply to the American work (Lock - Stančič 1995). Over the last decade, a growing awareness of the potential of GIS is evident from the publication of an increasing number of both practical case-studies and theoretical investigations. Important, especially for the critique of the use of some GIS routines, is the work of M. van Leusen (published in several articles and then combined into his PhD dissertation, Van Leusen 2002), as well as the manual 'Spatial Technology and Archaeology' by Wheatley and Gillings (2002). Also worthy of consideration are the CAA (Computer Applications in Archaeology) conference proceedings, published annually since 1973, and including GIS since the late 1980s, as well as the books edited by Lock (2000) and by Mehrer – Wescott (2006), and the recent book on Geographical Information Systems in archaeology by Conolly and Lake (2006).

[46] There are several examples of GIS-based survey studies in the Mediterranean area. Greece: Boeotia (Gillings and Sbonias 1999; Bintliff and Howard 2000), Aetolia (Bommeljè et al. 1987; Doorn 1993), Achaia (Simoni – Papaiannopoulos 1998; Petropoulos-Pontrandolfo-Rizakis 2004) and the Patras region (Rizakis et al. 2001), Kythera (Broodbank 1999; Bevan-Conolly 2002-2004), Corinth (Romano-Tolba 1996; Gregory 1998), Thessaly (Helly 1998), Crete (Phaistos – Watrous et al. 2004). Cyprus: Given – Knapp et al. 1999 and Given – Knapp 2003. Italy: Tuscania (Vullo and Barker 1997), Tiber Valley (Patterson - Millet 1998), Agro Pontino (Kamermans 1994; Attema 1993), Potenza Valley (Vermeulen 2002 and 2003), RPC project in Southern Italy (Attema et al. 2002; Van Leusen 2002 and 2004); Turkey: Vanhaverbeke - Waelkens 2003, Lycia (Hailer and Martin 1998).

[47] The use of modern terrain morphology to represent its past versions has been subjected to much constructive criticism, allowing it to proceed with more careful use. On DEM (digital elevation model) interpolation methods and biases see for instance: Beex 2004 and Farinetti – Sigalos 2002; on DEM quality for viewshed analysis see: Van Leusen 2002: chapter 6; on investigation for archaeological purposes of the 'topographic fabric' see: Duke 2003 - CAA2002.

[48] For an archaeological application see Bevan 2002 on Crete.

[49] See Renfrew and Zubrow 1994 for 'cognitive processualism'.

[50] Sometimes referred to as *desktop mapping* software.

definition of GIS will undoubtedly emphasise analytical capabilities, it must be recognised that in some respects a major strength of the software lies in its ability to integrate and manage large and diverse datasets (Gillings and Wise 1999, Vermeulen 2001). The risk is to create just a more sophisticated and attractive version of the old-fashioned 'Archaeology of dots'. On the contrary, the strength of GIS environment is that the dots can become meaningful within the system, being attached to incrementally higher levels of information.

Certainly analyses can differ according to the theoretical archaeological background followed, and also depending on the data-type involved in the GIS. In the case of Regional Archaeology, spatial and location analyses of settlement and landscape features can be carried out within a GIS system (see below), as well as analytical procedures helping the detection and interpretation of ancient landscape dynamics.

As stated earlier, much work identifies the risks of 'environmental determinism' (Limp 1997, Stančič et al. 1997, Given et al. 1999), and scholars either try to find a solution for it (Favory and Van der Leeuw 1998, Given et al. 1999), or they claim the urge of new GIS models designed explicitly for socio-cultural rather than geophysical processes (Verhagen et al. in press).[51] Some try to integrate current theoretical notions of landscape within GIS functionalities involving various ways of effectively humanising the landscape, stressing the perception of the landscape itself by people living in it in the past. As Gillings and Wise (1999) point out, these approaches initially attempted to comment on the perception and cognition of an individual situated in the landscape based on visibility and intervisibility studies involving line-of-sight and viewshed routines (for example, Gaffney *et al.* 1995; Wheatley 1995; Lock and Harris 1996; Van Leusen 2002: chapter 6). Further steps in this direction are various perception-based approaches (Llobera 1996), the reconstruction of landscape dynamics (Gillings 1995, Budja and Mlekuz 2001), and the integration of Virtual Reality into the GIS systems (Gillings – Goodrik 1996; Forte et al. 2003). In the last decade, several studies have also appeared concerning the application of agency concepts within the GIS analytical environment (Van Hove 2003; Llobera 1996). In contrast, the analysis of cognitive phenomena over a landscape, if overstating 'human' and cultural aspects of the landscape, such as symbolic or mental factors, can also lead to an abuse of phenomenological explanations.

Different kinds of locational and GIS analyses are often performed within GIS systems: thematic mapping, Thiessen Polygons, Nearest Neighbour Analysis, Quadratc analysis and Location Models, different kinds of statistical analysis, geostatistics,[52] Network analysis, Modelling techniques, buffers,[53] as well as predictive modelling.[54] I will describe here briefly two typical GIS analyses that can be useful for cultural considerations on past human behaviours and involve cognitive landscape parameters, such as landscape accessibility and visibility: Cost-surface analysis and Visibility/Viewshed/Line of Sight analysis.

Cost-surface analysis: as an improvement on buffers based on straight-line distance, catchment areas, usually investigated through circular or other-shaped buffers, can be weighted, within GIS, taking account of topography, therefore calculating not only straight-line distance but walking-distance or time-distance. A module to calculate such a relative cost-surface using the DEM is included in most GIS packages. It models the cost of moving through space, where costs are a function of both the standard costs associated with movement, and also of frictions and forces that impede or facilitate that movement (see, among others, Stančič et al. 1997; Llobera 2000; Van Leusen 2002: chapter 6; Gillings 2002: 151ff.; Pizziolo-De Silva 2004; Howard 2007). From the cost surface one can perform the shortest path analysis, leading to the simulation of ancient paths (among others De Silva-Pizziolo 2001: 284-5; Podobnikar et al. 2004). In some GIS work, the sites or activity *foci* employed in cost-surface analysis are differentially 'weighted' (Llobera 2000: 74; Van Leusen 1993 and 2002: 6.8). By means of these analyses, the relationship between man and the topographical environment can be examined, and physical and perceptive movement through the landscape investigated (Llobera 2000; Van Leusen 2002: chapter 6).

Visibility/Viewshed/Line of Sight analysis: marks a step in the direction of understanding the ancient perception of the landscape, through the identification of sites in viewsheds from both culturally and naturally defined points in the terrain. A viewshed is all the locations that are capable of being seen from a defined location or locations. The GIS uses elevation data (potentially with additional layers such as vegetation, structures, etc.) to compute line-of-sight. A refinement of this is in the distinction between the absolute visibility (probable viewshed) and the level of clarity (fuzzy viewshed) with which a site is seen. Another development is the interrelation between the visible and the accessible landscape. For reference see, among others, Van Leusen 2002: chapter 6; Gillings 2002: 151ff.. European archaeology has greatly developed this kind of analysis in order to explain complex social processes (relatively early examples are the work in Brač -Stančič et al. 1997- and in Kars -Novakovič 2001). For the concept and

[51] A session was organised at the EAA 2004 congress by J. Chapman and B. Gaydarska, with the title: *How GIS can contribute to a social archaeology?*

[52] A particular form of statistical analysis particularly performed within a GIS environment and in GIS-based landscape studies (see for instance Barceló - Pallarés 1998:65 and Van Leusen 2002: 5.15).

[53] Buffer analysis is used to investigate catchment areas or areas of interest around a point, polygon or along a line feature.

[54] For predictive modelling see Kvamme 1990, and, among others, Verhagen et al. 2000, Van Leusen and Kamermans 2005, Veljanovski and Stančič 2006. Although North American archaeology has the most experience with this technique, several recent European applications show its potential for regional archaeology (for the latest applications see Kamermans 2000; Verhagen and Gazenbeek 2006).

applications of cumulative viewshed analysis (viewsheds from a set of features), see among others: Wheatley 1995 and Van Leusen 2002 (chapter 6).

The GIS environment can help the refinement, re-thinking and re-application of traditional categories and thematics that have characterised Regional surveys and Settlement pattern studies: upland-lowland dynamics, agricultural practices and pastoralism, rural-urban interactions, nucleated vs. dispersed settlements, resource control, social stratification, development of complex societies, spatial patterning, recognition of boundaries between cultural groups, and recognition of the boundaries of large diffuse sites. An increasingly important role in Regional Studies is played by the anthropic and cultural variables of ancient settling dynamics, the perception that ancient people had of the surrounding natural environment and cultural *milieu*, and their relationship with the perception of time and space (spatio-temporal dynamics).
Many GIS practitioners and theorists argue that GIS technology has already began to reach maturity and archaeological GIS-based studies are now beginning to interrogate themselves, finding their own voice and identity and developing a reflexive body of theory. Moreover, quite a number of archaeologists believe that GIS is slowly revolutionising the way in which archaeology is performed, and that having used GIS it is impossible to return to other methods (e.g. Gaffney - Stančič 1991: 29-32). I agree with this: it forces one to rationalise data, to clarify one's thoughts and aims in order to be able to instruct the computer in a structured way, to collect data in an appropriate, adequate, organic and structured way. I would support an interaction, a continuation of old methods together with the new, as if they were two versions of the same reality. However, the use of GIS should not determine the nature of the research. The research questions of landscape evolution and transformation must come first, and GIS must at best be a tool to carry out such research.

The present research can be considered GIS-based in the sense that a GIS was the framework within which the landscape physical and cultural data have been collected and managed, and later analysed and presented. Within this framework, the construction of the archaeological dataset was, to a certain extent, led by the inner properties of the GIS system, which certainly informed, as a way of thinking, the information source critique process and helped in the structuring of the archaeological record in a coherent and consistent manner while maintaining the resulting archaeological datasets adequately flexible (see chapter I.2.2). On the other hand, the chosen approach to the landscape also plays a key role in the construction of the datasets, as well as in the analysis and interpretation processes. I would say that within the GIS environment, the approach to the landscape as described earlier in this chapter became practice, leading to the enucleation of location choices of past communities over the landscape, as well as to the association between landscape zones and activities, resulting in landscape characters, and to the exploration of cultural landscape meanings and settlement behaviours (see earlier in the chapter and chapter I.2.2).

I.2.1

The physical landscape datasets

The physical landscape analysis of Boeotia is meant to individuate landscape information, and on this basis perform further analysis concerning the past use of the land and the resources available to past communities, as well as the past communities' cognitive approach to their landscape. The methodology established, as discussed in detail in chapter I.2.2, dealing jointly with material culture, and environment, opts for the 'region' as the analytical unit. It aims mainly to assess the interface between human and social actions and landscape (Barker – Mattingly 1999: ix). The transformation of the landscape through time is also under study, in order to infer elements of human eco-dynamics.
The use of GIS can help in performing the analysis, being able to easily manage a large quantity of data and process them in a short time. In addition, it helps us to monitor the quality of the data, and to keep track of the processes employed and the calculations applied to the original records.

In this chapter, I will discuss the choices made and analyses performed on environmental datasets within the GIS system implemented. Physical landscape datasets inserted into the system will be described and examined.

Different physical datasets have been employed within the GIS system, and these come from different sources. They can be divided into:

- BASE data
- DERIVED data
- INTERPOLATED/managed data
- INTERPRETED and MODELLED data (results of analysis)

BASE data:
- Elevation data (manually digitised from georeferenced 1:50,000 topographical GYS[1] map): contour lines at 20m interval and spot heights.[2] The sheets used were: Athina-Eleusis, Bagia, Chalkis, Elateia, Erythrai, Kaparelion, Larymna, Libadia, Libanatai, Perachora, Thiva.
- Rivers and streams network (from 1:50,000 GYS map).
- Geological data (from 1:50,000 IGME[3] geological maps and surveys). Paper maps are based on the 1:50,000 GYS topographical maps.
- Aerial photographs for detail windows (Kephisos valley, Copais, Tanagra).

DERIVED and INTERPOLATED data:
- TIN interpolation from contours and spot heights (see BASE data) and the derived grid DEM (Digital Elevation Model) with a spatial resolution on the ground of 30x30m.
- Slope (derived from TIN model) – 30x30m grid.
- Aspect (derived from TIN model) – 30x30m grid.
- Hillshade (derived from TIN model) – 30x30m grid.
- Classified physiographical position from slope and elevation data (see dedicated section below) – 30x30m grid.

INTERPRETED data:
- Classified land capability (see dedicated section below) – 30x30m grid.
- Classified erodibility (tendency to erosion) (see dedicated section below) – 30x30m grid.

MODELLED data:
- Dynamic model of Copais lake fluctuations (realised according to topographic, environmental and historical data). For a detailed account of the model see chapter II.3.1 – PERILACUSTRINE MARSHY LANDSCAPE – and Farinetti 2008.

ACCURACY OF BASE DATA

Environmental layers are usually comprised of continuous variables or discrete variables that cover the whole surface of the area under study. Spatial resolution

[1] GYS stands for *Geographiki Ypiresia Stratou*: the Hellenic Military Geographical Service. The 1:50,000 topographical maps are based on photogrammetric compilations from vertical serial stereographs, scale 1:42,000 approximately (1945; field checked 1971 to 1973 -Chalkis sheet in 1955-; photogrammetrical ckecking from aerial stereophotographs 1986 to 1988).

[2] When I started this PhD research, there was no digital topographical data available for Greece, and even today there is not easy access to them, and the costs are, in any case, enormously high. Therefore, during the first stage of my research, I digitised manually 11 topographical paper maps at a scale of 1:50,000 provided by the GYS. The digitised layers are: contour lines (20m interval) and high spots, rivers and streams, modern villages and towns, the main road network.

[3] IGME stands for *Institouto Geologikon kai Metalleftikon Erevnon*: the Greek Institute of Geology and Mineral Exploration. Fieldwork was carried out during the 1960s and 1970s. Maps were published between 1965 and 1986.

of data can differ, as well as data quality, but environmental layers are usually characterised by a stronger linearity and consistency (at least in the way data have been collected and mapped, and in the way they have been inserted into the system).[4] On the other hand, as will be discussed in the next chapter (I.2.2), archaeologically meaningful entities as well as social variables marked by a high degree of non linearity are often represented as discrete rather than continuous, and it is much more difficult to give them a coherent spatial form or a consistent structure (Van Leusen 2002: 2.18-19).

Accuracy of environmental data sets used for subsequent analysis of the archaeological landscape depends on the accuracy of base/row data used.
Quality of results is linked with the scale at which data have been collected (or: with the scale and resolution of the base data used). As always when talking about accuracy of data, we refer to the quality of the data in relation to the problem investigated as well as the goal of the research. In our case, aims and scale of the work make adequate and acceptable for the majority of analyses the accuracy level given by the thematic cartography available for the region under study (1:50,000 GYS topographical maps, 1:50,000 IGME geological maps and hydrogeological map, Davidson's land suitability for arable agriculture map[5]).
Available maps and literature data on geology and land classification were verified and corrected as much as possible with fieldwork on the ground, both by means of an extensive and non-systematic field survey, and punctual sampling in sample areas,[6] while lacking a detailed study for soils over the whole region (not even at the same scale –1:50,000- of the geological maps available). At the local scale, in the area of ancient Thespiae (see chapter II.3.9) a study has been conducted by Shiel and Stewart (Bintliff-Howard-Snodgrass 2007: chapter 7) as for the reconstruction of land-use potential in the past, with attention to land degradation and soil dynamics. I used the work as general guidelines valid for the past land potential and land use in Boeotia, especially for areas with comparable soils and geomorphological characteristics (see below – GEOMORPHOLOGY AND LAND CLASSIFICATION).

[4] Physiographical/morphological and land capability layers are linear, based on the same base data (geological-topographical map georeferenced) and derived in the same way from the same base data sets (using the same classification, etc). On the other hand, we must take into account that topographical and geological maps are themselves results of sampling and interpolation processes, which the user cannot always easily detect and judge, especially if not a true expert in the discipline.

[5] A digitally drawn map result of D.A. Davidson's personal work carried out in Boeotia as for evaluation of land suitability for agriculture, with the use of core sampling (he focused especially on the plains and some hilly landscape areas). The map was made available to me through Prof. J.L. Bintliff. See also Davidson - Theocaropoulos 1992 and Davidson et al. 1994.

[6] See below for a discussion of the methodology applied.

The grid used for raster data and analysis has been set to 30x30m cell size. Resolution of base maps (topographical and environmental data, at the scale of 1:50,000), as well as a DEM based on elevation contour lines at 20m intervals, do not allow for a better resolution.
It could be noted that the environmental layers are supposed to be interrelated and cross-analyzed with archaeological entities, which are mainly represented as points within the system, and therefore, when interrelated with raster datasets, they would be represented as cells of 30x30m size (a reasonable minimum size for archaeological entities like our sites –see chapter I.2.2). If one considers the scale of the work with problems and goals at a regional level, and the fact that the environmental layers are supposed to be interrelated and cross-analysed with the archaeological entities, the resolution chosen could be considered as reasonable, being in accord with the base datasets (mainly environmental) as well as with the archaeological-cultural layers involved in the analysis.

As often happens in regional landscape studies carried out at a relatively small and therefore general scale, datasets involved are results of research on the present landscape and environment. Modern landscape datasets have thus been used, processed and created through interpolation and integration processes, in an attempt to define characteristics of the modern landscape to infer information and have a base of reasoning for a possible landscape of the past.

GEOMORPHOLOGY AND LAND CLASSIFICATION

Three main aspects of the geomorphology of the landscape have been taken into consideration, and raster GIS layers were consequently created:

I. physiographical position
II. soil erosion
III. land capability

Those aspects combined together for the definition of physio-geomorphological features allow us to define a series of landscapes and their stability.

I. Physiographical position

A class of physiographical position can be defined as follows:

> ***Physiographical Class****: a unit of geomorphological forms (landforms) with common characteristics that identify elements of a specific landscape. Physiographical classes are not defined absolutely, but according to the specific landscape under study.*

Classifying the landscape in physiographical units can help in the location of human settlements and activities more precisely by comparing them with the physical characteristics of a landscape, and can help to understand the type and degree of relationship between human settlement and physical environment.
For instance, in the case of the Copais basin, the central lake plain is differentiated from the foothills, even if they lie at low elevations comparable to that of the plain, since in relation to that particular landscape they are elements with different characters: the former permanently or semi-permanently flooded; the latter not. Likewise, foothill landscapes are different from the uplands or the gentle slopes of the narrower, inner valleys, even though they are characterised by similar slope ranges.

At a further stage, features can be differentiated according to their different soil characteristics also: in the example of the Copais basin, the plain presents clay and slightly permeable silt (more fertile where the lake was not permanently present but always susceptible to flooding), while in the foothills limestone appears and constitutes the active karstic zone (less fertile, marked by the presence of *katavothrai*,[7] but with a significant presence of seasonal or perennial water sources).[8]
The examination of physiographical position and of qualitative and quantitative presence of the different physiographical classes has been carried out in general terms for the whole of Boeotia,[9] and in detail in the analysis of the physical landscape of each *chora*.

Physiographical position can be obtained from the combination of slope and elevation data, using a suitable grid size. In our case, we should be aware that the low resolution of the base DEM (30x30m) does not allow one to appreciate in detail slope differentiations, which remain therefore only indicative, as they lack the intervening morphological complexity.

The following is the list of physiographical classes used for the characterisation of the Boeotian landscape:

- *uneven*: hillside steep or very steep (between 20% and 35% or more than 35%)

- *hilly*: hillside moderately steep (10-20%), sometimes steeper, but low in altitude

- *undulating plateau*: moderate inclination (3-10%), sometimes almost flat

- *plain*: slope <5%

Those four classes can be applied to the three main elevation ranges: mountain (upper), hills, plain (lower) areas.

[7] The *katavothrai* are natural swallow-holes, usually draining water from karstic basins.
[8] For details see chapter III.1.
[9] See chapter II.1.

Valley below 200 m asl [<200]

code	slope %	description
P1 P2	<5	lacustrine basin, valley
P3	≥5 <10	gentle slope
P4	≥10	foothill

Hill below 600 m asl [≥200 <600]

code	slope %	description
H1	<5	plateau
H2	≥5 <10	gentle slope
H3	≥10 <20	moderate slope
H4	≥20 <35	severe slope
H5	≥35	very severe slope

Mountain above 600 m asl [≥600]

code	slope %	description
M0	<5	plateau
M1	<10	plateau, gentle slope
M2	≥10 <25	moderate slope
M3	≥25 <45	severe slope
M4	≥45	very severe slope

A map of physiographical classes for Boeotia can be found in chapter II.1 (fig.2).

II. **Soil erosion**

Erosion is a natural process of surface soil sliding caused by various factors, and strongly influenced by human impact and in particular by land use (both agriculture and animal husbandry). Linked to the process of soil removal are the transportation and deposit of debris. Erosion is therefore a fairly complex phenomenon to analyse, because of the extreme variability of factors that can cause and influence it. Nevertheless, an attempt to classify it can be essential, considering the influence that erosion processes have on the actual soil potential (Gisotti 1983; Godard - Rapp 1987).
Generally speaking, for agriculture, erosion can be considered a negative phenomenon since its consequences are the alteration of the soil's superficial structure and thinning either of the 'arable soil' or, in the case of natural soil, of the superficial horizons which are the most fertile. Thus, natural erosion can cause the removal either of the surface stratum (A horizon) or, in extreme cases, of the whole soil (Comel 1972; Gisotti 1983; Castiglioni 1991). On the other hand, eroded soil often ends up in lower elevations where it can then be used for cultivation.
The speed of erosion processes is influenced by human factors as well as natural phenomena. When soil removal is more rapid than soil formation (*accelerated erosion*), complete removal of soil and appearance of bedrock can be reached (Castiglioni 1991: 412ff.). Accelerated erosion is favoured by human activities, in particular

reduction of and changes in vegetation coverage (deforestation to gain land for agriculture, general reduction of wild vegetation), overexploitation of soil both for agricultural or pastoral activities, increasing of slope (Butzer 1974). Accelerated erosion can also be influenced by natural factors, in particular abrupt environmental variations which create a strong imbalance, such as a long drought, damages to the vegetation cover, etc. (Megaw 1977; Mehemet 2002).

Accelerated erosion is therefore caused by incidental factors (both natural and anthropic), which often induce rapid phenomena of erosion and deposition. On the other hand, the so-called 'geologic' erosion (a natural dynamic process linked to the natural tendency to balance of environmental features, and surface soil also) is usually slow enough to be compensated by pedogenesis processes and to allow for the soil to regenerate (see, among others, Rice 1977 and Gary at al. 1973[10]).

The risk of soil erosion has for a long time been one of the geological hazards to which human activities are exposed (Gisotti 1983: 37ff. discusses this issue, which, however, is also examined by many others). For Greece, the majority of the country is marked by a hilly and mountainous (rarely high mountain) landscape, often constituted by easily erodible rocks, with an uneven morphology. Boeotia also presents this character, even if its landscape is marked by some wide valleys and the presently drained basin of Copais, the largest karstic basin (*polje*) of Greece (Dercourt 1977; Higgins and Higgins 1996). The erosion of these kinds of soils and morphologies is often strong, favoured by the Mediterranean climate, which is characterised by long dry and hot periods alternating with periods of short and heavy rains.[11] On the other hand, Shiel (2000) notes, quoting Davidson and Theocharopoulos 1992, how in general the Boeotia area appears to be mostly remarkably immune to serious erosion[12] which plagues other parts of the Balkan Peninsula (Shiel – Chapman 1988).

The present work employs the research work carried out by D.A. Davidson in Boeotia (Davidson - Theocaropoulos 1992; Davidson et al. 1994). Davidson correctly states that soil erosion is related to slope, soil type and texture, as he notes: "*As is to be expected, a close relationship was found between soil erosion and slope. Land with slopes of 3% or less is distinguished by the lack of erosion whilst at the other slope range, land in excess of 18% is dominated by erosion class 4...*" (1994: 373). A relation between erosion, soil type and texture indeed exists, but it can be far too complex to be taken into consideration here. In addition, the relationship between well developed soils (vertisols, alphisols) and lack of erosion, and the larger variability in degree of erosion of younger soils, once verified on the ground, could not attain the same degree of occurrence and reliability demonstrated by the erosion-slope relationship.[13]

Therefore, the classification made by Davidson (1994: 372) of four erosion classes is built substantially on the basis of the relation (considered as primary) between erosion and slope, and defines the percentage of subsurface horizon exposed, until the total erosion of soil A horizon:

1. None;
2. Subsurface horizon exposed in <30% area;
3. Subsurface horizon exposed in >30% area;
4. All of A horizon removed.

Davidson therefore determines the extreme classes:

≤ 3%	class 1
> 18%	class 4

from which the intermediate classes can be inferred:

>3 <10	class 2
>10 ≤ 18	class 3[14]

The combination of physiographical and erosion classes allows us to delineate 'morphological landscapes'. For instance, the colluvia that border the Theban plain towards Yliki lake can be classified as 'undulating plateaus' in valleys with high soil stability, while the Copais is a valley depression with a tendency to soil stability due to sedimentation, in various degrees from the central area to the border.

III. **Land capability** (in relation to agriculture)[15]

The employed methodology of land evaluation,[16] in an attempt to evaluate, at a general level, land potential and limitations for agricultural use, takes into account physical components of landscape, such as geology,

[10] Geologic erosion is the "*Erosion of rocks and soils under natural environmental conditions, undisturbed by human activity.*" (Gary et al. 1973: 484). Accelerated erosion is an erosion process "*...occurring in a given region at a greater rate than normal erosion, usually brought about by the influence of man's activities in disturbing or destroying the natural cover*" (Gary et al. 1973: 4).

[11] On the complexity of erosion in the Mediterranean, see also Bintliff 2000d and 2002.

[12] See also Shiel-Stewart 2007 for Thespiae area in particular.

[13] As a more general reference, Davidson's work employes the Universal Soil Loss Equation - USLE (Wischmeier-Smith 1965) commonly adopted in Europe (FAO-Unesco 1981).

[14] For the present work on Boeotia I have corrected Davidson's erosion classes (with multiple values) according to slope and physiographical classes.

[15] A definition is given by a Dictionary of Soil Science (Canarache et al. 2006): "*Land capability classification: the classification of lands according to their performances for various farming uses* (Soil Survey Staff 1993). *It is a qualitative land evaluation procedure* [...]". In this classification, lands are classified for the production of the more common field crops (in our case: arable crops more common in the past, olives and grapes), but not for crops requiring little cultivation or for flooded crops.

[16] "*Land Evaluation: a general term for any procedure used to estimate land quality.* [...] *The term often has a more or less synonymous sense to the term soil evaluation, most procedures making use, to a great extent, among other land characteristics, of soil properties. Qualitative, semiquantitative and quantitative land (soil) evaluation procedures are often differentiated* [...]" (Canarache et al. 2006).

geomorphology, pedology. The undertaken task was to combine the more relevant components of the environment in order to individuate the land capability of each area of landscape within the region, employing qualitative evaluation procedures. This examination has been interrelated with cultural factors; mainly the agricultural techniques and husbandry practice in use in different periods (plough agriculture, pre-ploughing agriculture, etc.). In order to define land capability, the geomorphological and pedological examination must be combined with an examination of climatic, vegetational and faunal factors, which, due to their complexity, especially when referred to the past, have been taken into consideration for the present work only in extremely general terms, being conscious of the limitations and risks of such a choice. On the other hand, the examination considers, at a simplified level, parameters influenced by different factors: rock erodibility, underground water circulation, morphology, slope (Gisotti 1983).

First step was the enucleation of geological formations and lithologies: more stable parameters, as landscape elements only slightly or not at all modifiable on a human scale. Once the formations were delineated through the examination of the available geological maps, different aspects have been analysed: technical characters, influence of surface morphology, phenomena to which they might be subjected, soils that they usually produce. Soils deriving from different rocks, delineated in this way, were classified. Afterwards, the results were verified and integrated with Davidson's and Shiel and Stewart's studies (see above), and with tests in the field. Finally, the resulting dataset was correlated with physiographical values, and the land capability classification scheme was set out.

The land capability values used in the present work result from the relationship between a basic soil classification, soil erosion and a general consideration of the degree of workability of soil related to low-technology agriculture.

A. *Soil classification*

Soil is the earth's surface layer, the contact environment between the inside and the outside of the earth (Duchaufour 1970; Comel 1972; Foth 1990; Arnoldus-Huyzendveld 2003), between rocks, sediments and subsurface water on the one hand, and air, pluvial water, vegetation, fauna, and human activities on the other. Soil is created by the alteration processes of the rocks once in contact with atmosphere and biosphere. In some respects, the soil has similar characteristics to the lithologies from which it derives, but differs from them since modifications occur related to the presence of vegetation, fauna, and micro-organisms, in addition to a whole series of transformations caused by human action. Soil is in fact sensitive to both natural and human influenced transformations of the environment (variation in climate and environment, settlement traces, land use, etc.), to which it reacts and of which it maintains traces, being at the same time both factor and consequence.

The formation and character of the soil are therefore influenced by several factors: the presence of biological activity (mainly produced by vegetation and animals), whose type and intensity strongly influence the character of the soil itself; man's action, that can strongly condition the nature and development of soil; the composition of the bedrock from whose alteration the soil is formed; climate, since rain, wind, sun and ice assist the physico-chemical rock alteration. Pedogenesis is the result of the integrated action of those diverse factors. The length of pedogenetic processes gives the age of a soil, whose formation starts at the moment when the alteration of the original bedrock begins. The development and age of a soil are the result of the combination of pedogenetic factors with morphogenetic processes (soil erosion and deposition) that model the surface and create landscape morphology (Gisotti 1983; Arnoldus-Huyzendveld 2003; Canarache et al. 2006).

Soil is considered a 'renewable natural resource', even if in fact it is quite fragile, considering that hundreds or even thousands of years might be necessary for its formation, while accelerated erosion or human action can destroy it in a few years or even a few hours. This is true not only today, but also in the past human action played a crucial role. Overexploitation of soil, due either to demographic pressure or to agricultural practices (in the ancient Mediterranean, for instance, with no use of crop rotation), could have led to the impoverishment of the soil in certain areas, with a consequent drop in soil fertility as well as renewal capacity.[17]

The subject of soil nature, characteristics and dynamics, is much more complex, and a detailed pedological study would be a basic tool in order to define both the nature and potential of different soils. However, considering the limitations of my own knowledge and work on the matter, a very simplified soil classification has been attempted. For the region of Boeotia, a soil map is not publicly available, and creating a map with adequate detail would certainly fall beyond the purpose of the work and the ability of the writer.

In order to produce an overall large scale mapping of Boeotian soils I used as a basis the available 1:50,000 geological maps, which cover almost the whole region[18] at a reasonably good qualitative level (approximate resolution 35/40 metres[19]). The diverse lithologies have been classified considering the soil type that would most probably correspond to each bedrock type. In table 1 is a catalogue of formations mostly present in the region of Boeotia, identified on the base of the lithologies derived from geological maps and grouped according to similar characters.

Soil types (to be related with land capability) were defined and mapped according to lithology, geomorphology and physiographical classes. The two factors that have been systematically employed in classification and analysis are slope and elevation.

[17] For mainland Greece see among others van Andel and Zangger 1990; Bintliff 2002.
[18] A sheet –*Chalkis*- is missing from the IGME archives.
[19] The resolution is indicated on the IGME paper maps.

Slope influences soil depth and development in particular. In plain areas, under the same pedogenetic conditions as for other factors, depth would be at its maximum value, while in areas with increasing slope, soil depth has a tendency to decrease due to erosion. Slope also influences external drainage, as well as surface water flow, and favours washing out processes on steep hillsides as well as stagnancy in flat areas or depressions. Elevation influences the leaching[20] processes of bedrocks and clay, due to the increase in rainfall at higher elevations, and, especially, a progressive slowing down of the mineralisation process of organic elements due to lower temperatures.

I believe that this method, despite its evident limitations (it does not directly consider, for instance, climate and vegetation coverage), can be considered suitable for the purpose of the present work, since:

a. Bedrock is one of the most important pedogenetic factors (see above), as it influences soil type, on one hand with regards to chemical characteristics/behaviour, producing soils more or less rich in nutritive elements, and on the other hand with regards to physical properties, since more or less compact or hard bedrocks can produce deeper or shallower soils accordingly, and more sandy or clayey soils, as well as soils of lesser or greater workability, especially with low-technology agricultural practices.
b. The greater part of Boeotia has fairly rough morphologies, with a discontinuous vegetation cover that easily allows for erosion processes and hydrogeological breakdown. Furthermore, those soils have been subjected to millennia of human actions (tilth, vegetation burning to gain land for cultivation, overexploitation by pasturage and agricultural activities soil is not capable enough). Thus, soils are affected by frequent renewal and are not greatly developed. All this assumed, one can consider the influence of bedrock quite prevalent compared to other pedogenetic factors.
c. Soils of Boeotian valleys, and the Copais basin in particular, follow different dynamics, and can be included in a general classification in which deposition factors are stronger than those of erosion. In the case of the Copais, a specific analysis has been carried out to estimate and clarify the state (either a large swamp or a proper lake) of the basin before the drainage, and consequently the type and availability of soils in different periods of history (see chapter II.3.1 – PERILACUSTRINE MARSHY LANDSCAPE – and Farinetti 2008).

[20] Alteration of rock surface due to chemical factors. Some elements dissolve in water, so the process is enhanced by rain, for instance.

A	**Alluvia**	Recent alluvial plains. Fluvial and Fluviolacustrine deposits. Clays, silt, fluvial sands and pebbles. Limno-palustrine as well as lacustrine sediments, recent and contemporary. Grey, brown, black/dark soils.
		Terraced ancient alluvia. River and lake deposits at elevations higher than the present river bed. Alluvial terraces. Talus, alluvial talus, alluvial pediments.
B	**Colluvia** Incoherent surface deposits that would cover mountain and hill slopes or are situated at their feet.	Deposits in foothill areas. Hillslope deposits. Elluvial and/or colluvial deposits. Glacis. Aquifer deposits, detrites, gravels. Red soils. Bauxites. Glacial deposits.
C	**Metamorphic Rocks**	Metamorphic rocks. 'Green rocks', 'ophiolites' (gabbros, serpentine, diabases).
D	**Clays**	Clays, tectonic heterogeneous complexes with clay as prevalent component. Schists; flysch; schistose clays and clayey schists.
E	**Marls** Marls are rocks whose composition is intermediate between limestone and clay.	Prevalently marl formations, marl-sand formations (when the marl component is prevalent).
F	**Carbonate Rocks** Carbonate rocks are constituted of calcium carbonate (CaCo3) and/or magnesium carbonate (MgCo3). Those constituted mainly by CaCo3 are limestones; those in which MgCo3 prevails are dolomites.	Compact and hard limestone. Mainly carbonate rocks, compact and hard, limestone, marls limestone, calcarenites (sandstone made of limestone grains rather than of quartz and pheldsphate grains), limestone breccia, crystalline and fossiliferous limestones, silex limestones, dolomite limestones, dolomites.
		Soft limestones, porous limestones, slightly coherent. Travertine deposits, calcarenites, limestone conglomerates, sandy limestones. To this category can be associated the slope breccia, comprised of deposits of cemented hard-break rock fragments, that often cover, with thin layers, limestone mountain and hillsides.
G	**Sandy Formations**	Siliceous sands more or less cemented, polygenic conglomerate, pudding stone, breccia. Ancient cemented talus. Sandy-marl flysch, sandy-clayey flysch. Sandstones with different components. Soft sandstones.
H	**Flysch** Flyschs are constituted by diverse rocks rhythmically alternated; they correspond to a series of various materials which regularly alternate in great thickness.	Flysch constituted by sandstone in alternation with marl, or limestone with clay, etc. The fertility of soils originating from flysch is strongly influenced by the component rocks.

Table 1. *List of formations mostly present in the region of Boeotia.*

B. *Soil erosion*

Determination of soil erosion has been carried out according to the procedure illustrated above (section II – SOIL EROSION). Soil erosion values are linked mainly to slope values but also to the degree of sustainability of agricultural exploitation and practice (mostly lacking, in ancient times, techniques of soil preservation[21]).

C. *Soil workability*

Soil workability can be directly influenced by slope, soil stoniness,[22] and terrain drainage properties. While dealing with past agriculture, soil workability must be considered in relation to low-technology agricultural practices (i.e. dry – non irrigated - agriculture, soil worked with hoes and/or wooden ploughs, the latter probably the exclusive instrument in Prehistoric periods – from Final Neolithic onwards in the Aegean - but still in use in the Classical period, along with the iron plough, which was, however, mostly in use in more recent periods (Roman / Late Roman[23]), often as a high value item,[24] which the majority of farmers could hardly afford.

Soil workability is a crucial factor, since we should always consider the agro-pastoral techniques presumably employed in the periods considered in order to determine land evaluation values.

The general assumption is that both for late Prehistory and Greco-Roman antiquity, we are dealing with low-technology agricultural practice, employing low-technology tools, as well as with soil preservation practice.[25]

As attested by several scholars (recently Van Joolen 2003; Kamermans 1993), land workability is the element that influences farmers' choices, rather than soil fertility. This is because, in the case of non-mechanised agriculture, land capability depends, as stated above, both on the possibilities offered by applied technology and on the quantity of labour required. Different technologies and practices, as well as the different integration between agriculture and husbandry (recently Van Joolen 2003; Nixon-Price 2001), would better exploit different areas and soils within the same region. This is true particularly in the Mediterranean area, which is characterised by a high degree of variability of soils, microclimate, etc.[26]

In the present work, the scale adopted to define land evaluation values, aiming at a starting point for a research framework for the definition and interpretation of the ancient landscape, allows us to consider two main general categories of agricultural practices:

- Hoe agriculture in the earliest periods (Neolithic)
- Plough agriculture, mainly with the wooden plough, with a progressive introduction of the iron plough in more recent periods (Final Neolithic to Late Roman)

Hoe agriculture was certainly accompanied by herding practices since Initial Neolithic (Perlès 2001: 73). Some scholars would suggest that along with farming and herding, gathering activities would maintain a role in the group's economy (Thomas 1990; Van Joolen 2003), but Perlès (2001: 16-17) for mainland Greece disagrees, noting how early farmers made little use of wild resources, as natural plant and animal resources do not seem to have influenced the location of Early Neolithic settlements. Regardless, landscapes less appropriate for cultivation (steep or moderately steep slopes with stony soil, or land with a marshy character – Perlès 2001: 10) were probably only partially protected and marked by rich vegetation, good for both gathering and grazing.

More suitable for hoe agriculture are light soils, not too deep (30cm), well drained on moderate slopes[27] and characterised by a low degree of stoniness. On the other hand, hoe agriculture is impracticable on heavy terrains (e.g. clayey textures) or soils either too well-drained or not drained at all, with too much or no water. It is also impracticable on completely rocky soils (extended surface rock, different from pebbles)[28] or on severe slopes, greater than 25-30%.

Many scholars seem to believe that hoe agriculture, probably along with the 'slash and burn' practice, was not accompanied by soil preservation measures (Neuenschwander and Peters 1988; Palmet al 2005; Bellwood 2005). If this was the case, very steep slopes, highly subject to erosion, would not be considered suitable for agriculture and were probably used for pasturage.

Among the classes of physiographical position defined and employed in this work, the most suitable for hoe agriculture would be P3, P4, H1, H2, H3, M0, M1 and partially M2.

Plough agriculture can be differentiated into i) a practice characterised by the use of the wooden plough, without fallow and with simple drainage technique, and ii) a practice characterised by the use of both wooden and iron ploughs and by short periods of field fallow, though

[21] We must consider, though, that terracing techniques for the retention of soil were in use since the Bronze Age. There is a long debate on this aspect for the Prehistoric period (Wagstaff 1992; Simpson 1992; Atherden 2000; Conolly 2002-4 among others) as well as dated examples from the Greco-Roman period (among others Lohmann 1992 –Atene Survey).

[22] A definition of stoniness can be found in FAO 1977 (55ff.) and is reported by Van Joolen 2003 (cap.2).

[23] See Forni -personal communication- in Van Joolen 2003:112.

[24] See White 1976 for the Medieval period, though many scholars do not agree.

[25] On the use of ploughing in earlier periods there is much literature available (Halstead 1995; Sherratt 1981; Pullen 1992). See also above, footnote 21, on terracing techniques.

[26] For the climate of Boeotia see chapter II.1.

[27] Slope classification for hoe agriculture: 0-25% (suitable); 25-55% (suitable with few limitations); >55% (severe limitations of use).

[28] For instance on upland slopes of central Helicon, above Evangelistria towards the Valley of the Muses (see chapter II.3.8).

without proper crop rotation (White Jr. 1976, Forni 1989a; Forni 1989b; Forni 1990; van Joolen 2003).

i) Without fallow and with a simple drainage technique (probably employed in prehistoric periods), the wooden plough has a limited potential. For this practice, as well as for hoe agriculture, light soils are more suitable, but deeper (at least 50cm) and resistant to pulverisation. The wooden plough 'crumbles' rather than ploughs. Generally speaking, wooden plough agriculture prefers the same terrain as that preferred by hoe agriculture, but with significant differences. It is characterised by a higher productivity potential, i.e. it allows for the cultivation of less favourable land (such as dry soils with simple irrigation systems), increasing cultivable areas, and increases soil productivity, favouring a better filling of the earth for seeds. The use of a wooden plough also allows one to cultivate the heavier soils of low river terraces,[29] while the use of the iron plough, certainly in use during the Archaic and Classical periods, would allow for a proper cultivation of alluvial valleys (see below). On the other hand, the best areas for the use of wooden ploughs without fallow practice are light soil land, marked by a low degree of slope[30] and stoniness. Unlike in the case of hoe agriculture, pebbles are in fact a strongly limiting factor for use of the wooden plough. Nevertheless, both practices tend to avoid extreme terrains, especially those too heavy or clayey, as well as steep slopes, difficult to cultivate with a plough drawn by man or animal, and also more subject to erosion processes.
Among the classes of physiographical position defined and employed in this work, the most suitable for wooden plough agriculture are P3, P4, H1, H2, H3, M0, M1 and M2.

ii) The use of iron ploughs (but together with wooden ploughs in the majority of cases) and practices of field fallow and rotation of crops, enlarges agricultural potential considerably, especially as far as the alluvial valleys and the lacustrine plains[31] are concerned, i.e. the most fertile areas and with the highest productivity potential (White 1976). Cereal cultivation, in particular emmer wheat (*triticum dicoccum*), can be extended. Heavy, wet soils can be worked by iron ploughs, which parallels a development of drainage and irrigation techniques, well described by Roman authors (i.e. Columella, *De Re Rustica* and Cato, *De Agricoltura*) and reported by Renaissance authors (i.e. Torello, *Ricordo di agricoltura*). At this stage, limitations to soil workability are linked to the socio-economic structures of the communities, rather than technology.
Among the classes of physiographical position defined and employed in this work, the most suitable for wooden/iron plough agriculture are P1-P2, P3, P4, H1, H2, H3, H4, M0, M1 and M2.

Within the present work, land classification processes have been applied using the first two criteria (soil classification for land capability and soil erosion) and taking into consideration the physiographical position. Proceeding to classification according to the third criterion (soil workability) would have required data collected on a different scale, and with a larger number of collected samples in different areas. Studying a large number of sample windows[32] in detail would probably be the best solution for this kind of research, but it is certainly time-consuming, and would probably need the support of a specialist soil scientist. Therefore, for the time being, information on soil workability remains at the level of an unclassified qualitative dataset which informed the general land classification (see below).

The resulting general land classification has been verified and corrected through fieldwork, through a more detailed study of some sample areas, of diverse lithologies, paying particular attention to the soils of the upland plateaus, hill slopes, foothills and talus areas, as well as of alluvial basins and major river valleys (Copais basin, plains of Thebes, Thisbi and Thespiae valleys, Tanagra basin).
The working hypothesis, checked, monitored and corrected this way, was compared with other dedicated studies, both on Boeotia (Davidson 1992; Davidson et al 1994; Shiel-Stewart 2007) and on other Mediterranean areas (Kamermans 1993; Van Joolen 2003; for similar work see also Wallace 2001 and Tomkins et al. in press).
For the present work, a first classification of land capability was created according to a simplified qualitative classification of soils, focused in particular on agricultural potential (FAO 1967; FAO 1968; FAO 1976; FAO 1977; FAO-UNESCO 1994; FAO 1998; Soil Surface Staff 1975; Soil Surface Staff 1975; see above for details). It is similar to that used by Shiel in its detailed study of Thespiae area (Shiel-Stewart 2007), developed by the Conservation Service of the US Department of Agriculture (Soil Surface Staff 1975; Soil Surface Staff 1975).[33] Its structure of eight land capability classes was simplified to four classes.

The defined four classes of land capability are:

F - Lands with few limitations restricting their uses (due to a tendency to wetness or dryness or very moderate erosion).

MF - Lands that have some limitations for agriculture (which reduce the choice of crops or

[29] See the middle/low Asopos terrace (appendix I.14 – THE TANAGRA SURVEY PROJECT).
[30] Slope classification for plough agriculture: 0-13% (suitable); 13-55% (suitable with few limitations); >55% (severe limitations of use).
[31] For instance the plains of Central and Eastern Boeotia.
[32] See as an example the work carried out in the Tanagra area (resumed in I.14 – THE TANAGRA SURVEY PROJECT and in Bintliff et al. 2006, and illustrated in detail in Farinetti-Sbonias in preparation).
[33] This classification adjusts land capability classes based on limitations due to erosion, wetness, rooting zone, and climate (Foth 1990).

require moderate conservation practices), due to wetness caused by different factors.[34]

LF - Lands that have more limitations for agriculture (which reduce the choice of crops or require moderate conservation practices), due to dryness and meagre soil depth which causes rooting problems or to high slope degree.

U - Lands that have severe limitations for agriculture (which reduce the choice of crops and require very careful management), due to very meagre soil depth and high slope degree, causing serious problems for access and rooting. In the majority of cases, they can be suitable for pasture, forestry, and grazing.[35]

Note: class MF and class LF produce the same effect in terms of limitations for agriculture, but should be separated due to the different strategies that can be used to overcome the problem (different conservation strategies). In a more simplified classification the two classes could also be merged.

A peculiar case is represented by FLYSCH formations, whose derived soils can be less or more fertile according mainly to the percentage of pebbles and rock components. For instance, flysch formations with a high schist component are characterised by a low degree of fertility (LF) and can be locally unfertile (U). In some areas flysch formations have been checked in the field during extensive fieldwork. These visits revealed that the fertility of flysch in the Boeotian landscape can also vary according to elevation. Therefore, the land evaluation layer was queried within the GIS according to elevation values.

Flysch below 600m: MF.
Flysch above 600m: LF (considering climate restrictions, i.e. slower soil formation as well as probable cold winters that might slow down vegetation growth).

Ultimately, the general classification of flysch as MF and LF according to elevation seems to be the most acceptable.

In some cases flysch land capability values were treated separately. In the peculiar micro-landscape of the Valley of the Muses, for instance, sampling in the field shows that flysch could be generally classified as fertile –F–, without exception. Through the attribution to this soil capability class, together with the attribution to a specific physiographical class (see above – PHYSIOGRAPHICAL POSITION; SOIL WORKABILITY) and the assignation of specific soil characteristics (soil stability for instance) one could obtain further information about a surface: for instance, soils on hill slopes are affected by differentiated erosion according to slope degree, etc. In the case of the Valley of the Muses, historical information supported the decision. This criterion cannot be extended though to other Boeotian landscapes, in some cases because flysch was checked in the field and proved to have a different behaviour, in other cases because soils could not be investigated adequately within the framework of the present research.

A map of land capability classes of Boeotia is shown in chapter II.1 (fig.3).
The classes in the map are applicable to both wooden and iron plough agriculture, though one should consider that the lower fertile areas, such as a bonified Copais basin or areas of the Theban plain, as well as the Kephissos basin and Asopos lower terrace in its lower course in the Tanagra basin, were ploughed with more difficulty with a wooden plough (see above – SOIL WORKABILITY).

Models based on land evaluation are *generic* (they can be applied to any area with a similar environment without reference to its archaeology) and *falsifiable* (Van Leusen 2002: 5.14), with archaeological fieldwork and/or core sampling. Certainly, I am aware that a proper land evaluation study would require a larger investment in palaeo-geographic reconstruction, also by means of systematic core-sampling programmes as well as palaeo-botanical analysis for palynological reconstructions. On the other hand, as already mentioned, for the purpose of assessing a landscape situation, the datasets that were available, and our limited fieldwork and core sampling, might perhaps constitute a methodology suitable for archaeologists involved in similar studies, which do not have available, as in this case, detailed pedological studies applied to so-called 'traditional' agriculture.
In any case, we should keep in mind that the landscape derived from modern maps and terrain visits is the landscape of today, a result of all the past transformations and interactions, and all the methods for reconstruction of past landscapes deserve more attention. Within the framework of the present research, an attempt has been made in this direction only in the case of peculiar landscapes, such as the Copais basin. For this, an attempt at reconstruction of the past landscape in different periods could be made (see chapter II.3.1 and Farinetti 2008). Also in the case of intensively studied windows, such as the Tanagra region, a much more detailed study could be made, which could lead to an analytical reconstruction of the recognisable element of past landscapes.

A definition and a map of land capability classes built in this way could therefore constitute a first degree of improvement of the interpretation of environment and soils simply related to the natural elements of quality of soil in relation to agriculture (pedogenetic and morphogenetic factors), which are often carried out in archaeological landscape work. This interpretation, even if basic and correct, carries the risk of simplifying the relationship soil type - most suitable cultivated crops. The

[34] Such as a superficial aquifer or a high clay percentage.
[35] Only few areas would correspond to the Conservation Service of the US Department of Agriculture's 8th class of land capability (rough land, even for woodland or grazing, and limited use to wild life food or cover and watershed) and therefore in the simplified classification used for this research categories have been merged. The U class is comprehensive of the 5th to 8th categories.

example reported in the chart below (table 2) may constitute an attempt in this direction, and could be used as a basis to examine the agricultural structure and potential of a past landscape, being conscious though, as its author notes, that it should be enhanced with a careful examination of cultural, technical and socio-economic factors involved.

In particular, for Boeotian soils, Shiel and Steward (2007) evaluated the agricultural potential of soils in the area of ancient Thespiae. The chart below (table 3), summarising Shiel's results in matching soils to crops, can complete the table above taking into account specific local soils. The categories used in table 3 can be applied to soils and topographical units of many areas of Boeotia, and generally correspond to our LF, MF and F land evaluation classes (see above in this chapter).[39]

EVALUATION	**high**	**medium**	**low**
CEREALS[36]	Soils with fine texture, deep or with limited depth, presenting low risk of hydric stress	Soils with medium or rough texture, deep or with limited depth, presenting a moderate risk of hydric stress	Soils with a meagre depth, or with limited depth and rough texture, presenting a high risk of hydric stress
OLIVE TREES[37]	Soils deep or with limited depth, not influenced by aquifer	Soils with a meagre depth, not influenced by aquifer	Soils influenced by aquifer (either permanent or temporary)
For grapes[38] the same criteria as for olive trees can be considered.			

Table 2. *Soil evaluation criteria for production of cereals (mainly wheat) and olive trees (after Arnoldus-Huyzendveld 2008).*

SOIL	**arable crops**	**olives/grapes**	**vegetables**	**fruit trees**
dry clay/calcareous shallow stony soils	moderate	high	poor	high
clay soil with silt and sand and good water capacity	high	high	poor	high
clay soil in foothill zones and streams with good water capacity	moderate	moderate	fairly high	low
clay/loam soil with good water capacity (flood plain)	low	low	high	low

Table 3. *The chart resumes the results of Shiel and Steward's work (2007) in the area of ancient Thespiae in Boeotia for soil evaluation criteria for production of different crops.*

[36] Mainly wheat, without differentiation.

[37] Tassinari (1976: 798): for olive trees, very compact, wet or very loose soils are not suitable.

[38] Tassinari (1976: 750): for grapes, soils of different natures can be suitable, and only excessively wet soils are not at all suitable.

[39] The F class is comprehensive of the two last soils in table 3, including foothills, lower plains and stream valleys.

I.2.2

The archaeological/cultural datasets and the research methodology

DEALING WITH THE ARCHAEOLOGICAL RECORD

The archaeological information available for the region of Boeotia comes from bibliographical records and archaeological reports, mainly concerning topographic research, excavations and rescue excavations carried out in the region, as well as intensive artefact surface survey results (for details see chapter II.2).

The starting point for the collection was Fossey's book (Fossey 1988), currently the richest source of information as far as extensive research on ancient settlements in Boeotia is concerned. It offers an excellent survey of the material published by the mid 1980s.

Following Fossey's general structure, data are gathered according to *chorai* (fig.1). This implies an approach to the areas of influence of the ancient *poleis* (as they appear in the Classical period) as a focus for analysing the settlement system. Due to the marked territorial character of the ancient Greek *polis* (see Snodgrass 1987/9 among others), and the 'village territory' character of the settlement in the majority of Greek regions and in the majority of historical periods (Kirsten 1956 and Bintliff 1994), for 1200 years of history the model of interaction between the city and its territory can be considered as effective and as having had a relatively strong influence on the natural settlement pattern. Therefore, there is a methodological reason for gathering information and structuring the research according to *chorai*, enhancing peculiarities and detecting similarities. This also facilitates the analysis of the lower rank sites mainly directly related to the *polis*. Furthermore, there are differences in intensity and quality of research in the various *chorai* affecting the available data and the knowledge of landscape and settlement history.

Generally speaking, according to the kind of archaeological evidence being dealt with here and its distribution in space, a suitable methodology for data recording is required. A system has been implemented to deal critically with the different levels of complexity of both the information and the archaeological evidence available from the bibliography (in a broader, regional perspective) and from more recent systematic artefact surface surveys, which intensively cover the landscape in search of all traces of human activity in smaller defined areas. The system should also be able to deal with a large chronological span (over 7000 years, from the Neolithic to the Late Roman period) as well as with multi-period sites.

The archaeological information encountered, which was recorded and collected by previous scholars in different periods and within different research frameworks, is characterised by a high level of inhomogeneity and by inner biases which are a challenge to uncover. Compared to environmental layers, whose spatial resolution as well as quality of data can differ, but which are usually characterised by stronger linearity and consistency (see chapter I.2.1), it is much more difficult to give to archaeologically-meaningful entities as well as social variables a coherent spatial form or a consistent structure. In particular, the archaeological record is marked by a high degree of non linearity.

Despite this, as if the problem did not exist, many of the current archaeological GIS projects often contribute to an indiscriminate production of maps simply corresponding to a database containing a discrete representation of archaeological data, often acritical, due either to a lack of prior theory or hypotheses about the problems the archaeologists want to resolve or especially to a lack of detailed monitoring of the available record. Thus, including the largest quantity of data and data layers into a map seems to be the main goal, without always paying adequate attention to the alphanumeric information attached.[1]

Archaeologists used to mechanically define social activity areas from evidence and/or artefact concentrations discovered on the ground, but I would suggest that in order to carry out meaningful analyses, the archaeological information and evidence would need to be evaluated in terms of what they really represent before associating evidence into landscape-meaningful social entities. Within GIS systems, the archaeological entities often do not carry with them information about their quality, precision, or even at the very least about how the information was collected and recorded. In other words, good metadata[2] information is lacking in many archaeological GIS projects, and this is sometimes due to a general disregard of this kind of information, or sometimes, as is often pointed out, it is due to the balance between quality and cost in terms of time, since any metadata collection operation is time consuming.

---

[1] We should be aware, though, as Gillings and Wheatley point out (2002: 16-17), that we "*are rarely interested only in where things are*"; on the other hand, we "*often have a wealth of information about the archaeological entities under study, and we should be able to deal with it in a suitable and coherent manner, according to our research aims*".

[2] On the concept of metadata see chapter I.1.1, and Gillings and Wheatley 2002: 86.

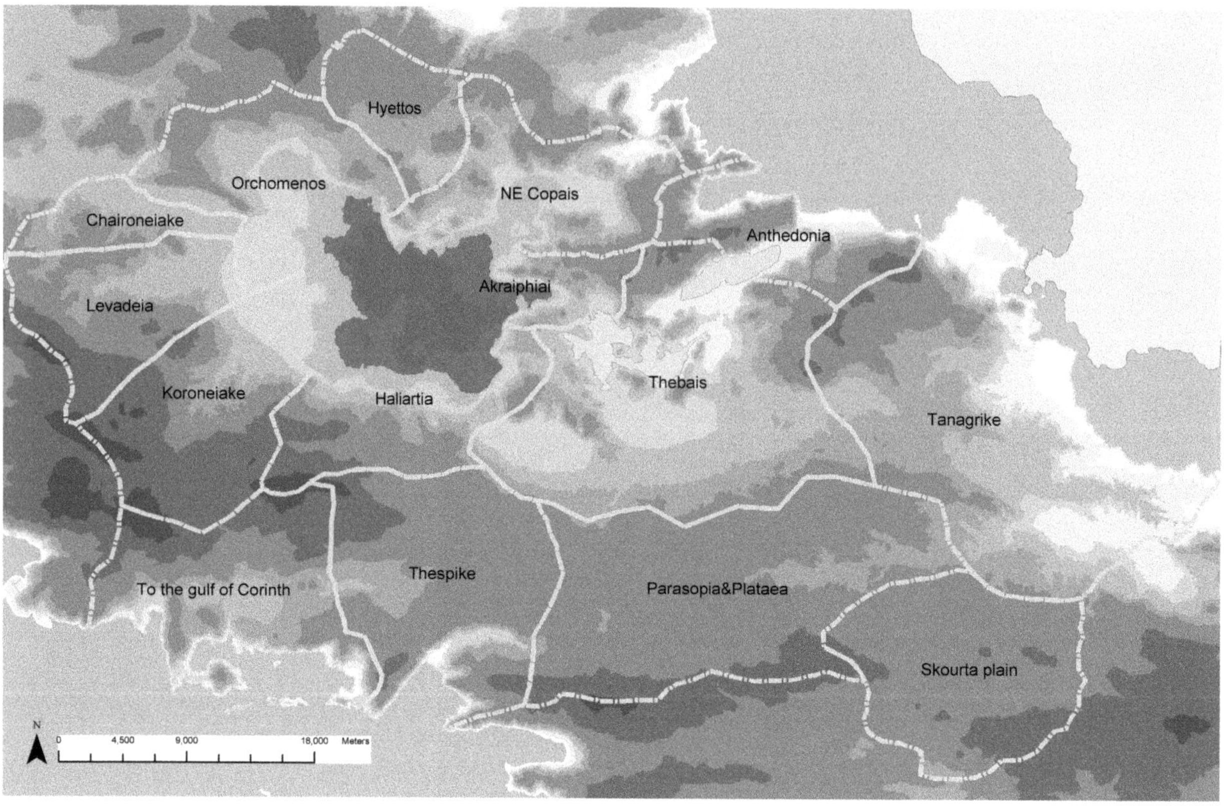

Fig.1. *Ancient Boeotia and the chorai (ancient polis territories)*

Thus, it is important to find a way to deal meaningfully with the uncertainty, fuzziness and inconsistency central to the archaeological record. Using GIS would enable researchers to move "*beyond the dots on the map*" (as in the title of Bintliff 2000b), monitoring the available archaeological record and its quality, as a starting point towards the performance of high quality spatial and landscape analyses aiming at a full understanding of the archaeological and social landscape. Consequently, the process should therefore begin at the phase of data collection and recording.

THE DATABASE STRUCTURE

Deconstructing information and reassessing known data

In order to deal with, and in an attempt to integrate, those inhomogeneous, non linear datasets, we would deconstruct the available archaeological record, through a process of strong source-critique, and reconstruct it in terms of meaningful units of social activities in the landscape, extracting the meaning from each individual component to best assess the huge quantity and variety of archaeological information available, taking into account the non-linearity and incoherence of the data set.

In the bibliographical record, the archaeological evidence is described, either in a general or in a more detailed form, clustered in different ways, but mainly according to geographical location and/or proximity to the main settlement in Greco-Roman antiquity. In archaeological reports the evidence is mentioned with an indication of its location, which is usually in strict relation to the modern settlement network. In this book, these general indications of location, and therefore groups of information, will be called *sites*, as they are in the majority of bibliographical records. Therefore, within the system, units of archaeological evidence, as noted in the field and reported in the bibliography, are collected according to '*sites*', as mentioned in the bibliographical record, which usually correspond to general modern indications of location (or traditional names). Following that, in the framework of a constructive critique, a process of deconstruction of the known archaeological information is carried out. This, therefore, also becomes a deconstruction of the notion of *site*, as conceived – in the majority of cases acritically – over years of archaeological research. Thus, a process of deconstruction of the archaeological information into the minimal '*units of archaeological evidence*' recognisable at this scale is carried out through a 'positive' critique. Afterwards, the known data managed this way are reassessed into a more coherent information structure, the system allowing for the formation of '*components*', i.e. 'archaeologically relevant spatial aggregates',[3] archaeological components of the landscape, which can then be used for further analysis (see scheme in fig.2).

In this way, information from the bibliography is deconstructed and the known data reassessed into a more coherent information structure, and special attention is

[3] According to Dunnell's definition (1992; see below in the text).

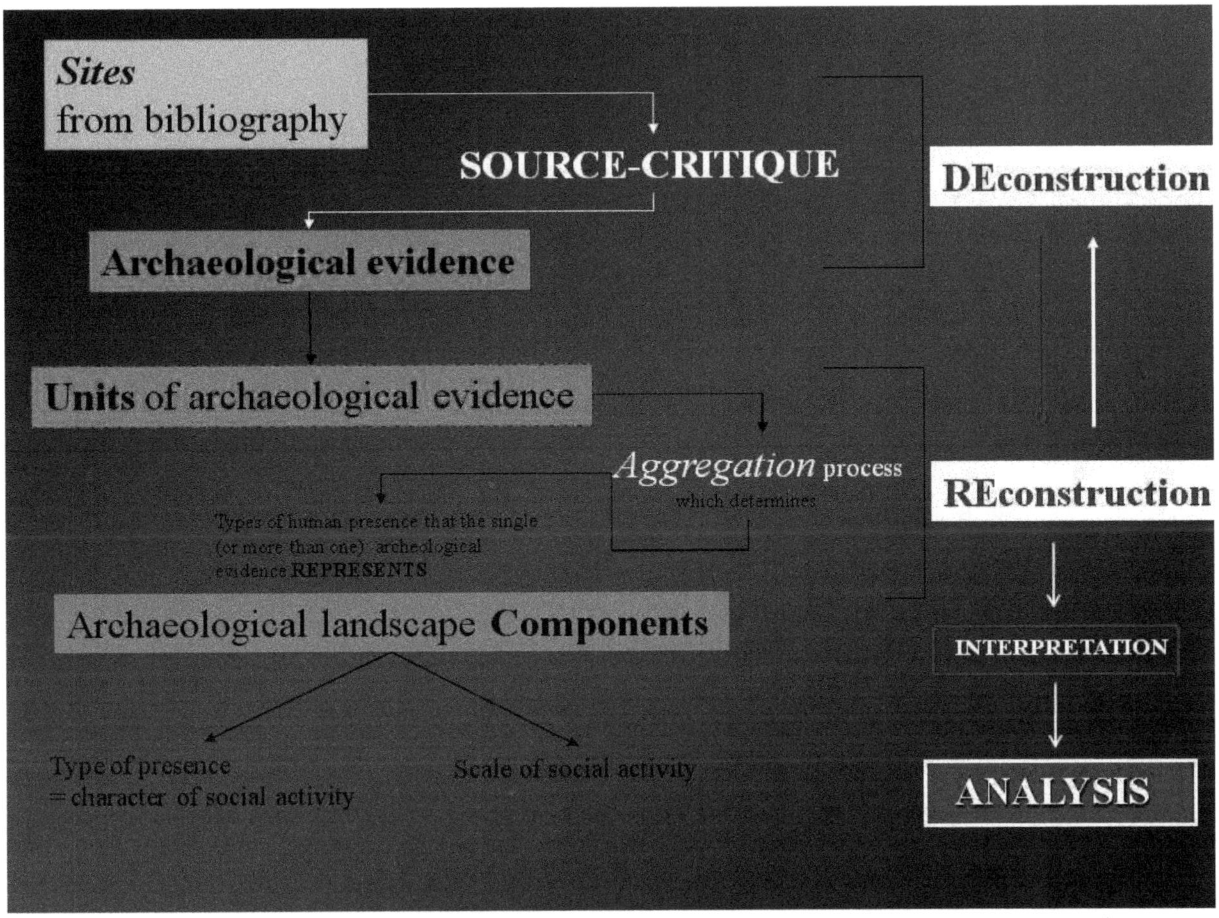

Fig.2. *Model of deconstruction and reassessing of the archaeological record*

paid to recording the type of source of each piece of information collected, in an attempt to carry out a coherent and structured source critique (see below). After the DEconstruction and REconstruction process, we would be able to better separate the '*sites*' (activity foci) in terms of the level of human activities they may represent, and in this way deal with the critical issue of *representativeness* and of valuing the information in terms of what it represents.

Unlike the traditional aggregations into sites, which, since the archaeological record is a non-renewable resource, may lead to the loss of information and biased destruction of the archaeological record,[4] our *components* are always renewable, updatable and rethinkable. The system allows us to re-compose the units into different meaningful entries, either once new variables are involved and our knowledge increases, or when different individuals deal with the interpretation of the same archaeological evidence. With credit due to the digital archives and opportunities for the updating of GIS systems, we can always modify and recreate constructed aggregations (as long as it is remembered that they are a construction), and the units of social associations can always be deconstructed (and destructured) again into the individual features (units of archaeological evidence), which can be, if needed, aggregated again in a different way.

By reinterpreting traditional sites, I try to deal with the archaeological evidence without feeling frustrated by the notion of site. On one hand, the siteless approach is critically applied while dealing with the archaeological evidence; recognising, collecting, recording and describing the single minimal units of archaeological evidence, known from the visit in the field, as part of an "*extensive and fluid social landscape*" (Bruck – Goodman 1999b: 11). Contrastively, by means of a critical examination of the archaeological information, the methodology allows for the aggregation of units of archaeological evidence into units of social associations (called '*components*'), which have a coherent archaeological meaning and can be made available for further analyses of the social landscape, in which topographical and natural features, animals and humans each play a role in the creation of cultural meaning (Bruck – Goodman 1999b: 11).

Through the deconstructing-reconstructing procedure illustrated above, we could get to the definition of '*site*' as a cluster formed by the composition of more social units (*components*) and their mutual relationship, keeping in mind that the aggregation process is, of course, entirely interpretative. Misinterpretations arise when archaeologists, talking about and employing 'sites' and archaeological aggregations, forget that they are simply

[4] Dunnell (1992: 36) states how the notion site not only biases our understanding of the human past, but it is also rapidly leading to biased destruction of the record, forever impairing our understanding of the human past.

archaeologists' constructions, a constructed reality, and consider the sites as archaeologically relevant, empirical units that exist independently of the archaeologists.
Conversely, in the process of our research, we would need 'objects' (although they will always be the partial - imprecise real) on the basis of which to carry out analyses, and make our analyses comparable with other projects or research works whose focus is the *site* (even though not always constructed in a critical way). As Dunnell writes (1992: 33): "*We require a series of units, not just a single unit, of association within which counts and patterns are archaeologically meaningful*". The construction of archaeologically relevant and interpretable aggregates is, therefore, required to some extent. As Dunnell continues (1992: 34), "*this view proceeds building up to, rather than dividing, spatial aggregates of interpretive significance*".

According to this 'constructed components' approach, the 'sites' cease to be considered as empirical (archaeologically relevant) units that exist independently from the archaeologists, but become, in the form of archaeological *components* of the landscape, the basic unit of the archaeological interpretation (or 'inferential association'). On the other hand, the *units of archaeological evidence* can be seen as the basic units of archaeological knowledge recording.
With reference to Dunnell's critical approach, whilst our units of archaeological evidence are existing features – empirical units which can be seen, discovered, described – our constructed components will not be existing empirical units which can be discovered and described (in that way, we would make the same mistake of some archaeologists with their sites – see, for instance, Dunnell's disagreement (1992: 29) with Goodyear *et al.* 1979: 39), but constructed features, whose critical interpretation can lead to the extraction of some archaeological meaning. For instance, as many years of research on off-site archaeology have shown, it cannot be ignored that only by linking total landscape sherd scatters and their chronology could one perform possible aggregations into sites and clarify other phenomena, such as a sherd on a site possibly belonging off-site from another era. Evidence comprising only one or a few sherds, therefore, could be interpreted either as constituting a unit of an archaeological component of the landscape, or as evidence of off-site landscape use (see below – SOURCE CRITIQUE, ATTRIBUTE FUZZINESS).

A relational database was implemented (in a Microsoft Access 2004 environment) in order to record the archaeological information available, and it was then included in the GIS system, realised in an ArcGIS 8.1 environment.
The final development of the reassessment of known data is at the mapping and spatial level, when environmental, geoarchaeological, geographical, and topographical landscape data help us in the interpretation of the components.

Phase I. *Deconstructing*: the Units of Archaeological Evidence.

A *unit of archaeological evidence* is anything that derives, through various processes, from past human behaviour and/or activities, either at the level of simple human presence or at different levels of activity focus, and it is characterised by a location; a result itself of human behaviour and human agency. It is, in Dunnell's words, "*a unit within which association is observational rather then inferential*" (1992: 33).[5]
The notion of a unit of archaeological evidence employed here is close to the definition of '*site*' given by Hole and Heizer (1973: 86), and followed by modern textbooks: "*Site is any place, large or small, where there are to be found traces of ancient occupation or activity. The usual clue is the presence of artefacts*". For us, though, the presence of artefacts does not immediately define a site, as they might, for instance, be related to off-site material. The units of archaeological evidence are therefore recorded first, and interpreted at a later stage (see below).
In the present study, in practice, a unit of archaeological evidence is a certain artefact (or a group of similar, homogeneous artefacts – see below) that exists today or could be seen by archaeologists or other scholars in a certain location. Any ancient feature that one may encounter constitutes archaeological evidence; it is what has remained to us from the past. Therefore, the recording of known archaeological evidence can be characterised as an observational and descriptive process (rather than inferential).[6]
Within the database, each unit of archaeological evidence is recorded as a record in the table [Arch_Evidence]. The system allows one to choose among different types of archaeological evidence, and therefore to enter different properties for each unit of archaeological evidence: type, location, chronology, detailed chronology, excavated yes/no, information on excavation, known by means of intensive survey only yes/no, 1st mention of the unit of archaeological evidence, type of presence.

Evidence type: the field registers the type of archaeological evidence, among a given list of possible types of evidence. The values included in the list, and recalled from the [Arch_Evidence_type] look-up (reference) table have been updated according to requirements during data entry for the present work – see table 1.[7]
Location: the field records a detailed location as reported in the bibliographical record.
Chronology: the field records the period to which the evidence can be attributed, among a given list of

[5] See Dunnell 1992 for a definition of artefact: "*anything that displays any attribute, including location, as a consequence of human agency*" – cfr. Spaulding 1960: 438; Dunnell 1971: 117.
[6] "*It is impossible to undertake archaeological analysis without grouping the data and so categories must be assigned. However, it would be naive to assume that the process of categorization can be undertaken in an entirely objective manner*" (Bruck – Goodman 1999b: 6)
[7] The question mark available for the record of each entry refers to the chronology assigned to the evidence rather than to the attribution of type.

agricultural production	Evidence that concerns agricultural production (e.g. grinding stones, beehives).
architectural fragments/building material	Architectural fragments (such as fragments of a frieze, columns, or building decoration) or building materials (building stones, bricks, or carved or squared blocks).
artefact concentration	Mainly pottery (but could also be lithic) from INT survey or from certain presence of human activity noticed by EXT survey. Not only quantitative but also qualitative.
scattered sherds/material	Surface sherds from EXT survey, mainly when density is unknown or cannot be quantified.
surface walls	Remains of walls or foundations visible on the surface.
isolated finds	Movable artefacts found on the surface but not associated (or immediately associable) with a context: e.g. a coin, a terracotta figurine.
epigraphical evidence	Inscriptional evidence.
excavated structures/layers	Unidentified excavated structures. If they can be identified then they become the corresponding Evidence type and the box Excavated yes/no will be ticked.
cemetery	Burial evidence, comprising more than one tomb or burial evidence.
isolated tomb/tumulus	Burial evidence, comprising only one tomb or a tumulus feature.
bath	Isolated bath buildings, visible on the surface or discovered during excavations.
boundary stone	*Horos*, inscribed boundary stones.
fortification	Remains of fortifications.
infrastructure	Infrastructure features, such as signs of a road, a bridge, an aqueduct, or a water channel.
mining	Signs of a special activity (mining).
quarry	Signs of a special activity (quarry).
reused	Architectural fragments/building material or inscriptions reused in later structures.
unknown	

Table 1. *Type of Archaeological Evidence.*

chronological periods. The values are recalled from the look-up (reference) table [Chronology] – see table 2.

Detailed chronology: the field records a more specific chronology than that recorded in the *Chronology* field, according to the most common divisions, as known from the bibliographical record (e.g. Early Hellenistic, 4th century BC).

Excavated yes/no: whether the evidence was discovered by excavation or not.

Information on excavation: information on the excavations which revealed or provided more detailed evidence.

INT only yes/no: whether the evidence is known from intensive field-by-field survey only, and not previously noticed by surveyors of the topographical tradition or during extensive survey work.

1st mention: the source which first reported the existence of the unit of archaeological evidence. This field is important for the source critique process, as will be clarified later (see below).

Type of presence: the field records the type of presence represented by the individual archaeological evidence itself – as much as our knowledge permits us – without considering any possible association with other units of archaeological evidence. The possible values are: 'human presence'; '? human presence'; 'activity focus'; '? activity focus' (see below and table 3). The question mark in front of the entry represents either an uncertain attribution of the evidence to a period, or an uncertainty inherent in the archaeological information.

Notes: the field allows the input of a discursive brief description of the evidence, with mention of the main bibliographical references.

Period ID	Periods	From	To
N	Neolithic	6800 BC	3300 BC
EH	Early Helladic	3300 BC	2100 BC
MH	Middle Helladic	2100 BC	1550 BC
LH	Late Helladic	1550 BC	1100 BC
SubM	Submycenaean	1100 BC	1050 BC
Preh	Prehistoric		
PG	Protogeometric	1050 BC	900 BC
G	Geometric	900 BC	700 BC
A	Archaic	700 BC	480 BC
C	Classical	480 BC	323 BC
H	Hellenistic	323 BC	31 BC
R	Roman	31 BC	330 AD
LR	Late Roman	330 AD	600 AD
gr-rom	Greco-Roman		
unknown			

Table 2. *Chronology*[8]

As seen above, the descriptive recording process of known data is accompanied by critical thoughts on the source(s) from which the information has been taken, and on the actual value of the evidence being recorded (see below also). Afterwards, yet within the framework of the archaeological evidence level, a further step can be taken which leads to the stage of the interpretation of the entries. The '*Type of Presence*' field, associated with each individual Archaeological Evidence entry, assigns to the evidence a general character which is the result of an interpretative process. The unit of archaeological evidence represents by itself a certain or possible human presence or activity focus. The known evidence is examined, weighted, and assigned a character (i.e. a scale of human presence that a certain piece of evidence might

[8] The chronological chart is based on general periodisation (Perlés 2001, Dickinson 1994, Alcock 1993). Absolute dates are only indicative.

represent if considered on its own) among the values listed in table 3. This will give a general impression of the area under study, which would show the overall distribution of human presence and human activities across time, at different levels of complexity.

human presence	From the evidence we cannot infer more than a human presence (e.g. 1 or 2 sherds).
? human presence	Either the chronology of the human presence is uncertain, or the evidence is comprised of isolated finds, characterised by a high level of mobility.
activity focus	Certain activity focus, whatever degree or level.
? activity focus	Either the chronology of the activity focus is uncertain, or we cannot distinguish, on the basis of the evidence itself, between a human presence and an activity focus level.

Table 3. *Type of presence.*

Phase II. *Reassessing*: Components and Activity Loci.
According to the approach discussed above, the aim was to develop a method of constructing units of archaeologically meaningful, inferential associations (*components* of the archaeological landscape) from smaller-scale observational units (*units of archaeological evidence*) – Fig.2. In the artificial formation of *components* (aggregation process), different units of archaeological evidence that can be spatially and logically related to each other are linked together. These would form a *component* of a certain type. The process is not characterised by a high level of automation, since it must be carried out critically and it must be revised many times, keeping in mind that the historical relatedness of the component's constituent elements (units of archaeological evidence) is highly variable and not always directly correlated with spatial proximity, and that "*objects found in spatial proximity may have* [...] *entirely unrelated histories that preclude a simple equation between spatial proximity and systemic relevance*" (Dunnell 1992:29).

Dewar (1986), discussing New England settlement, argues that 'sites' (concentrations of artefacts) should be distinguished from 'components' (artefacts from the same period) and from 'occupations' (artefacts from the same use). Our *components* were constructed taking into account, as we saw above, Dewar's components (archaeological evidence from the same period) and occupations (archaeological evidence from the same type). This is as far as the vertical dimension is concerned. As for the horizontal integrity, which for a site Dewar takes as a given, we used mainly a proximity criterion, but always keeping in mind what we stated above: two things/objects close to each other (in spatial proximity) are not necessarily assignable to the same component and may have entirely unrelated histories. Therefore, we also tried to establish historical connections between deposition events (even though this is much more difficult when dealing with extensively and not systematically collected data).

In practice, the units of archaeological evidence are all listed in the database form: 'aggregations'. They can be virtually picked up and located into a 'component in process'.[9]

A preliminary filter was created in order to select only the records of the [Arch_Evidence] table to which were assigned the same 'Chronology' value. This, together with other filters that one may apply (by 'Site', for example, or by 'Location' data on the form), helps the process of aggregation. Otherwise, the process is as much as possible independent from either any preconceived aggregation process or site identification.

In the process of aggregation, the mapping of each unit of archaeological evidence is very important. Some of the units are very mobile and can only be assigned with caution to a component (e.g. epigraphical evidence, coins, isolated finds, reused pieces, or architectural fragments).[10] Some other units of archaeological evidence simply belong off-site (see above), and they are not assigned to any specific archaeological component either.

Once a component is formed and the corresponding record created, the system allows the user to enter different properties for each component (chronology, detailed chronology, component type, excavated yes/no, known from intensive survey only yes/no, component status, details, other notes), including the research framework within which the discovery occurred.

Component ID: the ID code of each component, constituted by the code of the *chora* it belongs to spatially and a progressive number assigned by the system.

Component counter: a unique progressive number assigned by the system to each component.

Chora ID: the ID code of the *chora* within which the component is spatially included.

Toponym: usually the field records the traditional name, as known from the bibliography. When this is not available, the actual toponym of the geographical/topographical feature as appearing on maps (1:50,000 or 1:5,000) is recorded. In this way we can group the components according to the logic of association which would have been used in traditional studies to define a 'site', and we can recall a component with a text meaningful name referring to location, without necessarily referring solely to numeric codes, which are often confusing. Queries on the basis of this field would help in individuating the continuity or the recurring of occupation of a certain place. In this way, the toponym represents an *activity locus* (place of concentrated evidence), much closer to the traditional concept of 'site'. In each *chora* chapter, these *activity loci* are indicated by

[9] In the aggregation process different units of archaeological evidence that are logically or spatially related to each other and which correspond to the same component type are linked together. The network of spatial relationships would link social units, and would have a social meaning. We could analyse the social space on the basis of the different levels of human activities we can detect in it and their mutual relationships (spatial and logical/cultural) - see below in text (GIS and social landscape).

[10] Reused elements, as well as infrastructure evidence or inscriptional evidence, are not always associated with a archaeological component. Some of them remain simply units of archaeological evidence recorded.

a number on the archaeological map, and listed with their related *components* in the table linked to the map.
Type: the field registers the type of the component, from among a given list of possible types. The values included in the list, and recalled from the [Component_type] look-up (reference) table, have been continuously updated during data entry for the present work – see table 4. The question mark before an entry refers to the period of occupation rather than to the attribution of type.[11]
Details: details on location.
Position: the field gives an indication of the accuracy of position on the map. A list of choices is given, recalled from the [Position_type] look-up (reference) table: precise (GPS positioning, exact coordinates), certain (when the component has been positioned on the 1:50,000 map, and therefore the position carries the accuracy level of the map), approximate (when the exact location of the component is not known with precision), imprecise (see table 5 and *Spatial fuzziness*, below in the text).
Chronology: the field records the chronological period to which the component can be attributed, among a given list of chronological periods. The values are recalled from the [Chronology] look-up (reference) table – see table 2.
Detailed chronology: the field records a more specific chronology than that recorded in the *Chronology* field, according to the most common divisions, as known from the bibliographical record (e.g. Early Hellenistic, 4thC BC).
Discovery: the research framework within which the discovery occurred. A list of choices is given, recalled from the [Discovery_type] look-up (reference) table (see table 1), illustrated and discussed in chapter II.2. It constitutes the key field for the source critique process (see below).
Excavated yes/no: whether the evidence was discovered by excavation or not.
INT only yes/no: whether the evidence is known from intensive field-by-field survey only, and not previously noticed by surveyors of the topographical tradition or during extensive survey work.
Other notes: some remarks on the component that cannot be inserted in a structured way in the other fields. During the aggregation process this field is required for noting down thoughts and remarks.
Status: in the field can be noted, when available, information on the 'status' the component has within a certain component type. For instance, for settlement components the 'status' could be a *protopolis*, a *polis*, a village, a hamlet, etc.; for burial sites the 'status' could be a cemetery, a small rural burial site, etc.; while for cult places the 'status' could be a sanctuary, a small cult place, a simple cult installation (altar), etc.

settlement[12]
burial place
cult place
rural site/activity
fort
infrastructure[13]
special activity[14]
activity focus
human presence
? settlement
? burial place
? cult place
? rural site/activity
? fort
? Infrastructure
? special activity
? activity focus
? human presence

Table 4. *Type of Archaeological Landscape Component*

Position Type	Position Type Description
precise	GPS positioning or exact coordinates
certain	located on 1:50,000 or 1:5,000 map
approximate	imprecise possible location: a possible area is known, but without enough precision
imprecise	known existence but unknown location (ex: epigraphical evidence - probable provenance from the area of a museum object - unknown original findspot)

Table 5. *Type of position.*

Constructed as such, a *component*, unlike the sites in the bibliographical record, is no longer a place distinguished by artefacts (or a grouping of more places in spatial proximity and distinguished by the presence of artefacts), but it is comprised by the artefacts themselves and their spatial relationship. In a way, this approach takes its inspiration from the definition of site expressed by Binford (1964:431): 'The site is a spatial cluster of cultural features or items, or both. The formal characteristics of a site are defined by its formal content and the spatial and associational structure of the population's cultural items and features present'. Such a notion of site can be accepted though only as long as it is borne in mind that the spatial component is a feature that the archaeologists are constructing, and it does not –and cannot - exist independently of them, as discussed above (see Dunnell 1992).
We should now ask ourselves: what role should such a *component* play in our archaeological landscape research? The issue is discussed later in the section GIS AND THE SOCIAL LANDSCAPE. Here, I should remark that it was decided to use the term 'components' (rather than

[11] The question mark before the component type value is related to an uncertainty concerning the chronological attribution. In cases where I could not be sure about an interpretation of a component type (component character), I usually recorded the component as a simple 'activity focus', noting the possible character in the field Status (and mentioning the hypothesis in the Notes field and in the text).

[12] The meaning of the 'Settlement' component type is dependent on the chronological period considered. In the Greco-Roman period, it can mean a nucleated site (of either urban or rural vocation), while in the Prehistoric period, it can mean a habitation site (in both cases at different levels, specified in the '*status*' field, if known).
[13] a road, a bridge, an aqueduct, a water channel, etc.
[14] a working area, a production activity, a quarry, a pottery workshop, a limekiln, etc.

'sites') in order to make clear their relation to a wider context, as inferred activity foci of a particular kind in a particular period, as archaeological components of the landscape, as well as their being constitutive parts of more complex activity loci, the latter being much closer to the traditional concept of site, as seen earlier (*Toponym*). As we have seen above, I use the components as a "*merely synthetic construct created by the archaeologists to deal with varying spatial distributions of artefactual remains*" (Goodyear et al. 1979: 39, as definition of 'site').

Discovery_Type	Discovery_Type Description
Ext. topographical surveys	Site-oriented extensive research.
Int. topographical surveys	Site-oriented intensive research not involving artefact surface survey.
Int. and sys. topographical surveys	Research that covers the landscape looking for sites but without counting surface potsherds systematically.
Int. and sys. artefact surface survey	Intensive and systematic artefact surface surveys' reports and/or preliminary data.
Historical geography	Mainly research focused on the search for *poleis*, town and village settlements and cult places.
Rescue excavation	Archaeological reports are the main source of information on rescue excavations, as well as some conference proceedings.
Personal or group interest	Personal or group research interest: site-oriented / period-oriented extensive research; problem-oriented research; systematic excavation.
Travellers	Early travellers of the 18th and 19th (and beginning of 20th) C. (e.g. Leake, Gell, Ulrichs, etc.)
Other	Mainly cases in which evidence for certain periods is discovered by archaeologists interested in different periods or particular site types while investigation is in process.
accidental	Accidental discovery, mainly from field ploughing or tomb robbing/illegal excavations.

Table 6. *Type of discovery.*

Relationships

The database set up for this work is relational, in order to better fit the needs of the GIS system, built itself on a relational basis.

In a relational database a *relation* is a table which assigns a series of attributes (stored in *fields*) to each object of a particular class/group. Each recorded object with attributes assigned to it forms a *record*. For each record we can assign a value for N variables (the fields). The number of attributes assigned to an object in a table is called the *information degree* of the table. We have seen above, for instance, the attributes which have been assigned to each archaeological evidence unit and to each component.

Within the database we can have as many relations/tables as we want (in our case, [Arch_Evidence] and [Components] constitute the main tables, along with look-up/reference tables mentioned earlier). Each of them is linked to other relations by means of *relationships* of different kinds (1:1; 1:N; N:M), in order to deal with the data to maintain a 'referential integrity' of the data themselves (for instance, we are not allowed to add information containing an object into a table if there is no record concerning the same object in a table associated to the one we are working on) and to easily carry out cross queries.

The picture in fig.3 shows the relationships which link the main relations/tables within the database.

The table [Chorai] is linked to the table [Sites] with a 1:N (one to many) relationship: each site is contained spatially in a *chora*.

The table [Sites] is linked to the table [Arch_Evidence] with a 1:N relationship: the units of archaeological evidence are the logical results of the 'deconstructing' process of the archaeological record available and usually recorded by site in the bibliographical record.

The table [Components] is linked to the table [Arch_Evidence] with a N:M (many to many) relationship: in the reassessing process, each unit of archaeological evidence is assigned to one (in the majority of cases) or more components (see above in text). The table [Components&Arch_Evidence] is a connecting table which helps manage the N:M relationship between the two tables [Arch_Evidence] and [Components]. A component is formed by many (more than one) units of archaeological evidence, and the same unit of archaeological evidence can belong to more than one component (e.g. some surface walls can belong to a necropolis as well as to the associated settlement, at least as long as we cannot be sure to which component they actually belong). In the case of uncertainty of attribution, a question mark is ticked next to the attribution of the unit of archaeological evidence to the component.

The core of the whole archive is the [Arch_Evidence] table, which stores all the units of archaeological evidence itself. At the first stage of the process, these units are assigned to sites (according to bibliographical records or other sources); they then, metaphorically speaking, 'move' to the components. Ultimately, there will be four different archives stored in tables: the archive of 'bibliographical' sites; the archive of all the deconstructed units of archaeological evidence; the archive of the reconstructed components; and the archive of the wider activity loci, resulting from a clustering of the components. The four archives are linked to each other, but can also be read and queried independently.

Structured in this way, the system allows us to structure our information at more levels, each related to the others, simplifying the complexity and attributing metadata information in each stage of the process. In other words, it allows reasoning on collecting, recording, managing, analysing and presenting our archaeological spatial data

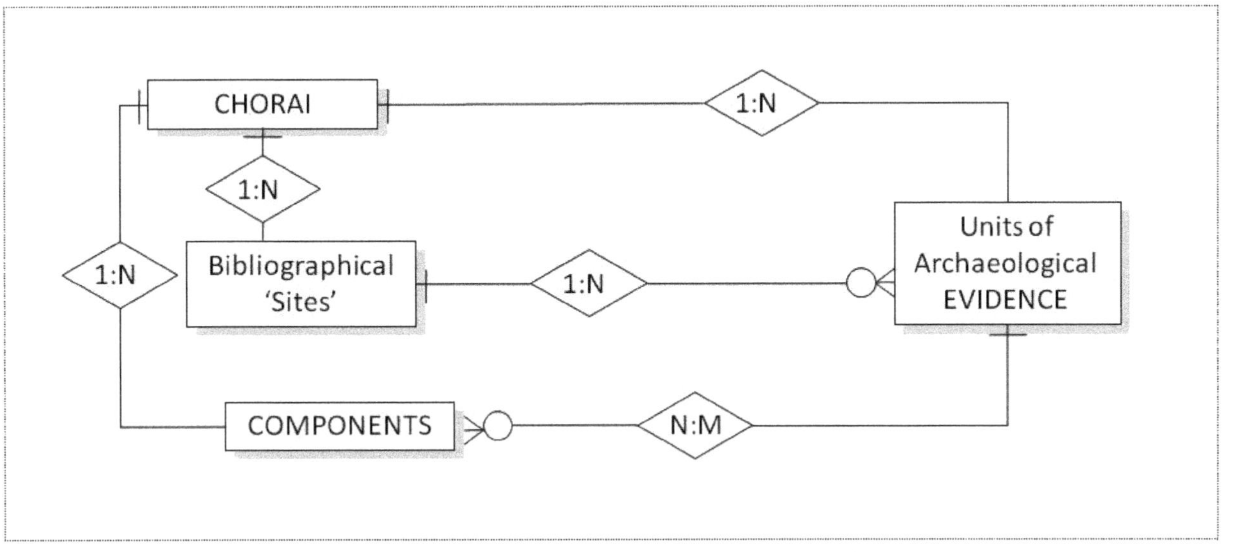

Fig.3. *The structure of the database with the logical relationships that link the main tables within it.*

from the textual/alphanumeric level, before entering into the mapping stage (see below).

SOURCE CRITIQUE

The deconstructing and reassessing processes therefore play an important role in building up the archaeological database for the Boeotia region. Behind these processes is a strong application of the so-called 'source critique' or 'info critique' (the '*Quellenkritik*' of the German tradition), which helps in the definition of metadata attached to the data within the system, and allows one to deal, to a certain extent, with the fuzziness inherent in the available archaeological data. Sometimes the fuzziness is expressed in the source (through indications of probability and uncertainty), but more often it remains implicit and must be detected through a critical examination of the information.

Within the database, fields are available to record and highlight the fuzziness at different levels: spatial, temporal and attribute (information content).

Temporal fuzziness: periodisation and chronological attribution

Diachronic studies are "*comparisons of settlement area structures in the same territory in various periods [..] as, among other things, they may bring a sophistication of our descriptive systems and profoundly influence the generation of archaeological structures and their interpretation*" (Neustupny 1994: 254, evoking diachronic studies). In order to proceed to diachronic analysis, a detailed chronology would be needed, while in the majority of cases it is not easily available in the archaeological record, and chronological attribution to the components remains at the level of larger temporal categories (Prehistoric or Greco-Roman, for instance).
Furthermore, issues of periodisation can heavily bias the archaeological record, as is known from previous reports and studies. Van Leusen, in his PhD thesis (2002), discusses these issues, showing how the chronological attributes of source observations come in a huge variety of terms and classifications, many of which need historical interpretation and may vary considerably from study to study. Moreover, chronological attribution to a site is often assigned on the basis of pottery types, creating typo-chronologies which "*tend to diverge as the distance between the observations increases, and a single period term may have different chronological significance depending on where an observation is made*" (Van Leusen 2002: 13.13). In addition, some periods can only be recognised if diagnostic forms or decorations are present, and the absence of those pieces may lead to an incorrect exclusion of the site from the record of that period; probably present but represented only by coarse ware (Attolini-Cambi et al. 1991; Alcock 1993).
Therefore, as Van Leusen states, "*interpreting the dating evidence presented by a source is often extremely difficult, and so various metadata fields are needed to record the nature and amount of uncertainty associated with any period assignment*" (Van Leusen 2002: 13.13). Sometimes old-fashioned terminology has to be adjusted to modern usage and temporal fuzziness is indicated by the source with expressions such as a generic 'Hellenic' or 'Greco-Roman' or other attributions, not always well determined. The identification of the source reliability is therefore also necessary for the determination of chronology.
In the present work, the attempt to deal with temporal fuzziness is represented by a field in the database assessing the validity of the chronological information, and by the application, for the chronological attribution too, of the same criteria of source critique applied to the content of the archaeological information.

Spatial fuzziness: location and positioning

The accuracy of spatial attribution to a piece of archaeological evidence or a component is strictly linked to the scale of the map on which the sites are positioned, or to the accuracy of the instrument used (e.g. GPS). In our case, the majority of sites have been positioned, according to indications given in the bibliographical record, on the 1:50,000 topographical GYS map,[15] and therefore the position carries the accuracy of the map. In the Raster format, resolution of the base grid is 30x30m, and this also gives the precision of the spatial attribution to the dot representing each component.

Within a regional compilation, all source observations have locations which are known with a greater or lesser precision and accuracy. Most of the archaeological-topographical work does not pay attention to *where* something was found, but only to *what* was found.[16] Sometimes, fuzziness in location is indicated by the source in an often informal and inconsistent manner, through the use of qualifications such as 'nearby', 'in the vicinity', 'not far from', etc. Within the GIS, even relatively small errors in position can have significant consequences at the stage of spatial analysis of the data, as pointed out by, among others, Van Leusen,[17] who also gives some examples and recommends "*[including] assessments of the precision and accuracy of such measurements wherever possible*" (Van Leusen 2002: 13.7).

In the present work, the attempt to deal with spatial fuzziness is represented in the database by a field (*Position*), which records an indication of the accuracy of position on the map (see above). At the stage of analysis, the presence of many components of uncertain position along with the low position accuracy due to the scale of the base maps, would bias the dataset too much for certain analyses. Nevertheless, when possible, I attempt some spatial analyses including only the features positioned with certainty (see chapter II.4).

Spatial fuzziness also concerns the extension of an activity focus, which is almost never specified in the source available, except for data from artefact surface survey, for which we have the extension of the surface evidence by phase, if not the actual original dimension of the component at a certain time. Therefore, it would be wrong to assign a more or less hypothesised (as never correctly documented) dimension to the component examined. In this work, when some indication is given for extension, it is documented in the Note field, and helps in the interpretation and assignment of a character to the component (*type* and *status* fields).

Attribute (information content) fuzziness

Source critique should always be applied in archaeological research. In Alcock's words, "*data must be examined to ensure that the patterns detected are not simply an artefact of the state of our archaeological knowledge – or ignorance*" (Alcock 1993: 49). Information cannot be taken for granted and not all information can be considered equal (with the same degree of quality or referring to the same level of occupation in the past). Fuzziness is inherent in the archaeological record (like non-linearity and incoherence), even if not explicitly expressed by the source, usually with the use of informal expressions such as '*probably* a site', or similar. Therefore, one should collect and work with metadata in order to extract meaningful entities from the archaeological record available.

As noted above, in the present work an attempt to deal with content fuzziness has been conducted throughout the whole process of archaeological data recording and structuring, by means of deconstructing and reassessing procedures. Within the database, for each unit of archaeological evidence, a field *1st mention* is available to record the first time the evidence is mentioned in the archaeological record, and the *Type of presence* field records a general attribution for each unit of archaeological evidence as representing a certain or possible human presence or activity focus (see above). At the component level, the *Discovery* field is available, in which information is recorded on the research tradition in which the information was first collected. Querying the database on the basis of this field would allow the discovery and investigation of biases in the archaeological map (see chapter II.2).

The multi-step procedure of data recording illustrated above, along with the recording of metadata, can be of some help in the source critique process, and allows the critical exploration of the available dataset.

Generally speaking, the critical issue is the question of *representativeness*. An effort should be made to assess the value of the information and to separate the 'sites' in terms of the level of human activities they may represent. One sherd from the Neolithic period might represent a Neolithic site (especially if found among Greco-Roman sherds, whose abundance creates a heavy background – Bintliff 1999; Bintliff et al. 2002), while in the case of Classical material scattered everywhere in the modern countryside, two Classical sherds in a certain location would not automatically mean a site.[18] Archaeologists dealing with settlement histories should pay attention to the critical issue of multiperiod sites, and especially to the level of complexity characterising the 'site' in the different periods of actual occupation recognised in a certain location. That is what the topographical tradition lacks; often recognising as real and, at the same level, the presence of some periods at a site mainly known for a significant occupation in other periods.

[15] GYS stands for *Geographiki Ypiresia Stratou*: the Hellenic Army Geographical Service.

[16] Fossey 1988: 20 writes: "*For the most part the important consideration is* what *is found, not* where". He remarks, correctly, that only if the entire surface area of all sites was covered in detail could the distribution of artefacts be possibly shown to have any significance. In this way, he himself stated the limit of extensive research, at least as far as location accuracy its concerned.

[17] Harris and Lock (1992:118-120) provide an early discussion of spatial error and fuzziness in the context of the representation of archaeological records in a GIS (Van Leusen 2002: 13.12).

[18] "*It seems likely that the prehistoric equivalent of a small historic farmstead with hundreds of sherds collectable today on the surface will be represented by a handful –maybe as little as 2-3 prehistoric sherds in a surface collection*" (Bintliff 2000b: 5) – see also Bintliff 1999.

Some evidence is less trustworthy than others with regards to representativeness. Some possible sites are represented by movable artefacts: for instance, inscriptions are very mobile, and the same could be said for coins. Sometimes only older finds are known from museum records as reported from an area (usually indicated by the modern toponym of the village, but with no clue to the precise location within the municipality – *koinotita*- boundaries). Moreover, in some cases we only know that 'something' was found at a location, without knowing the scale of human activity this 'something' may actually represent. Topographical tradition indistinctively makes up 'sites' as easily from a couple of sherds as from an actual artefact concentration, and considers that these sites could mean the same in terms of landscape. At the analysis stage, it would be wrong to mix together places so different in terms of nature of information, with the risk of comparing a settlement with a small farm, for example, as if they were identical in terms of landscape. This tends to be the problem with landscape studies of the topographical tradition (see below – THE MAPPING PROCESS WITHIN GIS).

How can one weight the data collected according to the most probable human presence which can be seen beyond each single dot? Obviously, it is impossible to do this quantitatively and on a general scale, but it is possible by considering, for each single dot on the map, what kind of archaeological evidence lies behind it. Were just a few sherds found there, or a concentration of artefacts? Who found the evidence? Did they find the site while looking for something else (e.g. human presence for other periods, in which they were much more interested)? What one can try to do is to consider each single dot in turn, and try to understand what the dot means in terms of human presence, and what is behind the material evidence representing it (Bintliff 2000b). This works at the level of data recording, and it should be performed during data processing in which a character is assigned to the human presence whose material remains were found at a certain place. Afterwards, at the level of interpretation, dots are analysed in terms of landscape, after the mapping process (see below).

Extensive/topographical surveys mainly revisit sites already known and larger centres and cult places, as stated by Fossey himself in his 1988 book, and give us an idea of what is present, but not of what is missing, in the landscape. On the other hand, intensive field-by-field surface fieldwalking ('intensive' meaning quantified observations and controlled / monitored artefact collections) give us reliable information also on the absence of presence, and on the character and significance of the actual archaeological presence. Furthermore, intensive and systematic surveys give a focus on the countryside and the rural segment of the landscape, for which we have insufficient information both from historical sources, epigraphical documents, and from extensive/topographical surveys.

In dealing with intensive field-by-field artefact surface survey, we can deconstruct the information until we obtain the smallest unit of archaeological evidence, since we have control of the information in the field. Conversely, in dealing with extensive/topographical survey data, we do not have control of the field record (if one never existed), and therefore we cannot always deconstruct information down to the smallest unit of archaeological evidence. Sometimes the smallest unit is what our source of information report as smallest. For example, often our sources report: "here is a LR site as I saw a few LR sherds". So we know at least that the smallest unit is 'a few sherds'. We will never know, though, how many sherds were actually there on the surface, and how large an area was explored. Thus, we are lacking the detailed *quantitative* information, but we do have, at least, the *qualitative* information, which can help. In other cases, though, a particular place is simply reported as 'Early Bronze Age site' without any information on the significance of archaeological evidence that was seen there. Nevertheless, we could infer the quality of the information from the background (theoretical and knowledge) and interest of the particular archaeologist who visited that place and gave us this (even general) information. We can always estimate, in some way, the qualitative information when dealing with extensive/topographical data, by applying to them a source critique process.

As noted above, a single unit of archaeological evidence can be part of more than one component. This is the case, for instance, of units with unknown chronology, e.g. a fortification dated to the Greco-Roman period may belong to the Classical settlement, to the Hellenistic settlement, to the Roman settlement and, probably, to the Archaic settlement.[19] The same is true also for a scatter of blocks and architectural fragments, without any given chronology.

During the aggregation (i.e. component formation) process, it is important to know what is behind each single activity mapped in order to assign a character, such as status, hierarchy, properties, etc., to the activity itself. Knowing its relationship with other units of archaeological evidence also helps. For instance, if there is a tomb and some inscriptional evidence, the single tomb could represent a cemetery, etc.

As S.E. Alcock has suggested in her book (1993), we should be able to establish the relative reliability of currently available survey results (source critique). We should ask ourselves if what we have is a genuine pattern representative of behaviour in the past, or whether it is rather a function of the relative 'visibility' of certain periods, and of our taxonomic schemes. For instance, for the Late Hellenistic – Early Roman period, only fine wares are well known and chronologically subdivided, and surveys examined by Alcock seem to have based the dating of sites on fine wares alone. This brings into question the visibility of all relevant sites, as the research is biased by the assumption that residents of all rural sites would have had equal access to all pottery types, and this was not the case. On the other hand, we cannot identify sites on the basis of imports alone either (Alcock 1993;

[19] We may get more precise information once a chronological study of the city wall is available.

Attolini-Cambi et al. 1991; Millet 1991), and the 'dissemination' time of specific types of high-quality ceramics must be considered as well.[20] I should point out though, that at least in the case of Boeotian cities, access to specific pots is unlikely to affect the picture, as nearly all the pottery was made within each *polis*. Rather, it is knowledge of the local wares and the sampling strategies for collecting the material during field survey which might affect the reliability of the pattern.

Once the data are plotted on a map (see below – THE MAPPING PROCESS WITHIN GIS), we may find that some areas have low amounts of data. We should ask ourselves whether the picture represents a reasonable paucity of sites, or whether areas may appear empty for topographical/geographical reasons, to geotaphonomic processes, or simply to lack of information. On the other hand, we can have a large amount of information about certain areas due to the interest of particular archaeologists (see chapter II.2 for further discussion on Boeotia) or to easier accessibility. Only a critical examination of the quality of data can help to weight an archaeological picture of the landscape having (very) good data in some areas and (very) poor in some others.

An acritical plotting of the known sites would result in maps full of dots which are, for the reasons illustrated above, often incomparable. Thus, any statistics derived from them would be totally inappropriate. Research projects carried out in the topographical tradition often run this risk. One can consider, for instance, Fossey's gazetteer and settlement history (1988) for Boeotia, and McInerney's (1999) work on Phokis (the region bordering Boeotia to the N); both very good studies with regards to the amount of information collected and discussed, but certainly lacking a critical approach. I will discuss Fossey's work in chapter II.2 and *passim* elsewhere in the present work, while I will comment briefly here on McInerney's work on Phokis.

In his book *The Folds of Parnassos* (1999) McInerney provides a good depiction of the historical landscape of Phokis, with a useful geographical description at the beginning, but in building up the archaeological map he lists and plots Phokian ancient sites on a map, without weighting them in terms of the archaeological information which lies behind them (chapter *Settlement and society*, pp.88ff). In analysing his data, he gives an interpretation of the peculiar increase in settlement in the Classical period, which in his opinion did not occur gradually in Phokis (unlike Boeotia, he says, according to data from the Durham-Cambridge Thespiae survey), but happened quite abruptly after the Persian Wars. McInerney compares data (densities and numbers) from intensive surveys with the data available to him, which come from extensive/topographical surveys. He does it, though, without a qualitative critique, and in this way falls into an acritical interpretation: the apparent lack of settlement in the Archaic period, which McInerney explains not by a process of depopulation but by a different way of settling (perishable dwellings, leather and wood tools, pastoralism mixed with intensive agriculture, leading to a lot of very small farmsteads in the landscape), could also be explained simply by the fact that sites cannot be found so easily because of the huge quantity of material coming from later periods (Classical and Hellenistic especially, with the peak of activity in the 4th C) which 'hides' material from other periods from the eyes of extensive surveyors.[21] Only a systematic and intensive survey could provide a proper answer to this question, and this brings to light the risk of comparing data sets so diverse and differently collected. The risk becomes even greater if this is done in numeric terms. It should be stressed here once again the lack of evaluation of archaeological evidence, emphasising how scholars still write books and assume settlement trends only on the basis of extensive survey, without applying any critical approach.

The general picture of ancient Boeotia that we can obtain from the site recognition carried out by means of archaeological, historical and extensive collection of data, has so far proved to be 'naturally' biased by a sort of city-centred or historical place-oriented or archaeologists' interests-oriented approach. This has resulted in a lack of information concerning lower rank settlement, such as small villages and farms, and the full range of occupation phases of well known sites.

On the other hand, intensive surveys provide information mainly concerning other levels of life in the past – small sites, rural areas, non elite elements – and allows a more detailed understanding of diachronic processes, for both rural and urban areas.[22]

Therefore, the general picture has to be somehow calibrated on the basis of this extra information. Since intensive surveys will hardly cover the whole area (so far ca 360 sqkm has been intensively surveyed), 'intensive

[20] Other issues could be: the effects of variations in the supply rate of ceramics; probable increased use of alternatives to pottery, such as glass and precious metals (this would have caused an exaggeration of the early imperial decline in site numbers); relationship of surface and subsurface remains; ability of survey to detect all strata of society.

[21] Material of less apparent periods will only be found through the careful examination of field-by-field surveyors, and with the correction and sampling techniques typical of their methods.

[22] Even within the framework of intensive surveys, we could only ever produce a near-precise picture of the ancient situation, since not all available kinds of investigation can be applied. Such an attempt has been made in England, for a small area, at Shapwick (Somerset), within the framework of a research project which is pioneering in many respects. There, across the whole territory of a parish of 1.284 hectares, a wide range of investigation methods and techniques have been applied: extensive survey, intensive survey, historical documentation, toponymic research, excavations, shovel-pit sampling, pollen and soil analysis, archaeological prospection (geophysics, aerial photography and remote sensing). Although we will never produce such highly-detailed information for an extended area (a *chora* or the whole of Boeotia), we will always be able to integrate information. As with the creation of a Digital Terrain Model from known individual spot heights (points with associated height values), we can, from the 'punctual data' given by surveys at different levels and research results, interpolate the surface of the development of the whole Boeotia province.

windows' represent only several small landscapes, as the picture (fig1 in chapter II.2) shows. Therefore, we cannot consider only the windows. What is true and proved for one area can be totally inappropriate for another.

The deconstruction of the traditional definitions of the 'site' illustrated above is due mainly to the development of intensive surveys, which were able to locate up to a hundred times as many sites per square km as earlier work at a less intensive level (Alcock et al 1994; Bintliff 2000a[23]).

The only way to compare intensive with extensive survey data seems to be the application of info- and source-critique to the information. This could be done, for instance, by reducing the evidence to the smallest unit. For example, artefact concentration would be the smallest unit of evidence in the case of artefact surface survey, which, with due caution, could somehow be compared with evidence reported from extensive survey, at least to partially overcome the qualitative gap, without disregarding, though, the quantitative gap, which cannot be overcome. This is why the comparison of values between the two levels is always dangerous (see discussion in Van Leusen 2002: chapters 4 and 13), and when producing maps presenting both intensively and extensively collected data, one should always mark the areas actually covered by intensive surveys (intensive windows). Although direct integration of the two data sets cannot be automatic, one should not become frustrated (Bintliff 2000a: 137), but should find a way to discuss the increasing developments in knowledge that intensive and systematic survey can bring, as well as taking into account older information, no longer visible on the surface but noted during extensive/topographical survey. (In the present work, this was carried out within the framework of the discussion of each single component – see *chorai* chapters).

We should keep in mind that even data collected by intensive and systematic artefact surface surveys should be processed with a strong 'critique'. The issue of visibility can greatly affect the representativeness of data. Geomorphological, pedological and taphonomical processes, as well as surface visibility (influenced by vegetation, agricultural use and other factors) and period visibility (discussed above), influence the recovery of archaeological sites (Barker-Symonds 1984: 287-288; Bintliff-Snodgrass 1985; Barker 1995: 49-50; Terrenato-Ammerman 1996; Terrenato 2000; Wells 2001; see Terrenato 2004: 38-42 for a review of the issue). Also, problematic is the case of multi-period sites, where the off-site background (heavier in certain periods) may bias the period site maps if not dealt with properly (see Gallant 1986; Bintliff 1999 and Bintliff-Howard-Snodgrass 2007). This can be overcome by comparing on-site data collected with off-site data in their immediate surroundings (see Bintliff and Howard 1999 for the methodology in use, for instance, in the framework of the Boeotia project, and some examples in the chapter II.3.14 and appendix I.14 - TANAGRA SURVEY). This moves beyond the purpose of the present work, however, and in some analyses we chose to include only those sites that were clearly datable (including only their main period of occupation), in order to avoid the risk of using data which could be dramatically changed following the in-depth examination of their composition. In the present work, I have inserted into the survey sites database only the surely dated artefact concentrations, mostly from preliminary reports (never fully and systematically published or currently being published). These are presented in each catalogue of sites related to the *chorai* in which they are included. Survey sites can be inserted into the database as 'artefact concentration' at the archaeological evidence level, and as settlement (hamlet, village) or rural site (according to the interpretation given) at the component level.

THE MAPPING PROCESS WITHIN GIS

Landscape archaeology deals with the interpretation and reconstruction of human behaviour over space, through time,[24] based on appropriate spatial units of analysis (see chapter I.1.2). Therefore, the final development of the proposed reassessment of known data is at the mapping and spatial level within GIS. In order to understand the development of the procedure, we should keep in mind the basic distinction, within a GIS system, between alphanumeric-textual and graphic-spatial archives, logically linked in order to perform selection and cross queries (see Wheatley and Gillings 2002: 12 and 60).

All textual information entered into the relational database has therefore been linked to the corresponding features mapped, and their spatial information, in order to create a system able to manage, process and analyse spatial and temporal complexity. The result is that behind each feature on the map, or on the screen, different degrees of information concerning it are stored. We should not forget that the mapping process accompanies each stage of the data recording and interpretation, and also acts as a critical reviser, especially in the aggregation (component formation) process.

As illustrated above, we have assigned each single feature (unit of archaeological evidence) to a different class/type, specified by the diverse character of the features. Classes represent the different types of archaeological evidence involved in the system. For instance, if the possible presence of an ancient occupation is known through epigraphical evidence only, this will be recorded simply as epigraphical evidence. The same would be true for the case of a few scattered sherds noted at a certain place. They will remain as such, scattered sherds, as long as we cannot find another indication of activity on that spot. If

[23] For Boeotia, Bintliff mentions a density of 1 site x 26 sqkm (extensive/topographical survey) compared to a density of 3-4 sites (or more) per 1sqkm (intensive artefact surface survey) – Bintliff 2000a.

[24] 'How' and 'why' human behaviour over space changes, the dynamics... (Van Leusen 2002: 5.12). "*Space, time and function can be thought of as different axes along which the available data can be differentiated*" (Van Leusen 2002: 5.13)

we do find other indications, then the presence of scattered sherds becomes evidence of sure and certain activity. Furthermore, if the scattered material can be associated with more distinctive evidence, this would enable us to assign a character to it. For example, if such scatters are associated with graves or with funerary stelae, then they would become a cemetery, while if we can associate them with walls, remains of buildings, fortification walls, then they could be assigned the character of a settlement. Low density scatters may represent small-scale or short term occupation sites, or even burials.
In this way we can monitor the quality, the suitability and the value of our data in qualitative terms (if not strictly quantitative) and proceed to a systematic 'info-critique' (see above).
Within the GIS system, which associates spatial with textual information, and through the aforementioned method of data collection and structuring, we are attempting to move beyond the dots on the map. In fact, in practice, the result is a graphical representation of the study area (a map) characterised by the presence of graphically diverse features, each one corresponding to different units of archaeological evidence associated at different levels of complexity and linked to the corresponding alphanumeric record in the database. In this way, paper sketches or maps of the individual sites are also digitised and inserted into the system, when available.

According to the same approach followed in the layout of the alphanumeric database, the first stage of recording of the archaeological evidence is followed by the mapping of the constructed components. Therefore, applying an interpretative filter, dots may become useful and helpful for visualisation and display purposes. The resulting map will be a map with dots. However, we should not forget that 'our' dots are not just floating points, but they remain within the system, and the information content, critically revised (see above) and stored in the [Components] table, remains attached to them.
Within GIS one could easily run the risk of returning to the traditional archaeological topographical tradition that approached landscape as an 'empty space' within which some points of particular interest, 'sites', are situated. This would be partially avoided if the dots are the result of a multi-step process of info-critique carried out on the archaeological information available, as illustrated above. On the other hand, we should keep in mind that we are dealing with data collected mainly through extensive topographical research and marked by a high degree of non-linearity. Therefore, the data do not have the character of intensive survey results which, if carried out in increasingly more sampling areas of the wider region, would fill in (with recorded presence and absence) the 'empty spaces', and would give us a better idea of the landscape trajectories.

Querying the database on the basis of the metadata fields (illustrated above in the discussion of temporal, spatial and content fuzziness) would allow the discovery and investigation of biases in the archaeological map (see chapter II.2), as would, for instance, investigation at the map level of the degree of correlation between the modern road network and the known units of archaeological evidence or components. The proximity to the modern road network can constitute a strong bias factor affecting the recovery of archaeological sites. In this work, a GIS analysis is carried out to define distance ranges from modern roads, in order to appreciate the possible increasing in presence of archaeological sites in the proximity of roads, due to easier accessibility or, in few cases, to road building or maintaining.[25] The results of the analysis are presented and commented in each *chora* chapter, taking into account the possibility of a real correlation between presence of sites and modern roads. The latter often trace the past road network by which some types of sites were located, or run along foothills and valleys, crossing areas where sites are much more likely to occur.
GIS can therefore be useful in visually representing biases in archaeological research or discovery of sites (density maps; geostatistical analysis to represent density of discoveries; relationships between hilltop sites and discovery, between modern road and discoveries, etc.) as well as helping in the evaluation of the cultural variable.

To sum up, our dots therefore represent the result of a multi-step process of info-critique carried out on the archaeological information available, as illustrated above. The importance of such a methodology can be seen more clearly if we critically consider some gazetteers and landscape studies that do not seem to carry out enough source critique, as well as some (recent) research projects carried out on the settlement history of Mainland Greece (McInerney 1999, examined above, and Fossey 1988, examined in chapter II.2). The aim is to produce, through the critical reanalysis of the quality of the available information, a map having a level of confidence in itself, exploring the less and more reliable possibilities. Most important is that we know the level of confidence that can be assigned to a map.

GIS AND THE SOCIAL LANDSCAPE

Social activities were, and are, performed in the landscape. In any single location, different social activities would have occurred through time, as well as different associations between them. Therefore, we would obtain archaeological evidence at different levels of complexity (Neustupny 1994: 248). For instance, some isolated finds may represent low level social activity areas, but could also represent a medium level social activity area on which taphonomic agents and post-depositional processes have buried or moved away some of the associated material culture. On the other hand, a few sherds from a certain period may represent activity at higher level if noted among a large concentration of later material. Very mobile items, such as inscriptions for

[25] A similar analysis was carried out by Rajala et al. 1999, who interpreted the correlation between site locations and distance to Roman roads as indicating discovery biases.

instance, can hardly have a meaning in themselves, unless they are found in association with other material. By looking at the spatial and cultural relationships occurring between social units we could try to detect associations between lower or higher levels of archaeological evidence, which could constitute meaningful entities in terms of activity areas at different degrees of complexity.
During the systematic analysis of the archaeological record to produce aggregations into culturally meaningful entities (i.e. components), a chronology value is attributed to each component. In other words, each of the diverse components represents different choices taken over the landscape in different periods. This is why the database is constructed to have a single period of occupation assigned to each component; each component representing a single 'choice' over the landscape in each different period,[26] while it has to be understood as a historically constituted entity.[27]

GIS archaeologists usually display environmental layers as representing potential taskscapes, and then analyse their dots or polygons, representing archaeological features, on top of the resulting layer. This is often done as if the dots (or polygons) represent the full picture of occupation of the landscape. Yet, as stated earlier, it would be wrong to perform the analysis without assessing limits and being conscious of the research biases. On the other hand, since, due to our limited archaeological knowledge, we do not have the complete picture, we should examine each single focus of activity, in the direction of an '*archaeology of place*'.[28] Once deconstructed and reconstructed, the archaeological record according to components, we would examine each single component/activity focus (GIS dot or polygon) on the map, trying to understand its meaning in terms of landscape and in terms of 'social use' of the landscape in the past, within the macro-region and sub-region (see chapter I.1).
Each single 'DOT' could provide information on the social use of the landscape if examined over a set of landscape data, such as raster environmental datasets for the recognition of suitable activity landscapes (land potential, communication routes, relationships with other components/activity foci, probable activity areas, etc.). Therefore, as a preliminary step to dividing the landscape into 'task-areas' (such as agricultural production, settlement/habitation, industrial activity; associating activities and landscape zones according to environmental and cultural factors), which could be represented by GIS polygons (or groups of pixels) associated with a specific function, we would analyse each single dot in terms of environmental landscape and of social use of the landscape, and then in relation with other activity foci in an attempt to consider the relationship between people, place and landscape.[29]

The aim of the research is to give each archaeological inferential component (the meaning of a dot in terms of human activity) a meaningful interpretive context. On one hand is the analytical and critical deconstruction of the archaeological record (of *what is* known in the landscape) described above; on the other hand is the consideration of bits of landscape as wholes,[30] considering the absence of presence and presence of absence (Van Leusen 2002) of remains of social activities, and detecting taskscapes in terms of connection between activities, objects and people (for the concept of *taskscape* see Ingold 1993 and 2000 and chapter I.1.2 for discussion).
By examining the relationship between landscape and activities (what/who influences what/whom; what/who informs what/whom; how they interrelate), we can derive the potential significance of the material from the landscape characteristics and, conversely, we can derive the potential significance of the landscape and landscape features from the social and spatial characteristics of the material culture.
According to an 'activity areas' dynamic approach, based on a meaningful interaction between the 'Community Area' and the taskscape theory (see chapter I.1.2), we can try to assess the shifting of use of landscape, in terms of the different social choices taken over an area of the landscape in different periods....[31] According to a 'settlement dynamic' approach, we can analyse the shifting of the main settlement, the shifting of satellite settlement location, the disappearance of settlement, and similar aspects regarding the main residential cores within a micro-region, according to the revised settlement chamber approach as illustrated in chapter I.1.2.[32]

Within this framework, the *chora* constitutes a micro-region/sub-region included in the wider region of Boeotia. Within a *chora* several 'community areas' can be active, one of which usually gravitates around the main settlement centre, the *polis*. Within each *chora* micro-landscapes can therefore be defined, marked by a distinctive landscape 'character', a concept which constitutes a bridge between landscape and region /

[26] Choices over the landscape can be seen as acts of agency. For the concept of 'agency' see M. Johnson 1999: 189, and Tilley and Shanks 1992: 122-6, as well as Dobres - Robb 2000, and Wheatley 2000: 123-31. See also Van Hove 2003: 201-207, Llobera 1996, and Corsin Jiménez 2003: 137-153.

[27] "[...] *understanding places, including settlements, as historically constituted entities. The act of settling may involve reference to the previous use of that place*" (Bruck – Goodman 1999b: 14).

[28] For a discussion on the distinction between the concept of space and place see, among others, Tuan 1977: 12; Thomas 1996: 31, as well as Tilley 1994: 14ff, and Knapp and Ashmore 1999: 20f.

[29] Settlement studies go beyond site-centred approaches to consider relationships between people, place and landscape (Bruck - Goodman 1999b: 12).

[30] As opposed to the individual archaeological evidence and its typology etc.

[31] Proximity to resources; landscape use; land suitability according to production type for the period; relation with other landscape elements and other components; activity areas within the settlement chamber; taskscapes.

[32] With the 'settlement dynamic approach' I refer both to the community area theory (Neustupny 1991 and 1994; and Kuna 1991), as well as to the dynamic settlement system approach (Dewar 1986 and 1992). See also chapter I.1.2 for details.

micro-region (see chapter I.1.2). There is a level of landscape characters that can be examined in the long-term perspective, i.e. the level which belongs to the interaction between environment and human activities, which, in other words, belongs to the taskscape. This is what is of interest in the present research, aiming to identify settlement hierarchies (the size and distribution of community areas, the relationship between central places and their satellites, etc.) and to explain them in economic and social terms, to describe the inner structure of the community areas (settlement patterns on a local, community level) including the relations between habitation, production, burial and ritual areas, to identify the main general as well as culturally specific factors (environmental, economic, social, symbolic etc.) in past site-locational preferences, to evaluate the reliability and explanatory power of individual datasets collected by different research methods. Following the trend given by the settlement chambers approach (revisited through the community area theory and the taskscape approach - see chapter I.1.2), one can obtain more detailed and larger clues on the social use of the landscape by looking at the location in the landscape of the units of social evidence, at the settlement and activities pattern, at the association between activities and landscape zones, and especially by looking, in the long term, into the movements and shifting of areas of use (activity areas) within a settlement chamber. The shifting of the major settlement, or the gradual transformation and modification of settlement structures through time, for instance, as well as (and especially) the shifting of activity areas in space through time, are all objects of the research.

Following the above statements, it is clear that the archaeological record to be used for archaeological landscape analysis, and for assessing the social use of the landscape, must be revisited in a 'GIS-based' (GIS-designed, GIS-informed) way in order to become suitable for more meaningful analyses. Therefore, in the interpretation of archaeological landscape data, the use of GIS becomes a GIS approach to landscape studies, which informs and drives data collection and management up to or down to the analytical level.

APPLICATION OF COST-DISTANCE ANALYSIS

Traditionally, site catchment analysis is aimed at individuating territories exploited by a settlement. Its first application in archaeology was the work of Vita-Finzi and Higgs (1970: 1-37).[33] The basic principle of the method is the acceptance of the relationship between the fruitful exploitation of a source by a population, and the amount of energy consumed by that population in order to exploit it. The limit of a site catchment area (i.e. the distance from the site exploiting it) differs according to the type of settlement or the period considered. For the Greek landscape, Bintliff (1977: 112) suggests two hours' travel time for a site related to animal husbandry, and one hour travel time for an agricultural community, while later he (Bintliff 1997c and 1999c) goes into more depth by recognising a dynamic development of territory size, and concludes (following Flannery 1976) that "*the 5-kilometre farming radius may operate in certain settlement scenarios, such as pioneer farming 'infill' situations, but not others, with mature 'filled' farming landscapes stabilizing into 2.5-kilometre radius catchments*" (Bintliff 1999c). Accordingly, the average cross cultural distance between villages is 4-5km (Bintliff 1999c).

In order to individuate and graphically present site catchment areas, different methods have been developed, especially within geographical studies. Allocation analysis, otherwise known as 'tessellation' of space, is based on the concept of division of space on the basis of the geometric properties of a set of points. It is based mainly on Euclidean distance and gravity based rules, and considers the space as 'flat'. The most common application of the method in geography and archaeology is the so-called Thiessen/Voronoi polygons,[34] a developed version of which is the X-tent model[35] (Hodder and Orton 1976; Wheatley and Gillings 2002: 149-151). Within a GIS environment, the application of algorithms allows identification of which cells belong to which source based on closest proximity (in a straight line). An output raster is produced that records the identity of the closest source cell for each cell. Each cell in an allocation raster receives the value of the source cell to which it will be allocated. The allocation analysis in itself is based on measurements of Euclidean (straight line) distances, but it can also be performed on the basis of a cost weighted distance, which considers the space as topographically variate (rather than 'flat') as well as taking into consideration the cost (in time or energy) of moving over the terrain within space (see below). By offering tools for calculating cost- (time- or energy-) distance, GIS can help to overcome an excessive and risky abstraction of space.

The series of GIS techniques that are based on the capacity to assign a cost value to each cell of a raster dataset/matrix and then to calculate the cumulative cost on the map (Van Leusen 2002: 6.1) is given the name *Cost-surface analysis*. The concept of cost-surface is the expression of energy or time cost of moving over the terrain. In order to calculate this, not only the Euclidean distance between the locations is taken into account, but also other variables, among which primarily the terrain

[33] The concept is based on the idea that the further an area is from a site, the lower the probability of that site's inhabitants exploiting that area.

[34] Thiessen/Voronoi polygons (a sort of tessellation) are based on a simple nearest-neighbour method for partitioning a featureless space, equivalent to a gravity model operating in 'flatland', and result in a complete tessellation of space (Van Leusen 2002 : 6.4).

[35] The X-tent model (used by Renfrew and Level 1979, and Grant 1986 among others) refers to the influence of the size (or rank) of a centre onto its territory. It is assumed that the degree of influence of a centre is proportional to a function of its dimension, which would diminish gradually as the distance increases (see also Van Leusen 2002: chapter 6).

slope (Wheatley and Gillings 2002: 151; 159-162).[36] The possibilities offered by this technique have led to some confusion as to the best way of calculating costs, as illustrated extensively by Van Leusen (2002: 6.5).[37] Some of these methods consider terrain slope as the key factor for the speed of the moving on the terrain. For the analysis carried out within the framework of the present study while exploring site catchments, boundaries between communities and settlement chambers, I calculated the accessibility of settlements on the basis of slope, according to a formula provided by Gorenflo and Gale 1990 (and re-proposed by Tobler 1993[38]). The relationship between terrain slope, velocity/speed and cost value is represented by the formula:

$$V = 6 \exp \{-3.5 * \text{abs}(S + 0.05)\}$$

where V is the walking speed in km/h, S the terrain slope in degrees, and 'exp' is the base for natural logarithms. This function is symmetrical but slightly offset from a slope of zero so the estimated velocity will be greatest when walking down a slight incline (Van Leusen 2002: 6.6). For each degree of slope[39] is assigned the time (min/km, approximated value) that a man employs to walk it, taking into account his speed. The result is a time-cost surface with values from 12 to 395, representing the effort to move from one cell to the other (expressed in min/km).
From the cost-surface one can derive the *cost-weighted distance*, i.e. the Euclidean distance corrected by the cost surface values (from one or multiple features). The distance obtained in this way can be considered temporal rather than geometric (expressed in time –minutes- rather than in metres), and constitutes a continuous raster map, of which any number of values may be used to provide 'cut-off points' or boundaries to the catchment or territory (Van Leusen 2002: 6.4-5). If the cost-distance is allowed to run until all available space has been used, the result is a tessellation of space similar to Thiessen/Voronoi polygons.[40]

The cost-surface analysis can be used to determine the potential catchment area of a site, by defining a buffer area around the settlement site based on a specific distance radius, taking into account the topography.
On the other hand, the continuous surface of topographically corrected distance values, which results from the analysis, can be explored in terms of movement in the landscape and of relative distances between settlement sites. In this way, settlement chambers can be appreciated, although they result from more qualitative issues than do site catchments. Therefore, the continuous surface resulting from cost-surface analysis constitutes the basis for the settlement chamber analysis carried out for each *chora*, and helps to better evaluate the impact of physical constraints on the creation of settlement areas, the spatial inter-relationships among settlement sites, as well as the degree of movement over the landscape of settlement foci (see *chorai* chapters and chapter II.4).

In the cost-surface analysis carried out within the framework of the present study and aiming at exploring site catchments, boundaries between communities and settlement chambers, the partially reconstructed ancient roads (mainly physical routes as they are mapped in this work - see fig.5) have not been added as a correcting factor of the analysis. The analysis would be too strongly biased by the incompleteness of the dataset, as a detailed and systematic study of the ancient road network would be needed. This can be demonstrated by a GIS distance analysis from reconstructed and potential routes performed on the archaeological components. The fact that 82% of the archaeological components listed and mapped in this work are further than a distance of 100m from the partially known and supposed physical routes network, would testify for how many routes and ancient paths used in the past we miss in the constructed dataset. Such a study is not available nor could have been done within the framework of this work as it would go beyond its scope. Furthermore, what is explored by means of cost-surface analysis in the present work is the relationship between the settlement and its territory for its economic use mainly, rather than the communication network between communities nor the chosen and preferred paths followed by main routes.

[36] Compared to other kinds of analysis, such as Thiessen/Voronoi polygons or circular catchment analysis, performed in a 'flat' space on the basis of Euclidean distance and gravity based rules, cost surface analysis allows the simple 'flat' geographical space to be supplanted by a set of complex cost surfaces incorporating many relevant properties of the terrain, taking into account the expenditure of time or energy (Van Leusen 2002: 6.1).

[37] For instance, travel cost surfaces can be isotropic (the same in all directions) or anisotropic. Because of the effect of both slope and terrain, the cost of traversing a particular location may differ depending on which direction it is being crossed in (Van Leusen 2002: 6.5). However, most authors agree that travel cost has both an isotropic and an anisotropic component; the former exemplified by costs relating to the type of terrain and ground characteristics (soil, vegetation, and wetness), the latter by costs relating to slope and streams.

[38] See Tobler Waldo, www.geodyssey.com/papers/tobler93.html (last visit April 2008). The formula has been employed for instance by Verhagen et al. (1999), who calculated the accessibility of settlements in the Vera Basin, Spain.

[39] Slope classified up to 45 degrees because above it the terrain can be considered unwalkable. The highest value in our Boeotia dataset is 85.28 degrees. In the slope model therefore slope values fluctuate between 0 and 85.28 degrees.

[40] This process is called 'allocation' in the ESRI ArcGIS Spatial Analyst extension. Performing allocation via the Straight Line Distance function allows one to find the cells that are to be allocated to which source based on closest proximity, in a straight line (=Thiessen/Voronoi polygons), whilst performing allocation via the Cost Weighted Distance function takes into consideration the cost of travelling over the land rather than the straight line distance.

CULTURAL-ANTHROPIC GIS LAYERS

In addition to the archaeological evidence and components, from the Neolithic to the Late Roman period, other GIS layers included into the system refer to cultural-anthropic landscape features. These are listed and described below.

Layer **COMPONENTS** [points]: the layer includes the components recorded in the archaeological database of ancient Boeotia built up within the framework of the present work, mapped according to information given in the bibliographical sources and built up according to the methodology stated above in this chapter.

Layer **CHORAI** [polygons]: the borders of ancient *chorai*, into which ancient Boeotia can be divided, represent approximately the limits of the *poleis* territories in the 4th century BC (at the time described by the *Hell. Oxy.).* They mainly follow geographically marked lines, building up topographical units, and sometimes they are known with higher precision due to some archaeological features discovered (e.g. a sanctuary situated at the border, as stated in ancient texts, or inscribed boundary markers– *horoi*). At other times it is much harder to identify the actual border line, and in those cases a 'guessed' line has been drawn as working hypothesis (fig1).

Layer **ROADS** [polylines]: the layer is constituted by the vectorisation of the main (i.e. paved) modern road network. Some roads, such as coastal or mountain roads, or the road running along the E side of Paralimni lake, are totally new (most of them are not even marked on the 1:50,000 Geological maps, and some are marked only on the most recent version of the 1:50,000 GYS topographical map). As basis for the vectorisation of roads I have used the 1:50,000 GYS topographical map drawn in the 1980s, which should have been the road network available to earlier researchers in the Boeotian landscape (Fossey, Lauffer, compilers of Prehistoric gazetteers, etc). The main purposes of this work are: i) to add a feature layer in order to make the maps much more comprehensible to users (to show the positions of sites in relation to modern landscape features, together with modern villages); ii) to create a layer on the basis of which to map possible biases in archaeological discovery according to the proximity of roads;[41] iii) to identify, when of interest, changes in communication patterns between past and present, and to compare with the physical routes much more used in past communication. Some mountain passes, for instance, disappeared (e.g. the pass through Helicon village or through Koukoura/Agia Anna), some others take different ways.[42]

Layer **ROUTES** [polylines]: physical ancient routes known or hypothesised, on the basis either of archaeological evidence or of ancient sources or modern travellers' reports.[43] Archaeological remains of ancient routes are recorded as units of archaeological evidence (AE) in the [Arch_Evidence] table of the archaeological database.

Layer **CHORIA** (modern villages) [points]: based on the 1:200,000 *nomos* map, recorded in the 1960s and updated in 1983 by the Hellenic Statistical Service - ESIE. The main villages and 2nd rank hamlets (*synoikismoi*) are included, constituting the *demoi* giving their name to the corresponding territories (see *Koinotites* layer).

Layer **CHORIA** (modern villages) [polygons]: based on the 1:50,000 GYS topographical map drawn in the 1980s. The polygons represent the built area of the village.

Layer **KOINOTITES** (modern village boundaries /administrative divisions) [polygons]: digitised from the 1:200,000 *nomos* map, recorded in the 1960s and updated in 1983, with attached information from the Hellenic Statistical Service (ESIE) concerning 1981 (population, number of hamlets, etc.)

Layer **FRANKISH TOWERS** [points]: based on Lock's 1896 map; the result of Lock's study on the remains of Frankish towers in Boeotia. In addition, the two main Frankish centres (castles of Thebes and Levadeia) have also been added, as well as some towers noted within the framework of the Boeotia Survey Project (Chaironeia and Tanagra/Ag.Thomas).[44]

Layer **OTTOMAN VILLAGES** [points]: based on research carried out by J.L. Bintliff and M. Kiel (Bintliff–Kiel in preparation) who positioned known and potential sites employing as reference old maps, including the

[41] The analysis does not really work for rescue excavation – most recent roads are not mapped, though recent rescue excavation work is discussed in the *chorai* chapters – nor for intensive systematic surveys which walk each field no matter the proximity to the easiest path, but for extensive survey work visiting the most accessible places.

[42] As far as roads not really belonging to the network but simply leading to churches/monasteries, I included that to Moni Panagias (by the Apollo Ptoion sanctuary), but omitted those to Proph.Ilias monastery above Vathy/Aulis –East end of Thebes sheet, W end of Chalkis- and to Moni Sagmatas (both not marked on the Geological maps, probably older than the topographical GYS).

[43] The main studies on routes in Boeotia are Burn 1949 (plan 42) and Pritchett 1985: 138-165 (*Routes over Helicon from Thespies towards Levadeia*), mainly on the Helicon area and central and W Boeotia. On routes in Boeotia see also Gomme 1911/12. Useful for the layer creation were also accidental discoveries of traces of ancient routes (mainly from rescue excavations) and Philippson's (1951) remarks on the physical passes. Sowerwein maps and time calculation, based on the 19th century postal times (given personally by the author to J.L. Bintliff and kindly passed onto me) were also helpful.

[44] The layer is probably incomplete because, since Lock's work, no other systematic study has been published on Frankish settlement in the region.

Carte de la Grèce (1852), and travellers' accounts, as well as recent place names.[45]

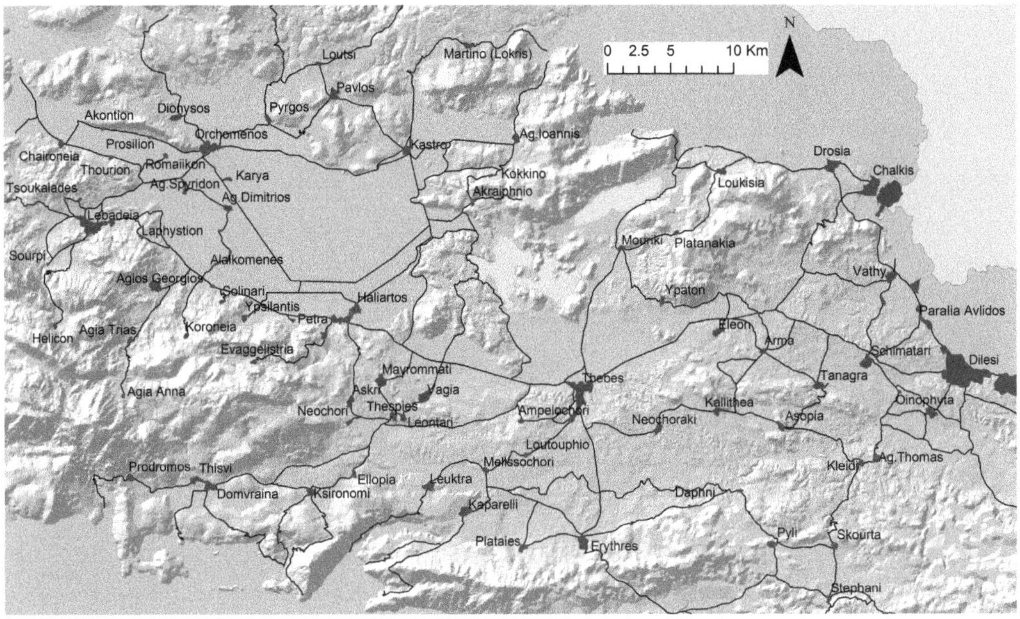

Fig.4. *Modern settlement (choria) and roads in Boeotia*

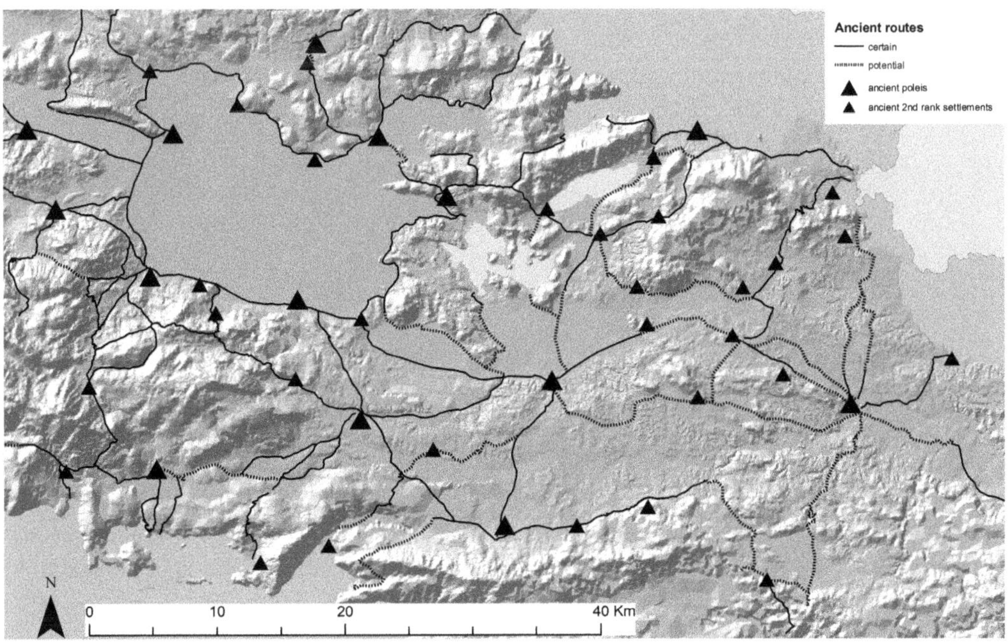

Fig.5. *Physical routes of ancient Boeotia. The archaeologically testified or most certain routes are represented with a continuous line, while dashed lines represent the potential routes.*

[45] See Farinetti – Sbonias 2004 for an example of the potential of GIS analysis in the study of Ottoman archive data from Boeotia, studied by M. Kiel (Bintliff-Kiel in preparation).

II.1

The Boeotian Landscape: Topography and Environment

THE ENVIRONMENT

Boeotia constitutes the middle *section* of Central Greece, a long strip of land between the Gulf of Corinth and Euboeoan straits. Boeotia extends, to use Strabo's term, as a '*tainia*'[1] from E to W, from the Euboean sea to the Crisean gulf, and in length it is almost identical to that of Attica (Strabo IX 2.1).

The rugged topography of Boeotia, and in general of mainland Greece, derives from the Alpine orogenic phase and the subsequent epi-orogenic subsidence events. The main topographic features are related to a system of ancient submarine ridges and furrows of predominant NW/SE orientation (Perlès 2001:9 summarising Bintliff 1977, Higgins and Higgins 1996, Jacobshagen 1986). During the epi-orogenic phase large basins were formed through subsidence, in direct relation with the pre-existing NW/SE ridge and furrow structure: the West Macedonian plain, the Thessalian Plain, the Saronic Gulf and the Copais basin, which constitute the Sub-Pelagonian Intermediate Zone, as well as the lowlands of Elis and Messenia. Despite subsidence and active erosion that filled the basins with flysch deposits, the result of this epi-orogenic phase is that Greece is a largely mountainous country.[2]

The largest part of Boeotia belongs to the Sub-Pelagonian Zone and the Parnassos Zone. The Sub-Pelagonian zone is defined by Higgins and Higgins (1996: 19) as "*the great belt of ophiolites and associated rocks (limestones, cherts), that were originally part of a continental margin, between the Pelagonian continent and the Pindos Ocean*". During the Tertiary, they add, "*a deep continental trough developed, which filled up with up to 5 km of continental and shallow-water sediments, called molasses, and deep water flysch sediments*" (Higgins and

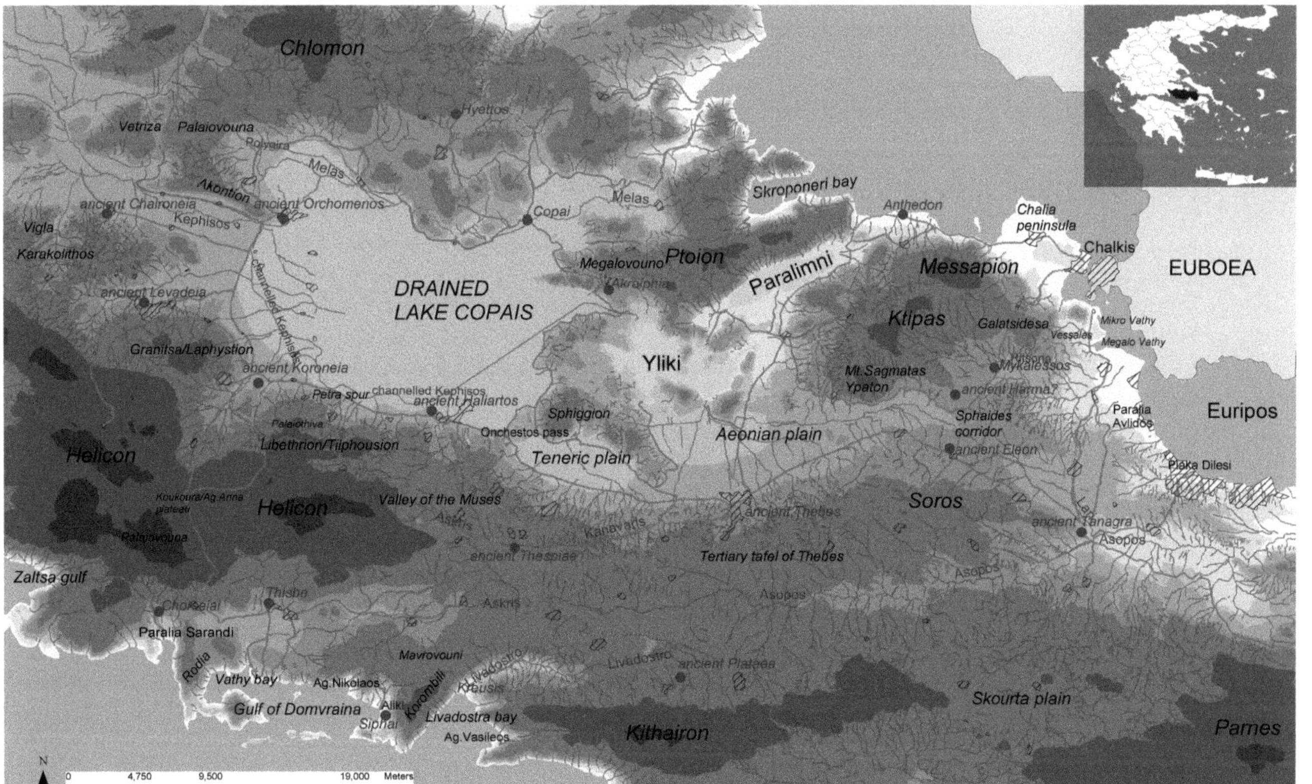

Fig.1. *Boeotia general map, with main topographical features and physical routes.*

[1] See Wallace 1979 for translation: 'ribbon-like stretches of country' / landstrip. Wallace (1979: 7) says that Strabo uses the word 'tainia' to describe land-strips dissecting large land-masses.

[2] "*More than two-thirds of Greece lies above 300m, and steep mountainous reliefs isolate the subsidence basins, creating constraints on inland communications*" (Perlès 2001: 9).

Higgins 1996: 19). In contrast, the Parnassos zone is characterised by a large carbonate series beginning with a dolomitic sequence of the Secondary Era, covered during the Tertiary by flysch.
During the Pleistocene, Boeotia, generally speaking, follows the evolution typical of the Mediterranean area. At the last peak of the last glacial period (22.000 – 12.000 years BP ca.) the glaciers on the high mountains of Parnassos and Helicon were not particularly extensive. The sea-level was lower than today and relief was deeply incised by rivers. The erosive action was significant and relatively rapid, helped by the nature of what we can assume was the most probable vegetation coverage in the period: extensive dry grasslands and limited woodlands.
Around 10,000 years BP, the end of the glacial period, marked by a general rise of the temperature, was accompanied by an increase in the sea-level. The rise of the sea formed inlets in the coastal valleys, which were progressively filled by alluvial sediments. A large debate exists today about these alluvial sediments in the Mediterranean region, regarding issues such as their nature, explanatory factors behind their development (general and local), deposition phases, chronology (the so-called 'older' and 'younger' infill) as well as overall dynamics (among others, see Bintliff 1987; Pope-Van Andel 1984; Bintliff 2002 and, for alluvia in general, Brown 1997).

The morphology of Boeotia is therefore characterised by a sequence of valleys, created by tectonic depressions generically oriented NW-SE, which are surrounded by two types of carbonate relief:

- high mountains of limestone and dolomitic limestone of Lias-Trias, that reach considerable elevation (Parnassos, Helicon, etc.) and are located mainly in the SW-S area;
- lower relief groups, rarely reaching 1000m high, formed by alternate and overlain layers: limestone, sandstone and chert formations of Jurassic and Triassic age; Pliocene marl, sandstone and conglomerate, as well as Pleistocenic sediments; these materials, subjected to erosion, form, along with the Pleistocene-Holocene fluvio-lacustrine sediments, the sedimentary deposits of valleys and inner plains.

These basins, or inner plains, belong to a series of basins (ellipsoidal in shape) that extend from Mount Oeta to the Euboea Gulf (on a line ca. 125km long) and run through Boeotia in a NW/SE direction, for ca. 80km E-W and ca. 40km N-S.
There are three main plains of Boeotia: the northernmost plain of Orchomenos, which derives its water mainly from the Kephisos river and was bordered by the former lake Copais; the Theban plains in the central area (namely the Teneric and Ionian plains) and the southernmost plains of eastern Boeotia, which derive their water mainly from the Asopos river.
Throughout the majority of the earlier periods, up until its drainage at the end of the 19th C AD, a large swampy lake lay among these plains; the Copais, of karstic origin, had a variable extension (see below) within a basin which extended ca 23km E-W and ca 13km N-S. Two other lakes characterise the landscape of central Boeotia: Yliki and Paralimni (see below).
Boeotia is in fact also characterised by extensive limestone karstic landscapes, with a number of both open and closed *poljes*. The karstic landscape is generally marked by a poor surface hydrography and by a well developed underground hydrography - factors which affect water availability as well as the fertility of the soil and land potential for agriculture. The principal examples of this kind of landscape in Boeotia are the Copais area and the Domvraina/Thisvi *polje*.[3]

Physical borders of the region.
The borders of the Boeotia region (see map in fig.1 and fig.1 in chapter I.2.2) are clearly marked by high mountain ranges to the W (with the Helicon massif that separates Boeotia from Phokis) and to the S (with the Kithairon-Parnes ridge that separates Boeotia from the Megarid and Attica), and also by the sea to the SW and the NE (by the Gulf of Corinth and the Euboic gulf respectively). Less defined borders are to the NW, where the border with Phokis opens through the Kephisos valley, marking a free passage to Boeotia (with clear consequences for the history of the region – see chapter II.3.3), to the E, where the frontier between the political regions of Boeotia and Attica has fluctuated throughout history (see chapter II.3.14) and is difficult to pin down, and to the N, where the relatively low pass towards Opuntian Lokris could constitute an easy way through the hills bordering Copais towards the sea.[4] Through the low hills to the E of Mount Klomon (see fig.1) a route has long existed leading from Orchomenos and the Copais basin to Larymna (a harbour probably controlled by the Boeotians in certain periods[5]).
Due to these less defined bordering features, the borders we have used for our analysis are conditioned to being real in certain periods only,[6] thus partly arbitrary for use

[3] For the Domvraina/Thisvi *polje* see chapter II.3.10. See Farinetti 2008 for a more detailed description of the karstic landscape and the Copais.
[4] The modern road running along the Paralimni lake was cut into the rock fairly recently and does not mark an ancient route, while the low water periods around 1990 onwards at Paralimni and Yliki revealed ancient roads running at lower levels along the lakesides.
[5] See Oldfather 1916; Fossey 1990a; Papachatzis 1981: 158; RE XII.1, s.v. *Larymna*: 880-1.
[6] Political historical borders of Boeotia were different in different periods, as we know from historical and epigraphical sources. In prehistoric periods the region can be seen as a whole, with the majority of cultural and topographical characteristics in common with one or other neighbouring regions. We should therefore define the borders historically, and we have often done so while examining individual *chorai*, but in the regional landscape description we have merely chosen a reasonable border.

all the time, but mostly have some geographic 'attractors' to make them come into use at certain times.

As mentioned above, the highest mountains in the area form an almost continuous mountain wall marking the SW and S borders of Boeotia: from the W, Helicon, Korombili, Kithairon, Pastra, and Parnes. The mountains are mainly Mesozoic limestone and form large blocks connected through mountain cols. The Kithairon-Parnes mountain chain appears as an impressive towering wall. As Munn (1989: 231) states: "*The ancient likewise thought of this mountainous region as a barrier between the lands of the Athenians and the lands of the Boeotians,*[7] *for this was a dividing line between ethnoi*", and the frontier was primarily a cultural frontier. On the other hand, it was not inevitable that nations were divided by it, since, as Munn (1989: 231) points out: "*the mountain region itself is by no means impenetrable, nor is it all rugged terrain. There are valleys and small plains with good land for cultivation, and still more land for grazing, between and among the ridges of Kithairon and Parnes. And it was inhabited in antiquity*".[8]

The main route joining Attica and Boeotia, Athens and Thebes, ran, in Pausanias' time[9] and still at the end of the 19th century - Frazer 1898 (and even, in the form of a secondary road, still today), immediately to the W of Eleutherai, a small fortified citadel on the southern slope of Kithairon (see fig.1).

The harbours.

Ephorus (indirectly quoted by Strabo IX 2.2) declares the superiority of Boeotia over its neighbours (Phokis and Attica included), not only due to the fertility of its soil (see below) but also because it is the only region that has three seas and a large number of good harbours; these advantages have allowed the development of contacts with Italy, Sicily, Libya, Egypt, Cyprus, Macedonia, Propontis, Hellespontus, etc.

In fact, along both shorelines of Boeotia numerous harbours were active in ancient times; they were more numerous and efficient on the E coast, up to the Gulf of Euboea (Larymna, Skroponeri[10] to the NW; Anthedon to the N; Aulis, Dilesi, and Oropos to the NE),[11] as compared to the S coast, facing the Gulf of Corinth (Tipha, Kreusis, Siphai, Hagios Ioannes, Vathy and Livadostro).[12] Heurtley's article (1923/25) is still a good, if slightly dated, study of Boeotian harbours.

Some of the natural landings were in use only or mainly in certain periods, as in the case of Anthedon and Dilesi, which saw their peak in Hellenistic and Roman times. Others functioned from Prehistoric times throughout antiquity, such as the harbours on the gulf of Corinth, probably already in use in the Mycenaean period (Heurtley 1923/25), and then flourishing in the Late Roman period, being apt for long-distance trade.

The shoreline.

The main factors involved in shoreline modifications are sea-level changes, tectonic activity and subsidence phenomena in general, and sedimentation or erosion processes.

After the extensive general modifications of the shoreline during the Pleistocene caused by changes in sea-level related to glacial phenomena, the shoreline may have further changed during the Holocene in the Neolithic period because of ongoing tectonic activity (Perlès 2001, quoting Morrison 1968; Stiros and Papageorgiou 1994). The tectonic nature of these transformations also seems to characterise overall changes in the Mediterranean area, though it is not so clear. Most probably, minor and local tectonic events, related to sedimentary phenomena, also had some considerable influence.

In the Volos gulf and the Franchthi area, for instance, detailed work has shown a rise in sea level of ca. 5m during the Neolithic (Perlès 2001, quoting van Andel 1987. See Bintliff 1977 for the Aegean picture in general, and Lambeck 1996). Detailed work on this coastline variation, focusing on the problem of the position of ancient sites in relation to the sea, has been carried out by Zangger (1993) for the Argolid, among others.[13]

The climate.

Climate has, naturally, an important influence on the environmental conditions of a region, while topography has strong consequences on the climate.

In the case of Boeotia, the diverse and rough nature of the ground, marked by a series of valleys or lowlands surrounded by relatively high mountains, small plateaus, and closed basins, along with a greatly varying amount of exposure to the influence of the sea, and finally characterised by large-scale karstic phenomena, helps create and determines several local microclimates within the overall climatic picture.

Osborne (1996: 54) for instance notes: "*so local are the variations, that it is by no means guaranteed that a wet year in one place will also be a wet year 20 miles away. Some local variations are the product of physical relief: Thebes, in the middle of a basically flat Central landscape, receives on average only 63% of the annual rainfall of Levadeia, which is 25 miles west of it and in the rain-excess belt of Mount Parnassos*".

[7] Herodotus VII 141.3; Xenophon *Mem.* III 5.25; Strabo IX 1.3.

[8] The Skourta plain is a nice example of this (see chapter II.3.14 and appendix I.14).

[9] Pausanias IX 1.1. Via this route Pausanias entered Boeotia from Attica.

[10] Fossey (in *Teiresias* Suppl. II 1979: 9-13) suggests that the Skroponeri bay was a key military harbour during the period of Epaminondas and the Theban hegemony.

[11] See Oldfather 1916, Heurtley 1923/25.

[12] See below, in the section with the description of the gulf of Corinth area.

[13] Sea level rose steeply 12,000-6,000 years BP, and this led to widespread submergence of old coastal plains; it then slowed down and in many basins alluviation kept place or even gained, thus over time many Aegean plains move forward into the sea, making ports landlocked (Bintliff 2000d and Bintliff 2002). See for instance the case of the Malian gulf (outside Boeotia to the N, constituting the northernmost part of the North Euboean Gulf) and of Thermopylae (Higgins and Higgins 1996: 81-83).

Included in the Mediterranean Zone, the climate of Boeotia is classified as 'meso-Mediterranean' (UNESCO-FAO 1963).

The mean annual temperature reaches 18.5°C and the rainfall trend, marked by a strong annual variability with short but heavy rainfalls, as is usual in the Mediterranean climatic zone, reaches its highest peak in November and December and its lowest trough in July, with rainfall concentrated in the period from October to March (75% of the total).

Inter-annual rainfall variability is even more important when the annual mean is low, particularly during the dry summer months. This element, associated with the high summer temperature, can cause dry stretches three months long, as often happens in the Mediterranean sub-semiarid zone. In the past, this fact often had an impact on the periodical risk of famine, related to the fall of climatic conditions below the minimum level required for cereal growth.

The mean annual rainfall in Greece, at the present day, fluctuates between 463mm (measured at Tanagra, eastern Boeotia) and 470mm (measured at Salonika – Shiel-Steward 2007), with a pluviometric gradient of ca. 40mm every 100m elevation gap. The suggested mean seems slightly higher than the one calculated for Central Greece, but many scholars note that both in Boeotia and Argolid the mean annual precipitation can drop under 400 mm (Perlès 2001: 14 quoting Greig and Turner 1974, Forbes 1989, Hansen 1991).

Seasonal fluctuation and inter-annual rainfall variability, even without regional or general climatic variations, along with the strength of rainfall, can be considered factors for erosion processes as well as soil weathering, recurrent in the whole region. C. Perlès notes: "*The effect of the Mediterranean climate on the mostly massive limestone elevations resulted in particularly steep slopes, which were conducive to important erosion. The sediments washed down by violent seasonal rains accumulated in the many deep depressions of tectonic and karstic origins, which are equally characteristic of the Greek countryside. Contrast between the rugged and steep mountains overlooking absolutely flat inner basins remains to this day a powerful experience*" (Perlès 2001:10).

The high variability and the violence of seasonal rainfalls also have a large influence on the systems of aquifers and springs, as well as on surface hydrology and the dynamics of wet zones. They can also be factors in the frequent flooding that still today affects some of the lower areas, which are also usually the most fertile areas. Finally they can explain the human effort since ancient times, to regulate and canalise water, which can be seen in high level hydraulic engineering work in the region.[14]

For thousands of years, most probably since the Final Neolithic / Early Bronze Age (from around 6000 BP onwards – Roberts 1998; Bintliff 1992b and 2002), the climate dynamic of Central Greece has been broadly comparable with the semi-arid Mediterranean climate today, characterised by winter rains and dry hot summers.

We can, however, trace some variation locally in different periods. Variations, such as increases and decreases in temperature, are recognisable on a general scale in different times from the Neolithic up to the so-called 'Little Ice Age' of Medieval-Early Modern times (from around the 16th to the mid 19th century AD). Generally speaking, Boeotia has (and most probably had) a less arid climate than neighbouring Attica, Euboea and the Argolid (Shiel 2000; Bintliff 2002).

As for Greece in general, average summer temperatures would have most probably been slightly lower around 7000BC and rainfall slightly heavier (Perlès 2001: 13 quoting Huntley and Prentice 1988 and Bottema 1994; Bintliff 2002).

In addition, as Osborne (1996: 55-6) notes, even though we have no ancient climatic statistics, we do have a number of descriptions of plant communities and of the agricultural year from ancient Greek writers of the Archaic and Classical period (for Boeotia, for instance, we can refer to Hesiod, living ca. 700 BC[15]), through which we can infer that the seasonal pattern in antiquity was much the same as the seasonal pattern today.

The hydrology.

Both the geo-morphological structure and the climate of Boeotia influenced and determined the surface and underground water system.

AQUIFERS

On a general scale, considering the main hydrogeological characteristics and geological formation of the area, three main aquifers can be recognised in the region (Papadopoulou – Gournellos 1996; Lambrakis – Kallergis 2001; Daskalaki – Voudouris 2008):

1) *calcareous and calcareous dolomite aquifers*, formed by limestone and dolomitic limestone complexes with a medium or high fissure and karstic permeability, and characterised by a relatively deep and powerful subterranean hydraulic circulation. The flow of the springs of these aquifers is about 100 to 1,000 litres per second;

2) *gravel-silt-sand aquifers*, formed by deposits of fluvio-lacustrine detritus complexes with a variable porous permeability, and characterised by a quite shallow subterranean hydraulic circulation, located primarily in the cones of dejection or in the lacustrine deltas. The flow of the springs is only a few litres per second, unless they are also fed by the water-bearing strata of the calcareous complex;

3) *sandstone-marl, limestone-marl and clayey-schistose aquifers* (limestone-marl and sandstone-marl-clayey systems), with a medium to low or absent fissure and porous permeability, characterised by a poor subterranean hydraulic circulation. The flow of the springs is a few litres per minute.

SURFACE HYDROLOGY

The modern landscape of central Boeotia is marked by two lakes: Yliki, formerly Likeri, and Paralimni (see below); it is also characterised by two karstic *poljes*,

[14] See, for instance, chapter II.3.10 for the Domvraina/Thisbi area, and Farinetti 2008 for Copais.

[15] Hesiodus *Op.* 609-17.

today drained: the Domvraina/Thisvi *polje* and the large Copais basin (see appendix II and Farinetti 2008).
The main rivers which cross the Boeotian landscape (see map in fig.1) are: the Kephisos (see appendix II and Farinetti 2008) and Melas (see appendix II and Farinetti 2008), the main rivers of the NW area; the Asopos (see appendix II - iBb), the main river of eastern Boeotia; smaller rivers and streams around Copais (see appendix II and III); and other smaller rivers and streams such as the Askris, Livadostro and Kanavaris within the area of the Tertiary *tafel* of Thebes (see appendix II - Tertiary *tafel* section and chapter II.3.9), and the Lefkos and Ismenos in the eastern (Aeonian) plain of Thebes (see appendix II - iA).

The vegetation.

According to Rackham (1983), within the panorama of vegetation in continental Greece, Boeotia is characterised by an extensive presence of low, woody and herbaceous vegetation (the so-called maquis, steppe and garrigue) with an important presence of wood coverage mainly limited to the N facing slopes and plateaus of the high mountains. Certainly a large part of the landscape, especially the fertile lowlands, the hill slopes and the gentle versants of the mountains, has been cultivated for millennia. The typical landscape of the Mediterranean polyculture is still today dominant in the larger part of the region. The core of the landscape seems to be still occupied by the original vegetation. Even these wild areas, however, show the vegetational characteristics of the well known group called "anthropic steppe" (grazing and forest cover alternated) – the result of biological migration encouraged by human vectors and ecological receptivity – which invaded the areas cleared, or at least occupied, by humans and then subdivided itself into *facies* according to the diverse microclimates.
As far as vegetational history and human influence are concerned, Perlès (2001:14-15) notes that the main distinctive character of Boeotia and central and eastern Greece compared with northern and western Greece, where the original oak forest coverage was very dense (due to more frequent rainfall), is that in the former the forest would have been naturally more open (at the beginning of the Neolithic era), especially from Boeotia southwards. The cutting of the forests due to agricultural requirements, along with the diffusion of grazing animals and the nature of the climate (marked by hot and dry summers), favoured the dispersal of steppe species and prevented the replacement of cut trees, causing a general diminishing of the forest. Low-growing shrubs and small trees (juniper and terebinth, usually under-represented in pollen analyses) would have been the dominant vegetation (Hansen 1991: 18).
The pollen diagrams for the former Copais Lake (Allen 1990; Greig and Turner 1974; Turner and Greig 1975; Turner 1978) indicate, for the first half of the Holocene, a deciduous oak woodland on the deeper soils, evergreen oaks on the thinner soils, and a steppe vegetation on the slopes (Perlès 2001: 14-15 quoting Rackam 1983). Pollen analyses and the study of the magnetic properties of the sediments from these cores, related to exposed sections of lacustrine sediments in the basin, allow for a first reconstruction of the climatic-vegetational evolution of the basin and constitute a useful interpretative guide for the valleys and lowlands of Boeotia. This research also gives indications of traces of human intervention from ca. 27,000 years BP up to 3,000/2,500 years BP (Allen 1990; Turner and Greig 1975).
In the period between the Pleistocene-Holocene transition and 3,000/2,500 BP, vegetation is typified mainly by an abundance of *Quercus sp.* (oak) up to ca. 5,000 BP, and after that by the decline of this and the gradual development of other tree species, such as *Fagus sp* and *Carpinus betulus*, and most especially graminaceous and herbaceous species, as well as by a small but significant growth in the presence of *Olea sp.* (which might indicate agricultural activities). According to Bottema (1994: 56), the presence of terebinth and juniper indicated a dry, open woodland around Copais, at about 8000 BP. The forest became denser afterwards (Bottema 1994: 57; Perlès 2001: 14-15).
As for magnetic properties, in the most recent part of the sequence the decline of *Quercus* corresponds to a series of peak values of magnetic susceptibility: "...*It must therefore be considered a possibility that the sediment at these levels comes from the inwash of magnetically enhanced eroded topsoil from catchments slopes*..." (Allen 1990: 177). Both pollen and magnetic analyses seem to indicate the decline of the humid postglacial phase during the period ca. 5,000 years BP, along with the spacing out of forest coverage (due to the decline of oak) in favour of a more arid climate,[16] bringing about the denudation and erosion of hill slopes. According to several authors, these processes of vegetational transformation and increasing erosion can be related to human impact, in association with climatic changes, either to fairly destructive agricultural activities (for example the 'slash and burn' practice), or rather to an over-exploitation of land related to socio-economical factors (see Zangger 1992, Roberts 1998 and Bintliff 2002).
Unlike N and W Greece where the original oak forest coverage was very dense (due to more frequent rain), in central eastern Greece the forest would have been naturally more open (at the beginning of the Neolithic era), especially from Boeotia southwards, and the climate would have been more pronouncedly Mediterranean (Bottema 1994: 57; Turner 1978). Once the forest was cleared, the hot and dry summers and the presence of grazing animals would have prevented the growth of trees in favour of steppe species (Perlès 2001:14-15). Osborne (1996:57-69) points out how areas of aromatic plants, characteristic still today of the so-called garrigue (evocative of the Greek countryside, and of sage and thyme for instance), which develop on soils too short of moisture for woodland plants, would have been brought into existence solely by the pasturing of animals. On the other hand, Rackham (1983: 304-5) argued that for Boeotia he could find no evidence to support the view that garrigue and steppe are results of original woody vegetation (macchia) successively degraded by browsing,

[16] i.e. the traditional dry climate in the Mediterranean.

burning, and woodcutting.[17] He suggests that the ratio of macchia, garrigue and steppe depend mainly on soil moisture and, when water is scarce, garrigue and steppe can exploit even the water stored beneath the areas occupied by patches of macchia. The spacing of macchia patches would depend mainly on how much ground is needed to supply each patch with water, often independently from browsing activities.

PHYSICAL LAND UNITS OF BOEOTIA

According to the methodology illustrated in chapter I.2.1, the study of the Boeotian landscape, which is the subject of the present work, also involves the examination of the physical landscape of the region and the evaluation of land potential (mainly for agriculture) by means of GIS.

Physiographical classification of landscape units.
Some results of the GIS analytical work carried out on environmental data and described in chapter I.2.1 can be commented on here, at a general regional level, in order to describe the physical landscape of the Boeotia region.

1	P1_P2	lacustrine basin, valley	26.6%
2	P3	gentle slope	3.5%
3	P4	Foothill	9.8%
4	H1	Plateau	18%
5	H2	gentle slope	4%
6	H3	moderate slope	8.9%
7	H4	severe slope	10.2%
8	H5	very severe slope	7.5%
9	M1	plateau	2.7%
10	M2	plateau / gentle slope	1.7%
11	M3	moderate slope	3.6%
12	M4	very severe slope	3.4%

Table 1. *Percentage of the different physiographical classes present in the region of Boeotia (P=plain; H=hill; M=mountain)*

The division of Boeotian landscape into three main classes according to elevation (plain, hill, mountain), and then into physiographical units (such as valleys, plateaus, watersheds, hill slopes, etc. – fig.2), allows us to analyse sections of the landscape (as we will do in the examination of each individual *chora* or micro-region) as well as the Boeotian landscape as a whole, which is our technique in the present chapter. We will address here several questions regarding the whole region.

Is Boeotia a mountainous area? GIS analyses of topographical data revealed that 11.5% of the ancient Boeotian region[18] lies above 600m asl. On the other hand, 88.5% of it lies below 600m asl, with 40% lying below 200m asl (see pie-chart in fig.2).

Despite these values (which are due especially to the fact that our dividing line between hill and mountain has been set at 600m asl), the presence of bordering mountains (such as Citaeron and Parnes) and the Helicon massif and the Ptoon-Messapion group marks and defines the character of the Boeotian landscape. The majority of the ancient *chorai* of Boeotia include mountainous terrain in their territory, which can be therefore considered almost as much a defining element as the fertile plains.[19]
Slope values higher than 45% are found in 30% of the mountain landscape.
Considering the statistics and the fact that 30% of the plain features are constituted by the lakes and the Copais lake, we should probably better describe Boeotia as a hilly landscape, since 48.5% lies between 600m and 200m asl. The hilly segment of the landscape (upland-plateau land), extended and widely fertile, actually

Hilly landscape (above 200m and below 600m asl)	48.5%
Mountainous landscape (above 600m asl)	11.5%
Plain (below 200m asl)	40%

constitutes a critical landscape key for Boeotia, as settlement choices in different periods of history nearly always concentrated there. Generally speaking, the most critical divisions of the Boeotian landscape can be defined as: crystalline limestone uplands; intermediate elevation hill and plateau land made up of flysch and soft limestones; lowlands of Plio-Pleistocene colluvia, alluvia and lake deposits. The middle area is nearly always the key one preferred for settlement in every period.
Looking at the variety within these landscapes, by separating the individual physiographical classes, we can differentiate between different morphologies, as can be seen in table 1.
By focusing on the pattern of the three main morphologies – hilltop, hillside and valley – we can see that as one would expect, the highest slope values and the largest frequency of high slope values can be seen in the mountainous zone, while the lowland is marked by the presence of the plains.
In the hilly landscape zones, both valleys and plateaus (H1) are recognisable throughout the whole region and actually constitute the second most represented physiographical class in such areas (see pie-chart in fig.2). The largest mid-land plateaus which mark the landscape are the large plateau south of Thebes running to Cithaeron (longitudinally crossed by the Asopos), the Skourta plain, the Hyettos area north of the Copais and the whole Thespiae area, as well as the low ridge separating the Tanagra and Thebes plains. Among the H1 class, hilltops can also be recognised, if larger than 30x30m, which is the grid size of the RASTER layer (see chapter I.2.1).

[17] Rackham (1983) points out that animals cannot stop woodland growing, as shepherds have to regularly burn it to allow them to graze.
[18] With the artificial borders marked according to that stated earlier in this chapter.

[19] See dedicated chapter for each *chora.*

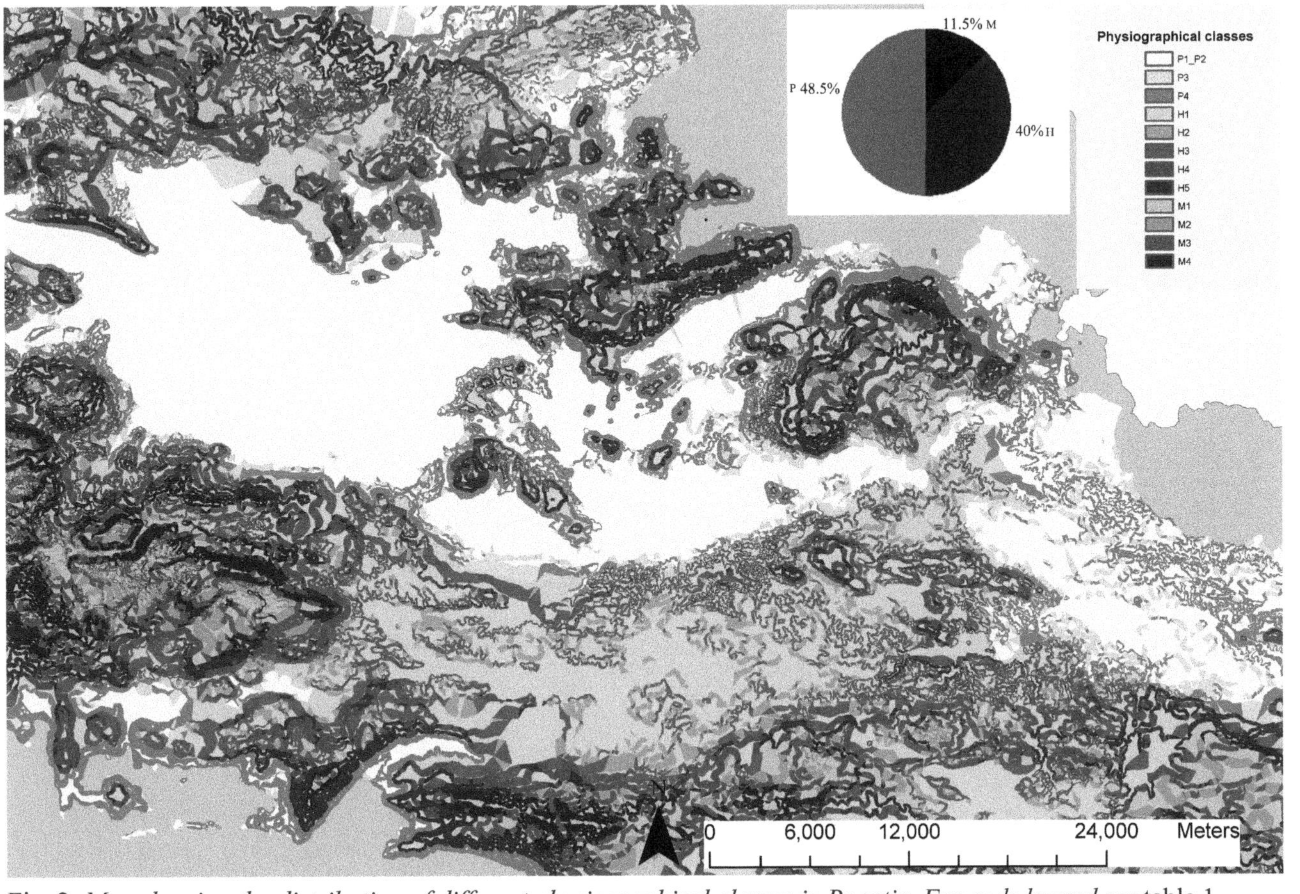

Fig. 2. *Map showing the distribution of different physiographical classes in Boeotia. For code legend see* table 1.

The 'foothill' class [P4] is clearly recognisable in several areas: at the edge of the lakes, along the coast to the Gulf of Corinth, around the Skroponeri bay, as well as corresponding to the dissected basins and plateaus round Copais, at Anthedon (modern Loukisia), or in the Valley of the Muses, or in a thin line at the edge of the Teneric plain west of Thebes. This morphology usually forms a line running along the base of the hills, or around basins and plateaus, but we can see areas with the same morphological characteristics (below 200m asl and slope ≥ 10%), though much more dispersed over the landscape, in the area E and NE of Tanagra up to the shoreline.

Within the hilly landscape zone, 37% is <5% slope (H1 plateau, see above). 21% is H4, 18% is H3, 16% is H5, and 8% H2. Therefore, if we exclude the upland plateaus (which I have pointed out as features characterising the Boeotian landscape), the rest of the hilly landscape is almost homogeneously distributed between the other classes, with a lower presence of H2 (gentle slope), which is limited to some fragmented areas mainly around the hill plateaus.

As mentioned above, the lowland is marked by the presence of the plains. 66% of the landscape below 200m asl is characterised by slope values lower than 5%, and they are, as we have seen, either lacustrine basins or valleys.

Soils and Land Potential

The dominance of limestone in the geological substrate of Boeotia and the presence of karstic processes have a direct influence on the nature of the soils of the region.

Soft limestone and calcareous sandstone are dominant in the uplands. Marls and flysch are extensively distributed in the mid-altitudes. Flysch presents great variability (see chapter I.2.1), but generally form deeper soils of good workability.

The bases of the relief frequently display cones of gravel and conglomerates of the Pleistocene age. On the gentle slopes of low hills at the borders of plains and valleys, the older fill can be found; the lowlands and the valleys are covered by Holocene silty alluvium and soil, as well as large areas of Pleistocene soil.[20] In general, within the Greek soil spectrum, the soils of the Boeotia lowlands are fertile and workable.[21]

[20] Large areas of Pleistocene soil are most of the Teneric and Aeonian Plains and some areas around the edges of Copais to the E, N and W of the lake.

[21] For soils see mainly Higgins and Higgins 1993. For a synthesis see Perlès 2001: 10-11. For Boeotia see Shiel and

Fig.3. *Map showing the distribution of land capability classes in Boeotia.*

F	Soils with few limitations restricting their uses because they demonstrate a slight a tendency to wetness or dryness or very moderate erosion.	41.9%
MF	Soils that have some limitations for agriculture, reducing the potential choice of crops or requiring moderate conservation practices, due to wetness caused by a variety of factors.	5.3%
LF	Soils that have more limitations for agriculture, reducing the potential choice of crops or requiring moderate conservation practices, due to dryness and limited soil depth, causing rooting problems.	17.4%
U	Soils that have severe limitations for agriculture, restricting the choice of crops and requiring very careful management, due to very limited soil depth and a high slope degree, causing acute difficulties for access and rooting.	35.4%

Table 2. *Land potential classes and their occurrence in the Boeotia region.*

A classification of land potential has been made for the Boeotia region, for the purpose of the present work (see chapter I.2.1). A summary of the classes and their occurrence in the region can be found in table 2 (see chapter I.2.1 for details).
As described above (see I.2.1), the land potential for agriculture has been evaluated mainly according to lithology and to some visits on the ground. Quantifying the results, a large part of the whole region can be considered suitable for agriculture with very little (almost no) limitation.

As we can see in table 2, the land best suited to agriculture (classes F and MF) accounts for 47.2%, while LF areas, where agriculture can be practiced if strictly necessary, occupy 17.4% of the whole region.[22]

Steward (2007). See the chart based on Shiel and Steward's work (2007) on Thespiae in chapter I.2.1 (table 3).

[22] Bintliff (1985: 15-17) suggests that in antiquity possibly half of the total land of ancient Boeotia could have been in cultivation. In 1961 the cultivable area was put at around one-third of the country. Allowing for the slightly different boundaries of ancient Boeotia and the varying size of the large Copais Lake, Bintliff calculated that this would be equivalent to some 800 sq.km. of cultivable land in antiquity, applying the same ratio. He suggested, on the basis of historic sources, that around 650sq.km. of arable land would have been required to

On the other hand, with regards to availability of pasturage areas, mountainous landscapes of Boeotia marked by sandstone and/or limestone can be devoted to husbandry, especially when the slope is steeper and scarce soil on the surface makes the land less suitable for cultivation. Quantitatively speaking, land suitable for pasturage, which can include LF and U classes, amounts to 52.8%. We should keep in mind, though, that F and MF land zones could also be used for grazing either during fallow or after harvesting, especially in the case of mixed economic systems or of agricultural systems characterised by culture rotation.

THE GEOGRAPHICAL SUB-REGIONS OF BOEOTIA

A good description of both the physical and historical landscape of Boeotia was given by Philippson in his book '*Die Griechischen Landschaften. Eine Landeskunde*'. The description of the Boeotian landscape is included in one of the five volumes published between 1950 and 1959, and in particular in volume I, part II (*Das östliche Mittelgriechenland und die Insel Euboea*), published in 1951.

In his very detailed yet discursive way of describing the Boeotian landscape, Philippson follows two different levels of landscape analysis. The first, which is much more detailed and discursively treated, approaches the region through various small landscapes and describes them with a very high level of detail in terms of physical and cultural landscape, and the second, which constitutes the top level of description (containing the other), and divides the region of Boeotia into five areas with distinctive geographical characteristics. These divisions, generally speaking, either correspond to the historical divisions into *chorai* around the main *poleis* of Boeotia (which are mainly defined by topographical boundaries), or combine together more of them into larger geographical units. The geographical areas are the plain and foothills of Chaironeia, the Helicon district (West Helicon, Laphistion-Levadeia-Koroneia, East Helicon, South foothills of Helicon and the Korombili area), the Copais basin, the Mountains to the E of Copais, and Eastern Boeotia.[23]

In the treatment of archaeological data (mainly at a descriptive level), I will follow a subdivision by micro-regions/*chorai*, defined both by topographical and political-historical characteristics.[24] On the other hand, at the analysis level, my study will follow Philippson's detailed level, in individuating small cultural landscapes with distinctive physical and cultural characters, which can coincide with our 'settlement chambers' (see chapter I.1.2).[25]

maintain hoplite and aristocratic classes alone. He inferred that probably in ancient times a far larger proportion of the landscape was under cultivation, or at least intensive herding, say 40 or even 50 per cent (in Classical times) rather than one-third (Bintliff 1985: 15-17 and 1997b: table 1).

[23] A brief general description of the geographical areas individuated by Philippson is included in appendix II.

[24] See chapter II.3 and *chorai* chapters.

[25] I approach the small landscapes detected by Philippson and apply within this framework the revised settlement chamber approach discussed earlier in chapters I.1.2 and I.2.2.

II.2

The Boeotian landscape: state of archaeological research

The archaeological landscape of Boeotia has been studied in different ways throughout the centuries of archaeological research, according to different perspectives and approaches applied. Many descriptive catalogues and monographs on archaeological findings and excavation, as well as synthetic studies and broader perspective research, some of them with a strong historical focus, have been published to date, making available to us a large amount of data. Within the framework of the present research, all archaeological information available for the region of ancient Boeotia has been collected and inserted into the database, according to the methodology described above (chapter I.2.2). This mainly included: archaeological reports, systematic and rescue excavation reports, extensive topographical research, intensive surveys – published and preliminary results, ancient travellers' reports (examined partially).

This series of information produces inhomogeneous, non-linear datasets, marked by different levels of complexity. For instance, different degrees of detail and complexity characterise the information and the archaeological evidence available from the bibliography (in a broader, regional perspective) in comparison with those resulting from more recent intensive and systematic artefact surface surveys, which intensively cover the landscape in search of all traces of human activity in smaller defined areas.
As seen earlier (chapter I.2.2), a field named 'discovery' is present in the database structure, in which is input an indication of the research framework within which the discovery of each individual cluster occurred. In order to maintain coherence, the choices are limited to: extensive topographical surveys, intensive topographical surveys, intensive and systematic topographical surveys, intensive and systematic artefact surface survey, historical geography, rescue excavation, personal or group interest, travellers, accidental, other (see table 6 in chapter I.2.2). In filling in the field for each cluster, attention has been paid to the first discovery (when known) of the particular cluster (in this way monitoring the knowledge actually available), while later researches are indicated in the bibliography (and indicate the quality of the information). This is why our research was primarily interested in the mapping and monitoring of available archaeological knowledge and its possible biases within each area under study.
By indicating the source type, the archaeological information can more easily be checked and monitored, and by querying the field, the database user can easily find all records (clusters) found through a precise discovery type, and detect biases in discovery. (This analysis has been performed, for instance, for each *chora* – see *chorai* chapters.)[1] Discovery type is indicated for each single cluster, also for multiperiod sites, in an attempt to indicate the discovery type for each single period of occupation or use, each of which constitutes a different cluster. With the value 'other', for instance, I indicate cases in which evidence for certain periods was discovered 'by chance' on a site whose main occupation is investigated by archaeologists interested in other periods or other site types.
In fact, the whole of the archaeological record available for Boeotia comes from one of those different kinds of research trends or discovery occasions, and every source of information available can be classified in one of these groups. Generally speaking, an overview of the cluster discoveries would give us a general overview of the state of research on the Boeotian archaeological landscape.

According to the classification available in the database (see above), the available data will be briefly presented in this chapter, in order to give a quick overview of the state of research on the Boeotian landscape.

DESCRIPTION OF AVAILABLE DATA

Archaeological information available for the region has been collected until 2006.
Primarily descriptive sources (such as archaeological reports or gazetteers) must be distinguished from analytical and synthetic works, which do not directly present the archaeological information but rather try to analyse the settlement and the landscape history (even if partial, concerning only certain areas, certain periods or a certain topic of research), and can therefore be included in the research and interpretation already carried out on Boeotian landscape history.[2] Here, I will present mainly descriptive sources, from which the archaeological record available is actually generated, aiming at the construction of the archaeological map. Synthetic works are quoted

[1] In an attempt at monitoring the biases in the archaeological knowledge available for the region, I also proceed, in each *chora* chapter, to other consideration of the data (metadata inquiries), e.g. GIS analysis to monitor the relationship between discoveries and proximity to modern roads.

[2] Some works present both aspects, e.g. Fossey 1988 (with gazetteers and settlement analysis), or Lauffer Kopais I (with comments on the settlement pattern attached to site descriptions).

within the single analytical and synthetic chapters of the present work.

Available sources of archaeological information include:

Extensive topographical survey reports: site-oriented extensive research, conducted over the whole region or in wide areas, including mainly Fossey's 1988 detailed gazetteer of Boeotian sites (certainly the starting point for the present work).

Intensive topographical survey reports: site-oriented research conducted intensively (but not systematically, and not involving artefact surface survey) over quite small areas, such as, for instance, the very detailed and useful work conducted by Lauffer in the Copais area (Lauffer Kopais I), or the one conducted later by Knauss et al. (Kopais 1, 2, 3).

Intensive and systematic topographical surveys: this category includes, for Boeotia, various research: part of the work done by Munn in the Skourta plain area, when fieldwork was conducted (apart from the areas where transecting was carried out – see below) by examining the landscape (25% of the whole area, especially in the locales of Pyli, Skourta and Stephani), looking for sites (in this sense I indicated the survey as topographical), but without counting sherds systematically. It was not a proper artefact surface survey but was nevertheless conducted in an intensive and systematic way. The category also includes Fossey's rural survey in the Chostia area (territory of ancient Chorseiai – Fossey Chostia I), as well as David French's Prehistoric site survey for Central Greece, carried out since the 1960s (French 1972).

Intensive and systematic artefact surface survey reports and/or preliminary data: in Boeotia, intensive and systematic artefact surface surveys have been carried out in different areas (mainly the areas of Thespiae, Haliartos, Hyettos, Tanagra and lately Koroneia) by the Boeotia Survey Project (since 1978, directed initially by Bintliff and Snodgrass, then by Bintliff alone, and currently by Bintliff and Slapšak – see bibliography for references),[3] in the ancient Thisbe area and in the islets in the gulf of Domvraina by Gregory (1980, 1986 and 1992), in the Skourta plain by Munn (limited to the transect survey - see bibliography for references), in the urban area of Plataea by an Austrian team led by Konecny, in collaboration with the local *ephoreia* (in the person of V. Aravantinos), all characterised by intensive and systematic fieldwork, albeit conducted according to slightly different methodologies (fig.1).

Historical geography: including mainly research focused on the quest for *poleis*, town and village settlements and cult places (Kirsten's and Pritchett's studies, for example).

Rescue excavation: archaeological reports, both Greek (AD; Praktika; AAA; Ergon) and foreign (mainly journals of the foreign archaeological schools in Athens: Archaeological Reports, BCH chronicles, etc.), are the main source of information on rescue excavations, as well as some conference papers (especially from the proceedings of the Boeotian international conference[4]) presenting results from new (often unpublished) rescue excavations.

Personal or group research interest: this category includes site-oriented and period-oriented research (Dickinson's work for instance, HS&D 1979), problem-oriented research (for example the work of Knauss et al. on Copais,[5] or other works published mainly in journals concerning Greek studies) as well as systematic excavations or test soundings carried out to address specific questions or to investigate further sites already known.
Gazetteers reporting mainly Prehistoric sites, or evidence from later periods only when they occur at Prehistoric (or possible Prehistoric) sites (for instance, Hope Simpson 1965, HS&D 1979, Syriopoulos 1968 and 1983-84), are usually included into the category of 'Personal or group interest', and sometimes in 'Ext. topographical surveys' according to the kind of research that led to the actual discovery of the site.[6]

Travellers: some sites were first discovered during the journeys carried out by early travellers in the 18th and 19th (and the beginning of the 20th) centuries. They especially visited the main sites known from historical sources. Among other travellers see especially Leake, Gell, Ulrichs, Ross (see bibliography for reference). Travellers were not read systematically or thoroughly, but were checked only on certain occasions with regard to specific problems (see quotation in text and footnotes).

Other: mainly cases in which evidences for certain periods were discovered by archaeologists interested in other periods or other site types while their investigation was in progress.

Accidental: Accidental discovery, mainly from field ploughing or tomb robbing/illegal excavations.

[3] For the present work, I have inserted into the survey sites database only the artefact concentrations securely dated, mostly from preliminary reports (never fully and systematically published or currently being published) – see each individual *chora* chapter.

[4] *A', B'* and *Γ' diethnes synedrio Boiotikon meleton* (1988, 1995 and 2000).

[5] Knauss et al. mainly KOPAIS 3, but also several articles.

[6] Gazetteers are obviously taken into consideration for this purpose only when scholars working on them first discovered the particular cluster.

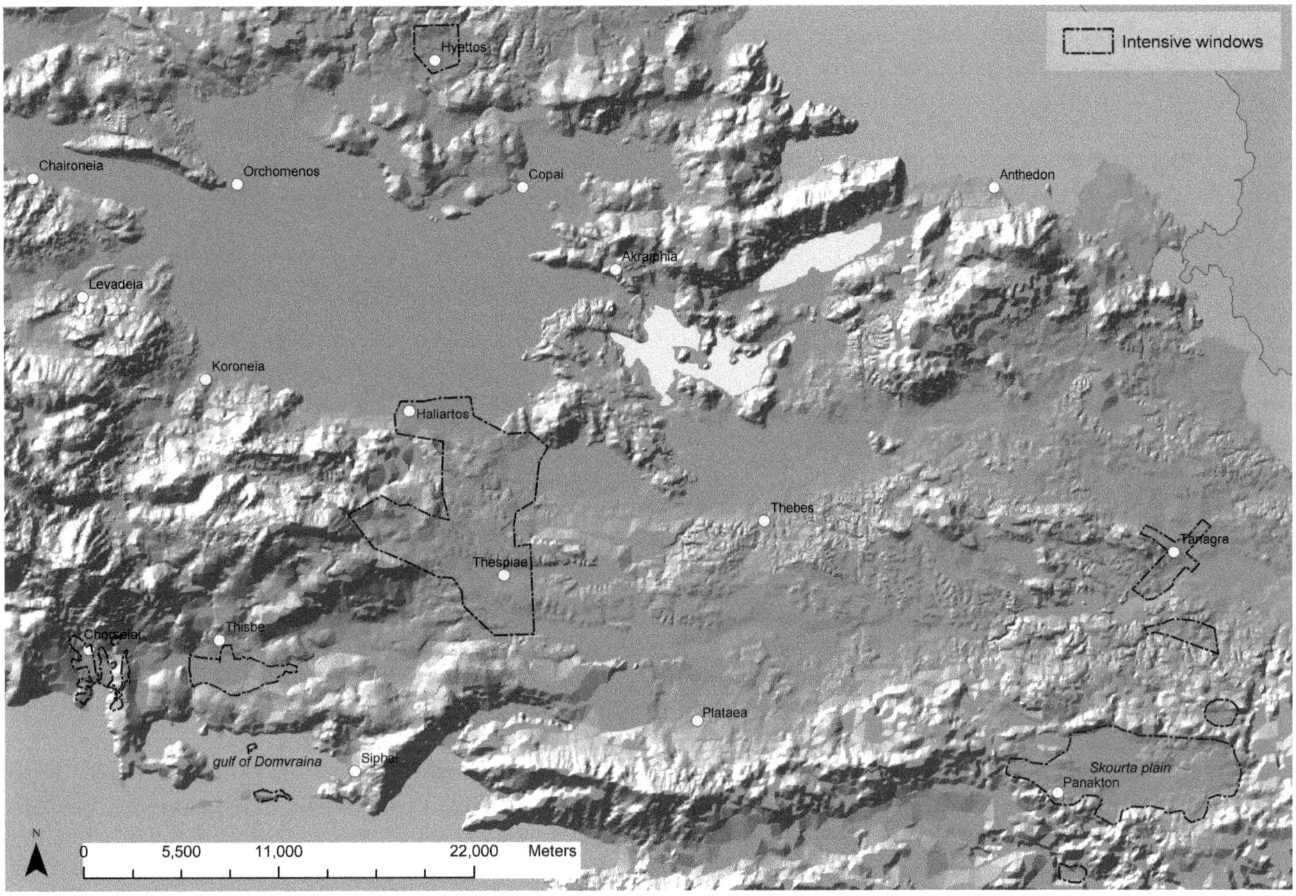

Fig. 1. *Areas of Boeotia intensively and systematically surveyed.*

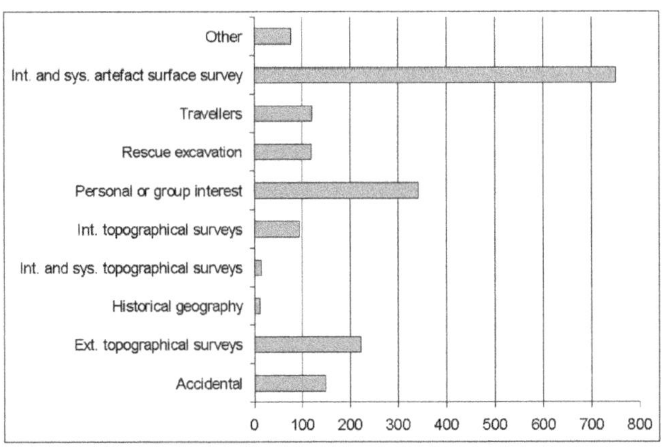

Fig.2. *Graph illustrating the proportion of components discovered in the whole ancient Boeotian region within different research frameworks.*

In addition, there are some sites known only from ancient texts which mention *polis* names, cult place dedications or distances between places, sometimes providing an indication of their position, but sometimes not. For the present work, only located sites, archaeologically known (whether mentioned by ancient sources or not), have been included in the database. Mention of the other sites (known only from ancient texts) has been made in the chapters for the individual *chorai*, in the majority of cases within the description of a cluster, in discussing the possible identification, or, in the case of settlements, also in the concluding discussion of village places and settlement chambers.

Among the ancient texts, Strabo's and Pausanias' itineraries certainly constitute the first, contemporary, travellers' reports, used by many scholars working on Boeotian topography.

The starting point for the collection of available archaeological information was Fossey's 1988 book - still the richest source of information as far as extensive research on ancient settlements in Boeotia. Following Fossey's general structure, data are also gathered according to *chorai* (ancient *poleis* territories) in the present work.

BIASES IN AVAILABLE SOURCES

As we have seen, some of the available sources focus on a particular period, sometimes also mentioning the presence of other periods (though unsystematically, as in the case of the Prehistoric Gazetteer for instance), others are diachronical, and take into account different periods (naturally scholars who compile the accounts are more familiar with certain periods and their data can therefore be slightly biased), others are concentrated on a specific area / on an area of specific interest (works on Copais for instance) and this results in areas better and less known (we know almost nothing about the Asopos valley, and we have little evidence from the Livadeia area, etc.).

Some of the sources give the precise location of the evidence they report, leading to an easy positioning on the available topographical base maps, while some others allow us only to guess just few details about the location.

Thus, in taking information from diverse bibliographical data sources, the resulting archaeological map will be quite variable in density of mapped clusters as well as detail of information (accuracy of the attribute, in GIS language) and accuracy of position. We therefore face problems of integration of data from different data sets; some more accurate and less biased than others.[7]

Generally speaking, as often happens in archaeological research, the information available is biased by personal research interests or research targets.
On the other hand, we can obtain a less biased picture from rescue excavation data rather than from data collected within the framework of site-oriented or period-oriented research. From rescue excavations, for instance, we might obtain information on the rural segment of the landscape, usually known almost only by artefact surface survey, and, both in the former and in the latter case, we might get 'unexpected' information, as neither type of investigation is focused on a specific and restrictive research aim.
Moreover, a basic distinction can be made between studies producing data qualitatively described (topographical work, some excavation reports, etc.) and studies producing data quantitatively collected (artefact surface survey, recent systematic excavations, etc.).[8]

Conversely, in the case also of intensive artefact surface surveys, even if realised in the same methodological and theoretical framework, the results obtained are not always easy to compare (see van Leusen 2002: chapters 2 and 13, Alcock – Cherry 2004, and chapter I.2.2 in the present work).
In Boeotia, several intensive and systematic artefact surface surveys have been carried out in different areas in the last three decades: the Boeotia Survey Project, mainly in the areas of Thespiae, Haliartos, Hyettos and Tanagra (since 1978; directed initially by Bintliff and Snodgrass, then by Bintliff alone, and currently by Bintliff and Slapšac);[9] Gregory's project in the Thisbe area (Gregory 1980 and 1992); Fossey's survey of the area of ancient Chorseiai (Khostia I); the Skourta plain survey by Munn and his team (limited to the transect survey[10]); the Plataea urban survey by an Austrian team led by A. Konečny in collaboration with the *ephoreia* (V. Aravantinos)[11] – see fig.1.[12]
Those projects are all characterised by intense and systematic fieldwork, albeit conducted according to slightly different methodologies.[13]

Within the tradition of gazetteers, either diachronical or period-oriented, we can detect different approaches and tendencies.

Among others, Fossey, especially in his topographical gazetteer published in 1988, shows a particular interest in some areas (especially the area to the Gulf of Corinth, where he also carried out an intensive survey project, at Chostia) and in some periods or particular types of archaeological evidence (fortifications and forts, for instance).
In general, he sometimes reports on places he visited personally (fieldwork carried out mainly in the late 1960s and 1970s), either known sites or places he seems to have discovered, giving fresh information from the ground, while other times he reports information from other sources (accounts up to ca.1980-1982).
As Bintliff (2000a) points out, Fossey's book is the natural, if not the *only* starting point for the construction of an archaeological map of Boeotia. "*It offers an excellent survey of the published material by the mid-1980s, and also provides a useful basis for propositions regarding settlement dynamics testable by further more intensive landscape research*" (Bintliff 2000a: 123).
His approach is extensive and site oriented, and biased by a sometimes acritical approach to the sources employed. The quantitative aspect of the evidence he reports on (the quantity of sherds he saw) is usually unclear, and needs to be questioned according to the issue of representitivity (see chapter I.1.2). This is not only the case for his own visits, but also an inner bias in earlier accounts that he uses (that often create settlement sites from a few sherds on the ground). Despite all this, his collection of information constitutes the main resource of our

[7] For a broad discussion of this issue see Van Leusen 2002, focused on the integration of data from intensive and extensive surveys, as well as from intensive surveys of different kinds carried out in different periods and within different research frameworks.

[8] In general, it must be pointed out that the archaeological record has suffered damage during the landscape's recent history (e.g. through deep ploughing, intensive building and industrial activities).

[9] Bintliff 1985; Bintliff – Snodgrass 1985; Bintliff 1992d; Bintliff – Howard 1999; Bintliff – Howard - Snodgrass 1999; Bintliff et al. 2000; Bintliff et al. 2004; Bintliff 2006; Bintliff - Howard – Snodgrass 2007; Bintliff – Slapšac 2007; Bintliff et al. 2008.

[10] An intensive and systematic artefact surface survey (fieldwork from 1985 to 1989) carried out in 3 areas: the Panakton area (around Panakton); an area NE of Stephani; an area to the NE of the latter and further from Stephani (marked on fig. 5.2 of Munn – Zimmermann-Munn 1989).

[11] See *ArchDelt* and *AReports* on the project, from 1997 onwards.

[12] One could add to the list the the first intensive sherd collecting, at modern Dritsa/Harma (ancient Eleon), in the 1930s (!), illustrated by a photograph in van Effenterre's book, Les Boeotiens (vanEffenterre 1989: 29). Recently, a project (The Eastern Boeotia Archaeological Project) has started in the area of on the plains surrounding the modern villages of Arma, Eleon, and Tanagra (first report in *Teiresias* 37 (2007) part 2).

[13] We should keep in mind that even data collected by intensive and systematic surveys should be processed with a strong 'critique', mainly comparing on-site data with off-site data collected in their immediate surroundings (see Bintliff – Howard 1999 for the methodology in use for instance within the framework of the Boeotia project).

archaeological database, and he usually provides, through his own visits or other sources of information available to him, more valuable information as far as the chronological aspect is concerned.
One could also note that surface pottery reported by Fossey seems to be, in the majority of cases, mainly fine ware, because he uses it as diagnostic for dating. In this way, though, we lack information on the actual presence of coarse ware on the sites, even if undatable. On the other hand, Fossey certainly recognises Greco-Roman wares better than does Lauffer or other scholars whose major interest is the Prehistoric period.

As for Prehistoric gazetteers, they note, when available or when noticed, evidence from later (Historical) periods, but since this is not their main interest (nor their specialisation), their information on historical periods can not be taken for granted, and most of the time is incomplete, even if useful (especially HS&D 1979). On the other hand, Fossey, as we saw, through his own visits or other sources of information available to him, provides more valuable information as far as the chronological aspect is concerned.
As for the majority of Prehistoric gazetteers, the main problem is the spatial fuzziness of their data, which are rarely precisely located in space. Though HS&D 1979 report geographical coordinates of sites (probably approximate central location), for instance, it is not always clear precisely where at the site something was noticed, especially for the more complex settlement sites.
A useful work is that of Syriopoulos, who gives useful information especially on the Geometric (and Protogeometric) period, sometimes overlooked by other studies. In his two gazetteers (1968 and 1983-4) of Prehistoric sites, he sometimes copies information from previous similar works (Hope Simpson 1965, for instance, and in his second publication also HS&D 1979), and he sometimes reports new information from personal visits or recent accounts.

Papachatzis' work (1981 - a topographical account on the basis of Pausanias' text) has also been used on some occasions. He combines Pausanias' text with the results of modern excavations and visible monuments, in a clearly narrative approach.

Pritchett's topographical work[14] has been used especially for marking ancient routes, and also for discussions on some particular sites and identifications.

As far as the theoretical background of their work is concerned, these studies are mainly extensive topographical research, also considering the periods when they were carried out.[15] The most recent general work (Fossey 1988) still tends in that direction.

On the other hand, Lauffer's work (summed up in the 1986 Kopais I book, but actually carried out largely before the Second World War) moves towards a much more intensive level of research, apparently less site-oriented. It is focused on the Copais area, with special interest in certain areas surrounding the lakes.
Examining his work considerably carefully, the impression was given that he visited some areas of the Copais basin and its surroundings more intensively then others. For instance, in the Koroneia and Haliartos area (as well as in NE Copais - Ptoon area), it seems that he actually walked (intensively even though not systematically in 'modern artefact surface surveys' terms) almost every metre of terrain, while in other areas his movements seem to have been rather 'site-oriented' or along main roads (in the Orchomenos or Hyettos area, for instance).[16]

Another good recent study is that of Knauss and his German team, who carried out major archaeological work with the German Institute led by H. Kalcyk (Knauss et al. Kopais 3). Though it is of interest mainly for the technical hydraulic section, the three volume work on Copais (Knauss et al. 1984, 1987 and 1990 – Kopais 1, 2, 3) also provides sound information on the archaeological evidence they noted (especially in Kopais 3).[17] The research is biased by the fact that they were following the 95m contour line around the basin in order to discover sites closely bordering the lake, and in particular the three settlements mentioned by the ancient authors (mainly Strabo) as 'swallowed up' by the lake in an early period (old Orchomenos, Eleusis, Athinai).[18] Knauss et al. attempt identifications of the activity foci they found with places mentioned in ancient texts. For the purposes of our work, it is very much more of interest, that traces of human activities and occupation, dating back to different periods, have been found in the Copais area, irrespective of their exact identification. Some of them show continuity of occupation, even if the character of this occupation still requires further study.[19]

As noted earlier, new and different information on the archaeological record of the region is also provided by recent rescue excavations, which are to some extent less biased by archaeologists' 'personal' and special research interests.
Furthermore, rescue excavations also help in increasing knowledge of some particular evidence. For instance, the

[14] Pritchett's topographical work is published in five volumes (from 1965 to 1982).
[15] Generally speaking, topographical studies are mainly lacking in distinguishing the level of complexity the 'site' is characterised in the different periods of actual occupation recognised in a certain location. In particular, small pot samples from large multiperiod sites often tend to be given equal credit with the large samples representing the major periods of occupation (for this see also Bintliff – Howard 1999 and chapter I.2.2 of the present work).
[16] See also chapter II.3.1 for comments on this work.
[17] The archaeological study was led by H. Kalcyk.
[18] See also chapter II.3.1 for comments on this work.
[19] In some cases an attempt was made to identify the site character, but my efforts are based only on other people's information and the topographical location of site and geomorphological processes that could have interested the formation of the archaeological records available to us, as well as human biases that could have contributed to the creation of the archaeological record.

ancient Boeotian cities that lie below modern towns are known better through excavations. Thebes is the most impressive, but Levadeia is also an example, and rescue excavations are almost our only means of knowledge of the ancient city. In fact, by means of rescue excavation in advance of modern construction projects, much more information can be obtained compared to that from other *poleis* now lying in the modern countryside (ancient Tanagra and Thespiae, for instance). On the other hand, the layout of ancient towns lying beneath modern ones is not so clear or easily definable (e.g. Thebes and Levadeia, but especially the latter); neither are the positions of sub-urban features (such as cult places or necropoleis).

For the Boeotian plains also, rescue excavations have been providing new information in recent years, especially with the construction of the new national road (highway) through the middle of the plains (Tanagra, Thebes and Copais), as well as work for the railway which, running through the foothills, has provided information on the areas where most ancient activities took place, and also the installation of a gas pipeline (*Agogos Physikou Aeriou* – DEPA 1993-94) which crosses Boeotia from NW (area of Davlias) to SE (area of Dervenochorion and Parnithas) - AD48 1993: 180.[20]

Therefore, as noted earlier, rescue excavations constitute an important chapter in the archaeological knowledge of the region (see fig.2), although the data must be collected under some limitations (especially time limits), as pointed out by the *ephor* of Boeotia, V. Aravantinos (2004: 82). Efforts have been made in the last decades to obtain useful and well-collected information from rescue excavations, while, as the *ephor* states: 'in the '50s, during work for the National Road, vases were found by workers of the Ministry of Public Works' while no archaeologist was present. The situation has now changed. Relevant examples of extended excavations for significant public works increasing archaeological knowledge are given from Akraiphnion (cemetery and other activities along the Athens-Lamia National Road) and Thebes (cemetery to the NE of the city, by the underground tunnel for the road Thebes-Mouriki/Thebes-National Road).

Some areas are also known better than others due to the specific interests of some scholars or of a group of scholars interested in a particular issue. For instance, the Dilesi area (and the area between Schimatari and Dilesi) is known quite well due to the intensive topographical work carried out by Brown while looking for the sanctuary of Apollo (Brown, BSA 1905/6: 93ff). Furthermore, the burial pattern in the area of ancient and modern Tanagra was quite well known due to the high level of interest in cemeteries of the Mycenaean and Classical/Hellenistic periods, due to the discovery of *larnakes* and terracotta figurines respectively. In the same way, the area of Copais has been widely and thoroughly researched due to interest in particular issues concerning the hydrological behaviour of the former lake and surrounding settlement (see Lauffer and Knauss' Munich team for instance, above). Furthermore, fortifications on the E Copais area are well known from Noack's research around Gla. He interpreted as LH fortifications or fortified settlements, all the spots that then, after more careful examination (mainly by Lauffer), were actually found to be Greco-Roman settlement sites and, usually, fort sites. Noack's travels and notes though did help the next researchers to directly go to these places and investigate them more carefully.

In several cases, research led by personal research interest (in particular chronological periods rather than in specific areas) allows for the 'accidental' discovery, at the same spot or in the immediate vicinity, of remains of other periods or other characters than the main research focus. In these cases, quite frequent and marked as 'other' in the 'discovery' field of the database, the archaeological record can be considered somewhat less biased, at least as far as the character of the evidence discovered is concerned, if not for the location of the discovery, which is not casual as it is led by an interest in other periods. Conversely, in the case of rescue excavation both discovery and location are casual.

Fossey, for instance, like other scholars with expertise in material of the historical period, recognises different phases in potsherds whilst other visitors to the site (mainly Prehistorians) do not. This often happens, for instance, with sherds from the Hellenistic period, for which the picture improves for number of known activity foci due to scholars who can recognise it among other material.

Therefore, an interesting research issue is represented by the multiperiod sites. When dealing with extensive/topographical surveys, we often have as a result a high presence of multiperiod sites, which could then, however, turn out to be a biased picture of continuous occupation trends. The majority of data in any case derive from substantial multi-period sites rather than small, shorter-lived sites that are usually identified only by intensive surveys or through rescue excavation or (less often) accidental discoveries. In addition, in the topographical tradition, small pot samples from large multiperiod sites often tend to be given equal credit with large samples representing the major periods of occupation (for this see also Bintliff – Howard 1999 and chapter I.2.2 of the present work).

Studies on settlement patterns and settlement histories

As mentioned previously, those descriptive/compiled works, which constitute the foundation of our archaeological data collection, often stand alongside topographical or landscape studies which perform a synthesis of the data available to them, in an attempt at reconstructing a settlement history of Boeotia, or parts of it. One of these is certainly Fossey (in the second volume of his 1988 work), who attempts to give a long-term history of settlement in the region, but also Buck, who tries to complement historical sources with archaeological data, from a non-archaeological point of

[20] For example: an EG cemetery near Solinari, remains of an ancient road near Thourion, etc.

view. Philippson and Kirsten (in Philippson 1950-9), through a historical geography approach (though less archaeological data, and in a minimal detail, were known by then[21]), also try to detect settlement trajectories.
Furthermore, some scholars offer synthetic work limited to small areas within the region, such as the Boeotia Survey project articles (in particular the Valley of the Muses and Thespiae), or the detailed work for the Skourta plain, etc. (see above). Moreover, as highlighted earlier, notes and issues on settlement patterns are also available in some mainly 'descriptive'works, which, however, also present attempts at synthesis, as in the case of Fossey 1988 (with gazetteers and settlement analysis) or Lauffer Kopais I (with comments on the settlement pattern along with site descriptions), or some archaeological reports.
We can note that in some of these works, attention is not paid to a critical examination of the actual archaeological evidence, and this often leads to the creation of settlement maps (and consequent settlement histories) on which it is not always clear what the dots represent in terms of actual evidence on the ground (just a few sherds, or complex evidence comprised of artefact concentration, remains of structures, excavated layers, etc.).[22]
As noted earlier, reference to and discussion of those synthetic works are made *passim* in each individual *chora* chapter, according to needs and occurrence, as well as in chapter III.2.

[21] For a discussion on the book see chapter II.1.
[22] As for the issue of representitivity see chapter I.2.2.

II.3

The *chorai* of Boeotia

As mentioned earlier, cultural and environmental landscape data concerning Boeotia are gathered, in the present work, according to *chorai* (fig.1 in chapter I.2.2). This implies an approach to ancient *polis*' territories as a focus for analysing settlement systems (see chapter I.2.2 for methodological implications).

In the following chapters, landscape data concerning each individual *chora* will be presented, illustrated and commented upon.
The *chorai* belonging to the Copais area will be presented first, in a geographically clockwise order; then the *chorai* of Central and Eastern Boeotia will follow.

Each individual *chora* chapter will be structured as follows:

- A narrative/qualitative illustration/topographical description of the physical topographical characteristics, along with a 'physical land units and land evaluation' section, in order to present and describe 'quantitatively' the landscape ('landscape in numbers') and its potential.
- A discussion of the boundaries of the *chora*, based on information available from ancient sources and epigraphical documents, as well as from other scholars' reasoning.
- A section 'Archaeological record', in which the archaeological evidence available is listed briefly according to toponyms and corresponding components. The archaeological record for each *chora* is presented in detail and discussed in appendix I, along with the description of the *activity loci* and *components*. Its final version can be defined as a sort of deconstructed-reconstructed version of what could traditionally be a 'catalogue of sites' (see chapter I.2.2). The description of each component group (*activity locus*/site – according to toponym) typically includes: a general description of location (or a brief topographical introduction when available); a brief history of research at the site along with a description of the process of component formation/'construction' and a brief 'source critique'; interpretive remarks and discussion. In some cases, some components are grouped in the presentation and discussion. This happens either when they share the same location (according to toponym), representing different phases and/or character of occupation of the same *activity locus*, or when they present common issues (in the latter case, the same component may also appear in different sections of the text, which discuss common problems). This grouping according to toponym allows for a description of archaeological evidence and issues which can be compared with the traditional gazetteer of sites, though conceptually different, as they are the result of the deconstructing-constructing process. In this way the cycle is closed from site to site again (see also chapter I.2.2). For each site, the main bibliographical references are mentioned in the text, and more can be found in the database under each component, or in the bibliographical reference quoted in the present work.[1] Components are listed in topographical order, starting from the main centre (usually the *polis*) of the *chora*.[2]
- Following the list of components, in each chapter's 'Archaeological record' section the archaeological record available is discussed as a whole according to the data sources and available information which produced it, and datasets' biases will be discussed.[3] Afterwards, landscape themes are investigated and landscape issues are discussed. Addressed questions could be: *what to do with the archaeological dataset available? What does the available dataset mean in terms of landscape?*

In the case of *chorai* which have been investigated (even partially) by intensive artefact surface survey, data are briefly presented in the corresponding appendix. In the present work, I have inserted into the survey sites database only the artefact concentrations securely dated,[4] mostly from preliminary reports (never fully and systematically published, or currently being published). A special case is represented by Tanagra *chora*, for which a

[1] In several cases, for instance, the reference to Fossey 1988, or other updated accounts, refers to the list of accounts reported there.
[2] Component numbering is given in an automated way by the software. It does not, therefore, strictly follow the topographical order.
[3] Mainly through statistical analysis which can be carried out on the research carried out in the area, such as queries and graphs on the database's '*Discovery*' field, and on the proportion of rescue excavation and excavation and surveys (extensive/topographical work), or queries on the amount of known components per period, etc. Proximity to the modern road network is also used to investigate biases.
[4] Only the sites that were clearly datable (including only their main period of occupation) were included, in order to avoid the risk of using data which could change dramatically following a more in-depth examination of their composition.

separate section has been created (in appendix I.14) in order to present data and results of the Tanagra Survey Project, for which freshly collected and critically monitored data are available to me as a member of the research project.

In the final section, some landscape issues will be discussed period by period, throwing light especially on some Prehistoric period (generally Neolithic and Bronze Age) and Historical trends. The *chora* by *chora* discussion could seem methodologically unsuitable for Prehistoric occupation, yet a critical examination and a general interpretation of the Prehistoric landscape within each *chora* facilitates the understanding of diachronical processes which took place at the micro-regional level. As for the Historical periods; different segments of the human landscape will be examined when information is available (town level, village level, cult places, rural occupation, etc.). Activity foci that are not clearly recognisable are also discussed, and questions are addressed. Addressed questions could be: *what can we do with this undetermined evidence? How can we look at them in terms of 'landscape'?*

Furthermore, some trajectories of the landscape in the long-term are commented upon, and actual or potential settlement chambers within the *chora* are highlighted, according to the regional and micro-regional approach to the landscape illustrated in chapter I.1.

II.3.1

Central Helicon: Koroneiake

TOPOGRAPHICAL SETTING

At the SW corner of the Copais, to the SE of the Granitsa or ancient Laphystion, lies a bay of the Copais basin, which runs 6/7km inside the Helicon relief and coincides with the *chora* of ancient Koroneia.[1] The Northern edge of the Koroneia bay is defined by the Granitsa/Laphystion, running in a N-W direction, while the Southern edge is defined by the rocky ridge of Libethrion/Tilphousion, running in a S-E direction.
The bay penetrates the Laphystion and the Northern and Central ridge of Helicon (see chapter II.1). The inner part reaches a height of 400m before the mountain slope rises to the Palaiovouna peak (1748m asl), while towards the Copais the level is ca 100m at the entrance to the valley, and the hills between range from 212 to 244m high, crossed longitudinally by the Pontsa valley and its tributaries. The fertile basin is not a plain (the elevation range of the Koroneia valley is higher than that of Levadeia) but a *Tafel* made of soft, young, tertiary deposits (Philippson 1951: 449). It is composed mainly by the alluvial valley of the Pontsa river[2] and its tributaries. The land is very fertile and easily cultivable, both in the river valleys and in the non-flooded area at the entrance of the basin (see below), with a very rich cultivable soil. A ridge (Megalo and Mikro Butsurati) runs along the central axis of the basin and separates the larger Pontsa valley from the narrower valley of the smaller stream, Kakaris, which runs parallel to it (see map in fig.1).[3]

Davidson's map[4] reports fertile alluvial deposits from a river and not from the lake, in the area where the Phalaros/Pontsa river and the Kyarios/Kakaris river[5] flow into the Copais. Whilst the Herkyna river used to flow from Levadeia into the Copais basin with a delta (therefore giving the area a marshy character), one might suppose that the Phalaros and Kyarios did not create a delta, so that the surrounding area was more fertile due to their regular deposits. During visits to the area it was noted that the Phalaros and Kyarios outlets were characterised by less marshy and more alluvial deposits than the Herkyna outlet into the basin. Today the Kephisos is channelled and takes all the water from springs and streams flowing into the Copais. Only a small marshy area is preserved by Rachi (Philippson 1951: 474).[6]

As for traces of ancient landscapes, in the area of Agia Triada there is still a forest of oaks - a residue of the original Holocene woodland on the Helicon mountain.

As Kirsten (in Philippson 1951: 670) noted, the landscape in the S and SE Copais is much more remote than other areas bordering the lake. Here the continuation of the projection of the Helicon mountain (Laphystion – Tilphousion) allows only narrow passes to the edge of the lake. The natural obstacles characteristic of the area made it a suitable choice for a battlefield in ancient times.[7]

The ancient *polis* of Koroneia was located close to the edge of the large plain, between the Pontsa valley to the NW and the valley of a parallel stream (Kakaris)[8] to the SE, and lay on the final spur of the low ridge Mikro and Megalo Butsurati (mentioned above), crossing the basin longitudinally in a NW direction for ca. 4km (fig.1 and fig.2).

---

[1] Kirsten (1951: 670) writes of '*polis landscapes*' for the Boeotian *chorai*, and then characterises as '*gulf landscapes*' the *poleis* gravitating on Copais.
[2] The Phalaros/Pontsa river was, according to Philippson (1951: 474), totally dry in summer. Today one can still see water running along its course in high summer.
[3] The general description of the area is based mainly on Philippson 1951: 449f and Lauffer Kopais I: 76, with personal observations added.
[4] For a reference to the map see chapter I.2.1.
[5] See reconstruction of the river in Knauss et al. (Kopais 3, plan 2.13).

[6] A marshy zone, formed mainly by the Herkina river outlet, extended to the W of the Copais lake water, by the modern villages of Ag.Dimitrios for ca. 5 km. up to the foot of Laphystion near Rachi (Philippson 1951: 473 – see fig.13). This marshy area formed a dangerous zone between the plain of Kephisos and the plain to the S edge of the basin, through which the main road ran. This marsh probably played host to the battles of 85 BC - Mitridates *versus* Silla - and 1311 - Frankish cavalry *versus* Catalan soldiers (Philippson 1951: 473). It is also marked in the GYS 1:50,000 map (*Lebadeia* sheet) and in the 1:200,000 Boeotia *nomos* map. It was probably semi-independent from the lake's fluctuations, and its permanent marshiness (even after the drainage) is due to other factors.
[7] Two important battles took place in front of ancient Koroneia in 447 and 394 BC.
[8] There is a question regarding the identification of the Pontsa or Kakaris river with the ancient Permessos (Lauffer Kopais I: 80-1), though the river Pontsa should probably be identified with the ancient Phalaros (Fossey 1988: 323 and others), and Kakaris with Kyarios. The Triton stream should probably be identified with the small stream by Solinari village (Papachatzis 1981: 214 n.6).

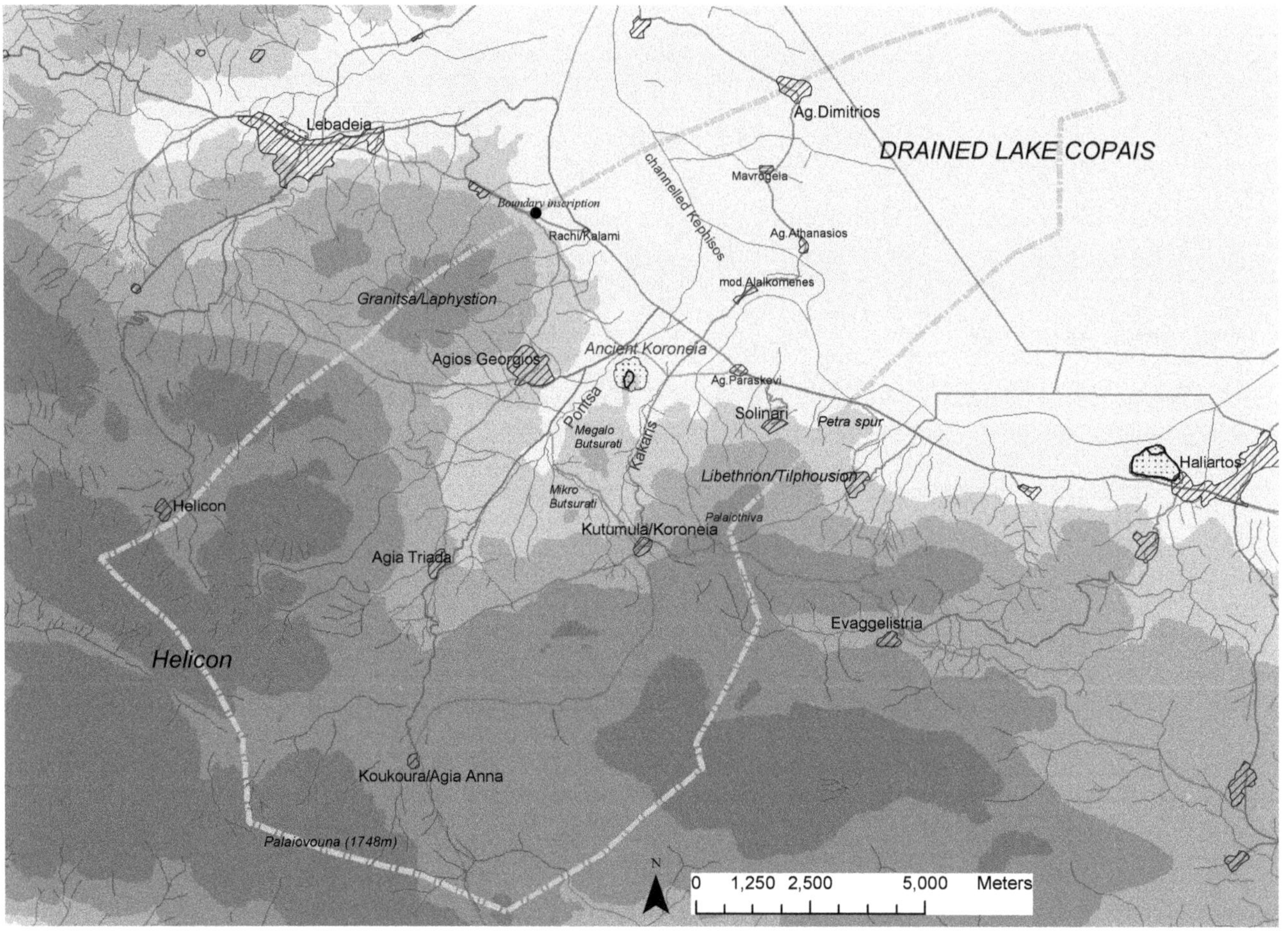

Fig.1. *Topographical setting of the chora of Koroneia.*

Boundaries

It is clear from the topographical introduction and a glance at the map (fig.1), that the *chora* is clearly defined by natural borders, being a lateral bay of the Copais basin surrounded by mountains to the W (NW – Granitsa/Laphystion ridge[9]), to the S (Helicon Palaiovouna ridge, rising up at the very end of the valley, beyond Stevenikon/Agia Triada), and to the E (SE – Libethrion/Tilphousion ridge or Palaiothiva/Vigla ridge).

A boundary inscription has been found at the NE foot of the Granitsa/Laphystion ridge,[10] not far from the church of Agios Ioannis.[11] It would have marked the boundary between Koroneia and Levadeia.

Another boundary inscription was found in an ancient dyke in the Copais basin.[12] According to Fossey (1988: 501), it would have marked the boundary between Koroneia and Orchomenos.

At some time probably in the Archaic period,[13] the natural Southern border, the narrowest part of the valley before the start of the slopes up to the peak of Palaiovouna (1748m), was extended beyond the Stevenikon area to include part of the Koukoura valley, when the small upland plateau of Koukoura/Agia Anna, considerably high up in the Helicon massif (attempts to identify it as the territory of ancient *Hippotai* have been made – Fossey 1988: 339; Roesch 1965: 53) was split between Koroneia and Thisbe (Fossey 1988: 337), as stated by an inscription (IG VII 1870).[14] This area is a natural crossroad for the main mountain routes through Helicon, as noted by Fossey (1988: 339 and 1990: 205-7), and earlier by Pritchett (V 1985: 156) – see fig.1 in chapter II.1.

To the N and E, the boundary was provided by the Copais lake and marshes. Examining the actual extension of the lake (see fig.1 and below), there was a wide non-flooded

---

[9] Fossey remarks (1988: 337) that Granitsa in the Slavic languages means 'frontier' (Kretschmer, Glotta 1914: 265).

[10] The inscription [AE1671] is reported in SEG XXIII 297; SEG XXXV 406; Vollgraff BCH 1902: 570; Roesch 1965:61-3. The find spot is marked on fig.4 in Camp II (see fig.1).

[11] for a discussion of the N boundary see Fossey 1988: appendix IV.3

[12] Inscription [AE1108] was found in ancient dyke at Dodekakleidi about 0.5km N of Xinos. Fossey (1988: 501) saw it in the garden of Charalambos E. Alambasis at Ag. Dimitrios. A tall, slender block of grey limestone, square in cross section, roughly shaped. 106.0 cm (h) x 34.0 cm (w) x 35.0 cm (th). A smoother band, 9.2 cm broad and containing the inscription, runs across the front of the stone: "OROS" (2nd-1st centuries BC).

[13] This would match with the interpretation of the Archaic period as age of expansion in Boeotia (cf. Bintliff 1994 and chapter III.2).

[14] Fossey considers the settlement area of Koukoura/Agia Anna separately (1988: 339ff).

area at the entrance to the Koroneia basin, whilst the marshes came closer to the edge at the Petra spur (SE), as noted by Leake (1835: 137 - see fig.1). To the N, at the foot of the Granitsa/Laphystion ridge, in the area of Rachi/Kalami, a permanent marshy area is known (see above, footnote 6), and reported by many travellers and Philippson (1951: 474).

The *polis* territory was called *Koroneia* (Thucydides I 113.2) or *Koroneiake* (Strabo IX 2.19, 28 - Hansen 1996).[15]

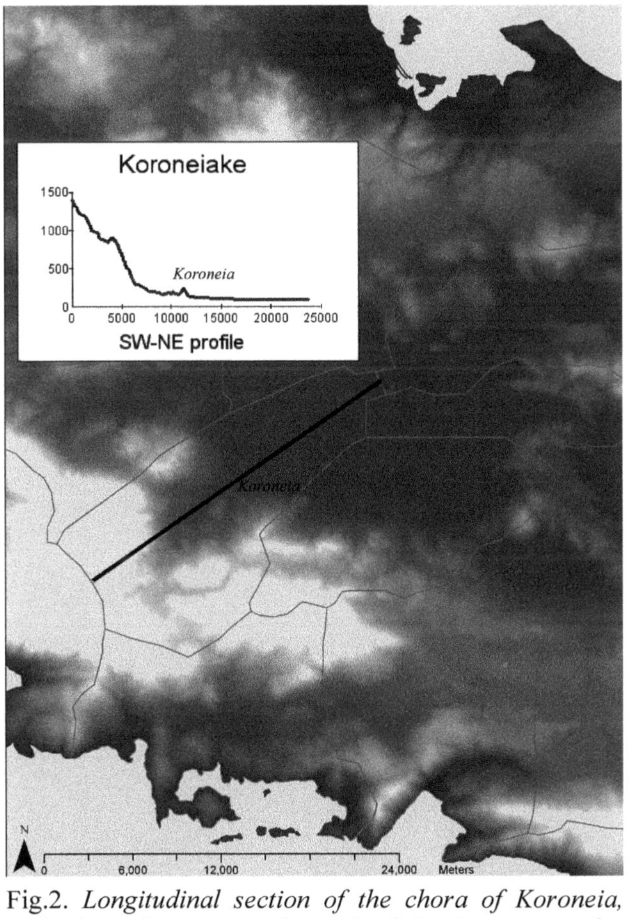

Fig.2. *Longitudinal section of the chora of Koroneia, with the polis spur at the end of the Butsurati ridge visible, in front of the flat area at the entrance to the valley.*

PHYSICAL LAND UNITS

Half of the *chora* can be described as mountainous (above 600m). This is due primarily to the presence of the Helicon massif in the SW part of the *chora*, as well as the higher part of the two ridges radial to Copais closing the Koroneia basin to the N and the S. The hilly landscape is composed predominantly of the slopes of these N and S ridges and the central ridge of Butsurati, at the end of which ancient Koroneia lies. The plain (lowland) of the valley itself is quite small, compared to the much more open Levadeia basin (see fig.2 in chapter II.1). On the other hand, the percentage of plain lanscape, including the portion of the Copais basin available, increases immediately in front of the ancient *polis*, and probably extending quite far into the basin itself (see digital reconstruction of lake fluctuations in fig.10).

Hilly landscape	20.7%
Mountainous landscape	44.9%
Plain	34.4%

1	P1_P2	lacustrine basin, valley	31%
2	P3	gentle slope	0.9%
3	P4	foothill	2.4%
4	H1	plateau	5.2%
5	H2	gentle slope	0.3%
6	H3	moderate slope	2.8%
7	H4	severe slope	6.9%
8	H5	very severe slope	5.6%
9	M1	plateau	9.5%
10	M2	plateau/gentle slope	5.6%
11	M3	moderate slope	14.6%
12	M4	very severe slope	15.2%

Table 1. *Percentage of the different physiographical classes present in the Koroneia area (P=plain; H=hill; M=mountain).*

Lacustrine basins and valleys (P1_P2) are the predominant features in the plain landscape, constituted mainly by the Copais basin and its edge, and only in small part by the end of the Pontsa and the Kakaris stream valleys and their outlets into the Copais basin.

Foothill landscape (P4) characterises the area immediately below and following the heights marking the shape of the valley.

The site of the ancient *polis* of Koroneia lies on the flattish top (H1) of quite a steep spur (P4) at the end of the ridge crossing the basin longitudinally and constituting part of the hilly landscape of the *chora*. The hilly segment is marked by quite a strong presence of plateaus (H1), considering the total area assigned to this section of the *chora* landscape. In particular, plateau features characterise the gently sloping slopes of the Northern ridge bordering the area, as well as the area at the end of the Phalaros/Pontsa river valley. On the other hand, the central ridge (at the end of which lies ancient Koroneia), the higher areas of elevation surrounding the basin, as well as the outer slopes of the rocky ridge of Libethrion/Tilphousion are marked by a steeper morphology (H4, H5).

The central Butsurati ridge ends at its SW, and leaves space for an opening constituted by the lateral valley of a tributary of the Phalaros/Pontsa, through which the two valleys (Pontsa and Kakaris) joined at the back of the valley, before the elevation increases following the slope of Helicon (as clearly visible in the topographical layer in fig.1).

[15] According to the sources, it comprised *Hermaion* (a valley), *Metachion* (a fortress between Koroneia and Orchomenos) (Ephorus [*FGrH* 70] fr. 94a), and the sanctuary of Athena Itonia (Strabo IX 2.29; *Scholia* Apollonius Rodius 1.551a) – Hansen 1996: 90.

Helicon is characterised by the presence of smaller or larger upland plateaus (M1). Within the Koroneia *chora*, a quite recognisable mountain plateau stands out towards the S: the Koukoura/Agia Anna plateau (see this chapter – *passim* and especially *component KO_13*).

RESOURCES

A map showing land capability (for agriculture – see chapter I.2.1) for the Koroneia area can be seen in fig.14. The *chora* landscape is characterised by fertile and arable land suitable for agriculture, in and at the edge of the Copais basin, as well as in the river valleys (class F). The Eastern Pontsa river valley covered by alluvia is larger, longer and more articulated (with its lateral tributaries) than the valley of the Kakaris stream, to the W side of the ancient city site. The site of the ancient *polis* of Koroneia occupies a neat boundary between two contrasted landscapes both fertile; the hilland and the Copais edge lowland. A fertile area covered with good arable soil can also be found on the upland plateau of Koukoura/Agia Anna. On the other hand, class MF covers almost the whole of the hilly landscape, creating a large cultivable area available up to a height of 600m. The plateau and gentle slope features recognised (see above) in the hilly segment of the landscape are therefore potentially suitable for agriculture. Areas belonging to class LF are mostly suitable for grazing, considering their elevation and slope.

As noted by Fossey (1990c), the concern to save agricultural land from flooding evidenced in the 2nd C AD (text from Hadrian, 125 AD – setting up of dikes to channel the rivers flowing into the SW Copais, in order to stop them flooding good agricultural land) suggests that crops were the main economic resource of this valley in Roman times, as they are today.

A resource also available to the *chora* is the lake itself, with the parallel forms of economy it can offer (see below and Farinetti 2008).

Quarrying was probably also available. Traces of an ancient quarry (*component KO_88*) for building materials have been noted on the ridge constituting a lower spur of the Eastern Laphystion ridge.

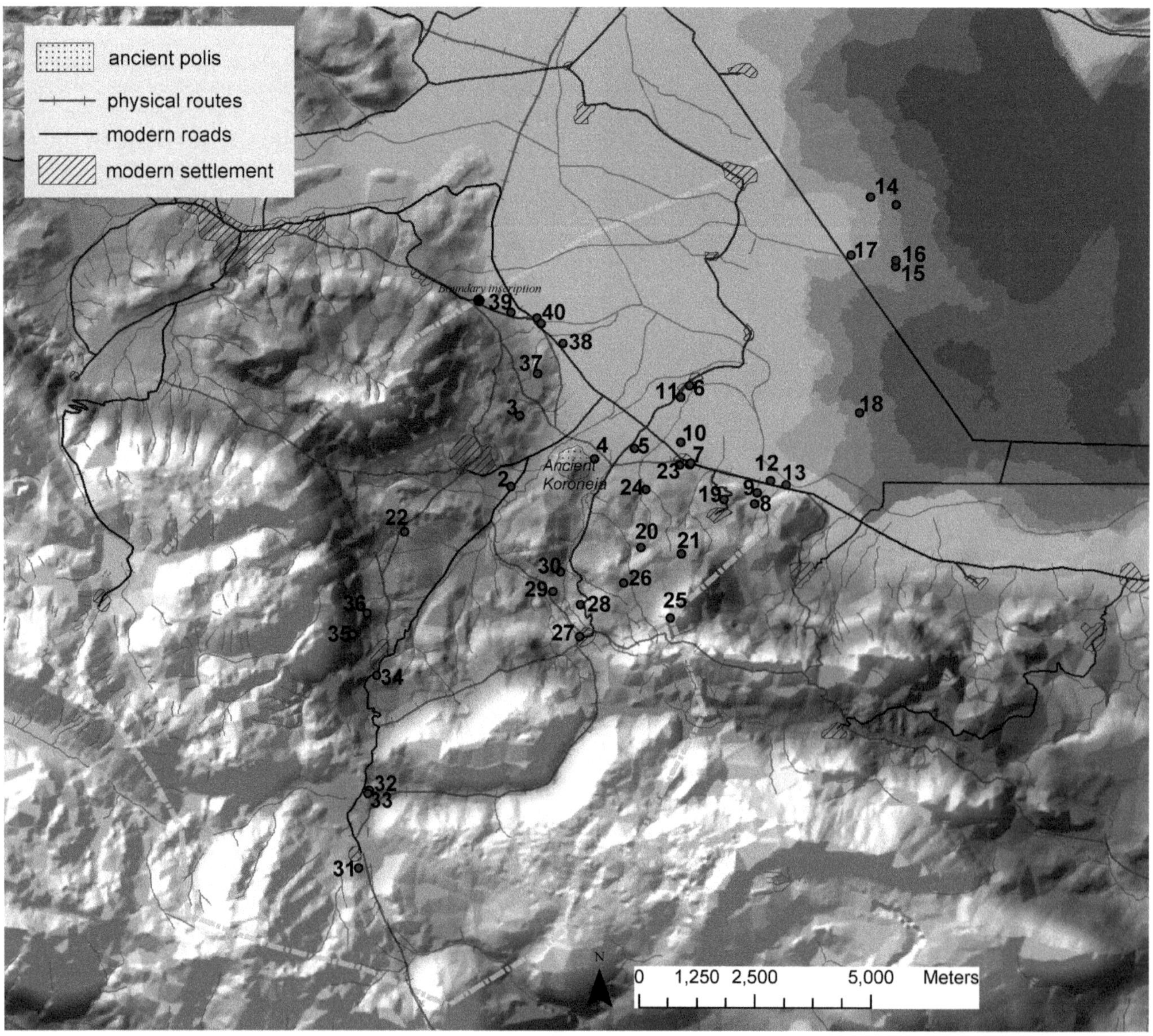

Fig.3. *Archaeological map of Koroneiake chora.*

THE ARCHAEOLOGICAL RECORD

1	**PALAIA KORONEIA**	*Components KO_1 to KO_10*
2	**Ag.Athanasios**	*Component KO_92*
3	**Ag.Georgios – Xerokamara**	*Component KO_87*
4/5/6/7	**ITONION**	(4) *Components KO_14 and KO_16 to KO_19;* (5) *KO_83;* (6) *KO_33 and KO_34;* (7) *KO_41*
7	**Agoriani/Ag.Paraskevi**	*Components KO_40 and KO_41*
8	**ALALKOMENAI**	*Components KO_21; KO_24 and KO_25*
9/10	**Sanctuary of ATHENA TRITONIA/ALALKO MENIA**	(9) *Components KO_63 and* (10) *KO_84*
11	**Mamoura/Alalkomenai**	*Component KO_35*
12	**Solinari – National Road**	*Components KO_93 and KO_94*
13	**Solinari Agios Ioannis**	*Components KO_27 KO_27 and KO_28* (Tilphousion?); *KO_29*
14/15	**Ag.Dimitrios East**	(14) *Components KO_11 and KO_20;* (15) *KO_70*
16	**Xinos**	*Components KO_65 to KO_72*
17	**Xinos**	*Components KO_64*
18	**Petra – Triton**	*Components KO_73, KO_74 and KO_85*
19	**Solinari**	*Components KO_22 and KO_26*
20	**Solinari - Agios Ioannis**	*Component KO_82*
21	**Solinari - Agia Paraskevi**	*Component KO_30*
22	**Pontsa - Agioi Taxiarchoi**	*Components KO_31, KO_32 and KO_61*
23	**Agoriani-Dekedes**	*Components KO_36 to KO_39 and KO_12*
24	**Solinari – Kaminia**	*Components KO_86*
25	**Paliothiva**	*Component KO_80*
26	**Tsoroko – Ag.Paraskevi**	*Component KO_81*
27	**Kutumula/Koroneia - Ag.Taxiarchoi**	*Component KO_75*
28	**Kutumula/Koroneia - Ag.Giorgios**	*Component KO_76*
29	**Koroneia – Kaminia**	*Component KO_77*
30	**Butsurati**	*Components KO_78 and KO_79*
31	**Koukoura/Agia Anna**	*Component KO_13*
32	**Milia**	*Components KO_49 and KO_50*
33	**Milia S**	*Components KO_95 and KO_96*
34	**Agia Triada – plateau**	*Component KO_59*
35	**Agia Triada cave**	*Components KO_51 to KO_58*
36	**Agia Triada**	*Components KO_60*
37	**Triapi**	*Component KO_88*
38	**Magoula Kalami**	*Components KO_89 and KO_90*
39	**Kalami – Lioma**	*Components KO_42 to KO_47*
40	**Kalami/Rachi**	*Components KO_48, KO_62 and KO_91*

Table 2. *List of archaeological components and activity loci mapped in fig.3.*

Rescue excavations were, on several occasions, an unbiased way of discovering archaeological remains, independent of personal research interests (see chapter II.2). For instance, in 1994, works for the *Agogos Physikou Aeriou* (gas pipeline), running from NW Boeotia to the area of Thebes, brought to light significant archaeological remains (Aravantinos, AD49 1994: 284-286). The same is true of road and railway constructions, especially at the foot of the Petra mountain and in the Kalami area.
The Koroneia *chora* is particularly rich in available archaeological information, especially considering the absence of intensive and systematic surface survey. The lower area, between the city site and the immediate border of the basin, and the basin itself in front of the Koroneia valley, was investigated by accurate survey work by the German team of Knauss et al. (thematically oriented to evidence by the edge of the former lake Copais in order to examine the water level fluctuations - Knauss et al Kopais 3).[16] In addition, the areas immediately bordering the basin, as well as the whole valley and the inner ridges, were intensively covered by Lauffer within the framework of his research work on the areas surrounding the Copais lake (Lauffer Kopais I).[17] The Koroneia *chora* probably belongs to the areas that Lauffer visited more carefully and actually walked through thoroughly. Consequently, not only has the archaeological record available for the area increased, but, even at the scale of our work, the evidence is much more accurately located, allowing us to proceed to further analysis involving different datasets.

The graph (fig.4) illustrates the proportion of components discovered within different research frameworks as far as the Koroneia *chora* is concerned.
The quite significant presence of 'Other' entries (see chapter II.2) is due mainly to the fact that due to personal

[16] For the limit and the excellent aspects of their work see also chapter II.2.
[17] For the limit and the excellent aspects of his work see chapter II.2.

interest in, and frequent visits to, the well known Prehistoric sites at the edge of the Copais basin, traces of activities from historical periods were also found.
The high percentage of 'Personal or group research interest' is due to the intensive research carried out in the Copais and at its edges by Knauss' German team (with a special interest in the area of Copais located in the

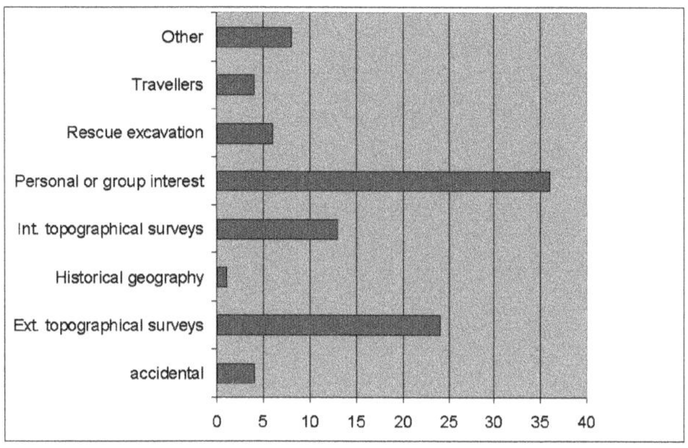

Fig.4. *Graph illustrating the proportion of components discovered within different research frameworks.*

Koroneia *chora*), and to the personal interest of scholars compiling Prehistoric gazetteers on the sites at the edge of Copais.
'Rescue excavation' entries are limited in presence, and are represented especially by the excavation work carried out along the road at the edge of the basin in the 1990s.

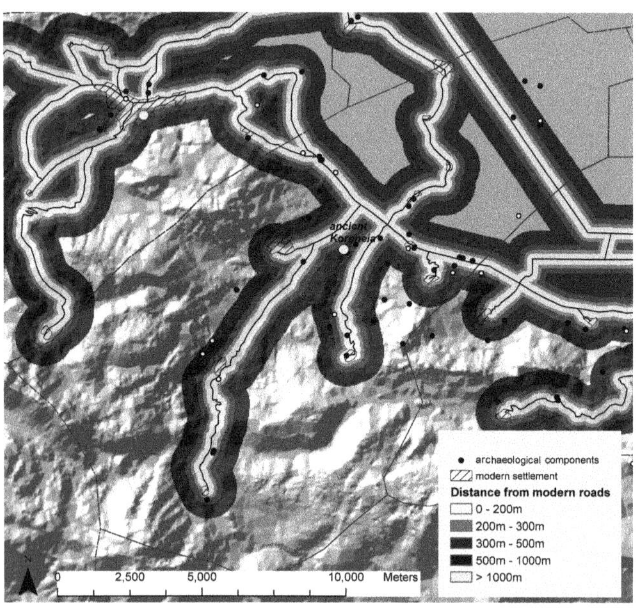

Fig.5. *Relationship between components and the modern road network.*

The ratio of Prehistoric to Greco-Roman components known is 25 to 66 (1:2.6), while among the historical periods, 43% are dated Archaic to Hellenistic, 27% Roman-Late Roman, and 30% are attributed to the general Greco-Roman period.

In fig.5 we can see the relationship between known archaeological sites and distance from the modern road network (30 of 96 components within 300m; 39 within 500m). Considering that of the remaining components (further from the road), almost 20 relate to ancient Koroneia and immediate surroundings. We may conclude that research has generally been quite biased by proximity to modern roads, especially as for the route along the edge of the Copais basin. Although this can be linked to the fact that the line was also used as a communication route in the past, we should be aware that in certain periods of history, probably the most preferred route was the inner one through the mountains (see fig.3).

For the Koroneia *chora*, we can observe an increase in the number of 'sites' when comparing the resulting map (fig.3) with Fossey's map of the area (1988: 322, fig.44). This is mainly due to the fact that there is a gap of almost 20 years between them; two decades of road construction and other infrastructure work, and therefore of rescue excavation, as well as years during which two major projects have been carried out: Lauffer (Kopais I), and Knauss et al. (Kopais 3), the first only mentioned briefly by Fossey 1988 in the *Addenda* section.

ANALYSIS OF THE *CHORA* LANDSCAPE

PREHISTORIC PERIOD

For the *chora* of Koroneia, we do have archaeological evidence reported for all the broad Prehistoric periods under consideration: Neolithic (fig.6), EH (fig.7), MH (fig.8) and LH (fig.9). The available archaeological record also includes three pieces of evidence generally datable to the Prehistoric period: a settlement (*component KO_89*), an unspecified but sure activity focus (*component KO_79*), and a single piece of evidence indicating human presence (a few flints noted during extensive survey work - *component KO_23*).
The location, physiographical character and type of evidence of this *component KO_23*, a very small ridge at the very foot of the hills below Solinari, is very close to that of the other surface concentration of flints found in the Levadeia *chora* (see chapter II.3.2 – *component LE_22*).
The knowledge on the prehistoric landscape of the area at the immediate edge of the Copais basin is constituted mainly by two known Early and Middle Helladic large settlement sites, Agoriani-Dekedes (*components KO_36 to KO_39*) and Kalami-Lioma (*components KO_42 to KO_45*). Apart from these Prehistoric sites gravitating towards the basin, other evidence from the Prehistoric period is known from the upland of the *chora*, by the modern village of Agia Triada, above the road towards the Koukoura/Agia Anna plateau and pass. There, apart from Neolithic evidence from the cave above the village of Agia Triada (*component KO_51*), EH activities are known from the cave (*component KO_52*), on a plateau within the village (*component KO_59*), and along the road (*component KO_60*), and MH is also reported. According to the excavation report on the work by the cave (AReports 2000-2001: 56), these activities should

probably be linked to the development of pastoral activities between the end of the EH and the beginning of the MH period. The so-called 'secondary product revolution' saw an increase in the number of flocks in order to produce wool and cheese/milk in addition to meat, and led to the use of and dwelling in a part of the landscape otherwise neglected because of its limited agricultural usefulness. It was probably used at an earlier stage for other, non-permanent activities / temporary purposes, such as hunting and gathering.

In fig.7, one can see the windows (lowland and upland) where the Early Helladic landscape appears. The concentration of knowledge of the EH period in the small area of Agia Triada is curious. It could either be the accident of a landscape window more carefully examined (due to the presence of the cave site), or it could be a landscape window actually more intensively occupied in a certain period (probably corresponding to the secondary product revolution),[18] and for which the occupation left traces still visible in the landscape (probably also because of a rather low level of activity during later periods). In the intermediate area (coinciding with the hilly segment of the landscape), the Prehistoric landscape is less known, and one cannot state with certainty that EH life was concentrated in these archaeologically better known areas. On the other hand, the presence of the lake and its resources (see chapter III.1) in the case of the lowland activity foci, as well as the available pasturage in the case of the upland foci, could give value to the archaeological record available, as well as a meaning in terms of landscape. Generally speaking, locations on low hills seem to be preferred in this period, as attested by the two settlement sites at Agoriani-Dekedes (*component KO_37*) and Kalami-Lioma (*component KO_43*) at the edge of the Copais. Though, this could also be due to a bias in knowledge affected by the proximity to roads and by prominent positions over the basin. Adding to the picture the components generally datable to the Prehistoric period, an activity focus (*component KO_79*) appears in the hilly intermediate area, just below the modern village of Koroneia, on the other side of the Kyarios/Kakaris stream valley; probably on an old river terrace.

A very partial knowledge of the archaeological landscape biases also the Middle Helladic period, though in this period the available archaeological picture is enriched by the German team's research within the Copais basin (mainly Knauss et al., Kopais 3). Evidence of MH occupation is also known within the Copais basin, with at least one activity focus that could be interpreted as an area of habitation (we cannot say if temporary or permanent on the basis of the available information). These were probably linked with the exploitation of the lake's resources as well as with vicinity to water (see chapter III.1).

From the available record, the LH picture of human occupation in the Koroneia *chora* would look similar to that of the MH period, though for the site of Kalami-Lioma the LH presence seems reduced and of less certain character compared to the previous periods. On the other hand, once again one ought to say that any comment can be only hypothetical, as we must be very cautious in the interpretation of these periods for which only poor datasets are available and which can be biased by several factors (proximity to modern roads, prominent locations), even stronger than data available for later periods.

Looking at the available Neolithic archaeological record, we should also consider that the resulting picture appears very similar to that from the mapping of evidence from later prehistoric periods, due to an evident bias in knowledge. Neolithic was found where research was carried out on sites with later occupation.[19] The activity foci known for the plains are related to Copais life (see chapter III.1), just as the habitation sites and activity foci of later Prehistoric periods. On the other hand, the signs of FN/EH occupation (excavated burnt layers) brought to light at the Agia Triada cave site (*component KO_51*) must be linked with activities in the upland landscape before the specialisation of pastoralism. These activities could be hunting, or the raising of flocks for meat, and occupation of the cave could have been either temporary or permanent, and of course because it is a natural shelter.

[18] The EH period is marked by an expansion of settlement into relatively marginal landscapes, almost certainly facilitated by the diversification of agricultural strategies and an increased reliance on pastoral production (Halstead 1987; Whitelaw 2000). In the Koroneia area we could imagine at this time an increased interest in the mountain areas, rich in water and grazing land.

[19] The appearance of Neolithic is usually in deeper strata of an excavated site, or on the surface together with later material in larger quantities.

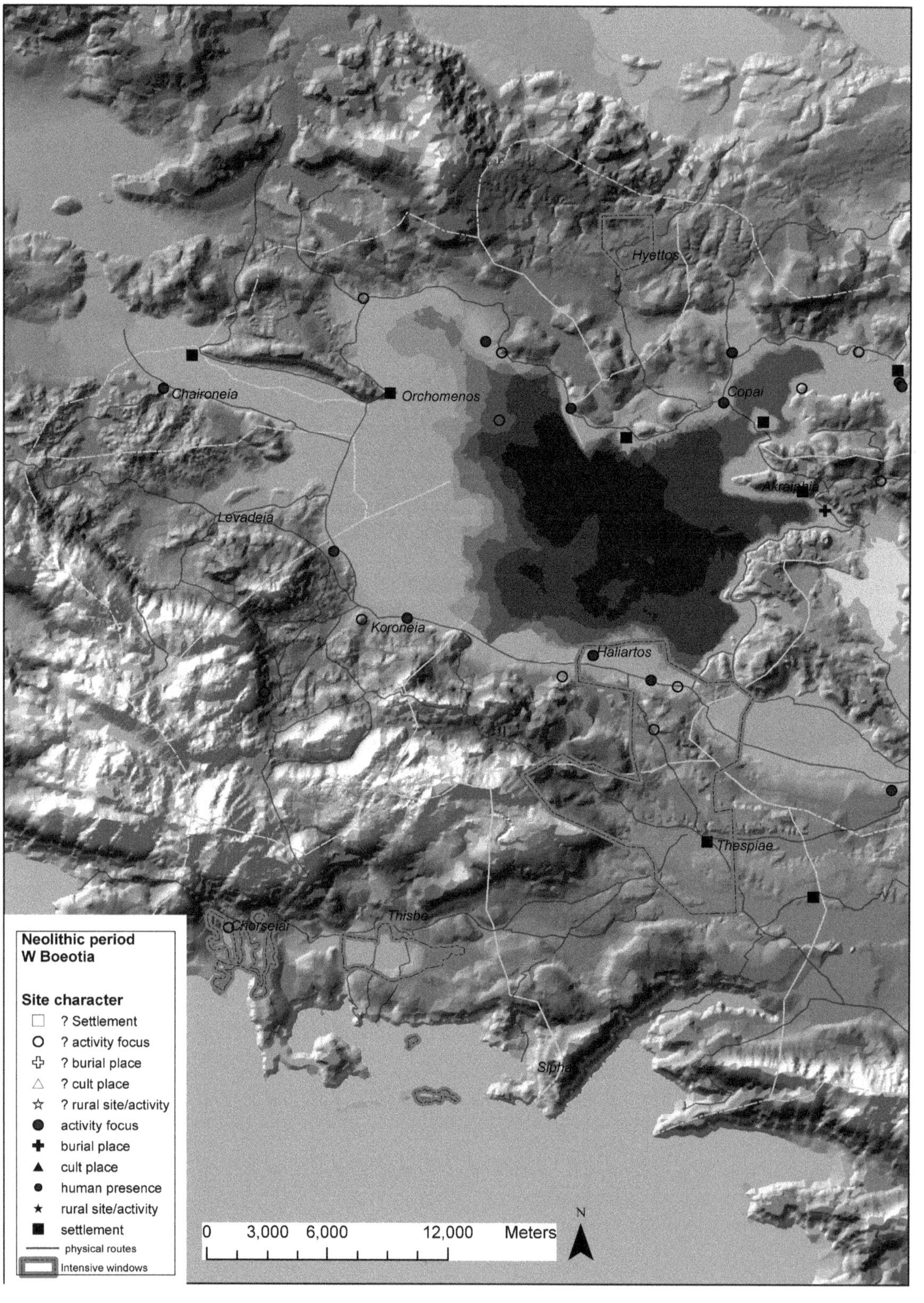

Fig.6. *Neolithic map of W Boeotia*

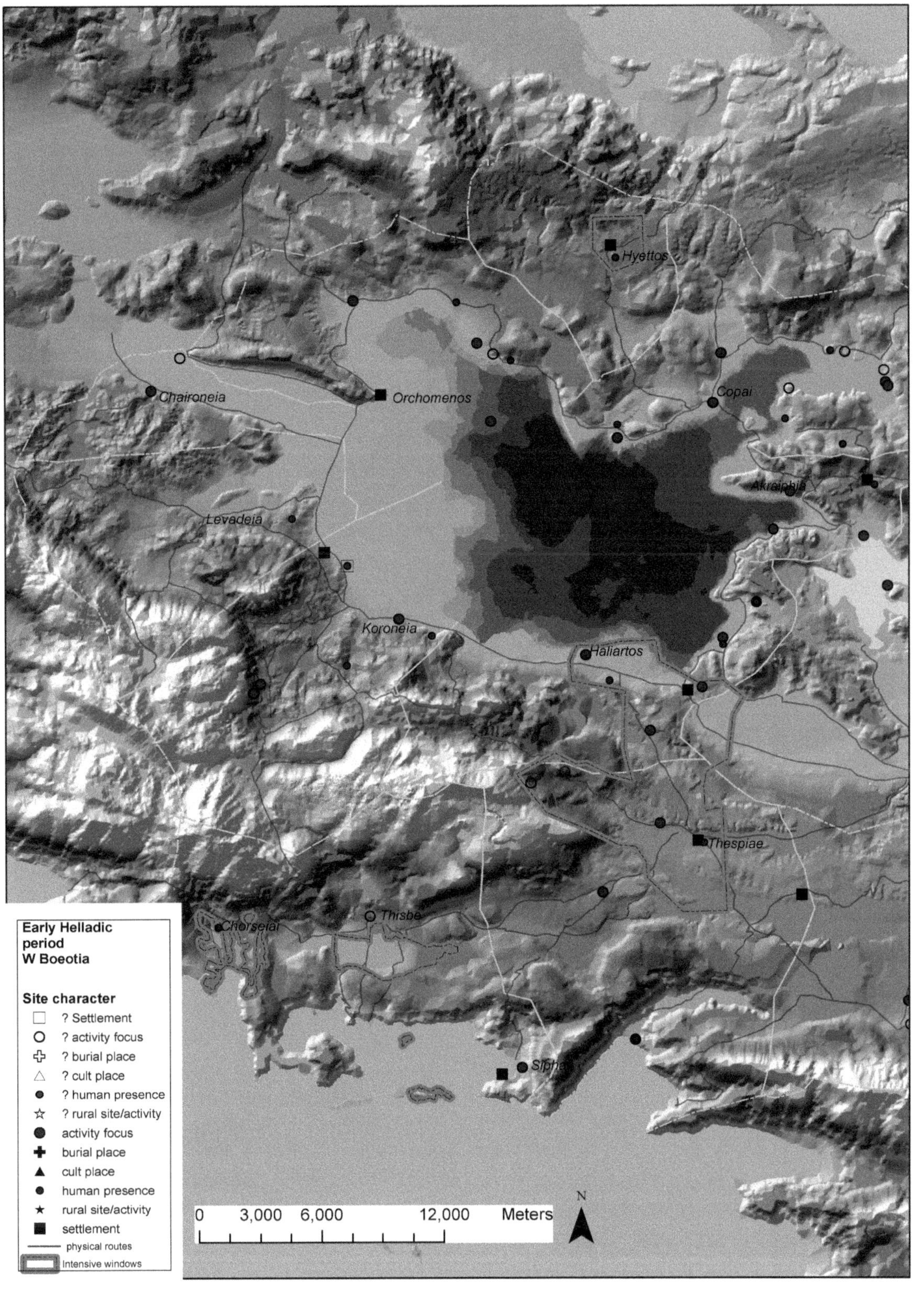

Fig.7. EH map of W Boeotia

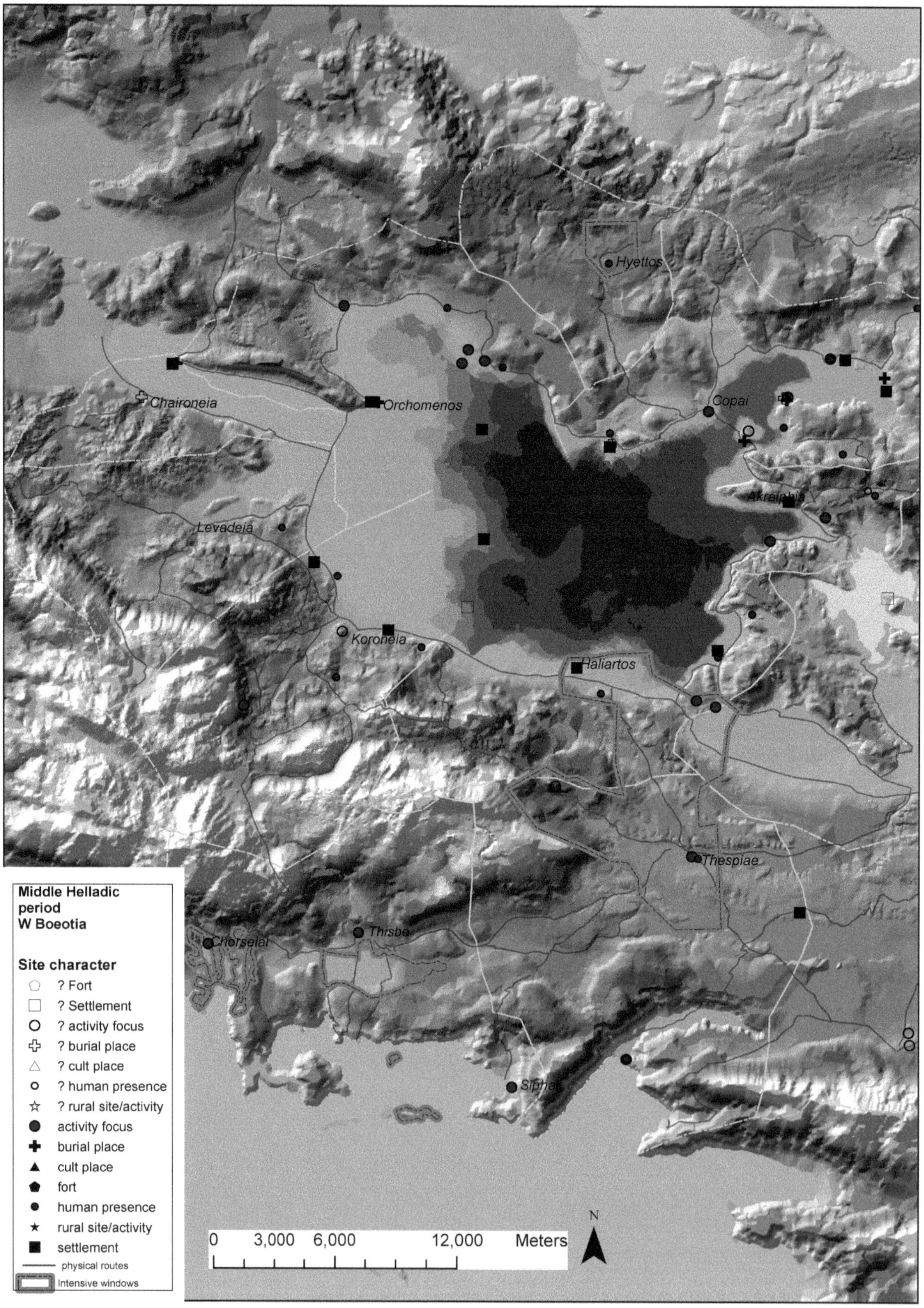

Fig.8. *MH map of W Boeotia*

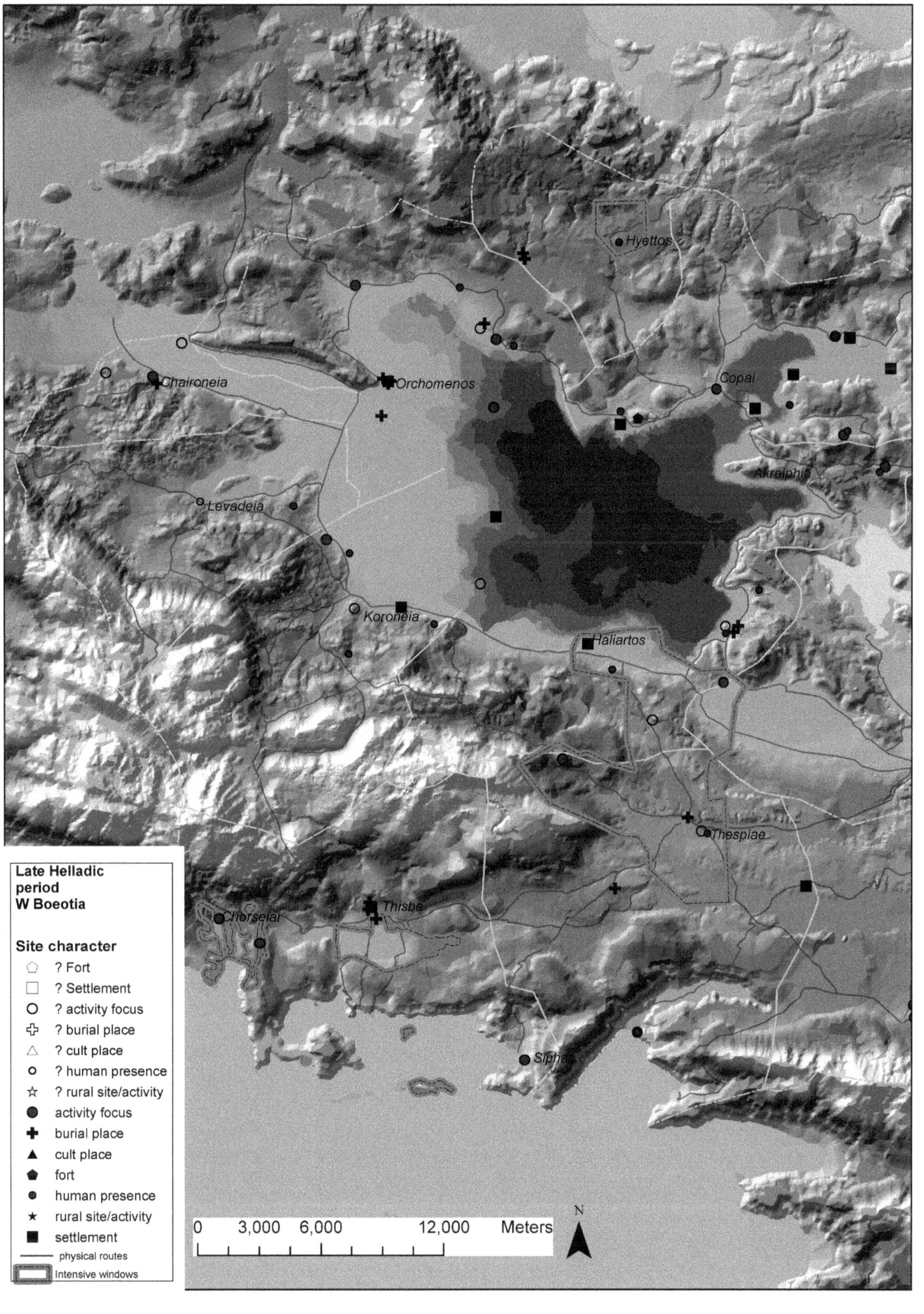

Fig.9. *LH map of W Boeotia*

GRECO-ROMAN ANTIQUITY[1]

Town level

The ancient *polis* of Koroneia was located, as seen above, in the centre of the opening of the valley. The ancient *polis* of Koroneia lay in a backward position compared to the passage of the perilacustrine road, but was visible from it and could control the crossroads of the main route crossing the whole of Boeotia (Thebes-Levadeia), following the Southern edge of the Copais basin, and the routes to inner Helicon (see fig.1 in chapter II.1). The ancient *polis* was located close to the edge of the large plain,[2] between the Phalaros/Pontsa valley to the NW, and a parallel stream (Kyarios/Kakaris)[3] to the SE, on the final spur of the low ridge Mikro and Megalo Butsurati (mentioned above), crossing the basin longitudinally in a NW direction for ca. 4km (fig.1 and fig.2). Surface sherds (whose density on the ground is incredibly high, especially in some areas) indicate the presence of an urban settlement there since the Geometric period (*components KO_1 and KO_5 to KO_8*), and excavations conducted on the acropolis' E slope attest the presence of the Roman agora as well as structures indicating a Late Roman settlement and cemetery (*components KO_9 and KO_10*). The city knew periods of favour (especially in the Roman Imperial period) and decline (VI and II C BC).[4]

Village level

At least one satellite settlement is known for the *chora* of Koroneia from ancient sources: ancient Alalkomenai (see *components KO_21, KO_24, KO_25*). There have been various attempts at identifying archaeological remains with this ancient *polichni*, as mentioned by Pausanias (see Alalkomenai components).[5] What is of interest here is that a settlement at the village level existed since at least the Classical period (and probably earlier), and probably continued its life until the Early Roman period. It was located to the E of ancient Koroneia at the Eastern edge of the *chora* (by its physical limit: the rocky spur of Petra), and its settlement chamber was comprised of a very small valley, which the site overlooks, between ancient Alalkomenai and Tilphousion (see dedicated components), and protected from flooding (see fig.3). Evidence has been found attesting the presence of a Classical, Hellenistic and Roman focus, probably a settlement site, at the entrance to the small valley. Alalkomenai could easily represent a case like Askra, of a *proto-polis* which never had the strength to become a *polis* (chapter II.3.9).

Another Greco-Roman settlement was probably located in the upland plateau of Koukoura/Agia Anna and attempts have been made to connect this with ancient *Hippotai*, probably occupied only in Classical-Hellenistic times (*component KO_13*).

An inner location at the back of the valley would be a place suitable for a potential village (or large rural settlement) site, and the grazing areas above were probably held by a rural occupation whose characteristics we still lack (see below - *rural segment*). Though Lauffer would suggest the presence of another Greco-Roman village settlement in the area of the modern village of Kutumula/Koroneia (mainly on the basis of *component KO_76* and *KO_77*), the evidence is still weak and could relate to another kind of activity focus rather than a settlement site (see below – LONG TERM SETTLEMENT TRENDS).

Rural segment

The parts of the fertile (F) areas covered by alluvia, and of the MF slopes covered by flysch closer to the city, were probably farmed by the inhabitants of ancient Koroneia, while areas further distant, as well as landscape zones suitable for agriculture in the inner valleys, would have been farmed through a system of farmsteads (some temporary, some permanent) as in other Boeotian rural areas (see examples in Thespiae, Tanagra[6]). In the outer *chora*, beyond a 2-2.5 km radius from the city,[7] the landscape would likely have seen the presence of hamlets and villages, in addition to farms.

Zones suitable for grazing were also available, especially in the inner parts of the valley and on the steeper slopes of Helicon, as well as on the rocky areas of the ridges (especially the Libethrion/Tilphousion ridge, towards its E spur), and these landscapes would therefore have been covered with sites linked to these activities.

Most of the information about this segment of the landscape is lacking, mainly due to a lack of intensive field-by-field surface research. Conversely, for the Koroneia *chora*, the work of Lauffer (Kopais I) provides information on a few rural or other activity (industrial) foci within the *chora* (*components KO_78, KO_86, KO_87?*). *Component KO_78* on Butsurati ridge can be considered with certainty to be a Classical farm in the environs of ancient Koroneia, exploiting the territory nearby (higher hill area, inner valley, good for slope agriculture and probably also a suitable starting point for taking sheep to graze on the upland plateau in the summer). *Component KO_86*, in the vicinity of Kaminia (by Solinari), immediately to the S of the confluence of two small creeks, could also be interpreted as a rural site (Roman?) in considering both the position of the site and

---

[1] Period maps are included in chapter II.4, figs.17-19-21-23-25-27.

[2] The city occupies a neat boundary between two contrasted landscapes both fertile, the hilland and the Copais edge lowland, as often happens for the Boeotian city sites (see also Levadeia, for instance – chapter II.3.2).

[3] For the identification of the streams see Lauffer Kopais I: 80-1; Fossey 1988: 323 and others (see above *passim*).

[4] For details see appendix I.1 (PALAIA KORONEIA).

[5] Reported under the site entry in appendix I.1 is the discussion of the various identifications.

[6] For this, see for instance Stissi 'The Archaic to Hellenistic Pottery' in Bintliff et al. 2008 for Tanagra, Bintliff-Howard-Snodgrass 2007 for Thespiae.

[7] In Boeotia, the land within 2km of the town is usually farmed by inhabitants of the city (see rural-urban model in Snodgrass 1987/9 and in this work chapter II.4).

the evidence.[8] Both components are located in the Mid-fertile flysch landscape of the valley hillslopes.
Burial evidence by Kutumula/Koroneia (*component KO_77*) might also be connected with the rural exploitation of the upland landscape (see below – LONG TERM SETTLEMENT TRENDS).

Burial areas

No burial places strictly related to the city seem to be known for the Classical-Hellenistic-Roman period, though, by comparison with other Boeotian *poleis*,[9] they must exist, and the usual burial location is around and along the road exiting a Greek city. Only some excavated bases and funerary *stelai* indicate the presence of a Greco-Roman cemetery area by a church between ancient Koroneia and Ag.Georgios (*component KO_92*). On the other hand, remains of burials from earlier (LG – *component KO_15*) and later (LR – *component KO_10*) periods were found by the city site.
Burial areas of the Geometric period have also been found elsewhere in the *chora*: an EG cemetery found by the road by the Petra spur during a recent rescue excavation (*component KO_93*, which could be linked to an earlier settlement at Alalkomenai – see above), and a LG cemetery found during the systematic excavation by the local ephorate of a building of Classical date S of the hamlet of Ag.Paraskevi (*component KO_40*).
In the majority of cases, for the Classical to Roman periods, the material culture available related to burial areas is provided by epigraphical evidence found in different places in the *chora*. Find spots have been approximately located where possible, but inscriptions are very mobile and cannot provide precise locations of burial areas; only that they existed somewhere in the vicinity (*components KO_68 and KO_70*, *KO_22* and *KO_48*). Only a necropolis of Imperial date has been noted (*component KO_20*) in an inner location within the Copais basin.
Two burial areas have been given a general date because there is inadequate information to assign a more precise chronology: a grave tumulus (isolated tomb/tumulus), generically dated to the period, in the area between the city of Koroneia and the normal extension of the Copais lake (*component KO_11)*, and a burial area found in the inner Kakaris valley, also dated to the Greco-Roman period, with tile-roofed tombs reported by Lauffer (*component KO_77*).

Religious areas

Our picture of the sacred landscape in the *chora* of Koroneia is quite rich and well represented in the archaeological record. It is also complicated by the different attempts at identification of the various archaeological evidence for cult places with sanctuaries mentioned in ancient texts.

A Boeotian federal sanctuary is known from the *chora*, the *Itonion* (probably to be identified with *component KO_84* – but see also *components KO_14, KO_16, KO_19 and KO_33* and *34* for identification hypothesis). Moreover, the available archaeological record includes extra-urban sanctuaries (or sanctuaries associated to a settlement), namely: by the Koroneia city site *component KO_9* – LR, *components KO_14, KO_16, KO_17, KO_18* and *KO_19* – A to LR; and the sanctuary of Athena Alalkomenia - Tritonia by ancient Alalkomenai (identified with *component KO_63*, or by others, less probably, with *component KO_84*).
The sanctuaries of Athena Itonia and Athena Alalkomenia were separate but dedicated to the same deity who, as daughter of Tritonas, was believed to be the goddess of fertility of flooding areas (Philippson 1951: 474).
Other known Greco-Roman cult places in the *chora* are: *components KO_27* and *KO_28*, by the rocky spur of Petra, identified as the Classical-Hellenistic sanctuary of Apollo Thilphosios; *components KO_31* and *KO_32*, the Pontsa Roman-Late Roman sanctuary by the church of Ag.Taxiarchis; *components KO_55* to *KO_57*, the Archaic to Hellenistic occupation and use as cult place of the cave site above Agia Triada village; only epigraphical evidence leads to the placement of a cult place (Apollo Paiaion?) to the East of the marshy area at the outlet of the Herkyna river, in an inner position within the Copais basin (*component KO_69*).
Other components have been interpreted as cult places by some scholars, but do not show enough clear evidence for it (and have been assigned a generic 'activity focus' character in the database (as in the case of *component KO_81*, Tsoroko - Ag.Paraskevi).

Forts and fortifications

The fort of Paliothiva (*component KO_80*), on the rocky ridge of Libethrion/Tilphousion, closing the Koroneia valley to the S, is the only fort known for the *chora* in the Greco-Roman period, apart from the city wall fortifying the *polis* of Koroneia itself, and a probable fort on Granitsa/Laphystion (between Koroneia and Levadeia *chorai*), listed in chapter II.3.2 (*component LE_26*). The fort lies just above the saddle carrying one of the two routes leading out of the Zagara valley (Haliartos *chora* – see chapter II.3.8), as marked on the 1:100,000 Greek General Staff map (1943) - Pritchett 1985: 155 (see fig.3). The Petra/Tilphousion fort itself, at the E rocky spur of the Libethrion/Tilphousion ridge, is listed under Haliartos (chapter II.3.8).

Other activities / unspecified activity areas

Some unspecified activity foci, of various characters, can be identified in the general landscape of the *Koroneiake* archaeological record. *Component KO_47*, for instance, is a Hellenistic activity focus noted at the site of Kalami – Lioma due to the presence of a large Prehistoric site there. The same is true of the site of Agoriani - Dekedes, where a few isolated Classical black glaze sherds were found when examining the Prehistoric surface material there at the large Prehistoric site; the sherds (*component KO_12*) could represent a rural burial place probably in

[8] See appendix – *component KO_86* for an interpretation of the place as a possible sanctuary (Lauffer Kopais I).
[9] See, for instance, examples outside Tanagra and Thespiae also known from intensive survey, and excavated examples by other Boeotian *poleis*, Tanagra included.

association with a farm, or slight traces at the edge of one of the Koroneia burial areas.
The same is true for the Palaio Koroneia site itself in the Geometric period (*component KO_5*), when we do not know exactly the extent of the occupation at the site (better known for other periods), where only few surface sherds have been recognized as Geometric.
On the other hand, some other components are known to a larger extent, but we still cannot specify a function with certainty. At the site of Agoriani/Ag.Paraskevi (*component KO_41*), for instance, excavation has discovered a building of Classical date, containing a hoard of 5th C BC coins, with no clear function.
Other components, given the character of 'activity focus', could also be assigned with a question mark to other categories, and have therefore been discussed in the different sections above: *component KO_81*, a probable cult place; *component KO_86*, a possible rural site; *component KO_87*, a possible rural site or workshop area near Koroneia.

Three main habitation areas are quite evident in the distribution of settlement and known activities in the Koroneia area: perilacustrine, mid-height, and upland.[10] The *polis* itself, and the major settlements, as discussed below, mainly occupy the hillslopes of the fertile mid-height zone (MF in the land capability classification – fig.14). I will consider in detail the evidence for the upland and perilacustrine marshy landscape.

The upland landscape

A particular type of landscape to be studied for the Koroneiake is the uplands, mainly inhabited, as the data suggest, in the Prehistoric period when the uplands become valuable in terms of exploitation for specialised pasturage, as well as in medieval periods, mainly for security reasons.
Pastoral activities and settling related to upland areas can be considered as crucial for the Koroneia area, marked by the presence of mountains (see above) and of land suitable for grazing (U and LF zones, as well as MF and F zones when fields are fallow – fig.14) within the Helicon landscape.
For Helicon, Burn (1949: 317) notes that the "*Boiotoi of the Early Iron Age, who were essentially plain-dwellers, seem to have taken little interest in their mountains*" (Hesiod's father, a foreigner, was able to take land there). Conversely, in the Final Neolithic and Early Bronze Age, economic trends seem to have brought people to the mountains (and to Helicon as well) – see the Agia Triada cave and plateau site (*components KO_51, KO_58 and KO_59*).

[10] This can also be noted elsewhere around the Copais basin. Especially from the EH period, we note a presence of settlement at the edge of the basin and in upland areas. During Greco-Roman times, mid-height zones also became occupied, mainly by rural activities, while the economic use of both perilacustrine and mid-height zones, as well as the uplands in particular, is evident in the medieval Ottoman period.

The Koroneia area and its archaeological record, (even if poorly known), may be suitable for the examination of pastoralism issues in the Helicon mountain. The same may be true for Levadeia *chora*, though there we have no archaeological record available. In our case, even if we do not have a great deal of archaeological information available for the mountains,[11] we do have an idea of the extension of the *chora* and how much of it is mountainous area suitable for pasturage, and therefore we can hypothesise on the probable economy and way of living. In addition, the marsh area bordering Copais might also have been a good area for grazing, as would be attested by the fact that the majority of inscriptional evidence concerning right of pasturage comes from Copais (Migeotte 1994).
Research on mountain landscapes is directly linked with research on pastoralism and on people and communities dedicated to pastoralism, either because of tradition,[12] or because of economic exploitation.

Products from pastoralism, in the Mediterranean area mainly husbandry of sheep and goats, are: meat, milk (and cheese), and wool. Manure can also be considered a product. Initial exploitation was for meat as a primary product. Later, between the Neolithic and EBA period, with the so-called 'secondary products revolution' (Sherrat 1981[13]), people began the exploitation of animals not only for breeding but for intensive production of milk and wool. Wool is a primary product only for a specialised economy. In the Koroneia *chora*, the Agia Triada cave and plateau are good examples of an increased use of relatively marginal landscapes, and increased reliance on pastoral production (Halstead 1989) within the EH period.

For the historical period, one could investigate whether a *polis* like Koroneia (provided with so much available upland and lowland land for grazing) had an interest in controlling these upland areas, or whether it rather exploited them because they were available and could not be made any other suitable use of. In our case, the fact that the Helicon watershed was important to Koroneia for grazing and some farming is shown by its border dispute with Thisbe mentioned in an inscription.[14]

[11] Also, from the point of view of material culture, we usually do not have much remaining from pastoralists. They have been often called 'invisible people'.
[12] See for instance, in Boeotia of the Ottoman period, the case of Arvanitic communities mostly dedicated to husbandry activities due to their economic tradition, while Greek villages were initially still linked mostly to agricultural practice (see Farinetti – Sbonias 2004).
[13] According to Sherrat 1981, in the Near East, in the 4th millennium, we assist at the so-called 'secondary products revolution', practised mostly at large settlements, and perhaps only by the richest households in these, and by the ruling elite in the palace economies. Dairy production is probably taken on throughout Europe from the Final Neolithic period, as well as the use (or sharing with others) of the plough by most farmers.
[14] See above in text and IG VII 1870 (Roesch 1965: 53; Fossey 1988: 337).

Since we know (from historical and epigraphical evidence) how large a role was played in the Greco-Roman period by the relationship between pastoralism and agriculture (Nixon-Price 2001), we might suppose an exchange of products and land within the community, based also on gender division of labour.

Nixon and Price (2001: 405) distinguish three varieties of pastoralism:
i. stationary pastoralism.
ii. mobile economy within a restricted area.
iii. long distance transumance.
In Koroneia we probably deal mainly with the second case: a mobile economy within a restricted area, within which the flocks might go up to summer pastures approximately a day's walk from the permanent habitation site.
Certainly, though, there were also forms of stationary pastoralism,[15] probably conducted mainly along with agricultural activities, on the marshy lowland at the edge of the Copais as well as in the agricultural fields after the harvests, and in less fertile areas in higher locations within the valley (the Butsurati ridge for instance, or the lower slopes of the Petra/Tilphousion ridge.) Some evidence of quite upland farms, which could be used for agricultural purposes and contemporaneously be bases for upland pasturage (one day or short periods or seasonal), are known from the Koroneia area (see, for instance, Butsurati –*component KO_78*). A special case could be represented by the high plateau of Koukoura/Agia Anna (see above).
As for long distance transhumance, we do know, for instance, that in the 20th century, and to some extent even today, people from the Zagora/Evangelistria village (up on Helicon in the parallel valley) took their flocks to Aliki, to the sea in the Gulf of Corinth, during winter (see Nixon-Price 2001: 405 - inverse transhumance). Something similar could have happened in the upland settlement within the Koroneia area, when inhabitants dedicated to husbandry could exploit the upland pasturage in the spring and summer, while exploiting lower areas (either by the sea or even by the Copais) in the winter. Invisible people might then become visible on the mountains in summer and in the plain and on the coast in winter. Their families, however, remain visible all year round in the lowlands, plains and villages. Transhumant flocks may coexist with stationary flocks during the locally fertile period (winter-spring in the Koroneia lowlands).

Regarding the cultural aspects of pastoralism, it is worth mentioning Hesiod (Hesiodus *Th.* 22-35), who was taught his song by the muses while he was shepherding his lambs. The useful slopes of mountains like Helicon provided to the plains below both pasturage and the opportunity to receive divine wisdom and inspiration, in the Near Eastern tradition (West 1997). Koroneia is on the W face of Helicon, but the sacred landscape linked to pastoralism is an aspect of research worth a special investigation (see for instance the cave sanctuary of Agia Triada[16] – *components KO_51 and KO_58*). The cultural attitude to pastoralism was therefore quite positive in Archaic-Classical-Hellenistic times, giving value to the shepherd as a social figure. In the Roman period, areas of pasturage were open to everybody, as they were under central control (especially border areas often of critical use in earlier times[17]), but the attitude towards them changes slightly. In the Late Roman period the attitude to pastoralism was different: the pastoral slopes were 'outside' and dangerous.

As far as the environment is concerned, the *chora* itself, as seen above, can be considered as marked by a high physical and natural attitude to pastoralism/pastoral landscape (mountains, marshy grazing land). On the other hand, the territory of the majority of the *chora* is quite fertile, and inhabitants did not need to concentrate on grazing activities, which was probably *tout-court* complementary.

The perilacustrine marshy landscape
Marshes, swamps and fens represent at best the ambivalence which characterises marginal areas (*eschatai*), external to the cultivated space but certainly not passive and inert elements of the rural landscape.[18] In fact, such areas are not simply a pure obstacle to the expansion of cultivation, or potential sources of diseases and poor climate, but in the majority of cases are ideal places for a parallel economy (fishing, plants, ribs, husbandry, etc.). Therefore, apart from the attempts carried out to drain, at least partially, some areas of the Copais basin (see fig.11 and Farinetti 2008), one might suppose the existence, sometimes confirmed by the available archaeological record, both of rural sites exploiting sources of parallel economy from the border of the lake, as well as cult places linked to the water[19] which would populate and give active life to the marshy landscape.

[15] See Halstead 1987, who would support the hypothesis of stationary pastoralism as the most suitable form of pastoralism in Classical Greece.

[16] Immediately downslope of the cave sanctuary today is a small church, dedicated to the Virgin. Behind it, a holy tree can be seen, with icons in the central break. The tree, as well as the surrounding forest, is an oak, probably belonging to the ancient vegetation (forest of wild oaks). The cave sanctuary, as well as the modern church close to the holy tree, is near a creek. They can all be linked to the use of upland areas (in the proximity of the creek I also found a chopped flint, probably Paleolithic).

[17] see Daverio Rocchi1988, Alcock 1993.

[18] As Traina (1988) and Fantasia (1999 after Traina) point out. See Farinetti 2008.

[19] See Pappadakis AD 2 (1916): 217ff for a list of cults of Herakles in the Copais.

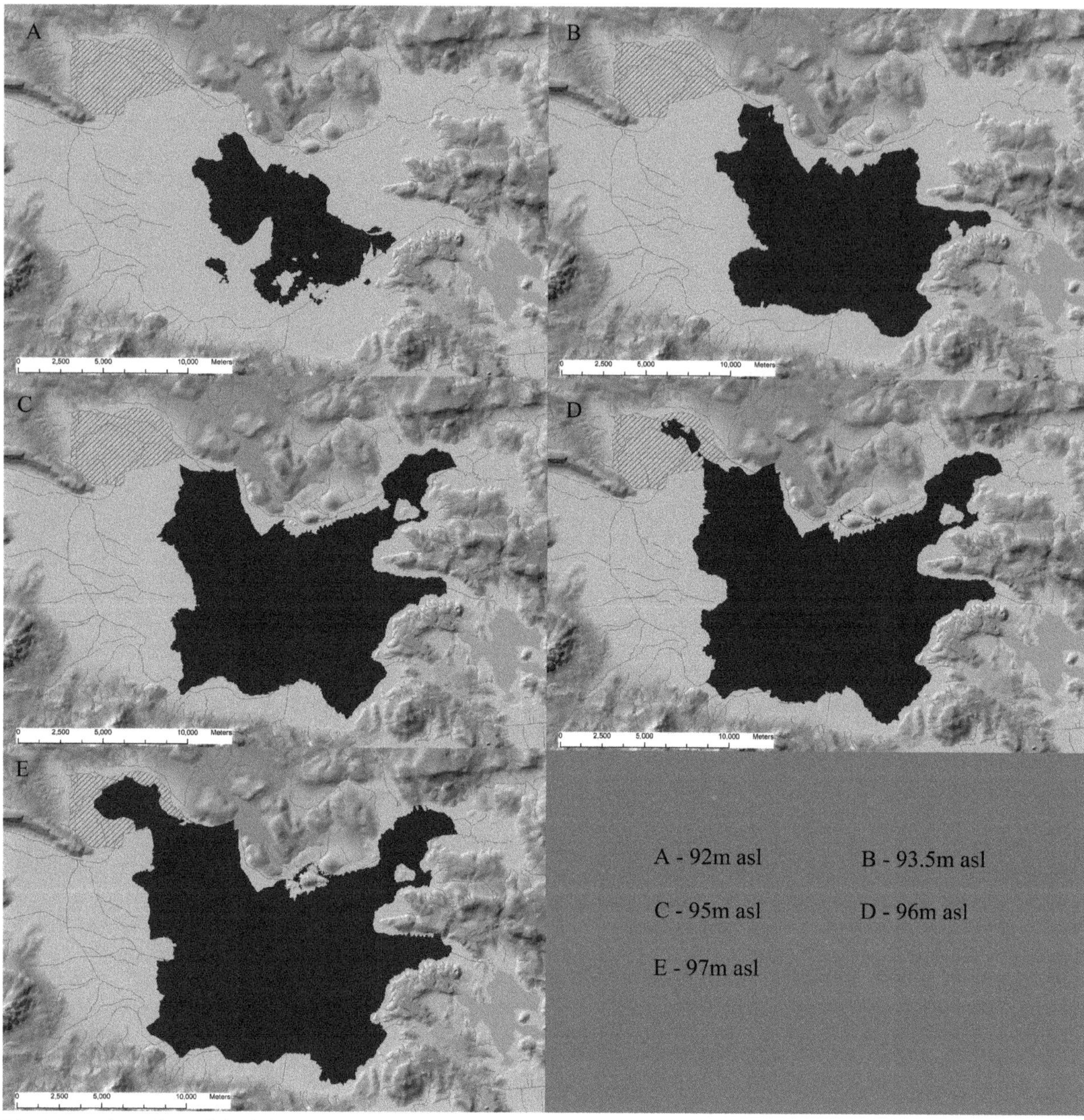

Fig.10. *A digital model of the water fluctuations of Copais lake.*

The digitally reconstructed model of lake fluctuations[20] can help in evaluating the relationship between *poleis*, perilacustrine activities and the marshy lake. The model is illustrated in detail in Farinetti 2008 and it is briefly presented here below:

1. Annual model

1a. DRY periods (One or more DRY years). MIN extension of the lake: up to 92m *a.s.l.*. Relation lake/marsh: marsh progressively covers the whole wet surface and in some periods the lake is replaced by marsh. (fig.10 - A)

1b. WET periods (One or more WET years). MAX extension of the lake: up to 96m (or more –97m for instance- in catastrophic years) and of the marsh. Relation lake/marsh: the lake increases and marsh decreases; marsh progressively becomes lake. (fig.10 - D and E)

2. Seasonal model

2a. SUMMER. MIN extension of the lake: up to 93.5m (or even smaller –up to 92m *a.s.l.*- in dry periods or in very dry summers). Relation lake/marsh: large marsh and small lake. (fig.10 - B)

2b. WINTER. MAX extension of the lake: up to 95m (or larger –up to 96m *a.s.l.*- in wet periods or in very wet winters). Relation lake/marsh: more balanced; the lake increases but usually does not reach the 1b situation. (fig.10 - C)

[20] A dynamic digital reconstruction of lake fluctuations has been implemented, combining the results of the work of the German team led by engineer J. Knauss and the Munich University group (Knauss et al. Kopais 1-2-3; see Farinetti 2008 for further details), the geomorphological and palaeoenvironmental studies carried out on the area, the examination of aerial photographs and the known archaeological evidence and historical data.

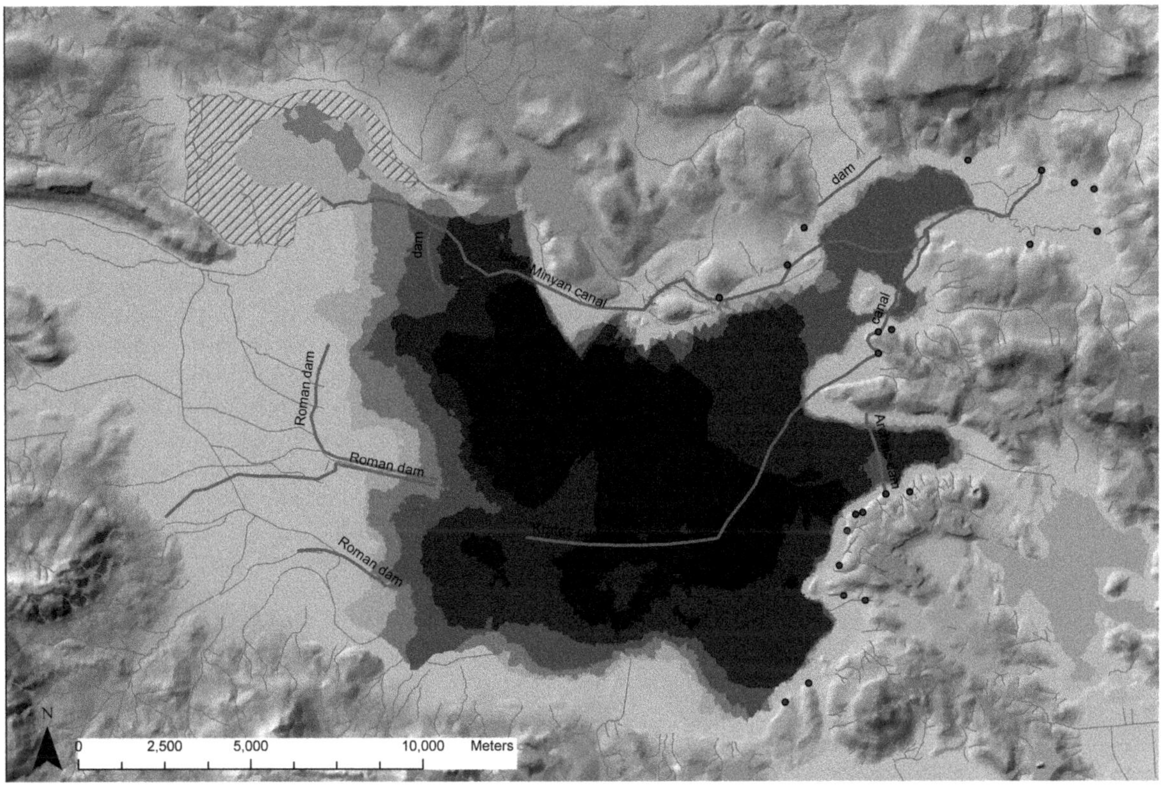

Fig.11. *Map of the main recognised features related to the various attempts at drainage of the basin over time (after Knauss et al. Kopais 3).*[21] *The dots represent the katavothrai.*

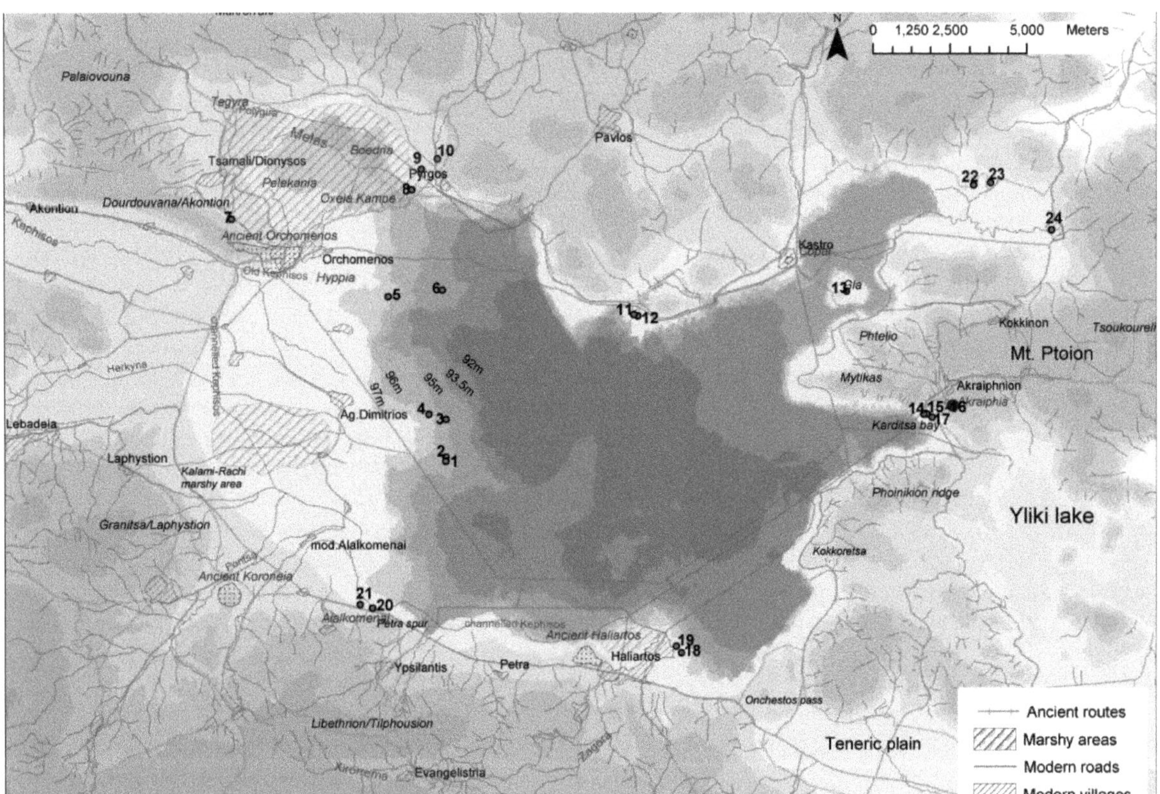

Fig.12. *Greco-Roman sites along the reconstructed lake edge.*

[21] The map does not include the whole system of channels, dams and polders that totally drained the basin in the Mycenaean period (for details see the reconstruction made by Knauss et al. Kopais 1).

As already pointed out by Knauss and his team (Kopais 3) while conducting researches at the edge of the lake to investigate its historical level(s), sites are mainly attested along the 95m asl contour. This level is hypothesised by the Munich Kopais-Project group as the maximum level reached in antiquity in normal wet conditions (see above, 2b; fig.12 and Farinetti 2008).

In the Koroneia area, the archaeological record available would attest mainly the existence of peripheral activities by the edge of the lake in the Greco-Roman period since Geometric times (burial areas -*KO_70, KO_11, KO_20, KO_93* in map fig.12 nos.1-3-4-21 - and cult places – *KO_69* and *KO_27* in map fig.12 nos.2 and 20- in the area of the modern village of Ag.Dimitrios[22]), while a rural exploitation is not directly attested. Nevertheless, the aforementioned activity foci do attest some use of the marshy landscape, along with the burials, possibly marking rural plots along a road exiting the *chora*, known at the very edge of the basin by a marshy area (Kalami/Rachi *components KO_48 and KO_62*, along with other signs of habitation activities – *component KO_91*[23] – map fig.3 no.40). The results of intensive research in other areas of the Copais basin might shed light on activities and life in this perilacustrine and marshy segment of the landscape: by ancient Haliartos, for instance, in a landscape similar to that of the Koroneia area, at the S edge of the basin, a string of Archaic-Classical rural sites (chapter II.3.8 and map fig.12 no.18 – survey site HAL B4) and a large Late Roman estate (Kahrstedt site – map fig.12 no.19) attest the rural exploitation of the edges of the lake in the Greco-Roman period. The same kind of landscape use is known for the area of Orchomenos (*component O_42* – a Classical and Roman possible rural activity; *O_66* – a Classical possible rural site; *O_60* – a Roman settlement area – map fig.12 nos.6-9-8) and along the E edge of the basin (*NC_7, NC_73* of uncertain character; *NC_45* – a Geometric to Classical probable hamlet site; *NC_56* – a Classical probable farmstead; *AK_53* – a Late Roman villa rustica – map fig.12 nos.12-13-23-24-17).

LONG TERM SETTLEMENT TRENDS IN THE CHORA *LANDSCAPE*

Though a careful settlement history and a detailed analysis of the shifting of settlements (main and satellite) would be possible only after intensive research into every segment of the landscape, data available for the Koroneia *chora* are indeed enough to attempt a general understanding of the settlement history in the area.

The Greco-Roman *polis* of Koroneia was located, as noted above, in the centre of the opening of the valley, in a position from where it could control the passage and also the road going to the inner Helicon and then to the Thisbe area, and the sea, through the Koukoura/Agia Anna pass. The city-site was abandoned at some point at the end of or after the Late Roman period, and the main settlement of the area moved elsewhere. The remains of the Frankish tower still visible on the slope occupied once by the greek *polis* provides evidence of a Frankish settlement there, overlooking the main routes but also in an inner enough position to protect itself from exposure to attacks. Both the Late Roman occupation and the Frankish tower on the lower acropolis[24] indicate the survival of medieval and probably ancient Greek communities, and continuity into Ottoman and Modern times namely at Ag.Georgios, to the W side of ancient Koroneia and on the SE slope of the Laphystion ridge, immediately within the entrance to the valley, where the main settlement moved from the city further into the hills, also exploiting part of the lake. Thus Ag.Georgios, today the largest village of the area, is possibly the replacement of the ancient city from the 14th/15th century AD. It is recorded in the Ottoman archives as a large Greek community already in 1466. During the process of renaming villages in modern times, the name of Koroneia was given to the smaller upland village on the slope of the Tilphousion ridge. This does not, however, play a central role within the area.
Knauss et al. (Kopais 3: 66) seem to imply that the Prehistoric centre of the area would have been at Agoriani-Dekedes, which would then have moved to Koroneia in historical times. Knauss and his team seem to have been looking for the Homeric Koroneia, and that is why they were searching for one central settlement. In my opinion, the fact that at Koroneia little Prehistoric pottery has been found (see appendix I.1) should lead us to ask whether we should look for an 'old-Koroneia' elsewhere at all. Prehistoric occupation could also have been different, without a central settlement site, and the few prehistoric sherds seen at Koroneia may mark the presence of a small settlement among others (Agoriani-Dekedes; Kalami-Lioma). Intensive and systematic field-by-field survey, begun at the city site in 2007 (within the framework of the Boeotia City Survey Project), should certainly provide some answers.

Within the wider *chora*, apart from evident small landscapes supporting a settlement site in antiquity (such as ancient Alalkomenai and Koukoura/Ag.Anna), we can identify a number of long term potential areas by looking at the traditional settlement pattern. Even if the settlement and other activity foci may have shifted in place within those settlement areas, people living in it exploited and used the same land, the same territory, and experienced the same landscape. In order to identify potential

[22] In the whole area of the Copais basin, see cult places: *O_35, O_66*? (fig.12 nos.7-9), *AK_55* (fig.12 no.14); burials: *O_25*?, *O_58* (fig.12 nos.5-10) *NC_37*, NC_87, *AK_57* (fig.12 nos.22-11-15); and a pottery kiln installation *AK_58* (fig.12 no.16).

[23] Signs of possible rural occupation are also known on the Kalami-Lioma hill, known as a Prehistoric settlement site (see appendix I.1), immediately above Kalami/Rachi and the marshy area.

[24] The ancient city probably became a refuge site in the Early Byzantine period, then a village in Middle Byzantine times and in the Late Byzantine period was still there and controlled by the tower.

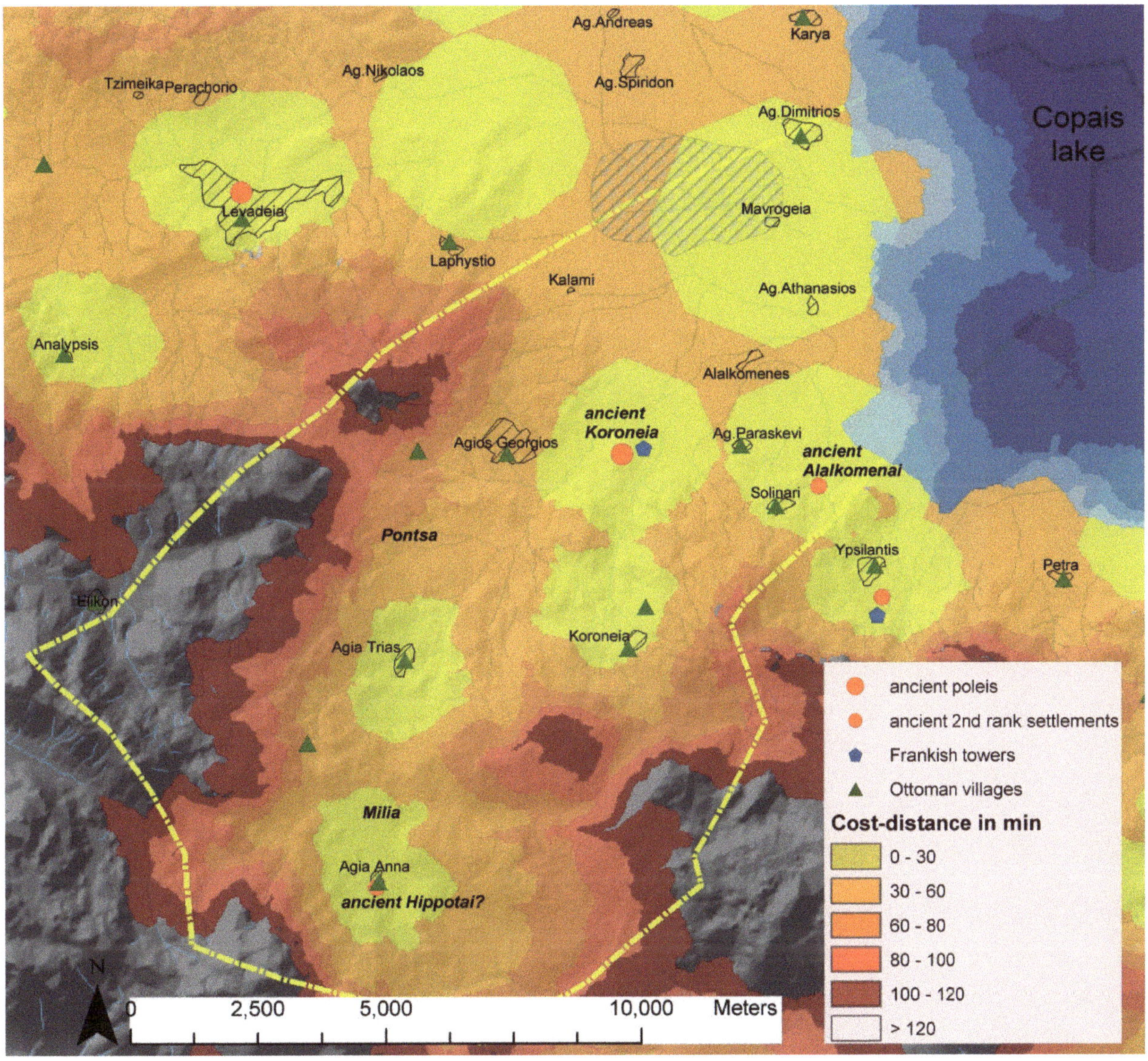

Fig.13. *Classified surface representing the cost-weighted distance (1/2 h walking and further ranges) from recognised 1st and 2nd rank ancient settlements (represented by larger and smaller dots). Areas without dots indicate potential settlement chambers. Ottoman villages and Frankish towers have also been added to the map to show their spatial relationship with the Greco-Roman settlement network and to appreciate potential settlement chambers.*

settlement chambers I applied a GIS cost-distance analysis based on the Greco-Roman 1st (*poleis*) and 2nd rank settlements (both certainly positioned and potential). Critical radius is the half an hour walking time distance from the settlement, defining territories potentially and probably exploited by it (see chapter I.2.2 for details[25]). The settlements included in the dataset used for the cost-distance analysis are *poleis* and satellite settlements from the Greek antiquity (Archaic-Classical-Hellenistic periods) over the whole Boeotia region, and can be considered coeval for the purposes of the analysis on potential long-term settlement chambers. Even if in the majority of cases the settlement rank is known for mature Greek antiquity, I preferred to test the model without differentiating *a priori* the settlements according to their rank, as the model is supposed to be working as a comparison layer for a long-term analysis and is based on the subdivision of the territory applicable also to early periods of history. Then, protourban nucleated settlements (*komai*), defined also as *proto-poleis*, were sharing the territory fairly regularly. Afterwards, only some of them would have reached the status of *poleis* and would have created differentiation between 1st and 2nd rank settlements. The model itself, result of the analyses, shows a relatively larger territory for the proper *poleis* (see chapter II.4). The map in fig.13 displays the results of this analysis, commented upon below. Marked are the ancient hypothesised settlement network on which the cost-distance analysis is based, known Frankish towers,

[25] For a similar analysis using Thiessen polygons and straight line-distance of 2.5km radius applied to towns and villages see Bintliff 1994b (fig.20 and p.219: "*in the Classical 5th-4th C BC there is a dominant tendency for the entire cultivable land of Boeotia to be divided into modular territories of some 2.5 km radius, each exploited by a polis or village*"). For the methods employed in performing the analysis see chapter I.2.2.

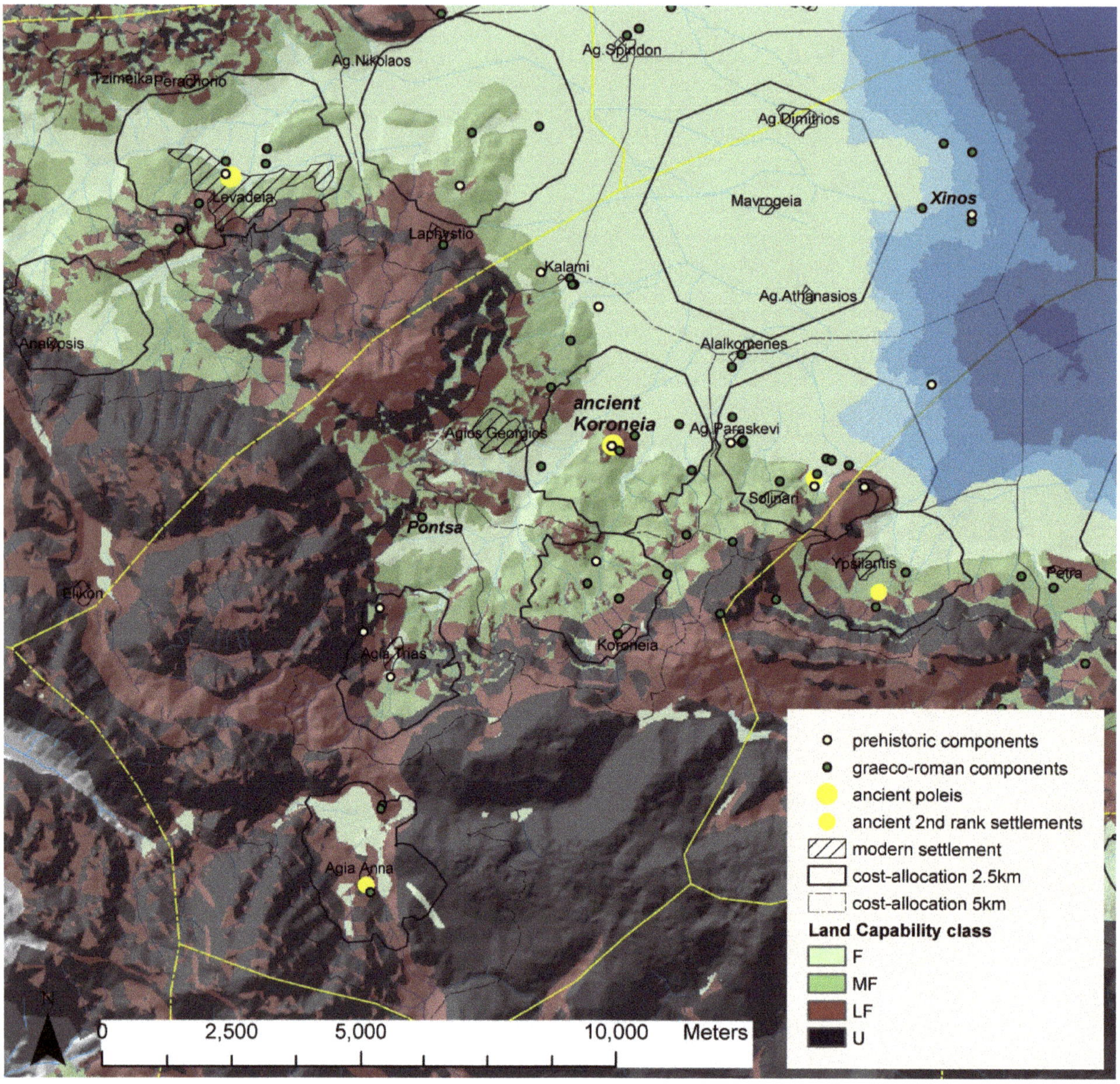

Fig.14. *Map showing the Greco-Roman settlement network, the polygons resulting from the cost-distance analysis (marking half an hour and one hour walking time distance) and dots representing the known archaeological components (same as in fig.3), with land capability information underlain.*

identifiable Ottoman villages, and present day villages. The map in fig.14 shows the Greco-Roman settlement network, the polygons resulting from the cost-distance analysis (marking half an hour and one hour walking time distance) and dots representing the known archaeological components (same as in fig.3), with land capability information underlain.

In the plain area along the edge of the former lake[26] a series of settlement chambers open up. These are occupied today by the modern villages of Ag. Dimitrios, Alalkomenai, Ag. Paraskevi and Solinari, which exploit the fertile alluvial soils (fig.13-14). Some of these have predecessors in the Ottoman period. Ag. Dimitrios is recorded in the Ottoman archives from 1466 to 1570, then in 1642-1688 as a Greek village. The large size of this village even in 1466, along with its lakeside position, as well as its survival through the troubled 17th and 19th centuries, indicate a settlement chamber on the immediate W outskirts of Copais, probably affected by seasonal flooding.[27] The archaeological evidence at Xinos (*components KO_64 to KO_72,* as well as *KO_11* and *KO_20*) would support this. To the S of it the second rank villages of Mavrogeia and Ag. Athanasios occupy part of the plain today). Further S, Mamoura/the modern village of Alalkomenai, in the middle of the Phalaros/Pontsa's former outlet area, does not find any possible attribution among the villages recorded in the Ottoman archives,[28]

[26] For the reconstruction of the lake fluctuations in the past see above in the text – PERILACUSTRINE MARSHY LANDSCAPE and Farinetti 2008.

[27] Seasonal floading is shown in the lake fluctuation model (fig.10 and see above in the text – PERILACUSTRINE MARSHY LANDSCAPE and Farinetti 2008 for details).

[28] Its existence has probably to be linked with the drainage of the lake.

while only uncertain evidence might indicate the presence of a sanctuary and/or burial site in antiquity (*components KO_33* to *KO_35*). On the other hand, the area of the settlement of ancient Alalkomenai, in a small valley facing Copais, at the E entrance to the Koroneia valley, certainly constitutes a small settlement chamber. A settlement site (*components KO_21, KO_24, KO_25*) was located there in Greco-Roman times (a small *polis* – defined as *polichni* in Roman times – a satellite of Koroneia[29]), but Byzantine and Ottoman sherds were also noticed by Lauffer (Kopais I: 75) to the W of Solinari village, down the slope in the Dichalorevma locality, probably marking the location of a small village/hamlet in medieval times. Solinari (formerly Sulinari) has probably to be identified with the Ottoman village of Seta/Sela or Suli recorded in the Ottoman archives. It could represent a case of a shifting of settlement with the village of ancient Alalkomenai.

Other possible small settlement areas might have existed at the edge of the Copais and towards the border with the *chora* of Levadeia. The two Prehistoric settlement sites of Agoriani-Dekedes (*components KO_36 to KO_39* by Ag. Paraskevi) and Kalami-Lioma (*components KO_42 to KO_45*) may mark two small landscapes available at the edge of Copais, exploited since the Neolithic period and working as such especially in Prehistoric periods (in parallel with a probable Prehistoric settlement site at palaio-Koroneia – *components KO_2 to KO_4*), while later included into larger settlement chambers (at least for the Agoriani area). In the area of Agoriani-Dekedes and the modern settlement of Agoriani/Ag.Paraskevi, Lauffer (Kopais I: 110) seems to suggest a shifting in location of the Prehistoric settlement, that would have moved in historical times to the SE foot of the hill.[30] I would rather suggest a shifting in use of the place (see Agoriani-Dekedes entries: *components KO_36 to KO_39, KO_9, KO_12*): settlement site on the flat top of the hill in the Prehistoric period, abandoned in the Dark Age; cemetery site SE of the hill in the Geometric (LG) period linked to a habitation site which was probably nearby; probable Classical rural cemetery on the flat top of the Prehistoric mound, probably associated with a Classical farm nearby or just at the edge of a cemetery of the *polis* of Koroneia along the road.[31] In modern times a hamlet (Ag.Paraskevi, recorded already in the 16th-century archives) occupied the area at the foot of the mound, by the road. At Kalami-Lioma later occupation (Classical-Hellenistic, *components KO_46 and KO_47*), probably a rural site, is attested at the site of the Pre-Mycenaean settlement itself, with possible Hellenistic-Roman burials nearby (*components KO_48 and KO_62*). The cost-distance analysis shows a potential gap in the area of Kalami (fig.13), which could however be interpreted as an empty area marking the border between the *chorai* of Koroneia and Levadeia. In an inner location, the Late Roman sanctuary in the locality of Pontsa, situated in an inner valley of the Pontsa river tributaries running parallel to the S to the Ag. Georgios basin, might mark the focus of a potential small settlement chamber, and the cost distance analysis 2.5km / 1/2h walking distance radius might confirm the space for a potential small settlement area (fig.13).

In conclusion, both the modern and Ottoman settlement pattern, as well as the available traces of Prehistoric and ancient occupation, show the existence of villages occupying and exploiting the edge of the lake.

In the inland of the *chora* of ancient Koroneia there is today a series of villages which date back to the medieval and Ottoman periods[32] and show the economic use of the uplands in medieval and postmedieval times.

Kutumula/Koroneia is an Albanian hamlet recorded in the Ottoman archives in 1466-1570 and 1646-1688. Higher up on Mt. Helikon is Steveniko/Ag. Triada, a Greek village recorded in 1466-1570 and Koukoura/Agia Anna, in the upland plateau opening up beyond and above Steveniko/Ag. Triada, on an inner route to Helicon, a Greek village recorded in 1466-1570 and 1646-1688. An additional village in between, Milea, is recorded in the Ottoman archives in 1521-1570 and again in 1642. The availability of land and resources for settlement activities in the wider area of Milea/Milia is confirmed by two possible rural sites (Milia – *components KO_49-50* as well as *KO_95-96*), which in Greco-Roman times, though, might have gravitated around the settlement area of Koukoura/Agia Anna. The area of Koukoura/Agia Anna constitutes a well topographically-defined upland settlement chamber in inner Helicon, also occupied in antiquity, and attempts have been made to locate ancient *Hippotai* at the edge of this wide upland plateau - fig.14 (see *component KO_13* and Fossey 1990e: 205-7). In the deepest, narrowest part of the Koroneia *chora*, the Agia Triada area can also be considered a small upland landscape, even if not so well defined as the Koukoura/Agia Anna case. The area is occupied in the Prehistoric period (especially EH – see above), but its high position within the valley, by the passage of the upland road, gives it a long-term potential. The secure, high position, overlooking all of the *chora* of Koroneia, would constitute suitable characteristics for a Dark-Age refuge settlement,[33] certainly inhabited in the Late Frankish / Early Ottoman period. By that time, the lower area of Kutumula/modern Koroneia village, on the slope of the Libethrion/Tilphousion ridge, was also chosen for settlement. The cost-distance analysis indicates that there is room for a settlement site there, as the Ottoman and

[29] In the Roman period, the only other settlement existing in the area of Koroneia seems to have been the village of Alalkomenai (Pausanias IX 33.5-7). As noted by Fossey (1990c), the village still existed in the imperial period while the sanctuary nearby (*components KO_63*) had fallen in disuse.

[30] The same situation, as Lauffer (Kopais I: 110) notes, occurred at Pyrgos, in the NW part of the Copais basin (by the temple of Apollo – see chapter II.3.4) and in other places.

[31] The building with a hoard found just down the slope (*component KO_41*) still has to be explained.

[32] They are mainly arvanitic but probably refound Byzantine settlements.

[33] There is no evidence for it, but the village really looks, even today, like a fortified settlement (see for a parallel the case of Agios Konstantinos – Tanagra, appendix I.14).

present settlement pattern would attest (fig.13). Lauffer would suggest the presence of a Greco-Roman village settlement in the area of the modern village of Kutumula/Koroneia (mainly on the basis of *component KO_76*[34]), but the evidence is weak and could relate to another kind of activity focus rather than a settlement site. The settlement of modern Kutumula/Koroneia was probably created in the medieval period,[35] with the intense economic use of uplands. Like the village of Zagora/Evaggelistria to which it is connected by a mountain road, it is clearly a medieval upland settlement located in the inner valley. Those upland areas in antiquity could have been exploited solely by farms or as a base for husbandry activities. The location of an Ottoman village does not directly mark a settlement chamber used as such in Greco-Roman times also, as is attested by the case of Ginosati, in the area of ancient Tanagra, where the Ottoman village (Vionis 2006) exploited an area occupied in antiquity by farmsheds and rural sites (see chapter II.3.14 and appendix I.14 – THE TANAGRA SURVEY PROJECT).

[34] See also the cemetery location for the supposed village at Kutumula/Koroneia - *component KO_77*.

[35] An Albanian hamlet recorded for the first time in 1466, it could have refounded a Byzantine settlement.

II.3.2

Northern Helicon: Levadeia

TOPOGRAPHICAL SETTING

The valley of Levadeia is separated to its N from the Chaironeia valley by the ridge terminating in the Kourpeiko, while to its S it is separated from the Koroneia valley by the Laphystion ridge. It constitutes a bay which enters into the Copais basin on its Western side, as well as the Koroneia valley to its S.

The fertile basin of Levadeia is ca 2km wide and ca 6km long. The modern town of Levadeia lies at its SW corner. The valley is wider to the town and narrows rapidly thereafter; in its wider part it is divided along its axis by a low schist ridge (Tryphiolithari) which elevates the level of the plain. The plain begins at 100m asl (Kephisos) and rises to 160m asl (Levadeia town). The town of Levadeia controls three small plains of which it marks the central point (fig.1). The city controls the routes that lead from Thebes both up to Delphi and down to the Antikyra bay through the Tsoukalades pass, as well as the old inner route from Thebes to Chaironeia through the Kerata pass (fig.1), and the entrance to the route towards the inner Helicon, through the pass behind Surpi.[1]
From the ridge behind the town of Levadeia springs the Herkina river (called Kria at present[2]) through a system of karstic springs.[3] The springs are still visible today in the upper part of the modern town, where they have been transformed into a small park acting as a fresh refuge on hot summer days.

The Herkina river has been well channelled, in connection with the drainage work in the Copais.[4] In the past it was characterised by a delta outlet, and therefore a marshy area marked the landscape (see chapter III.1).[5]

Boundaries
The limits of the *chora* are marked geographically.[6] They are comprised of a valley running radially NE to SE from the Copais basin towards Helikon mountain. Along the N side the Kourpeiko ridge separates it from the area of Chaironeia, while along the S side the Laphystion ridge separates it from the area of Koroneia.
A boundary inscription has been found below the Eastern peak of Mt.Granitsa/Laphystion, to the E of the church of Agios Ioannis, built beside an abundant water source, E of the modern Laphystion village. The inscription[7] mentions the border between Koroneia and Levadeia, which ran over Mt. Laphystion, apparently from some spring, up a ravine, to an altar of Zeus Laphystion (Pausanias IX 34.5). The latter may have been a simple ash altar of the type often used for the worship of Zeus on Greek mountains (Camp II 1991 quotes Langdon 1976). The springs might have been those by the church of Ag.Ioannis (Camp II 1991: 195). Significantly, Granitsa (the former name of Laphystion village[8] and mountain) is the Slavic word for 'border' (Kretschmer 1914: 265).
The W boundary runs through the uplands of Helicon and separates Boeotia and Phokis. According to Dasios (1995), two natural boundaries are clearly recognisable: one to the E of the Vathyrema river which belonged to Boeotia, and another to the W which, from the Karakolithos heights, follows the line Karakolithos - Lophos Viglas – Panopeus acropolis, to get to the N route out of the valley, where it meets the valley of the lower Kephisos.

---

[1] The description of the landscape of the Levadeia *chora* is based on Philippson 1951: 445ff, quoted also by Fossey 1988: 343, and on personal observation.
[2] The Herkyna river has been identified with the ancient Probasia river, which according to Theofrastus (*HP* IV 11.8) ran from Levadeia towards the Copais. Probasia river is mentioned in an inscription also (IG VII 3170).
[3] Philippson 1951: 446/7 gives a detailed geological description.
[4] See also Knauss et al. Kopais 3 1990: 17.
[5] The Herkyna, S of the Hyppia plain turned and ended (together with the water of the Kalamaki spring) in a marshy zone that extended to the W of the Copais lake water, by the modern villages of Ag.Dimitrios for ca. 5 km. up to the foot of Laphystion near Rachi (Philippson 1951: 473). This marshy area, probably semi-independent from the lake's fluctuations, formed a dangerous zone between the plain of Kephisos and the plain to the S edge of the basin, through which the main road ran. In IG VII 3170 some words (l.5 and l.16) refer to dams which were probably linked to attempts at drainage of the marshy area corresponding to the outlet of the Herkyna river into the lake (the inscription refers to public land to be divided and rent, in the area between Orchomenos, Chaironeia and Levadeia). Today, only a small marshy area is preserved by Rachi (Philippson 1951: 474; the area is marked in the GYS 1:50,000 topographical map -*Lebadeia* sheet- and in the 1:200,000 Boeotia *nomos* map).
[6] Philippson: 445ff; Fossey 1988: 349.
[7] The inscription is recorded as AE1671 in this work. See SEG XXIII 297; SEG XXXV 406; Vollgraff 1902: 570; Roesch 1965: 61-3. See also Dasios 1995.
[8] Village history in Camp II 1991: 196, footnote n. 10.

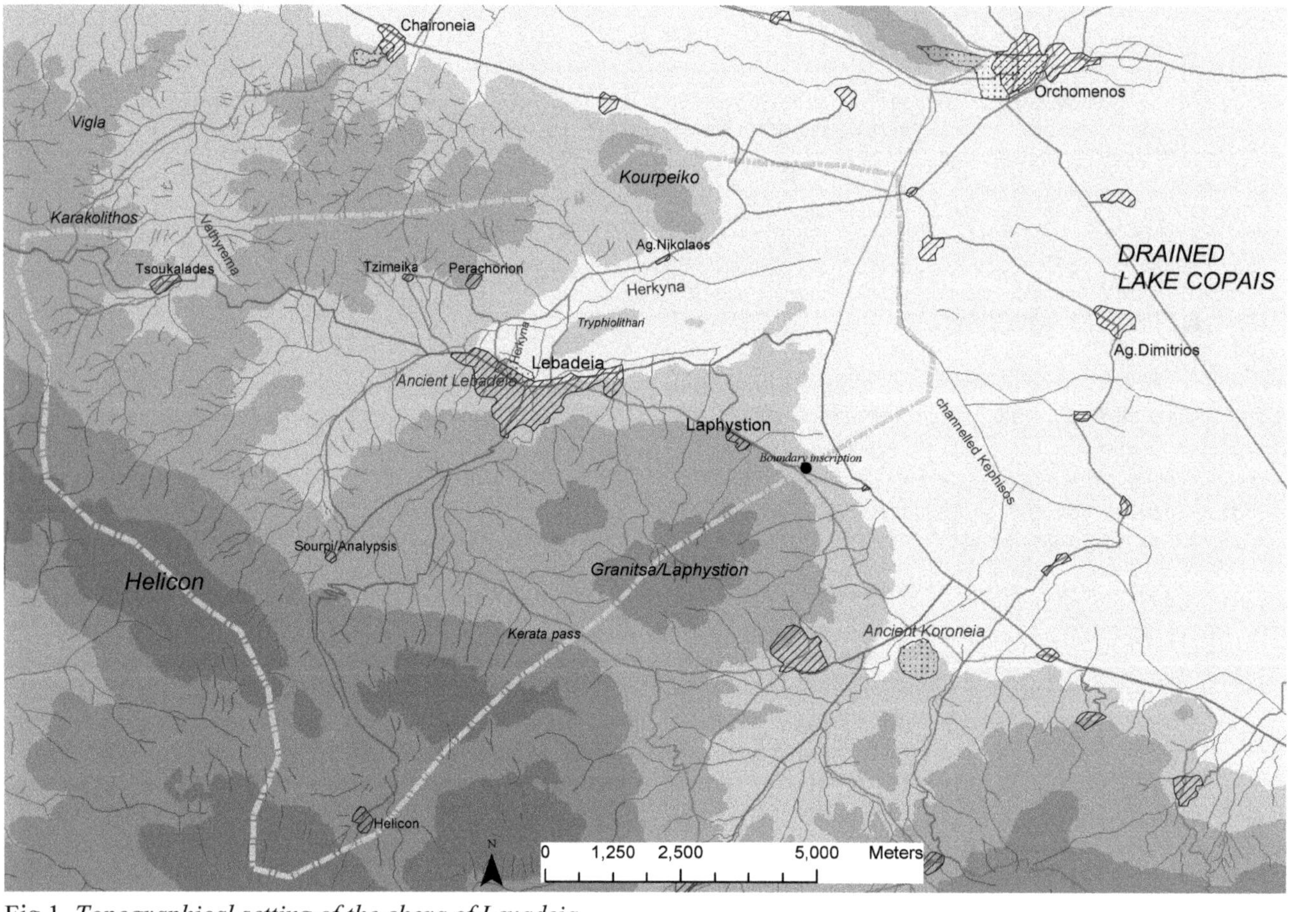

Fig.1. *Topographical setting of the chora of Levadeia.*

Fossey (1988: 501 and 497-500) notes, on the basis of ancient texts,[9] that the *chorai* of Levadeia and Chaironeia did not border the lake and therefore locates the limit of the *chorai* very much lying back in relation to the Copais basin. He proves his argument with a boundary stone, found among the remains of the Herkyna dike,[10] in the middle of an area which constitutes the Western part of the basin. It carries simply the word *ὅρος*. Fossey (1988: 497-500) believes it marked, in Early Roman times at least,[11] the boundary between Koroneia and Orchomenos since, as he believes, the territories of Levadeia and Chaironeia did not reach the lake and therefore Koroneian and Orchomenian *chorai* were neighbouring. Kirsten (1951: 670) also assumes that the territories of Chaironeia and Levadeia did not reach the lake.[12] I think that this should be reconsidered, since the lake in this Western part of the basin was really lying back (i.e. the water of the lake did not fill in the 'bays', unlike Karditsa bay, or the NE bay), forming a real plain (see fig.3 and fig.5). Perhaps the *chorai* of the two *poleis* (Levadeia and Chaironeia) extended as far as the beginning of the plain, with the plain itself belonging mainly to Orchomenos (this, probably, is the location of the city plain, mentioned as Hyppia in ancient texts[13]). A division of the land based on a cost-distance analysis[14] from the *poleis* surrounding the lake would support this hypothesis (fig.2). This might also help to explain the inner location of Chaironeia and Levadeia within their valleys.

---

[9] Pausanias IX 39.1; Theophrastus *HP* IV 11.8; Strabo IX 407; IG VII 3170.

[10] The inscription [AE1108] was found in Dodekakleidi locality, about 0.5km N of Xinos (Fossey 1988: 501). The Roman dam in which the inscription was found is mapped in chapter II.3.1 fig.11 after Knauss et al. Kopais 2.

[11] The boundary stone is dated to the 2nd-1st C BC (Fossey 1988: 501).

[12] Kirsten states (1951: 670): "*Because the territory of Orchomenos extended until the highland (until Probatia and Kephisos), which was also the border with Levadeia and Chaironeia, Orchomenos was the richest polis of the Copais basin and that's why it was predestined to be the leader, the leading centre of the Copais area*".

[13] The Hyppia plain is known, in association with Orchomenos, from ancient sources (Theophrastus *HP* IV 11.8) as the largest and nicest plain of Boeotia (Plutarchus *Sull.* 20.4-5), where the infamous Minyan cavalry was trained. See Knauss et al. Kopais 3 fig.6.8 and in this article fig.4.

[14] A similar attempt at exploring territorial divisions in the area was carried out by Fossey (1988, fig.64), using Thiessen polygons (based on Euclidean straight-line distance not weighted by topography). Here the cost-distance analysis (performed within the GIS system) takes into account, in calculating the distance between locations, the effort (cost) of moving (walking) through the land. The algorithm used is mainly based on distance and slope (see for instance Stančič et al. 1997, Gillings and Wheatley 2002: 151ff, Howard 2007), therefore taking into account the terrain morphology (see chapter II.3.1 – LONG TERM SETTLEMENT TRENDS).

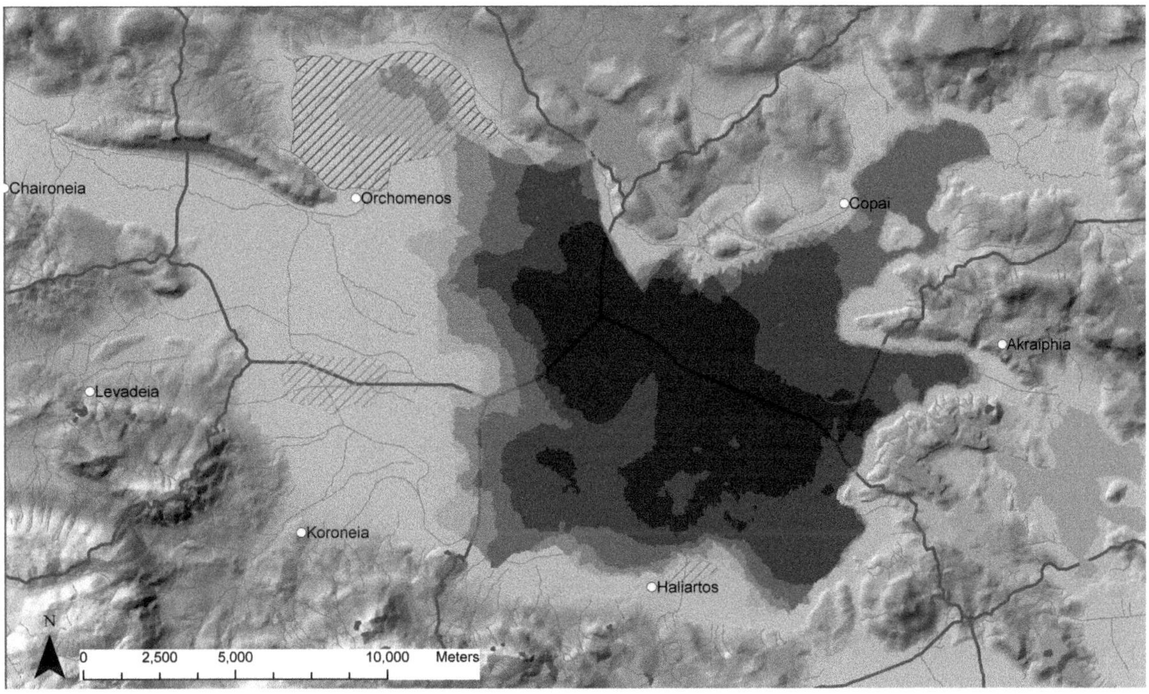

Fig.2. *A division of the land based on the results of a cost-distance analysis from the poleis surrounding the lake. For the reconstruction of the lake fluctuations shown in the map see chapter II.3.1 – PERILACUSTRINE MARSHY LANDSCAPE.*

PHYSICAL LAND UNITS

The physical landscape of the *chora* can be described as mainly between 200 and 600m elevation. A quite large part of the *chora* is mountainous (see fig.2 in chapter II.1). This is due mainly to the presence of the Helicon massif in the Eastern part of the *chora*, as well as of the two ridges radial to Copais closing the Levadeia basin to the N and S. The mountainous part of the landscape can be set at 21.5% of the total area, while the hilly segment is at 51.4%. Areas below 200m constitute 27.2% of the *chora*. Since the limit of the *chora* belonging to the ancient *polis* of Levadeia cannot be precisely outlined towards the Copais basin, the amount of available lowland could change considerably.

Hilly landscape	51.4%
Mountainous landscape	21.4%
Plain	27.2%

1	P1_P2	lacustrine basin, valley	18.1%
2	P3	gentle slope	2.2%
3	P4	foothill	6.9%
4	H1	plateau	13.6%
5	H2	gentle slope	1.4%
6	H3	moderate slope	10.2%
7	H4	severe slope	13.6%
8	H5	very severe slope	12.6%
9	M1	plateau	3.9%
10	M2	plateau, gentle slope	1.8%
11	M3	moderate slope	7.3%
12	M4	very severe slope	8.5%

Table 1. *Percentage of the different physiographical classes present in the Levadeia area (P=plain; H=hill; M=mountain)*

Lacustrine basins and valleys (P1_P2) are predominant in the plain landscape, constituted by the Copais basin and the Herkina river valley. Foothill landscape (P4) marks the area immediately below and following the heights surrounding the plains. Within the foothill landscape the ancient *polis* (as well as the modern town) of Levadeia also lies. Plateaus seem to be the main features of the hilly segment of the landscape. As mentioned by Fossey (1988: 343), a larger, gently sloping area lies to the South West of Levadeia, enclosed by the Helicon range to the South and SW. The open upland area, crossed by two streams, is quite well recognisable in the physiographical map (fig.2 in chapter II.1), as well as in the topographical background of fig.3.

RESOURCES

The *chora* landscape is characterised by fertile and arable land suitable for agriculture, at the edge of the Copais basin (though some areas are, as seen above, marshy, and therefore suitable for other kinds of activity) and in the Herkina river valley.
A map showing land capability (for agriculture – see chapter I.2.1) for the Levadeia area can be seen in fig.7. While the most fertile areas are constituted mainly by the Copais plain and the Herkina river valley (together with some upland river valleys), class MF (Mid Fertile) is predominant, constituted principally by the flysch zone of the slightly elevated areas.[15] The combination of upland plateau or gentle slope (within the hilly landscape especially H1 and H2, but also M1 and M2, though less

[15] For attribution of the flysch to class MF in this landscape, see chapter I.2.1.

represented) with MF land constitute upland areas suitable for cultivation and the existence of a human rural landscape.

Land for pasturage is also available in the Helicon upland landscape, as well as behind the city site, where the valley narrows and elevation increases (yellow in the physiographical class map), and on the slope of the ridges (Kourpeiko and Laphystion) separating the Levadeia basin from the Chaironeia and Koroneia *chorai.*

As Fossey (1988: 349) points out, a κοινόν των λατόμων is recorded at Levadeia,[16] referring to local stone and building material that could be mined in the area (on Helicon as well as on the Laphystion ridge). Epigraphical evidence also indicates the presence of one or more quarries in the area. The temple of Zeus Basileus (*components LE_7* to *LE_9*) was made in local stone, local limestone and black granite from Helicon.[17]

THE ARCHAEOLOGICAL RECORD

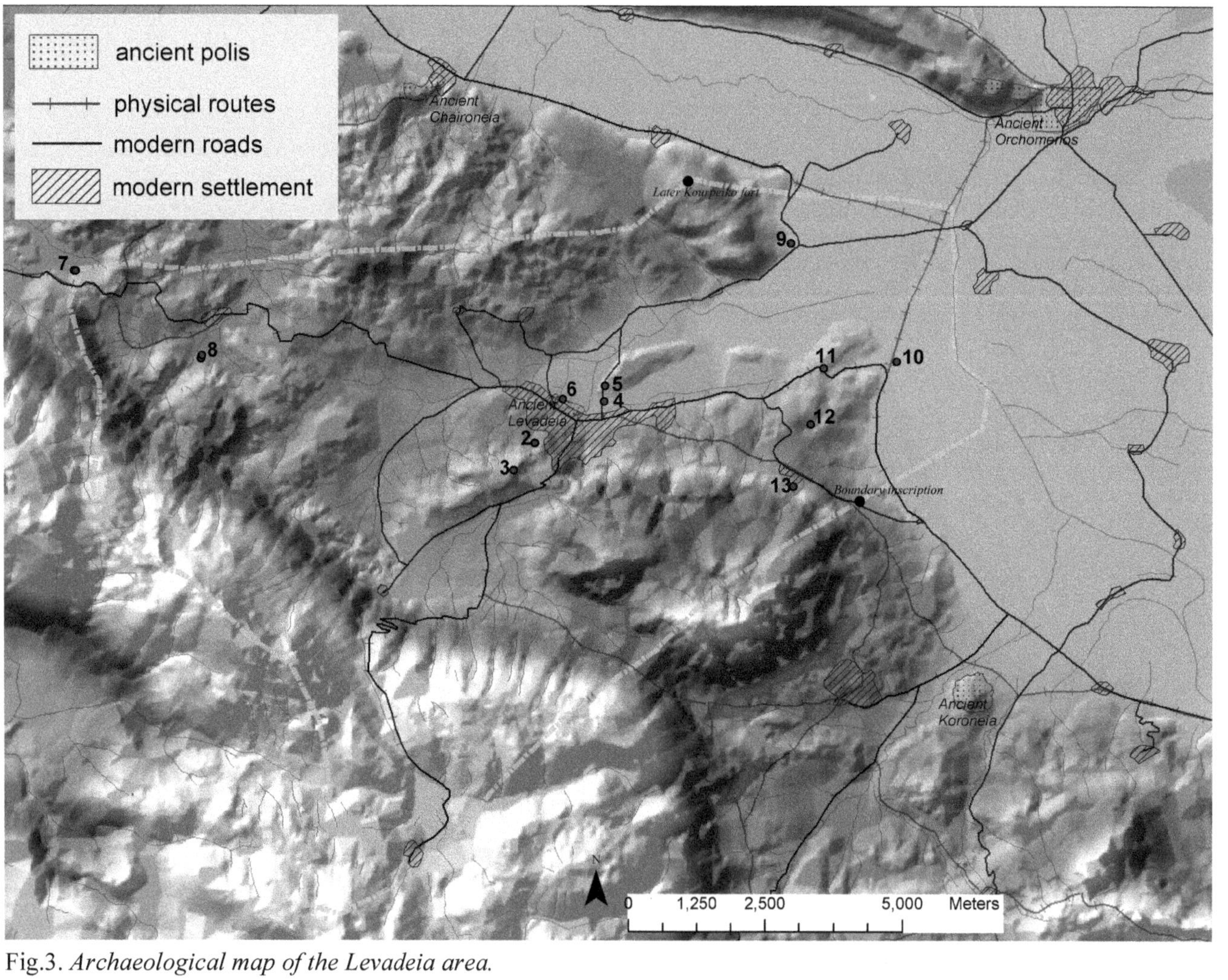

Fig.3. *Archaeological map of the Levadeia area.*

[16] AD 1917: 421-3, n.2 no.5; Roesch 1982: 182-3.

[17] Fossey (1988: 349) reports as another resource available to the ancient *polis* of Levadeia income from the prophetic shrine of Trophonios, undoubtedly relevant in the period of prosperity of the healing hero sanctuary.

1	**LEVADEIA**	*Components LE_1 to LE_6*
2	**Levadeia – Herkyna springs**	*Component LE_10*
3	**Levadeia - Proph.Ilias**	*Components LE_7 to LE_9*
4	**Levadeia NE**	*Components LE_14 and LE_15*
5	**Levadeia – Ag.Vlassios**	*Components LE_11, LE_12 and LE_13*
6	**Levadeia N**	*Component LE_24*
7	**Karakolithos**	*Components LE_16 to LE_18*
8	**Levadeia-Delphi National Road – Tsoukalades**	*Components LE_19 and LE_25*
9	**Thourion-Ag.Ioannis**	*Component LE_23*
10	**Mavrogeia**	*Components LE_20*
11	**Neraidogorna-Ag.Taxiarchoi**	*Component LE_21*
12	**Laphystion-Kazakines**	*Component LE_22*
13	**Laphystion**	*Component LE_26*

Table 2. *List of archaeological components and activity loci mapped in fig.3.*

The ancient *polis* of Levadeia is, as seen above, known only by means of rescue excavation, carried out for infrastructure or building within the modern town which lies above the ancient town. For this reason, the archaeological record is poor, though probably less biased by personal research interests, and the evidence is known in quite some detail as results of excavation, while other Boeotian *poleis*, such as ancient Tanagra and Thespiae, lying today in open fields, are known only from surface remains. Since the ancient *polis* lies beneath the modern one, the layout of the ancient town of Levadeia, as well as the position of close extra-urban features (such as cult places or necropolis), are not so clear or easily definable.

Personal interest in the city site[18] and its sanctuaries mentioned by Pausanias (the famous Trophonios *in primis*) add information at the city level. Comparatively, available archaeological datasets for the rest of the *chora* are not very rich in information about ancient occupation, being mainly results of rescue excavations for road construction, extensive topographical survey, and local information, while the non systematic intensive survey carried out personally by Lauffer[19] enriches the picture of lower level activities in the landscape.

The graph (fig.4) illustrates the proportion of components discovered within different research frameworks. In the case of the Levadeia *chora*, the large presence of rescue excavation on the city site considerably increases this type of discoveries.

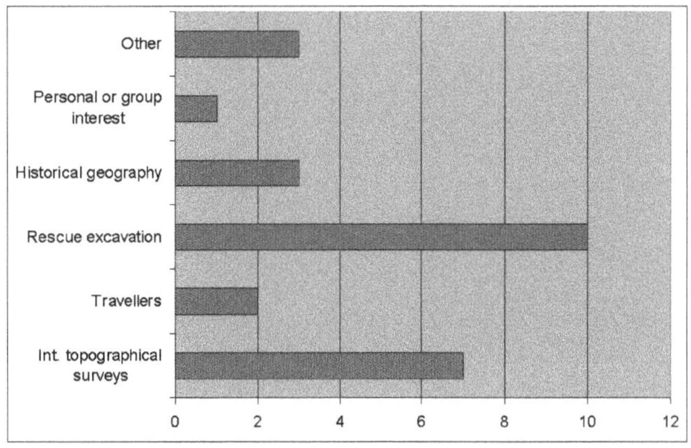

Fig.4. *Graph illustrating the proportion of components discovered within different research frameworks.*

No Prehistoric evidence seems to have been recovered, apart from two indications of uncertain attribution. The ratio of known Prehistoric to Greco-Roman components is 2 to 26 (1:13), while among the historical periods, 54% is dated Archaic to Hellenistic, 25% Roman-Late Roman, and 21% is attributed to the general Greco-Roman period.

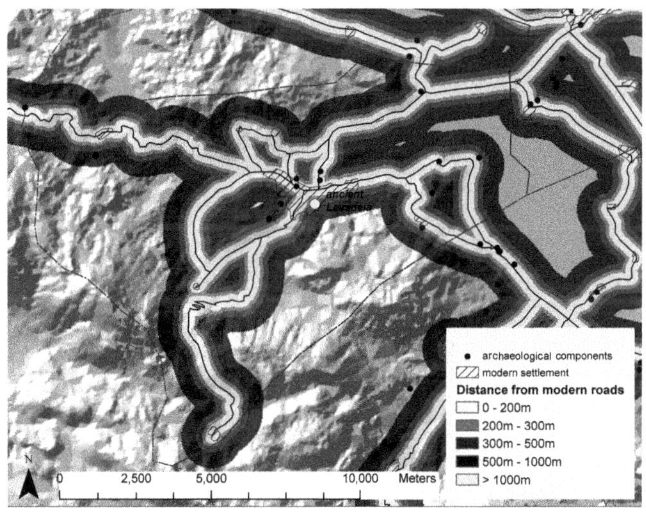

Fig.5. *Relationship between components and modern road network.*

In fig.5 we can see the relationship between known archaeological sites and the distance from the modern road network. 12 components out of 26 are within a distance of 300m from modern roads, and 22 are within a distance of 500m from the road. In this case, the proximity to the road certainly biased the discovery of archaeological data, and this could correlate with modern construction levels by roads (see the high percentage of discoveries by means of rescue excavations, which increase accidental discoveries and therefore to some degree the reliability of the archaeological picture[20]).

[18] Especially by L.A. Turner for her dissertation (Turner 1995).
[19] See mainly Lauffer Kopais I.

[20] In the case of burials by Tsoukalades, for instance (*components LE_19 and LE_25*), we would never have known of a possible ancient settlement there without the accidental discovery of burials through rescue excavation for the construction of a road (far from any personal research interest or presence of activity focus of other periods or types).

ANALYSIS OF THE *CHORA* LANDSCAPE

PREHISTORIC PERIOD

For the Prehistoric period, almost nothing seems to have been recovered in the landscape for the time being, apart from two pieces of evidence (of uncertain attribution) to which the character of 'human presence' has been assigned: AE787 (LH) which corresponds to *component LE_6*, comprised a potsherd dated to the LH period reported as from the area of historical Levadeia, and AE1246 (generally Prehistoric), a concentration of obsidian flints noticed by Lauffer on the NE slope of Laphystion (*component LE_22*). Lauffer says the presence of a Prehistoric site here is notable in such an inner location compared to the better known and studied Prehistoric sites around Copais. As suggested above, I would rather say that the evidence cannot immediately be identified as a 'site' but requires further examination. On the other hand, one could also point out that the presence on the surface of highly visible lithic material is notable. The experience that intensive surface survey has gained in the Boeotian landscape leads it to be characterised mainly as a pottery landscape rather than a stone tool/flint landscape. While the dense carpet of pottery marking Boeotian terrain is well known, only in some specific areas of quite large or important prehistoric sites can one find lithic material on the surface. As far as this evidence is concerned, the soil and the possible taphonomic processes this could have brought the material to the surface at that particular spot should be investigated.
A much more detailed survey of the upland areas would probably detect a situation similar to that in the mountainous segment of the Koroneia area (see chapter II.3.1). The physiographical characteristics, especially of the wider area which opens up in the mid-elevation landscape SW of Levadeia, together with the presence of fresh water provided by the two streams crossing the area, suggests probable Prehistoric occupation there (EBA and MBA).
Even if in a different physical landscape, the same could hold for the area of the modern village of Tsoukalades, a smaller area marked by the presence of fertile and easy workable alluvia formed by a stream, workable also in earlier periods, surrounded by the much steeper slopes of Helicon, suitable for pasturage.
The picture of the life in the Levadeia area in the Prehistoric period can be clarified only by looking at the wider area of the Copais basin (figs. 6 to 9 in chapter II.3.1), within which some areas are better known than others and can provide more detailed information to fill in the picture and attempt an interpretation of the dynamics of the Prehistoric occupation in the area.[21]

[21] see chapter III.1 and the chorai gravitating towards the Copais basin, such as Chaironeia and Koroneia, bordering Levadeia territory.

GRECO-ROMAN ANTIQUITY[22]

Town level

The ancient *polis* of Levadeia was located, as seen above, at a crossroad between different routes. The earliest possible archaeological evidence for the occupation of the town itself dates to the 8th C BC.[23] The *polis* lived until the Late Roman period. Although not very substantial in Hellenistic times, (its muster-lists show a population smaller than Akraiphia, Thespiae and certainly Orchomenos), it is presented by Pausanias (IX 39.1 to 40.2) as among the most prosperous cities of Greece by his time.
The location of the city, at the NE end of its *chora*, and at the SE edge of Boeotia on the border with Phokis, marked the border and was along main routes (see above). The presence of a prosperous Pan-Boeotian sanctuary there,[24] which brought pilgrims and hence some extra income, might be linked to this location, while otherwise Levadeia was a small town with its own fertile land[25] and a rural based economy. Today, the modern centre of Levadeia still plays a crucial role, though its character has changed, being the administrative centre of the *nomos* of Boeotia (due to its centrality during the Ottoman period when it was the seat of the Turkish governor of the province) and having preserved its ancient position along the route towards Delphi (even though the modern route does not run through the city but bypasses it, running along its edges).[26]

Village level

Neither hamlet nor village seem to have been represented by the known archaeological evidence in the area of Levadeia until Lauffer surveyed the area (during his own intensive, though not systematic fieldwork/survey). *Component LE_21* (see description above) represents a very probable site of a small settlement in the Greco-Roman period.[27] The location also would fit the hypothesis, since satellite settlements in these lateral basins of Copais seem to often lie towards the exit of the valleys into the larger central Copais basin, such as the satellite settlement of Alalkomenai in the Koroneia *chora* (see chapter). On the other hand, if the material

[22] Period maps are included in chapter II.4, figs.17-19-21-23-25-27.
[23] Slight is the evidence for the occupation of the town in the Geometric period (Turner 1995). Scarce are also the archaeological finds from the Archaic period: a sculpted grave stele dated to the Archaic period and some epichoric stelae which may also date to this period - see *component LE_1*.
[24] see Trophonion, below and in appendix I.2 (*components LE_7 to LE_10*).
[25] In a location between mixed land types, as often happens for the Boeotian city sites (see also Koroneia, for instance – chapter II.3.1).
[26] For centuries it was also the centre of a flourishing cotton industry (Philippson 1951: 449), but today few people work the land, and the majority of inhabitants are involved with administration of the province and, marginally, tourism, due to the route to Delphi (Fossey 1988: 343).
[27] I have not visited the place personally, but the given indications lead to this interpretation.

constituting the evidence (sherds and tile) is dated to the Roman period, the component could be interpreted as a villa site, and would add information on the rural segment of the landscape.

A place suitable for a potential village (or large rural settlement) site would be the gently sloping upland area exploited today by the village of Surpi (see below – LONG TERM SETTLEMENT TRENDS), as well as the area of the modern village of Tsoukalades (see burials - *components LE_19 and LE_25* and see also *components LE_16 to LE_18* at Karakolithos).[28]

Rural segment

We do not have any specific information on rural sites in the *chora*, due to a lack of research on this segment of the landscape. A few components to which discoverers assigned other characters or did not assign any character, could be interpreted in terms of landscape as rural sites (see for instance *components LE_20*; *LE_23*; *LE_21 etc.*). Furthermore, both according to information gained through intensive surveys carried out elsewhere in the Boeotia region, and considering the landscape character of the *chora*, one can observe here the availability of enough land for agriculture and pasturage (see above) to expect a rural occupation of the landscape similar to other nearby areas (e.g. Koroneia *chora*).

Burial areas

Burial areas have been detected to the N and NE of Levadeia town, and were most probably strictly related to the city (*components LE_11 and LE_12*; *LE_14 and LE_15; LE_24*). They might, however, be estate markers for community farmers (especially in locations further from the city), as happens in the surroundings of other Boeotian *poleis* (see Tanagra for instance[29]). Other burials have been found by means of rescue excavation on the mountain along the route to Phokis, dated to the Roman period by Tsoukalades (*components LE_19 and LE_25*) and to Classical/Hellenistic times in the Karakolithos area (*components LE_16 and LE_17*). Those burials are not related directly to the city, and for a possible interpretation see the components' descriptions (in appendix I.2).

Cult places/Religious areas

Apart from sanctuaries and cult places known from the city site (in particular the sanctuary of Trophonios – see Levadeia town section) and its outskirts, some extra-urban sanctuaries are known, some of which are identifiable with places mentioned by Pausanias (*components LE_7* to *LE_9* (Temple of Zeus Basileus); *component LE_21* (Temple of Zeus Laphystion at Neraidogorna-Ag. Taxiarchis), as well as a cult place site in a cave (*component LE_10*).

[28] For further considerations see below – LONG TERM SETTLEMENT TRENDS.

[29] Chapter II.3.14 and appendix I.14 –the Tanagra survey project.

Forts and fortifications

The presence of forts marking passes or borders can be observed in the landscape. The 4thC BC fort on Laphystion (*component LE_26*) probably overlooked the pass of the main road along the border of the Copais basin. The site of a fort (most probably Hellenistic) has been noticed at Karakolithos, on a mountain pass over Helicon (*component LE_18*), probably overlooking the border between Boeotia and Phokis, and marking a junction at the end of a probable boundary also between the areas of Levadeia and Chaironeia (see fig.3 and in chapter II.4 fig.21 and figs.29-31). Also the later (probably medieval) fort on Kourpeiko marks the border between the Levadeia and Chaironeia *chorai* (fig.3) and Lauffer (Kopais I: 145-6) would suggest it could mark the location of an earlier fort.

Other activities / unspecified activity areas

At the edge of the Copais basin, and at the entrance of the Levadeia valley, two undetermined activity foci were noted. One is *component LE_20* , which I have tried to interpret in terms of landscape (see appendix I.2), the other is *component LE_23*, for which only a few sherds are known together with later (medieval) material. Both could probably be interpreted as rural sites, considering archaeological remains and landscape location, and would therefore help our poor knowledge of the rural segment of the *chora* (see above).

Routes

Levadeia is still today in a strategic position. The city controls the main road from Thebes up to Delphi and down to the Antikyra bay through a number of small plains and then the Tsoukalades pass, where it crosses the watershed and enters ancient Phokis (see fig.1 – Philippson 1951 : 445; Fossey 1988: 343). A road would have followed the same route in ancient times also, as attested by archaeological evidence recovered along it (see appendix I.2 – Karakolithos: *components LE_16 to LE_18*) or a slightly shifted path at some points, as seems proven by the archaeological remains found 700m ca to the S of the modern road in the uplands (see above – Tsoukalades: *components LE_19 to LE_25*). Levadeia also controls the old inner route from Thebes to Chaironeia through the Kerata pass (fig.1 in chapter II.1), as well as the entrance to the route towards inner Helicon, SW of the city, through the pass behind the village of Surpi (Philippson 1951: 445; Burn 1949: plate XLIV).

LONG TERM SETTLEMENT TRENDS IN THE CHORA *LANDSCAPE*

The available data, coming mainly from extensive survey as far as the extra-urban landscape is concerned (see above), are not enough to allow for a meaningful analysis of the long-term life of the *chora* landscape. Thus, only a few reflections can be made, in the light of experience gained through the results of intensive surveys in other areas of Boeotia, and through considerations of the traditional settlement pattern.

Our impressions of the landscape by period (Geometric to Hellenistic) are quite poor (figs. 17-19-21-23 in chapter II.4), while if components dated generally to the Greco-Roman period are added, one can visualise a more 'archaeologically populated' landscape (fig.17 in chapter II.4). The denser occupation of the landscape, known for the Classical period elsewhere in Boeotia and in Greece, could be attested here by the presence of a probable satellite settlement site growing up to the NE of the main city site, by the entrance of the valley,[30] and the presence of some known or probable rural sites (*components LE_20, LE_21, LE_23*). Known burial sites are mainly related to the city, except for an isolated one on the mountain pass by the Karakolithos fort (*component LE_18*).
Conversely, for the Roman period we only have evidence of burial sites certainly datable to the period (once again, by the city and on the mountain along the modern road). Once again, adding the Greco-Roman sites of no better specified chronology, we can see other possible features populating the landscape, such as the settlement site (settlement or villa site?) at the entrance to the valley (see above).

The whole landscape section comprising the *chora* is even spatially centred on the city site of Levadeia. Accordingly, the modern town of Levadeia takes a central position in the landscape, administratively controlling almost all the territory of the ancient *chora*. It is notable that apart from modern Laphystion there are today no other first rank villages with the status of *koinotis* in the territory of the ancient *chora*. This is due to the large size of the city of Levadia and its status as administrative centre, which the town has carried through the Ottoman, Frankish and medieval period, as the administrative centre of the *Kaza of Livadia* in the Ottoman period and as the base for a castle in the Frankish period.
On the other hand, the dimensions of the *chora* seem to allow for the presence of other settlements, as noted already by Bintliff (1994b: fig.20) as the result of a Thiessen polygons analysis, where two empty potential village areas are located to the E and the W of the town of Levadeia. Considering the traditional settlement pattern, one notices that in the higher elevations on the NW, W, SW and E of the city a series of settlement chambers (working as such in the medieval and modern periods) open up, which control passes and alluvial valleys in the inland.
At a high elevation to the E of the city is situated modern Laphystion (formerly Granitsa), which is also recorded in the Ottoman defters in 1466-1570 (fig.6). The village of Granitsa/Laphystion controls, from an elevated position on the slopes (H3) of the Laphystion ridge, the area at the entrance to the valley. As seen above, for the Greco-Roman period we have evidence for a large rural site or a village (*component LE_21*) immediately below the slopes of Laphystion, not far from the Granitsa village but in a much lower position at the edge of a small rise (fig.6 - Neraidogorna). It may well be associated with *component LE_20*, gravitating towards the same bit of landscape.

In the interior W of the city there is a series of second rank villages, situated next to alluvial valleys with high fertility soils, which also go back to the Ottoman period: modern Tsoukalades (recorded in 1466-1570 and 1646-1688), Mikari/Makresi (ruined today but recorded in 1466-1570) and modern Analypsi (formerly Surpi, recorded 1506-1688). We cannot be sure that these areas were not exploited by the inhabitants of the city of Levadeia itself in Greco-Roman periods, but they would easily support rural sites if not hamlet/villages as in later periods.
As is clear from fig.6, showing the results of a cost-distance analysis based on ancient 1st and 2nd rank settlement,[31] a potential settlement chamber in antiquity could be recognised in the area of the modern village of Tsoukalades, along the road towards Helicon to Delphi. The area also supported a village in the Ottoman period (Ottoman defters –1466, 1506 and 1570- as Tsoukalades Yorgi). In this area the only archaeological evidence available is remains of Roman burials known by means of rescue excavations (*components LE_19 and LE_25*), which could be connected to the presence of a settlement or a large rural site (linked with pasturage) nearby, in that period at least (fig.7 and fig.3 no.8). There the landscape is marked by the presence of H2 class (becoming H1 in some areas), and we therefore still have a similar situation of an opening up of the landscape in this case associated with a river valley with fertile and easy workable alluvia. The route to Phokis would probably have run more to the E than the modern national road, crossing the village of Tsoukalades and following the natural passage (see above - TOPOGRAPHICAL SETTING, fig.1 and fig.3).[32]
The traditional village of Mikari, to the SE, is located where a stream valley constitutes a gently sloping area (H1), at the contact point between fertile alluvia and MF flysch and shist-sand soils. We have no archaeological information available for the area in antiquity, but we might suppose the existence of a settlement (hamlet or even village) at least in some periods of history, either there or in the stream valley immediately to the S. A cost-distance analysis seems to allow for such a hypothesis (fig.6), though in the Greco-Roman period the use of the uplands is not always as intensive as in later periods.
Further S, the wide upland area / hilly landscape plateau to the SW of Levadeia, on the slopes of Helicon which enclose the valley to the S, is characterised by the presence of H1 (plateau) and H2 (gentle slope) physiographical classes (see table 1 and fig.2 in chapter II.1, and fig.3) and crossed by two streams. The area can

[30] The settlement site's continuity into the Roman period is not sure, and even the site itself could have known a Roman phase of occupation only, probably as a villa site (see above and appendix I.2, *component LE_21*).

[31] See chapter II.3.1 – LONG TERM SETTLEMENT TRENDS.

[32] Evidence of burials to be linked to settlement activities are also attested in the Karakolithos area (*components LE_16 and LE_17*), along the same route to the NW. Also, according to Dasios (1995: 245-255), working on the border between Boeotia and Phokis, these remains would testify to the presence of upland settlement in the area.

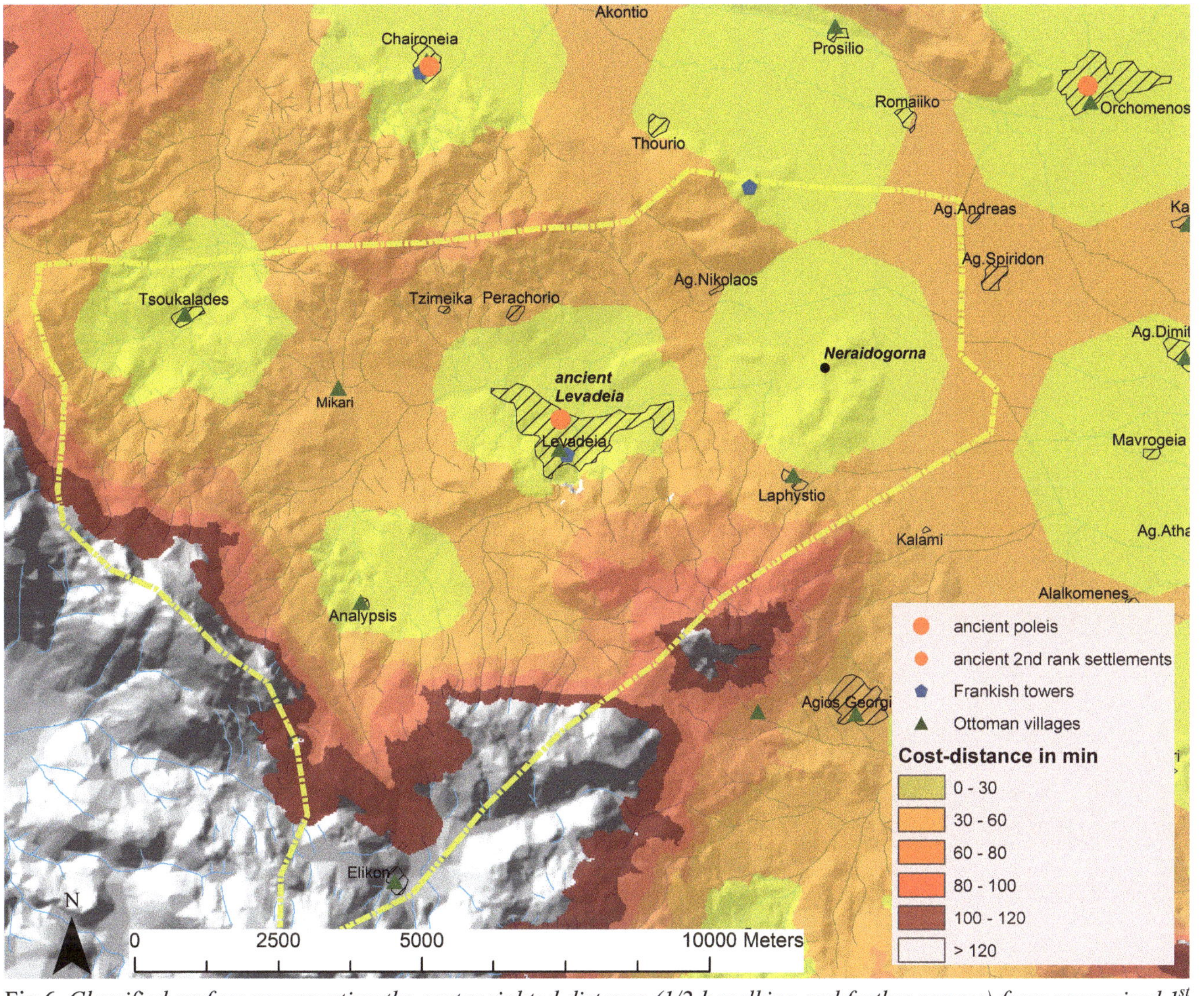

Fig.6. *Classified surface representing the cost-weighted distance (1/2 h walking and further ranges) from recognised 1st and 2nd rank ancient settlements (represented by larger and smaller dots). Areas without dots indicate potential settlement chambers. Ottoman villages as well as Frankish towers and the Frankish main centre of Levadeia have also been added to the map to show their spatial relationship with the Greco-Roman settlement network and to appreciate potential settlement chambers.*

be considered of MF (mid fertile) land (MF flysch in this case) crossed by small alluvial areas along the streams. It is located just below the steeper slopes of Helicon, suitable for pasturage. Today it is overlooked by the modern villages of Surpi/Analypsi and Ano Surpi (present in the Ottoman archives -1506 to 1688- as Surpi), and it would constitute a potential settlement chamber also for occupation (at the village or rural level) in antiquity.

Finally, Elikon (formerly Zerikia) is a high upland village on Mt. Helicon, also recorded in the 1466-1570 Ottoman archives. It certainly goes back to the infill and strong economic use of the upland landscape in later medieval periods, and one should not look for parallels in Greco-Roman antiquity.

On the slopes N of the city there are modern second rank hamlets which do not go back to the Ottoman period: Tzimeika, Perachorion and Ag. Nikolaos (probably exploiting the hills and part of the central plain). As for antiquity, although the available record does not include any considerable trace of activity in that section of the landscape, it could have been mainly exploited by rural sites gravitating towards the city. The existence of one hour walking distance radius around this N/NW part of the territory of the ancient city of Levadeia suggests also the possible exploitation of the area by city dwellers or by farms related to it. Even today all the hamlets belong to the *koinotita* of Levadeia.

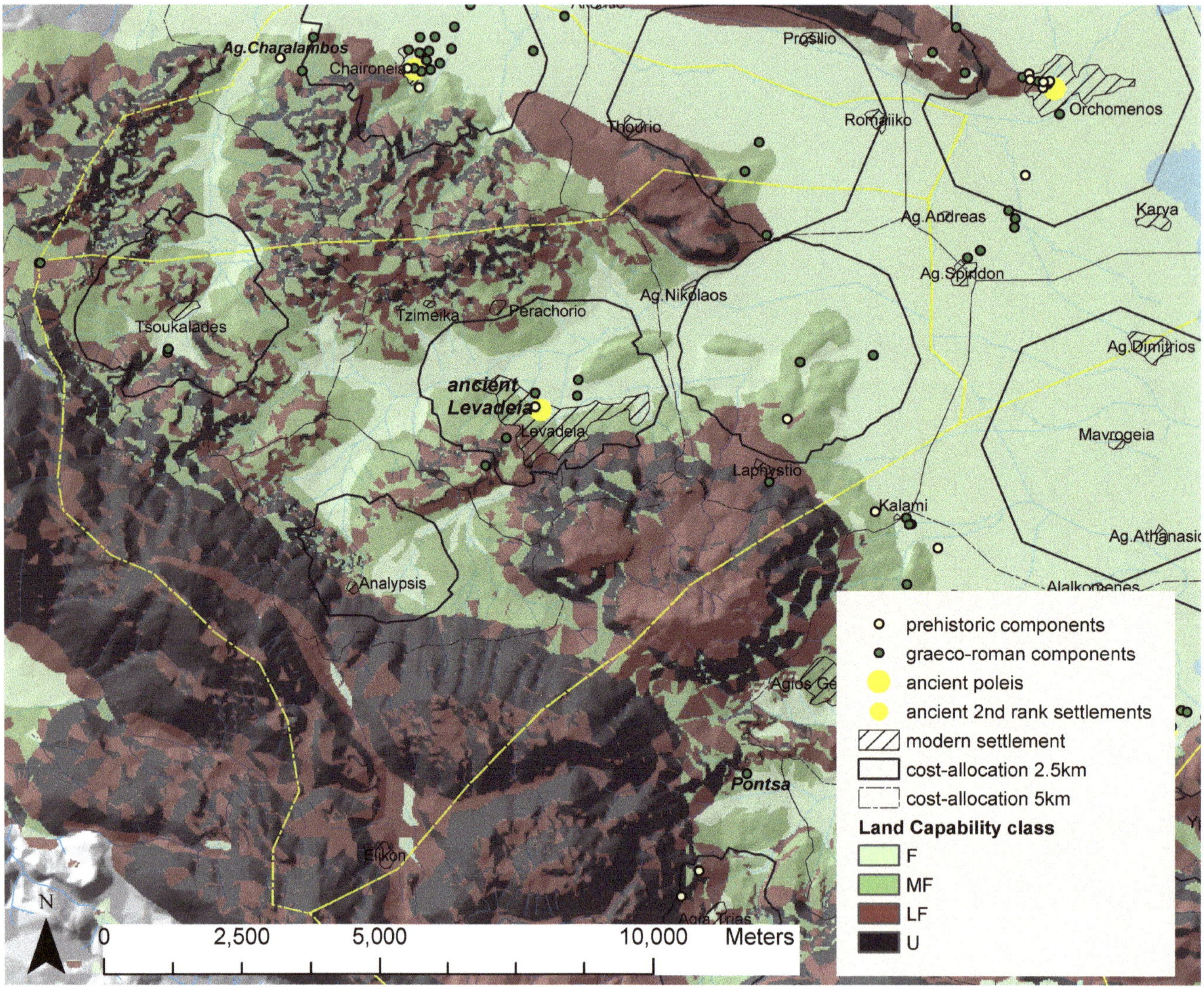

Fig.7. *Map showing the Greco-Roman settlement network, the polygons resulting from the cost-distance analysis (marking half an hour and one hour walking time distance) and dots representing the known archaeological components (same as in fig.3), with land capability information underlain.*

II.3.3

Chaironeia valley: Chaironeiake

TOPOGRAPHICAL SETTING

The plain of Chaironeia corresponds to the lower Kephisos valley (of ca 4km length) before it enters the Copais (fig.1). As Philippson (1951: 430) notes, it has the form of a depression rather than a basin, which runs in an E-SE direction and opens into the Copais plain.
Today, the whole of the basin lies in Boeotia, and is occupied by a few villages around it (Thourion, Chaironeia, Agios Vlasios, Davleia, Mavroneri, Akontion), but in ancient times the border was somewhere between the two communities/*chorai* of Panopeus (Phokian *polis* – today in the area of Agios Vlasios) and Chaironeia (Boeotian *polis*) - see below.
The width of the plain is almost uniformly 3km. To the N, the plain is bordered by the long wall of Akontion/Dourdouvana, a long low ridge (ca 20km long) which runs along the plain until it enters the Copais basin at Orchomenos.
The territory of the ancient *polis* extended on the right bank of the river to the southern edge of the plain, which is marked by a wide extended low mountain lying (in an E-W direction) below the Helicon mountain, and constituting its northern foothills. The southern mountain border of the Chaironeia plain is shorter (ends earlier in the W) than that of the N (marked by the Akontion), and ends at the Kourpeiko promontory, next to the Levadeia railway station, at the crossroads to Orchomenos. The Kourpeiko promontory divides the plain of Chaironeia from the Levadeia valley, which belongs to the Copais basin system and runs into the mountains surrounding Copais from that location.[1]
The landscape of the area is, and was in the past, marked by the presence of the river Kephisos and the variations of its meanders. Even today, after the drainage of the Copais, much of the Chaironeia plain has a tendency to be marshy.[2]

Boundaries

Chaironeia is the last Boeotian city before Phokis (Thucydides IV 76; Pausanias IX 40.2; Pausanias X 4.1). As noted earlier, the *chora* has geographically-marked boundaries,[3] being constituted by the lower Kephisos valley before it enters the Copais. Nevertheless, the borders are not always so clear. The valley is marked to the S by the ridge of Kourpeiko (which divides the territory of Chaironeia from the Levadeia area, and is the only clear border), while to the N the only potential physical boundary with the territory of Orchomenos is the line of the Kephisos.[4] To the E, the area of Chaironeia probably reached the edges of the Copais basin, but not the flooded area of the lake itself (see chapter II.3.2 – BOUNDARIES– for a discussion on the matter). To the W, the boundary between Boeotia and Phokis is still not recognisable with certainty, but on the S side of the valley, somewhere between the ancient cities of Chaironeia and Panopeus, a Phokian city was located at Agios Vlasios, almost 4km W of Chaironeia.[5] A detailed study of the matter has been published quite recently (Dasios 1995a), concerning especially the border line on the N side of the valley. As Fossey reports (1988: 384) "*the territory of Chaironeia could hardly have reached Mt. Hedilion on the N side of the plain as Meyer suggested* (Theopompus, *Hellenika* – Halle 1909), *for the territory of Orchomenos bounded upon that of Panopeus* (Strabo IX 416)".[6] Dasios (1995a), however, he comments that along the border between Phokis and Boeotia, one of the weakest points is precisely that of Agios Vlasios, at the N end of Vathyrema river valley, between Phokian Panopeus and Boeotian Chaironeia. He also notes as problematic the attribution of the area in the open Kephisos valley, as well as of the small valley between the Akontion/Dourdouvana, the Idilion/Vetriza and the Yphanteion/Paliovouna.[7] The latter could have belonged to Chaironeia, but Dasios suggests it belonged

[1] The description of the landscape of the Chaironeia *chora* is based on Philippson 1951: 430ff, quoted also by Fossey 1988: 375, on Lauffer (Kopais I: 131), and some personal observations.
[2] Fossey (1988: 384) reports from Pausanias (IX 42.7) that, in antiquity, inhabitants of Chaironeia distilled perfumes from various marsh flowers.
[3] Philippson 1951: 430; Fossey 1988: 384.
[4] Historically speaking, the relationship between Chaironeia and its stronger neighbour Orchomenos was always marked either by slight conflict or open tension (see for instance in the Archaic period –chapter II.3.4). For the border see Fossey 1988: 384 and also below in the text (LONG TERM SETTLEMENT TRENDS) and chapter II.3.4 – TOPOGRAPHICAL SETTING.
[5] Fossey (1988: 384) suggests that the boundary might be marked by the stream (ancient Molos?) which flows from the valley immediately to the E of Agios Vlassios.
[6] Fossey concludes that, for this reason, Orchomenos must have had land in the Kephisos valley (Fossey 1988: 384). Certainly in the Archaic period, Chaironeian territory was reduced to Orchomenian control (Roesch 1965).
[7] For the archaeological record available for the area, see below (*components CH_31 to CH_33*).

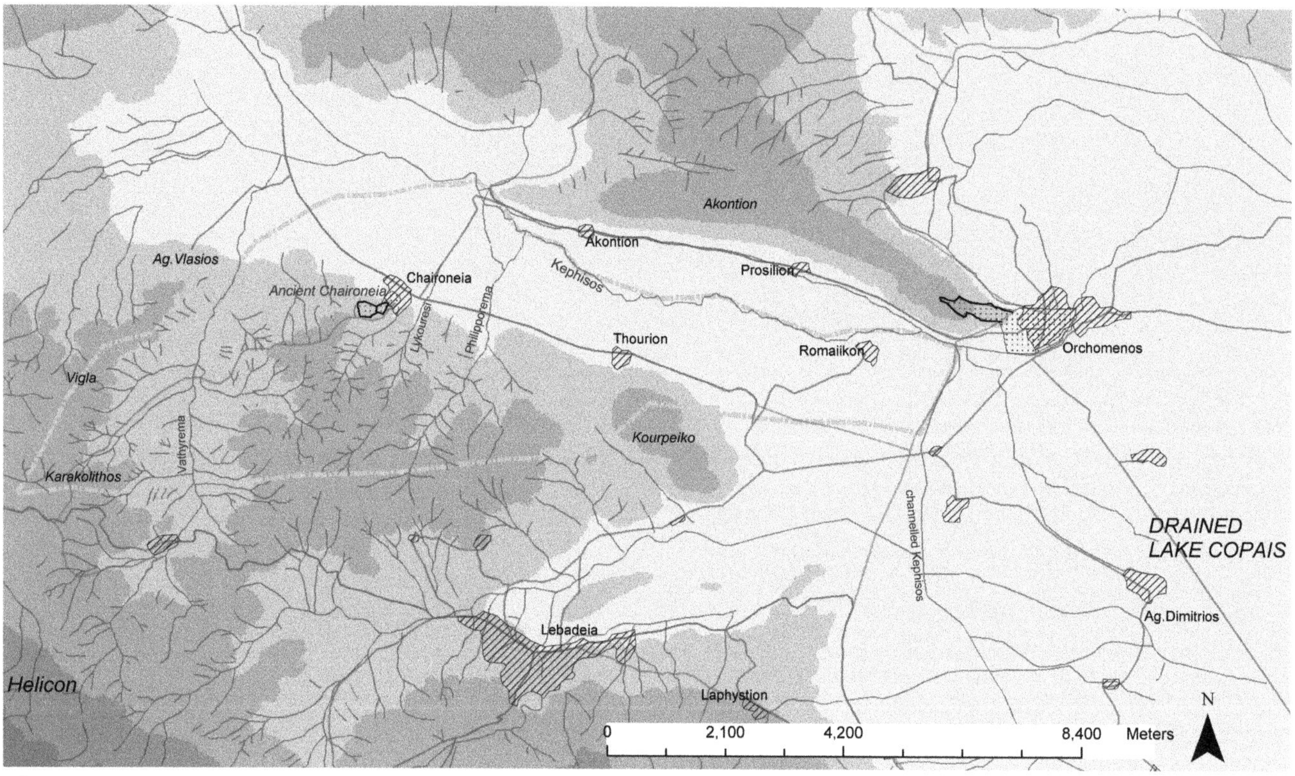

Fig.1. *Topographical setting of the chora of Chaironeia.*

to Panopeus, following Pausanias (IX 39.1) and Strabo (VIII 2.42), according to which Panopeus' territory bordered that of Orchomenos.[8]

PHYSICAL LAND UNITS

The physical landscape of the *chora* can be subdivided between the landscape of a plain (ca 50%), constituted by the Kephisos basin, and the landscape of the hills bordering it at its southern edge, which also separated the Chaironeia *chora* from that of Levadeia (ca 50%). Unlike the basin of Levadeia and Koroneia, the Chaironeia area does not reach the Helicon massif, and therefore the mountainous (>600m) segment is not present in the landscape (see fig.2 in chapter II.1).

The physical border with the *chora* of Orchomenos has been located along the Kephisos river itself, though this cannot be stated with certainty for every historical period (see above). Consequently, the lowland area belonging to Chaironeia may have changed considerably, also with regard to the uncertain border towards Copais[9] (see above and chapter III.1).

Hilly landscape	48.3%
Mountainous landscape	0%
Plain	51.7%

1	P1_P2	lacustrine basin, valley	42.1%
2	P3	gentle slope	0.9%
3	P4	foothill	8.6%
4	H1	plateau	19.2%
5	H2	gentle slope	1%
6	H3	moderate slope	3.8%
7	H4	severe slope	12.7%
8	H5	very severe slope	11.6%
9	M1	Plateau	0%
10	M2	plateau/gentle slope	0%
11	M3	moderate slope	0%
12	M4	very severe slope	0%

Table 1. *Percentage of the different physiographical classes present in the Chaironeia area (P=plain; H=hill; M=mountain)*

The borders of the plain rise quite steeply from it (P4 class), and the P3 class is almost absent.

Three large upland plateaus (H1), and several smaller ones, characterise the landscape of the hilly segment. Furthermore, in the upland of the *chora*, within the hilly zone, a quite fragmented landscape can be noted, ranging from H1 to H5, which corresponds to the lowest and the highest values of slope within the hilly landscape.

There is no mountainous segment (classes M1 to M4 are absent). The landscape is almost equally subdivided between the plain and the hilly zone (see above).

RESOURCES

As is visible in fig.6, the *chora* include very few unfertile areas (U). Low fertile areas (LF) are represented by the rocky area above the Chaironeia city site, and by the limestone ridge of Kourpeiko above Thourion village.

[8] See previous note: Fossey (1988: 384) uses the same historical texts to validate his hypothesis. As for the border between Boeotia and Phokis, see also chapter II.3.4 (*boundaries*).

[9] Did the territory of Chaironeia reach the water of the Copais lake? See above (*boundaries*) and chapter III.1 for discussion.

For this area, flysch soil formations have been assigned the land capability value of LF, because of the high presence of pebbles. In gently sloping areas they have been assigned a value of MF as, under special conditions and using adequate techniques, these soils can also be cultivated (see chapter I.2.1). On the other hand, the soil of the plain, characterised by heavy river deposits and with a high tendency to marshiness, especially in certain areas, also requires specific evaluation concerning its suitability and workability. Like Koroneia and Levadeia, the ancient city sits at the interface of fertile hilland soils and valley lowland soils.

THE ARCHAEOLOGICAL RECORD

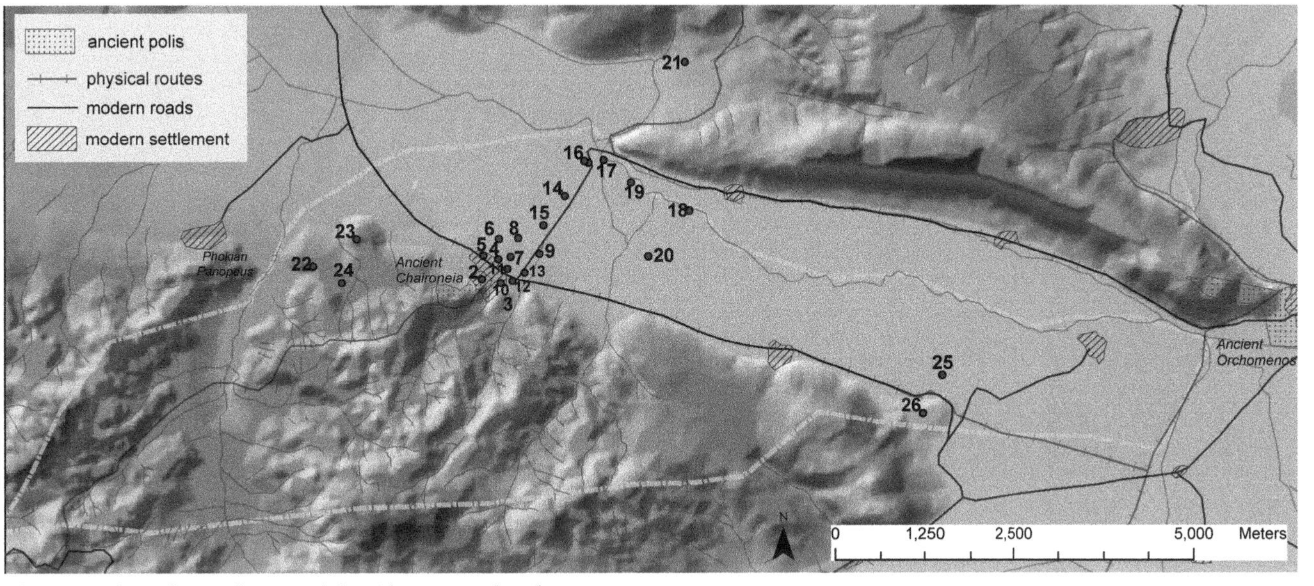

Fig.2. *Archaeological map of the Chaironeiake chora.*

1	**CHAIRONEIA**	*Components CH_1 to CH_9 and CH_38*
2/3	**CHAIRONEIA area**	(2) *Components CH_10 and* (3) *CH_11*
4	**Chaironeia – dioikitirio**	*Components CH_26 and CH_45*
5	**Chaironeia N**	*Component CH_12*
6	**Chaironeia – Plateia Scala and Vourlies**	*Components CH_46 to CH_49*
7/8	**Chaironeia – Rema**	*Components CH_34 and CH_35* (7); *CH_36* (8)
9	**Chaironeia - Kovras/Agortses**	*Component CH_50*
10/11	**Chaironeia museum**	*Components* (10) *CH_15, CH_16 and* (11) *CH_37*
12	**Chaironeia East**	*Component CH_19*
13/14	**Chaironeia North and Chaironeia bridge**	(13) *Component CH_14;* (14) *components CH_13 and CH_17*
15	**Varka**	*Component CH_18*
16/17	**Magoula Balomenou**	(16) *Components CH_21 to CH_24 and CH_13;* (17) *component CH_25*
18/19	**Akontion – Kephisos**	(18) *Components CH_40, CH_41 and* (19) *component CH_42*
20	**Chaironeia plain**	*Component CH_20*
21	**Agioi Apostoloi - Agios Georgios**	*Components CH_31 to CH_33*
22	**Agios Charalambos**	*Components CH_27 to CH_29*
23/24	**Isoma**	*Components* (23) *CH_43 and* (24) *CH_44*
25	**Thourion – Koumoulia**	*Component CH_39*
26	**Thourion East**	*Component CH_30*

Table 2. *List of archaeological components and activity loci mapped in fig.2.*

As for the city site, being the modern village of Chaironeia located above the site of the ancient *polis*, accidental finds and rescue excavations constitute the main provenance of our archaeological knowledge of the ancient city site, as for Levadeia or Thebes.

Most of the information concerning the archaeological record in the Chaironeia *chora* seems to have come from the most recent decades. This is mainly due to rescue excavations and to the attention paid by the local ephorate to any invasive intervention in the landscape. Rescue excavations in the area have taken place for the widening of the Kephisos river bank, for the construction of channels controlling the flooding of the area, and for the installation of a gas pipeline (1993-94).[10]

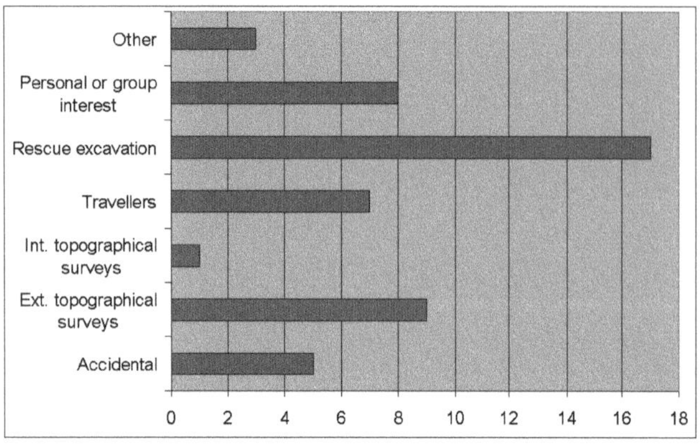

Fig.3. *Graph illustrating the proportion of components discovered within different research frameworks.*

especially the most recent ones, greatly contribute to shed light on the transformation of the ancient landscape of the *chora*, in particular on the rural segment (see below), which is still poorly known in the majority of the Boeotian areas which have not been covered by intensive surveys.

In fig.4 we can see the relationship between known archaeological sites and the distance from the modern road network. The fact that the majority of known evidence is located by the road exiting the city and crossing the plain is due to a combination of all the factors illustrated above (lack of intensive survey, river deposits burying sites, rescue excavation activity for channeling).

The ratio of known Prehistoric to components is 11 to 39 (1:3.5), while among the historical periods, 54% is dated Archaic to Hellenistic, 33% Roman-Late Roman, and 13% is attributed to the general Greco-Roman period.

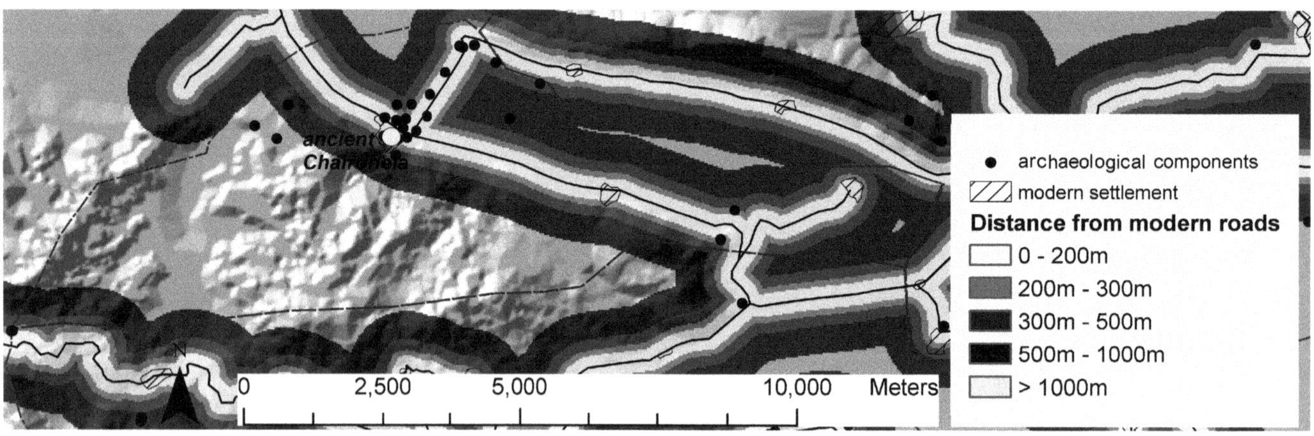

Fig.4. *Relationship between components and modern road network.*

The graph (fig.3) illustrates the proportion of components discovered within different research frameworks. In the case of the Chaironeia *chora*, whilst the large portion of the chart connected with travellers mainly concerns the city site of Chaironeia itself, the majority of archaeological discoveries have been obtained by rescue excavation (for urban and rural evidence), followed by a large amount of evidence uncovered by research led by personal interest, in particular the Prehistoric Magoula Balomenou site and the Agioi Apostoloi - Agios Georgios site at the border with Phokis.

Considering that the landscape of the area is greatly affected by the Kephisos river deposits, in addition to the fact that no intensive surveys (not even non-systematic) have been carried out in the *chora*, we have quite a rich archaeological dataset available.[11] Rescue excavations,

ANALYSIS OF THE *CHORA* LANDSCAPE

PREHISTORIC PERIOD

For the Prehistoric period, it seems that the lower Kephisos valley (which corresponds to the *chora* of Chaironeia) has similar characteristics to the upper valley, which is, however, in Phokis (Lauffer Kopais I: 131), and this is probably due to the presence of the river clearly marking the landscape and its geomorphological characteristics. On the other hand, low mounds corresponding to Prehistoric habitation sites or activity foci are typical in general of the NW area of Copais (see chapter II.3.4). Lauffer (Kopais I: 131) mentions the low *magoulas*, typical of the upper valley in Phokis but also visible in the Chaironeia area and the W part of the Copais basin, unlike the rest of the Copais area.

[10] See above within the single site descriptions.

[11] Extensive survey to locate sites known from the bibliography and from rescue excavation (on the GYS 1:5,000 map) has recently been carried out by the local ephoreia. A map is published in AD 51 (1996): 284 (fig.14), on which sites are marked with black dots. Despite the fact that in the article there is no clear association between these dots and the information on the sites, correspondence can be found between some known sites and the black dots represented on the map, and therefore the sites could be mapped within the GIS.

The only places with traces of Neolithic occupation known for the area are Chaironeia itself (acropolis and N slopes - *component CH_7*), and the site of Magoula Balomenou (Middle and Late Neolithic – *component CH_21*); the former intensively investigated because of the presence of a major later site; the latter discovered through excavation due to interest in Prehistoric material found in previous excavations at the site. The resulting picture is therefore very far from representative of the Neolithic human landscape in the area, which was certainly suitable for Neolithic life (fig.6 in chapter II.3.1). The presence of a large river could be a factor appropriate to Prehistoric settlement, for archaeological presence, but also a potential burying factor, for archaeological absence (the river deposits potentially burying earlier evidence).

The same is true for the Early Helladic period, for which the archaeological record available again presents a similarly limited picture: a probable activity focus of an undetermined kind at the Magoula Balomenou site (*component CH_22*), and at Chaironeia (*component CH_8*) – see fig.7 in chapter II.3.1.

For the Middle Helladic period, the extended settlement at the site of Magoula Balomenou provides a parallel with MH Orchomenos, both at the foot of Akontion. Their relationship, as Tzavella-Evjen points out (1988: 37-41), was probably close, and the two settlement sites should be studied together. MH is unknown elsewhere within the Chaironeia *chora*, with the exception of the uncertain *component CH_10* (only one MH pot from a burial) – see fig.8 in chapter II.3.1.

For Late Helladic, the only certain presence are chamber tombs and burials known from the surrounding of the Chaironeia city site, probably related to a settlement on the later acropolis of historical Chaironeia (*components CH_9 and CH_11* see above – Chaironeia section) – see fig.9 in chapter II.3.1. Otherwise, evidence from the period is uncertain (*components CH_24 and CH_29*).

Generally speaking, one could say that the Chaironeia area is known to any scholar of Greek Prehistory due to the presence of the impressive and well known, even if never systematically published, Magoula Balomenou site.

GRECO-ROMAN ANTIQUITY[12]

Town level

The ancient *polis* of Chaironeia controls the lower part of the Kephissos valley before it enters the Copais basin. It is located along the road from Northern Greece, through Phokis to Boeotia, Attica and the Peloponnese. The *chora* of Chaironeia is quite small, and gravitated around the *polis* itself, which was occupied probably since early times (Geometric and Archaic[13]), and until the Late Roman period. Even today, the main settlement (the other is Bramaga/Thourion) of the area is the modern village of Chaironeia, situated just above the ancient city site.

Above the ancient *polis* a fortified acropolis was built (probably since A period, see appendix I.3), providing the city a fortified area, quite well protected, above the settlement itself. It was apparently fortified since the Late Bronze Age (see above), which testifies to the weak location of the city site, at the only open entrance to Boeotia from the N.

Village level

If we consider as belonging to the *chora* of ancient Chaironeia only the right (S) bank of the Kephissos (without the left, N, bank), it could seem that there are no suitable locations for a potential village. There are two villages in this area today: Kapraina/Chaironeia and Bramaga/Thourion compared to which Chaironeia is a little larger. In ancient times, when Chaironeia was a larger *polis*, there would have been no need for a satellite village settlement. The position of the Thourion settlement, though, at the entrance of the valley, is similar to the situation of other W Copais areas, such as Koroneia or Levadeia,[14] where the ancient satellite settlements seemed to be located towards the exit of the valleys into the larger central Copais basin (see below - LONG TERM SETTLEMENT TRENDS).

If we take into account the small valley to the N side of the Kephisos valley (an area which, however, could have belonged to Phokis – see above Agioi Apostoloi), the existence of a Classical-Hellenistic and Roman? settlement has been hypothesised and linked with archaeological evidence by Dasios (1995a), on the other side of the probably formerly marshy area (*Loutsa*), from the Magoula Balomenou site (see appendix I.3 and below – LONG TERM SETTLEMENT TRENDS).

Rural segment

The archaeological record available for the area is quantitatively not very rich, as seen above. We have all periods represented but in low quantity as for number of sites. As is often the case, the resulting picture lacks a well-documented rural segment. In particular, no information on Classical (Archaic and Hellenistic) rural sites seems to be available. Conversely, investigated villa sites (or potential villa sites) from the Roman period are relatively quite well represented.

Roman (or LR) villa sites are reported for the area along the Kephisos river in ancient texts (Kahrstedt 1954: 109). Pausanias mentions a cemetery with flowers (especially iris) all around a villa, and archaeological data from mainland Greece tell us that such villa sites also often had burial areas nearby, or by roads nearby (*Mesogaia* 2001). Plutarch had a friend (Soklaros) who had a villa by the Kephisos and was an expert in conifers. As evident from the examples, the archaeological dataset from the *chora* offers us *components CH_13* and *CH_14*, with the cemetery *component CH_17* probably in association with *component CH_13*.[15]

[12] Period maps are included in chapter II.4, figs.17-19-21-23-25-27.

[13] On the territory of Chaironeia reduced to Orchomenian control in the Archaic period see Roesch 1965.

[14] see relevant chapters.

[15] See also Agios Charalambos site, for another possible villa site in the area, as well as Agioi Apostoloi - Agios Georgios site.

Burial areas

Known particularly from rescue excavations, the burial landscape of the *chora* is mainly characterised by burial areas linked with the *polis* of Chaironeia (*components CH_46* to *CH_49*), mainly to the N of the city site. Monumental burials are also known from the area, whose landscape is marked by the presence of two monumental tombs (polyandrion type) - *components CH_19* and *CH_20. Component CH_15* (Roman?) probably testifies to the presence also of a burial area linked to the city site to the E, but it could also represent a rural cemetery.[16]

The archaeological record available for the *chora* also includes burials associated with villa sites (or near to them) in the Roman period (see above – *component CH_17*, and potentially also others similar to this), and rural burial areas associated with possible isolated farms or rural plots.[17] Cist-tombs (mainly) in groups related to a farm have been noted in different localities in the area (see for instance *component CH_50* and probably also *components CH_34, CH_35* and *CH_36*), mainly to the NE of the city site.

Moreover, the discovery of the cemetery area by the Kephisos (Late Geometric-Archaic and some Classical graves – component *CH_40, CH_41* and *CH_42*) could mark the character of the river as a border with the territory of Orchomenos, especially in an early period (during the period of expansion of Orchomenian control over Chaironeian territory - Roesch 1965).

Cult places/Religious areas

Apart from known or probable cult places within the city site, such as the probably urban cult place dedicated to Asklepios, recognised by the ancient theatre, beneath the church of the Panagia (*component CH_38*), few other places have been reported by scholars as potential extra-urban sanctuaries or cult place sites (see Ag.Charalambos - *components CH_27 and CH_28*).

Attempts at the identification of the Apollon Thourios temple mentioned by Plutarch (*Sull.* 17.6-8) have been made (see *components CH_27, CH_28 and CH_44*). A certain location for a cult place, however, independent of the identification, is not reported.

Forts and fortifications

According to the available archaeological record, there are no fortifications overlooking the wide passage from Phokis through the Kephisos valley. On the other hand, the city of ancient Chaironeia itself, as noted above, had a quite well protected fortified area above the settlement itself (a fortified acropolis, with a defensive purpose). Probably already in the LH period, the site of the later acropolis of ancient Chaironeia was a fortified settlement (*component CH_9* and fortification AE792).

Recently the picture of the fortified landscape of the *chora* has been enriched by the discovery made by Camp et al (1992) of the fortification identified with Mt. Thourion mentioned by Plutarch (*component CH_43* – Isoma).

Other activities / unspecified activity areas

Some unspecified activity foci, of differing characters, can be pointed out in the general panorama of the *Chaironeiake* archaeological record. For some a possible character can be suggested (discussed under each individual component[18]), but for others not.

LONG TERM SETTLEMENT TRENDS IN THE CHORA *LANDSCAPE*

The landscape of the area is, and was in the past, marked by the presence of the Kephisos river.

The river meanders greatly, which means that it was and is subjected to a high degree of variability in its 'path'. Therefore, we cannot simply consider the current river bed as if it were that of the past, but, while trying to understand past landscape dynamics, we must also bear in mind that even the past, and also during the periods considered in the present work, the river course shifted through time (see under Magoula Balomenou site), with consequent changes in the landscape.

Landscape life was characterised by the presence of regular swamp creation and flooding, and the alternation of wet and dry periods. The settlement pattern in the area ought therefore to be studied accordingly, by taking into account any possible variation. It would be interesting to investigate how much the presence of the river, the shifting of its course, as well as the presence of large marshy areas, would have influenced human life and the human activities pattern in the valley.

Only a field campaign of intensive and systematic core sampling could provide a more certain reconstruction of flood dynamics, and could help to build up a realistic model. In the meantime, I simply consider the different possibilities of water presence in different periods, also according to the flooding dynamics of the Copais lake (known in much more detail mainly due to the work of Knauss' team). In fact, water balance in the area is strictly linked not only with the behaviour of the Kephisos, but also with the water level variations of the Copais (see chapter III.1).

As seen above, in the Late Prehistoric period the landscape was settled, and the area was probably considered suitable for early agriculture from the Neolithic period (judging by the presence of an important settlement site at Magoula Balomenou). The Early Helladic is less well represented, while in the Middle

[16] The doubtful interpretation of *component CH_15* (see appendix I.3) is linked with the still uncertain border line of the lower town, lying somewhere below the modern village (see appendix I.3).

[17] Evidence of small rural burial areas from A-C-H period associated with isolated farms are also found quite often in intensive systematic surveys in the Boeotia region, and are visible on the surface through a few freshly-broken fineware sherds. See, for instance, the Tanagra Survey (chapter II.3.14 and appendix I.14 – THE TANAGRA SURVEY PROJECT).

[18] For instance, both the LR and the H component at Agios Charalambos, marked generally as 'activity focus', might represent a rural site rather than a cult place (see above in text, under the entries).

Helladic, Magoula Balomenou again expanded, and the occupation known for that period gives us an idea of the use of the area.

For the LH period, from an examination of the dimensions of the Cyclopean fortifications, Fossey (1990b: 100ff) suggests that the site of Chaironeia might belong to the series of fortifications controlled by Mycenaean Orchomenos. Thus, according to Fossey, the site might have been an important site in the LH landscape, but we have only his report and hypothesis on this (see above – Chaironeia *component CH_9*), without any other clear indication of LH occupation in the area – see above.

If Fossey is right in dating the acropolis as a large Mycenaean site, then it must imply that the Kephissos Valley was cultivable over large areas and was not a marsh, either because it was managed by local drainage or just better naturally drained than in recent times. The Magoula Balomenou site also implies a better drained valley in prehistory, as a major settlement surrounded by wetland and marshes till recently.

In the Geometric and Archaic periods, the archaeological record available offers us the case of the Chaironeia city site itself, and the large necropolis along the Kephisos (*components CH_34 to CH_36*), which might be related to the city itself, marking a presumable border (see above).

With Chaironeia always being the main habitation focus of the area (always at a city level), in the Classical and Hellenistic periods, the available picture is that of an empty landscape as far as habitation sites are concerned, though we might suppose that this was not the case, especially considering the suitability of the landscape (fig.6) and the good amount of evidence available for later periods (R and LR) for living in the countryside (figs.23-25-27 in chapter II.4). Despite this apparent lack, data on the rural landscape are available through evidence concerning burials probably marking rural estates (*components CH_50*; *CH_34 to CH_36*) for the Archaic-Classical-Hellenistic periods. They are located mainly E/NE of the city towards the edge of the fertile plain (fig.2 nos.7,8,9 and fig.6). The landscape of burials therefore gives us an idea on the transformation in the rural settlement in the area, less represented in the archaeological record.

In the Archaic-Classical-Hellenistic periods we have evidence for burials (*components CH_46* to *CH_49*) related to the main city site, Chaironeia,[19] along with the above mentioned 'rural burials' related to a process of progressive infill of the landscape with individual farm sites linked to the system of small property, located in different areas of the landscape. City burials in the area N of the city continue until the mid 2nd C AD, when graves begin to appear in the area formerly occupied by the city site,[20] then abandoned. This can be read as proof of a reduction in the size of the city itself in the course of the Roman period. In the same period, evidence is available for the rural segment of the landscape, which is marked by the presence of villa sites (see above – RURAL SEGMENT) and large estates. At the beginning of the Late Roman period (mid 3rdC AD), the Chaironeia archaeological record seems to testify to the presence of burials (*component CH_45*) in the location of a large villa of the previous period (see *component CH_26*), which could be associated with the labour force for the villa and a probable different organization of functional spaces rather than to a decay of the property system of large estates.[21]

In the *chora* of Chaironeia, therefore, we obtain, mainly from excavation data, the picture of a settlement pattern similar to that we usually find in surveys. Since the information comes from rescue excavations we do not get an idea of the density of sites; information that we obtain usually through intensive surveys. Rescue excavations, though, certainly offer less biased information on the archaeological landscape than that from 'thematic' excavation (see chapter II.2), and provide information on the settlement pattern which produces a picture similar to that from intensive systematic surveys in other areas of Boeotia (see for instance the results of the Tanagra survey[22]).

Examining the settlement dynamics in the area (fig.5), in modern times just below the ancient acropolis lies the modern village of Chaironeia (former Kapraina). It is also recorded in the Ottoman archives (1466-1688) as a large Greek village, indicating continuity of Greek population, presumably from ancient Chaironeia.[23] To this village belongs the main area of the territory of the ancient *chora*. The ancient *chora* can also be described as a large unique settlement chamber, controlled by the city of Chaironeia. If we consider as belonging to the *chora* only the right (S) bank of the Kephissos (without the left, N, bank), as noted above,[24] there was no apparent place for another settlement, even a satellite, when the *polis* of Chaironeia was larger.[25]

[19] Burials start just outside the modern village, testifying the expansion of the ancient *polis* in that period (fig.2 no.4).

[20] For instance, burials dated to the imperial period have been found in the area of the museum of Chaironeia (information from the 2002 poster exhibition in the Thebes museum). One example recorded in our database could be *component CH_15*, though published as of uncertain date (see appendix I.3).

[21] As suggested in the panel of the exhibition at the Museum of Thebes. The villa / large estate system seems usually to begin in Boeotia in mid Roman times (200-400 AD), and they reach their maximum size and extent in Late Roman (400-600 AD).

[22] See above footnote 17, chapter II.3.14 and appendix I.14 – THE TANAGRA SURVEY PROJECT.

[23] There is also Byzantine and Frankish evidence in the area.

[24] See section VILLAGE LEVEL.

[25] Fossey notes that the territory of ancient Chaironeia covered virtually the same area as that of the two modern villages of Kapraina/Chaironeia and Bramaga/Thourion (Fossey 1988: 385), and see also Bintliff's Thiessen polygons analysis for ancient Boeotia (Bintliff 1994b fig.20), where only Chaironeia appears as exploiting the whole area.

On the other hand, two more modern villages are situated in the wider *chora* area today: Thourion (formerly Bramaga) and Romeiko (formerly Arapochori), in the central and east side of the *chora* territory, marking settlement chambers in this central and eastern half of the *chora*. In the case of other *chorai* on the same side of the Copais basin (Koroneia and Levadeia, closest neighbours), we noticed the presence in antiquity of satellite settlements towards the Copais plain, at the entrance to the 'valleys' (Alalkomenai for the *Koroneiake*, a possible settlement in the Laphystion area for the Levadeia *chora*). We could hypothesise a similar situation for the *chora* of Chaironeia, in which a second rank settlement could have been located in the Thourion area, where a Frankish tower and settlement is located, as well as the modern village of Thourion. Actually, the presence of a Frankish tower and settlement SE of modern Thourion strengthens the potential gap in this area, although no archaeological evidence is available for earlier periods. The results of a cost distance analysis[26] indicate adequate space for a settlement in the area. On the other hand, one should also consider the possibility that the Kourpeiko/Thourion area could have worked as border area for Levadeia instead, occupied also in ancient times by control posts and watch towers (see Lauffer Kopais I: 145ff and *component CH_30*).
In addition, just behind Chaironeia, a topographically defined small landscape, potentially suitable for a sanctuary with its 'catchment area', can be located in the area of Ag. Charalambos (*components CH_27 to CH_29*), a flysch mid fertile area overlooking the Vathyrema river valley – figs. 2, 5 and 6.

The discussion above considers as the northern border of the *chora* the course of the Kephisos river. If we also include in the territory of ancient Chaironeia the area N of the river, additional settlement chambers open up along the Akontion ridge. The Akontion mountain, an area also suitable for a pastoral economy (fig.6), is occupied today by two villages, Akontion (formerly Visvardi) and Prosilion (formerly Velis), both probably going back to the Ottoman period. Therefore, the traditional and Ottoman settlement pattern allows for the possible existence of a settlement on the N side of the Kephisos valley opposite Chaironeia, in order to understand the settlement network possibilities in antiquity also. Akontion village, in its position and its vicinity to Chaironeia, seems rather to be linked to the shrinking of the medieval/post-medieval Chaironeia village and the exploitation of Akontion slopes for pastoral activity, while Prosilion village could give us a clue to the placement of a settlement in the area at the entrance to the valley. Such a settlement could have exploited the area corresponding to the entrance of the Kephisos river into the Copais basin, and could have been located somewhere in the area of Thourion, Prosilion and Romaiikon. The evidence provided by *component CH_39* is too weak and undetermined to be meaningful so far, although its position at the entrance of the valley towards the Southern border would fit. The cost distance analysis allows for a potential settlement chamber that would also have respected the border towards Levadeia marked by the Kourpeiko (fig.5).
Finally, a potential small settlement chamber, potentially suitable for a rural settlement (at hamlet level) could be located in the area of the Agioi Apostoloi-Agios Georgios site (*components CH_31 to CH_33*), a small valley with alluvial fertile deposits. The cost distance analysis defines fair room for it (fig.5), but the area might however have been in Phokis for most of its history – Dasios 1995a.

[26] See chapter II.3.1 – LONG TERM SETTLEMENT TRENDS.

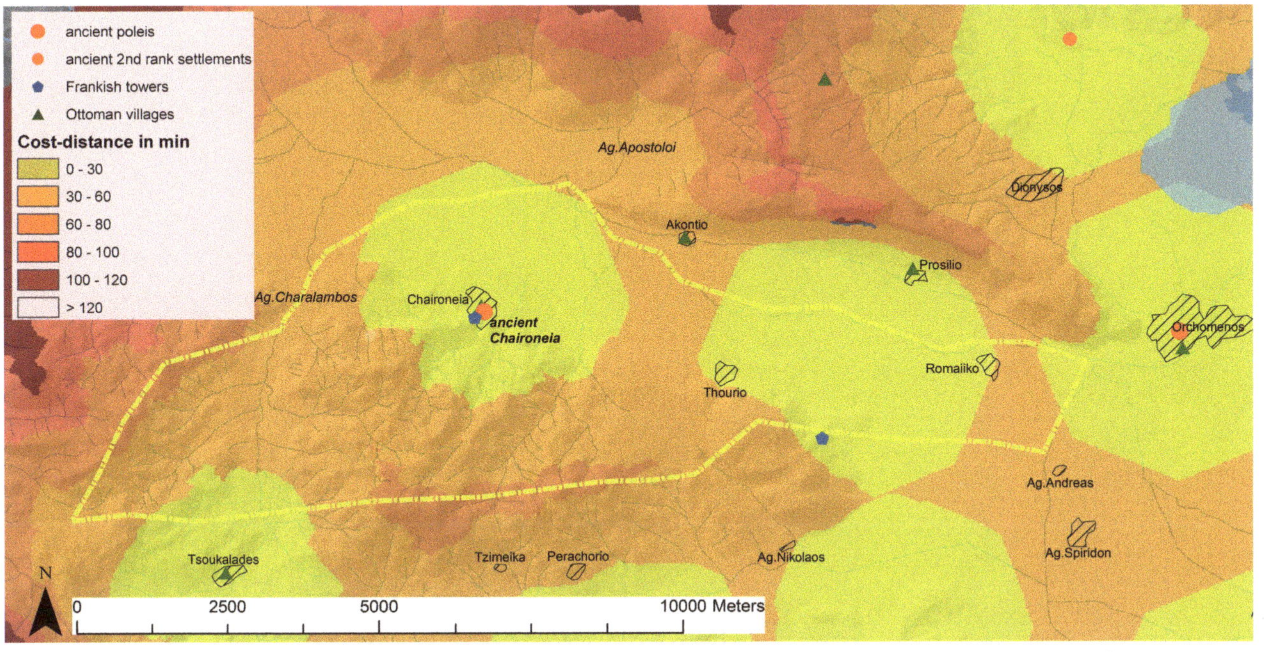

Fig.5. *Classified surface representing the cost-weighted distance (1/2 h walking and further ranges) from recognised 1st and 2nd rank ancient settlements (represented by larger and smaller dots). Areas without dots indicate potential settlement chambers. Ottoman villages and Frankish towers have also been added to the map to show their spatial relationship with the Greco-Roman settlement network and to appreciate potential settlement chambers.*

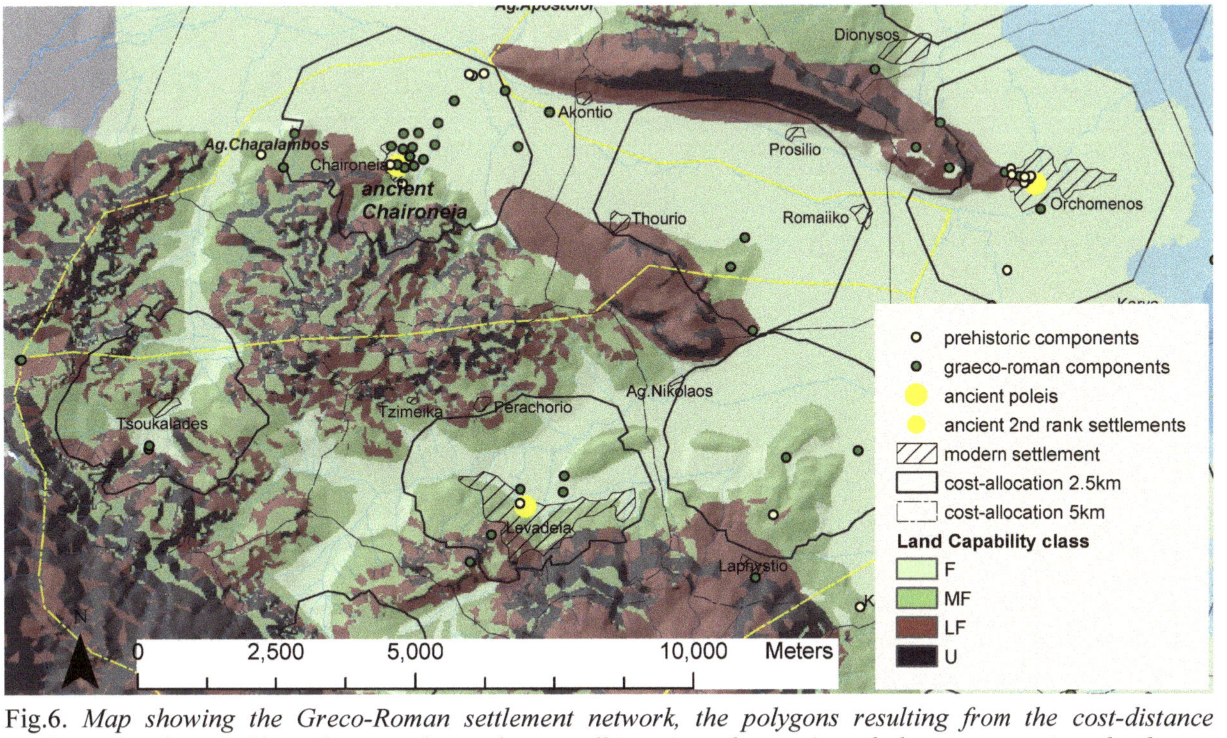

Fig.6. *Map showing the Greco-Roman settlement network, the polygons resulting from the cost-distance analysis (marking half an hour and one hour walking time distance) and dots representing the known archaeological components (same as in fig.2), with land capability information underlain.*

II.3.4

The Copais area: Orchomenos

TOPOGRAPHICAL SETTING

The *chora* of Orchomenos generally included the area of the Dourdouvana/Akontion ridge (at the E end of which is ancient Orchomenos, and along which to the S runs the Kephisos river) and the area before it towards the Copais, as well as the NW bay of the Copais (otherwise called Tsamali bay, after the village of Tsamali/Dionysos along its W edge).

The gulf of Tsamali, to the NW of Copais, opens up just behind the Akontion ridge, and reaches a distance of 4km from Orchomenos. To the N side of the Akontion, the Tsamali bay runs into the mountains (Philippson 1951: 466). The bay is bordered to the SW by Mt. Akontion, to the W by Palaiovouna ridge, to the NW by Makrorrachi, and to the N by Mt. Chlomon (1080m asl), of which the others are spurs (Fossey 1988: 351). The plain along the edge ends to the E with a Frankish tower by the village of Pyrgos, from where the higher mountains forming the basin edge begin.
The NW bay is characterised by a hydrological balance, and by a different behaviour not strictly related to the fluctuations of Copais (see chapter III.1). It was an area mainly characterised by the presence of a permanent marsh, due to the flooding of the Polygira stream (Philippson 1951: 474) and the presence of the springs of ancient Melas - one of the two main rivers feeding the Copais.[1] It is probably this small gulf that Theophrastus names 'Orchomenian lake'. In winter, the marsh joins with the lake, but it does not constitute part of it. The marshy character of the Melas basin is described by Plutarch (*Sull.* 20). This area of Tsamali, which would have looked the same in Plutarch's time (*Pel.* 16; *Sull.* 4-5), should be the swamp known as Pelekania, which produced reeds for flutes and created some islets (Theophrastus *HP* IV 11.8 - fig.12 in chapter II.3.1).[2]
Lauffer (Kopais I: 149) notes that there was probably a communication route within the basin that crossed it (from Pyrgos to Orchomenos),[3] though the basin was often entirely swamp. As Philippson (1951: 476) notes, the roads coming from the W avoided the swamp, going to Orchomenos or to Copais along the N edge[4] (right around the bay, from Pyrgos to Orchomenos, passing by the site of Polygira – Lauffer Kopais I: 149 and see below), while in modern times works have deepened the Melas bed, therefore collecting the swamp water and creating fertile land. In summer, the arm of the lake along the N edge of the gulf created a swamp collecting the water from the Kephisos and Melas (see above).
Today, like the rest of the basin, the NW bay has been drained, and, as Fossey notes (1988: 351), "*only the marshes around the springs and the head-waters of the slow-moving Melas remind one of the area's former water-logged state*" (see below).

The Akontion is a long, narrow ridge running in an E-W direction. It is bounded to the N by the Melas river, which emerges through several springs at the N foot of the ridge, and to the S by the Kephisos river, which 'wimples like a snake' (Strabo IX 3.16) across the valley.
The Kephisos runs along the S edge of the Akontion, through Chaironeia basin, and reaches Orchomenos before draining into Copais. In Philippson's time (and today), in the tract where the river crosses the former Petromagoula settlement,[5] the riverbed is dry, while in earlier times it probably was not (Philippson 1951: 472).[6]

Boundaries

To the N and the NW, the natural limits of Orchomenian territory are the mountain ridges separating the NW Copais bay from the territories of the ancient East Lokrian town of Kyrthones (=Kokkala, see Oldfather AJA 1916: 154-172), and the Phokian towns of Abai (=Palaiochori Exarchou, see Yorke JHS 1896: 291-312; Fossey 1986: 78-81) and Hyampolis (=Bogdanou, see Fossey 1986: 72-76).

---

[1] The Kephisos is the other.

[2] The area was visited also by Ulrichs 1840: 192; Philippson 1951: 474 – see fig.12 in chapter II.3.1 and Farinetti 2008. On the normally waterlogged nature of the NW bay in Classical times see Plutarchus *Sull.* 4-5; *Pel.* 16; for the fluctuations in the extent of the lake cf. Theophrastus *HP* IV 11. 3.

[3] Mentioned also in Ulrichs 1840: 193f. and Vischer 1875: 583f.

[4] For instance, Plutarch (*Pel.* 16), mentions that, in order to go from Tegyra (identified by Lauffer with modern Pyrgos, but see below) to Orchomenos, Pelopidas and his army had to use a path over the mountains (see also Pausanias IX 38.6). Whether Tegyra is located at the site of Polygira or Pyrgos, Plutarch's description of Pelopidas' path over the mountains is suitable for both identifications (see below).

[5] The former settlement of Petromagoula when joined with the former settlement of Skripou to its NE formed the modern village of Orchomenos.

[6] Before it dried up, different arms of the river flowed towards the villages of Karya and Ag.Dimitrios, while its main course flowed NE, via the gulf of Tsamali, towards Pyrgos, and then drained into the lake when the water level was high, or into the Melas in dry periods (Philippson 1951: 472).

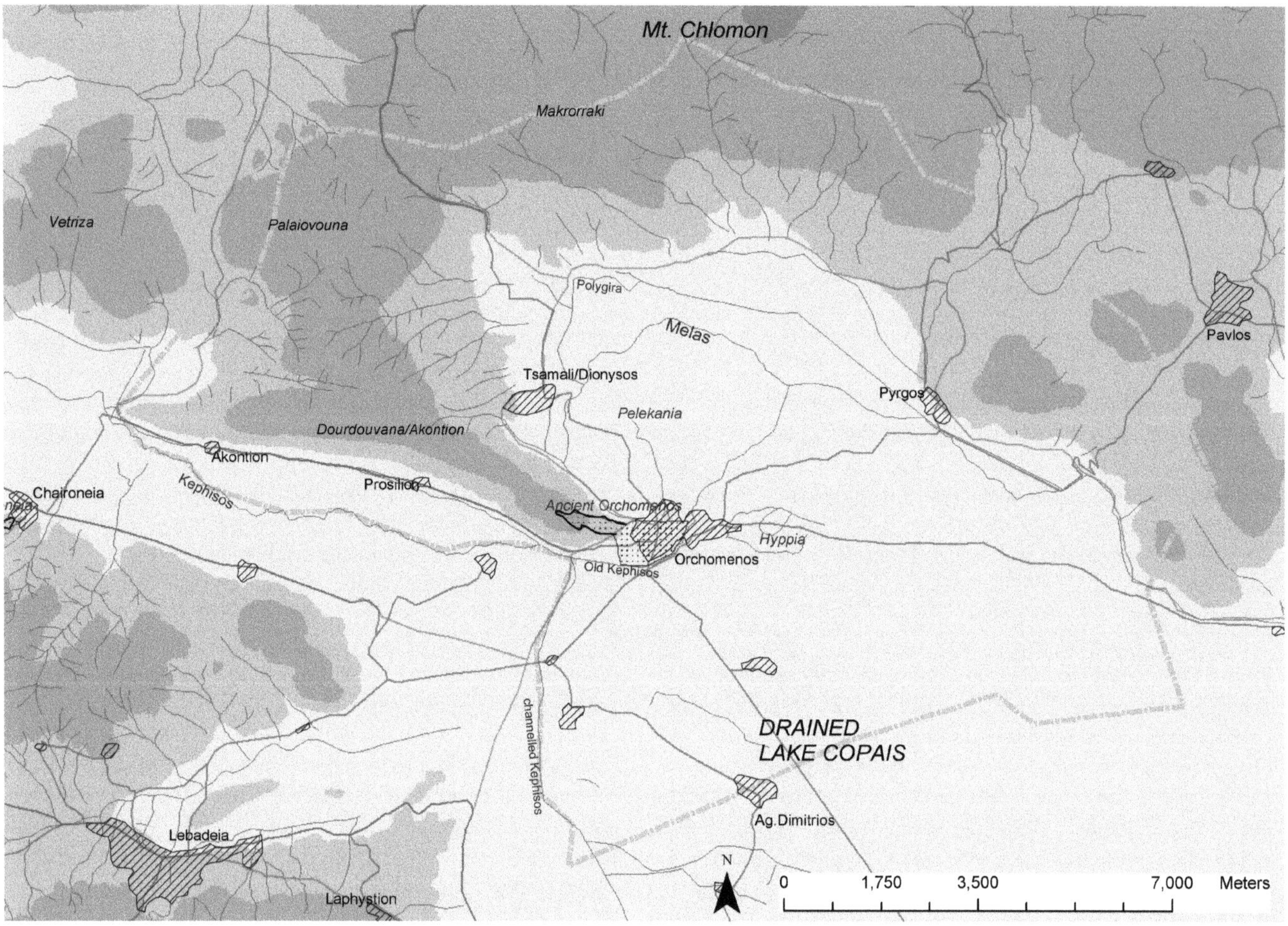

Fig.1. *Topographical setting of the chora of Orchomenos.*

The extent of Orchomenian territory along the N shore of the Copais is unclear (Fossey 1988: 364). As Fossey notes (1988: 364-5), to the E, the lake and the peripheral marshes would probably provide the limits; to the W, Orchomenos' territory would border the Chaironeia *chora*[7] and in some periods of history that of Phokian Panopeus[8] (the border running probably along the Kephisos river flowing S of and parallel to the Akontion, and along a line from the river towards the Paliovouna and Mavrokastro peaks to the NW of Akontion itself, marking the border with Phokis[9]); to the S boundaries are more difficult to establish (see below – RESOURCES- and Farinetti 2008).

PHYSICAL LAND UNITS

The physical landscape of the *chora* can be described as mainly below 200m asl, as is evident from the figure (fig.2 in chapter II.1) and the percentage chart below, mainly due to the presence of the Copais basin.

The mountain segment (> 600m) is not represented, except for a very small area to the N, marking the border of the *chora* with Phokis. The border with the surrounding *chorai* of Boeotia and other regions are therefore marked by lower or higher hills up to 600m.

Hilly landscape	36%
Mountainous landscape	0.3%
Plain	63.7%

1	P1_P2	lacustrine basin, valley	53.3%
2	P3	gentle slope	3.8%
3	P4	foothill	6.7%
4	H1	plateau	12.4%
5	H2	gentle slope	1%
6	H3	moderate slope	6.8%
7	H4	severe slope	11.2%
8	H5	very severe slope	4.6%
9	M1	plateau	0.3%
10	M2	plateau/gentle slope	0%
11	M3	moderate slope	0%
12	M4	very severe slope	0%

Table 1. *Percentage of the different physiographical classes present in the Orchomenos area (P=plain; H=hill; M=mountain)*

[7] The Chaironeia *chora* was reduced to Orchomenian control in the Archaic period (Roesch 1965).

[8] Strabo IX 416.

[9] Dasios (1995) is certain that Panopeus possessed the whole Idilion/Vetrisa mountain and probably at least a part of the small valley to its E, to get the NW hillsides of the Yphanteio/Paliovouna, where the border at Mavrokastro (480m) was controlled by a fort, whose remains can still be seen – see Dasios 1995b.

Lacustrine basins and valleys (P1_P2) are predominant in the plain landscape, constituted mainly by the Copais basin and the Tsamali bay.
Foothill landscape (P4) marks the area immediately below and following the heights surrounding the plains. Within the foothill landscape also lies the ancient *polis* (as well as part of the modern village) of Orchomenos.
No mountainous segment is represented, since the *chora* landscape morphologies do not reach the 600m asl elevation (apart from a very small area -M1- on the heights at the border with Lokris).
A lot of H4 and quite wide areas occupied by upland plateaus can be recognised within the hilly landscape (H1 – 12.4%).

RESOURCES

The *chora* landscape is characterised by fertile and arable land, suitable for agriculture, in the empty space at the edge of the Copais basin (even though some areas are, as seen above, marshy, and therefore suitable for other kinds of activity), while the fertile soil in the Tzamali area should be removed from the picture because it is permanently marshy (see above).
At the time that Fossey was recording his gazetteer (late 1970s – early 1980s), the NW bay (Tsamali bay), like the rest of the basin, was cultivated for cereal and tobacco (Fossey 1988: 351), but since then cultivation has changed, and today the bay has mainly cotton and other irrigated crops. Looking at the soil of the bay,[10] one notices that after drainage and with intensive irrigation agriculture, as well as water pumps working continuously, it has gradually become less calcareous, with a slight sandy component, which would both be considered unusual in a former marshy area. This is due to the intense water catchment carried out by farmers which has lowered the water table.
Due to the flooding of the Kephisos river, the promontory of Akontion is surrounded by a large fertile area which was usually not flooded, extending 3km to the NE and 6km to the E and SE of Orchomenos. Philippson (1951: 472) noted that the soil here is fat, with red clay soil at some points. It was cultivated with cotton, cereals and vineyards, and in 1928, between Levadeia and Skripou, some olive groves were burned by frost. It seems that due to the drainage of the lake, which used to make the climate milder, the winter became colder. In 1934, to the E of Skripou, Philippson saw herds of cows and horses.

A map showing land capability (for agriculture – see chapter I.2.1) for the Orchomenos area can be seen in fig.6. While the most fertile areas are constituted by the Copais plain, the part of Kephisos valley probably belonging to Orchomenos, as well as the part of the inner upland area to the E marked by stream deposits, is also fertile. The extended wide plateau area to the E should, however, be considered partially unfertile (U), and partially low fertile (LF), due to the presence of limestone lithology.

[10] Personal observation.

For the probable extension of the *chora* of Orchomenos in the Copais basin, the flood situation of the W area of the basin itself and consequent availability of fertile land to the *polis chora*,[11] a few comments can be made. In the Archaic period,[12] Orchomenian coins carried on the reverse an ear of corn (Babelon 1907: 938 and Hill 1906: 7). Fossey comments (1988: 364) that this is 'somewhat ironical', since it is doubtful that the land of Orchomenos was ever very productive of crops, since the northwestern bay was marshy in Classical times.[13] In the light of Knauss' information and of our attempt at modelling the hydrological situation and the fluctuations of the lake's extents (chapter II.3.1 and fig.10; fig.2 in chapter II.3.2), we might more carefully examine the interpretation of the relationship of the *polis* of Orchomenos with the lake, and the character of its economy. We know that variations in the lake's water level did not influence the area of Tsamali (NW corner – Theophrastus' Orchomenos lake), whose permanent marshiness was due to the water of the Melas and the springs at the foot of Akontion (see above). We may consider, for instance, the presence of a very large plain to the S of the *polis*, which probably belonged mainly to it,[14] and was perhaps also very fertile, rich in thick soil (limus deposits), though probably slightly marshy most of the time (see Philippson, who discusses the area of Agios Dimitrios – see above), taking into account the actual hydro-geological conditions of the plain. Theophrastus informs us of the density of Boeotian wheat (*HP* VIII 4.5) due to thick soil and cold climate (*CP* IV 9.5). Despite all this, in front of Orchomenos lay a large cultivable area, fertile and quite suitable for cereal cultivation, even if slightly or temporarily marshy, as in the case of the modern village of Mouriki, at the E edge of Yliki lake.

As for other possible resources available in the *chora* territory, Fossey (1988: 365) mentions: reeds[15] and pasturage.[16] Grazing land was available on the slopes of Akontion and the other ridges surrounding the NW Copais, as well as in the marshy areas (especially used for horses). Grazing activities are attested in the marshy area of the Tsamali bay: in 1928 the 214 inhabitants of Tsamali village were all nomad shepherds by the Polygira spring (Philippson 1951: 476). For a parallel economy available from the lake and swampy areas, see Farinetti 2008.

[11] Fossey 1988: 365 says: "*It is perhaps doubtful whether the land of Orchomenos can ever have been very productive of crops; this makes the reverse type of her coinage, an ear of corn, somewhat ironical*".
[12] First Orchomenian coinage, around 550 B.C.
[13] Fossey refers to Plutarchus *Sull.* 4-5 and Theophrastus *HP* IV 10-11.9 who talk about the 'Orchomenian lake', which, however, probably refers to the Tsamali bay (see above).
[14] See discussion in chapter II.3.2 –BOUNDARIES- and fig.2.
[15] See Plutarchus *Sull.* 4-5; Theophrastus *HP* IV 11.3.
[16] IG VII 3171 indicates the importance of pasturage to Orchomenos'economy, and the strict control of pasturage rights (Roesch 1965: 60 and 212-3). The inscription mentions cattle, horses, sheep and goats.

THE ARCHAEOLOGICAL RECORD

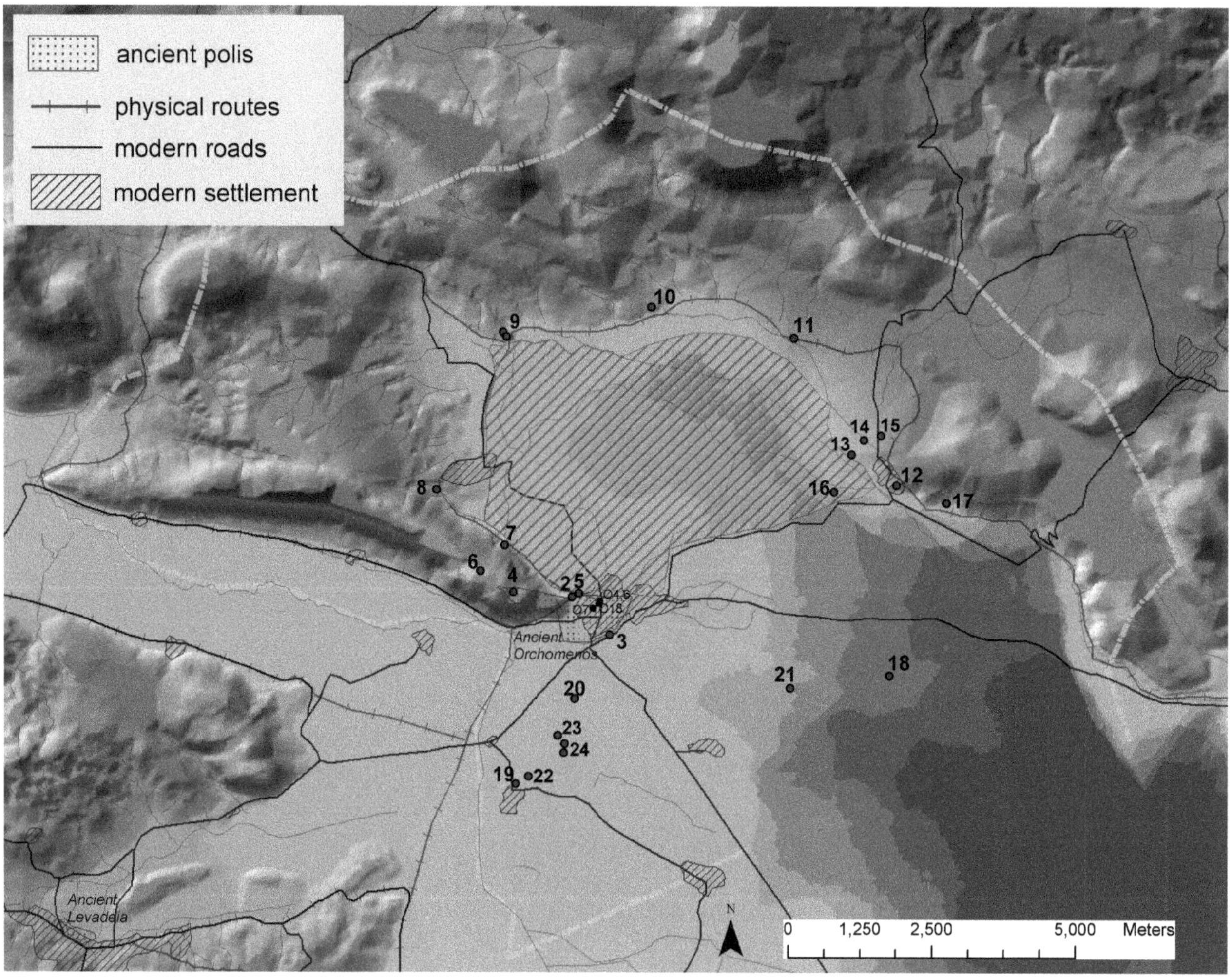

Fig.2. *Archaeological map of the Orchomenos chora.*

1	**ORCHOMENOS**	*Components O_1 to O_11; O_13 and O_14; O_16 to O_19*
2	**Orchomenos**	*Component O_12*
3	**Orchomenos**	*Component O_15*
4	**Orchomenos**	*Component O_16*
5	**Orchomenos neighborhood**	*Component O_24*
6	**Orchomenos quarry**	*Component O_21*
7	**Melas source**	*Components O_35 and O_36*
8	**Tsamali**	*Component O_37*
9	**Polygira**	*Components O_28 to O_34*
10	**Avrokastro–Mavromandili**	*Component O_44 to O_46*
11	**Erima Kalyvia**	*Component O_47*
12	**Pyrgos**	*Components O_48 to O_56*
13	**Pyrgos *magoula***	*Components O_62 to O_66*
14	**Pyrgos NW**	*Component O_57*
15	**Pyrgos-Trassa**	*Component O_58*
16	**Pyrgos-Melas dam**	*Components O_59 to O_61*
17	**Mavrospilia**	*Component O_67*
18	**Old Orchomenos**	*Components O_23, O_38 to O_43*
19	**Vranesi tumulus**	*Components O_26 and O_27*
20	**Orchomenos tumulus**	*Components O_22*
21	**Orchomenos Magoula SE**	*Component O_25*
22/23/24	**Vranesi**	*Components O_68* (22), *O_20* (23) *and O_69 to O_70* (24)

Table 2. *List of archaeological components and activity loci mapped in fig.2.*

All studies concerning Copais covered within their research the area of Orchomenos to some degree, meaning that it has therefore been quite intensively investigated (though not by means of systematic intensive artefact surface survey).
The majority of the studies, especially the earlier ones, have been focused on Bronze Age periods and material. The landscape of W Copais is characterised by the presence of low Prehistoric *magoula* sites similar to those typical of Phokis, which attracted (and attract) the interest of many archaeologists as well as of travellers in the past, and as such, they have been thoroughly visited.
A large quantity of archaeological information is available for the city itself; in a less clear form from early excavations, and in a more systematic way from recent excavations (mainly rescue) within the modern settlement.

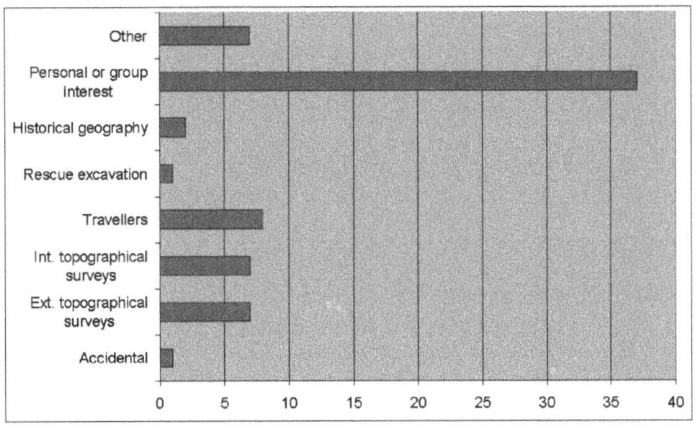

Fig.3. *Graph illustrating the proportion of components discovered in different research frameworks.*

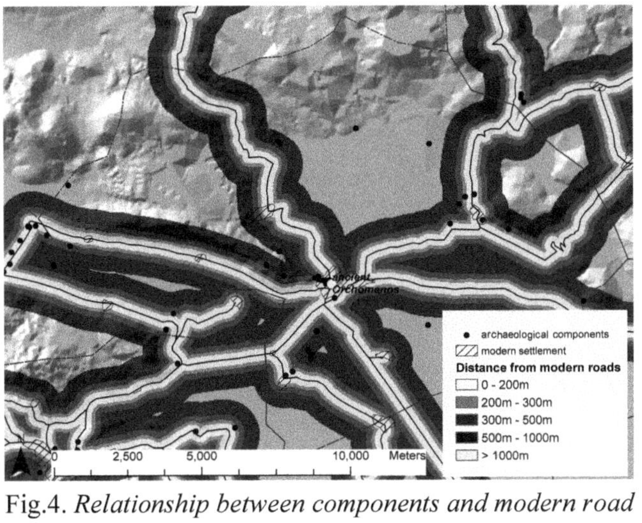

Fig.4. *Relationship between components and modern road network.*

The graph (fig.3) illustrates the proportion of components discovered within different research frameworks. In the case of the Orchomenos *chora*, the large presence of 'Personal or group interest' entries is due to many studies of the area having a personal interest on a specific topic, mainly related to the Prehistoric occupation as well as the behaviour of the lake and the settlement dynamics around it. Compared to the Levadeia *chora* (chapter II.3.2), for instance, there is a notable decrease in the occurrence of discoveries through rescue excavations.

The ratio of prehistoric to Greco-Roman components known is 30 to 33 (1:1.1), while among the historical periods, 63.3% is dated Archaic to Hellenistic, 23.3% Roman-Late Roman, and 13.3% is attributed generally to the Greco-Roman period.
The difference in knowledge of the prehistoric landscape compared, for instance, to the Levadeia *chora*, is probably not actually due to a less intense occupation of the Levadeia area in the prehistoric period (although the area of Levadeia was not characterised by the presence of a large nucleated centre, as was Orchomenos), but to a bias in archaeological knowledge due to strongly prehistoric-oriented research in the Orchomenos area, due to the importance of the main settlement site (Orchomenos itself) during the Bronze Age.
Compared to the extension of the *chora*, there is an extremely low number of known Greco-Roman sites (a much different behaviour pattern in the data as compared to the Chaironeia *chora* -chapter II.3.3-, with much fewer data for the extension).

In fig.4 we can see the relationship between known archaeological sites and the distance from the modern road network. Only 16 components (excluding those related to the city site) of 70 are within a distance of 300m from modern roads. Unlike the Levadeia *chora*, for instance, in the area of Orchomenos, proximity to the modern road network apparently did not bias the discovery of archaeological data (though it is biased by other factors, and in particular by personal interest oriented research, as seen above).

ANALYSIS OF THE *CHORA* LANDSCAPE

PREHISTORIC PERIOD
The analysis of Prehistoric settlement in this area is in tandem with consideration of the whole Copais basin (or at least its Northern edge) for the Prehistoric period (see figs. 6 to 9 in chapter II.3.1).
Lauffer (Kopais I: 163-4) remarks that only two other sites in the Copais are similar to the *magoula* site by Pyrgos (see appendix I.4 - *components O_62 to O_66*) in shape and height: Nisi by Stroviki (*components NC_1 to NC_4*) and Gla (*components NC_to to NC_72*). Other places in N Copais, such as the 'isle' E of Spitia-Katavothra (see chapter II.3.6), are linked to the border by land, or, like the large *magoulas* close to Kokkinon, Topolia-Kopai and Tourlogiannis, are spurs of higher mountains.
The shifting of settlement along the N edge of Copais, between the sites of Pyrgos and Magoula Pyrgos can be noted (*magoula*: Neolithic and EH, less strong MH and LH - Pyrgos: MH and LH stronger).

MH settlement sites are usually known in a slightly elevated location (see MH material, in a low location that could represent a burial area - *component O_61* as well as *component O_4*, MH burial at Orchomenos). MH

habitation sites known are: Orchomenos at the foot of the Akontion (*component O_3*); Old Orchomenos (*component O_40*); Pyrgos (*component O_50*) and Pyrgos Magoula (*component O_64*), which finds a parallel in the Magoula Balomenou site in Chaironeia area (*component CH_23*). Others are simply activity foci of unspecified character (*components O_30 and O_61*).
The MH settlement at Orchomenos (*component O_3*) can be connected, as Tzavella-Evjen did, with the MH settlement found at the other extremity of the Akontion ridge, at Magoula Balomenou site (*component CH_23*).
Lauffer (Kopais I: 131ff) suggests that the power and strength of Orchomenos over the Western Copais in the MH period is due to the fact that it controlled part of the Kephisos valley that lies higher (protected by lake water). He says that the other settlements in the MH period (in the NW gulf (Pyrgos and Polygira, and Stroviki) and in the SW (Kalami, Koroneia and Agoriani)) had a much smaller area available for agriculture than did Orchomenos (the Kephisos lower valley), and that they were in a less secure position, therefore never reaching the level of power of Orchomenos. Lauffer therefore gives greater importance to the fact that Orchomenos controlled the lower Kephisos valley, also giving it the strength to build up Minyan power and control over the rest of Copais (dikes, katavothrai, acropolis of Ptoion, etc.), unlike Kahrstedt (AA 1937: 9) who identifies Orchomenos' power with its control of the NW gulf (Pyrgos, Polygira, Stroviki, etc.). With regards to Lauffer's point; no other Copais *polis*, not even in later periods, had all this land available; they always had to recover land from water.[17]

The LH fortified sites known in the area could be considered part of the Mycenaean fortress system controlling Copais (Lauffer Kopais I, Fossey 1990d), located mainly in the NE bay (see chapter II.3.6).
Many cave sites are known from the area of Copais, some of which are published, some others simply reported without details (in this *chora*, Mavrospilia – *component O_67*), and others known from personal communication only and not yet investigated (but we would prefer not to include them in the archive for heritage protection reasons, as requested also by local authorities).
As noted by many scholars (Lauffer *in primis*), this area of the Copais basin is characterised by the presence of low *magoulas*, mainly of antropic origin, typical of the Thessalian landscape and usually marking Prehistoric habitation sites or tumuli. They form what we could call a distinctive '*magoula* landscape'. Different *magoula* sites have been visited by archaeologists, and we thus have many sites called *magoula*, most of the time difficult to identify and to distinguish one from the other (see *components O_62 to O_65, O_22, O_68 to O_70, O_25*). Sotiriadis had noticed a lot of small tumuli all around the area, the tallest being the Orchomenos tumulus, followed by the Vranesi tumulus, slightly lower, and a lot of other small 'bumps', protruding slightly from the surface. Within them he noticed the presence of Hellenistic (Classical?) and Roman sherds on the surface, probably to be related with burial areas (apparently the lower areas are Historical period burials, the higher could be real Prehistoric *magoula* sites). Some *magoula* sites were also noted and visited by Knauss et al. (Kopais 3). Some of the components representing *magoula* sites were given an 'unknown' chronology,[18] so do not appear in the maps per period. In the absence of further research, we could add them either to the Prehistoric landscape picture, or rather to the Historical one, if they mark tumuli/burial sites.

GRECO-ROMAN ANTIQUITY[19]

Town level

The *polis* of Orchomenos lies in a strategic position with regards to routes (fig.1) and the control of the whole Copais basin.[20] It is also located in a position dominating the wider area of the basin (to its S) that is free of water most of the year. Historical sources tell us that the city tried to gain control over the whole Copais area during the Archaic period. At that time, Orchomenos dominated Chaironeia[21] and Hyettos,[22] Olmones,[23] Hyria,[24] Levadeia,[25] and probably Koroneia[26] as well as Copais. Links with Haliartos are also probably implied in the connection of the mythical genealogies of the two *poleis*.[27]

Village level

There are indications of early occupation sites: Orchomenos, Pyrgos, Polygira (Aspledon and Tegyra of ancient sources – see above for identification issues), which could have been acting as *proto-poleis* (Bintliff 1994b, see chapter III.2). From the late Archaic - early Classical period onwards, apart from the *polis* of Orchomenos, settlement sites might have been located at Pyrgos (*components O_54 and O_55*), Polygira (*components O_33 and O_34*), and only probably at Avrokastro–Mavromandili (*components O_46 to O_49*).

[17] In addition, Orchomenos probably also had the whole area in front of it, to its S, part of the Copais basin but not wetland (probably only seasonally marshy).

[18] Since sherds are not reported and these magoula sites were noted only as characteristic topographical features.
[19] Period maps are included in chapter II.4, figs.17-19-21-23-25-27.
[20] See appendix I.4 –ORCHOMENOS- for *polis* history and details.
[21] See chapter II.3.3.
[22] Hesiodus *frg.144*; Pausanias IX 36.6. See Wilamowitz 1922: 19.
[23] Pausanias IX 24.3.
[24] Hesiodus *frg.143*.
[25] Wilamowitz 1922: 18f.
[26] See chapter II.3.1.
[27] We know from Pausanias (IX 34.5) that the mythical Orchomenian tradition (dating back probably to the Archaic period) claimed under its jurisdiction areas including the Laphystion mountain, Koroneia and Haliartos. The Tilphousion oracle, for instance, was probably under Orchomenian control in Archaic times – see Schachter 1994 s.v. *Teiresias (Orchomenos)*.

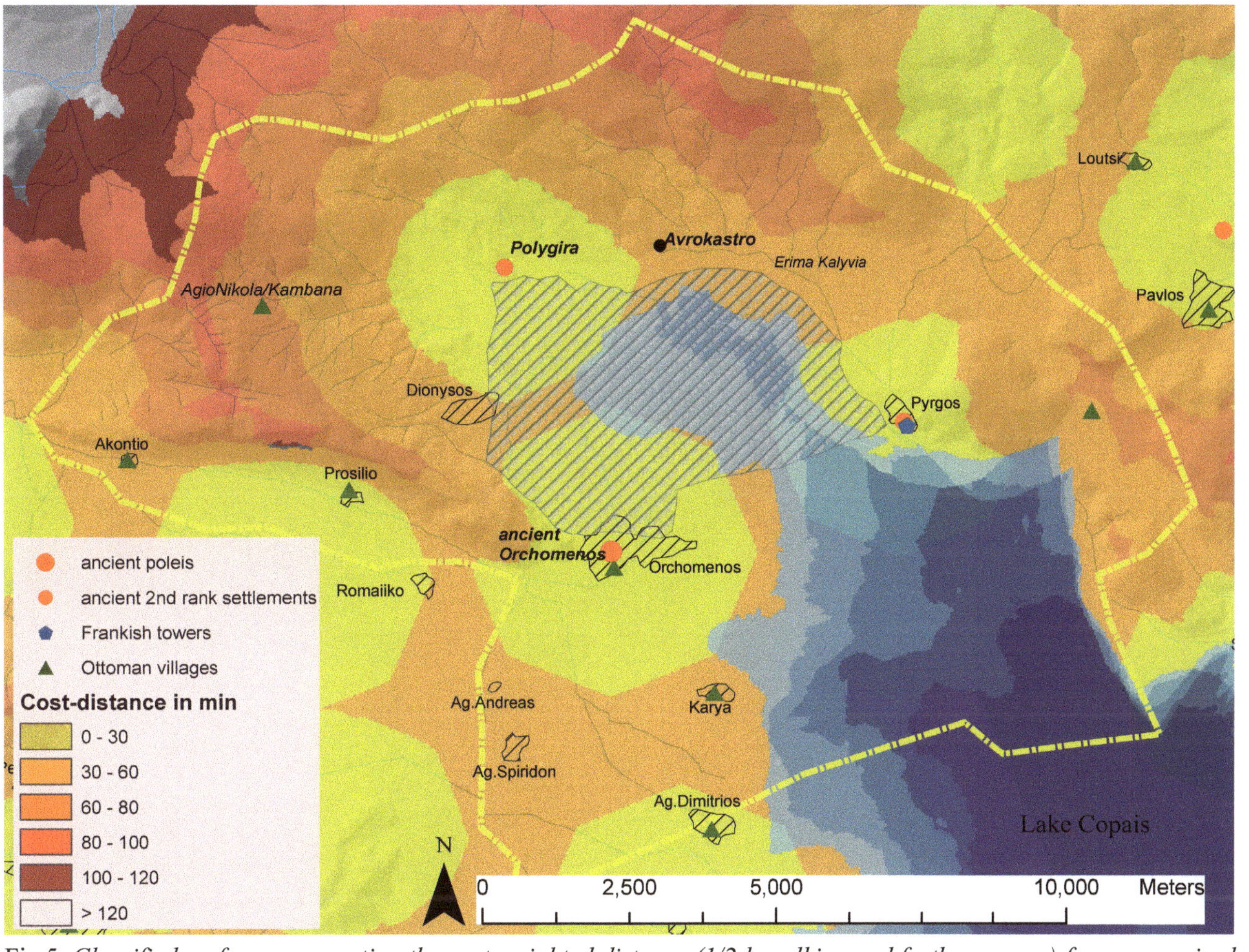

Fig.5. *Classified surface representing the cost-weighted distance (1/2 h walking and further ranges) from recognised 1st and 2nd rank ancient settlements (represented by larger and smaller dots). Areas without dots indicate potential settlement chambers. Ottoman villages and Frankish towers have also been added to the map to show their spatial relationship with the Greco-Roman settlement network and to appreciate potential settlement chambers.*

Aspledon and Tegyra, whatever sites they would have to be identified with,[28] never reached the status of *polis*.

Rural segment

The rural segment is poorly known if one excludes the evidence at the so-called site of Old-Orchomenos by the edge of the lake (*components O_23 and O_42, O_43*) and the little other historical evidence known from sites with mainly Prehistoric occupation.

Burial areas

Burial areas related to the *polis* of Orchomenos are known in its immediate surroundings, towards the plain (in the area where the village extends today). The burial landscape is then enriched with the tumuli common in this area of Copais, some of which are Prehistoric *magoula* sites (see above PREHISTORIC PERIOD), while some others are (or could be) recognised as burials of Historic (Greco-Roman) date.

Cult places/Religious Areas

Apart from sanctuaries and cult places known from the Orchomenos city site (in particular the sanctuary of Asklepios – *Asklepieion* – see above under Orchomenos) and its outskirts (a cult place attributed to the cult of Herakles at the source of the Melas), few extra-urban sanctuaries are known (at the site of Polygira - Apollon Tegyraios? -, at Tsamali) or hypothesised (at Magoula Pyrgos and Old Orchomenos). The two supposed cult places at the source of the Melas and by Tsamali are said to be connected by the road going around the bay at a higher level (see above for the cult places and below for the road).

Forts and Fortifications

The only fortifications datable with certainty to the historical period are the Hellenistic defensive walls with towers above Orchomenos (on the E summit of the Akontion ridge – *component O_16*). Probable fortifications of historical date have been noted at the sites of Avrokastro-Mavromandili [AE458] and Pyrgos [AE605].

[28] See appendix I.4 for discussion on identification

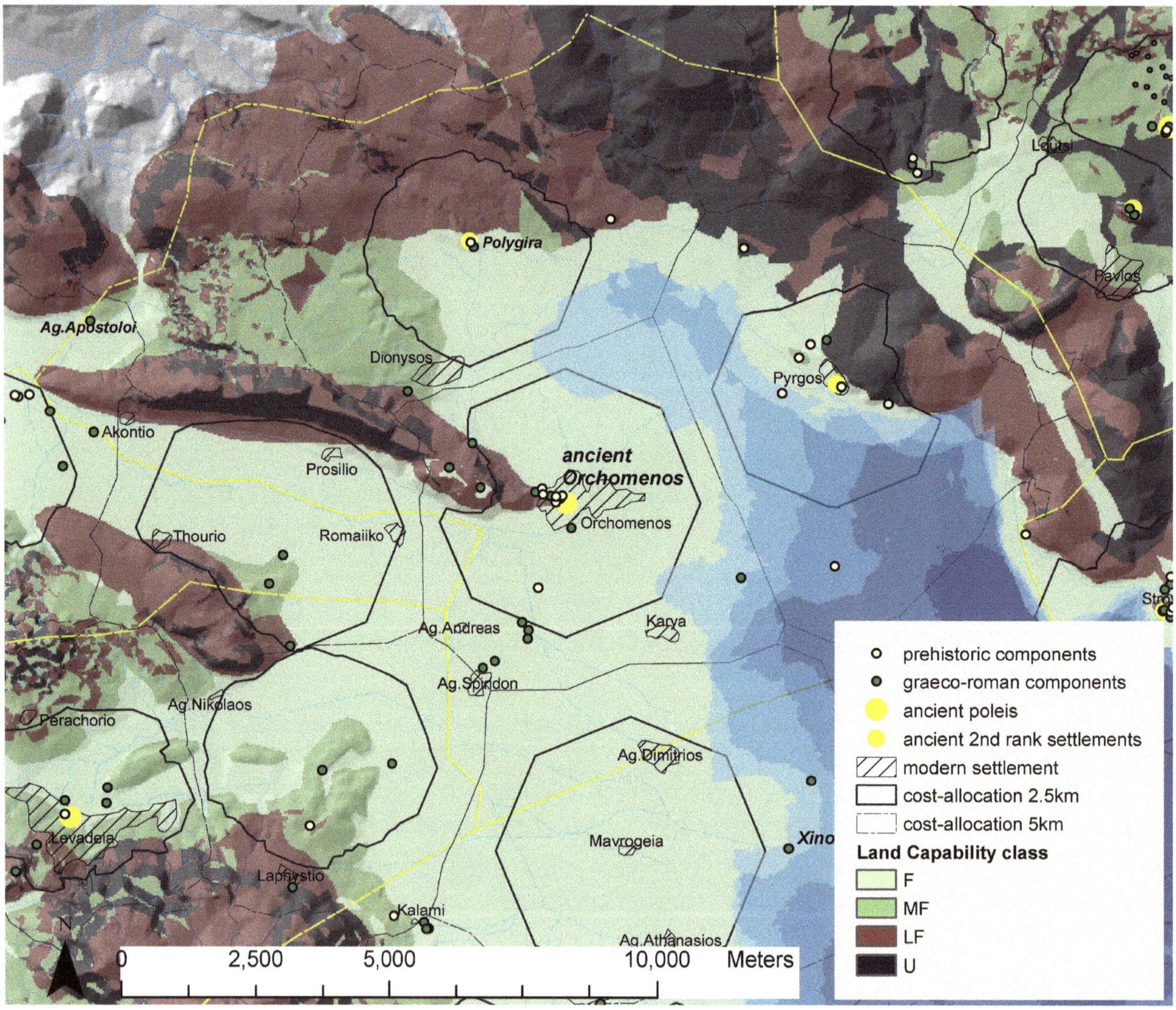

Fig.6. *Map showing the Greco-Roman settlement network, the polygons resulting from the cost-distance analysis (marking half an hour and one hour walking time distance) and dots representing the known archaeological components (same as in fig.2), with land capability information underlain.*

Other activities / unspecified activity areas
Undetermined activity foci were noted, which can not be assigned any character at the present stage of research. The majority of these might constitute part of the less-known rural landscape. Unspecified activity foci are also known at the edge of the Copais basin (sites detected by the Munich team led by Knauss), probably linked to the exploitation of the swampy lake resources also in the Greco-Roman periods.

LONG TERM SETTLEMENT TRENDS IN THE CHORA *LANDSCAPE*

Just as in Chaironeiake the landscape was marked by the presence of a river (the Kephisos), the *chora* of Orchomenos is characterised by water in the form of swamps, marshy river flows and the Copais lake.
The *chora* of Orchomenos extends on the two sides of the Tsamali bay, and yet it is constituted also by the area to the S of it, bordering the lake, the so-called Hyppia plain,[29] as well as by a wide (probably dry[30]) area to the S of Orchomenos, extending probably as far as the village of Vranesi/Ag.Spyridon, or even further S, to the village of Ag.Dimitrios (see also above, discussion for the Prehistoric period). The ancient city faced S, towards an area probably free from water and potentially highly fertile, although occupied by seasonal and/or local marshes in wet periods (see above –RESOURCES- and Farinetti 2008).
The modern centre in the geographical area of the ancient *chora* is also Orchomenos, formerly Skripou (one of the two original constituents of modern Orchomenos, the other being Petromagoula). The importance of Skripou in the medieval period as a centre of Byzantine Greek population (the Middle Byzantine monastery of Skripou)

[29] See map in Knauss et al. Kopais 3: 183 fig.6.8 and fig.1 in this chapter.
[30] See chapter III.1.

shows a possible continuation from antiquity,[31] abandonment and recolonisation by an Albanian *katun* (Iskres) which continued into the modern village (Koder – Hild 1976).

The area in the Copais plain is occupied today by a series of villages S of Orchomenos (Ag. Spyridon, Karya, Ag. Dimitrios). Although this area and its settlement is partly associated with the total drainage of the lake (at the end of the 19th century), it was also dry land in antiquity, forming a plain extending S of the ancient city (see above). This is also indicated by the existence of post-medieval villages in the area, predating the lake drainage. Ag. Dimitrios is recorded in the Ottoman archives from 1466 to 1570, then 1642-1688, as a Greek village. The large size of the village of Ag. Dimitrios even in 1466, and its lakeside position, as well as its survival through the troubled 17th and 19th centuries, indicate a settlement chamber at the W edge of lake Copais even before the drainage.[32] The village of Karya could also have an Ottoman predecessor (Kiver/Kirnoz) but the attribution is not sure (Bintliff-Kiel in preparation). Also in antiquity, the area immediately S of Orchomenos was most probably controlled by the city, and one might suppose the existence of one or more small hamlets, exactly like those that exist in the traditional settlement pattern (Ag. Spyridon, Karya). In the PG and G periods, the existence of a tumulus by the modern village of Ag.Spyridon (Vranesi tumulus – *components O_26 and O*_27 and no.19 in fig.2), which could be related to another small settlement (Fossey 1990d) rather than directly to the growing settlement of Orchomenos, might be linked to this hypothesis. This potential settlement could in any case fill the settlement gap to the S of Orchomenos visible in fig.5.

The edges of the Tsamali bay are shared today between the modern villages of Pyrgos and Dionysos (formerly Tsamali). For both villages it seems that there is no Ottoman predecessor (or that it has not yet been identified). The Frankish tower by the contemporary village at Pyrgos, however, indicates a settlement chamber in this area. As for the territory of the modern village of Tsamali/Dionysos, the only identified Ottoman village in the area is Agio Nikola/Kambana, associated with the famous Byzantine church of Ag. Nikolaos sta Kampia (12th and 13th centuries) (Bintliff-Kiel in preparation). The Byzantine and Ottoman occupation in this area, but also the possible continuity of the Greek population (reflected in the Greek name of the village), indicate the existence of settlement chambers and potential gaps in antiquity. Bintliff (1994b: fig.20), as the result of a Thiessen polygons analysis, already noted possible settlement chambers available around the bay.

The settlement site of Pyrgos, situated opposite Orchomenos on the axis of the Tsamali bay, was occupied in ancient periods (*components O_48 to O_56*). This Eastern area of the bay constituted a settlement chamber, since the Late Neolithic and Bronze Age (with the sites, both contemporary and alternate, of Pyrgos and Pyrgos *Magoula*, at the E edge of the Tsamali bay), also providing certain evidence of occupation at least in the Classical period. The NW edge of the bay would likely constitute a settlement chamber (fig.5), with relevant changes in settlement location though, if one considers the modern village of Dionysos, the Ottoman village of Agio Nikola/Kambana further up on the mountain, and the earlier Polygira (Tegyra?) settlement, archaeologically attested since Prehistoric times (*components O_28 to O_34*).

In the upland areas above the Tsamali bay to its E, one can probably recognise another small landscape in the area of *Erima Kalyvia* (fig.5). As Fossey notes (1988: 361), *Erima Kalyvia* was probably an offshoot of the village of Kolaka on Mt.Chlomon to the N, and the people of that village cultivate all the land in this area. A settlement might also have existed in this area in the Prehistoric period,[33] and even later, since the area could constitute a suitable small settlement chamber with fertile land available as well as grazing zones (see RESOURCES above and fig.6). In some periods of ancient history, a settlement (or at least a fortified site) was located at Avrokastro-Mavromandili (*components O_44 to O_46*), further to the W. Although the two locations (Erima Kalyvia and Avrokastro) are quite distant from each other, they could constitute an indication of the possible existence of a further settlement area in the upland NW of Tsamali bay, as the gap, albeit small, left between the two potential settlement areas centred on the locations of Polygira and Pyrgos seems to attest (see fig.5 showing the application of a cost-distance analysis[34]). Cautiously though, I prefer not to locate a potential 2nd rank settlement in this area, since Avrokastro could even be part of the Polygira settlement chamber (being 35 minutes walking distance). Actually, Fossey suggests the possibility of a shifting in occupation between Poligyra and Avrokastro in Hellenistic times, though the data are too poor to support this (see appendix I.4).

Finally, the Akontion mountain behind Orchomenos, suitable for a pastoral economy, is occupied today by two villages, Akontion (formerly Visvardi) and Prosilion (formerly Velis), both probably going back to the Ottoman period. Being situated along the Kephisos valley, this area belongs geographically to the Chaironeia area (and its settlement history has been discussed in chapter II.3.3), although it might have been controlled by Orchomenos in some periods of history. In fact, a strategic reason might have led Orchomenos to seek for control of the left (N) bank of the Kephisos river: the *leophoros* mentioned by Pausanias (X 34.7 – X 35.5) and leading to Opus in Lokris would probably cross that area, running along the Akontion ridge (Kopanias 2008: fig.5 – see here fig.6).[35]

[31] Though almost dead by Roman imperial times, and deserted in the Early Byzantine period (see Koder - Hild 1976).

[32] For Ag.Dimitrios see also chapter II.3.1.

[33] A prehistoric burial place is known in the area (*component O_47*).

[34] See chapter II.3.1 – LONG TERM SETTLEMENT TRENDS.

[35] The road was still in use in the 19th century, leading to the village of Bogdanou (ancient Hyampolis), as it is marked on *La Carte de la Grèce* (1852).

II.3.5

The Northern mountains of the Copais: Hyettia

TOPOGRAPHICAL SETTING

The *chora* is constituted by the mountains and uplands to the N of the Copais basin, on the Eastern shoulder of Mt.Chlomon (1080m asl). It is a real geographical landscape unit, comprised of a small series of linked plains, which deserves to be considered by itself in order to understand upland settlement chambers. Hyettos was probably the largest centre in the area in historical times, with at least one other quite large settlement coexisting (Olmones?[1]).

The area is marked by the presence of a series of upland

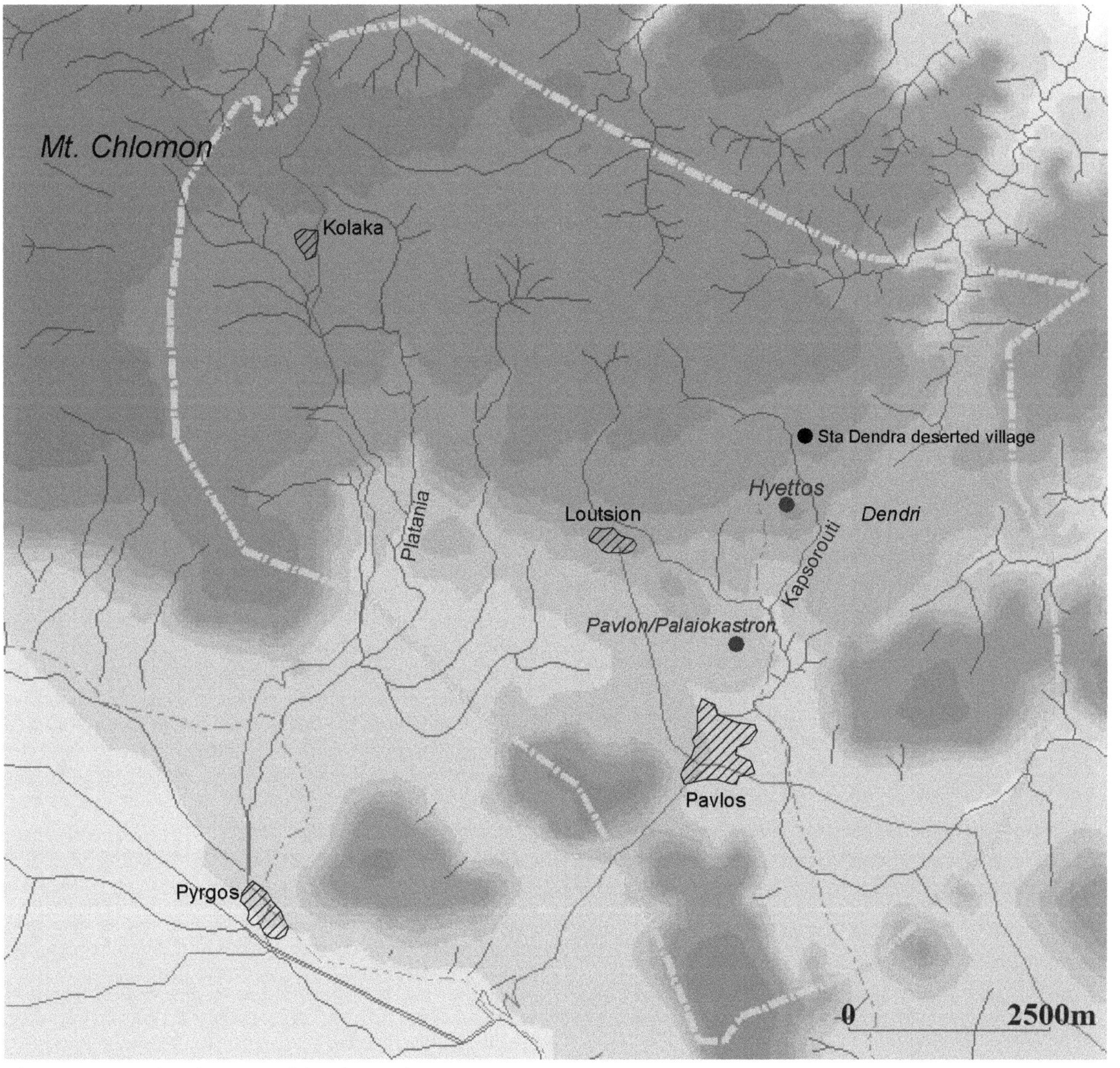

Fig.1. *Topographical setting of the chora of Hyettos.*

[1] See appendix I.5 for identification problems.

plains crossed by rivers and streams. The Kapsorouti stream flows to the SE of the site of Hyettos (and is further augmented by the spring there – see below) and runs to the S, flowing to the E of the site of Pavlon-Palaiokastron. The river crosses almost the entire Hyettos upland plateau.

Lauffer (Kopais I: 176, sketch 184) draws the stream as coming to the site of Hyettos from the E, while it is not visible as such in the 1:50,000 map, where it seems to emerge at the foot of the acropolis of Hyettos, by the water source, being a small stream beforehand. We should also consider the marshy character of the Dendri area, the flattish area to the E of Hyettos, marked as having a marshy character on the 1:50,000 GYS map,[2] though probably not swampy in antiquity.[3]

Boundaries

As reported in Fossey 1988: 293, "*Boeotia, in ancient as in modern times, included only the Southern part of the mountainous area which forms the North boundary of the Copais basin. These mountains are an extended Eastern shoulder of mt. Khlomon (1080m) and the Boeotian area lies to the South of the main West-East watershed*". The Boeotian area which constituted the *chora* as presented here is marked by the presence of a small series of linked plains. The nature of the boundaries is discussed by Étienne - Knoepfler (1976: 189-197).

PHYSICAL LAND UNITS and RESOURCES

Though the area looks mountainous in terms of landscape characters (as well as for landscape life – see below), the elevation in fact does not reach values higher than 600m, and therefore, according to our classification of elevation ranges, the mountain segment is not represented (see fig.2 in chapter II.1). The elevation range between 200m and 600m (namely the hilly landscape) is characterised by steep morphologies (H3, H4 and H5 are quite well represented) and stream valleys, which open up into a series of smaller and larger plateaus (class H1, clearly marking a large part of the landscape) with a remarkable difference in elevation compared to the Copais basin not far below, a fact that gives the landscape a mountainous character. The 9% represented by areas below 200m is constituted merely by the plain approaching Pavlos to its SE and the quite steep surrounding foothills (class P4).

Hilly landscape	91%
Mountainous landscape	0%
Plain	9%

1	P1_P2	lacustrine basin, valley	4%
2	P3	gentle slope	1.4%
3	P4	foothill	3.7%
4	H1	plateau	32.8%
5	H2	gentle slope	10.4%
6	H3	moderate slope	21%
7	H4	severe slope	19.3%
8	H5	very severe slope	7.2%
9	M1	plateau	0%
10	M2	plateau/gentle slope	0%
11	M3	moderate slope	0%
12	M4	very severe slope	0%

Table 1. *Percentage of the different physiographical classes present in the Hyettos area (P=plain; H=hill; M=mountain)*

Approaching the area from the SE (from Kastro) one gets the idea of the wide fertile plateaus available in this upland area. The first is the area called *Vargia* on the 1:50,000 GYS map, crossed longitudinally by the road approaching the modern village of Pavlos from the S. Having left Pavlos, and heading to the modern village of Loutsi, another wide upland plateau opens up, crossed by the modern road and marked towards the E by the outstanding rise that was the acropolis of a small ancient *polis* (identified as Olmones, see below). To the NE of 'Olmones', at a distance of about 1.7km, is another rise (identified as the acropolis of ancient Hyettos, see below), closer to the higher hills bordering this system of plateaus. Between the two rises, is another wide flat area (H1), with a gently sloping section (H2), which seems in most part unsuitable for agriculture, while the soil in the Northern part seems less workable (personal visit to the area, and intensive survey results report soil with a lot of pebbles - currently used for olive trees - and no rural sites discovered), while to the N of the Hyettos city site, on a foothill small platform and on the slopes,[4] a large number of rural sites were discovered (see fig.2).

To the SE of the acropolis of Hyettos is a wide area where minerals are mined. Black-green soil is still visible today, with a lot of open mines (on the city site itself, along the E slopes of the acropolis) for iron and magnesium (on the nature of these iron sources cf. Petraschek 1954, Bakhuizen 1979).

As noted above, the area is only partially available for agriculture (on part of the plateaus[5] - see fig.7), and is mostly suitable for pasturage. Exploitation of this particular landscape must have been similar through the centuries,[6] although the Dendri plain (see above) was

[2] The Dendri basin area takes its name from a deserted village called Sta Dendra to the N of Hyettos – marked as Metochi on the 1:50,000 GYS map.

[3] The plain to the E of the city was probably not swampy in antiquity to judge from heavy offsite and sites on and by it of Greco-Roman date, discovered by the Intensive Survey work carried out by the Boeotia project (see below in text).

[4] Mainly comprised of very fertile Neogen and flysch formations.

[5] Étienne - Knoepfler 1976: 198 note that, at the time of their study, cereals occupied more than half of the cultivable land, and probably, they say, the same was true in Greco-Roman times. Corn and tobacco are the crops in the plain of Pavlon. Wine production had a limited presence, but probably more than in modern times (Étienne - Knoepfler 1976: 198), and the same would be true for olives.

[6] We have evidence of pirates stealing θρέμματα (Étienne - Knoepfler 1976: 198-9; Fossey 1988: 298). Étienne - Knoepfler 1976: 198 note that people from the area were not probably only shepherds, but husbandry must have played a predominant role

probably not swampy in antiquity and would thus have allowed for good land which could offer rich cereal production. The hills to the N and W were probably cultivated at wider crops with olives and wine.

As Fossey notes (Papers 1990e: 208), in antiquity the population of the plain was perhaps not supported only by agriculture and herding, as today, but also by the exploitation of the local iron sources, essentially the only ones in Boeotia which may have been mined then. Traces of ancient iron working at Pavlon/Palaiokastron were noted (Bakhuizen 1979), and mentioned by ancient sources (for instance Plinius *HN* 128, who mentions magnesium occurring at Hyettos).[7] Certainly mines were operational during the first half of the 20th century.[8] As pointed out by Étienne & Knoepfler (1976: 200), the Hyettos area is known as one of the regions richest in iron resources in Central Greece.

THE ARCHAEOLOGICAL RECORD

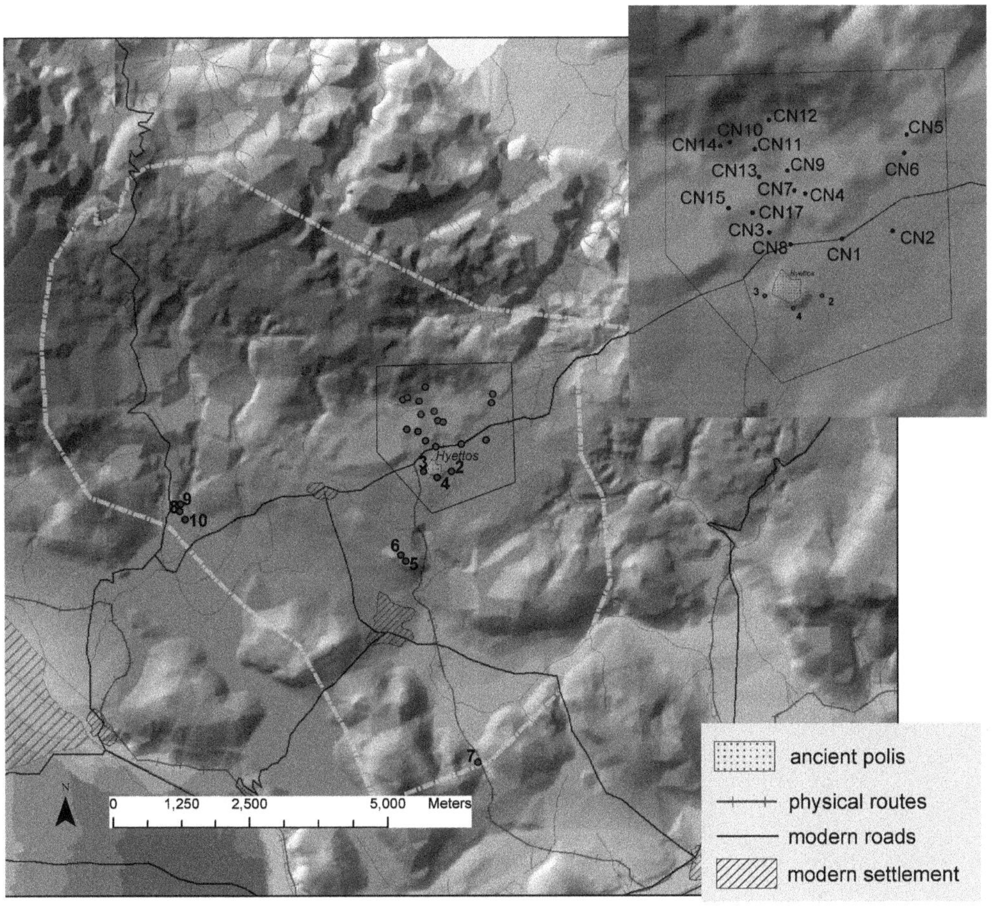

Fig.2. *Archaeological map of the Hyettos chora. The box indicates the area intensively and systematically surveyed.*

in the economy. Sheep-folds were probably quite far from and invisible from the city site, and therefore susceptible to pirate incursions (Étienne - Knoepfler 1976: 199).

[7] Étienne - Knoepfler 1976: 200-1; Fossey 1988: 299. No slag has been found in the area, but if magnetite was exploited, the ore was probably not smelted (Davies 1935: 246).

[8] A small village existed in the first half of the 20th century by one of the mining areas (Mine Tsouka, at the E edge of the territory of Hyettos). It had 68 inhabitants in 1928, but none by 1971 (Étienne - Knoepfler 1976: 200).

1	HYETTOS	*Components HO_1 to HO_6*
2	Hyettos East	*Component HO_7*
3	Hyettos West	*Component HO_8*
4	Hyettos South	*Component HO_9*
5	Pavlon-Palaiokastron	*Components HO_10 to HO_13* [Olmones?]
6	Pavlon-Palaiokastron	*Component HO_14*
7	Megalovouna-Kiapha	*Component HO_17*
8/9/10	Kolaka-Agios Ioannis	*Components HO_15* (8) *and HO_16* (9) *and HO_18* (10)

Table 2. *List of archaeological components and activity loci mapped in fig.2.*

Knowledge of the archaeological record concerning the area relates mainly to research on the city site of Hyettos and few sites discovered accidentally, mainly through illegal excavations. Additional information on the urban and rural landscape has been added by the intensive systematic artefact surface survey carried out in the area. The known data are therefore quite clustered (see fig.2) around the major settlement sites (namely Hyettos and Pavlon-Pavlokastron), the Mavrovounion mountain pass, and the area S of Kolaka (*components HO_15, HO_16 and HO_18*), which we could consider as belonging topographically to the area of interest.
The graph (fig.3) illustrates the proportion of components discovered within different research frameworks. In comparison to other *chorai*, rescue excavations are not significantly present in the Hyettos *chora*. This is due to the fact that no modern town is in the area, nor have infrastructure works been carried out in the area.

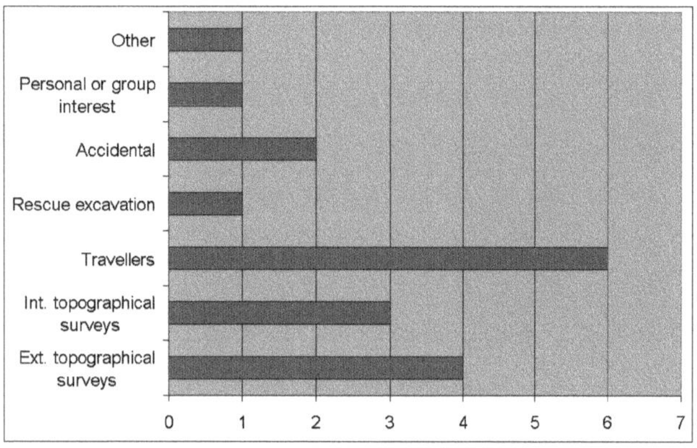

Fig.3. *Graph illustrating the proportion of components discovered within different research frameworks.*

A Systematic Intensive Surface Survey carried out in the area around Hyettos (area surveyed mapped in fig.2) from 1989-1991 by J.L. Bintliff and his team (mainly from the University of Durham, UK) discovered sites listed in appendix I.5 (table 1 – *Survey Sites*) and mapped in fig.2 (Bintliff 1992d).

If we include in the statistics the results of the intensive and systematic surface survey that concerned a small area of the *chora*, then the picture changes, as one may expect (fig.4).

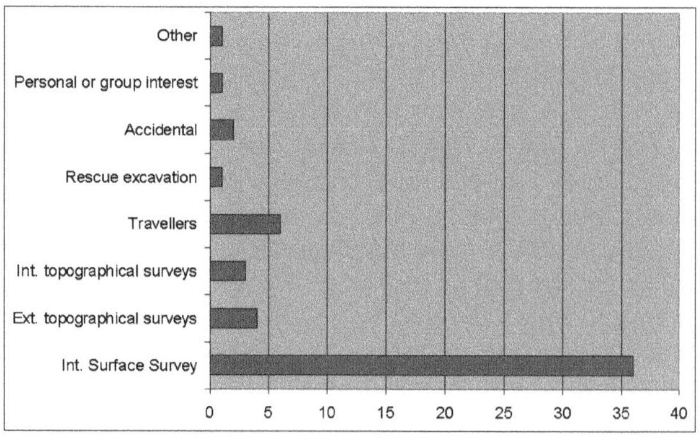

Fig.4. *Graph illustrating the proportion of components discovered within different research frameworks, including Intensive and Systematic Artefact Surface Survey.*

In fig.5 we can see the relationship between known archaeological sites and the distance from the modern road network. No strict correlation is visible because the discoveries of sites came about from different factors (see above).

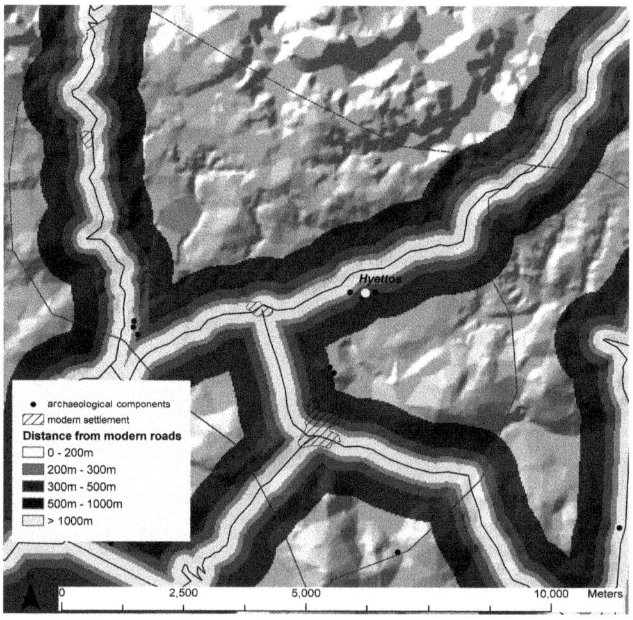

Fig.5. *Relationship between components and modern road network.*

The ratio of known Prehistoric to Greco-Roman components is 3 to 15 (1:5), while among the historical periods, 43% are dated Archaic to Hellenistic, 28.5% Roman-Late Roman, and 28.5% are attributed to the general Greco-Roman period. Adding the results of the intensive artefact surface survey, the ratio of Prehistoric to Greco-Roman components increases only slightly to 7 (4+3) to 48 (33+15) (1:6.8), while among the historical periods, 64% are dated Archaic to Hellenistic (since the number of known rural sites rises), 27% Roman-Late Roman, and only 10% are attributed to the general Greco-Roman period, the chronological attribution being more precise.

ANALYSIS OF THE *CHORA* LANDSCAPE

PREHISTORIC PERIOD

An EH settlement has been recognised through intensive artefact surface survey, as well as cist graves (SurveySite CN3), probably Prehistoric (MH?), to the immediate N of the Hyettos city site (see fig.2), whilst robbing activities in the area brought to light the existence of a large LH burial area to the W of Loutsi Village, S of the hamlet of Kolaka, probably indicating a large Prehistoric centre in the area.

The picture of prehistoric landscape available for the *chora* is, generally speaking, quite poor, considering the probable exploitation of this kind of landscape, suitable for husbandry, especially in the EH period, in conjunction with the secondary product revolution (see chapter II.3.1 and figs. 6 to 9).

GRECO-ROMAN ANTIQUITY[9]

Town level

The town level of the *chora* is represented by the Hyettos site. The only ancient source which provides information about Hyettos is Pausanias (IX 24. 3-4, 36.6), who mentions Hyettos as a κώμη[10] (jointly with Olmones) which never reached the town level and was dependent on the *polis* of Orchomenos. On the other hand, as pointed out by Étienne - Knoepfler 1976, epigraphical evidence shows that Hyettos was independent in the Archaic period and until the Late Roman period, with probably a short period (447-387) during which it was annexed by its strong neighbour.[11] Moreover, intensive and systematic urban survey (Bintliff 1992d) determined the maximum extension of the Classical–Early Hellenistic city as ca. 20ha, showing a quite large town in the Boeotian panorama, which needed and had good land (mid fertile) in its surroundings. Since it had control of this upland area, historical and political factors take it also to the *polis* level. Hyettos provides evidence in its inscriptions for Hellenistic population decline (the later Hellenistic ephebic lists from the acropolis wall) and financial and economic crises (Roesch 1965b: 256-61; Étienne-Knoepfler 1976: 201-210), observed also in the intensive survey results (Bintliff 1999d: 27-29). The reductive situation continues in the Mid Roman/Late Roman period and is also referred to by Pausanias (see below – LONG TERM SETTLEMENT TRENDS).

Village level

Archaeologically, the site of Pavlon/Palaiokastron (identifiable with Olmones - see *components HO_10 to HO_13*) shows characters of a dependent settlement. As long as we cannot prove its identification with ancient Olmones,[12] we should consider it as a smaller second rank dependent settlement. Apart from the possible settlement at Pavlon-Palaiokastron no other nucleated settlement site is known from the area.

Rural segment

The rural segment does not seem to be represented in the archaeological record available in the bibliography – including mainly rescue excavation and extensive surveys (with the exception of the evidence of unknown date S of Kolaka village – *component HO_15*). On the other hand, from the picture resulting from the intensive and systematic artefact surface survey carried out in the area, one can obtain much more information concerning rural sites, from isolated farms and sheds in the Classical-Hellenistic period to villa sites in the Roman and Late Roman periods (see below – LONG TERM SETTLEMENT TRENDS). In the Classical-Early Hellenistic period the finds in the area of ancient Hyettos conform to the 'picture of relatively low rural density and a concentration in small to medium farms' noticed also at Tanagra and Haliartos but also in other Greek surveys, in contrast to the picture of high rural density, including large Classical rural sites found in the survey sector S of Thespiae city (Bintliff-Howard-Snodgrass 2007: 146).

Since husbandry played an important role in the economy of Hyettos, one must hypothesise the presence of a number of sheep-folds (αυλαί) in the area, especially in the elevated areas surrounding the plains. They were probably quite far from and invisible from the city site, as Étienne - Knoepfler (1976: 199) point out, and therefore susceptible to pirate incursions (as testified by Hyettos' decrees mentioning pirates stealing θρέμματα).

Burial areas

Components HO_7, *HO_8* (and *HO_9*) are burial areas related to the city site and immediately at its outskirts below the acropolis hill.

There is apparently no evidence for rural burials, not even from intensive artefact surface surveys.

Cult places/Religious areas

No evidence is known for the area.

Forts and fortifications

In addition to the fortified acropolis of Hyettos, and the fortified site identified by some scholars with ancient Olmones (see above), the only known fort is *component HO_17*, along the upland route corresponding to a mountain pass.

LONG TERM SETTLEMENT TRENDS IN THE CHORA *LANDSCAPE*

As seen above, the two prominent sites of Hyettos and the supposed Olmones lie along the central axis of an upland plain with fairly good land. Considering the territory of ancient Hyettos, the most fertile areas extend

[9] Period maps are included in chapter II.4, figs.17-19-21-23-25-27.

[10] Pausanias describes a second rank settlement as Askra, for instance, in the same way (κώμη), while he describes the second rank settlement of Alalkomenai as πολύχνη.

[11] Otherwise Orchomenos and Hyettos used to constitute two separate 'districts'.

[12] See appendix I.5 for the question of identification.

W and NW of the city, while the area to the E and S, towards the Pavlon-Palaiokastron (Olmones?) site is full of stones and not suitable for agriculture (fig.7). Thus, the territory available to the Pavlon-Palaiokastron site mainly extended to its W and S. If the identification with Olmones is accepted, an inscription from Hyettos (IG VII 1808), stating that the Hyettos settlement chamber was bordered by the territory of Olmones to the W and S, would fit this picture.[13]

If the site at Pavlon-Palaiokastron is to be identified with ancient Olmones, then one may ask why the two small *poleis* (*komai*) according to Pausanias, or the main *polis* of Hyettos and the dependent settlement at Pavlon/Palaiokastron according to archaeological and epigraphical information, were both long-lived in this relatively small area, from early history to Late Roman times (this is confirmed for Hyettos, still unproved for the supposed Olmones, though still present in Pausanias' time at least). The results of the cost-distance analysis[14] show that there is not enough space in the area for two fully extending ½-hour-walking-distance-radius potential settlement chambers. The existence of both the two ancient settlements could be related to the peculiar economic exploitation of the area, based, as seen earlier, on farming, herding, and also mining, as well to its upland location, both of which also allow in later periods for the existence of two or more settlements.

In fact, there are three Ottoman villages in the wider *chora* of ancient Hyettos, all appearing in the 1466 archives (Andrea Loutsi, Pavlo Muzak and Gjin Vendre) and concentrated in this open upland section of the landscape. The first two continue till today (Lutsi and Pavlo) and are associated with alluvial soils in their vicinity (fertile soils on fig.7), in the area of the possible territory of the second settlement of the *chora* of Hyettos (Pavlon/Palaiokastron), which is situated just N of modern Pavlo. Conversely, in the territory strictly exploited by the ancient city of Hyettos there is no Ottoman or modern village. Although there is probably a lack of continuity of the Greek population of the Greco-Roman city, the deserted village of Sta Dendra, identified by the Boeotia survey half a kilometre N of the city, indicates a probable continuity of the Greco-Roman city through the earlier Middle Ages. It is also mentioned in the Ottoman archives of 1466 (Albanian hamlet Gjin Vendre) which continues in Early Turkish times (Bintliff-Kiel in preparation). The village is associated with available mid fertile soil; see above for the town of Hyettos.[15]

Further to the NW is the modern small upland hamlet of Kolaka. In ancient times, the ancient *polis* of *Kyrtone* (not yet identified) should probably also have been in this area. Kyrtone is an ancient *kome* mentioned by Pausanias (as *polichni*) as on a high mountain, close to a water source from the rock. Forchhammer (1857: 17) places Kyrtone where we know that Hyettos is, while Oldfather (1916: 163ff, followed by Papachatzis 1981: 166) identifies as Kyrtone the area of the modern village of Kolaka. The village belongs today to the province (*nomos*) of Lokris, as well as having a much lower location (Ag. Ioannis), to the S of Kolaka and to the W of Loutsion village and not very far from it (ca 2.5km), where some remains of ancient[16] structures are visible above the Platania torrent (*component HO_15*). The area is on the Chlomon mountain, close to a water source and with patches of mid fertile land available (see figs. 1 and 7 and Étienne - Knoepfler 1976: 211), and might indicate a possible settlement area, working as such at least in some periods of history and exploiting upland midland zones (see fig.6). The LH burials (*components HO_16 and HO_18*) known from the area of Kolaka indicate also a possible Mycenaean centre in the area.

Therefore, this upland plain and its surrounding mountainous areas may support many larger and smaller settlements, at times very close to each other. This may be linked to the life system in a mountainous landscape, which usually produces a dispersed settlement pattern, with often no need of a very large central hub.

As for the rural landscape in ancient times, intensive surveys in the area surrounding Hyettos (Bintliff 1992d) testify to the presence of rural sites, concentrated to the N of the city, in the basin and hills towards the mountain slopes, and associated with the mid-fertile zone (fig.7). The rural occupation recognised through intensive survey seems to be mainly Classical and Hellenistic, with small rural sites and farmsteads.[17] Classical-Hellenistic sites seem to be much closer to the city site than in other cases in Boeotia (see Thespiae and Tanagra for instance, in chapters II.3.9 and II.3.14). The finds in this area of ancient Hyettos conform to the 'picture of relatively low rural density and a concentration in small to medium farms' noticed also at Tanagra and Haliartos but also in other Greek surveys, in contrast to the picture of high rural density, including large Classical rural sites found in the survey sector S of the city of Thespiae (Bintliff-Howard-Snodgrass 2007: 146).

A few considerable sites seem to be attested in the Roman/Late Roman period (fig.2 - CN 5,6,7,8,14; fig.27 in chapter II.4), along with the city itself progressively shrinking to a small village level on the acropolis (see appendix I.5). A substantial Late Roman villa complex (CN7) associated with a highly fertile secluded basin of deep soils was located 1km N of the ancient city (fig.2 and fig.7). An olive press found here, constructed in secondary use from a monumental base brought from the declined city, indicates both the intensified land use and

[13] από \ δε δύσεως οι Αρέσκοντος Ολμωνί \ ου κληρονόμοι, από δε μεσημβρίας Σύμφορος Ολμώνιος (IG VII 1808, line 2), mentioned by Lauffer (Kopais I: 178).

[14] See chapter II.3.1 –LONG TERM SETTLEMENT TRENDS.

[15] The village can be associated with the fertile area immediately to the N of Hyettos and exploited in antiquity by the ancient city (fig.7).

[16] Unfortunately unspecified whether Greco-Roman or Medieval (see *component HO_15*).

[17] Inferences from the contemporary ephebic lists for the Boeotian Confederacy (Roesch 1965b: 256-61; Étienne-Knoepfler 1976: 201-210) suggest a total population of some 3,500 people in town and country for the state of Hyettos, of which around 70% lived in the city and 30% in rural hamlets and farms (Bintliff 1992d).

economic specialisation taking place in the Late Roman period, as well as the abandonment of parts of the city to the advantage of Roman villa-owners (see fig.27 in chapter II.4).

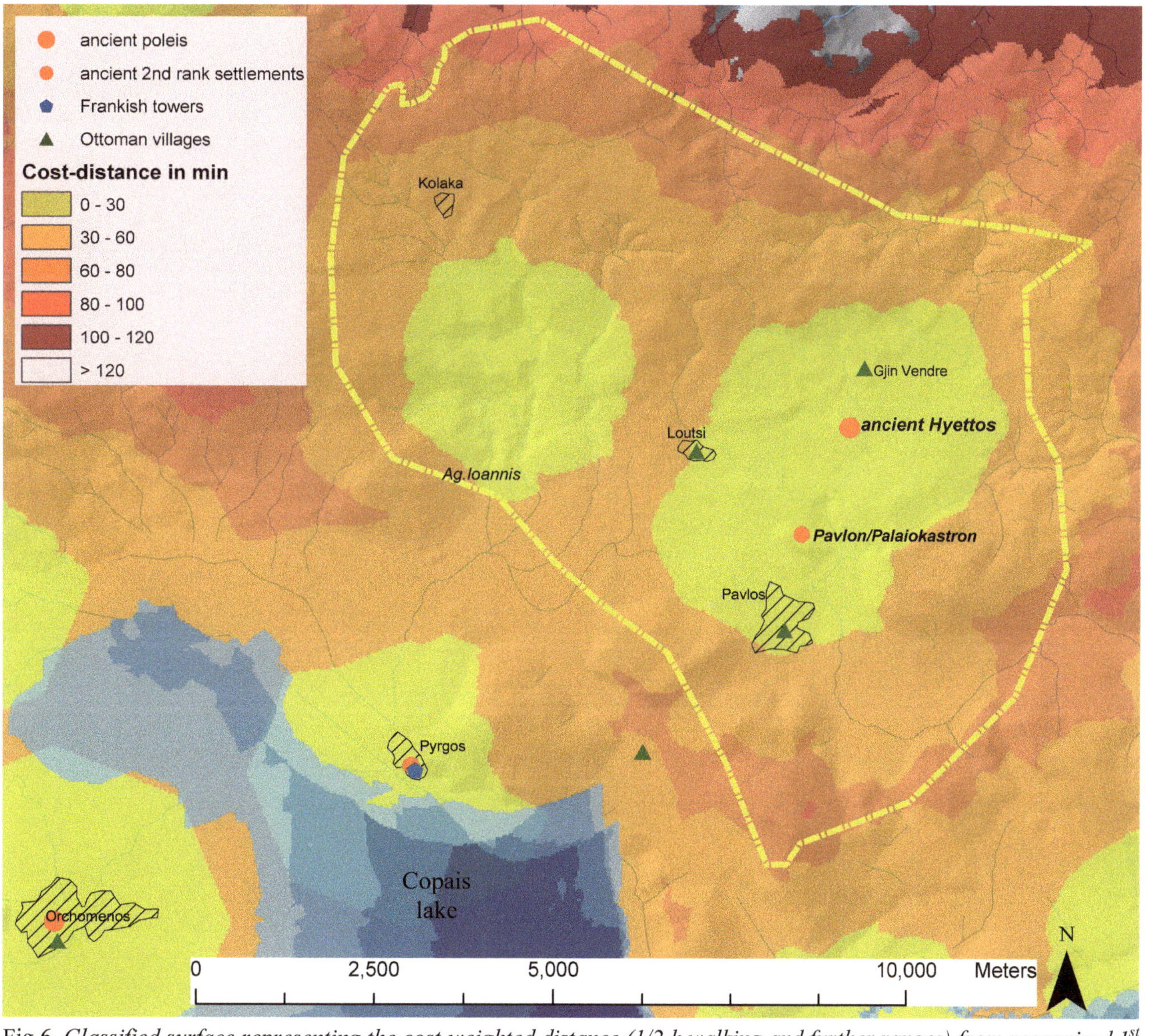

Fig.6. *Classified surface representing the cost-weighted distance (1/2 h walking and further ranges) from recognised 1st and 2nd rank ancient settlements (represented by larger and smaller dots). Areas without dots indicate potential settlement chambers. Ottoman villages and Frankish towers have also been added to the map to show their spatial relationship with the Greco-Roman settlement network and to appreciate potential settlement chambers.*

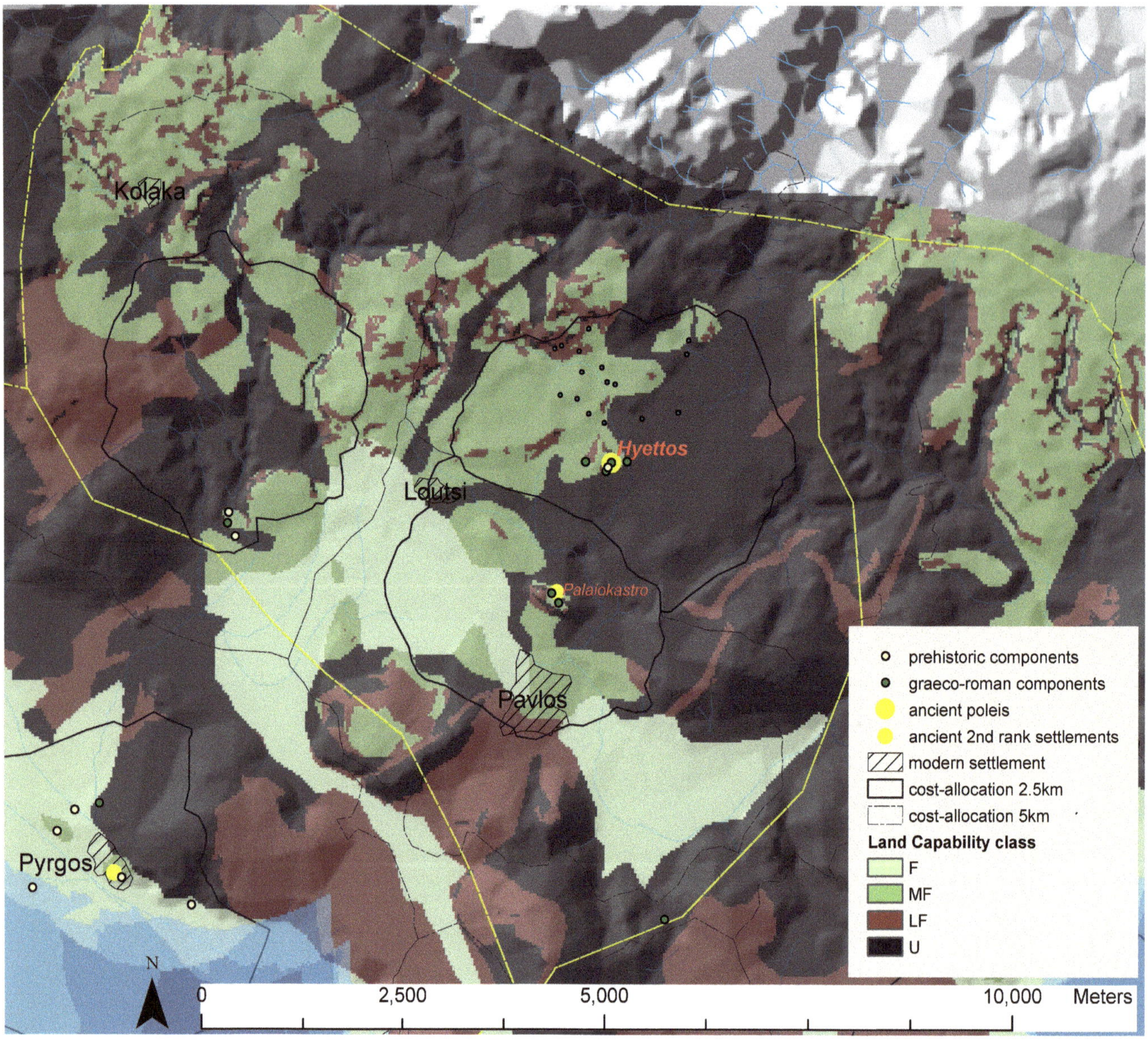

Fig.7. *Map showing the Greco-Roman settlement network, the polygons resulting from the cost-distance analysis (marking half an hour and one hour walking time distance) and dots representing the known archaeological components (same as in fig.2), with land capability information underlain.*

II.3.6

The Copais area: Copai and the North-East bay

TOPOGRAPHICAL SETTING

The North East Copais area is constituted mainly by the NE bay of the Copais basin and the surrounding elevated areas, namely the central and Eastern part of the Copais' North fringe (Fossey 1988: 277), as well as the limestone Megalovouno ridge, towards the E.

The landscape of the bay today is marked by the presence of the partially canalised and meandering Melas river, visible from afar also within the plain because of the high and thick reeds than grow all along its course. The Melas runs along the N edge of the basin and afterwards, after Kastro, and especially after the promontory of Pyrgos-Agia Marina, crosses the plain transversally. It used to flow into the *Megali Katavothra*[1] at its far E end until 1893, when the river was diverted elsewhere with the drainage of the lake.

The main settlement of the area today, as in the past, is Topolia/Kastro. It is situated on a hill/islet, isolated from the edge of the basin, which, when the water is low, is connected to the land by an isthmus. In the modern period it was, together with Pyrgos, the large settlement by the lake, and gave the lake its name (*Topolia limni*), as it did in ancient times (*Copais* after Copai, the ancient town on the site of the modern settlement). Topolia/Kastro is situated at the entrance to the North-Easternmost bay of Copais, which, immediately after Topolia/Kastro, opens up towards the N into a flat inner area, through which runs the road towards Martinon, whose E side is constituted by a flat area with vineyards (called Ampelia),[2] which opens towards Martinon and is overlooked from the E by the Proph.Ilias mountain. To the E of this opening, the *Topolia* bay forms a tongue-shaped feature whose edge is marked by promontories

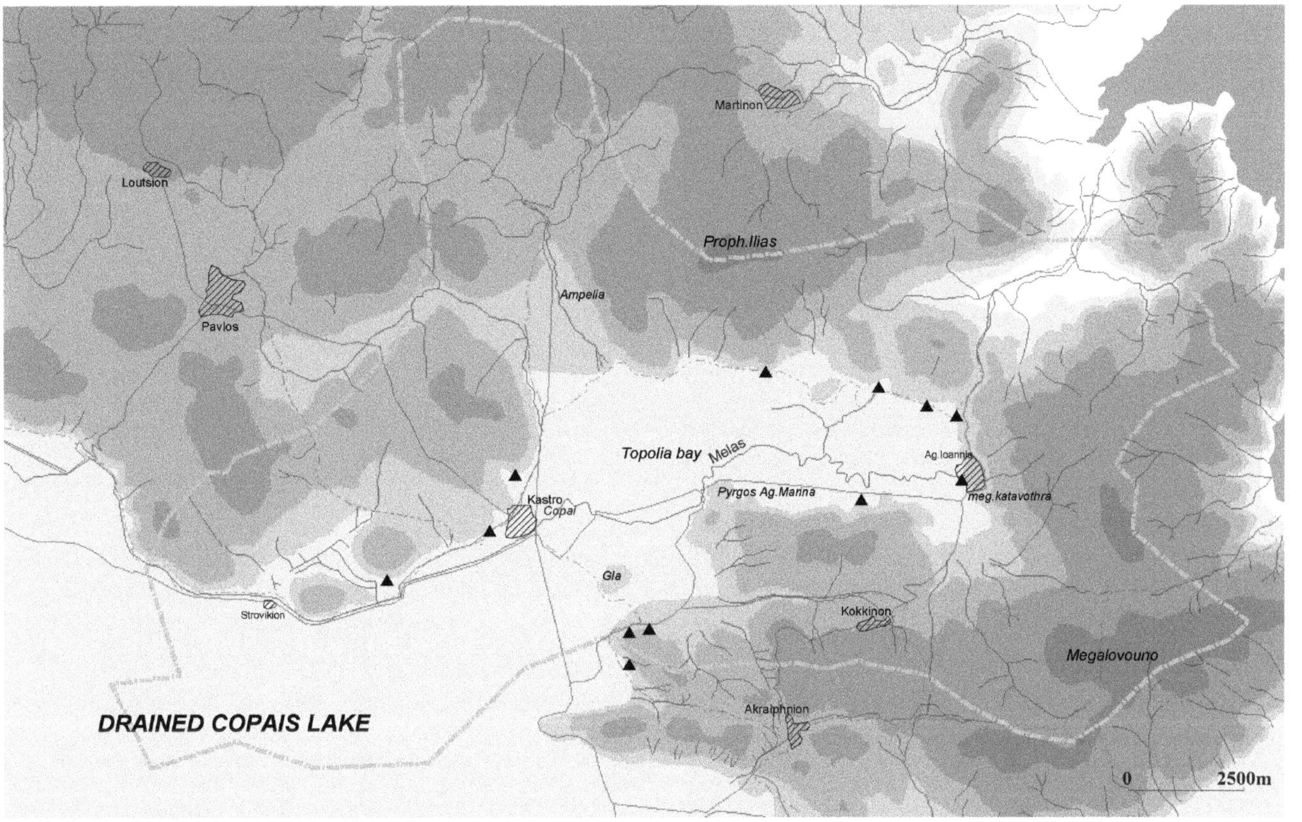

Fig.1. *Topographical setting of the chora. Triangles mark the katavothrai.*

[1] A *katavothra* is a karstic swallow-hole (see chapter III.1).

[2] Philippson 1951: 486. The area is crossed today by the Athens-Lamia highway.

and small gulfs, where many *katavothrai*[3] (marked in fig.1) open up.
The land of the Topolia bay is partially cultivated, though the majority of the area is covered with meadows and reeds, especially on the E edge. On the other hand, in the area of Kokkino, to the NW slopes of the Megalovouno ridge, as well as in the valley running E-W down to the Copais, vine cultivation marks the landscape (as at the time of Lauffer's visit, around the mid 20th century).

Boundaries
The boundaries are defined by the shape of the NE bay itself, and the mountain fringe around it. The border between Copai and Akraiphnion is still uncertain and, as Fossey points out (1988: 277), the inclusion of the Eastern end of the NE Copais bay in the territory of ancient Copai is considerably less than certain, though, in considering the boundary stone (*horos* - IG VII 2792), found at Phtelio point, it seems reasonable to see the land N of the Megalovouno ridge, at whose end it lies, as that of Copai, and to the S as belonging to Akraiphia.[4] Another inscription was interpreted as boundary stone, once again between Copai and Akraiphia, datable to the 6th – 5th C BC. The text suggested by Lauffer reads: [hópoς A]κραι[φιέον] [κ]αί Κοπ[αίον.[5]

PHYSICAL LAND UNITS

As one might expect from looking at the landscape, the greater part of the area is comprised of a flat landscape, partially covered by the Copais (lake or marshes) – table 1. On the other hand, the mountainous segment (>600m) is almost completely absent, since heights are not characterised by high elevation (see the Hyettos *chora*, belonging to the same North Copais upland landscape, and quite different from the South Copais upland landscape), while the hilly segment is marked by a significant presence of rough morphologies (high slope values, for instance) - see fig.2 in chapter II.1.

RESOURCES

With the exception of some lower areas in the NE bay itself, which with drainage could be exploited for agriculture (still probably always with a permanent marshy character), some areas are suitable for vineyards (the bay to the N of Topolia/Kastro and upland areas of Kokkino). Mostly, though, the carbonate rocks of the elevated areas around the bay are suitable for husbandry activities, while the slopes of Megalovouno were probably covered by forest.[6] Land capability values for the area are visible in fig.6. Fishing could also take place,[7] and a resource, part of the parallel economy offered by the lake (Farinetti 2008), was also the eels of Copais.[8] Iron and nickel deposits are known around the E and S edges of the NE Copais bay. They are mined today (crossing the area of Ag.Ioannis, it appears as a 'red landscape', though no evidence for exploration in antiquity seems to be attested[9]).

Hilly landscape	37%
Mountainous landscape	2%
Plain	61%

1	P1_P2	lacustrine basin, valley	39%
2	P3	gentle slope	3.8%
3	P4	foothill	17.5%
4	H1	plateau	7%
5	H2	gentle slope	1%
6	H3	moderate slope	7.3%
7	H4	severe slope	13.4%
8	H5	very severe slope	8.5%
9	M1	plateau	0.8%
10	M2	plateau/gentle slope	0.6%
11	M3	moderate slope	0.7%
12	M4	very severe slope	0.2%

Table 1. *Percentage of the different physiographical classes present in the NE Copais area (P=plain; H=hill; M=mountain).*

[3] Among others, Spitia Katavothra, Binia Katavothra, which is divided into two parts and might be artificial, and to the S of it the Megali Katavothra.
[4] Traces of a necropolis found at the NW and N foot of the ridge (*components NC_84 to NC_86* - see appendix I.6) could be burial areas connected with the town of Akraiphia, but could also be interpreted as rural burials.
[5] SEG XXX 440; Teiresias, 10 [1980] AE/2. Lauffer *Chiron* 10 (1980: 161-2) has no doubt that the inscription (dated to the 6th C BC) was found *in situ* in the plain, 500m NW of Phtelio cape and 500m SW of Gla.
[6] In the whole of the valley from Karditsa, the W slope of Megalovouno to the Copais basin, a process of deforestation has taken place. Nowadays the area is full of undergrowth and brambles (Lauffer Kopais I: 254).
[7] Topolia = τόπος ἀλεῖα, as reported in Fossey 1988: 288.
[8] The NE bay produced the best eels, see Frazer 1913: 132-3, quoted by Fossey 1988: 288. Schliemann (1882: 162) reports on the eels of the bay as a resource for the village of Topolia/Kastro, already renowned for their bulk and fatness in antiquity, and which Pausanias commends.
[9] Bakhuizen 1976: 51-7, reported by Fossey 1988: 288.

THE ARCHAEOLOGICAL RECORD

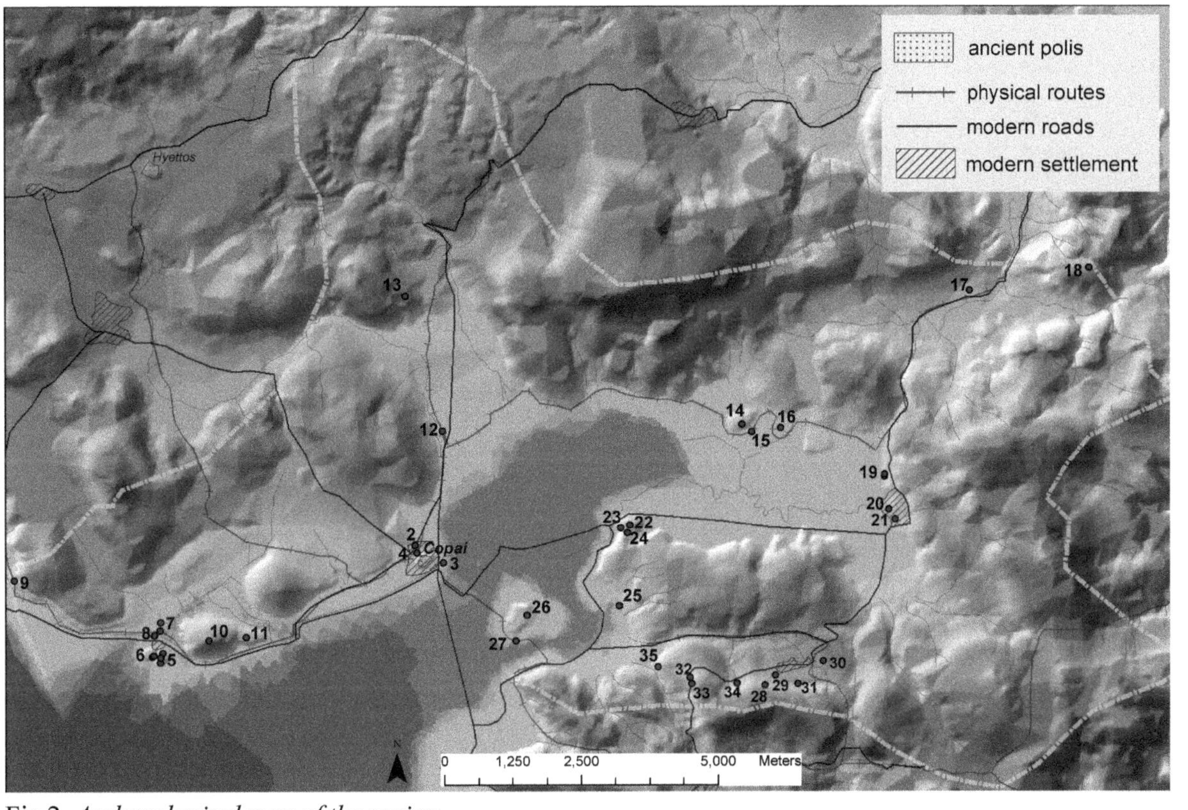

Fig.2. *Archaeological map of the region.*

1	**Topolia/Kastro**	*Components NC_15 to NC_24* (ancient *Copai*)
2/3/4	**Topolia/Kastro**	Topolia/Kastro NW (2): *Component NC_25;* Topolia/Kastro East (3): *Components NC_26 to NC_28* and Topolia/Kastro village (4): *Component NC_29*
5/6	**Stroviki-Nisi**	E hill (5) *Components NC_1 to NC_4 and NC_7 to NC_10, NC_5* (South), *NC_6* (North); W hill (6) *Components NC_11 and NC_87*
7	**Stroviki-Ag.Georgios**	*Components NC_88 and NC_89*
8	**Stroviki-Ag.Georgios**	*Component NC_90*
9	**Baroutospilia**	*Component NC_14*
10/11	**Tourlogianni**	Tourlogianni (10) *Component NC_12* and Tourlogianni-Bazaraki (11): *Component NC_13*
12	**Magoula Kavkala**	*Components NC_30 and NC_31*
13	**Alogopatisia-Kaphkala**	*Component NC_32*
14/15	**Chantza**	Chantza (14) and Chantza nearby (15): *Components NC_33 to NC_37*
16	**Agios Ioannis-Spitia Katavothra**	*Components NC_38 to NC_47*
17	**Souvliki**	*Component NC_80*
18	**Bazaraki**	*Component NC_81*
19	**Spilia Tsoutso**	*Components NC_48 to NC_50*
20	**Agios Ioannis-Megali Katavothra**	*Components NC_51 to NC_56*
21	**Megali Katavothra SE**	*Components NC_57 to NC_58*
22/23/24	**Pyrgos-Agia Marina**	Pyrgos-Agia Marina *Components NC_59 to NC_66* (22), *NC_67* (23) and Pyrgos-Agia Marina-Chuntiklissa: *Component NC_68* (24)
25	**Agia Marina SE**	*Component NC_69*
26	**Gla**	*Components NC_70 to NC_74 and NC_91*
27	**Gla-Talantouseza**	*Component NC_75*
28/29/30/31	**Kokkino**	Kokkino near (28): *Component NC_76 –* Kokkino W (29): *Component NC_77 -* Kokkino NE (30): *Component NC_78 -* Kokkino S (31): *Component NC_79*
32	**Megalovouno NE**	*Component NC_84*
33/34	**Megalovouno NE**	*Components NC_85 and NC_86*
35	**Megalovouno-Blaitesa**	*Components NC_82 and NC_83*

Table 2. *List of archaeological components and activity loci mapped in fig.2.*

Knowledge of the archaeological record concerning the area is largely comprised of the personal interest-oriented research on Prehistoric settlement in the area, and Prehistoric sites constitute the majority of known components in the area (58%). Lauffer, as well as Knauss' équipe, have intensively investigated the landscape around the gulf, also providing useful information on periods of occupation other than the main Prehistoric foci. In this way, many historical settlement or small activity foci are known because of their location on top of or close to Prehistoric foci (they are represented by the discovery type entry 'other'– see pie chart below).

The graph (fig.3) illustrates the proportion of components discovered within different research frameworks. For the North East Copais area, discovery frameworks are almost equally represented, with the exception of rescue excavation, with only slight representation, and accidental discovery (no robbery activities taking place; no presence of a large *polis* centre). The significant number of entries discovered through 'Int. topographical surveys' is due to the very good research work in the area carried out by Lauffer (and mainly published in Kopais I, 1986). A large number of the entries for 'Personal or group interest' are due, as in other *chorai* bordering Copais, to the interest, by groups and people compiling gazetteers, in Prehistoric settlement around the lake, as well as to the research carried out by Knauss' team (Kopais 2), which threw light also on the infrastructures (dams and polders) related to the late Prehistoric drainage of the lake. The entry/value 'Other' represents, in the NE Copais case, especially components of historical periods known mainly from research carried out on sites displaying Prehistoric occupation (see chapter II.2).

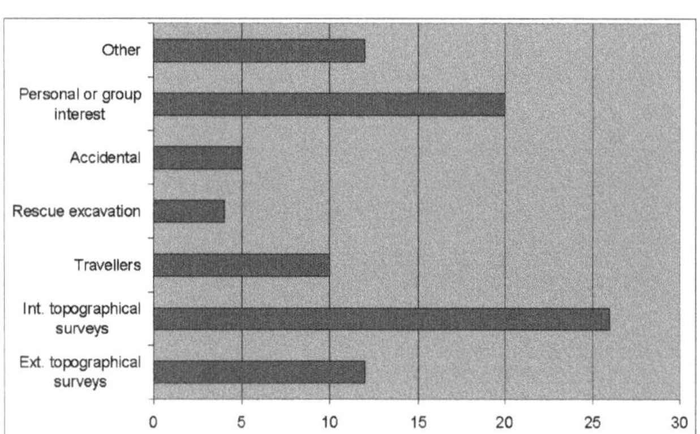

Fig.3. *Graph illustrating the proportion of components discovered within different research frameworks.*

In fig.4 the relationship between known archaeological sites and the distance from the modern road network can be seen. A strict correlation between the discovery of sites and the presence of a road (which makes discovery easier and archaeological remains more accessible) is clearly visible in the case of the components around Kokkino and in the Megalovouno area, while the apparent correlation in the case of Copai and of Stroviki-Tourloianni is mainly due to the position of the road running along the edge of the basin, where the ancient settlement was.

The ratio of known Prehistoric to Greco-Roman

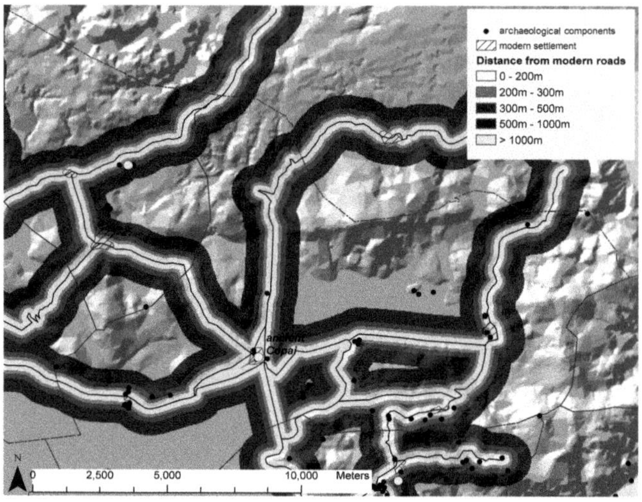

Fig.4. *Relationship between components and modern road network.*

components is 48 to 36 (4:3), with an unusual preponderance of known Prehistoric sites for the reasons discussed above, while among the historical periods, 64% are dated Archaic to Hellenistic, 18% Roman-Late Roman, and 18% are attributed to the general Greco-Roman period.

ANALYSIS OF THE *CHORA* LANDSCAPE

PREHISTORIC PERIOD

The picture of the Prehistoric landscape of the NE Copais bay is quite rich (see figs. 6 to 9 in chapter II.3.1).

With regards to the Neolithic, a lot of sites are known for the area, sometimes clearly datable to the Neolithic period (fig.6 in chapter II.3.1). This is due to the special interest in Prehistoric periods which led people working on the Copais and its surroundings, as well as to the probable strong Neolithic occupation in the areas bordering the lake, either in caves or in open settlement. Some of the known Neolithic activity foci correspond to sites occupied in later periods, and are therefore known due to the more or less intensive research carried out on them.

The Neolithic landscape of the area is also characterised by cave sites. The Baroutospilia cave has been entered into the database (*component NC_14*), but others have also been reported with Prehistoric finds (Bouka, Ag.Nikolaos, and other caves by the connection between Copais and the Yliki lake, in the area of Vourka, and on both sides of the Athens-Lamia National Road: Saracino, Baroutospilia itself, and others. For a brief summary see AD 1971: 219-220.[10]

[10] Spyropoulos reports on a visit to all the caves around Copais (AD 1971: 219-220).

The Early Helladic period is also quite well represented in the available picture of the Prehistoric landscape (fig.7 in chapter II.3.1). There are a few clear activity foci, probably settlement sites, such as Magoula Kavkala, north of Copai (where Neolithic was also found) – *component NC_31*, or the site at Agios Ioannis above the Megali Katavothra – *component NC_52*, or at Megali Katavothra SE - *component NC_58*, and probably at Topolia-Kastro and Stroviki-Nisi (*components NC_16 and NC_2*). They are all situated along the edge of the lake, in fairly elevated positions, with the exception of the site at Magoula Kavkala, which is on a low mound and occupies an area back from the lake, overlooking an inner bay. In the case of the area around the Megali Katavothra, foci of apparently EH date have been found in several places (Spilia Tsoutso, a spur above Megali Katavothra, and a depression to the SE).

For the Middle Helladic, a number of major sites of the period are known, as in other areas surrounding Copais (see, for instance, Koroneiake to the S) – fig.8 in chapter II.3.1. Clear MH occupation of a certain extent is known at Stroviki-Nisi - *component NC_3* with cist graves associated, on the peninsula of Agios Ioannis at Spitia Katavothra – *component NC_40* with cist graves associated, and on the spur above the Megali Katavothra – *component NC_53*. They can all be considered settlement sites. Pyrgos-Agia Marina (*component NC_61*) was probably also a settlement site in the MH period (though the principal occupation is later – LH), as well as Topolia-Kastro (while Gla probably was not). As for Chantza, see discussion under the site. As noted earlier, several burial places can be associated with settlement sites in the MH period. According to Lauffer (Kopais I: 36 and 233), in the Eastern part of Copais, we see the recurrent situation of MH necropoleis situated slightly away from the settlements, quite often facing it, such as at Agios Ioannis-Megali Katavothra and Spilia Tsoutso (see above – *component NC_50*), Topolia-Kastro, Pyrgos-Agia Marina (but see above - *component NC_67*), as well as, in another *chora*, at Medeon (Haliartia).

The NE bay of Copais is quite well known for the presence of Late Helladic / Mycenaean forts and fortified sites, examined in relation to each other by Fossey (1980),[11] and many scholars link them to the drainage of the Copais in the Late MH?/LH period.[12]

Generally speaking, light can be thrown on the Prehistoric landscape of the NE Copais area if considered in connection with the drainage works and dams throughout the area in the Late MH?/LH period. See, for instance, the case of Pyrgos-Agia Marina and possible relationships with Gla (see above, under the site), Stroviki with dams below the fortified acropolis, and the Agios Ioannis peninsula at Spitia Katavothra, also by a Prehistoric dam (fig.9 in chapter II.3.1).

LH forts have been found at Tourlogianni and probably at Bazaraki, while most of them were probably fortified sites (settled even if only for short periods or for specific defence and/or control purposes): Stroviki-Nisi, Agios Ioannis-Spitia Katavothra, Pyrgos-Agia Marina as well as the most widely known Gla. Also, a LH settlement site with no traces of fortification has been noticed at Agios Ioannis-Megali Katavothra (the absence of fortification, if real, would give the settlement a peculiar character within the *panorama* of the region). As for other activity foci known for the LH period, a circuit wall has been noted by travellers at the site of Topolia/Kastro, but no indication of date is available, nor of the kind of occupation of the site in the LH period. Contradictory, also, is the LH occupation at Chantza (see above, under the site).

The NE bay of Copais is the area where the edges of the basin are closer to the border of the lake (flooding area), and it may be seen also as more isolated (Lauffer Kopais I: 211). On the other hand, since the palatial powers of the LH period, the area was probably under the influence of a strong central power, probably Orchomenos, which might have also controlled Gla (Lauffer Kopais I: 211[13] and Iakovidis 2001).

The question is whether those fortified sites were actual settlement sites or only inhabited either for short periods or for specific purposes (control or defence). For Gla, Stroviki-Nisi, as well as Agios Ioannis-Spitia Katavothra, the hypothesis has been advanced that they were proper settlement sites. The same can be said for apparently unfortified sites (Agios Ioannis-Megali Katavothra). Traces of a possible Mycenaean-Minyan settlement were found below Gla in a polder area (Kopais 2: 216 – *component NC_91*). No burials have been found from this period though, according to the available archaeological record.

According to Lauffer (Kopais I: 233), the Bronze Age settlement in the area is characterised by settlement sites situated on the mountain spur into the Copais basin in positions quite well protected from the rear by high mountain ridges and peaks. For cultivation, they would also have used the immediate area of the basin itself. Lauffer proves this theory, noting the absence of any evidence for the period in locations less well protected from behind (such as at Nisi Ridge to the E of Ag. Ioannis/Spitia Katavothra, and the Magoula S of the Megali Katavothra (Lauffer Kopais I: sketch 217).

When considering settlement patterns in the area, one should always bear in mind that in the late MH? / LH period the bay was drained in its entirety, so the lower areas were also dry and available for agriculture, presenting a situation probably very similar to that of today.

[11] On the fortifications of the Mycenaean Copais, see Loader 1998.

[12] See Spyropoulos AAA 1973: 201-210 (for the date of the drainage); Lauffer Kopais I: 131ff ; Knauss et al. Kopais 1; Knauss 1995: 83-95. See also Iakovidis 2001.

[13] Lauffer (Kopais I: 208) notes that the LH forts in the NE bay are missing a settlement site controlling it, and he locates this settlement at the site of Stroviki-Nisi, which he identifies with the Homeric *Mideia*. This, however, probably does not correspond to the real situation.

GRECO-ROMAN ANTIQUITY[14]

Town level
The main centre of the area in Greco-Roman times, Copai, was never a large city site. Therefore, we might suppose a quite peculiar settlement situation in this remote NE area of Copais, also considering its peculiar landscape character, marked by the presence of the lake marshes. The development of Copai as the main centre, even if small, in historical times (it participated in the 1st Confederation with one beotarch - Larsen 1968), can be due to its position at the entrance of the bay, along the route at the edge of the Copais (coming from Skripou and running along the right side of the Melas river, fig.1), and from there upland to Lokris and the sea. A Frankish bridge over the Melas[15] at Topolia constitutes a clue to an ancient road from Atalanti which would cross the lake towards the S and would be in use only in the spring and summer (Pilippson 1951: 485).

Village level
By comparison with other micro-regions around the Copais (along the S edge, for instance), the landscape of this area is peculiar in terms of environmental character due to the behaviour of the water, and also with regards to topography, since it is composed of a series of small geographical units of differing character all around the Copais, both immediately at the edge of the bay and in more upland areas, rather than of a single valley. Considering the archaeological record available, there is little evidence that may represent small villages/hamlets of the historical period, although there was room for them in several micro-landscapes, as the structure of the Prehistoric landscape seems to show (see above – PREHISTORIC PERIOD and below – LONG TERM SETTLEMENT TRENDS).
Two settlement areas can perhaps be archaeologically recognised: the presence of a settlement area on the N and W hangs of Megalovouno can be traced through burials and other archaeological evidence (*components NC_84 to NC_86* and *component NC_78*, by the modern village of Kokkino), while traces of habitation noted at Stroviki-Nisi (*component NC_7*) may indicate the presence of a settlement chamber in that area.

Rural segment
The site above Megali Katavothra (*component NC_56* – activity focus) can be interpreted as a rural activity focus or a farmstead, linked to agricultural activities in the area on top of the spur jutting into the Copais. The evidence known from Agios Ioannis-Spitia Katavothra (*components NC_43 to NC_45*) could represent a small hamlet linked to a cult place, but most probably it was just a long-lived (Geometric to Classical) small group of houses. Rural occupation can also be traced at Alogopatisia-Kaphkala (*component NC_32),* linked to the farming of the flat area of Ampelia nearby.

Burial areas
A few burial areas are known for the period. Some are near and related to the *polis* of Copai (*components NC_25 to NC_28*), with the bulk of the evidence being represented by the extensive cemetery area discovered in the vicinity of Gla through rescue excavation (*components NC_26 to NC_28*). Others could be linked to hamlets/villages further from the *polis* site, though the attribution is not clear (*components NC_84 to NC_86*, for instance).

Cult places/Religious areas
There are several component entries characterised generally as 'activity focus' that have been interpreted as cult place by some scholars (e.g. *components NC_32, NC_7, NC_43 to NC_45*). The interpretation, however, is usually based on evidence too vague to allow a secure assignment of 'cult place' character.
In presenting the site at Alogopatisia-Kaphkala (*component NC_32* - along a long-distance road) as a possible location of a cult place, Lauffer (Kopais I: 248) discusses the pattern of cult places around the Copais, noting how, in the Copais area, they are either in a significant location in the mountains, close to a fresh water spring, on a terrace above a plain or plateau, or along roads at points of a certain importance (Onchestos, Petra, Alogopatisia).
Apart from these hypothesised locations suggested by Lauffer (supported by too little archaeological evidence), there are no cult places known for the NE Copais area.

Other activities / unspecified activity areas
Undetermined activity foci were noted which cannot be assigned any particular character at the present stage of research. Some of the entries with a generic 'activity focus' character might be interpreted as rural site/activities. Furthermore, as noted above, there are several component entries characterised generally as 'activity focus' that have been interpreted as a cult place.

LONG TERM SETTLEMENT TRENDS IN THE CHORA *LANDSCAPE*

The landscape of the NE bay can certainly be considered as a 'lake landscape' since, in this area, the water level was always higher and settlement is really marked by a perilacustrine character (see digital reconstruction of the lake fluctuations in fig.2, chapter II.3.1 and Farinetti 2008).
The landscape of the area is characterised by a series of small geographical units of differing character all around Copais, both at the immediate edge of the bay and in more upland areas. These were occupied by different settlement sites in the Late Prehistoric period, within which shifting of settlement and activities within these small areas (settlement chambers, as we would define them) can be noted. The same holds for historical times also, though differently. Small settlement chambers are recognisable as such mainly in Prehistory:

[14] Period maps are included in chapter II.4, figs.17-19-21-23-25-27.
[15] Ruins of an ancient bridge – Miller 1921: 132-3

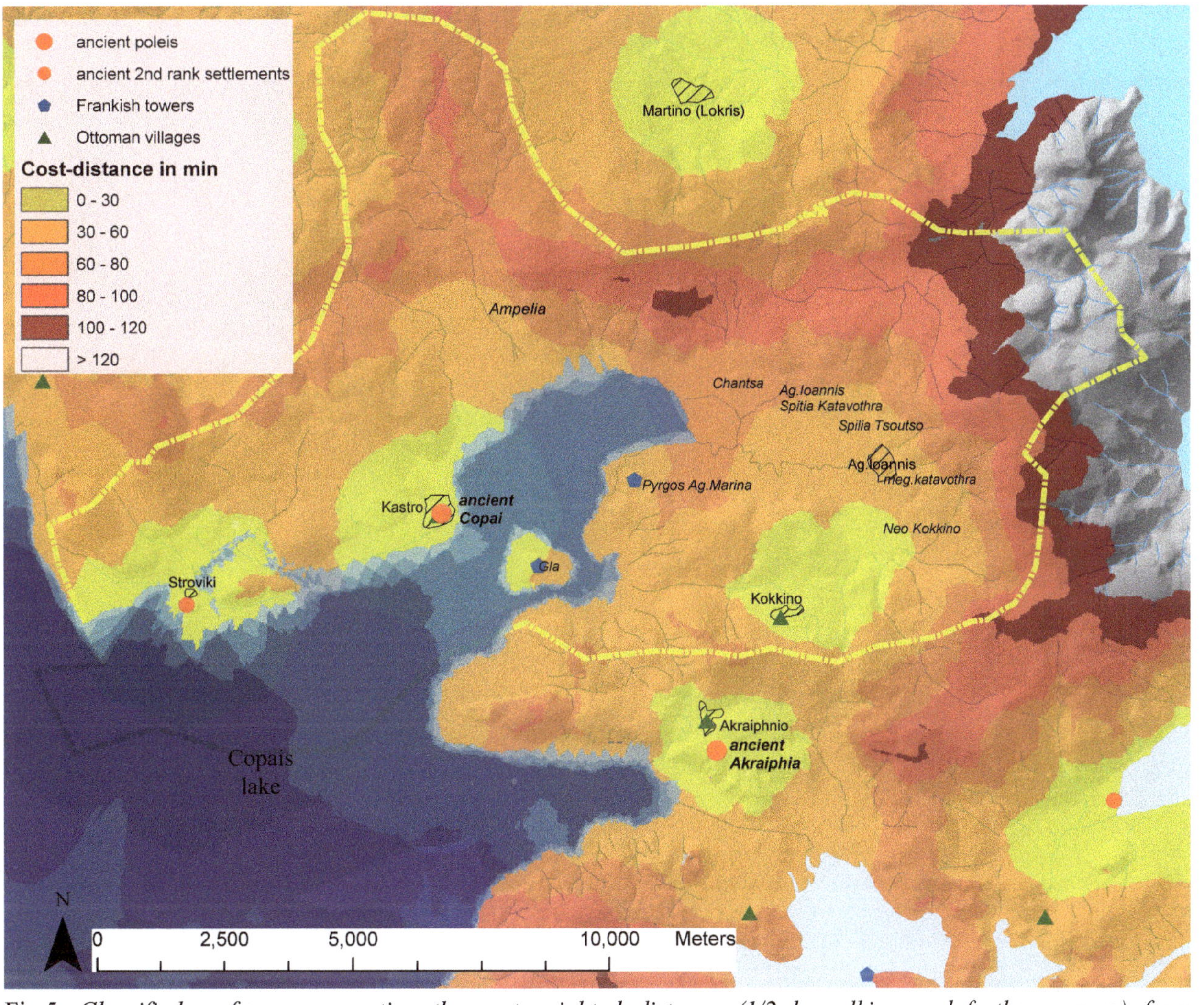

Fig.5. *Classified surface representing the cost-weighted distance (1/2 h walking and further ranges) from recognised 1st and 2nd rank ancient settlements (represented by larger and smaller dots). Areas without dots indicate potential settlement chambers. Ottoman villages and Frankish towers have also been added to the map to show their spatial relationship with the Greco-Roman settlement network and to appreciate potential settlement chambers.*

Meg.Katavothra - Spilia Tsoutso and Meg.Katavothra S, Ag.Ioannis - Chantza, Copai, Pyrgos - Agia Marina, Gla.

Examining, for instance, the relationship between Ag.Ioannis-Megali Katavothra and the Tsoutso cave site, in the Prehistoric period, the small settlement chamber presents a settlement site by the Tsoutso cave (in the N and EH periods – *components NC_48 and NC_49*), and a contemporary settlement site (or at least a clear activity focus – *components NC_57 and NC_58*) by the depression to the S of Megali Katavothra. Traces of EH were also found above Megali Katavothra itself. Afterwards, in the MH period, occupation is recognised at the Megali Katavothra site, with a cemetery site above the Tsoutso cave (*components NC_53 and NC_50*), while in the LH period the settlement site (apparently unfortified – *component NC_54*) is at the spur above Megali Katavothra. According to Lauffer (Kopais I: 232), the site above Megali Katavothra replaced in importance the earlier settlement in the depression to the SE (*components NC_57 and NC_58*) because its position was exposed and it was therefore smaller in extension. The Megali Katavothra settlement is bigger, better protected and has a larger area available for cultivation (on the flat slopes at the site and behind). When the draining of the lake took place, the site on the spur above the Katavothra would have increased its importance.
On the peninsulas by Spitia Katavothra (Agios Ioannis and Chantza), as well as at Topolia/Kastro, there is also occupation from Neolithic to LH, and at Pyrgos-Agia Marina, certainly from MH to LH. The inner bay beyond Topolia-Kastro, called Ampelia, where Magoula Kavkala is located (*components NC_30* (Neolithic) *and NC_31* (EH)), may also constitute a settlement chamber.

The consideration of the Prehistoric settlement pattern can help us infer useful information for the understanding of settlement strategies in the Greco-Roman period. In historical times, with the development of the town of Copai (located at Topolia/Kastro) as a nucleated settlement in the area, the landscape changed. The presence in Greco-Roman times of the larger settlement of Copai and of other satellite settlements (see below) would incorporate most of the small settlement chambers occupies the location of the ancient city of Copai, controlling parts of the lake. It is possibly also a pre-

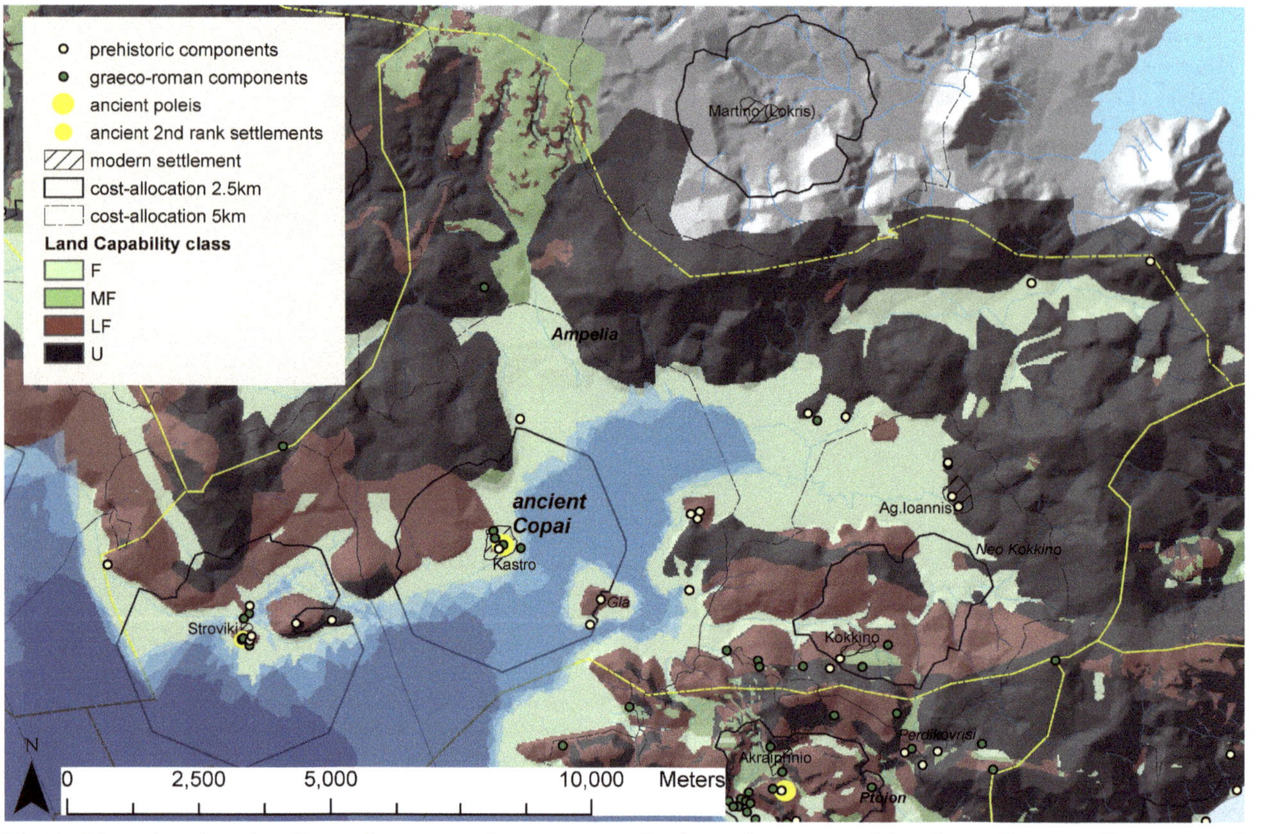

Fig.6. *Map showing the Greco-Roman settlement network, the polygons resulting from the cost-distance analysis (marking half an hour and one hour walking time distance) and dots representing the known archaeological components (same as in fig.2), with land capability information underlain.*

functioning as such in the Prehistoric period, especially those at the immediate edge of the bay. There, the archaeological evidence represents rural activities or farmsteads (*components NC_43 to NC_45, NC_56*) which, in areas with not much cultivable land available, were probably related to activities in the marshy[16] plain below or to pasturage/husbandry. The flat inner area called Ampelia (cultivated as vineyards) to the N of Topolia/Kastro (a half to one hour walking distance[17]), occupied in the Prehistoric period (see above), and where traces of Greco-Roman occupation have also been found (Alogopatisia-Kaphkala - *component NC_32*), can probably also be placed under the direct control of Copai in Greco-Roman times (with one or more small hamlets to exploit the land), as the absence of a settlement in more recent periods also might confirm.[18]

Thus, during the Greco-Roman period, Copai was probably the main settlement in the area. The central character of this location is also visible in the Medieval /post-Medieval and traditional settlement pattern. As in the past, the main settlement of the area today is Topolia/Kastro. The village is recorded in the Ottoman archives from 1466 onwards with the name Topolya, and Ottoman community as the Byzantine sherds on the hill indicate (Koder - Hild 1976). The Ottoman archives refer to it as a Greek village, indicating a probable survival of the Greek population.

Despite this, due to the peculiar character of the topography and environment in the area (see above), small settlement chambers can still be identified (as actual or potential ones), and the town of Copai probably never become more than a large village. The traditional settlement pattern helps us recognise these potential settlement areas.

Located in the Eastern part of the *chora* today is one more village, Kokkino, which is identified as the Albanian village of Dima Kokkino, which appears in the archives from 1466 onwards and holds up well in the 17th century. It is located in the inland mountainous country associated with MF and LF soils, suitable for grazing, but also with the alluvial bed of the streams to the N. Also included in this area is the area of Neo Kokkino, closer to Meg.Katavothra, which clearly exploits the MF (mid fertile flysch) upland area (see map in fig.6) nearby, and also had the river valley below (see figs.1 and 2). A potential settlement chamber is drawn to include both these areas, as well as the modern second rank settlement of Ag. Ioannis further N, as the results of a cost-distance

[16] Marshy only in certain periods and to a certain degree (see Farinetti 2008).

[17] See the cost-distance analysis in fig.5.

[18] The area is included in the modern *koinotita* of Kastro.

analysis[19] show in fig.5. The cost-distance analysis shows enough room for a settlement in this far Eastern edge of the bay, as also suggested by Bintliff (Thiessen polygons analysis in Bintliff 1994b: 219 and fig.20), although large parts of the lower areas are potential marshes. The settlement could either be located in the upland country by the modern village of Kokkino (see above), or in a lower location by Neo Kokkino. The existing archaeological evidence comes from the upland zone (*components NC_84 to NC_86* and *component NC_78*, by the modern village of Kokkino). The fact that this evidence dates back both to the Classical and Roman periods strengthens the possibility that this area was exploited in antiquity.[20] Even today the area is rich in vineyards, and includes the valley running E-W down to the Copais basin. With regards to the area of the modern second rank settlement of Ag. Ioannis, the available archaeological record does not provide enough evidence for interpretation as a settlement site.

Another modern 2nd rank village, Stroviki, is situated in the far Eastern part of the *chora*, indicating a possible area for settlement activities. In fact, at the higher eastern hill at Stroviki-Nisi the signs of occupation for the Classical period (*component NC_7*) might represent a rural hamlet of considerable relevance (with corresponding cemetery site on the smaller western hill – *component NC_11*).[21]

---

[19] For the application of this analysis see chapter II.3.1 – LONG TERM SETTLEMENT TRENDS.

[20] Lauffer Kopais I: 225 also suggested that those traces of a necropolis found to the NW and N foot of the Megalovouno ridge could indicate burial areas connected to a village settlement there.

[21] The evidence is obviously known because of the heavy presence of Prehistoric signs of occupation, representing a large Prehistoric settlement (*components NC_1 to NC_6*).

II.3.7

The Eastern Copais area: Akraiphiai

TOPOGRAPHICAL SETTING

The *chora* of Akraiphia is part of the Copais area, but the bridge of land between the two lakes (Copais and Yliki) as well as the Ptoion upland areas also belong to it (see fig.1).

Lauffer (Kopais I: 251) considers the area bordering Copais, between the NE bay area and the Southern Onchestos area, as the E Copais area. In antiquity the area probably belonged to Akraiphia (see borders below), and for the data recording, therefore, I consider it as a whole together with the surroundings of Akraiphia and the Ptoion mountain, accessible only from the N and NW, i.e. from the Akraiphia area.[1]

Below ancient Akraiphia the main Copais basin forms a bay (named Karditsa after the former name of the modern village of Akraiphnio),[2] which, at its deepest end, gets very close to the Yliki lake, from which it is separated by a low pass.[3] Water naturally occupies part of the bay,[4] which was protected by a dam crossing the bay in a N-S direction, probably built in the Mycenaean period and certainly working in the Archaic period (Knauss Kopais 2). Remains of a channel were found, which are probably to be identified with the Akraiphia internal channel (Aravantinos 2004); see fig.11 in chapter II.3.1.

According to the digital model of the fluctuations of the Copais lake (see fig.2; chapter II.3.1 and fig.10; Farinetti 2008), the water used to fill the bay to a certain point, but would never have reached the area occupied by the cemeteries (which would lie along the road in a marshy area) and the Late Roman rural activities (exploiting the resource of the peripheral foothill area), where a road also ran (see below: *components AK_7 to AK_18 and AK_53*).

The Ptoion ridge dominates the landscape of the area. Lauffer (Kopais I: 263) quotes Pindar fragment 51b (Hymn to Apollo Ptoios) which describes the Ptoion mountain. The Ptoion was also called *Skroponeri* or *Struzina*, as mentioned in Philippson 1951: 491, and it is described as an abrupt wall of rudist limestone (Megalovouno, 558m; Pelagia, 724m; Tsoukoureli, 698m; Tabouria peak, 838m) which descends towards the S to the Skroponeri gulf. The rudist limestone alternated with a series of flysch formations, a line of which runs along the N side of the Pelagia peak (see below: *component AK_46*) (Philippson 1951: 491-2). The N side of the limestone block is characterised by the flysch depression where the monastery of Ag.Pelagia lies. At the W end of the slope, on the same flysch area, in a valley which opens N-S (370m elevation), lies the sanctuary of Apollo (see below: *components AK_32 to AK_37*). It is a very beautiful landscape with limestone rock above soft schist, with a few olive trees and water sources. Further to the W lies the village Karditsa/Akraiphnio, on schist, which is separated by a line of limestone from the gulf of Copais to its S. W of Karditsa the schist continues to another small Copais gulf (between the Mytikas and Phtelio capes), while to the N and the S of the village the limestone ends in the Mytikas and Phtelio capes themselves.

The landscape of the S edge of the gulf is characterised by light coloured soils of tertiary formation (Philippson 1951: 493), while between Copais and Yliki lake lies a limestone plateau with a karst basin (Asprokampos – 137m elevation) in the middle of the pass (which probably belongs to the Haliartos *chora*).

Boundaries

The natural boundaries of the *chora* of Akraiphia are considered to be to the E the highest parts of Ptoion, probably Anemomilos (543m) and Pelagia (724m), but possibly also including Tsoukoureli (698m), as noted by Fossey (1988: 265). At its SE corner it should reach lake Yliki. The N and S borders are less certain. Towards the N, the boundary between Akraiphia and Copai would probably run from the Phtelio cape (where a boundary inscription - IG VII 2792 [AE1212] - was found in the 19th century; dated to the Hellenistic period – end of the

[1] Lauffer (Kopais I: 252) implies that the Ptoion area belongs to the Akraiphnio and Copais area, rather than to the Yliki lake and Thebes area, and supports the hypothesis of strong links between the Ptoion area and Akraiphnio and the Copais rather than with Yliki or Paralimni (Kopais I: 281). According to Lauffer, settlement is known only on Ptoion's E side, and *Hyle*, the only settlement known from ancient texts, is connected with Eastern Copais (Lauffer Kopais I referring to Homer *Il.* V 708).

[2] Philippson (1951: 466) notes how the E edge of the main Copais basin widens out twice: firstly corresponding to the Akraiphnio (formerly Karditsa) bay (which opens up S of the Mytikas cape and runs towards the E), and further to the S, corresponding to the small gulf of Davlosis (see Haliartos *chora* – chapter II.3.8). The Karditsa bay has been identified with the *Atamanische* plain by some scholars (among others Gomme 1911/12, Frazer 1913, Guillon 1943).

[3] The pass is constituted of limestone, and its E part produces serpentine marble (Philippson 1951: 488). Today the pass constitutes the exit of all the Copais water (towards lake Yliki by means of a tunnel).

[4] A swampy area still exists today before the Karditsa/Mytikas promontory, as noted by Philippson (1951: 488).

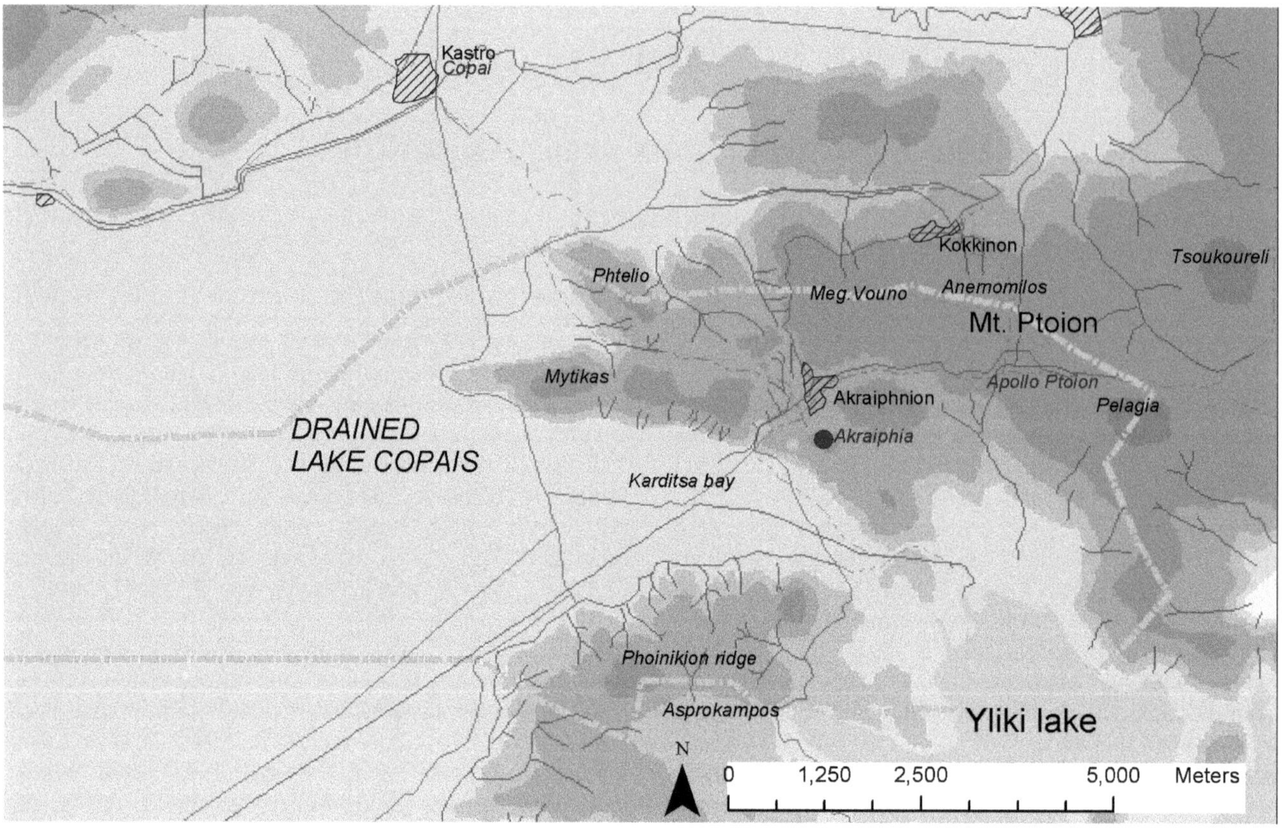

Fig.1. *Topographical setting of the chora of Akraiphia.*

4th C BC[5]), along the Megalovouno watershed, leaving the N side of Ptoion to the area of ancient Copai (see also chapter II.3.6).
Another inscription was found and interpreted as boundary stone between Copai and Akraiphia, datable to the 6th – 5th C BC. The text suggested by Lauffer reads: [hόρος Α]κραι[φιέον] [κ]αί Κοπ[αίον.[6] To the S, the border was probably marked by the first watershed within the fringe of heights defining the S edge of the Karditsa bay (Phoinikion ridge), on which two small fortifications are known (*components AK_23 and AK_24*). As stated by Müller (1995: 660), failing detailed written information, three types of 'monuments' would permit the identification of the possible frontiers of the region: the boundary marker at the Phtelio cape, the sanctuary of Apollo at Perdikovrisi marking the E frontier, and the small defensive forts located at different points in the territory, even if they cannot be dated with precision.[7] The absence of any access to the sea for ancient Akraiphia is confirmed (Müller 1995: 660).

PHYSICAL LAND UNITS

The edges of the Copais basin rise quite abruptly, as is usual, all around the Karditsa bay. While the upland area is mainly characterised by smaller or larger plateau areas, mainly cultivated, the modern village of Akraiphnio is located on a protected slope, facing the Copais lake, and the ancient acropolis is to the S of it, on a rocky spur overlooking the surrounding plateaus.
Mountainous elevations are present only in the Easternmost corner of the *chora*, on the Ptoion ridge (M2 class is not represented), while in general the landscape is marked mainly by the usual mid range elevation which characterises the fringe N and E of Copais, with steep slope features (H5) alternating with gentle slopes and upland plateaus. The high percentage of plain landscape is due to the artificial subdivision of the Copais area among the bordering *chorai*, and could be even considerably less, according to the actual portion of lake included (see fig.2 in chapter II.1).

Hilly landscape	23.5%
Mountainous landscape	0.5%
Plain	76%

[5] See AD 1970: 229. See also Daverio Rocchi 1988: 120-122.
[6] SEG XXX 440; *Teiresias*, 10 [1980] AE/2; Lauffer *Chiron* 10 (1980: 161-2). The inscription (dated to the 6th C BC) was found in the plain, 500m NW of Phtelio cape and 500m SW of Gla.
[7] As Müller reports in Müller 1995: 660. Some forts (Pelagia, Mytikas – see components below in the text) have been identified, while others (Vathy Spitari, Kori) have not.

1	P1_P2	lacustrine basin, valley	63.5%
2	P3	gentle slope	1.2%
3	P4	foothill	11.3%
4	H1	plateau	6.9%
5	H2	gentle slope	0.2%
6	H3	moderate slope	2%
7	H4	severe slope	6.3%
8	H5	very severe slope	8.1%
9	M1	plateau	0.3%
10	M2	plateau/gentle slope	0%
11	M3	moderate slope	0.1%
12	M4	very severe slope	0.1%

Table 1. *Percentage of the different physiographical classes present in the Akraiphia area (P=plain; H=hill; M=mountain).*

RESOURCES

Three types of land characterise the landscape of the small Akraiphia *chora* (Müller 1995: 659): the limestone mountain chain, reaching a height of 700m, formed of extensive karst surfaces; a dry but cultivable region at the foot or on the slopes of this; a lacustrine or semilacustrine region (the lakes of Copais, Yliki and Paralimni) drained in antiquity by the employment of a series of artificial outflows using the *katavothrai*.

Apart from the areas of carbonate rock, the Akraiphia *chora* also presents areas potentially suitable for agriculture. To the E of ancient Akraiphia, as well as in between the ridges, are upland fertile areas, while to the SE of the Ptoion ridge towards Yliki, an area of lake deposits creates a considerable amount of cultivable land (fig.6). Though, due to its scale (1:50,000), not visible on the map (fig.6), a few small areas of cultivable soil can also be found on the limestone areas, with soil deposits due to slope erosion. The presence of the boundary marker, establishing the frontier between Akraiphia and Copai at the end of the 4th C BC, is probably proof of the existence of disputed agricultural land in the small cultivable pockets between the Mytikas and Phtelio capes (Müller 1995: 659).

It would seem that the Mycenaean dike across the mouth of the bay may have remained in use until the early imperial period when, despite attempts to repair it, it gave way to the lake's waters (from Fossey 1988: 275: Fossey 1973; Kenny 1935: 193-4; Kahrstedt 1954: 84-5; Lauffer, AD 1971: 243-5; Fossey 1979: 554-560; see also Knauss et al. Kopais 2).

Ptoion, as well as ridges of Megalovouno, offers land for pasturage. Olives and vines are also cultivated in the flysch formation (valued here as MF –class 2) to the N of the *polis* site. Fishing was also a valuable occupation in the area, a fact that is proved by a Hellenistic inscription (list of fish prices - Vatin 1971) found in the *agora* of the ancient *polis*. Iron deposits are also known on Mt. Ptoion and its SW outskirts (though it is uncertain whether they were exploited in antiquity – Bakhuizen 1976: 56-7[8]).

[8] The geology and iron sources are treated by Petraschek 1954, *passim*.

THE ARCHAEOLOGICAL RECORD

1	**Akraiphnio Skopià**	*Components AK_1 to AK_6*
2	**Skopià**	*Components AK_28 and AK_60*
3 to 10	**Akraiphnio burial areas**	*Components AK_54 to AK_59* (3); *AK_7 and AK_8* (4); *AK_11* (5); *AK_12 to AK_17* (6); *AK_13 and AK_14* (7); *AK_48 to AK_52* (8); *AK_18* (9); *AK_9 and AK_10* (10)
11	**Akraiphnio Grava**	*Component AK_53*
12	**Akraiphnio-Nat.Road junction**	*Components AK_55, AK_57 and AK_58*
13	**Akraiphnio-Nat.Road S**	*Component AK_56*
14	**Phtelio-Vristika**	*Component AK_21*
15	**Mytikas**	*Component AK_22*
16	**Yliki NE - Vathy Spithari**	*Component AK_23*
17	**Yliki NE**	*Component AK_24*
18	**Spilia tou Sarakinou**	*Components AK_25 to AK_27*
19	**Megalovouno S slope**	*Component AK_19*
20	**Megalovouno E slope**	*Component AK_20*
21	**Ptoion Kastraki**	*Components AK_29 to AK_31*
22	**Ptoion Perdikovrysi**	*Components AK_32 to AK_37*
23	**Ptoion Perdikovrysi - W slope**	*Components AK_38 to AK_40*
24	**Perdykovrisi valley – plateau**	*Components AK_41 to AK_43*
25	**Perdykovrisi valley**	*Component AK_44*
26	**Ptoion Pelagia**	*Component AK_45*
27	**Ptoion Pelagia Monastery**	*Component AK_46*
28	**Tsekoureli**	*Component AK_47*
29	**Ag. Nikolaos**	*Components AK_61 to AK_65*

Table 2. *List of archaeological components and activity loci mapped in fig.2.*

*Sites on the N edge of Yliki lake, though probably belonging to Akraiphia, are presented and discussed under the *chora* of Thebes, where the Yliki lake and the surrounding settlement will be discussed as a topographical unit.

As for the Akraiphia *chora*, knowledge of the archaeological record concerning the area is due in large part to rescue excavation; mainly work in the modern village and especially along the Athens-Lamia highway, where extensive excavations have also been carried out recently for the construction of a new junction and enlargement of the highway, leading to further discoveries in the necropolis area (see above).

The graph (fig.3) illustrates the proportion of components discovered within different research frameworks. For the Akraiphia area, accidental discoveries are apparently also quite common. This is due to the discovery of tombs (which constitute one of the main features of the known

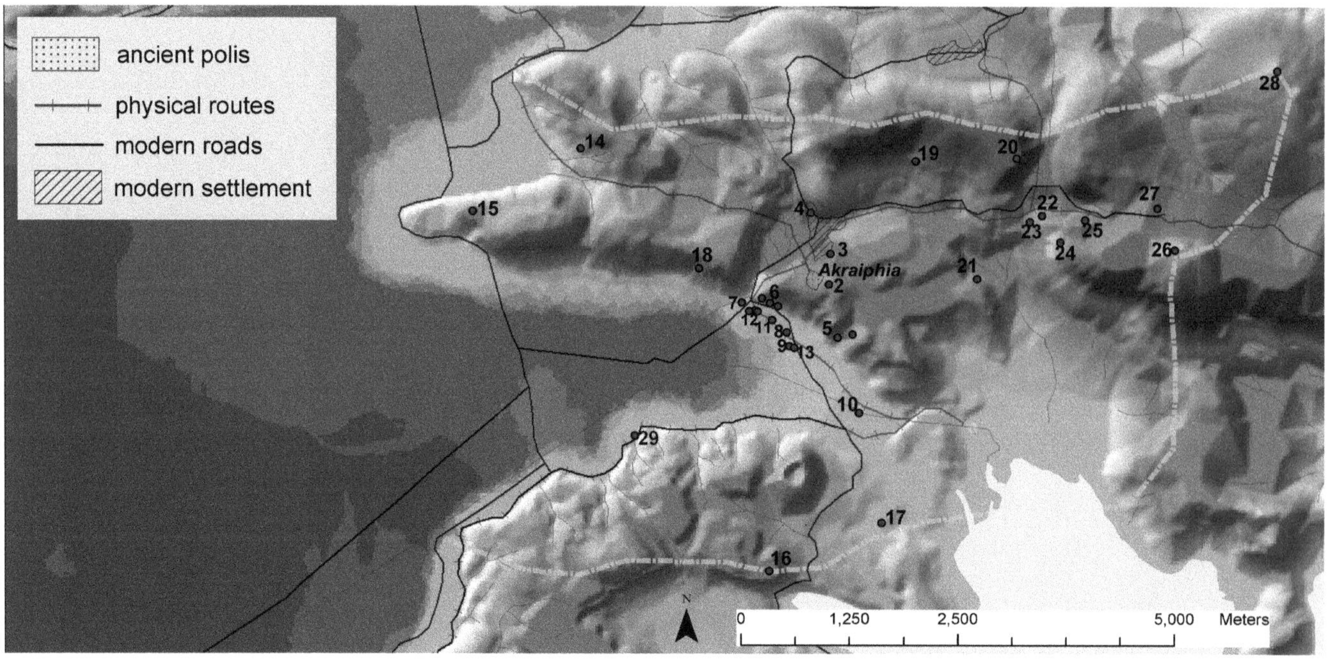

Fig.2. *Archaeological map of Akraiphia chora.*

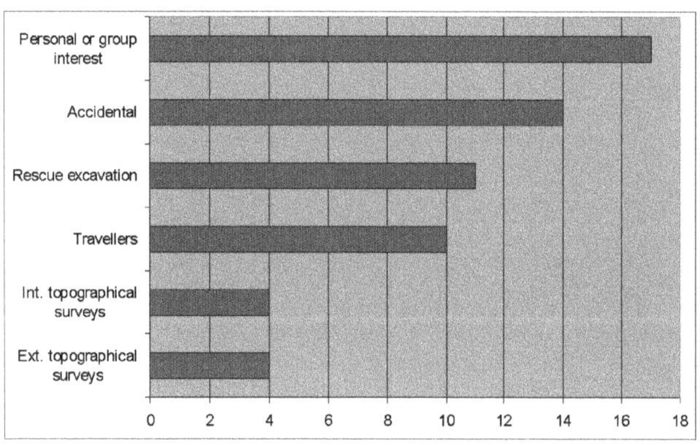

Fig.3. *Graph illustrating the proportion of components discovered within different research frameworks.*

archaeological landscape of the area) mainly through ploughing or illegal excavations.

Personal or group interest, which characterises the majority of discoveries, includes the systematic excavations at the Ptoion sanctuary (French School) and at *Spilia tou Sarakinou* (Speleological Service), as well as the discoveries made by Knauss' équipe interested in the areas at the edges of the Copais lake. Furthermore, personal interest in forts led to personal visits to the highest and most prominent peaks in the area, and in general the whole area of Ptoion.[9]

The 'Other' value is not represented within the discovery panorama of the *chora*, perhaps mainly because the archaeological record seldom offers us the case of multiperiod sites.

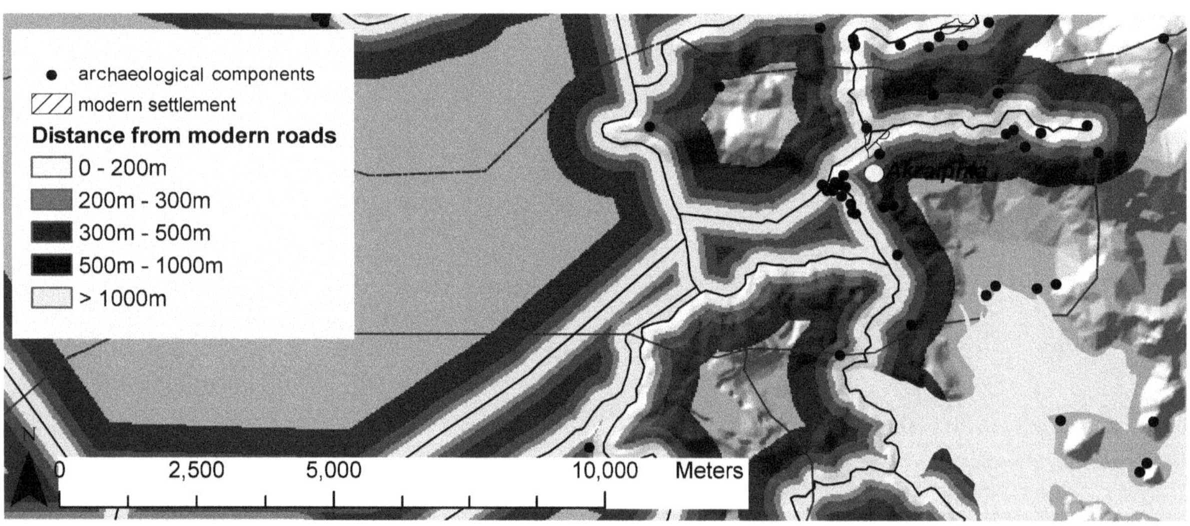

Fig.4. *Relationship between components and modern road network*

[9] Good maps of the archaeological area are given by Guillon 1943 vol. II, pl. V and Lauffer 1959, while more recent discoveries have been published in various archaeological reports.

Interestingly, all the available archaeological record comes from the N and NE side of the Karditsa bay. No archaeological remains (with the exception of two hilltop forts) are apparently known along the Southern edge of the bay, comprised of rocky limestone formations, which separates the Copais basin from the Yliki lake.

In fig.4 we can see the relationship between known archaeological sites and the distance from the modern road network. Known sites are in a quite strict correlation with the proximity of roads, with the exception especially of the known hilltop fort sites, whose knowledge is mainly due to a special interest in fortification (see below – Noacks's research, for instance, as well as that of the French School). Also distant from the modern road network are the known sites at the N edge of Yliki, discovered due to the interest in perilacustris areas and oriented research. In the lower area by the bay, occupied by burials, the strict correlation with the modern road network is due to the road construction work, which in the majority of cases allowed for the archaeological discoveries. In the upland area, discoveries along the upland path on the Ptoion can be related to the existence of an old route.

The ratio of known Prehistoric to Greco-Roman components is 11 to 48 (1:4.4), while among the historical periods, 50% are dated Archaic to Hellenistic, 31% Roman-Late Roman, and 19% are attributed to the general Greco-Roman period.

ANALYSIS OF THE *CHORA* LANDSCAPE

PREHISTORIC PERIOD

The area of Akraiphia being part of the E region of the Copais basin, settlement is much influenced by the presence of the marshy lake, and marked by characteristics which typify other areas around Copais, such as the NE bay in particular (see chapter II.3.6).

As elsewhere in the Copais area, for the earliest period we have evidence of occupation in caves, as at *Spilia tou Sarakinou* (*components AK_25 to AK_27*), where major Neolithic occupation was excavated. The Sarakinou cave lies at a higher elevation than other cave sites in the region, compared to the basin base level, a fact that gives it a special character. Neolithic occupation is also known in the plain within the Karditsa/Akraiphnio bay, below the ancient city site (fig.6 in chapter II.3.1), where, in deep strata below the Archaic and Classical cemetery, Andreiomenou found two complete vases similar to those of the Final Neolithic periods found fragmented in the aforementioned cave. Their probable burial use, as well as their discovery in thick lake deposits, could mean that in the Final Neolithic period the lake was some distance from the cave location (at least 1000m), as pointed out by Sampson 2000 – see the digital reconstruction of the lake fluctuations in the present work (chapter II.3.1 and Farinetti 2008).

There are few sites known for the EH and MH periods (figs.7-8 in chapter II.3.1). *Components AK_38 and AK_39* (Perdikovrisi site) could be linked to the occupation of upland segments of the landscape during the transition between the Neolithic and Early Bronze Age, due to the so-called Secondary Product Revolution. The *Spilia tou Sarakinou* cave site also displays traces of occupation in the Early and Middle Helladic (*components AK_26 and AK_27*).

As Lauffer (1940: 187) noted, the city site was probably not occupied during the Neolithic nor during the Bronze Age, when the main settlement of the area has probably to be found elsewhere (on the Ptoion, or down the Copais). On the other hand, Knauss (Kopais 2: 247ff) reports traces of a Prehistoric fortification within the Archaic circuit of the city.

A network of lookout towers (and roads?) must have existed in Mycenaean times – cf. Hope Simpson 1965: no. 415 (see *components AK_43 and AK_44*), as well as Lauffer's (Kopais I: 285) reasoning that the area of Ptoion (on the upland pass between E and W Boeotia) must have been strategic in the Late Prehistoric period. Despite this, few outposts have been noted which can actually be dated to the LH period (fig.9 in chapter II.3.1); the majority are datable to the Greco-Roman period (and probably to the 4th C BC) – see below FORTS AND FORTIFICATIONS.

GRECO-ROMAN ANTIQUITY[10]

Town level

The town of Akraiphia was located in a prominent position almost in the centre of what was presumably its *chora* (on Skopia hill, to the S of modern Akraiphnio village). The fortification of the town as well as the city landscape has been investigated in particular by the work of Guillon and Feyel (BCH 1936: 461 and Guillon 1943). The *polis* of Akraiphia had a certain importance from the earliest historical periods (Geometric-Archaic), considering the richness and abundance of tombs found dated to these periods. In the Geometric period the settlement probably had a *protopolis* character, although Akraiphia seems to be much more developed in earlier historical periods than other Boeotian *poleis* (since the area was soon controlled by Thebes – Hansen 1996: 79). The city seems to flourish at the beginning of the Hellenistic period also, after the decline of Thebes (many burials related to the city site are known from the lower area of the bay; the main core of the circuit wall would date to that period). The presence of a Late Roman burial (*component AK_6*) in the city area would indicate the decline of the city in the mature Roman and Late Roman period, known also for other Boeotian cities.

It was a middle-sized *polis*, which would probably have concentrated all the population of the small area, and probably most of the rural segment as well (see below – LONG TERM SETTLEMENT TRENDS).

[10] Period maps are included in chapter II.4, figs.17-19-21-23-25-27.

Village level
If we do not consider the N side of the Megalovouno ridge (occupied by the modern village of Kokkino – see chapter II.3.6) as part of the Akraiphia *chora*, the sole area of possible suitability for a settlement site is the Perdikovrysi area, at a crossroad between upland plateaus and mountain pathways, and characterised by wide fertile terraces and fresh water springs. The establishment of the Apollo sanctuary and related habitation and activities from early historical times, however, would fill in this 'available' space and exploit its resources. Otherwise, the settlement shows a strong nucleation into the town of Akraiphia (see below – LONG TERM SETTLEMENT TRENDS).

Rural segment
The rural occupation of the landscape of the area in Greco-Roman antiquity is almost completely unknown, with the exception of a Late Roman *villa rustica* site (*component AK_53*) excavated in the foothills below ancient Akraiphia and an evidence from burials (*component AK_19*, see BURIAL AREAS), probably reflecting the decline of the city in the R-LR periods. As suggested earlier, in earlier periods the land available for agriculture was probably farmed by the inhabitants of the city site (and in the upland areas the landscape probably would have seen the presence of some sheep-folds at the most). On the other hand, probably in the Roman, and certainly in the Late Roman period, other areas of the landscape would have been occupied and exploited intensively according to the large property system recognisable elsewhere in Boeotia and mainland Greece during the Roman period.[11] The location of *component AK_53*, low in the bay not far from the water is similar to one of the villa sites recorded in the Kephissos valley by Chaironeia (see chapter II.3.3), and presents a lower position, not unusual in Boeotia for rural sites of the Late Roman period (information which can also be inferred from the results of the intensive artefact surface surveys carried out in the region – mainly Thespiae and Tanagra – see chapter II.3.14 and appendix I.14 – THE TANAGRA SURVEY PROJECT).

Burial areas
The burial landscape of the area is archaeologically very well known, especially as far as the cemeteries linked to the *polis* centre are concerned. The main cemetery of Akraiphia lay at the S foothill of the acropolis hill (Geometric to Hellenistic period), in the Karditsa bay, while other burial foci have been noted (see above for detailed locations, but the main one, especially in Hellenistic and Roman times, is at the N foot of the hill, where the modern village lies). The necropoleis probably lay along the road towards Thebes (those in the area of Grava as well as that found further E in the Gouni area – *component AK_9*). The Roman burial area represented by *component AK_19* and the possible Classical burials (*component AK_63*) are not connected with the city site but probably with the exploitation of the rural landscape.

Cult places/Religious areas
The sacred landscape of the area is dominated by the presence of the upland sanctuary of Apollo Ptoios (known as *Ptoion* - sanctuary and *heroon* at Kastraki and Perdikovrisi on the Ptoion – *components AK_29 to AK_37*), which, located in a strategic position at a crossroad of upland routes, gives life and prosperity in the Greco-Roman period to an otherwise remote upland area.[12]
In addition to these major cult places, a small religious building (*component AK_55*), interpreted as a cult place dating back to the Geometric period, has recently been excavated in the area of the Karditsa bay immediately below the acropolis hill, in an area known as Grava because of the presence of cemeteries (see above). It would probably have been connected with the burial area, occupied as such since the Geometric period.

Forts and fortifications
Fortifications in the E Copais area (mainly on W Ptoion) are well known due to Noack's research in the environs of Gla. Although some of the forts in the area could also be dated to the LH/Mycenaean period, all the spots which he interpreted as LH fortifications or fortified settlements, after more careful research (mainly by Lauffer) turned out to be Greco-Roman sites and, most of the time, fort sites. Evidence from layout, technique, orientation and pottery make them later (probably 4th C BC, the period of Theban hegemony), and having no connection with Gla. Nevertheless, Noack's journey and notes led later research and helped scholars to go directly to these places and examine them more closely (see discussion under the sites above).
Several forts of historical date are therefore known from the area of ancient Akraiphia, and mark its landscape, especially in the upland area of Ptoion, controlling crucial upland and lowland passages. Apart from the circuit wall of the city site (enclosing quadrangularly the acropolis on Skopia hill), several isolated forts are known at strategic points in the area, overlooking passes or passages both towards the Copais and the Yliki lake. These probably belonged to the strong fortification system controlled by Thebes and set up in the 4th C BC (see *components AK_21 to AK_24; AK_41-AK_42 (probably); AK_45 and AK_47*), to control the passes towards W Boeotia.[13] They are clearly located along the boundary between the strictly Theban territory and the area of Akraiphia, which Thebes wished to control strictly for its central location within the region and its consequent clear strategic importance (fig.2 – nos. 16, 17, 26, 28).
Several forts were noted on the heights above the Ptoion-Perdikovrisi sanctuary: a fort on the highest peak of Ptoion (*component AK_47* – no.28 in fig.2), and two

[11] For a summary see Alcock 1993 and reviews. For southern Argolid see Jameson– Runnels– van Andel 1994, for Laconia Cavanagh et al. 1996 and Cavanagh-Mee-James 2005.

[12] The area of the Ptoion mountain is also crucial for control of upland as well as lowland passages, and the landscape is marked by the presence of forts (see below in the text).

[13] One of the first *poleis* that Thebes secured under its control was Akraiphia (Lauffer Kopais I: 285; Hansen 1996: 79).

other forts noted in the area, one on each side of the plateau above the temple of Apollo Ptoios (*components AK_42 and AK_45*– nos. 24 and 26 in fig.2 - Kenny 1935).
The forts in the area would guard the approaches from Anthedon (via the Paralimni lake), from Thebes, and from Larymna. As Lauffer (Kopais I: 281) suggests, it seems, though, that they would also guard the side of the Copais area.[14]

LONG TERM SETTLEMENT TRENDS IN THE CHORA *LANDSCAPE*
As seen above, the *polis* of Akraiphia had a certain importance from the earliest historical periods (Geometric-Archaic), considering the richness and abundance of tombs found dated to these periods. The city flourished again during the final phases of the Classical period and at the beginning of the Hellenistic period, after the decline of Thebes, with signs of decline in the Late Roman period.
In the wider territory evidence for settlement is lacking (see above – *Town level* and *Rural segment*). This nucleated pattern is also supported by the Medieval/Post Medieval and traditional settlement pattern. In the *chora* of ancient Akraiphia, controlling a middle to low fertility area, with small high fertility zones by the lake, is situated only one modern village, Akraiphnio, former Karditsa, which is recorded in the Ottoman archives from 1466-1688. The medieval church in the village, with the inscription of a Frankish lord dating back to the 14th century, might indicate continuity from antiquity into Byzantine and Frankish times and recolonisation by Albanians in the 15th century (Bintliff-Kiel in preparation).
By considering the three types of land which define the landscape of the small Akraiphia *chora* (see Müller 1995: 659 – limestone mountain, dry foothills region, Copais lake and marshes) one can obtain a picture of the settlement pattern accordingly. In the foothills and plain area of the bay we have burial areas connected to the *polis* of Akraiphia above, and the passage of the road, as well as the rural sites known at least for the Roman period, exploiting the bay (probably free from water due to the dam at the entrance to the bay). In the upland areas we do not have traces of habitation in the historical period (at most, the area would have been populated by sheep-folds and other buildings suitable for pastoral activities), but life in the area was probably active due to the presence of the sanctuary and its related activities.

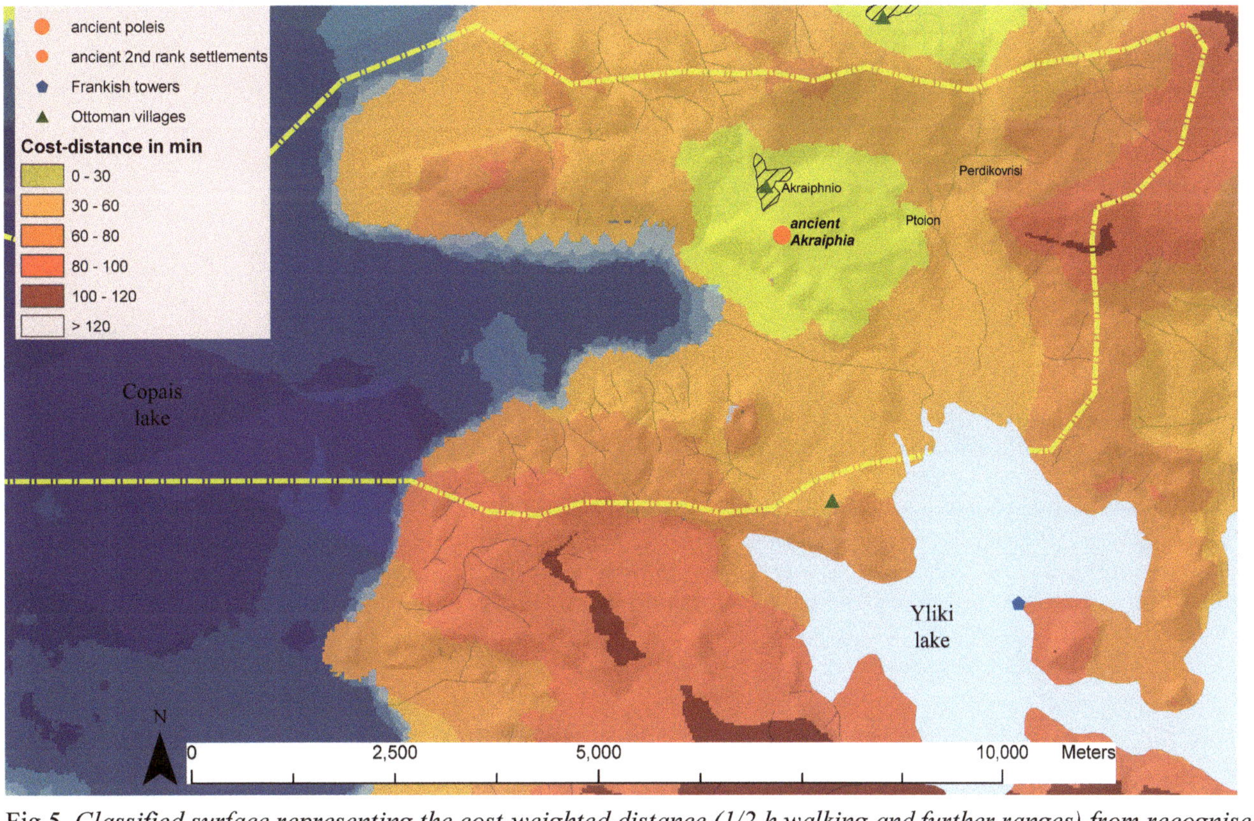

Fig.5. *Classified surface representing the cost-weighted distance (1/2 h walking and further ranges) from recognised 1st and 2nd rank ancient settlements (represented by larger and smaller dots). Areas without dots indicate potential settlement chambers. Ottoman villages and Frankish towers have also been added to the map to show their spatial relationship with the Greco-Roman settlement network and to appreciate potential settlement chambers.*

[14] The watchtower on Tsikoureli (*component AK_47*) seems to overlook only the W area and Copais, and the forts *AK_20* and *AK_45* would probably also have controlled that area (see chapter II.4).

The upland zone was also populated with several small forts (most of which are still partially standing) marking the mountainous landscape of this area as central with regards to position and strategic control of passes in Boeotia.
In the landscape nothing is known of classical rural occupation, and this is probably due to the few bits of cultivable land in the small (with regards to extension) *chora* having been farmed directly by people from the city. In the Late Hellenistic and Roman periods, the burial area moved to the slope of the Skopia acropolis, and a villa estate can be found in the lower area within the bay (*component AK_53*, a villa rustica site of Late Roman date) testifying a different use of the landscape (intensive export-oriented land use by large estates).

If we consider the area of Kokkino as belonging to the NE Copais area, than in the proper Akraiphia *chora*, even if we consider the upland Ptoion area to be a part of it, there is inadequate space for a second settlement chamber, especially in the Greco-Roman period,[15] as the cost-distance analysis[16] seems to confirm (fig.5). As already mentioned, the traditional settlement pattern confirms a diachronically nucleated settlement pattern in the region.[17] Only for the Roman period have we some evidence of further exploitation and settling in other areas (*components AK_19 and AK_53*).
The existence of a large and known sanctuary, such as that of Apollo at Ptoion-Perdikovrisi, however, would have created an area of influence of the sanctuary, rich in activities in the upland areas, which would otherwise be 'occupied' only by watchtowers in elevated positions. In fact, examining the available archaeological record, in the Perdikovrisi valley we find, as Lauffer also notes (Kopais I: 275), Prehistoric activities (with no certain focus – *components AK_38 to AK_40 and AK_43*, fig.2 nos.23 and 24), a Greco-Roman cult place (with habitation and connected activities – *components AK_32 to AK_37,* fig.2 nos.22), and a medieval/Byzantine? monastery (Ag.Pelagia) – fig.5 and fig.6.

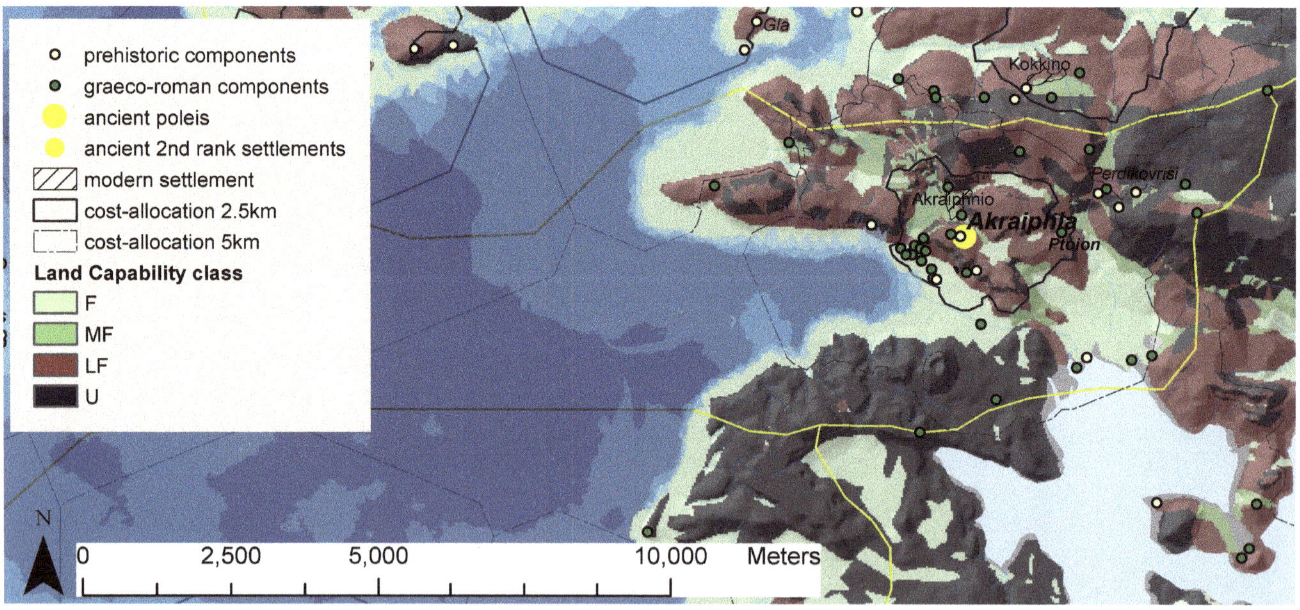

Fig.6. *Map showing the Greco-Roman settlement network, the polygons resulting from the cost-distance analysis (marking half an hour and one hour walking time distance) and dots representing the known archaeological components (same as in fig.2), with land capability information underlain.*

[15] See also Thiessen polygons analysis on Boeotian ancient settlement network in Bintliff 1994b: fig.20.
[16] See chapter II.3.1 – LONG TERM SETTLEMENT TRENDS.
[17] The village of Sengaina-Sence, reported in the Ottoman archives and located at the very S end of the *chora*, has to be considered as geographically belonging to the landscape of the Yliki lake, whose settlement history is discussed in chapter II.3.12.

II.3.8

The Copais area: Haliartia

TOPOGRAPHICAL SETTING

The area of Haliartos comprises the S and SE area of the Copais, from the rocky spur of Petra and the area of Vrastamites/Ypsilantis (to the SW) to the peninsula of Kokkoretsa to the NE, as well as inner valleys within Mt.Helicon (Fossey 1988: 301).

The Libethrion/Tilphousion ridge (N section of Mt.Helicon) juts into the Copais with three promontories: the Western one is Petra, a limestone spur with a flattish top (288m) at the foot of which is the *Tilphoussa* spring; the central one is a schist promontory that encloses to the N the small bay with the village of Vrastamites/Ypsilantis and the Frankish tower; the Eastern one is the spur occupied by the acropolis of ancient Haliartos, which closes to the E the small basin with the village of Siachon/Petra (Philippson 1951: 451).

According to Philippson (1951: 451) the Copais lake formally reached the Petra spur, while today the channelled Kephisos frees up the passage for the road and the railway. In fact, according to the fluctuation model (based mainly on Knauss et al. work - see fig.10 in chapter II.3.1 and Farinetti 2008; on the archaeological Classical and Roman remains before the spring at the foot of the spur, see chapter II.3.1), the lake water did not reach the spur even in the past, probably not even in the Late Medieval/Early Modern period, when the level of the lake was higher (considering the passage of the Turks during the War of Independence – see below and see chapter III.1). Therefore, presumably also in antiquity the road ran here[1] (fig.2).

As Philippson (1951: 474) reports, beyond the Petra pass begins a line of marshes formed by the streams. This line runs along the foothills of the hills below Ypsilantis village. There, along the S edge of Copais, runs a small terrace, 150m asl (55m above the main level of the Copais plain), to which also belongs the limestone acropolis of Haliartos (Philippson 1951: 467). The modern village of Haliartos is situated in the area to the E of the ancient acropolis. To the immediate E of the acropolis is a turn of the Zagora stream (now channelled), descending from the Evangelistria valley, and completely dry in summer. The road leading to Mazi (Philippson 1951: 474) lies above Haliartos to its S, while the village of Zagora/Evangelistria is at the very end of the valley.[2]

The territory of Haliartos also includes the end of the E edge of the Copais basin, from the Phoinikion ridge bordering the Karditsa/Akraiphnio bay to the Onchestos pass. The edge of the basin here is characterised by the presence of abrupt slopes (20-30m high) as well as *katavothrai* (swallow-holes) behind which begin the low hills characteristic of this part of the mountain border of Copais. There are promontories and small gulfs and islets before the promontories. In Philippson's time, the land in the bays was cultivated while the land of the lake was deserted (Philippson 1951: 489). Today the whole area is cultivated. In this area of Copais, remains of Minyan channels and dikes are visible. Along this Eastern edge, between the Copais and Yliki lakes, lies a limestone plateau with a karst basin in it (Asprokampos – 137m asl – Philippson 1951: 493-4), which could be included in the Haliartos *chora* (see below – BOUNDARIES). The limestone plateau continues to the S and comes close to the edge of the Theban *Tafel* (Philippson 1951: 494). In fact, only a low col separates the Copais basin, at its E side, from the NW edge of the Theban plain (Philippson 1951: 468). This col consists of rudist limestone and is crossed by two ways (20 / 30 m above the Copais) – the Southern one used today by the road, the Northern one by the railway (fig.1).[3]

Boundaries

The natural boundaries are quite clear: to the N was the lake; to the W the Palaiothiva-Petra (or Libethrion/Tilphousion) ridge marks a natural border which continues through the saddle between the Evangelistria valley and the plain of Koroneia; the S border line is marked by the Helicon and the lower ridge which bounds the Valley of the Muses to the N; while the E border is marked by Mt. Phagas as far as the Kokkoretsa peninsula (Fossey 1988: 320) and, probably, the Asprokampos plateau. As for the area around the strategic pass between the Teneric plain and the Copais,

[1] It should be examined whether the rise in elevation is natural or rather artificial to support the road and prevent it from flooding.

[2] Zagora in Slavic language means 'behind the mountain'; probably because of the route that came from the N valley (see below and fig.2).

[3] The latter lies between limestone blocks, while at the former, today more incised because of the cutting for the new road, the enlargement of an older one, schist with conglomerate as well as serpentine and pebbles are visible (Mesozoic or Young Tertiary - Philippson 1951: 468).

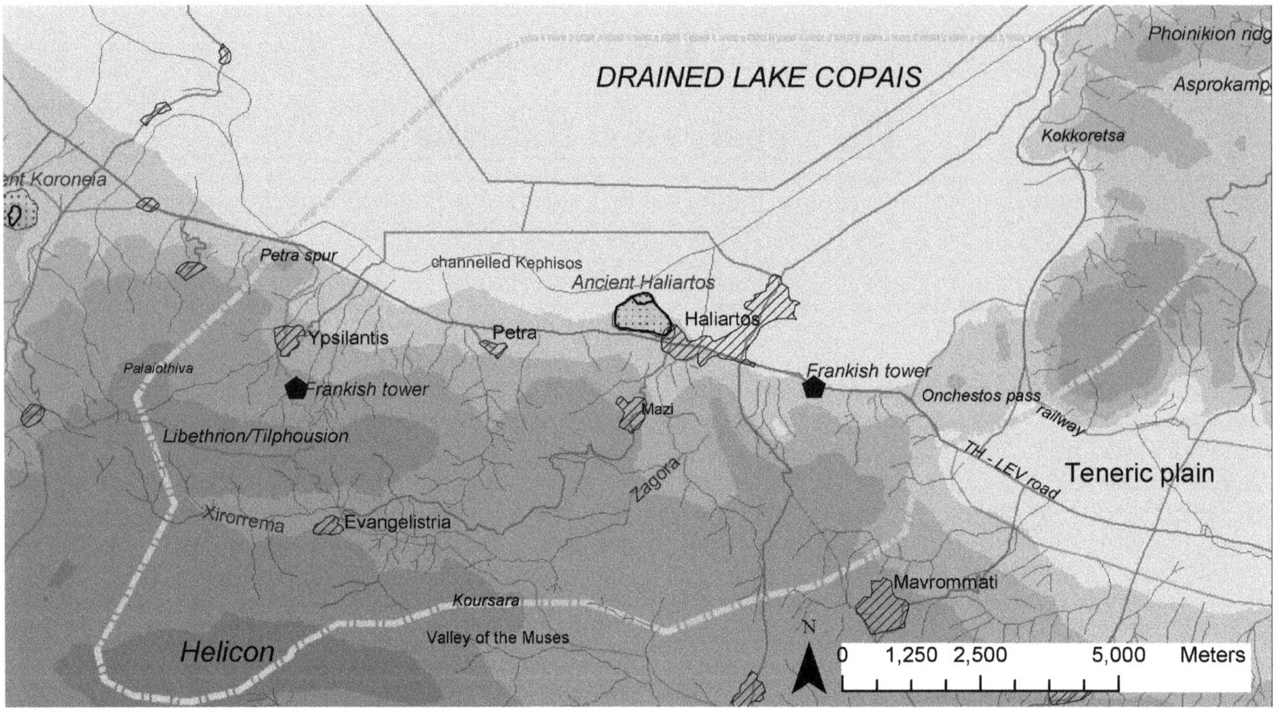

Fig.1. *Topographical setting of the chora of Haliartos.*

Onchestos was probably included into the territory of Haliartos (Fossey 1988: 320).

The border between the territories of Haliartos and Thespiae, on the light of intensive survey data, seems to lie in antiquity S of Mavrommati rather than N of the village, judging from the loss of small sites after Haliartos was destroyed by the Romans in 171 BC (J.L.Bintliff pers.comm. – see fig.19 in chapter II.4).

PHYSICAL LAND UNITS

Mountainous landscape is quite well represented because unlike the bordering N and E Copais area, behind Haliartos lies the Helicon massif, making the S area of the *chora* comparable in landscape character to the *chorai* bordering Copais to the S (along Copais' S edge), such as the areas of Koroneiake and Levadeia (see fig.2 in chapter II.1). On the other hand, the NE side of Haliartos belongs to the lower limestone landscape which characterised N and E Copais; in particular it is comprised of a low relief/ridge separating Copais from Yliki (see Akraiphiai *chora* – chapter II.3.7).

Plain areas would include the Copais basin, so the plain landscape percentage would increase according to the actual portion of the Copais basin free from water (see below and digital reconstruction of lake fluctuations in chapter III.1).

Hilly landscape	30.2%
Mountainous landscape	14.4%
Plain	55.4%

1	P1_P2	lacustrine basin, valley	45%
2	P3	gentle slope	2.3%
3	P4	Foothill	8%
4	H1	Plateau	8.4%
5	H2	gentle slope	1%
6	H3	moderate slope	5.7%
7	H4	severe slope	8.9%
8	H5	very severe slope	6.2%
9	M1	plateau	3.3%
10	M2	plateau/gentle slope	0.8%
11	M3	moderate slope	4.7%
12	M4	very severe slope	5.7%

Table 1. *Percentage of the different physiographical classes present in the Haliartos area (P=plain; H=hill; M=mountain)*

RESOURCES

Hence, as one might expect, the amount of available land would have varied considerably according to the fluctuations of the lake. Conversely, one can clearly see from fig.8 (including the digital reconstruction of the lake model) how much land bordering the Copais could have been covered by the lake. This land was probably seasonal marshland, but available for agriculture in certain seasons and years (see chapter II.3.1 and III for the digital model of the lake fluctuations; see also below – LONG TERM SETTLEMENT TRENDS). As also noted by Fossey (1988: 320), the basin slopes quite gently in this area, and different seasons could easily have provided quite different amounts of marginal pastures or marshes (mentioned in ancient texts: Strabo IX 407 and 411; Plinius *HN* XVI 66; Eustathius on *Il* II 503). Much of the land along the foothills was probably not even marshy, since it lies at a slightly higher level (see chapter III.1).

Certainly, resources from the marshy lake and the surrounding marshes were exploited by the local economy.[4] Reeds from the Haliartos marshes, used especially for flutes, are mentioned in Strabo IX 407 and 411, and Plinius *HN* XVI 66 (Fossey 1988: 320). Hunting activities were also practised (IG VII 2850 mentions a hunting club in the Roman period; IG VII 1828 – quoted in Fossey 1988: 320).

THE ARCHAEOLOGICAL RECORD

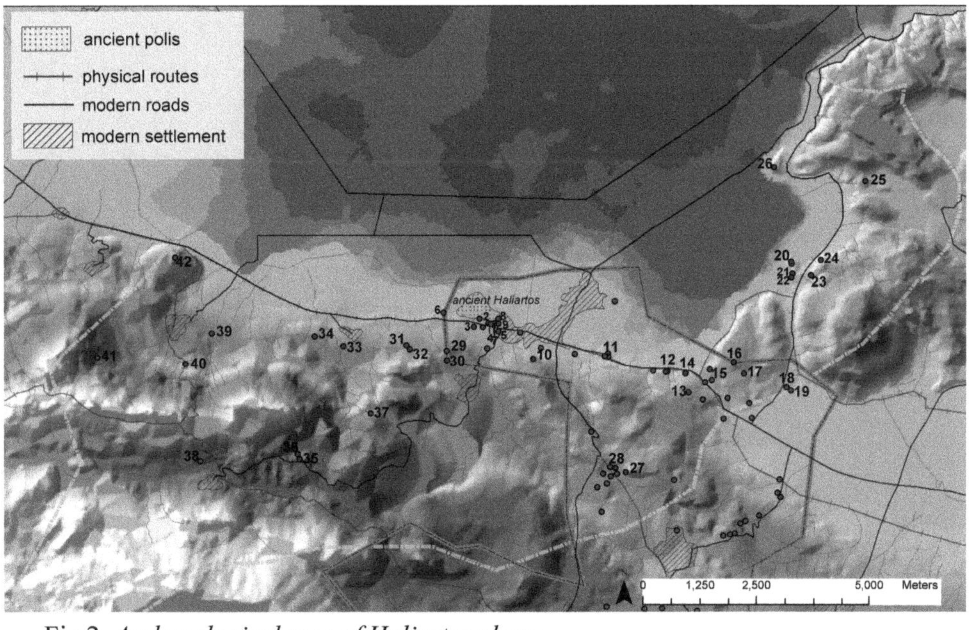

Fig.2. *Archaeological map of Haliartos chora.*

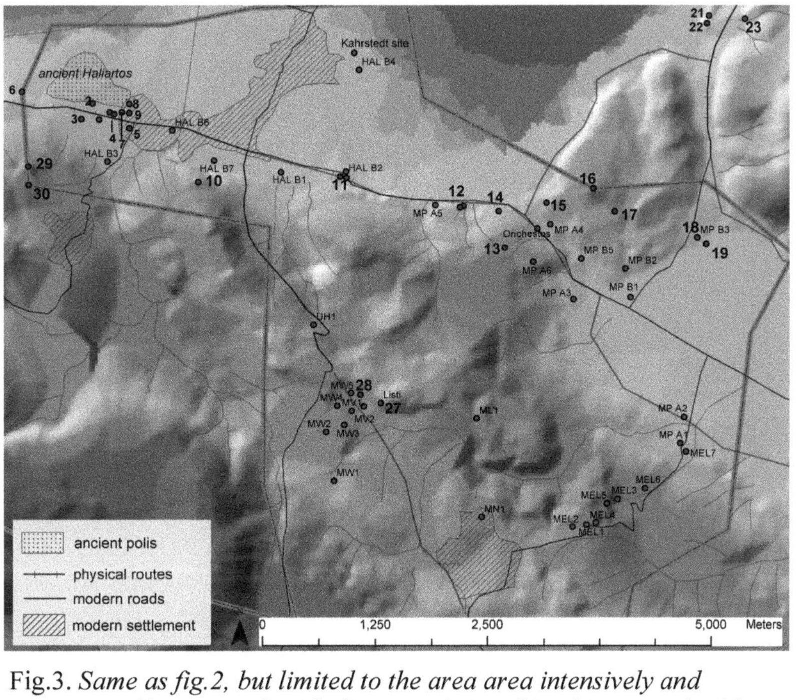

Fig.3. *Same as fig.2, but limited to the area area intensively and systematically surveyed, with discovered sites (listed in appendix I.8: table1- Survey Sites) marked.*

[4] Etymologicum Magnum: Ἁλίαρτος· ἡ πόλης. Ἢ ὅτι ἐκ τοῦ ἁλιέων καὶ τοῦ ἄρτου, reported by Fossey 1988: 320.

1	HALIARTOS	*Components HA_1 to HA_9*
2 to 6	Haliartos burial places	*Components HA_10, HA_11, HA_12, HA_13* (2); *HA_23* (3), *HA_24 and HA_80* (4); *HA_26* (5); *HA_27* (6)
7	Haliartos Kastron – SE	*Component HA_25*
8/9	Haliartos Kastron – East	*Components HA_81* (8) *and HA_82* (9)
10	Mazi Pyrgakos	*Components HA_14 and HA_15*
11	Pyrgos Haliartos	*Components HA_16 and HA_17*
12	Seidi	*Components HA_18 to HA_22*
13	Steni South	*Components HA_28 to HA_32* (ONCHESTOS sanctuary)
14	Steni SW	*Component HA_35*
15	Steni North – Tsumbitses West	*Components HA_33 to HA_34 and HA_85*
16	Tsumbitses-Kazarma	*Component HA_36*
17	Steni North - Tsumbitses East	*Components HA_37 and HA_38*
18/19	Tsumbitses Ag.Panagia	*Components HA_39, HA_40 and HA_76* (18); *HA_41* (19)
20	Megalo Kastraki	*Components HA_44 to HA_49* and *HA_83*
21	Kastraki area	*Component HA_50*
22	Mikro Kastraki	*Component HA_51*
23	Sphingion W	*Component HA_52*
24	Davlosis	*Component HA_53*
25	Kokkoretsa South	*Components HA_54 to HA_59*
26	Kokkoretsa - Proph.Ilias	*Component HA_60*
27	Listi	*Components HA_77 and HA_78*
28	Listi nearby	*Component HA_79*
29/30	Kariopoula	*Components HA_61* (29) *and HA_62* (30)
31/32	Stokthi Keramidissa	*Components HA_63* (31) *and HA_64* (32)
33	Petra	*Component HA_65*
34	Vigla	*Component HA_66*
35/36	Linopyrgo	*Components HA_67* (35) *and HA_68* (36)
37	Lykophio	*Component HA_84*
38	Mavri	*Components HA_69 and HA_70*
39	Vrastamites/Ypsilantis	*Component HA_71*
40	Ypsilantis Pyrgos	*Component HA_72*
41	Ypsilantis Ag. Nikolaos	*Component HA_73*
42	Petra/Tilphousion	*Components HA_74 and HA_75*

Table 2. *List of archaeological components and activity loci mapped in fig.2.*

A major part of our archaeological knowledge of the area is the result of the intensive topographical work carried out by Lauffer during his personal research on Copais (particularly focused on Koroneia and Haliartos *chorai*) – see graph in fig.4.

Extensive topographical surveys are also well represented within the *panorama* of discoveries, especially concerning actual and probable fort sites and cult places, along with travellers' discoveries.
Rescue excavation discoveries are mainly associated with building excavations in and by the modern village of Haliartos, as well as along the Thebes-Levadeia road.

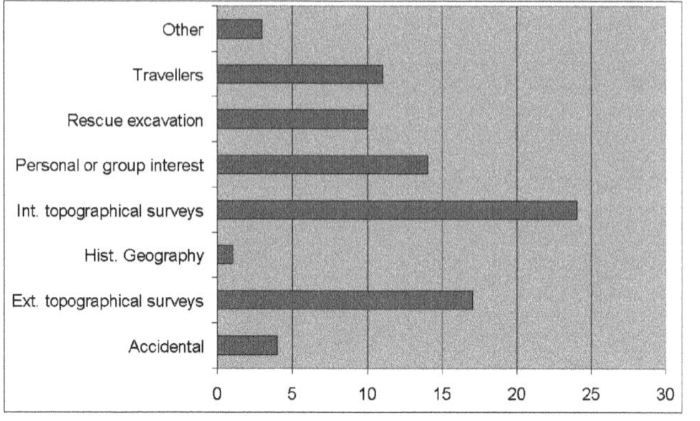

Fig.4. *Graph illustrating the proportion of components discovered within different research frameworks.*

A Systematic Intensive Surface Survey carried out in the area around Haliartos (area surveyed mapped in fig.2) by J.L. Bintliff and A.M.Snodgrass and their team (1986-1987) discovered sites listed in appendix I.8 (table 1 – *Survey Sites*) and mapped in fig.3.
Intensive surveys have contributed to knowledge of the landscape of the region. In fact, if we include the results of the intensive and systematic surface survey that focused on the Eastern part of *the chora*, mainly the lowland area towards Thespiae, the picture changes, as one may expect.

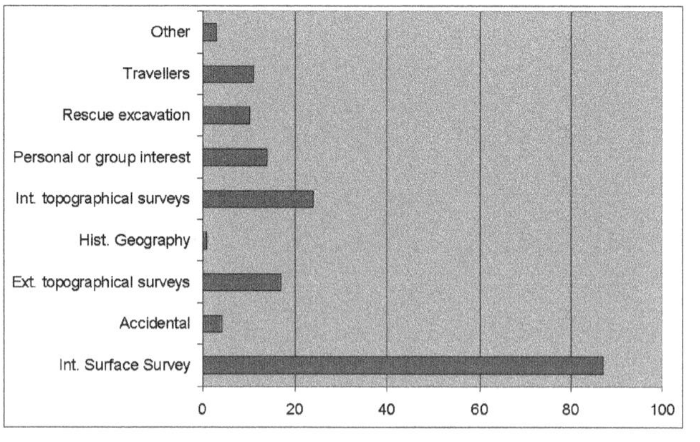

Fig.5. *Graph illustrating the proportion of components discovered within different research frameworks, including Intensive and Systematic Artefact Surface Survey.*

In fig.6 we can see the relationship between known archaeological sites and the distance from the modern road network. The apparent correspondence is weakened when considering the physical routes map (fig.2), since the known sites are along the roads corresponding to the probable ancient routes (mainly at the edge of the basin).

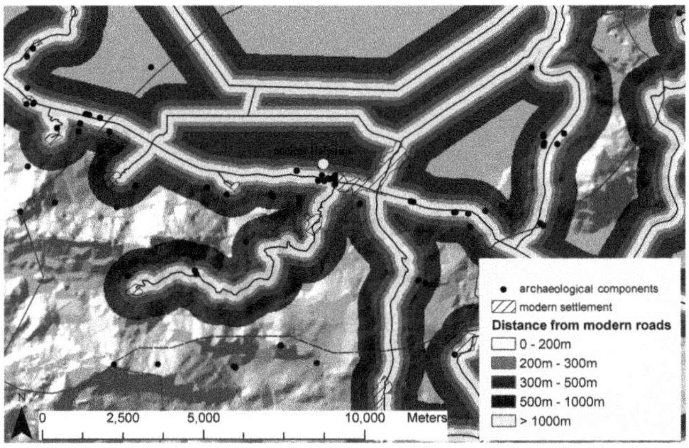

Fig.6. *Relationship between components and modern road network.*

The ratio of known Prehistoric to Greco-Roman components is 26 to 56 (1:2.15), while among the historical periods, 50% are dated Archaic to Hellenistic, 11% Roman-Late Roman, and 39% are attributed to the general Greco-Roman period.

ANALYSIS OF THE *CHORA* LANDSCAPE

PREHISTORIC PERIOD

The archaeological record available for the *chora* offers a fairly good picture as far as the Prehistoric periods are concerned (see figs. 6 to 9 in chapter II.3.1).
According to Lauffer (Kopais I), Neolithic sites are usually located around the Copais in rather open, fairly low, and not naturally defended locations compared to later Prehistoric sites. For the Haliartos area, *components HA_16 and HA_64* are known.
The site at Haliartos Pyrgos (*component HA_16* - Haliartos Pyrgos cave) is quite an important site in Neolithic Copais (occupied until Late Neolithic), together with Orchomenos and the Pyrgos *magoula* site in the NW bay. Abandonment after the Neolithic period at the Megali Katavothra site (NE Copais) and at Magoula Balomenou (Chaironeia), as noted by Lauffer (Kopais I: 45-6), who points out that in all three cases the abandonment was probably due to such locations being vulnerable to the effects of water and marshes. Bronze Age settlements in the Copais area are all better defended (usually closer to the lake but in higher locations and usually with a strong connection to the areas behind them). In the case of *component HA_16*, the site in the EH period had already moved to a higher position, on the plateau above the cave, as attested by intensive survey results (see above). The position is much better protected, though still by the spring and the lake. The Neolithic settlement at the S edge of the Copais area can therefore be compared with the settlement site in NW Copais. Likely EH sites are the EH activity focus found by the intensive survey by the Frankish tower at Haliartos (see appendix I.8, table 1 – *Survey Sites*: HAL B2, for which only Neolithic was known from the cave below), as well as at the site of Haliartos itself (*components HA_2 and HA_3*). These follow the pattern recognisable in neighbouring areas of the SW Copais basin for EH and MH sites, with elevated positions overlooking the edge of the basin (see Kalami-Lioma, Agoriani-Dekedes in the Koroneia *chora*).
In the LH period, the picture of settlement in the area is less well known. LH is known with certainty only at Haliartos Kastron (*component HA_4*) and at Tsumbitses Ag.Panagia (*component HA_76*), both quite elevated locations at the edge of the basin, while LH is only probably attested at Megalo Kastraki (*component HA_46*), with burials before it at Davlosis and Sphingion (*components HA_52 and HA_53*), and at Listi (HA_78), as well as at Kokkoretsa South and at Petra/Tilphousion (*components HA_55 and HA_75*), where attempts to date the remains of the fort to the LH period have been made (see appendix I.8). As Lauffer points out (Kopais I: 35), the typical relationship that seems to occur in the Copais area is that of Prehistoric (LH mainly) burial areas usually located in the foothills or on the slopes of the land opposite the (fortified) settlement to which they were attached, usually located by the lake (on a spur, promontory, islet). According to Lauffer, this would be the case for Megalo Kastraki (*component HA_46*) and the burial area by Davlosis (*components HA_52 and HA_53*). In this case, the settlement site would be Megalo Kastraki (but LH occupation is very doubtful - see appendix I.8, under the individual components).

GRECO-ROMAN ANTIQUITY[5]

Town level

Haliartos attests occupation from Protogeometric to Late Geometric times on the acropolis of the hill,[6] later to become the *polis* of Haliartos. It flourished in the Classical-Hellenistic period, reaching its greatest extent (around 40ha – Bintliff 1999d: fig.8) in the 4th C BC, and was destroyed by the Romans in 171 BC. The town of Haliartos used to lie in a strategic position, on a terraced promontory jutting into the lake and overlooking it at its SE corner. It lay along the route at the edge of the Copais which joined E and W Boeotia and crossed central Greece. In particular, when the Copais' water was at its largest extent, the *polis* of Haliartos dominated N to S access. Austin (BSA 1927: 206) notes *"the strategic importance of the position, to which the city doubtless owed its foundation"*.
Between the acropolis and the Southern hills constituting the lower Northern spurs of Helicon lies a quite flat area which is, however, higher than the main basin (Lauffer Kopais I: 47). The area was available to the city

[5] Period maps are included in chapter II.4, figs.17-19-21-23-25-27.
[6] Together with two or three very small rural sites in Haliartos immediate vicinity (survey sites HAL B4, HAL B5, HAL B6, HAL B7).

inhabitants, since, being above the 120m contour line, it was always dry and free of marshes (according to the digital fluctuation model – see fig.2; chapter II.3.1 – PERILACUSTRINE MARSHY LANDSCAPE- fig.10; Farinetti 2008). The area was crossed by streams coming down from the Helicon massif, creating a situation very similar to that of Koroneia, where, however, the ancient *polis* site was much more backwards).

Village level

Within the *chora*, in addition to the main *polis* of Haliartos, there is space for more settlement sites at a village level. The settlement attested at Onchestos would fill a gap. Initially, in earlier periods, the settlement was a village strictly linked with the sanctuary, and a satellite of Haliartos. It then probably became much more, having the status of a small town,[7] especially after the fall of Haliartos, when the SE Copais area was controlled by Onchestos, as mentioned in Diodorus XVII 10.4 (see above –*components HA_33, HA_34 and HA_85,* and below – LONG TERM SETTLEMENT TRENDS).

Another village is mentioned by ancient sources - ancient *Okalea* (in particular, Strabo IX 410). Since Leake (1835: 205f), attempts have been made to identify ancient Okalea with archaeologically more or less known sites. The various hypotheses suggested for the identification are: the Linopyrgo site (*component HA_68*), the Vrastamites-Ypsilantis site (Ross 1851: 31 - *component HA_71 and HA_72*), and Petra (Lauffer Kopais I – *component HA_65*), as well as Mazi Pyrgakos (Leake 1835: 205 – *component HA_15*). Up to now, however, attempts at identifying the site would seem to have been unsuccessful.[8] Among the many possibilities advanced, Lauffer suggests as most probable the Vrastamites-Ypsilantis site,[9] 6km distant from Haliartos and Alalkomenai, in accordance with Strabo IX 240, who reports (IX 410) Okalea as lying 30 stadia from Haliartos and Alalkomenai, between the two *poleis*.[10] During a personal visit to the site (after J. L. Bintliff, pers. comm.), Greco-Roman remains were noted, which would suggest the existence of a possible village below the Frankish tower in Ypsilantis valley (see below – LONG TERM SETTLEMENT TRENDS).

Rural segment

Our picture of the rural segment is enriched by the results of the Boeotia intensive survey, which discovered new sites and also clarified to some extent the occupation of the few previously known rural sites. For instance, at Seidi (*components HA_19 to HA_22*) the rural character of the site is attested by intensive surface survey carried out at the place, otherwise known only as an undetermined activity focus.

For the Protogeometric to Late Geometric period, the survey attested a nucleated focus on the acropolis hill of Haliartos, along with some very small rural sites in its immediate vicinity (survey sites HAL B4, HAL B5, HAL B6, HAL B7 – fig.3).

For the Classical period, intensive survey results indicate a picture of relatively low rural density and a concentration in small to medium farms (figs. 21 and 23 in chapter II.4) noted also at Hyettos and Tanagra, in contrast to the picture of high rural density, including large Classical rural sites found in the survey sector S of the city of Thespiae (Bintliff-Howard-Snodgrass 2007: 146).

The Classical florescence of the town of Haliartos is followed by a spectacular collapse, which also affects the rural landscape, after the destruction of the city by the Romans. At the city site a few isolated buildings indicate the existence of a small settlement focus of rural character (*components HA_25, HA_81, HA_82*). In the wider *chora* landscape, rural farms of Late Hellenistic – Early Roman date are almost non-existent in the Haliartos *chora* and very few in the overall results of the Boeotia Survey. Around the late 4th C AD, a radical change occurs in other parts of Boeotia, with the population rising again and the appearance of Roman villa and hamlet sites. It seems that the *chora* of the former city of Haliartos did not participate in this recovery (Bintliff 1999d: 29), although a few sites confirm the rural occupation of the landscape (*components HA_25, HA_81, HA_82,* and the Kahrstedt site, a large Late Roman villa estate by the very edge of the lake – see appendix I.8 table 1 – *Survey Sites*) – fig.25 and fig.27 in chapter II.4. Characteristic in this respect is that in the Late Roman period the area of the Onchestos sanctuary was turned into a villa establishment (*component HA_32*).

In higher areas in the Zagora valley one may expect sites linked with pastoral activities (*component HA_68*, for instance, could be interpreted as such, as well as *component HA_70*), as is usual on the slopes of Helicon, as also on the fringes descending towards the Copais basin (see Koroneia, Levadeia *chorai*).

Burial areas

The burial areas immediately related to the *polis* of Haliartos, by it and along the main road exiting the city towards Thebes and towards Koroneia and Orchomenos, are well known. No other burials are known from the *chora*, apart from some grave material which could be linked to a (large?) rural site in the vicinity (*component HA_69*), and the burial evidence (*component HA_62*) linked to the fort (Kariopoula - *component HA_61*), for which a strict link with a particular historical fact has been hypothesised.[11] *Component HA_35* is probably linked to the settlement at Onchestos.

---

[7] The town of Onchestos is mentioned in Pausanias IX 26.5 as well as in Strabo IX 410, 412, as well as by Diodorus XVII 10.4.

[8] Mylonas and Kirsten (RE 17 (1937) s.v. *Okalea* 2302-3) give a useful summary of the ancient testimonia. Mylonas and Kirsten even suggested that Okalea was not to be located in Copais, but between Thebes and Thespiae (following Apollodorus *FGrH* 244, F 197 and Plinius *HN* IV 7.26 - Fossey 1988: 314). Buck (1979: 9) also gives a brief account of the problems of identification. For a summary of the identification dispute see Fossey 1988: 316 note 55.

[9] This identification had already been suggested by Ross 1851: 31.

[10] Okalea is called *polypyrgos* in the Homeric *h.Ap.* 242.

[11] Lisandros' military camp before the attack on Haliartos (395 BC), mentioned by Plutarch (*Lys.* 29.11).

Cult places/Religious areas

The sacred landscape of the area is dominated by the presence of the sanctuary of Poseidon at Onchestos (*components HA_28 to HA_31*), at the passage between the Teneric and Copais plains. The sanctuary, whose life starts at the end of the 6th C BC and continues until the R period, lay in a strategical position along the border between Copais and the Teneric plain, and played an important active and symbolic role in Boeotian history as the meeting point of the Boeotian Amphictiony (see among others Kirsten, RE (1939) s.v. *Onchestos* 412-7; Buck 1979: 10; Schachter 1986 s.v. *Poseidon (Onchestos)*).

Other cult places are not recognisable in the picture offered by the available archaeological evidence.

Forts and fortifications

Several places are known which, considering archaeological evidence and location, might be interpreted as forts. Some of these are uncertain in date, and could be Prehistoric or Greco-Roman (4th C BC?), or both. Fortifications and watch towers indicate the strategic importance of the line Paliothiva-Petra as border line.[12] In this study, the Paliothiva fort is included in the Koroneia *chora*, while that of Vigla/Petra (the Tilphousion?) is recorded within this *chora*, not for historical political reasons, but just for simplicity and to avoid duplication.

The South and West forts of Kokkoretsa, if interpreted as such (see the individual components - *HA_54* and *HA_60*; uncertain character and uncertain chronological attribution), would mark the N border of the SW Copais bay, which belonged to Haliartos. The forts, if Greco-Roman in date, might have been controlled by the city of Haliartos.

The fortification at Kariopoula (*component HA_61*) represents a special case. It has been connected with a particular historical event and, if the interpretation can be considered valid, probably did not belong to the Haliartos *polis*, but was built for an 'independent' military need (see footnote 11). This could be meaningful in terms of landscape. Some of the fortifications built on the mountains of Boeotia, whose remains are visible today, can be seen as permanent sites (over at least a century, or even longer) for reasons of control or defence, but some other evidence may correspond to precise choices over the landscape for a specific meaning.

Other activities / unspecified activity areas

Some unspecified activity foci, of differing characters, can be pointed out in the general panorama of the archaeological record in the area. A possible character can be suggested for some (discussed under each individual component); but not for others.

Some of these should probably to be characterised as rural in character, linked either to agricultural or to pasturage activities (for instance, the historical site of Megalo Kastraki– *components HA_47 to HA_49*; component *HA_68* or *HA_70*).

LONG TERM SETTLEMENT TRENDS IN THE CHORA *LANDSCAPE*

The Southern side of the Copais basin (namely from Onchestos to Petra and Koroneia) is, according to Lauffer (Kopais I: 39), the most favourable for settling of the whole Copais area; rich in small streams running down Mt. Helicon and in fresh water springs. The area is defined by the Copais lake from the lower, Northern side, and by the Helicon uplands from the Southern. In the case of Haliartos, in particular, the long, fertile valley of Evangelistria marks the upland landscape. The upland valley of Evangelistria (crossed by the deeply incised Kalamaki/Ksirorrema river), also has an exit, through the uplands, to the area of Koroneia.

The plain below the hilly landscape between Helicon and the Copais was also dry in antiquity, because it lies a few metres above the flooding area and outside the marshy zone (see above and Farinetti 2008). It is a 2-3km wide zone, very fertile (marked to its N by the channelled Kephisos, and in antiquity probably by the natural course of the Kephisos, and crossed by the road joining Thebes to Levadeia-Orchomenos). Therefore, Haliartos and the settlements here had a large area available for cultivation (unlike the communities to the N and E of Copais). This is supported by evidence from the Boeotia Survey project. As reported by Bintliff and Snodgrass (AR 1986-87: 23ff.), a large Late Roman estate centre (called the Kahrstedt site) was found well into the lake N of the city of Haliartos, along with a few small Greco-Roman sites found between the 95 and 94.5 contours level, just above the rim of the basin (see fig.3 and appendix I.8, table 1 – *Survey Sites* and figs.21, 23, 25 and 27 in chapter II.4).[13]

This S edge of the Copais basin is marked by a series of settlement chambers[14] which run E-W and are associated with inland valleys and medium fertility hilly country of the northern slopes of Helicon and partially the fertile edge of the lake (fig.8). Modern villages, which go back to the Ottoman period, exploit these settlement chambers: Haliartos plays a prominent role both in antiquity and in the modern period.

Starting from the W we find the modern village of Vrastamites/Ypsilantis recorded in 1466-1688. Vrastamites is a Greek village, very important for its clear signs of continuity from earlier medieval times. The Frankish tower (13th century) above the modern village, and remains of houses, indicate a large Frankish feudal

[12] Kallet-Marx 1989: 301-311, provides a good consideration of visibility from towers (Palaiothiva and Evagelistria watch-towers), as well as a description of the sites of Paliothiva and of Evagelistria watch-tower as well as the routes (taken from Pritchett 1980).

[13] This information was used by Knauss and Kalcyck for the lake edge reconstructions, as well as for the construction of the digital model of the lake fluctuations (fig.10 in chapter II.3.1 and Farinetti 2008).

[14] These settlement areas are marked by bays defined by the Libethrion/Tilphousion spurs (see above – *topographical setting*). See fig.7.

village that continues and flourishes in the early Ottoman period (Bintliff-Kiel in preparation). Below the tower, remains of a possible Greco-Roman village are reported (see above – *village level*), and the presence of a settlement of second-rank status would fill in a gap at the W end of the *chora*,[15] as Onchestos did at the E end. A possible structure in the settlement organisation is visible in the results of a cost-distance analysis[16] having as a base a 2.5km average settlement territory from 1st and 2nd rank settlements (fig.7).

W of Ypsilantis the modern village of Petra (formerly Siakho) is also recorded in the Ottoman archives in 1540 and 1570. The area provides some evidence for ancient occupation (*components HA_65 and HA_66*), and Lauffer has suggested a possible identification with ancient Okalea (see above – VILLAGE LEVEL). In my opinion, the archaeological evidence is not enough to support the existence of a 1st or 2nd rank settlement in this area, although the results of the cost distance analysis show enough room for a possible hamlet (figs. 7-8).

Further to the W follows the area of ancient Haliartos, which is more open to the exploitation of the alluvial soils of the edge of the lake. The modern settlement there (immediately E of the ancient acropolis) has a central character today, as in the past, and offers an interesting case of settlement history. After the destruction of the ancient *polis* in 171 BC there seems to be some Roman-Late Roman agricultural activity at the site (see above – RURAL SEGMENT and Bintliff 1991b: 126). Medieval Haliartos goes back to Middle Byzantine times, if not earlier, and flourishes in Frankish times.[17] In the Ottoman archives a Greek village named Harmena appears (1466-1570), identified with modern Haliartos. The early appearance in the defters as well as the presence at the site of Early Ottoman pottery (and not later - Bintliff 2000a) would indicate the flourishing of the site in that period as well as the abandonment of the entire location[18] until a late 19th century refoundation.[19] Therefore, the site of Haliartos provides strong evidence for a nucleated settlement in several periods of the past. The ancient city appears as a small settlement focus in the Geometric period, along with some small sites in its immediate vicinity (see above – RURAL SEGMENT), exploiting the fertile area along the S side of the Copais. In the Classical period, the *polis* of Haliartos flourished, as did the rural landscape, with a moderate density of rural sites, as elsewhere in Boeotia (see above – RURAL SEGMENT). In the Roman period, after Haliartos' defeat, in 171 BC (see appendix I.8, under the city components), the evidence known from the city area and immediate surroundings is mainly to be associated with a rural occupation of the landscape, maybe hamlet-like, while the town-level settlement probably moved to Onchestos (see below). In the Late Roman period the *chora* of the former city of Haliartos did not participate in the general recovery of the landscape noted in other areas of Boeotia (notice though the LR Kahrstedt villa site by the edge of the lake). - figs.25 and 27 in chapter II.4.

At the eastern edge of the ancient *chora*, a final settlement chamber opens up, which would naturally include the Onchestos pass and the alluvial plains on both sides of it. The fact that this area belongs today to the modern village of Mavrommati, whose area also in the past could have been included in the ancient *chora* of Haliartos (see above – *boundaries*), would explain the apparent lack of a modern village in the strict Onchestos surroundings. In the Classical period the area is occupied by the important sanctuary of Poseidon (*components HA_33 and HA_34*[20]), associated with a habitation area, filling in the available settlement chamber in the area between the *polis* of Haliartos and the Teneric plain,[21] controlled by Thebes. The sanctuary overlooked the road joining Haliartos to Akraiphia.[22] During the 4th C BC, the meetings of the confederation (*koinon*) used to take place in Onchestos, and the sanctuary grew with a series of activities related to it as well as a small settlement, a satellite of Haliartos (once Thebes lost control over it). Later, as Theban hegemony weakened, there was a reduction in activities at the sanctuary, but the settlement probably attracted some of the population of the area. Progressively, with the development of civic activities, and especially after the fall of Haliartos (destroyed in 171 BC by the Romans), the settlement of Onchestos became a proper small *polis* site (*component HA_85*), being in Roman times the only nucleated settlement within the same settlement chamber that Haliartos had once controlled. In this sense, the town of Onchestos is said to have replaced Haliartos, in controlling the SE area of Copais, as mentioned in Diodorus XVII, 10.4 and proved

---

[15] This was already suggested by Bintliff in his Thiessen polygons analysis (Bintliff 1994b - fig.20).

[16] See chapter II.3.1 – LONG TERM SETTLEMENT TRENDS.

[17] The Frankish tower, by which Prehistoric occupation as well as a Greco-Roman activity focus have been noted (see appendix I.8, the Pyrgos Haliartos site), marks the location of another Frankish site, almost 3km to the E of the Haliartos *polis* site (fig.1 and fig.7), always by the edge of the basin.

[18] "*We have made a plausible case that the location was deserted in the troubled 17th century for an upland village called Mavrommati Harmena (a kilometre to its south), which ought then to have been Greco-Slav in original ethnicity. Yet this village in turn is abandoned by the 19th century and its population moves to merge with the Albanian in origin and still currently occupied village of Mavrommati (Yorgi), further away in a different village niche*" (Bintliff 2000a: 145-6). See fig.7.

[19] Modern Haliartos village was founded as a residence for the workers employed by the British Lake
Copais Company, "*but owing to its excellent communications and intermediate location between the two main provincial towns of Thebes and Livadeia, has gradually developed into a secondary service centre for surrounding villages*" (Bintliff 2000a: 146).

[20] For a similar case of a sanctuary filling in the gap for a settlement site, see the Ptoion sanctuary in the Akraiphiai *chora*.

[21] See also Bintliff's Thiessen polygons analysis (Bintliff 1994b - fig.20).

[22] Stephanus Byzantius, s.v. *Onchestos*, says that Onchestos was between the two *poleis*. The road between Haliartos and Akraiphia ran along the SE edge of the Copais basin.

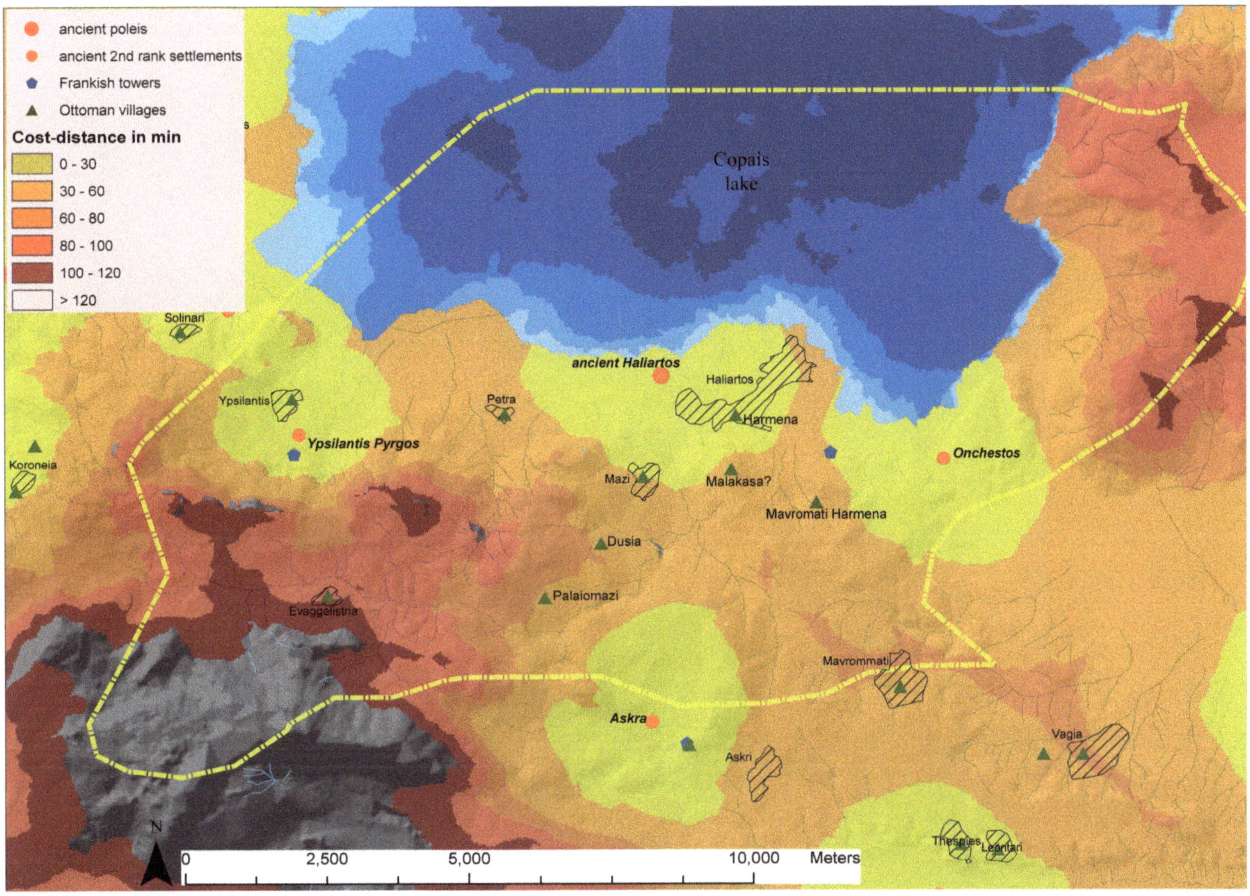

Fig.7. *Classified surface representing the cost-weighted distance (1/2 h walking and further ranges) from recognised 1st and 2nd rank ancient settlements (represented by larger and smaller dots). Areas without dots indicate potential settlement chambers. Ottoman villages and Frankish towers have also been added to the map to show their spatial relationship with the Greco-Roman settlement network and to appreciate potential settlement chambers.*

by archaeological evidence[23] (see above and Bintliff 1991b:124). In the Late Roman period the area of the Onchestos sanctuary was turned into a villa establishment (component *HA_32*), testifying to the transformation of the landscape during this period.

The southern part of the *chora* is occupied by part of Mt. Helicon. The Ottoman defters of 1642 record for this area a spatial sequence of villages, indicating also a route through the mountains. The villages are Mavrommati, Malakasa, Dusa, Espanos, Zagara/ Evangelistria and Rastamit/Ipsilanti (Bintliff-Kiel in preparation). On the map we can locate these Ottoman villages, of which only a few survived to the modern period (fig.7). Starting from E to W we find Mavrommati Harmena (recorded 1642, 1646, 1655),[24] the village of Malakasa (recorded 1466-1688, probably identified with the modern toponym Malaki in this area), the modern village of Mazi (also recorded in the Ottoman archives in 1466-1570), the now deserted village of Dusia (1466-1688, identified with the modern church of Zoodochos Pigi), the ruined village of Paleomazi[25] (possibly identified with the Ottoman village of Espanos which is recorded in 1466-1688 on the route between Dusia and Zagora), and modern Evangelistria (formerly Zagara, recorded in the 17th-century Ottoman lists (1642-1687)[26] (see Bintliff-Kiel in preparation for the attributions). The large number of villages in this upland area is remarkable, and shows the existence of small settlement chambers across a demarcated zone of small valleys and slopes of middle fertility in this part of the mountains.

Were these settlement chambers occupied by second order villages in antiquity? The cost-distance analysis shows an empty zone in the elevated area beyond the immediate surrounding of the town of Haliartos to its S and SE (fig.7), within one hour walk ranges from nearest settlements. Within it, the archaeological record provides

[23] Dates and chronologies actually coincide, as regards development at the Onchestos settlement site and evolution of the rural landscape (see above in text, rural sites by Haliartos).

[24] The deserted village, in the upland area to the SE of Haliartos, was recognised through the intensive survey conducted within the framework of the Boeotia Survey project. See above, footnote 18.

[25] The ruined village lies by a perennial spring in the Evangelistria upland valley, ca 4km SW of and above Mazi on the SE slopes of the Goulas mountain, today crossed by the Mazi-Evangelistria road. The deserted village lies on its left side.

[26] During the Ottoman period the settlement moved from the N slopes of the upland valley to the S slopes with springs, due to the increase in the number of inhabitants.

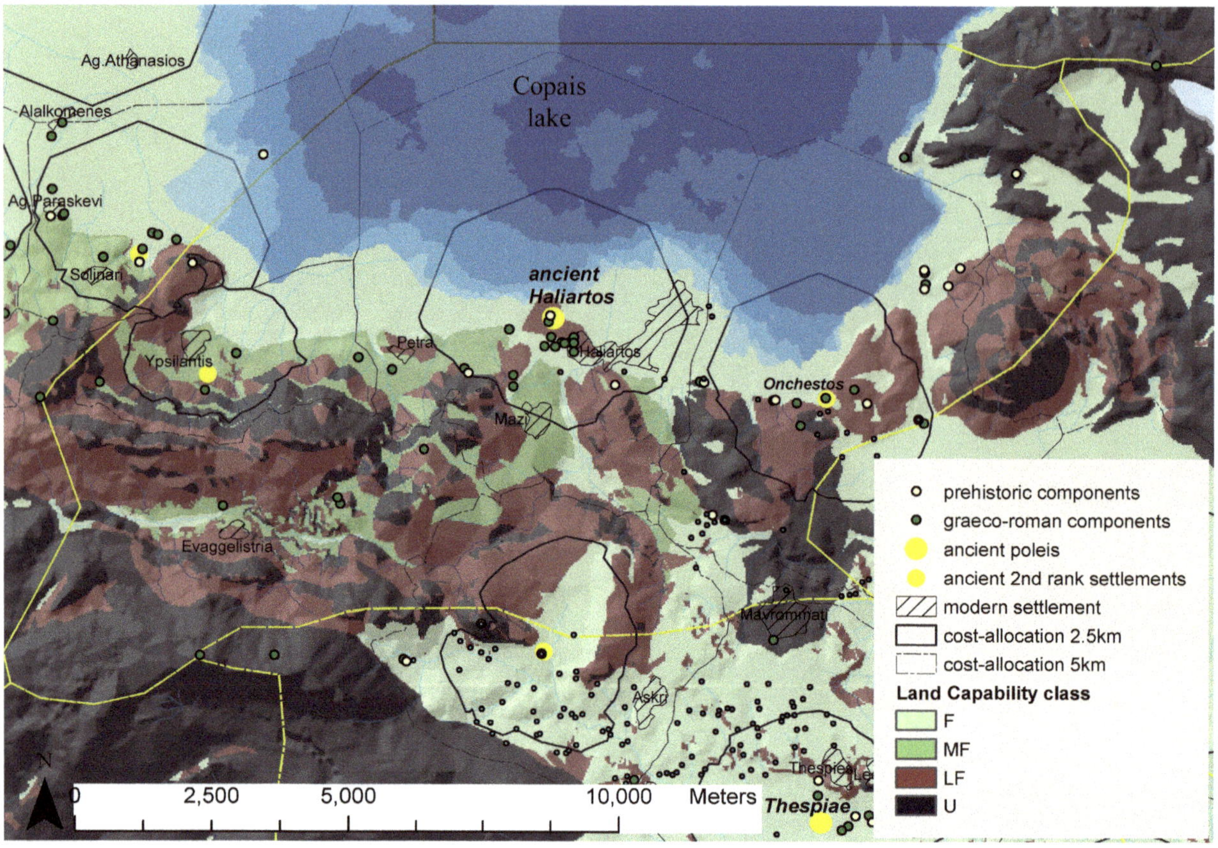

Fig.8. *Map showing the Greco-Roman settlement network, the polygons resulting from the cost-distance analysis (marking half an hour and one hour walking time distance) and dots representing the known archaeological components (same as in fig.2), with land capability information underlain.*

some evidence, but neither village nor hamlet sites have been recognised. Reused blocks in the medieval buildings of Palaiomazi,[27] which lies by a perennial spring in the Evangelistria upland valley, could be signs of Greco-Roman activity here.[28] To the NW of the village of Evangelistria, on a river terrace, remains from tile graves have also been reported (*component HA_69*). This evidence, along with the presence of watchtowers (see above), may indicate activities associated with upland uses (as better recorded in the *chora* of Koroneia – chapter II.3.1), but no substantial ancient settlement in this upland zone. This gap could possibly reflect a real pattern, as in general at the rural level the Boeotia survey seems to indicate thin settlement beyond the immediate surrounding of the town of Haliartos (see above).

[27] The reused blocs are recorded in the database as AE1396.

[28] See also Lauffer Kopais I: 49.

II.3.9

Eastern Helicon: Thespike

TOPOGRAPHICAL SETTING

The area under the control of the ancient city of Thespiae is not a unified and uniform plain, but rather a mixture of river valleys and intervening hills. It was heavily settled in modern times.
The ancient city of Thespiae and the surrounding areas (including the modern villages of Neochori and Ellopia) belong to what has been called 'the Tertiary *tafel* of Thebes' (see chapter II.1 and Philippson 1951: 502). The proper basin of Thespiae (constituting the 'central basin' of the Tertiary *tafel*[1]) extends from the turn of the Askris river (the abrupt turn that the river makes around a spur of Helicon running E-W – Palaiokarandas) to the W to where the Asopos river begins towards the E. It seems possible that the river Askris once constituted the upper part of Asopos (see chapter II.1).
To the S the basin is bordered by the Livadostra/Stavropotamos river, which has its source in a basin (2-3km; 280m asl) by the modern village of Kaparelli, at the foot of Kithairon (see chapter II.3.11). The basin is bordered to the E by a Tertiary *tafel* divided into several valleys (360m asl). The tafel comes out from the mountains at the village of Kriekouki/Erithrai and extends to the S towards the Asopos (Philippson 1951: 504). The Livadostra river valley extends towards the E, and flows along the foot of Kithairon in a wide cultivated valley (200m asl) by the village of Kaparelli. Further to the W the valley becomes narrower and deep, with banks covered with trees, and it turns to the SW between Korombili and Kithairon to reach the sea in a small plain by the few houses of Livadostra.[2] The W coast of the gulf of Livadostra comes in a direct line to Cape Patima, closing the wider bay to the SW. More complex and varied is the E coast reached by stream valleys from Kithairon, which today belong to the village of Kaparelli,[3] until Cape Punta, which closes the wider gulf to the SE (Philippson 1951: 505). The Livadostra gulf is 5km wide and cuts deeply into the land. It is open to the S wind, with the exception of Ag.Vassileos bay (Philippson 1951: 206).

The Northern uplands of the Tertiary *Tafel* of Thebes (Philippson 1951: 506-7 – iBc), running with continuous steep edges W to SE from behind Haliartos, at the edge of Helicon, up to Thebes and the Soros, separates the Parasopia and Thespiae areas from the central plains of Boeotia. On the *Tafel* are the villages of Mavromati and Vagia. To the W of Thebes up to Vagia this upland ridge goes further into the basin forming promontories and spurs. Behind the ridge, at the E edge of Helicon, is the Askris river, which, after emerging from the Valley of the Muses, runs in a SE direction across a soft, wide and very fertile tertiary valley, at the edge of which is the village of Palaiopanagia/Askri. At the foot of the edges of Helicon, where it turns to the W, lies the village of Neochori. After Neochori, the Askris river crosses a plain similar to that below Parapungia/Leuktra on the other edge of the basin. The village of Neochori lies at the starting point of a ridge jutting out from Helicon and ending in the Thespiae basin with the Palaiokarandas spur, around which the Askris river turns (Philippson 1951: 453), changing the direction of its course from NW-SE to NE-SW. Flowing towards the Domvraina basin, the river meets a rounded limestone ridge which leads the river into a small valley before entering (by Ellopia village) the Xironomi basin and crossing a series of small valleys opening to the N (Philippson 1951: 459).
There is a slightly marked low watershed between the rivers Askra and Kanavaris to the W of ancient Thespiae. The Kanavaris flows towards the E and becomes larger, taking water from several springs in the plain (see Bintliff-Howard-Snodgrass 2007), where the ancient city of Thespiae lies; afterwards it flows in its narrow valley through the hills, and emerges into the Aeonian plain ca 4km W of Thebes. The road from Thebes to Thespiae and Domvraina follows the Kanavaris valley as probably the ancient route did (fig.1). Two villages (Erimokastro/Thespies and Kaskavelli/Leondari) are situated on the edge of the Tertiary *tafel*, before the site of the ancient *polis* of Thespiae.
The so-called Valley of the Muses can be considered to be a small self-contained landscape within the wider area, opening into the Helicon massif in a SE-NW direction to the NW of ancient Thespiae, including at its SE the modern village of Askri.

Boundaries

The boundaries of the *chora* are clearly definable to the W and the S: to the W Mt.Helicon, to the S the Kithairon,

[1] See appendix II: iBb and Philippson 1951: 506 (iBb).
[2] In this plain are the remains of ancient Kreusis, see below in text.
[3] The harbour used by the inhabitants of Kaparelli today is not Livadostro but Ag.Vassileos.

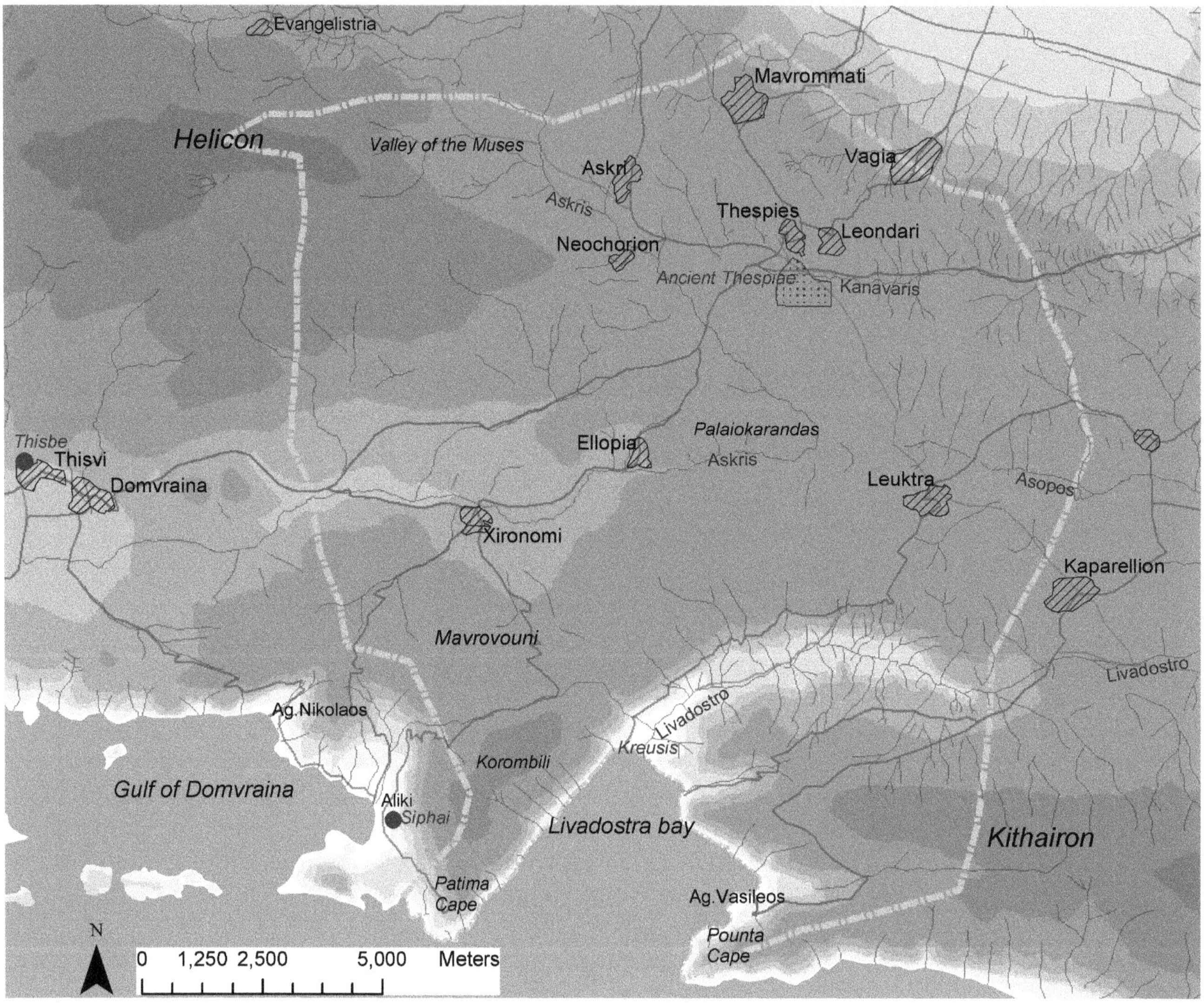

Fig.1. *Topographical setting of the chora of Thespiae.*

the Corinthian gulf and the Western ridge of Mount Korombili (Fossey 1988: 164). Conversely, to the E the boundary with Plataea and Thebes is not easily definable, as no clear topographical feature marks it, and only a sequence of low hills separates the bulk of the Asopos valley (belonging to Plataea and the Parasopia *chora*) from the Thespian plains proper; to the N the border with the *chora* of Thebes is even less clear; as Fossey (1988: 165) points out, it seems logical to include in the Thespike area the area of the Vagia and Mavromati villages, i.e. the N slopes of the chain of hills on which these villages stand, above the Teneric plain which Roesch (1965: 52 n.5) would give to Thebes and partly to Haliartos (its W swampy land). On the other hand, to judge from the loss of small sites after Haliartos'defeat by the Romans (171 BC), the border to Haliartos seems to be S rather than N of Mavrommati village.[4]

In antiquity, the boundary between the *chorai* of Thespiae and Thisbe[5] could have been marked by a dam, as suggested by Knauss (1992), who writes: "*The small prehistoric dam may have marked the boundary between the territories of Thespiai and Thisbe in later antiquity just as did the old barrages at Tegea, Mantinea and Orkhomenos in Arkadia. The Donakon, the reed-bed in the reservoir of the small dam, was situated in the territory of Thespiai as Pausania says*".

PHYSICAL LAND UNITS

Hilly landscape is predominant, and this is mainly due to the elevation range within which the majority of the *chora* lies, marking the landscape with more or less flat inner upland basins and valleys, where often no visible river flows. Within the hilly landscape, the predominant physiographical class is therefore constituted by plateau and gentle slope features (classes H1 and H2). H3 and H4

[4] This is the picture which seems to emerge from the systematic survey carried out in the area (Boeotia Survey Project) – see fig.3.

[5] The territory of Thisbe is included in the present work in the discussion on the Boeotian areas to the Gulf of Corinth (see chapter).

features mark the lateral slopes of the valleys and watersheds.
The mountainous segment is represented by the Helicon slopes and peaks, which mark the W end of the *chora* landscape, while plain landscape is represented only in the Livadostro valley and bay and by the lower valley of the Askris river towards the Domvraina basin, when the elevation range drops (fig.2 in chapter II.1).

Hilly landscape	79%
Mountainous landscape	11.7%
Plain	9.3%

1	P1_P2	lacustrine basin, valley	3.5%
2	P3	gentle slope	0.5%
3	P4	foothill	5.3%
4	H1	plateau	29%
5	H2	gentle slope	8.7%
6	H3	moderate slope	16.8%
7	H4	severe slope	12.9%
8	H5	very severe slope	11.5%
9	M1	plateau	2.3%
10	M2	plateau/gentle slope	1.5%
11	M3	moderate slope	2.7%
12	M4	very severe slope	5.2%

Table 1. *Percentage of the different physiographical classes present in the Levadeia area (P=plain; H=hill; M=mountain).*

RESOURCES

The basin of Thespiae, along with other small basins in the area, is covered with a brown, thick and muddy soil, which can be considered one of the most fertile soils of Greece, and is cultivated to a great extent. Generally speaking, the Tertiary *tafel* of Thebes of which most of the area is constituted (see above) is a formation marked by fertile soil (see in fig.8 the F areas). Some marshy areas are present, and some areas are apparently not characterised by the presence of surface streams, but water sources are not rare, and the villages of the area (Ellopia, Vagia, Neochori, Erimokastro/Thespies, etc.) still take their water from them (Philippson 1951: 502). The Livadostra river valley also constitutes a fertile area, to the SE, with fluvial deposits. On the other hand, the ridge that separates the Thespiae basin from the Livadostra valley constituted of dolomites and carbonate rocks, as well as the carbonate Helicon massif overlooking the area from the NW, do not allow agricultural production but create a landscape suitable for grazing (Shiel-Stewart 2007). Hunting activities are also attested on Mt.Helicon (IG VII 1828, quoted in Fossey 1988: 165).
A detailed study of the soil potential of the immediate surroundings of the city of Thespiae has been carried out by Shiel and Stewart (2007) and offers guidelines for understanding the agricultural landscape of the region (see table 2 in chapter I.2.1).

THE ARCHAEOLOGICAL RECORD

1	**THESPIAE**	*Components TH_1 to TH_12*
2	**Thespiae NE**	*Component TH_6*
3	**Erimokastro W**	*Component TH_13*
4	**Thespiae vicinity**	*Component TH_15*
5	**Thespiae S**	*Components TH_17 to TH_19*
6	**Thespiae E (poliandrion)**	*Component TH_14*
7	**Palaiokarandas**	*Components TH_ 29 to TH_32*
8	**Leondari NE**	*Component TH_20*
9	**Vagia**	*Components TH_21 and TH_22*
10	**Mavromati**	*Components TH_26 to TH_28*
11	**Askra Episkopi (ancient ASKRA)**	*Components TH_ 33 to TH_39*
12	**Askra Pyrgaki**	*Components TH_ 40 to TH_44*
13	**Askra - Sanctuary of the Muses**	*Components TH_ 45 to TH_49 and TH_23*
14	**Askra Helicon**	*Custers TH_50 and TH_51*
15	**Askra Hippokrene**	*Component TH_52*
16/17	**Neochorion**	*Components TH_96* (16) *and TH_97* (17)
18	**Ellopia – Koraki**	*Component TH_94*
19	**Ellopia – Agia Kyriaki**	*Component TH_95*
20 – 23	**Tatiza**	*Components TH_53* (20), *TH_54* (21), TH_55 to TH_57 (22), *TH_58* (23)
24	**Xironomi**	*Components TH_59 and TH_60*
25	**Livadostro E – Kastro**	*Components TH_61 to TH_69 and TH_24*
26	**Livadostro W**	*Components TH_70 to TH_74*
27 – 30	**Parapoungia/Leuktra**	*Components TH_75* (27), *TH_76* (28), *TH_77* (29) *and TH_25* (30)
31/32/33	**Arkopodi – Eutresis**	*Components TH_ 78 to TH_86, TH_84 and TH_89* (31)*; TH_87, TH_88 TH_90 to TH_92* (32); *TH_93* (33)

Table 2. *List of archaeological components and activity loci mapped in fig.2.*

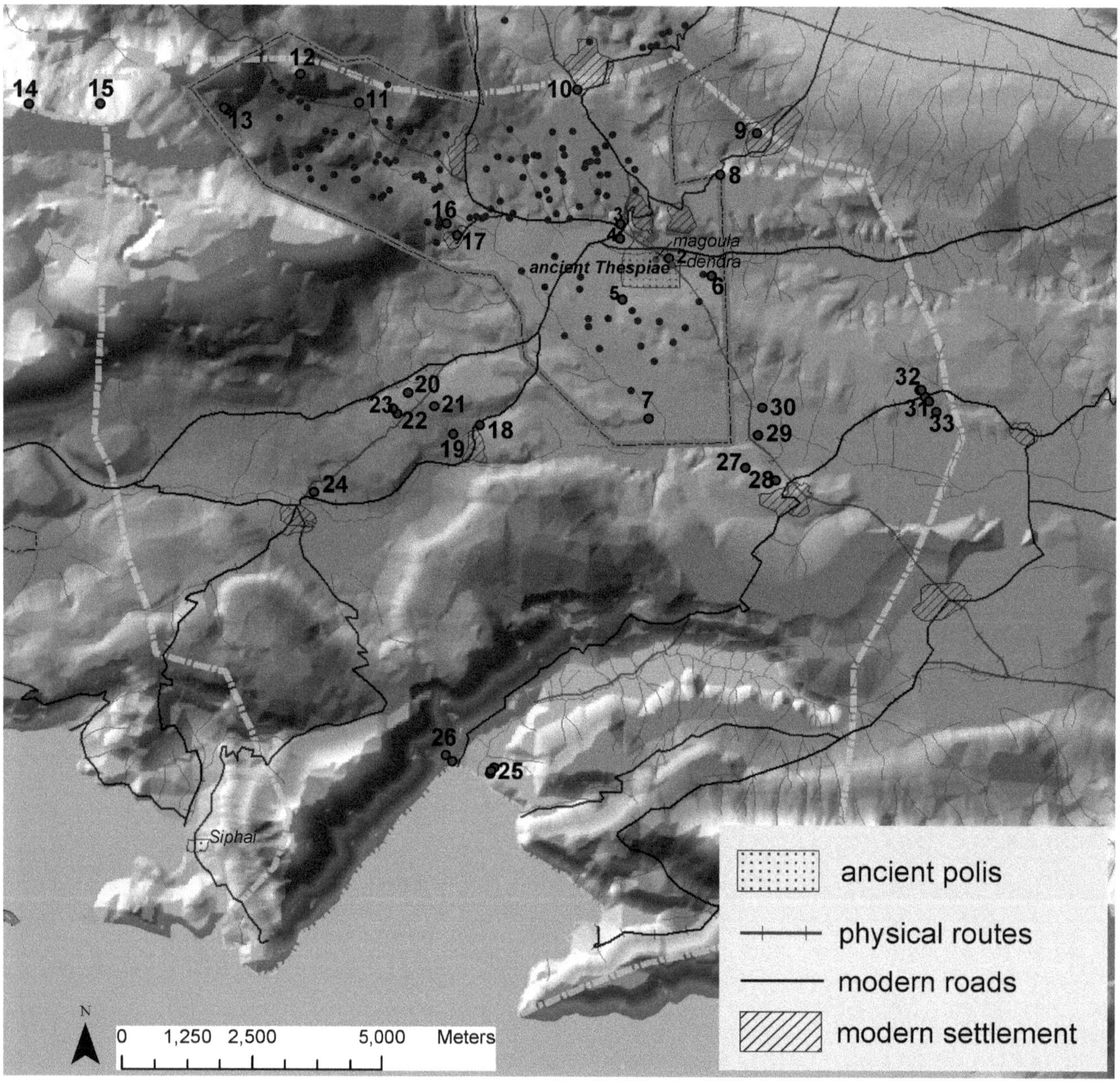

Fig.2. *Archaeological map of Thespiae chora.*

In the general picture of archaeological discoveries in the *chora* of ancient Thespiae, rescue excavations are little represented, probably because of the inner position of the area compared to the main Boeotian roads and urban centre.

Conversely, intensive and systematic artefact surface surveys carried out in several areas of the *chora* (in particular in the immediate surroundings of the city site and further to the N, as well as in the Valley of the Muses – see map in fig.3), covered different types of landscapes within the micro-region and give us a more reliable picture of the diachronic occupation of the landscape (see below and chapter III.2).

Discoveries led by personal or group interest are greatly represented, and this is due largely to interest in the main outstanding sites within the *chora*, such as the sanctuary of the Muses and the Hellenistic fortification in the Valley of the Muses, or the evidence at Livadostro/Kreusis and Eutresis. Accidental discoveries are also quite significantly represented, and correspond mainly to cemetery and burial areas discovered by means of inscriptional evidence.

A Systematic Intensive Surface Survey carried out in the area around Thespiae (area surveyed mapped in fig.2) in the 1980s by J.L. Bintliff, A.M. Snodgrass and their team discovered sites listed in appendix I.9 (table 1 – *Survey Sites*) and mapped in fig.3.

In the graph (fig.4) the segment 'intensive and systematic artifact surface survey' refers only to components representing periods of occupation discovered only by means of intensive survey at sites whose occupation in other periods was known already by extensive surveys or other kind of research. It does not refer therefore to the large numbers of activity foci reported in appendix I.9 (table 1 – *Survey Sites*). If we add these components to the general picture, the chart would change considerably (fig.5).

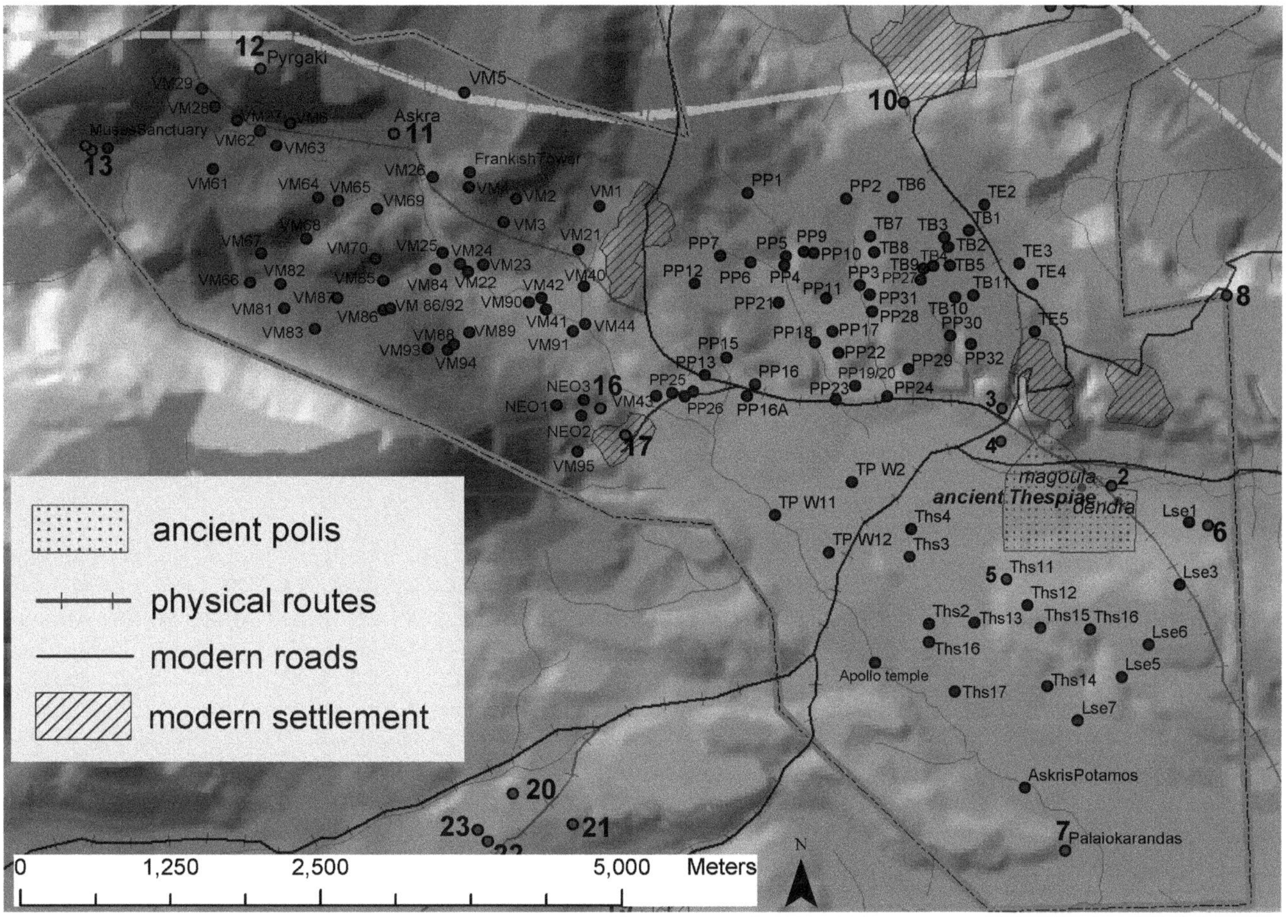

Fig.3. *Same as fig.2, but limited to the area intensively surveyed, with discovered sites (listed in appendix I.9: table 1 – Survey Sites) marked.*

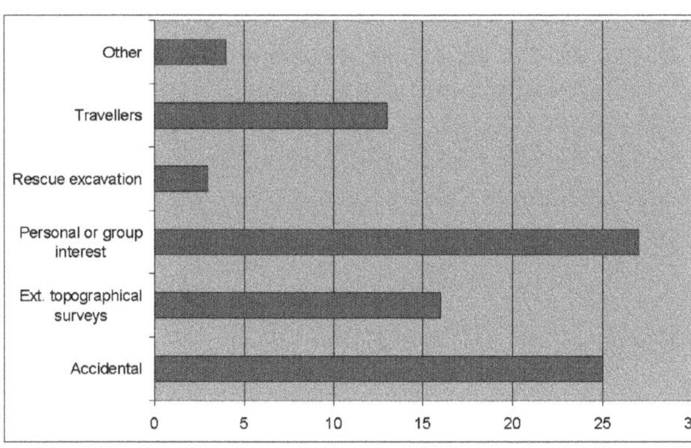

Fig.4. *Graph illustrating the proportion of components discovered within different research frameworks.*

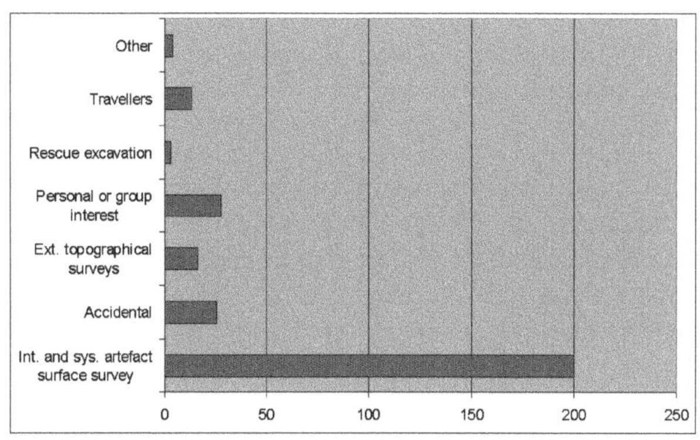

Fig.5. *Graph illustrating the proportion of components discovered within different research frameworks, including Intensive and Systematic Artefact Surface Survey.*

In fig.6 we can see the relationship between known archaeological sites and the distance from the modern road network. With the exception of the sites in the upper valley of the Muses and the prominent Palaiokarandas ridge, quite far from the main roads, the correspondence is quite striking. This is mainly due to research focused on the main settlements, which were along main routes that also correspond to the modern road network. The map with the results of the intensive surface surveys carried out in the area helps us to understand to what extent the archaeological record deduced from bibliographical sources mainly on extensive research could be incomplete.

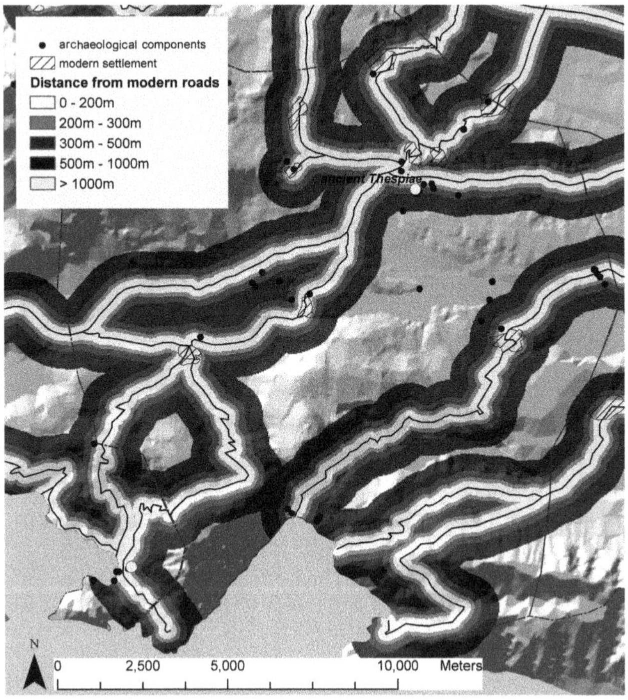

Fig.6. *Relationship between components and modern road network.*

The ratio of known Prehistoric to Greco-Roman components is 18 to 79 (1:4.4), while among the historical periods, 66% are dated Archaic to Hellenistic, 27% Roman-Late Roman, and 7% are attributed to the general Greco-Roman period. The Prehistoric landscape is comparatively quite poorly known, and this is probably mainly due to biases in research (generally less interest in the area shown by Prehistorians), as intensive survey conducted in parts of the area demonstrates, with a considerable number of Prehistoric sites discovered. The ratio between historical periods shows the general picture known elsewhere in other *chorai* of Boeotia and from intensive surface surveys for the rural landscape.

ANALYSIS OF THE *CHORA* LANDSCAPE

PREHISTORIC PERIOD

The Prehistoric landscape is quite well represented in the Thespiae *chora*, though mainly from sites which also show later occupation.

The Neolithic landscape picture is highlighted by the important sites at Thespiae Magoula (*component TH_9*) and the occupation at Arkopodi-Eutresis (*component TH_78*). EH occupation is known at Thespiae Magoula (*component TH_10*) and at the Livadostro East site (*component TH_61*), as well as at the Prehistoric settlement of Arkopodi-Eutresis (*component TH_79*). Signs of probable occupation are attested at the historical period site of the Sanctuary of the Muses (*component TH_49*), while traces of an EH activity focus have been found by Tatiza (*component TH_53*) – see figs.6-7 in chapter II.3.1.

For the MH period, the most important site known in the area is the settlement at Arkopodi-Eutresis (*component TH_80*). MH occupation is attested also at the Thespiae Magoula site (*component TH_11*), as well as at the Livadostro East site (*component TH_62*) along with possible MH tombs (*component TH_64*) – see fig.8 in chapter II.3.1.

The LH period, with a probable presence at the Thespiae Magoula site (*component TH_12*), and presence of uncertain character at Askra Pyrgaki (*component TH_40*), is mainly represented by the burial place W of the modern village of Erimokastro/Thespies (chamber tomb – *component TH_13*), and the rich Mycenaean tombs by Tatiza (*component TH_54*), as well as by the fortified settlement site at Arkopodi-Eutresis (*component TH_81*), and the elevated site at the E end of the Livadostro bay (*component TH_63*). The prehistoric occupation at the E end of the Livadostro bay, in an elevated and strategic position, attests a use of the coastal location in the prehistoric period, which corresponds with the prehistoric presence in other Boeotian bays to the gulf of Corinth (see chapter II.3.10) - see fig.9 in chapter II.3.1.

Prehistoric human presence is also attested at the Dendra locality, by the Thespiae city site (*component TH_16*), and elsewhere in the *chora* as shown by the intensive survey work carried out in the area (see appendix I.9 – table 1 – *Survey Sites*). In fact, intensive survey work in the area slightly increases our knowledge of the Prehistoric human landscape in the area, with the discovery of small rural sites in the plain.

GRECO-ROMAN ANTIQUITY[6]

Town level

The city of Thespiae (*components TH_1 to TH_8*) was situated at the formation of the Kanavaris river, at the crossroad of the routes Thebes-Thisvi (-Gulf of Corinth),[7] Koroneia-Plataia (-Athens), Haliartos-Kreusis, and in the centre of a very fertile area. Thespiae was a quite large city, which on many occasions was a strong adversary of Thebes. Developed in the Geometric and Archaic periods, in high Classical times the city of Thespiae reached an estimated area of ca. 100ha, as attested by the intensive

[6] Period maps are included in chapter II.4, figs.17-19-21-23-25-27.

[7] Following the Kanavaris valley until Thespiae and then crossing Thespiae plain.

and systematic artefact surface survey carried out on both the urban area and the immediate surroundings (Bintliff 1994a; Bintliff-Howard-Snodgrass 2007). During Late Hellenistic to mid-Roman times (ca 200 BC – 400 AD) the city shrank, though in the *chora* there were localised districts little affected by the general trend, in accordance with Strabo's statement (IX 2.25) that Tanagra and Thespiae stood out as relatively prosperous in his time in comparison with the rest of Boeotia.[8] In the Late Roman period the territory of Thespiae revived in both town and country (see below – RURAL SEGMENT), although its re-expansion seemed to extend only to its confined Early Roman size and never begins to approach its Classical maximum (as it happens, for instance, in its satellite settlement of Askra - see below and appendix I.9; Bintliff 1996b).

Thespiae had access to the sea through the Livadostro valley, at the end of which was the harbour of the city, identified as ancient *Kreusis* (see *components TH_70 to TH_74*).

Village level

Askra (*components TH_33 to TH_39*) is definitely one of the second-rank settlements present in the Thespiae *chora*, a *proto-polis*, exploiting a well-defined settlement chamber, which would have become a *polis* if it were not for the presence of stronger Thespiae. Hesiod's account seems to indicate that by the close of the Geometric era Askra was already a dependent village of the *polis* of Thespiae with its corrupt nobles or *basileis*.[9] Furthermore, a village settlement site could be at Arkopodi-Eutresis, if the evidence indeed supports such an identification (see *components TH_85* and below – LONG TERM SETTLEMENT TRENDS). A settlement was probably associated to the harbour at Livadostro. Other second-rank settlements can be hypothesised in the area of Xironomi and Ellopia and to the N of Thespiae (see below - LONG TERM SETTLEMENT TRENDS).

Rural segment

From the picture given to us by extensive surveys and excavation data only, the rural segment is certainly under-represented. With the exception of the Late Roman rural site at Tatiza (*component TH_58*), and the large wealthy Hellenistic-Roman rural site at Arkopodi-Eutresis (discovered due to the presence of occupation of other periods), only burial sites might allow the localisation of some rural sites (see below - BURIAL AREAS). On the other hand, if we add to the picture the sites discovered through intensive survey, we can note in the Classical-Hellenistic period the filling in of the landscape with rural settlement, especially of medium-sized farms and hamlets rather than small farms (hamlets are known in the area through intensive survey work carried out by the Boeotia project – LS3, LS6 and Askris potamos sites – fig.3 and fig.15 in chapter II.4). A densely scattered landscape was recorded especially N of the city of Thespiae, and in the valley of the Muses, while on the uplands N of the Thespiae plain, the survey did not record any signs of ancient hamlets and only scanty evidence of rural sites. Thus, imbalances are observed in the frequency of rural habitation in different parts of the *chora*, with high density of rural settlement in the inner zone of rural and urban nucleations (Thespiae city and Askra), and with lower rural densities in the outer zone of the *chora*, to the N and NW (Bintliff-Howard-Snodgrass 2007: 146-148).

The sites of the Late Hellenistic / Early Roman period, show a retraction of settlement intensity, in accordance with the major decline in overall activity in the *chora*. In the Mid to Late Roman period, the landscape undergoes a radical transformation with the appearance of villa complexes associated with hamlets of dependent labour. The peak is observed in Late Roman times, when the total surface area of rural settlement is twice the Classical-Hellenistic component, redirected into a new pattern of land use, associated with valley soils and marketable and transportable crops as a major product (Bintliff-Howard-Snodgrass 2007: 154-158).

Burial areas

In addition to the monumental Classical *poliandrion* (Thespiae E – *component TH_14*), by the city there are cemeteries linked to the city site: *components TH_17, TH_18* and *TH_19*, *component TH_15*, and probably also *component TH_6*. Furthermore, the multi-period (Classical to Roman) burial place at Palaiokarandas (*components TH_30 to TH_32*), with a probable Hellenistic shrine associated, should be linked to the city site.

Burial sites might suggest the presence of small villages or rural sites nearby, such as *component TH_20*, *components TH_21* and *TH_22*, probable *components TH_26 to TH_28*, *component TH_94* (and *TH_95*), *components TH_55 to TH_57* (Tatiza).

Cult places/Religious areas

The sacred landscape of the region is certainly marked by the religious character assigned to Mt.Helicon, with the presence of the well-known Archaic to Roman Sanctuary of the Muses (*components TH_45 to TH_48*). Still on Helicon, at the Askra Hippokrene site, a cult-use of the spring has been suggested (*component TH_52*).

Cult places are known at Arkopodi-Eutresis (Archaic-Classical cult place, which could be linked to the settlement – see *components TH_82* to *TH_84*), as well as in connection with burials (the Classical *poliandrion* – *component TH_14*, *component TH_29* as well as *component TH_6*).

Forts and fortifications

The landscape of the area is marked by the Geometric to Hellenistic fort at Askra Pyrgaki (*components TH_41 to TH_44*), overlooking the valley of the Muses and the N-S

[8] See the results of the survey conducted by the Boeotia Project in a district S of Thespiae city (Bintliff-Howard-Snodgrass 2007).

[9] Hesiodus *Op.* 38f., 248f. and 639. See also Strabo IX 2.25 for Askra being part of the Thespiae *chora* in the Roman period. The intensive and systematic artefact surface survey carried out at the site showed how by high Classical times Askra can be seen to have expanded to its mazimum size of 11ha. In Late-Hellenistic to Mid-Roman times the settlement of Askra, along with Thespiae, shrank, to recover its full Classical extent in Late Roman times (Bintliff 1996b).

passages. It was part of the network of fortifications of the 4th C controlled by Thebes, but was already in use earlier. Another probable Archaic-Classical watch-tower is known on Helicon (*components TH_50 and TH_51*).
In the Late Roman period, there was a fortified settlement at the city of Thespiae, as well as probably in the Livadostro bay.

Other activities / unspecified activity areas
Harbour activities are known at Livadostro, identified as ancient Kreusis, the port of Thespiae in Greco-Roman times (Paus IX 32.1).[10]

LONG TERM SETTLEMENT TRENDS IN THE CHORA *LANDSCAPE*

Central to the whole area is the large central basin of Thespiae. In antiquity this basin was under the immediate control of the city of Thespiae, while today this region is divided into large village territories, centred on fairly regularly spaced settlements mainly located at the edge of the Tertiary *tafel* of Thebes. This pattern goes back to the Ottoman and medieval periods (see Bintliff-Howard-Snodgrass 2007: 166-167 for the hamlets of Erimokastro, Palaeoneochori and Askra VM4 site). As Bintliff et al note: "*this dominance of nucleated community life may have begun in late antiquity, with the joint foci of the towns (such as Thespiae and Askra) and the large hamlet estates of the Later Roman countryside*" (Bintliff-Howard-Snodgrass 2007: 179).
The modern villages of Thespies (formerly Erimokastro, registered in the 17th century lists) and Leondari (formerly Kaskaveli, registered in the Ottoman defters 1466-1570) on the edge of the Tertiary *tafel*, before the site of the ancient *polis* of Thespiae, control the central part of the basin, replacing the plain Greek village of Erimokastro, which occupied the plain in the medieval period in the location of the ancient city.
To the W, Neochori, at the foot of the Helicon spur jutting to the E in line with the location of the ancient city, occupies the western plain. It was a Byzantine and Frankish era community continuing in the Ottoman (registered in 1506-1570) and modern period, moving downhill to the present village location. The ancient occupation of this settlement chamber is attested by the important Prehistoric site (survey site Neo1) and signs of Greco-Roman occupation (see survey site Neo 2 and 3; and PP25 and PP26 in table 1 – *Survey Sites* in appendix I.9).
To the NW of the central basin of Thespiae, the so-called Valley of the Muses can be considered a small self-contained landscape within the wider area, opening into the Helicon massif in a SE-NW direction. It belongs today to the modern village of Askri (formerly Panagia). To its W the Frankish and Ottoman site of Paleopanagia (VM4 - recorded as a Greek village in 1466-1646) indicates continuity of population from the ancient town of Askra (*components TH_33 to TH_39*) to the Frankish tower location and finally to the present village which moved to its new position in the 17th century (modern Askri).[11]
N of Thespiae plain, on the foothills S of the Teneric plain, are situtated the modern villages of Mavromati (registered 1466-1570) and Vagia (recorded 1506-1642 and in the 17th century lists), exploiting both the hilly country that belonged in antiquity to the territory of Thespiae, and the plains further N that belonged to the territory of ancient Haliartos and Thebes. The Boeotia survey did not record any signs of ancient hamlets in this hilly landscape, and even the evidence for small and medium rural sites is thin, in contrast to the densely scattered landscape N of the city of Thespiae and the valley of the Muses (Bintliff-Howard-Snodgrass 2007: 146).
To the S, the Livadostro valley today forms a distinct settlement chamber with the village of Leuktra in the inland (formerly Parapungia, registered in the Ottoman defters from 1466 to 1646) as a diachronic *focus* of settlement (the prehistoric site of Eutresis is nearby, also a *focus* of Byzantine and Frankish settlement). Here lies a possible settlement gap in antiquity, which can only partially be filled in by archaeological evidence (see above and Bintliff 1994b: fig.20). By the coast, Livadostro, the harbour site of modern Leuktra, appears as an active harbour in the 1540 register but without permanent population, as probable continuation from the Frankish site at this location (Bintliff-Kiel in preparation). In antiquity this potential settlement chamber, through which Thespiae had access to the sea, had been controlled by the city, through its harbour at Livadostro, identified as ancient *Kreusis*.
To the S of the ancient city of Thespiae also opens up the settlement chamber of the modern villages of Xironomi and Ellopia. The territory of Xironomi today contains part of the inland country and extends to the Gulf of Corinth, to the harbour site of Aliki (which formed a separate small *polis* in antiquity - see chapter II.3.10). In the Ottoman period the Albanian village Tatiza[12] (registered in 1466-1646, with possible remains of a Frankish tower also) occupied the settlement chamber just N of Xironomi, a small alluvial plain on the route from Thespies to Xironomi and Thisvi. Modern Ellopia occupies a separate settlement chamber in the interior, which forms part of the Tertiary *tafel* of Thebes.

In the Thespiae region we face the problem of the mid-land areas behind Thebes (on and to the S of the Tertiary *tafel* – the Thespike and Plataea-Parasopia area) where no physical constraints define settlement chambers and the settlements create culturally imposed divisions of the land. The settlement chambers examined above, which

[10] The Admiralty Handbook describes the port as too exposed. A coastal change would probably explain its use in antiquity. Heurtley locates at Livadostro (Kreusis) an important S-coast port in the Mycenaean period, as he locates remains of a road in use in the Mycenaean period on the W slope of the Livadostro valley (Heurtley 1923/25).

[11] See Bintliff's article on the settlement history in the valley from antiquity to the present (Bintliff 1996b).
[12] House walls and the church of Agia Triada are the visible remains of the deserted village.

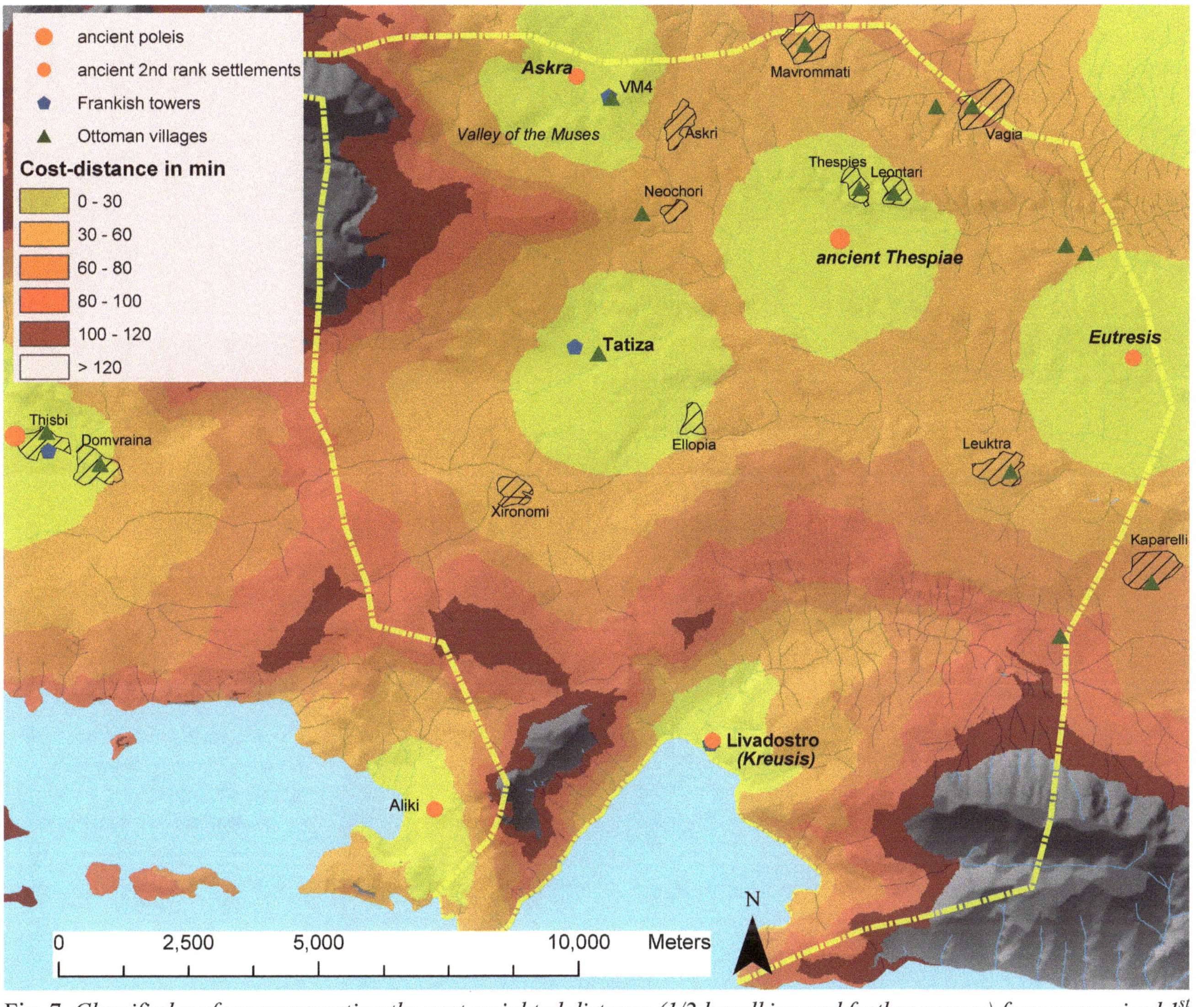

Fig. 7. *Classified surface representing the cost-weighted distance (1/2 h walking and further ranges) from recognised 1st and 2nd rank ancient settlements (represented by larger and smaller dots). Areas without dots indicate potential settlement chambers. Ottoman villages and Frankish towers have also been added to the map to show their spatial relationship with the Greco-Roman settlement network and to appreciate potential settlement chambers.*

are occupied by modern villages which to a large extent go back to the medieval and Ottoman period, indicate possible niches for ancient settlement and can be helpful in the analysis, representing possible locations of second-order settlements in the Greco-Roman period. For example, from the aforementioned modern and pre-modern settlement chambers, evidence for habitation exists for the Thespies/Leondari area with the ancient main *polis* of Thespiae, for the valley of the Muses, with the ancient *kome* of Askra, and for the area of Xironomi and Ellopia.[13] In those cases, the archaeological evidence of uncertain attribution can be supported by a cost-distance analysis[14] based on ancient settlement which shows enough room for one or two small settlements there (fig.7). Nucleated habitation can only be guessed for the area of Eutresis (see above) and Livadostro. As for the latter, a small scale settlement was probably associated with harbour and fort, as there was probably enough cultivable land to support it (fig.8). Apart from the small alluvial plain by the coast, today the landscape

[13] In the area of Tatiza, by modern Ellopia, a small village is hypothesised (AD21 1966: 202), according to surface archaeological evidence and excavated burials (Spyropoulos AD 1971: 221) – *components TH_53 to TH_57* (for the Classical period, see also *components TH_94 and TH_95*). Also by Xironomi, burial evidence could testify to the presence of a small community in the nearby area (see *components TH_59 and TH_60*). In the absence of a more detailed archaeological record, we can consider the more recent settlement pattern. The fact that today the area support two villages could be some support for the hypothesis of two small settlements, but the apparent absence of more than one Ottoman village in the area might also lead to the interpretation of the available archaeological evidence as corresponding to a village settlement in the area of Ottoman Tatize (much closer to Ellopia but in fact the predecessor of Xironomi in controlling the wider area down to the sea at Aliki) and some sort of hamlet by Xironomi (see also chapter II.3.10). The presence of one village in this area has already been suggested by Bintliff (1994b fig.20).

[14] See chapter II.3.1 – LONG TERM SETTLEMENT TRENDS.

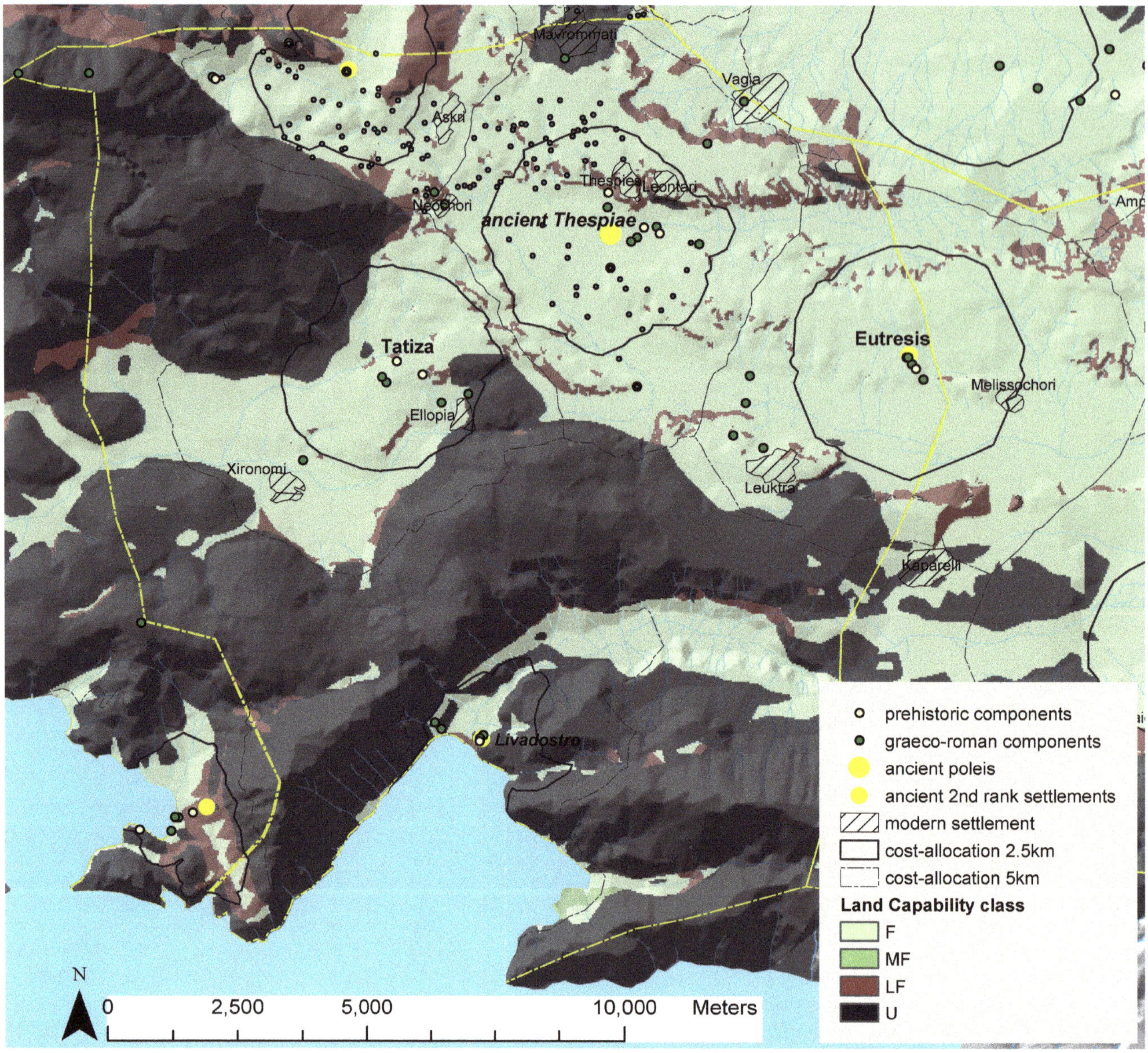

Fig.8. *Map showing the Greco-Roman settlement network, the polygons resulting from the cost-distance analysis (marking half an hour and one hour walking time distance) and dots representing the known archaeological components (same as in fig.2), with land capability information underlain.*

is marked by much hilly olive land on the lower slopes of the gorge, in stretches for several kilometers, though one should ask whether the cultivation is due to motor access for farmers, as it is so remote and previously was hard to access. As for Eutresis, the settlement nucleation gravitates onto the wider area of the modern village of Leuktra (the two locations are within a one hour walk of each other), where a settlement would have been in antiquity. The cost-distance analysis (fig.7) would also confirm the space for a village slot in the area, as already suggested by Bintliff (1994b fig.20) in considering the fact, also, that Thespiae and Eutresis lie in similar locations, on fertile uplands and near water sources. Eutresis fits the Plataea-Thespiae village gap, and dominates fertile land in all directions. The area was also crossed by the route from Thebes to the port of Livadostro, which was not direct but ran through the area of the modern village of Parapungia/Leuktra, where it was joint to the road from Orchomenos to Tespies (Philippson 1951: 505).

For other present-day occupied settlement chambers there seems to be scant evidence of antiquity. A characteristic example is offered by the settlement chambers of modern Mavrommati and Vagia, where, as mentioned above, there is slim evidence from the Boeotia survey for rural settlement in this area, also partially probably belonging to Haliartos (see chapter II.3.8).

Two places mentioned in ancient texts and probably in Thespike *chora* remain unidentified: Ellopia (Stephanus Byzantius s.v. *Ellopia*) and Leontarne ("below Helicon" on Schol. Venet. B on *Ilias* II 507), for which a location at Tatiza has been suggested (A.Snodgrass, pers.comm.).

Moving from nucleations to the rural level, our evidence from excavation data is accidental, and it is information from systematic surveys which sheds light on the picture (fig.3). In the sector of the Thespiae *chora* surveyed, for the Classical-Hellenistic period, the Boeotia survey project's team noted imbalances in frequency of small rural sites, hamlets and farms in different parts of the *chora* (Bintliff-Howard-Snodgrass 2007: 150-51). The inner northern zone, together with the valley of the Muses, produced evidence for high density of rural settlement, with a predominance of medium-sized farms and hamlets rather than small farms. According to Bintliff et al. "*they must be either hamlets composed of several farming families, or the centres of wealthy estates*" related to the issue of oligarchic class in the countryside (Bintliff-Howard-Snodgrass 2007: 134-35, 150). Some of these certainly mark hamlets or small villages (as the Askris Potamos site). Yet, despite the presence of these sites, most of the landscape in the THS/LSE sector was farmed by citizens resident in the city, leaving a small sector to be farmed from the rural sites. Conversely, the outer zone to the N and NW shows very low densities, noted also in the immediate countryside of the *chora* of Haliartos and Hyettos, marking regional differences in the urban-rural population in different parts of Boeotia (Bintliff-Howard-Snodgrass 2007: 146).

This exaggerated density of Classical Hellenistic sites, which begins about 1km N of the city of Thespiae, runs also across in the Valley of the Muses. In the Valley of the Muses, apart from the settlement of Askra,[15] there is a dense scatter of small and medium Classical-Hellenistic sites, but also some quite large hamlet size sites (Bintliff-Howard-Snodgrass 2007: 146, 148). This evidence indicates an enhancement of the rural population in the close vicinity of rural and urban nucleations, e.g. the city of Thespiae and Askra, with lower rural densities in outer segments of the *chora*, which were occupied in later periods (see discussion above for the medieval and post-medieval settlement pattern).

In the Late Hellenistic and Early Roman period a decline of settlement occurs in the eastern and southeastern hinterland with a reduction in the number of hamlets and farms. Also Askra may have been abandoned in Late Hellenistic – Early Roman times. Whatever was the reason (either an overall declining population and economy, or the swallowing up of small properties into larger estates), the sites of the Late Hellenistic to Early Roman transition "*show clearly a retraction of settlement intensity, rather than a loss of variety of niches exploited*" (Bintliff-Howard-Snodgrass 2007: 151-5). On the other hand, the SW hinterland sees an expansion of rural occupation, which marks a putative beginning of a new farming regime that would become fully visible in Mid and Late Roman times. By mature Roman times there are clear signs of revival of rural settlement, along with a new emphasis in land use. Villa complexes appear in association with hamlets of dependent labour. "*By Late Roman times, the total surface area of rural settlement [...] is twice the Classical-Hellenistic complement. Although the later Roman city appears only marginally larger than in Early Roman times [...], activity in selected parts of the* chora *has vastly increased*" (Bintliff-Howard-Snodgrass 2007: 157). Instead of a variety of catchment types typical of the Classical period, we have the dominance of a single type of environmental niche (the most fertile and moist valley colluvial-alluvial soil). Extensive estates around large rural sites can be reconstructed, meant to produce the entire range of crops and animal products to which the different parts of the landscape were suited, along with a more market-orientated economy for the wealthy land-owning class (Bintliff-Howard-Snodgrass 2007: 158).

[15] Which according to the calculations of J.L.Bintliff housed a population of 1,250 people within an area of 11ha of Classical occupation (Bintliff 1996b).

II.3.10

Three small *chorai* to the Gulf of Corinth: Siphai, Thisbe, Chorseiai

TOPOGRAPHICAL SETTING

From a geographical and topographical point of view, we can talk about the whole area including the mountains behind and the small gulfs facing the Gulf of Corinth and belonging to ancient Boeotia. On the other hand, if we are interested in the area under the control of a particular *polis* in the Greco-Roman period, we should examine separately the individual territory belonging to a single community.
Philippson (1951: 455ff mainly) offers a good detailed topographical – geological description of the area as a whole. Conversely, Fossey (1988: 167-196) gives a description of each separate *chora*, since in the area we have in fact three small *poleis*, each with their own area. From E to W, they are Siphai, Thisbe, and Chorseiai.

As described by Philippson (1951: 455f), the promontory between the Zaltsa gulf and Sarandi bay creates an area 6km wide between the sea and Mt. Helicon. In antiquity the area belonged to the Phokian *polis* of Voulis, while the hinterland of Sarandi bay belonged to Boeotia (with the area of ancient Chorseiai and modern Chostia/Prodromos). In this area, on a mountain spine, was the fort of **Chorseiai** (today Kastro) 250m asl, below the monastery of Ag.Taxiarchoi. The passes through Helicon to Sarandi bay in the area are quite difficult, and they were probably not used for trade,[1] while from the Eastern Domvraina basin a route is open through an incision[2] (at a height of 320m as opposed to the average elevation of the surroundings of 420m), at the Eastern end of which lies the modern village of Chostia/Prodromos.
The whole S slope of the Palaiovouna ridge (which marks the border of Boeotia with Phokis and constituted the S ridge of W Helicon – see appendix II) has water and forest coverage but also some cultivated terraces. This W part of the Helicon fringes to the E is bordered by a long

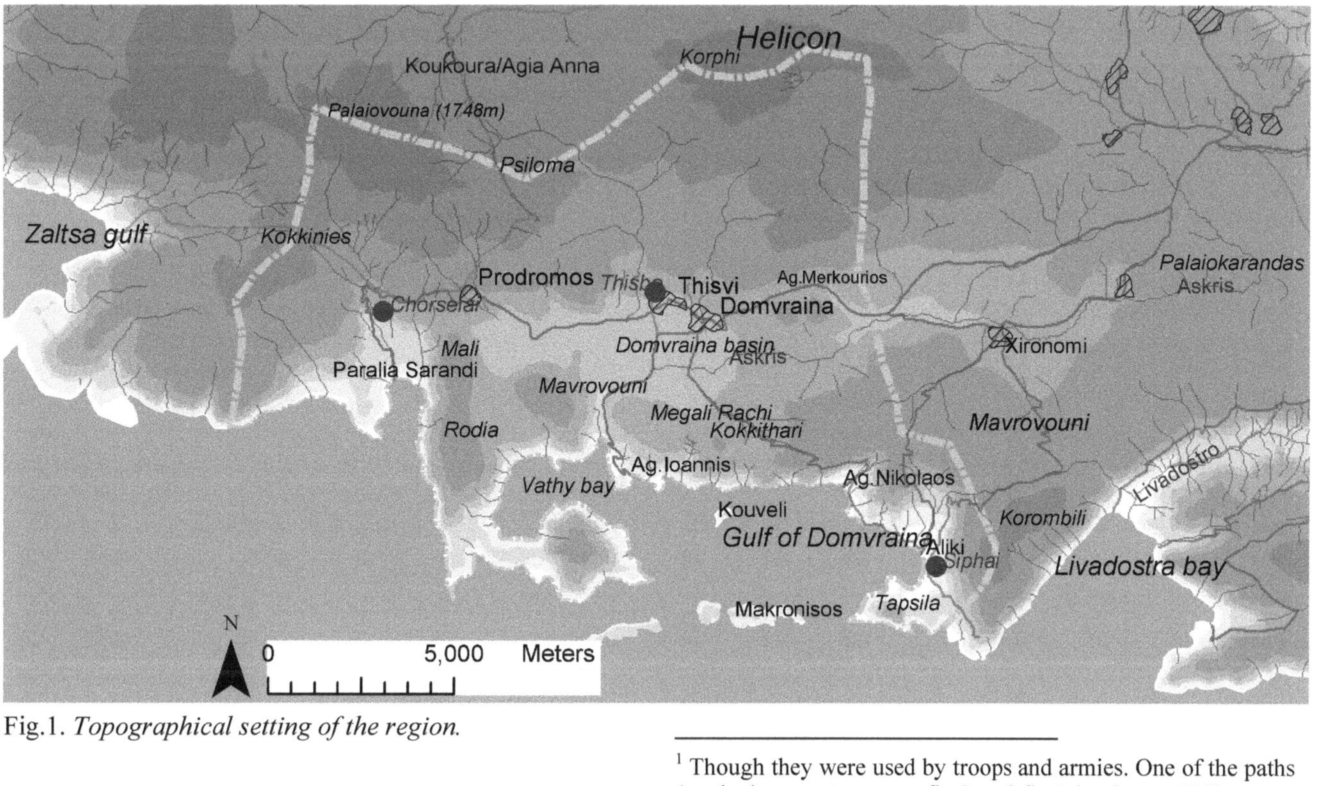

Fig.1. *Topographical setting of the region.*

[1] Though they were used by troops and armies. One of the paths (marked on route map – fig.2 and fig.1 in chapter II.1) comes from the upland plateau of Koukoura/Agia Anna (see Koroneia *chora* – chapter II.3.1).
[2] The incision through which the route runs is at a height of 320m as opposed to the average elevation of the surroundings of 420m (Philippson 1951: 455).

extended N-S hill, from 643 to 420m in elevation, with an incision to 320m. In this incision, at the foot of the mountain, lies the village of Chostia/Prodromos, which exploits the plain to its E.

To the E of the Chorseiai area, and separated from it by the L-shaped Rodia/Agiomachos promontory, is the mountainous coastal area forming the wide so-called gulf of Domvraina. In the middle of this wide gulf is the islet of Kouveli.

At the W end of the gulf is the Vathy bay that was never used as harbour due to the difficult access. The gulf, throughout which the depth of water is greater than 30m, is divided in smaller gulfs, among which the smallest is that of Agios Ioannis, the harbour of Domvraina, and the most Southern harbour of Boeotia – see Heurtley (1923/25). On this tiny gulf is a small hamlet with an Albanian population (fishermen with a few boats; in Philippson's time busy with the commerce of corn and cereals from the Copais to Corinth). The port is narrow and unsafe. A road connects the village with Domvraina and Thebes. In antiquity the bay was the port of the ancient *polis* of **Thisbe**, lying in an inner karstic basin (partially occupied by a marshy lake - see below) separated from the sea, the wide gulf of Domvraina, by a chain of coastal mountains crossed by the road from Agios Ioannis.

At the Eastern end of the wide gulf is the bay of Aliki, with a small plain where until a few decades ago were salt installations (seen by Philippson 1951: 456f), and in antiquity was the polis of **Siphai**. The small gulf, a potential sea-lake (from which the name of the modern village derives), is overlooked by the Korombili ridge, on whose slopes are areas with plantations producing olives and oil. Philippson (1951: 456f) notes that the harbour is more suitable for pirates and illegal commerce than for legitimate trading activities because the connection with the interior is difficult.[3]
Apart from a path around the coast, at the foot of Korombili (Roesch 1967: 292), there are only three ways to access Aliki from the hinterland (Tomlinson-Fossey 1970: 244): one, followed by a very recent graded road, leads through the high col between Korombili and the chief peak of Mavrovouni, and descends into the NE corner of the small coastal plain; another main route, possibly that used in antiquity, passes over the W shoulder of Mavrovouni and, beyond the Spartan military base (*component GC_17*), comes into the N side of the plain (marked on fig.2); a third route crosses over the very W part of the Mavrovouni ridge, descends to the coast and thence into the NW corner of the plain.

The forests of the high mountains, the cultivated plains, and the rocky limestone areas, along with the sea, with gulfs and small islets, though very close to each other, constitute several different landscapes whose proximity in a relatively small area is typical of the Greek landscape (Philippson 1951: 461).

The Korombili ridge, whose highest peak is 907m asl, lies between the Gulf of Domvraina and the Livadostra bay. The ridge, to the N, runs into a col which forms (see 1:200,000 *nomos* map) a flat surface (645 to 614m asl) lying between the Xironomi village basin and the Livadostro river valley (Philippson 1951: 461). These SE slopes of Korombili are mostly bare, with the exception of an area above the small gulf of Ag.Nikolaos with a small Aleppo pine forest.
The basin of Xironomi (360m long) forms the entrance to the small landscape of Domvraina, and it is, in fact, a sort of depression between Helicon and Korombili (constituting one of the systems of depressions along the gulf of Corinth).

The Domvraina valley
Due to the marked karstic character of the region, especially in the Western part of Boeotia we note the phenomenon of the formation of *poljes*. Some of them have been dried in recent times, but humans always had to fight against the emergence of water on the surface, or sometimes to make the best use of it. The Copais basin (examined in chapter III.1) is the major case in the region, and in the Balkan area, but we also have other cases of formation of small poljes, such as in the Domvraina valley.
The Domvraina valley is situated in the ancient *chora* of Thisbe, which extended between Mt.Helicon and the bay of Domvraina (see Pausanias IX 32.3[4]). It is a closed *polje*, surrounded by gentle slopes, with a basin of ca. 220 sq.km, crossed by the Askris river (ancient Permessos or Termessos) which flows into the deepest part of the valley, not far from ancient Thisbe. The valley is dominated by the Helicon massif, whose Mesozoic limestones are mainly covered by Pleistocene deposits. As Shiel and Steward (2007) state, "*...in the lowest-lying areas, deposition of material eroded from the limestone massif and from the upstanding areas of Pleistocene deposits continue to the present; the existence of relatively unconsolidated sediments ensured that the landform is still evolving*". The unconsolidated superficial deposits of the alluvial plain are poorly provided with free water, and the fertile soils are scarcely developed.
The lowest part of the valley, marked by the 120m contour line and 6 sq.km in area (Knauss 1992), is occupied by marshes which still today, in the winter, become flooded and form a lake of variable depth and extension. Such a phenomenon is due to the karstic subterranean and superficial circulation of the valley, apparently without open *katavothrai*.[5] As Knauss (1992: 35) points out, the valley of Dhòmvraina is the only polje of Greece with no visible, open *katavothra*. We cannot exclude the possibility, however, that some *katavothrai* might be buried under the Holocene sediments of the valley, and that in some historical periods they might

[3] For the harbours on the gulf of Corinth see Heurtley 1923/25.

[4] πρῶτα μεν ὄρος ἐστὶ πρὸς θαλάσσῃ τοῦτο δὲ ὑπερβαλόντα πεδίον σε ἐκδέξεται καὶ μετὰ τοῦτο ἄλλο ὄρος· ἐν δὲ ταῖς ὑπωρείαις ἐστὶν ἡ πόλις (Pausanias IX 32.3)

[5] Natural subterranean channels, swallow-holes, typical of karstic basin landscapes.

have been active. In antiquity, a system had been implemented in order to alternately free from excessive water the two halves (E and W) of the marshy *polje* (E and W) of the marsh, by means of a dam crossing the basin, as attested by Pausanias (IX 32.3). This would allow the inhabitants of Thisbe to cultivate the free half. In prehistory also, the water behaviour of the polje was controlled (see fig.2 and Knauss 1992 map in fig. 4.3).

The hydrological basin of the valley corresponds only partially to the hydrogeological basin, because a portion of the effective precipitation which penetrates underground as far as the water-bearing strata, reaches water sources situated outside the basin and feeding also, for instance, the contiguous Asopos basin. This kind of phenomenon is also quite common in karstic systems. The main stream of the valley, the Askris, originates in the Valley of the Muses, from which it has water most of the year. The river runs initially to the SE towards Thespiae and the steep-sided promontory of Palaiokarandas (chapter II.3.9, fig.1). In the Thespiae basin the river loses, most probably especially in periods of heavier rainfall, part of its water to the Asopos, from which it is separated by a low watershed (Knauss 1992; Sheil and Steward 2007). At this spot, the Askris turns to the W-SW, collects water from some of the seasonal creeks fed by the Helicon and ultimately drains into the lowest part of the valley, near ancient Thisbe. The typical Mediterranean rainfall, characterised by a strong annual variability and with short but heavy rainfall, alongside the absence of any superficial or subterranean effluent, is a basic factor in the erosion phenomena, and in the frequent floods that still today affect the lowest part of the valley, which is also the most fertile. This may explain the long-term/continuous human intervention in order to regulate the water, by means of drainage works, many of which are of considerable size. The valley has been a centre of inland drainage for centuries; since Prehistoric times.[6] Banks, channels, and basins for water collection, have been described by Pausanias during his visit to Thisbe, and remains of these works have been noted since the 19th century (Knauss 1992).

Pausanias (IX 32.3) mentions the flooding of alternate portions,[7] and the fact that inhabitants of Thisbe built up a bank, to deviate the watercourse 'in alternate years' to one side of the bank and cultivate the land which is on the other side.[8] Unfortunately, he does not give us either a date or a chronological reference, even reported by a saying, for this work, as is the case for many other drainage works in Greece mentioned by the sources (Fantasia 1999), from which we can almost never argue a date of construction and creation of such measures. This may tell us something about the perception of the ancients in relation to this kind of human intervention on the water landscape. At least in Pausanias' time, what was important was probably that such works were there, functioning, and probably also that people made them at some point, no matter when.

PHYSICAL LAND UNITS

The three types of landscape are all represented in the wider area, and also within the three small individual *chorai*. The physiographical classes are also all represented (see table 1 and fig.2 in chapter II.1).

Hilly landscape	42.3%
Mountainous landscape	26.9%
Plain	30.8%

1	P1_P2	lacustrine basin, valley	9%
2	P3	gentle slope	1.9%
3	P4	foothill	19.9%
4	H1	plateau	9.5%
5	H2	gentle slope	1.4%
6	H3	moderate slope	8.4%
7	H4	severe slope	12.6%
8	H5	very severe slope	10.4%
9	M1	plateau	4.6%
10	M2	plateau/gentle slope	4.9%
11	M3	moderate slope	8.7%
12	M4	very severe slope	8.6%

Table 1. *Percentage of the different physiographical classes present in the Gulf of Corinth area (P=plain; H=hill; M=mountain)*

The plain landscape is represented by the inner basin and depressions (P1_P2 and P3 - namely, for instance, the larger Domvraina basin, the area from ancient Chorseiai towards the Sarandi bay, and a few smaller areas behind and within the so-called limestone plateau following the shore-line[9]), and their sloping foothills (P4).

Quite steep slopes characterise the hilly part of the landscape (steep slope features -H4- mark the fringes of Helicon as well as the elongated ridge enclosing the Sarandi bay –S of ancient Chorseiai- to its W), combined with small plateaus (the only larger one is the end of the Permessos/Askris valley before it descends into the Domvraina basin). The physiographical situation characterising the end of the Permessos/Askris valley and the Domvraina basin into which the river flows is interesting; one can clearly see how the Domvraina basin constitutes a lower depression (P1_P2 class) separated by a steep slope (P4) from the Permessos/Askris valley, which is in the hilly segment, forming a higher plateau (H1).

The mountainous segment (>600m asl) is represented almost exclusively by the Helicon massif, overlooking the whole area from the N and NW.

[6] For the archaeological evidence, on drainage and related works, of the various periods, see Knauss 1992.

[7] τὸ δὲ πεδίον τὸ μεταξὺ τῶν ὀρῶν ἐκώλυεν οὐδὲν ἂν λίμνην ὑπὸ πλήθους εἶναι τοῦ ὕδατος εἰ μὴ διὰ μέσου χῶμά σφισιν ἐπεποίητο ἰσχυρόν· καὶ οὕτω παρὰ ἔτος ἐς μὲν τὰ ἐπέκεινα τοῦ χώματος ἐκτρέπουσι τὸ ὕδωρ τὸ δὲ ἐπὶ τὰ ἕτερα αὐτοῦ γεωργοῦσι (Pausanias IX 32.3).

[8] Kirsten 1951: 459, n.1. According to some scholars (see Fantasia 1999), Knauss 1992: 43ff is probably wrong in interpreting the phenomenon as a non-functioning polder to the W side of the dam, in the year of Pausanias' visit.

[9] see appendix II – b4.

RESOURCES

Generally speaking, the area is mainly characterised by unfertile areas (U) for agricultural activities, with few exceptions (F class constituted mainly by the Domvraina basin, which, however, is occupied seasonally by marshes – see below, and the valley to the S of ancient Chorseiai, with a few other alluvial pockets). The low fertile (LF) area to the NE is part of a very steep slope landscape on the Helicon massif and therefore probably only used for grazing (see fig.6).
In the following paragraphs, we will look more carefully at the land characteristics and resources of each single community chamber.

The area of ancient **Siphai** comprises the small plain around the Aliki bay and the mountain slopes that enclose it: Korombili to the E, and to the N its large projection, the ridge of Mavrovouni; to the S the small ridge of Tapsila, another outcrop of Korombili, which forms the arm of land enclosing the bay to its S (Fossey 1988: 167). No permanent watercourses flow through the area, which is, however, crossed by many seasonal rivulets. Part of the plain supports cereal crops, but it is mainly cultivated with olive trees (Fossey 1988: 167). In the past there was a large salt pan in the NE corner of the bay (seen by Philippson 1951: 456f). In the past, fishing activities would be also form a part of the economy of the area,[10] as well as grazing on the coastal plain and surrounding slopes.[11]

The area of ancient **Thisbe** comprises the area between Mt.Helicon to the N[12] and the bay of Domvraina to the S (Fossey 1988: 178), and therefore includes the large karstic basin of Domvraina (a *polje*), the centre of inland drainage, whose landscape was very much different in antiquity, marked by fluctuating marshes (see above).[13] To the E, the border of the area is denoted by the southward projections from the Helicon massif which divide the *polje* from the plain of Thespiae, with the pass of Ag.Merkourios as the real only break (Roesch 1965: 52 and plan XIV.1 quoted by Fossey 1988: 177); to the W by the southward projection of Helicon known as Malia Senga[14] (with the peak of Mali), which separates the area of ancient Thisbe from that of ancient Chorseiai. To the S, separating the inner basin (*polje*) from the gulf of Domvraina is a lower ridge of mountains (namely, from E to W, Kokkithari – Chimadio/Kivadio/Megali Rachi – Mavrovouni) – Gregory 1992: 19. From the inner area, there are a few routes to the sea: one at Kreusis, the ancient port of Thespiae (Pausanias IX 32.1; Pieske 1922), in the Thespiae *chora* (see chapter II.3.9); another in the area of Thisbe, at the SW corner of the Domvraina basin, towards Vathy/Ag.Ioannis, the ancient port of Thisbe. The latter runs through a low pass between the Chimadio/Kivadio/Megali Rachi and Mavrovouni ridges.[15] Only the narrow valley of the Permessos/Askris river easily connected Thisbe to the rest of Boeotia.
The W part of the *polje* is farmed today by the inhabitants of the village of Chostia/Prodromos (which, however, mainly exploits the area of ancient Chorseiai, to the W). The main farmland available to ancient Thisbe was the basin of Domvraina (according to the condition of the *polje*) and its surrounding slopes, as well as the small littoral of the gulf of Domvraina (Fossey 1988: 184).
The W part of the N edge of the basin is formed by the hillsides of the Palaiovouna ridge (Helicon), sloping abruptly, while the E is a much more gentle karstic landscape with limestone depressions (dolines) among red soils, along with fields and their boundaries. In Philippson's time, there were olives and vineyards at the S edge of the basin, while cereals and corn were cultivated in the fertile areas (Philippson 1951: 458). The other three sides of the basin (E-S-W) are bordered by the limestone plateau (constituting the littoral string of mountains – see appendix II).
As Gregory (1992: 21) points out, the territory controlled by Thisbe was naturally limited in extent and quality. In the inscription, IG VII 2225, reporting a *Senatus Consultum* of the Imperial period, the nature of Thisbe's terrain is threefold: land, harbour, and mountains (ll.18-19). The production of crops is mentioned in l.53 (though the agricultural capability of the Thisbe basin was certainly limited to the rich and deep alluvial soil of the basin,[16] yet always threatened by flooding, contrasting with the poverty and bareness of the surrounding hills and mountains – Gregory 1992: 21). Wine production is not mentioned, despite the land producing good wine today (Fossey 1988: 184). Grazing was also an exploited resource but, certainly, a large part of the city economy was based on the port of Vathy and the exploitation of long-distance trade.[17]
The control of the area of Koukoura/Agia Anna was therefore crucial to ancient Thisbe both due to the need for more cultivable land to support its population, and for

---

[10] A particular type of fish was caught there: Aristoteles *HA* II 13 (from Fossey 1988: 167).

[11] Especially in the past, but still today, flocks are taken from the Zagora/Evangelistria village on Helicon to these lowland pasturages during the winter.

[12] Peaks of Korphi 1265m asl and Psiloma 1105m asl. The upland plateau of Koukoura to the N of this line was disputed between Koroneia and Thisbe, as documented by IG VII 1870 (in this study, the area has been included in the Koroneia *chora*).

[13] S and SW of the modern village of Kakosi/Thisbi is a karstic basin whose lower part (100m) is a winter lake which collects the water from the snows of Helicon (larger in March). The basin also collects water from the river Askris, which ends here (Philippson 1951: 458). The Permessos/Askris river probably had a swampy bed in this area in ancient times (Gregory 1992: 29). For a description of the physical characteristics and nature of the Thisbe karstic basin (*polje*) see above in text and Knauss 1992.

[14] Stamalios (to the N = Mali) and Rodia (to the S) on the 1:50,000 GYS map.

[15] For ports and communication routes in the area see Heurtley (BSA 1923/5).

[16] Deposits of the Permessos/Askris river.

[17] Since Mycenaean times, when a considerable N-S trade through Boeotia has been argued (Heurtley 1923/25, vs Gomme 1911/12 who disagreed).

control of the upland routes connecting the N of Greece and W Boeotia to the harbour on the Gulf of Corinth. Both these reasons would explain the dispute with Koroneia over control of the upland plateau (see above, footnote 12).

The area of ancient **Chorseiai** is comprised of a small valley in the S flank of Helicon, ending at the Corinthian gulf in the coastal plain of Sarandi. It is crossed by seasonal rivulets (Fossey 1988: 187). The settlement chamber of the small ancient *polis* is geographically well defined by mountains belonging to the Helicon massif to the E-N-W, and by the sea of the Gulf of Corinth to the S (Fossey 1988: 196).
Even today access to the area is limited to the route over the pass of Malia Senga[18] from the modern village of Chostia/Prodromos, and to another much more difficult pass over the higher ridge to the W of the valley (Rachi Mada - Kokkinies on the 1:50,000 GYS map) leading to the Zaltsa bay and the ancient town of Voulis in Phokis, which the territory of ancient Chorseiai bordered (Roesch 1965: 56-8, remarking on the isolated nature of the area; Dasios 1995[19]).
Only a very small part of the territory is strictly cultivable, with olive groves and vineyards.[20] Pasturage land and fishing are also available resources (Fossey 1988: 196).

THE ARCHAEOLOGICAL RECORD

Siphai *chora*

1	**Aliki (ancient SIPHAI)**	*Components GC_1 to GC_8*
2	**Aliki SW**	*Components GC_9 and GC_10*
3	**Aliki S**	*Component GC_11*
4	**Alonnisi Alikis**	*Component GC_12*
5	**Mavrovouni Kastron**	*Components GC_13 to GC_18 and GC_79*

Thisbe *chora*

6	**Kakosi-Thisvi (ancient THISBE)**	*Components GC_19 to GC_25, GC_77 and GC_78*
7/8/9	**Kakosi-Thisvi**	*Components GC_26 (7), GC_27 (8) and GC_28 (9)*
10	**Kakosi-Thisvi area**	*Component GC_29*
11	**Ormos Ag. Ioannis**	*Component GC_30*
12	**Vathy (gulf of Domvraina)**	*Components GC_31 to GC_36*
13	**Kivadio**	*Component GC_37*

THISBE BASIN survey sites	Thisbe basin D-3	*Component GC_38*
	Thisbe basin C-1	*Components GC_39 to GC_41*
	Thisbe basin C-3, D-1, D-2, E-2, E-1	*Components GC_42 and GC_43, GC_45 to GC_54*
	Thisbe basin C-4	*Component GC_44*
KOUVELI island survey sites	Kouveli island K-1	*Component GC_55*
	Kouveli island K-3	*Component GC_56*
	Kouveli island K-4	*Component GC_86*
MAKRONISOS survey sites	Makronisos M-1 (Diporto)	*Components GC_87 and GC_90*
	Makronisos M-2	*Component GC_89*
	Makronisos M-3	*Component GC_88*

Chorseiai *chora*

14	**Prodromos Kastron (ancient CHORSEIAI)**	*Components GC_57 to GC_66*
15	**Prodromos Kastron S**	*Component GC_67 and GC_68*
16	**Prodromos Kastron N**	*Component GC_69*
17	**Prodromos Kastron NW**	*Component GC_70*
18	**Mali**	*Components GC_71 to GC_74*
CHORSEIAI COUNTRYSIDE survey sites	Chostia WR-1	*Component GC_80*
	Chostia WR-2	*Components GC_81, GC_82 and GC_83*
	Chostia WR-3	*Component GC_84*
	Chostia M-2	*Component GC_85*
	Chostia WP-1 (Sarandi Peninsula)	*Component GC_75*
	Chostia ER (Eastern Ridge)	*Component GC_76*

Table 2. *List of archaeological components and activity loci mapped in fig.2.*

[18] Stamalios (to the N = Mali) and Rodia (to the S) on the 1:50,000 GYS map.
[19] In the past, as today, an upland road very twining led from Voulis to Helicon, through a pass (between Palaiovounas - 1748m asl- to the E and Tsiveri -1560m asl- to the W.) The same pass probably also followed the Phokis – Boeotia border (Dasios 1995, who refers to Pritchett 1985: 144, n.9.
[20] As reported in Fossey 1988: 196, Thisbe helped Chorseiai during a shortage of grain (probably in the late 2nd C BC) IG VII 2383.

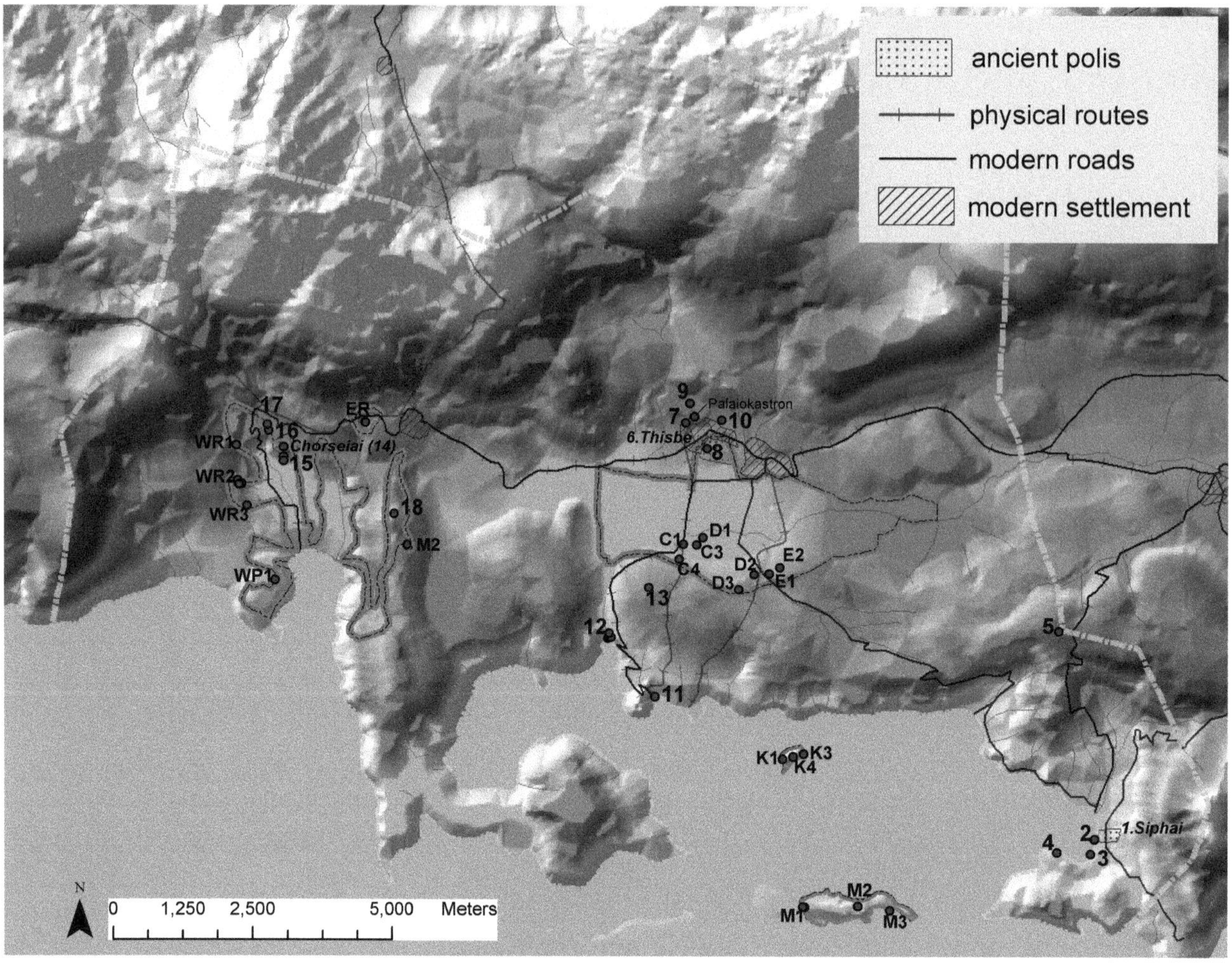

Fig.2. *Archaeological map of Gulf of Corinth area*

The wider area consisting of the three ancient *polis* chambers is archaeologically known by means of extensive and topographical surveys; in particular the topographical work of Tomlinson and Fossey (BSA 1970: 243-263).

More systematic research (but always marked by a topographical imprint) was carried out by Fossey and his team in the area of ancient Chorseiai. Their results are presented in several articles and conference papers, as well as in the monograph Khostia I (1986). Systematic research was focused on the site of ancient Chorseiai, at Mali in 1980, and in 1983 on the so-called Western ridge (Kokkinies on the 1:50,000 GYS map) and on the so-called Western peninsula (Gauvin Khostia I: 51),[21] and concentrated on structural visible remains (see below – RURAL SEGMENT).

Intensive and systematic survey has been carried out by T.E.Gregory in the area of ancient Thisbe[22] (in 1979 and following years – results presented in Gregory 1992) and on the small offshore islands of Kouveli and Makronisos in the gulf of Domvraina (1981, 1982 and 1984)[23] – see above, Thisbe basin – Kouveli - Makronisos.

Problems-oriented research has also been carried out in the area, e.g. by Heurtley, working on Boeotian harbours (Heurtley 1923/25), and focusing especially on Bronze Age evidence.

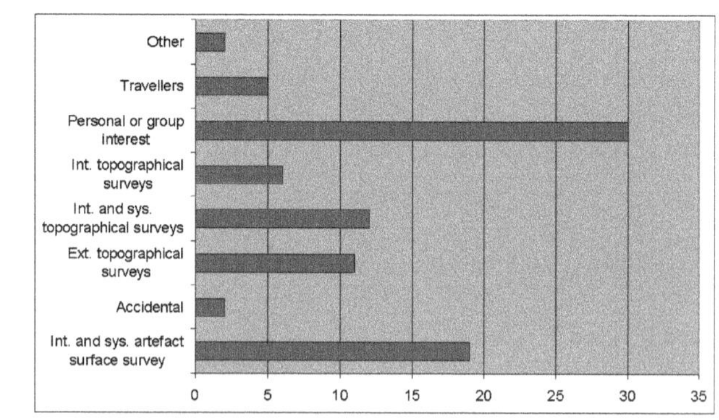

Fig.3. *Graph illustrating the proportion of components discovered within different research frameworks.*

[21] The surveyed areas are marked in Fossey - Gauvin 1983.

[22] The surveyed areas are marked on Gregory 1992: 24, fig.3.4. This was an intensive and systematic field-by-field artefact surface survey with fieldwalkers 10m apart. The aim was the investigation of the relationship between the urban area of Thisbe and its rural hinterland, and the collection of evidence of settlement patterns in the Thisbe basin over time (Gregory 1992: 22).

[23] Intensive and systematic surface survey, mainly architectural, with fieldwalkers 5m apart.

In fig.4 we can see the relationship between known archaeological sites and the distance from the modern road network. Correspondence seems quite strict, but this is probably due to the modern road circuit corresponding to the old one as topographically conditioned, rather than to a bias in research strongly influencing the resulting pattern.

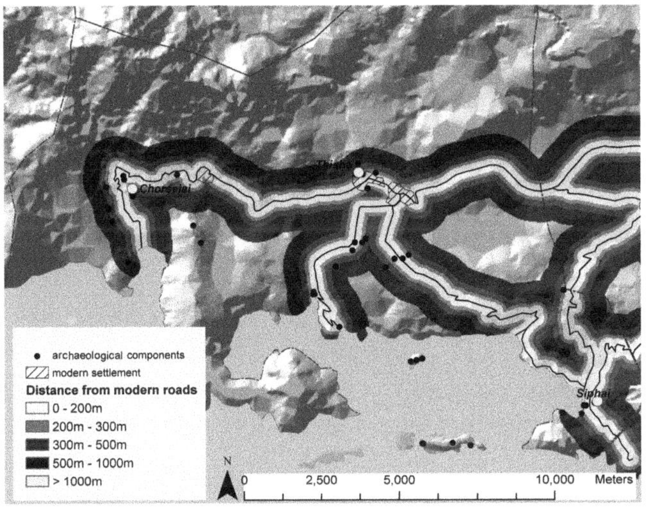

Fig.4. *Relationship between components and modern road network.*

The ratio of known Prehistoric to Greco-Roman components is 13 to 67 (1:5.1), while among the historical periods, 49.3% are dated Archaic to Hellenistic, 39.1% Roman-Late Roman, and 11.6% are attributed to the general Greco-Roman period.

The quite unusual abundance of Roman-Late Roman (mainly Late Roman) sites is probably due to the success of intensive and systematic surveys in filling in the rural segment of the landscape, as well as to the flourishing of the area in that period (see below).

ANALYSIS OF THE *CHORA* LANDSCAPE

PREHISTORIC PERIOD

Prehistoric occupation in the area (from EH through to LH, it seems[24]) is known at the three historical *polis* sites, identified mainly because of careful investigation at the Greco-Roman sites, but also due to their outstanding location within the relatively small settlement chambers (see *passim* in the chapter).

An important EH settlement has been identified at Alonnisi Alikis, in a suitable location for a habitation site of that period, in a fairly elevated location overlooking a sea-lake[25] (see fig.7 in chapter II.3.1).

MH activity foci are known from the three *polis* sites of the Greco-Roman period (see fig.8 in chapter II.3.1).

LH activity foci are located at the historical city sites of both Siphai and Chorseiai (*components GC_3 and GC_60*), while LH burials are known from the outskirts of the important LH site identified at Palaiokastro hill (historical Thisbe - *components GC_26 and GC_28*). An activity focus in the LH period (which could represent a refuge place – Fossey 1988: 196) has been identified also at Mali (ridge to the E of historical Chosteiai city site) – *component GC_71* (see fig.9 in chapter II.3.1).

Apparently the area was strategic for N-S long-distance commerce from the Mycenaean period, as suggested by Heurtley 1923/25 (though Gomme 1911/12 disagrees). In fact, traces of ancient roads, for which a probable use in Mycenaean times has also been suggested, leading to the natural harbours (certainly in use in historical times) of Aliki and Vathy, have been noted, as well as the possibly fortified LH settlement on the hilltop by the sea at Aliki (*component GC_3*). Furthermore, it is self evident that Thisbe (carrying a name of pre-Hellenic origin), must have played an important part in the economy of Mycenaean Greece with its harbour town of Vathy, constituting the natural haven on the N shore of the gulf of Domvraina for routes between Mycenae and Kenchreae, and a Boeotian route via Thespiae to Thebes (Evans *JHS* 1925: 1-42, with reference to the so-called 'Treasury of Thisbe'[26]).

Interestingly, no Prehistoric activity focus has been noted by either the intensive topographical or the field-by-field artefact surface survey work carried out in the area, with the exception of some scattered obsidian flints mentioned briefly by Fossey and Gauvin when discussing the 'unfortunate condition', with regards to visibility of archaeological remains, of the alluvial plain below ancient Chorseiai (Fossey and Gauvin 1983: 237; see below footnote 28).

GRECO-ROMAN ANTIQUITY[27]

Town level

The town level is represented by the three *polis* sites which constitute the main settlements of the territories they controlled, at least from the Archaic to the Late Roman period (see above for description and borders of those geographically quite well marked settlement chambers; see appendix I.10 for detailed chronology). Chorseiai, and especially Siphai, can be considered as belonging to the category elsewhere called *proto-poleis*, which appeared in early periods, and developed into towns but never into proper *poleis*, while Thisbe, far larger in extension, is a good example of a second-rank

[24] Less sure is the EH presence at Kakosi/Thisbi Palaiokastro site (*component GC_19*) and Prodromos Kastron (*component GC_58*).

[25] The location is comparable to that of the majority of EH sites around the Copais basin.

[26] See appendix I.10, *component GC_28*.

[27] Period maps are included in chapter II.4, figs.17-19-21-23-25-27.

real historical *polis*, as pointed out also by Gregory (1922: 19ff).

Village level

Secondary level (satellite) settlements are less easily recognisable. In the small *chora* of Siphai there was probably no place for a second settlement site (apart from a settlement grown up in the Roman and Late Roman periods around the harbour installations in Aliki bay); nor in the Chorseiai area, where the hilltop settlement recognised at Mali seems to have been linked to specific historical periods and defensive urges (see above, Mali site). Even the *chora* of ancient Thisbe, larger in extension and today supporting two villages (separated by historical accident, but for the most part one large village), did not allow for the presence of a second rank settlement site (see above – resources), if one excludes the settlement that grew up in the Roman and Late Roman periods around the harbour of Vathy. *Thisbe basin C-1* (*components GC_39 to GC_41*) was probably a large rural site (hamlet-like) rather than a proper village.

Rural segment

The rural landscape of the chora is represented mainly by the results of the two intensive surveys carried out in the Domvraina basin (Gregory 1992, primarily) and in the area of ancient Chorseiai (Gauvin Khostia I: 51-81, mainly).

For the area of ancient Chorseiai, the picture of the rural landscape given by the survey work is that of a quite heavily populated countryside in the Late Classical – Early Hellenistic period, as attested by other survey projects in the wider region (mainly the Boeotia survey project). On the other hand, while the wider Boeotia survey project is dealing especially with potsherd concentrations and scatters on the surface (forming various assemblages), archaeological evidence available here is mainly comprised of architectural and structural features (single buildings or complexes with or without courts or other adjacent structures, sometimes with tower buildings). This is partially due to the personal research interest of the scholars involved in the survey,[28] but certainly also to post-depositional and taphonomic factors, mainly erosion, which would allow structures to be visible on the surface, while soil with associated material would mostly have been washed away. This offers a favourable situation for the study of architectural remains, not always easily available elsewhere.[29] The rural sites found are relatively small, and some of them are probably to be interpreted as '*fermes d'élevage*' (Gauvin Khostia I: 80), linked to pastoral activities and to the keeping of the flocks within the enclosure (see for instance site WP-1 *component GC_80*).[30]

As for the Domvraina basin, rural occupation of the landscape seems to have been concentrated in a broad band around the N, S and E of the plain at the edge of the large marshy *polje* area partially controlled in antiquity (Knauss 1992),[31] at elevations between 100 and 140m above sea level, and the population appears to have followed the course of the water-control device S of Thisbe (see above). Apart from the urban settlement of Thisbe and the Vathy port, the rural segment is represented by a large rural agglomeration (*components GC_39 to 41*: Classical, Hellenistic and Late Roman) and a network of small isolated farmsteads (Gregory 1992: 33), mainly occupied in Late Classical-early Hellenistic and Late Roman times (and giving therefore the usual picture of the Boeotian countryside).

Burial areas

Burial areas are apparently not known. This might be due to a bias in research, mostly concentrated on ridges and outstanding locations, rather than to the plain and by the lowland routes, where necropoleis were probably placed. Even at the outskirts of the detailed-survey city sites (Chorseiai and Thisbe), few signs of burials seem to have been noted, with the exception of some Roman burials noted at various points near the ancient roadway descending from the S gate of ancient Chorseiai (*component GC_68*), and some accidental finds probably belonging to Roman burials reported from around the modern village of Kakosi/Thisbi (*component GC_29*).

Cult places/Religious areas

Few cult places are known, and some are only mentioned by ancient sources or epigraphical evidence. Archaeologically known are *component GC_69* (remains of a temple to the W of ancient Chosteiai), *component GC_11* (a Roman extramural cult place at the very Southern end of Aliki beach), and, on the W slopes of Mavrovouni, *component GC_13* (a small sanctuary from Geometric to Hellenistic times).

[28] The scholars involved in the survey are much more interested in standing structures or building remains, in their plan, and employed walling techniques, rather than to pottery assemblages (which they use to date the structures). Moreover, they concentrated on the ridge slopes and would not look in the lower alluvial valley since "*il serait presque miraculeux de pouvoir retrouver aujord'hui trace*"(meaning 'structural trace') of any of the rural buildings (which one might suppose must have exploited the valley for cereal cultivation, olive as well as wine and fruit trees), because of the heavy alluviation which would have filled the Sarandi plain with recent soils. They note that a careful examination of the area has given occasional scatters of black glaze or Late Roman sherds –as well as some obsidian flints- in freshly ploughed fields (Gauvin Khostia I: 80 and Fossey and Gauvin 1983: 237). Gregory's work in the nearby Thisbe plain (not to mention other areas of Boeotia or elsewhere in Greece) would prove the opposite, giving meaning to those artefact scatters mostly not associated with visible structural remains (see above in text – Gregory 1992).

[29] The same phenomenon happens in Attica. See the Athene survey, carried out by Lohmann in Attica (Lohmann 1985 and 1992), and also the earlier work of Young on South Attica farms (1956: 122-126).

[30] Husbandry, probably in combination with bee-keeping and oil production (Gauvin Khostia I: 80).

[31] In Pausanias'time the two halves (E and W) of the basin were alternately drained and exploited (Pausanias IX 32.3; for details see Knauss 1992, map in fig.4.3).

Forts and fortifications

In general, the known fortifications are linked to the *polis* settlement sites, which were fortified in most periods of history. Other forts (namely watchtowers) are related to the control of passes, mainly connecting the inner areas with the coastal plains and ports, through the coastal ridges (see above, land and resources) – *component GC_37* (Kivadio) and *component GC_79* (Mavrovouni), as well as the forts above Vathy harbour (*component GC_35*) and at the N end of the Vathy bay (*component GC_36*).

LONG TERM SETTLEMENT TRENDS IN THE CHORA *LANDSCAPE*

Several modern villages (Prodromos, Thisvi, Domvraina) in the territory of the ancient towns occupy the inland fertile plains and extend their territory to the coast to the S, where their small harbour sites are located, and to the N, to include part of the upland mountainous country of the southern ridges of Helicon (Paliovouna, Motsara). The villages go back to the Ottoman period, and in some cases show continuity from antiquity through the medieval period. The small harbour sites also appear active in the medieval and Ottoman period, creating a pattern comparable with the *polis*/harbour pattern of the ancient settlement system.

The modern village of Prodromos (formerly Chostia), recorded in the Ottoman archives from 1466 to 1570 and then 1642 to 1688, forms an example of a possible surviving Greek community from the small ancient city of Chorseiai to its W, certainly active in Late Roman times (Fossey Khostia I – see appendix I.10), hence the survival of the name (Bintiff-Kiel in preparation). Sarandi (modern Paralia), the harbour of Chostia, appears in medieval portulans, indicating the existence of an 'inland village / harbour (*skala*)' pattern in the medieval and

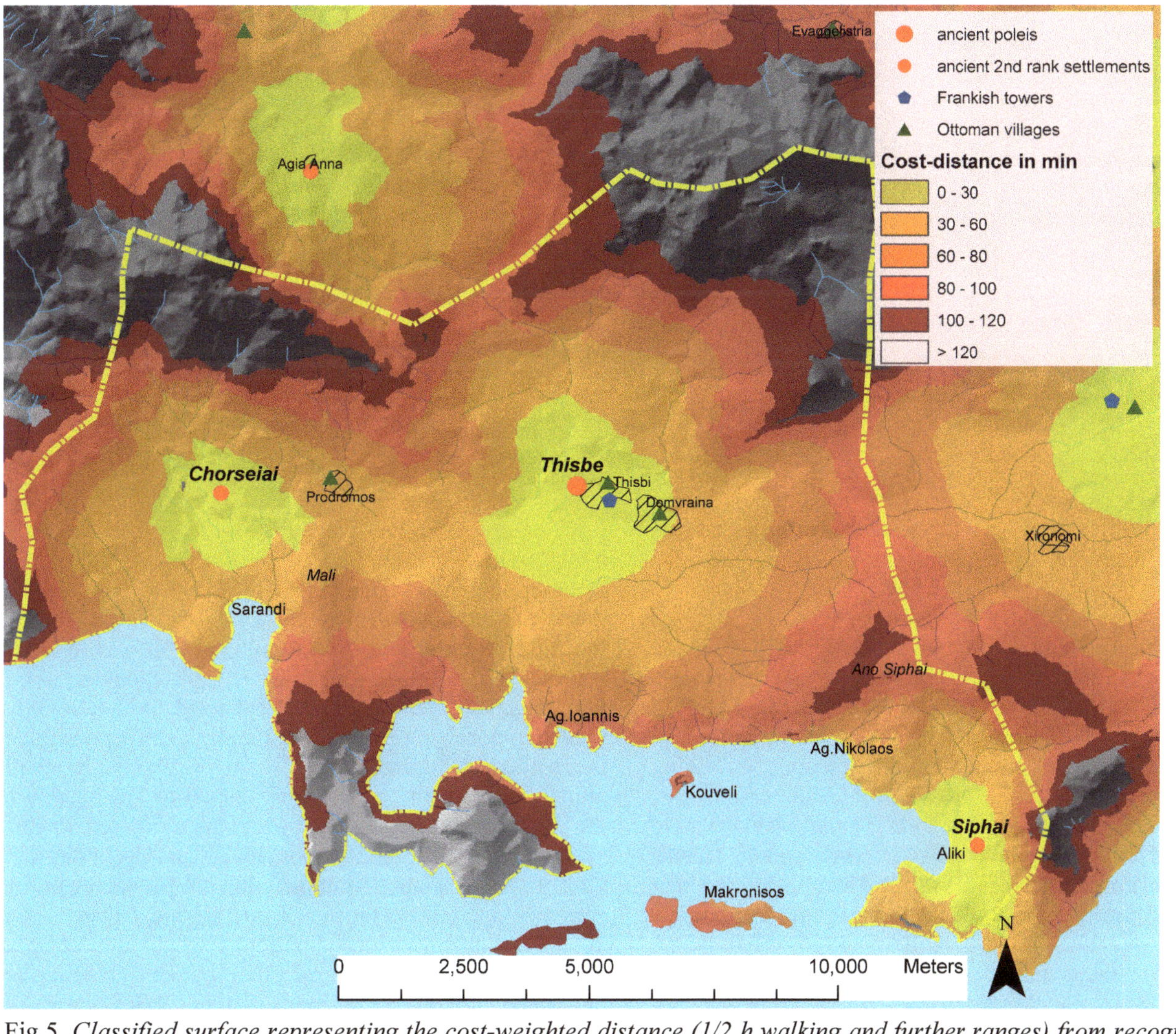

Fig.5. *Classified surface representing the cost-weighted distance (1/2 h walking and further ranges) from recognised 1st and 2nd rank ancient settlements (represented by larger and smaller dots). Areas without dots indicate potential settlement chambers. Ottoman villages and Frankish towers have also been added to the map to show their spatial relationship with the Greco-Roman settlement network and to appreciate potential settlement chambers.*

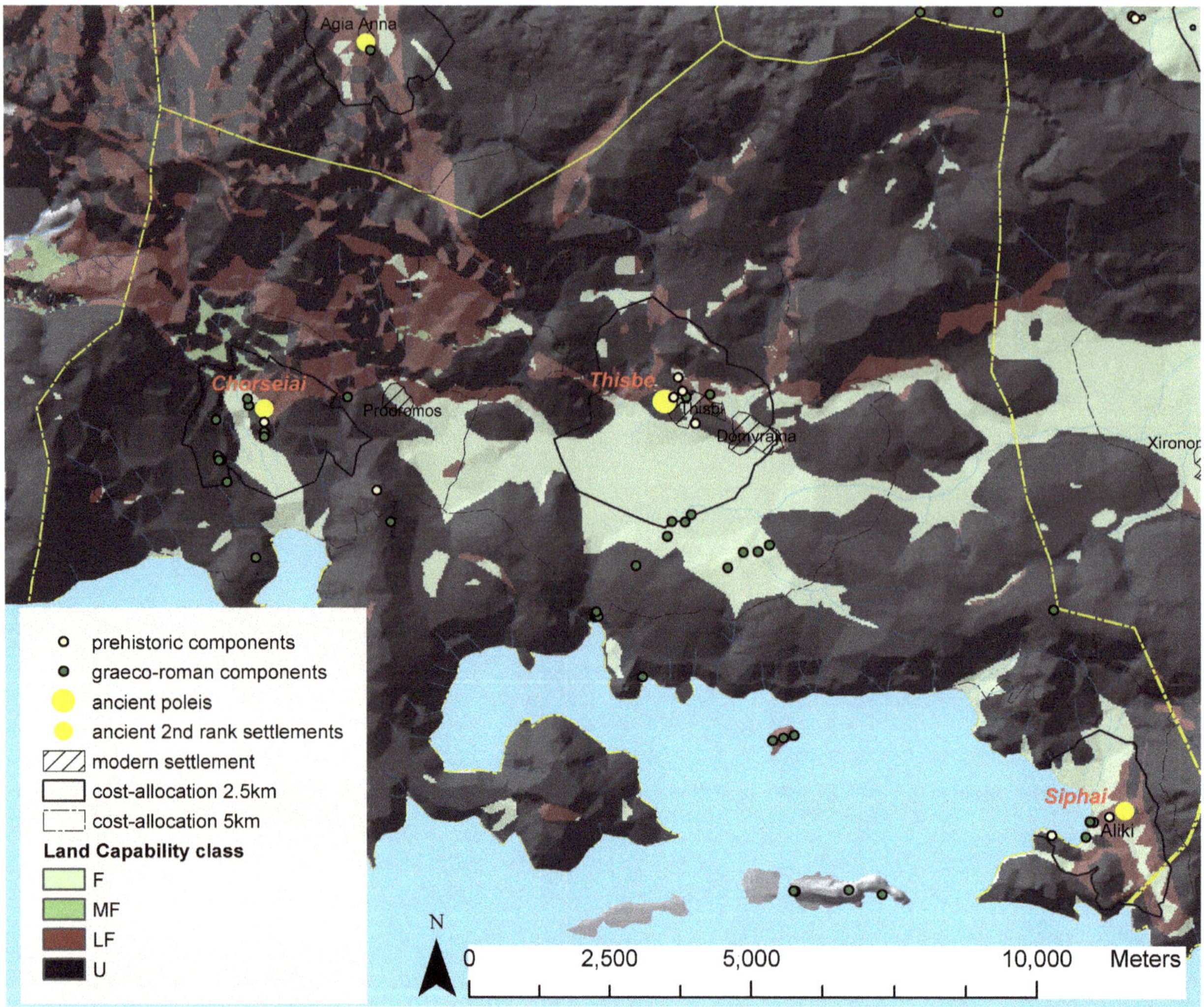

Fig.6. *Map showing the Greco-Roman settlement network, the polygons resulting from the cost-distance analysis (marking half an hour and one hour walking time distance) and dots representing the known archaeological components (same as in fig.2), with land capability information underlain.*

post-medieval period (see the results of a cost-distance analysis[32] performed for the ancient settlement network - fig.5).

Thisvi (formerly Kakosi) is the next modern village to the E, occupying the territory of the ancient *polis* of Thisbe. This small Greek village is recorded in the Ottoman defters (1466-1688) and also shows evidence of important Frankish and Byzantine remains, indicating descent from ancient Thisbe. The settlement chamber of the ancient *polis* with its wide fertile plain is also exploited today by another village next to modern Thisbe, Domvraina (modern Korini). While Thisvi is the remnant of the Byzantine village, the people of Domvraina, recorded in the Ottoman register (1466-1570) as an Albanian village, probably settled here to enlarge the reduced Greek population (Bintliff-Kiel in preparation).[33] Small harbours to the S (Ag. Ioannis and Ag. Nikolaos) respectively serve the two villages,[34] which in antiquity belonged to a single settlement chamber (see fig.5).

Another settlement chamber exists today to the E along the inland alluvial valley of the Askris river, occupied by the modern village of Xironomi in the interior and its harbour site, Aliki, on the coast. In the Ottoman period the Albanian village Tatiza (registered in 1466-1646, with also possible remains of a Frankish tower) occupied the settlement chamber just N of Xironomi, a small alluvial plain on the route from Thespies to Thisbe. Aliki, the harbour site of modern Xironomi, is linked to this village through the 1871 toponym Tatiza Aliki (Bintliff-Kiel in preparation). Although this settlement chamber seems comparable to those of Chorsiae and Thisbe to the W, the settlement history in antiquity seems to have been different. The harbour site of Aliki was cut off from its hinterland and forms a separate small *polis*, Siphai (fig.5). The inland interior and the fertile alluvial valleys seem to be connected to Thespiae, which appears to have

[32] For the application of this GIS analysis see chapter II.3.1 – LONG TERM SETTLEMENT TRENDS.

[33] Yet the Slavic origin of the name suggests a deserted medieval site at this location (Bintliff –Kiel in preparation).

[34] Ag. Ioannis is possibly to be identified with the location of an Ottoman village.

controlled this area of natural passes.[35] We should note, though, that this separation may also have a topographical reason, since the Mavrovouni ridge (higher and wider than the limestone coastal plateau separating the Thisbe basin from the sea) cuts the plain of Xironomi from the small coastal plain of Aliki (Siphai). The latter is larger that the other small harbour sites to the W, and is therefore able to support a small city or *kome*.

Therefore, in the Boeotian area facing the Gulf of Corinth, the medieval/post-medieval and traditional settlement pattern is comparable with the ancient settlement system (fig.5). In antiquity, the area was divided among three *chorai* with similar characteristics: the main centre in an inner position (except Siphai), an occupation of the countryside which differs amongst the different periods of the Greco-Roman, and a harbour site (or more) on the coast, more or less easily reachable from the main settlement. Communication routes were land-routes connecting the settlements, but certainly also sea-routes connecting the harbours.[36] The known sites are concentrated around the small fertile inland basins and coastal plains (fig.6), while they are absent in the peripheral border zones which are dominated by limestone ridges (with the exception of forts[37] and the early settlement of Mali, *components GC_72* to *GC_74*). The topography therefore creates clear borders between the three community areas.

The Gulf of Corinth linked the cities to, rather than separated them from, the outer world. As stated by Gregory (1986: 21), both Kouveli and Makronisos islands[38] seem to have been part of a widespread long-distance commercial network, as attested by the harbour sites and the occupation on the islands[39] in the Late Roman period (see appendix I.10, Kouveli and Makronisos), contemporary to the harbour sites at Vathy and Aliki (see above *passim*), all part of a larger commercial settlement pattern in the Gulf of Domvraina and beyond. In fact, the area (mainland and gulf) was interested by long-distance trade routes, considering that the products from Copais (and from all central Greece) also used to be traded beyond Boeotia through the harbours on the gulf of Corinth, and from these ports goods could be transported with relative ease across the coastal ridge and into the interior of Boeotia.[40]

Chorseiai, and especially Siphai, can be considered as belonging to the category elsewhere called *proto-poleis*, which appeared in early periods, and developed into towns but never into proper *poleis*, while Thisbe, far larger in extension, is a good example of a second-rank real historical *polis*. Besides, a long-term settlement approach would give emphasis to the occupation of the area especially in the Late Classical – Early Hellenistic period, as elsewhere in Boeotia, for rural infill of the countryside, as well as in the Late Roman period, which saw the flourishing of the harbour sites and of the long-distance trade routes, giving a central role to an apparently only marginal landscape (and probably from the LH period).

A few more words could be given to the area of Thisbe, where intensive field-by-field artefact surface survey has also been carried out (in 1979-80 – see above).

Thisbe, as noted above, is a good example of second-rank polis. "*It played only a minor role in political and military developments but could, in favorable circumstances, assert its independence*" (Gregory 1992: 19), as in the Hellenistic period for instance. Thisbe appears in the historical record in the Hellenistic period and in Late Roman times (Gregory 1992: 19). Its prosperity in the latter period has probably to be connected with the flourishing of the Vathy harbour site, belonging to the city and important for long-distance trade (see above).

As seen above, the Domvraina (or Thisbe) basin (another main economic resource of the city site) has been subject to inundation and alluviation throughout the centuries, and into modern times.[41] During their survey, T.E.Gregory and his team carefully observed natural and man-made cuttings in different spots in the plain and they point out that the central part of the plain was apparently probably never densely inhabited, "*either because it was too swampy or [..] because the land there was rich and had to be exploited for its agricultural potential.*" (Gregory 1992: 32) Habitation, they note, seems to have concentrated in a broad band around the N, S and E of the plain, at elevations between 100 and 140m above sea level, and the population appears to have followed the course of the water-control device S of Thisbe (see above). A threefold settlement hierarchy seems to work in the Thisbe Basin: the urban area of Thisbe, the port of Vathy, the large rural agglomerations (*components GC_39 to GC_41*; Thisbe basin C-1), and isolated farmhouses or outbuildings (Gregory 1992: 33 – see corresponding components above), providing the usual picture of the Boeotian countryside, mainly occupied in Late Classical-Early Hellenistic and Late Roman times.

Within the same settlement chamber, the village of Kakosi/Thisvi continues its central role through time (since the Byzantine period – see above) preserving its

[35] See chapter II.3.9 for a discussion of settlement in the area.

[36] As Dasios (1995: 245) notes, communication between ancient Voulis in Phokis (Chorseiai's neighbour to the W) and other Phokian *poleis* were by sea. We might suppose it would be the same as for the Boetian *poleis* to the Gulf of Corinth.

[37] For instance, Ano Siphai, *component GC_13*.

[38] Fossey (1988: 174) suggests that some of the islands in the bay of Domvraina may have belonged to ancient Siphai.

[39] The inner rural sites (see above, Kouveli and Makronisos sites) would probably be sparsely inhabited, and their primary function would have been to supply the larger commercial settlements, where the population would concentrate.

[40] The road from inner Boeotia came through the Steveniko pass (Agia Anna/Koukoura basin is therefore a crucial area, as well as ancient Thisbe and the Domvraina basin) – Philippson 1951: 458.

[41] This could be the reason why the central area, on either side of the Permessos/Askris river, presents very low densities of surface material, although Gregory wonders how much the absence of settlement evidence is due to taphonomic reasons and how much corresponds to the actual situation (Gregory 1992: 32).

Greek population. At the beginning of the 20th century, the village still had several thousand inhabitants and the administrative status of a *dimos* (Gregory 1992: 19), while the plain, Gregory notes, was by that time largely devoted to the growing of wheat, and a customs house, at Ag.Ioannis testifies the use of this secondary port (more easily reached in modern times than Vathy, but already in use in antiquity – *component GC_30*).[42] Today the village has shrunk considerably (300 people reported by Gregory), as the harbour lost importance (because of improvements to overland transportation and the opening of the Corinth canal).

[42] By the village of Kakosi are the remains of a stone-built dam that used to lead the marsh water towards the S in order to facilitate communication with the Ag Ioannis harbour even in periods of significant rainfall (Philippson 1951: 458).

II.3.11

The upper Asopos basin: Parasopia and Plataea

TOPOGRAPHICAL SETTING

Parasopia is the ancient name for the upper basin of the Asopos. It constitutes a geographical unit, mostly comprised of a Tertiary plateau, rich in cultivable soil, and it can be considered one of the most fertile areas of Boeotia.
The Asopos river marks the landscape of the *chora* in its entirety, crossing it transversally from W to E. In its upper part, the Asopos is a river with not much water flowing in a large basin, which falls from W to E from 300 to 240m a.s.l. (Philippson 1951: 502[1]). During its course, the river sometimes flows within narrow banks marked by reeds, while sometimes it opens up (Philippson 1951: 506).[2] To the N of the river the area is characterised by small valleys and basins.
The Livadostra/Stavropotamos river marks the landscape of the SW corner of the *chora*. The river has its source in a basin (2-3km; 280m asl) to the E of the Kotroni hill[3] at the foot of Kithairon (Philippson 1951: 504). The basin, namely the open area before Plataea,[4] is bordered to the E by a Tertiary *tafel* divided into several valleys. The *tafel* outcrops from the mountains at the village of Kriekouki/Erithres, and extends to the S towards the Asopos. To the N, the Livadostra valley is separated from the Asopos plain by a low rise, atop which is the deserted village of Pyrgos,[5] concerned mainly with cattle in Philippson's days (Philippson 1951: 504).
An elevated limestone zone from the area N of Korombili[6] runs towards the E with an average elevation of 320m asl. At its end lies the large village of Kaparelli, and Eastwards the isolated limestone hill of Kotroni (385m asl). To the N, parallel to the limestone ridge, is a cultivated slope covered with soil from the Tertiary period, on the N slope of which is the Parapoungia/Leuktra village, formed by three settlements which look over the wide valley basin (Philippson 1951: 505).
The ancient city of Plataea, the largest settlement in the area in historical times, with the status of *polis*, was situated S of the Asopos, on a low terrace (320m asl – today named Kastro) on the foothills by the plain, E and below the village of Koukkla/Plataies (390m asl), located at the upper end of a talus slope.
Further to the E, along the foothills of Kithairon, is the village of Kriekouki/Erythres, to the E of which the mountain range turns slightly to the N with the Pastra ridge, below which ancient Erythrai lay[7] (Philippson 1951: 506).

Boundaries

The borders of the *chora* are: to the S, Mt. Kithairon and Pastra, to the N the Soros range, to the E the line across the Asopos valley at the point where the foothills of Pastra approach the Soros range and there is a narrow ravine of the Asopos.[8] To the W the boundaries are less defined geographically, and the border would probably lie somewhere between ancient Plataea and the Thespiae area, probably corresponding to the rolling, barren hills W of Thebes and the very slight rise between the Asopos river system and the Askris system to the W,[9] as Fossey (1988: 132) suggests. Since geographical limits are not so clear in this area, political limits would also fluctuate. Much of the area, in fact, lay very close to the great power of Thebes, and precisely in the direction where natural divisions are least clear. As Herodotus (VI 108) and Pausanias (IX 4.4) inform us, the S boundary of the territory of Thebes was at some point in history fixed by arbitration at the Asopos river itself.

PHYSICAL LAND UNITS

As can be seen from the chart above, the plain landscape is practically absent, and the hilly segment constitutes the majority of the landscape. As in the case of Thespiae *chora* (see chapter II.3.9), this is due to the elevation range within which the *chora* landscape is included, rather than to the morphological character. In fact, the majority of the hilly landscape is comprised of features belonging to class H1, constituting a plain landscape as far as morphology is concerned, but simply located

[1] A detailed geological description of the Asopos valley can be found in Philippson 1951: 506. See also in appendix I.14 - THE TANAGRA SURVEY PROJECT.
[2] There are several bridges over the river (Philippson 1951: 506).
[3] Kotroni hill, an isolated limestone hill 385m asl, E of Kaparelli village.
[4] In the open area before Plataea the famous battle of 479 BC took place; the battlefield lies between the *tafel* of Kriekouki/Erythrai and the upper part of the Livadostro basin (Philippson 1951: 504).
[5] A Frankish tower stands by the village, and is named after it. The location has a nice view towards the Plataea plain where the battle of 479 BC took place.
[6] See chapter II.3.10 and map in fig.1 (II.3.10).
[7] Ancient Erythrai has been identified with the Daphni plateau location (see *components PP_50 to PP_54*).
[8] The E border would run not far from ancient Tanagra to its W.
[9] For the Askris river, see chapter II.3.9.

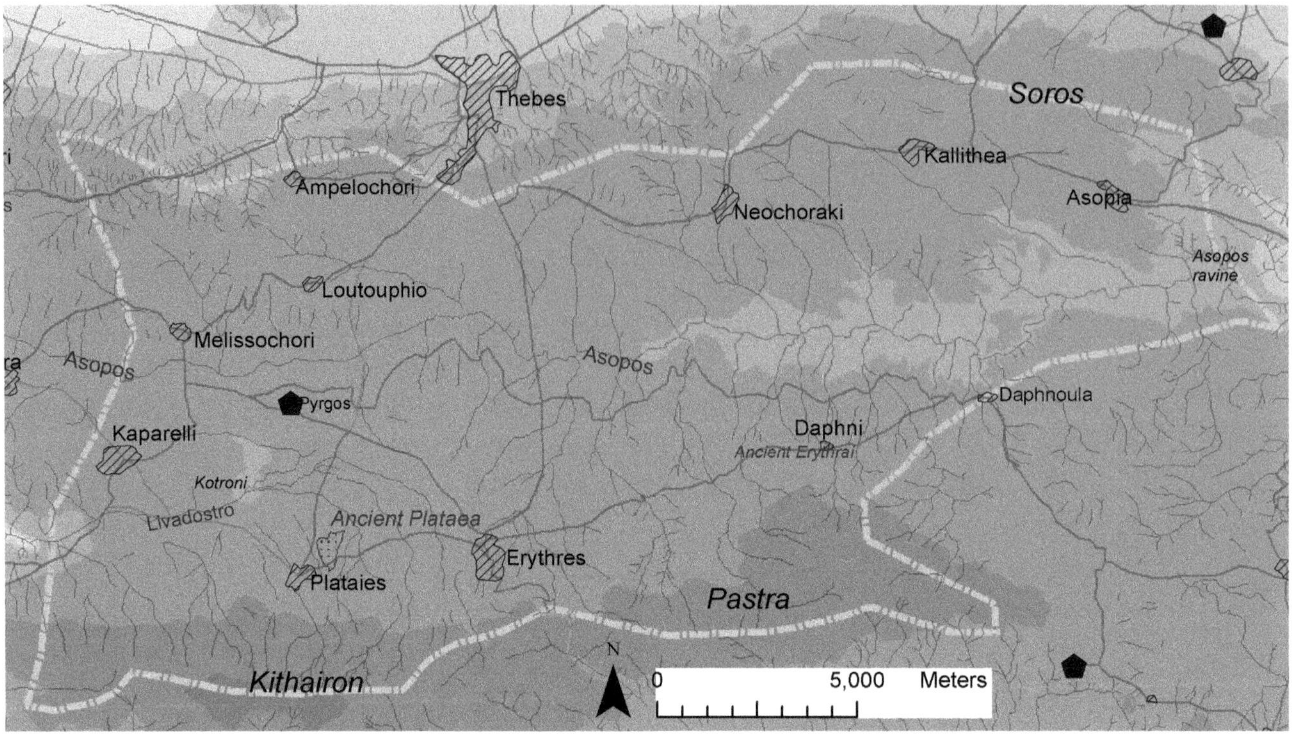

Fig.1. *Topographical setting of the Parasopia and Plataea chora.*

entirely above a 200m elevation (which we have taken as the elevation border between plain and hilly landscape – see chapter I.2.1). As is visible in the map fig.2 in chapter II.1, H1 features constitute the Asopos valley (comprising a Tertiary plateau) as well as the basins marking the landscape of the N side of Asopos (see above, general description). Steeper slopes characterise the E area of the *chora*, marked by a much more differentiated landscape and a narrower Asopos valley.
Mountain landscape is represented by the slopes of Kithairon above 600m asl, along the S edge of the *chora*.

Hilly landscape	87.1%
Mountainous landscape	12.8%
Plain	0.1%

1	P1_P2	lacustrine basin, valley	0.0005%
2	P3	gentle slope	0.005%
3	P4	foothill	0.002%
4	H1	plateau	46.3%
5	H2	gentle slope	12.8%
6	H3	moderate slope	15.3%
7	H4	severe slope	9.2%
8	H5	very severe slope	3.5%
9	M1	plateau	3.4%
10	M2	plateau/gentle slope	1.8%
11	M3	moderate slope	4.1%
12	M4	very severe slope	3.4%

Table 1. *Percentage of the different physiographical classes present within the Parasopia and Plataea area (P=plain; H=hill; M=mountain)*

RESOURCES

The plain between Plataea and Thebes is marked by high fertile Neogen sediments and older pre-Holocene colluvia. The streams, including the longer Asopos, are embedded deep in their valley, allowing mainly for narrow patches of alluvia.
As stated above, the area is rich in cultivable soil, with intense crop production. The map in fig.10 clearly shows the predominance of fertile land (class F) in the area. The value MF is totally absent, while the U value is represented by the highest slopes of the Soros range (to the NE) and the limestone heights on the slopes of Kithairon along the S border, as well as the few dolomite formations showing on the surface in the area where the modern village of Kaparelli lies (W of the map).

THE ARCHAEOLOGICAL RECORD

1	**PLATAEA**	*Components PP_1 to PP_10; PP_76 and PP_77*
2	**Plataea SW frindges**	*Components PP_1, PP_2, PP_76 and PP_77*
3 – 6	**Plataea burial places**	*Components PP_12 and PP_13, PP_14* (3); *PP_17* (4), *PP_18* (5) *and PP_19* (6)
7	**Plataea N – Kotronaki**	*Components PP_15 and PP_16*
8/9	**Neochorakion**	*Components PP_20 to PP_22* (8); *PP_33 and PP_75* (9)
10	**Moustaphades/Kallithea – Pyrgaki**	*Components PP_23 and PP_24*

11	**Kallithea Metochi**	*Components PP_71 to PP_73*
12	**Soros**	*Components PP_25 to PP_28*
13	**Soros**	*Components PP_32 and PP_74*
14	**Soros SE**	*Components PP_29 to PP_31*
15/16	**Asopia**	*Components PP_56 and PP_57* (15); *PP_58 to PP_64* (16);
17	**Asopia vicinity**	*Components PP_65 and PP_66*
18	**Asopia E**	*Components PP_67 to PP_69*
19	**Asopia S**	*Component PP_70*
20-22	**Asopia S**	*Components PP_78* (20); *PP_79* (21); *PP_80* (22)
23	**Daphni Kastron**	*Components PP_47 to PP_49*
24	**Daphni plateau (ancient ERYTHRAI?)**	*Components PP_50 to PP_54*
25	**Daphni**	*Component PP_55*
26	**Erythrai – Pantanassa**	*Components PP_34 to PP_38*
27	**Erythrai - Pantanassa foothill**	*Components PP_ 39 to PP_40*
28	**Erythrai - Pantanassa area**	*Components PP_41 and PP_42*
29	**Erythrai – Pyrgos**	*Component PP_43*
30	**Erythrai - Pantanassa NE**	*Components PP_44 to PP_46*

Table 2. *List of archaeological components and activity loci mapped in fig.2.*

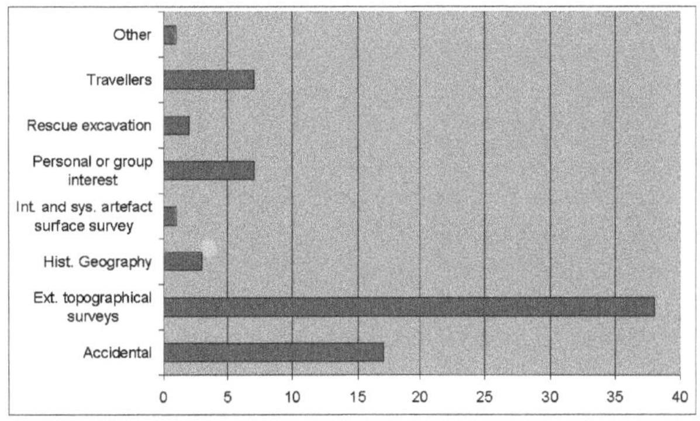

Fig.3. *Graph illustrating the proportion of components discovered within different research frameworks.*

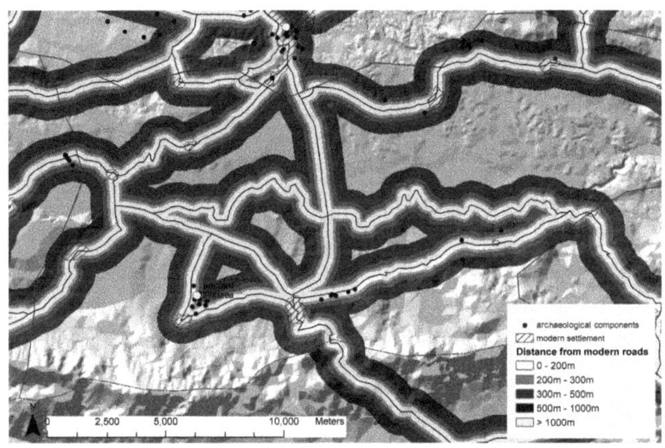

Fig.4. *Relationship between components and modern road network.*

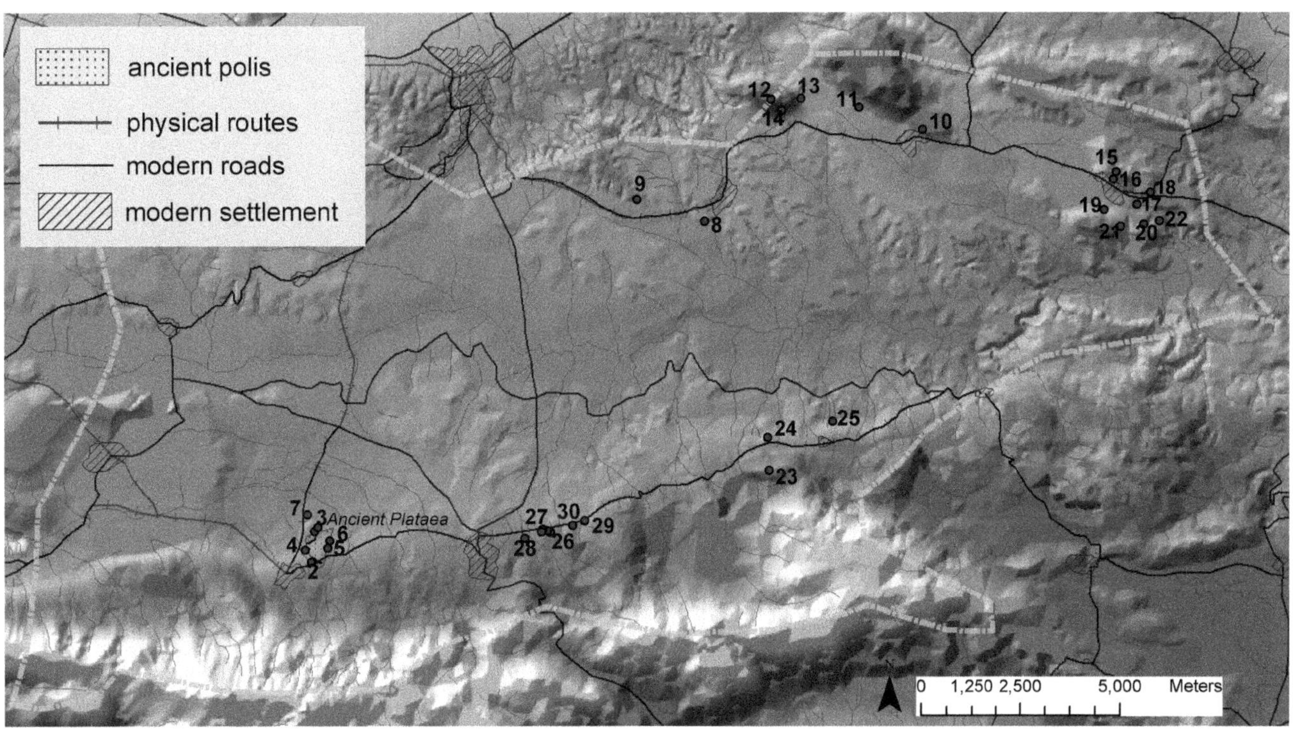

Fig.2. *Archaeological map of Parasopia and Plataea region.*

The archaeological record available for the area is concentrated along the foothills of Kithairon and Soros, respectively the S and the N edges of the *chora*, and archaeological research has focused mainly on hilltop locations, possible acropolis sites for the settlement known for the area from historical sources.
Conversely, the middle Asopos valley and the wide surrounding plain, constituting the central part of the area crossing the Plataea-Parasopia *chora* transversally, is archaeologically less known. This is due mainly to the absence of an intensive and systematic artefact surface survey carried out in the area.
The location of the known archaeological record (as evident in the maps) is due mainly to research focused on the settlement site (actual or probable/possible) in the area. As so often happens, the rural segment of the landscape is missing.
Rescue excavation is also poorly represented in the discovery panorama (see graph in fig.3); its presence being due only to works for the gas pipe-line to the S of the village of Asopia, which brought to light, apart from burials, also a rural site.

In fig.4 we can see the relationship between known archaeological sites and the distance from the modern road network. The correspondence seems striking, and the archaeological record available would indeed appear to be biased by proximity to the modern road network. This is mainly due to the absence of research carried out in the wide plain crossed by the Asopos river, and to interest in prominent sites at the edge of the basin, as well as the fact that the modern road through the foothills most probably follows an ancient route, and this could also, albeit partially, explain the location of known sites along it.

The ratio of known Prehistoric to Greco-Roman components is 14 to 60 (1:4.3), while among the historical periods, 31% are dated Archaic to Hellenistic, 14% Roman-Late Roman, and 12% are attributed to the general Greco-Roman period.

ANALYSIS OF THE *CHORA* LANDSCAPE

PREHISTORIC PERIOD

For the Prehistoric period, we are missing a potentially large quantity of data, considering the suitable landscape that the Asopos river would offer to Prehistoric settlement, especially in its middle course. Sites which may have been by the tributaries of the Asopos, or on upper terraces of the river itself (see the Prehistoric settlement pattern for Tanagra) elude our knowledge.
Only one earlier site, Neolithic and Early Helladic, has been identified N of Plataea in the Kotronaki locality (*components PP_15 and PP_16*). There is a LH activity focus at the site of historical Plataea, while MH is only probable; LH activity foci can also be located at Erythrai – Pantanassa (*component PP_34*) and at the Daphni plateau (*component PP_50*) sites. Prehistoric occupation is attested also at the Asopia site (*components PP_61 to PP_63*), since Prehistoric potsherds were found on the slopes, probably washed down from the top of the Prophitis Ilias hill (see figs.5 to 8).
LH burials are known from the areas of Soros (*component PP_74*) and Moustaphades/Kallithea (*component PP_23*), linked to a settlement nearby. An attempt has been made to identify it with the Soros Easter peak (see Soros –*component PP_25*- and also the discussion of the Moustaphades/Kallithea site).

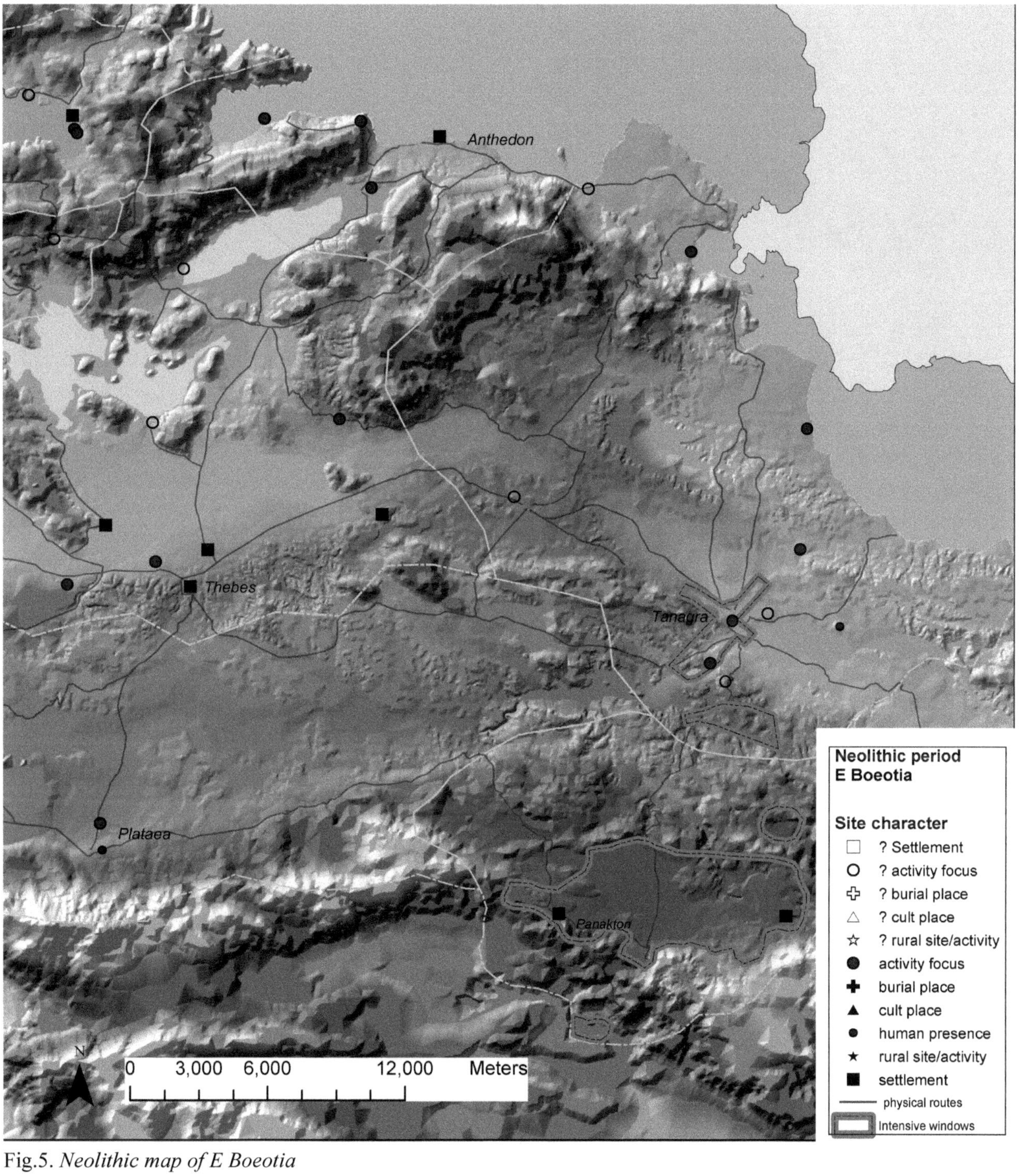

Fig.5. *Neolithic map of E Boeotia*

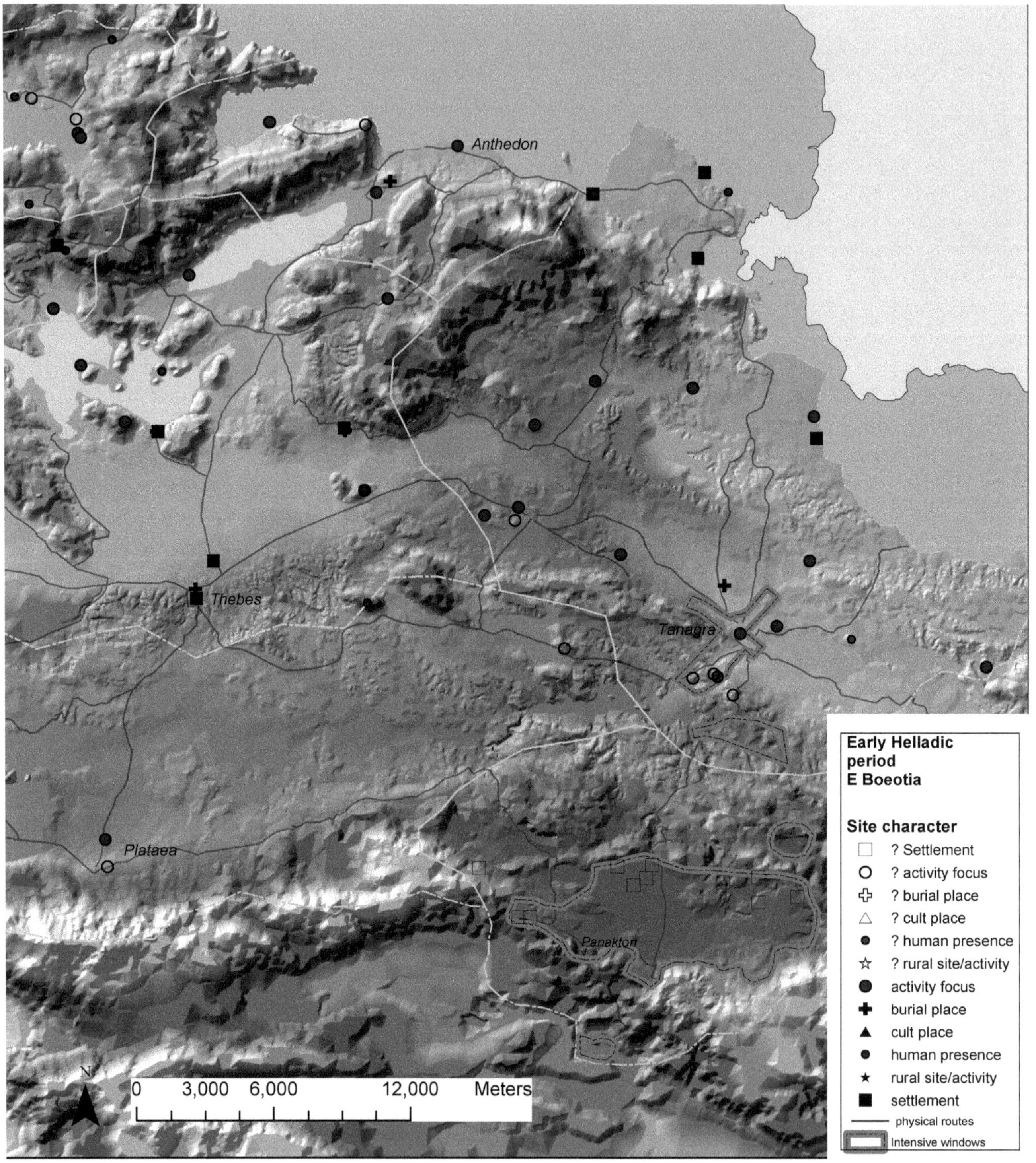

Fig.6. *EH map of E Boeotia*

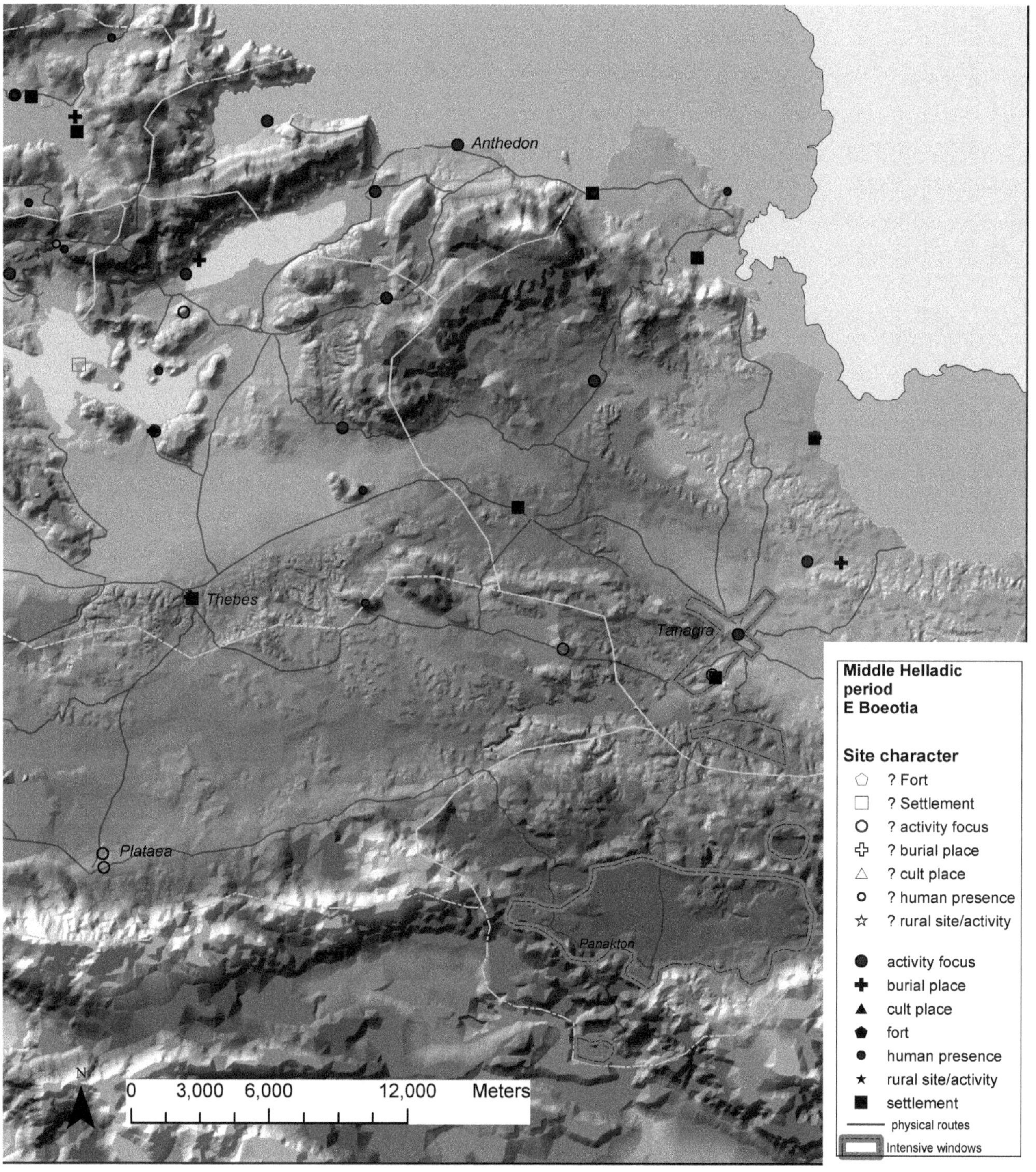

Fig.7. *MH map of E Boeotia*

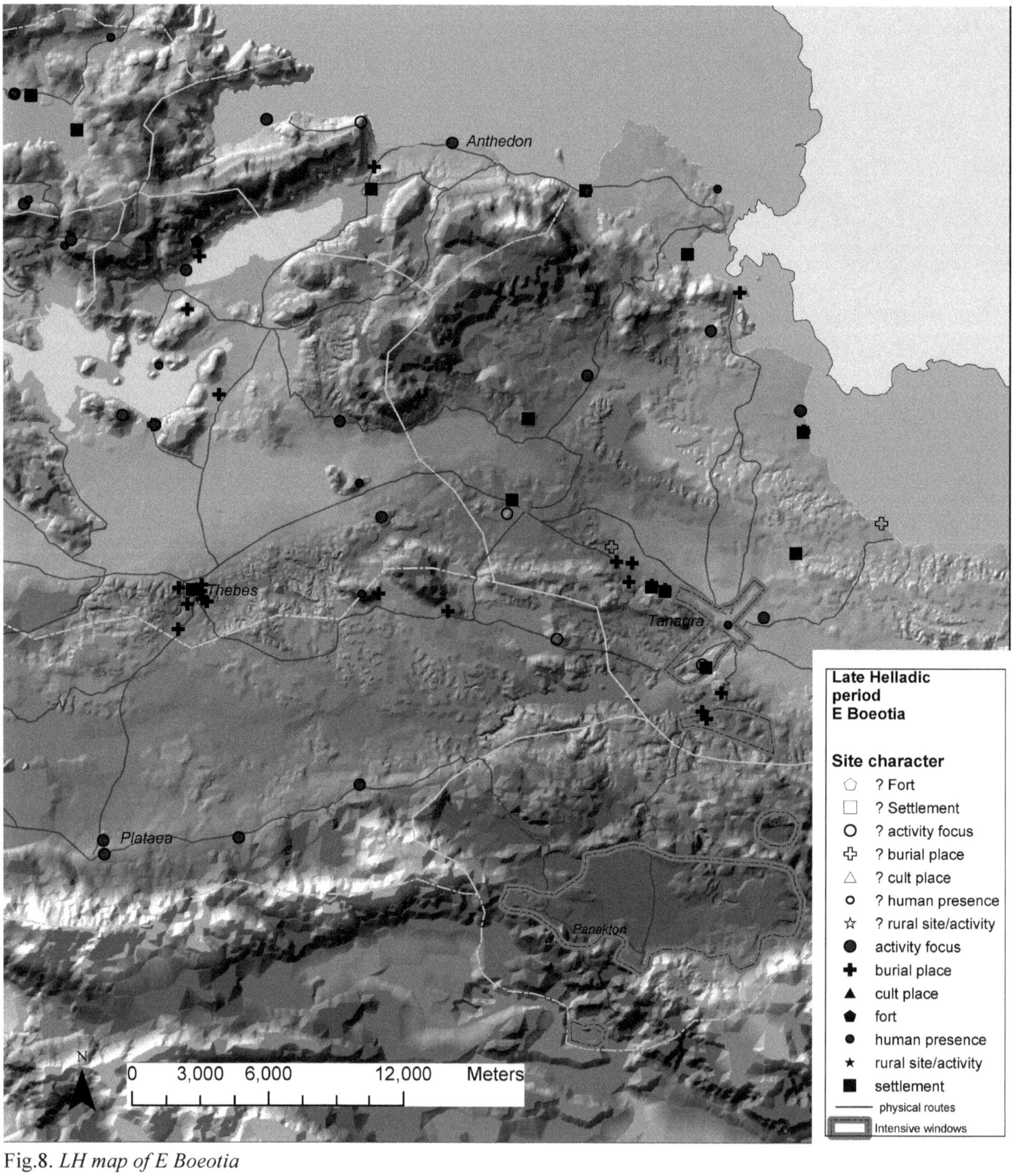

Fig.8. *LH map of E Boeotia*

GRECO-ROMAN ANTIQUITY[1]

Town level
The town of Plataea is the main centre in the area, with the status of *polis* and the right of participation in the Boeotian confederation. The larger walled area of Plataea corresponds to the 5th C BC city. Occupation in early periods of history is poorly known in the area, with the exception of the intensively studied city site of Plataea, where Geometric and Archaic occupation is also attested with certainty and in detail. Its position close to Thebes (without physically well-defined borders with Theban territory – see above) and the wide fertile land available to it (see above – RESOURCES), made it and its area an object of interest to Thebes. Despite Thebes taking over half its land after its victory over the Plataeans in 426 BC, the large size of Plataea during the 4th C can be explained by the fertility of the still available land. Also in Late Roman-Medieval times the place had a status, as attested archaeologically (see appendix I.11).
The importance of the city of Plataea is also based on the central position between the valley road to the port of Kreusis/Livadostra and the important route through the pass above Kithairon to Eleusis, Megara and Athens (=the modern road to Athens and Vilia), and Thebes was certainly interested in this area of crucial strategic importance against Athens.

Village level
The existence of two villages in Greco-Roman times might be supposed for the area including the modern villages of Moustaphades/Kallithea and Neochoraki, and the Soros hilltop area, along the N edge of the wider Parasopia region, based on the existence of fortified hilltops and burial areas (see appendix I.11 for discussion). Another village might have been located in the Klembotsari/Asopia area, mainly because of the existence of burials in the surroundings (see appendix I.11, under Asopia). A potential gap can also be located in the area of Loutouphio.[2]
Along the S edge, at the foot of Kithairon, three settlements are mentioned by ancient sources, from E to W: Erythrai (located at the site of Daphni), Hysiai (identified with the Erythrai-Pantanassa site) and Plataea. Only for Plataea can a clear nucleated settlement status be archaeologically defined with certainty. For the others, only hilltop fortifications are known, which we can only guess were in connection with a settlement (see above under Daphni Kastron and Daphni plateau, and Erythrai-Pantanassa sites).

Rural segment
As noted earlier, the rural segment is poorly represented in the archaeological record of the *chora*. *Components PP_44 to PP_46* might be interpreted as a probable rural site, occupied from the Classical to Roman period, while recent discoveries during gas pipeline works to the S of the village of Klembotsari/Asopiae indicate a rural site probably linked to pastoral activities (*component PP_78,* with probable rural burials – *components PP_79 and PP_80*).

Burial areas
Several burial areas are known in connection with the city site (*components PP_11 to PP_14 and PP_17 to PP_19*), from all periods of Greco-Roman history. Within the city area, outside the NW 'acropolis', Late Roman tombs are reported (*component PP_11*). They could be an indication of the shrinking of the city at that time (probably to the NW), as happens elsewhere in Boeotia.
Other burial areas are known in the vicinity of possible settlement sites: *component PP_32*, *components PP_20 to PP_22, PP_33 and PP_75*, *component PP_55*, *components PP_65 to PP_70*. Some burials might be associated with agro-pastoral activities and habitation (*components PP_79 and PP_80*).

Cult places/Religious areas
An attempt has been made to identify *component PP_29* (Soros SE) with a cult place, though there is too little evidence. A similar case is that of *component PP_40* (Erythrai - Pantanassa foothill).

Forts and fortifications
Several forts and/or fortified sites are known from the area. In the majority of cases, attempts have been made to identify the fortifications reported as connected with a settlement or as containing a small refuge settlement within (see *components PP_26 to PP_28*; *PP_ 35 to PP_37; PP_47; PP_59 and PP_60*), while in a few cases they could be single forts or watchtowers (*component PP_47*, *component PP_43* of uncertain date though, *components PP_50 and PP_60*).
The fortification on the Soros easterly higher peak (*components PP_26 to PP_28*) is probably to be considered as just a fort (and not the acropolis of a settlement) in historical periods.

LONG TERM SETTLEMENT TRENDS IN THE CHORA *LANDSCAPE*

Settlement chambers are difficult to recognise in this long, wide area, where there are no clear physical borders to define topographical units available for a settlement. The great availability of water sources make the area easily settled anywhere one might choose, and the high land potential characterising the area in its entirety (see above – RESOURCES) would not allow the identification of local patches available for potential settlements. In order to enucleate a pattern feasible for antiquity, an examination of the modern and traditional settlement pattern could help.
In the area of Parasopia there are today a series of village territories, centred on fairly regularly spaced settlements on both sides of the Asopos valley. The community borders fall on the Asopos river, so that settlement chambers are created on both sides of the Asopos valley,

[1] Period maps are included in chapter II.4, figs.18-20-22-24-26-28.
[2] For a discussion of the settlement history in the area see below in the text – LONG TERM SETTLEMENT TRENDS.

containing part of the plain and the hilly country N and S of it.[3]

N of the Asopos river we find six modern villages, some of which go back to the Ottoman period. In many cases additional villages are recorded in these territories in the Ottoman period, indicating the existence of several settlement chambers in this area (see fig.9). Modern Asopia (formerly Chlebotsari) goes back to the Ottoman period (1466-1646). An additional village is registered in the Ottoman archives in the southern part of the modern village territory. Modern Kallithea (formerly Moustaphades) has no predecessor in the Ottoman period. Yet nearby, by a torrent W of Kallithea, is located the deserted village of Likouresi registered in 1506-1646. The settlement chamber of modern Neochoraki was occupied probably in the Ottoman period by the village of Golemi, recorded in the 16th and 17th centuries. Modern Loutouphio has also an Ottoman predecessor (Luta, recorded 1466-1688). In the settlement chamber of modern Ampelochori to its N, the deserted village of Samari goes back to the Ottoman period (16th-17th centuries). A second Ottoman village (Kokinaris Likores, founded 1506) is possibly located SE, by the northern border of the settlement territory of modern Melissochori. Modern Melissochori (formerly Baltsa/Baltsia, recorded 1521-1688) occupies the last settlement chamber N of the Asopos river in the W part of the territory of ancient Parasopia. Its territory includes a large part of the Asopos plain, also exploited in the past by the deserted village of Khalki, E of Melissochori (recorded 1506-1688).
On the southern side of the Asopos river, settlement today is comprised of three villages in the foothills, from W to E Plataies (formerly Kokla – no more than a village), Erythrai (formerly Kriekouki, recorded 1466-1646 and today administratively within Attica) and Daphni (formerly Darimari, recorded in 1646) to the SE.[4] To the W of Plataea is the large village of Kaparelli (recorded in 1466-1688), located at the end of the S edge of the tertiary *tafel* of Thebes and controlling part of the plain and the nearby slopes of Kithairon, with a smaller village/hamlet, Melissochori, to its N, in the middle of the wide Asopos plain.

All these modern villages, the majority of which go back to the Ottoman period, contain in their territories part of the fertile plain of Asopos, the Livadostra basin and the southern slopes of Kithairon, by the foot of which they are situated. The most favourable is the settlement chamber of Plataies, which contains a large plain. This settlement chamber, occupied in antiquity by the settlement of Plataea, with the status of a city, continued to be an important focus of habitation in the medieval and postmedieval period (considerable Byzantine and Frankish remains outside the city walls[5]), recolonised by Albanians after a gap in the Frankish period (Bintliff-Kiel in preparation). A pair of Ottoman villages named Kokla Kebir and Sagir is recorded in this area in 1466-1688 and 1540-1680 respectively (Bintliff-Kiel in preparation), continuing in the Early Modern and Modern periods (modern Plataies, formerly Kokkla). North of Plataies (within the village territory of modern Kaparelli) the 19th century village of Pyrgos, with a Frankish and a possible Ottoman village, constitutes a further settlement chamber in this area.

A cost-distance analysis[6] performed for the *chora* might help to indicate the potential areas gravitating onto a settlement focus potentially shifting through time. As there are no physical constraints to the forming of settlement chambers, they could be determined arbitrarily by proximity relationships, and within them settlements would seem to have been very mobile through time, without firm constraints on the location of the centre and yet always within the same wider area and probably without changing much, with regards to the number of nucleated settlements exploiting the large amount of available fertile land and resources.

The *polis* settlement of Plataea is the major centre in the area, though the actual limits of its territory against other possible nucleated centres within the wider area, which would certainly have existed, are not easy to define. The wider Tertiary plateau which mostly constitutes the *chora* landscape has always been densely populated, with its potential able to support a large number of inhabitants.
One might suppose, for instance, the existence of two villages in the area of Neochoraki-Soros-Kallithea (see appendix I.11 and above - *village level*). In the modern settlement pattern, the distance between the two modern villages of Neochoraki and Kallithea is 5 km. Kallithea's physical landscape comprises mostly excellent land in Neogen, and the village has plenty of fresh water too. Certainly this area would have suited a settlement in antiquity. On the other hand, as there is space today for two villages whose economy is based mainly on agricultural activities, there must have been space for two settlements in the wider area of Neochoraki and Kallithea

[3] For the boundaries of modern *koinotites* and settlement structure in the area see also discussion and comparison of the modern settlement pattern and the supposed ancient one in chapter II.4 for the wider Boeotia region and figs. 1 and 14 (chapter II.4). The *koinotita* of Thebes goes behind the low series of hill to the S of the town, and exploits part of the Asopos valley, to the Asopos itself. It can be noted also that the Asopos river actually works as border for the other *koinotites* gravitating towards its valley as well (fig.1 in chapter II.4).

[4] In SE Parasopia, the only sizeable village, as noted by Fossey (1988: 116), is the modern village of Darimari/Daphni. In the recent past in SE Parasopia, there was also the hamlet of Katsoula/Rapendosa (Daphnoula on the map), with only 52 inhabitants by 1961, and having disappeared from the statistical book of *demoi* by 1981 (in 1967 Fossey (1988: 116) saw only 2 people resident). The majority of the inhabitants seem to have moved to the Skourta plain, while still farming the Parasopia land (Fossey 1988: 116).

[5] See also recent systematic work on the city by the Austrian team lead by A.Konečny (preliminary report in Konečny et al. 2008).

[6] See chapter II.3.1 – LONG TERM SETTLEMENT TRENDS.

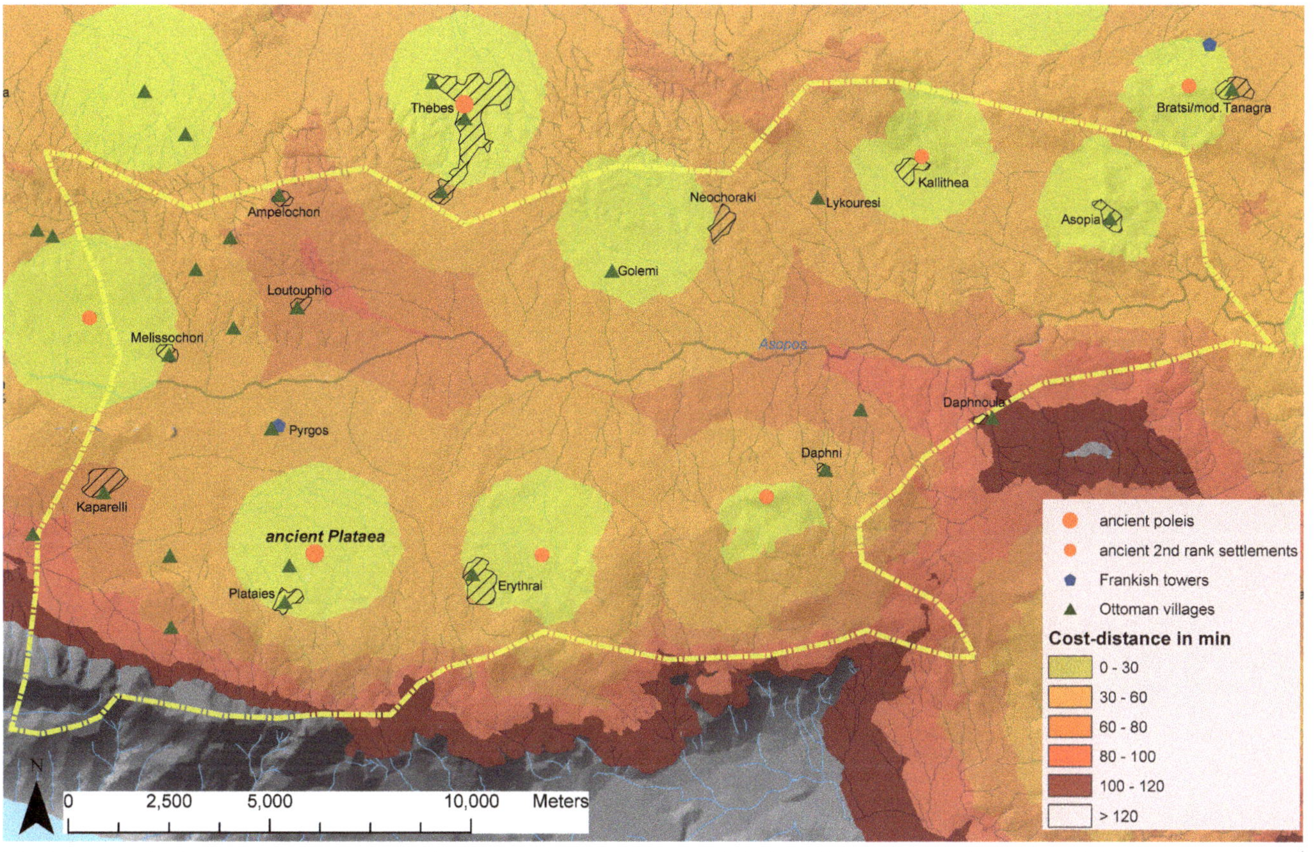

Fig.9. *Classified surface representing the cost-weighted distance (1/2 h walking and further ranges) from recognised 1st and 2nd rank ancient settlements (represented by larger and smaller dots). Areas without dots indicate potential settlement chambers. Ottoman villages and Frankish towers have also been added to the map to show their spatial relationship with the Greco-Roman settlement network and to appreciate potential settlement chambers.*

in the past also.[7] This can be supported by the existence of village settlements recorded in the Ottoman defters, within the settlement chambers occupied by the modern villages (Lykouresi by Kallithea and Golemi, not far from Neochoraki). The weakness of a natural logic of space due to the lack of physical constraints in the area determining settlement chambers has as a result a much freer shifting in location of the settlement focus within zones delimited only by the existence of and proximity to neighbouring settlements. There is no need, therefore, to suppose the sites of ancient village settlements in the same locations as the modern ones, as later periods' settlement histories seem to suggest. In the cases of Kallithea and Neochoraki, the Greco-Roman occupation (attested mainly by burial evidence[8]) could have been elsewhere within the wider area.

Today along the N limit of the *chora*, to the E, is also the village of Klembotsari/Asopia, also 5km ca from Kallithea. There would have been space here for another settlement in the past as well, whose existence seems to be attested by burials among other slight evidence. In the NW part of the *chora* are situated today the small hamlets of Loutouphio and Ampelochori. Here once again the existence of modern and Ottoman settlement foci might help to identify at least one potential settlement niche in the past, although in this case the archaeological evidence is completely lacking. The gap can easily be evaluated on a map showing a cost-distance allocation analysis from the recognisable or supposed (on the basis of evidence) Greco-Roman settlement sites. On the other hand, one should also consider and evaluate the proximity of Thebes and its satellite villages, expanding onto and exploiting the Tertiary *tafel*. As pointed out earlier, the definition of the settlement structure of the landscape in this area, without physical constraints and very close to Thebes, powerful in the Greco-Roman period, can not be easily performed.

On the S side of the *chora*, along the foothills of Kithairon, as seen above (*village level*), there would have been at least two other settlements to the E of Plataea (ancient Hysiai and ancient Erythrai, according to historical sources, which could be represented by the archaeological evidence at Erythres and Daphni[9]). An examination of the landscape and land availability would fit with the indication of the historical sources, as there is enough land and resources for two villages in addition to

[7] See also Bintliff 1994b (fig.20), who locates a village settlement in antiquity in the area of Kallithea and evaluates a potential gap in the Neochoraki area, as well as in the Asopia and Loutouphio areas (see below in the text).
[8] See appendix I.11 for details.

[9] See appendix I.11 for details; table 2 and the map in fig.2 for location.

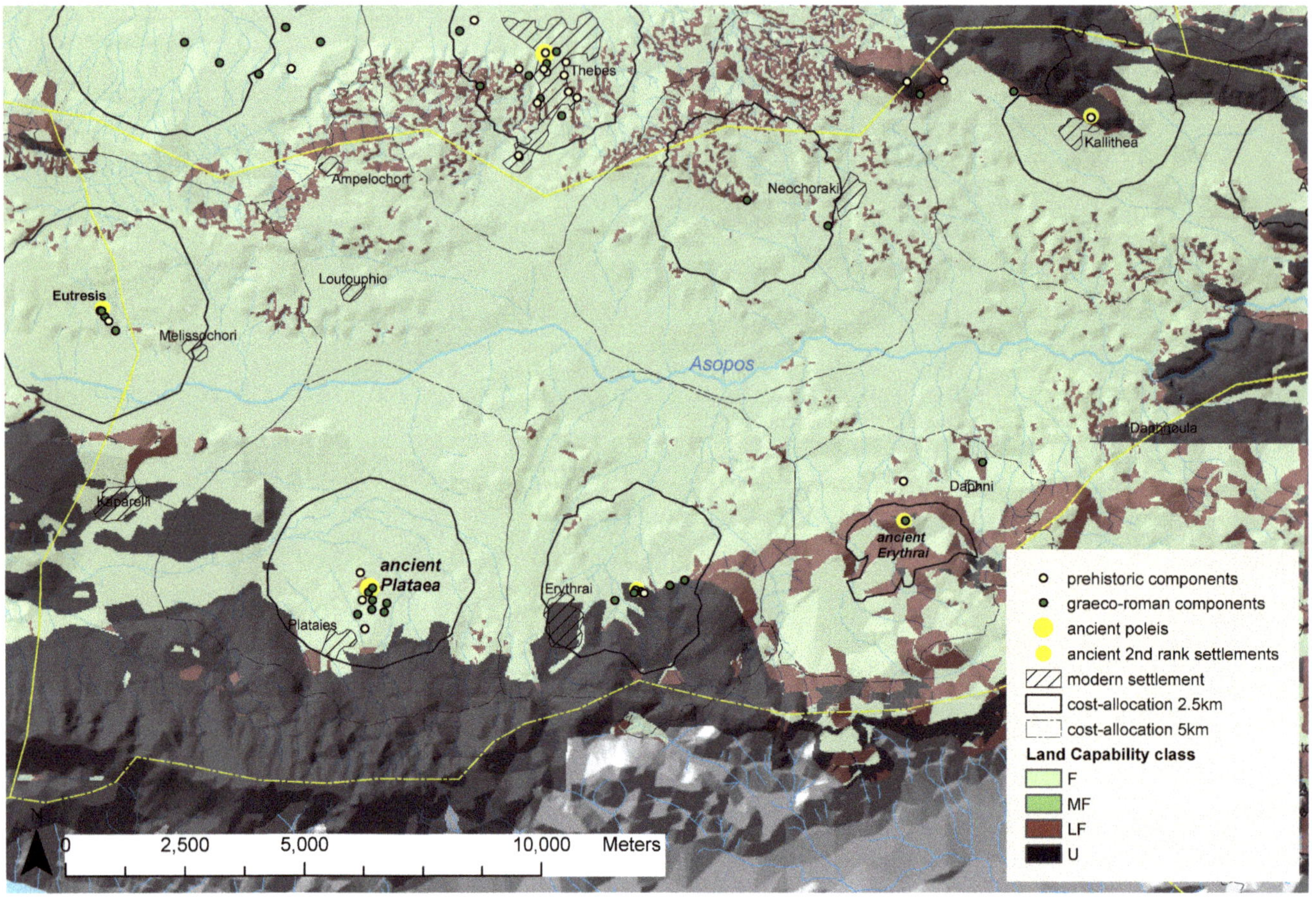

Fig.10. *Map showing the Greco-Roman settlement network, the polygons resulting from the cost-distance analysis (marking half an hour and one hour walking time distance) and dots representing the known archaeological components (same as in fig.2), with land capability information underlain* (the very E part of the *chora* is missing as there is no available geological map at the 1:50,000 scale – the Chalkis sheet is missing from the available IGME maps).

the *polis* of Plataea, as the traditional and Ottoman settlement pattern (examined earlier) would also testify.

Yet, as we have already noted, as there are no physical constraints to the forming of settlement chambers, they could be determined arbitrarily by proximity relations. For example, as Philippson (1951: 502) notes, in the interior part of the Tertiary *tafel* of Thebes, in the immediate hinterland of the city, are only four small villages: Ampelochori, Loutouphio, Melissochori, Neochoraki.[10] The presence of Thebes influences the surroundings, whose landscape is marked by a few small villages.[11] On the other hand, in the Ottoman period, it seems that a more dispersed settlement pattern existed, with many small villages especially in the W part of the *chora*. The existing archaeological evidence would indicate for antiquity a more nucleated pattern consisting of Plataea and a number of villages at regular distances, more or less comparable to the modern settlement pattern.

Yet, future work at the rural level might change this picture. The Thespiae survey, for example, recorded regional differences in the urban/rural population in different parts of the *chora*, e.g. presence of hamlets and small villages in the inner zone by the city and very low densities in the outer zone of the *chora*. As for the Parasopia area, the gaps at its W and E edges, occupied by the modern village of Kaparelli to the SW, Ampelochori to the NW, and Daphni to the SE, might reflect real low densities in the outer zone of the *chora*.

[10] The villages record inhabitants in 1889 – 625 (all together), and in 1928 - 1544 (383 each village), while, e.g., Kaparelli to the S had in 1889 – 883 and in 1928 – 1752 (Philippson 1951: 502).

[11] In general, as Philippson (1951: 502) notes, the settlement in the landscape gives a primitive and backward impression.

II.3.12

The Theban plain and the area of the lakes: Thebais

TOPOGRAPHICAL SETTING

The area of Thebes is mainly comprised of the Aeonian (properly the 'Theban') plain, though the Teneric plain (at least partially if not all[1]) and the area of lakes Yliki and Paralimni to the N also form a part of it, though the latter constitutes a topographical unit on its own.
The W part of the plain of Thebes, once named the Teneric plain (*Tenerikon pedion*), extends up to 10km up to the hills of Onchestos (see chapter II.3.8) and is 4km wide. It has no river flowing through, and a *katavothra* at the N end. The Eastern part is 100m asl while the W is 91m asl. Only a low col separates the E side of the Copais basin from the NW edge of the Theban (Teneric) plain, which, in this NW corner, is lower than the Copais basin (91m asl – appendix II–iB; Philippson 1951: 468) and in winter often becomes marshy, as surface water collects in the large Variko marsh. Further E towards Thebes are also lower flat marshy areas: just W of the outlet of the Kanavaris river into the plain of Thebes (see below), immediately below and N of the Tertiary *tafel* of Thebes (see appendix II -iB- and chapter II.3.9; Philippson 1951: 502ff). The S border of the Teneric plain is constituted by the rises of the Tertiary *tafel*, whose N slopes are marked by the presence of a fertile talus formation made of colluvial deposits.
The E part of the plain of Thebes, named the Aeonian plain and much larger (the 'Theban' plain proper), extends 11km E-W and 6km N-S (from Yliki lake to Thebes). From Thebes to Yliki the elevation range varies from 130m to 100m asl. It is very fertile, with no marshes,[2] since the water of the Kanavaris and Lefkos streams, which respectively cross the W part of the plain in a N direction and the E part in a W direction, flow into the Yliki lake through one of the two gaps between the hills bordering the lake (Philippson 1951: 408). The S border of the Aeonian plain is constituted by the rises of the Tertiary *tafel*, whose N slopes are marked by the presence of fertile marls.
The landscape of the NE edge of the Theban plain is marked by the presence of Mt.Sagmatas (746m asl), which extends into the plain of Thebes from the Messapion/Ktipas chain and stops abruptly, creating a passage from the area of Thebes towards Eastern Boeotia (see chapter II.3.14). The mountain, named Ypaton in ancient times, indeed marks the landscape picture, so that its ancient name, meaning 'the highest (mountain)', makes sense (Philippson 1951: 497). Below the W slope is the modern village of Sirtzi/Ypaton, recently renamed after the mountain, with a large spring at its edge.
From Ypaton to the plain of Mouriki (by lake Yliki, see below) is an area where serpentine runs between gentle slopes, fertile red soils and *elluvia* formations, which the eye can easily distinguish from the surrounding grey limestone. Towards the W, the serpentine forms the pass (211m) between the plain of Mouriki and Thebes, as far as the E bank of Yliki lake where limestone again appears (Philippson 1951: 497).

The Yliki (otherwise called Likeri) lake has an articulated perimeter and follows an E direction. The promontories around the lake form a group of limestone hills, such as Stroulongo and Kokkinovrachos, which rise as islets in the Theban plain, and the majority of which are visible from Thebes (Philippson 1951: 494). On the S side of the lake, between the rocky hills, are two gaps through which the plain of Thebes reaches the lake. The lake is separated to the NW from the Karditsa/Akraiphnion bay (see chapter II.3.7) and the Copais basin by a low pass,[3] reaching the Sengaina bay (Philippson 1951: 495). Theban access to the Yliki lake is either slow, from the E side of the lake through the hills and the hamlet of Yliki, or more direct from the S shore.
The elongated Paralimni lake, extending in a SW-NE direction, lies to the NE of Yliki. The continuation from one lake to the other is not straight, and between them is a distance of approximately 2km. There are two passes from Yliki lake to the SW side of Paralimni lake: the W

[1] See below and chapter II.3.8. The Eastern part belonged to Thebes with more certainty.

[2] The Aeonian plain is very fertile but with no marshes, according to Philippson (1951: 408). On the other hand, Fossey (1988: 199) notes "*a tendency of the Aeonian plain to marshiness has been controlled in modern times by the cutting of a few drainage channels*", and Ulrichs (1840) reports some ancient dams in the E part of the plain of Thebes. In the 1:50,000 GYS map, channels are clearly visible, especially the channeled Kanavaris river crossing S-N the W part of the Aeonian plain. These channels could, however, also be linked to the great withdrawal of water for irrigation, greatly increasing in the last few decades, when irrigated cultures have progressively taken the place of wheat and cereals, still marking the agricultural landscape in Philippson's time (Philippson 1951: 508). Philippson (1951: 508) notes that springs, which today allow for vegetable garden cultivation, can be found in the vicinity of Thebes. Otherwise, he says, the plain is mostly cultivated with cereals.

[3] Along the W coast opens up a road that leads to Karditsa/Akraiphnion (Philippson 1951: 495).

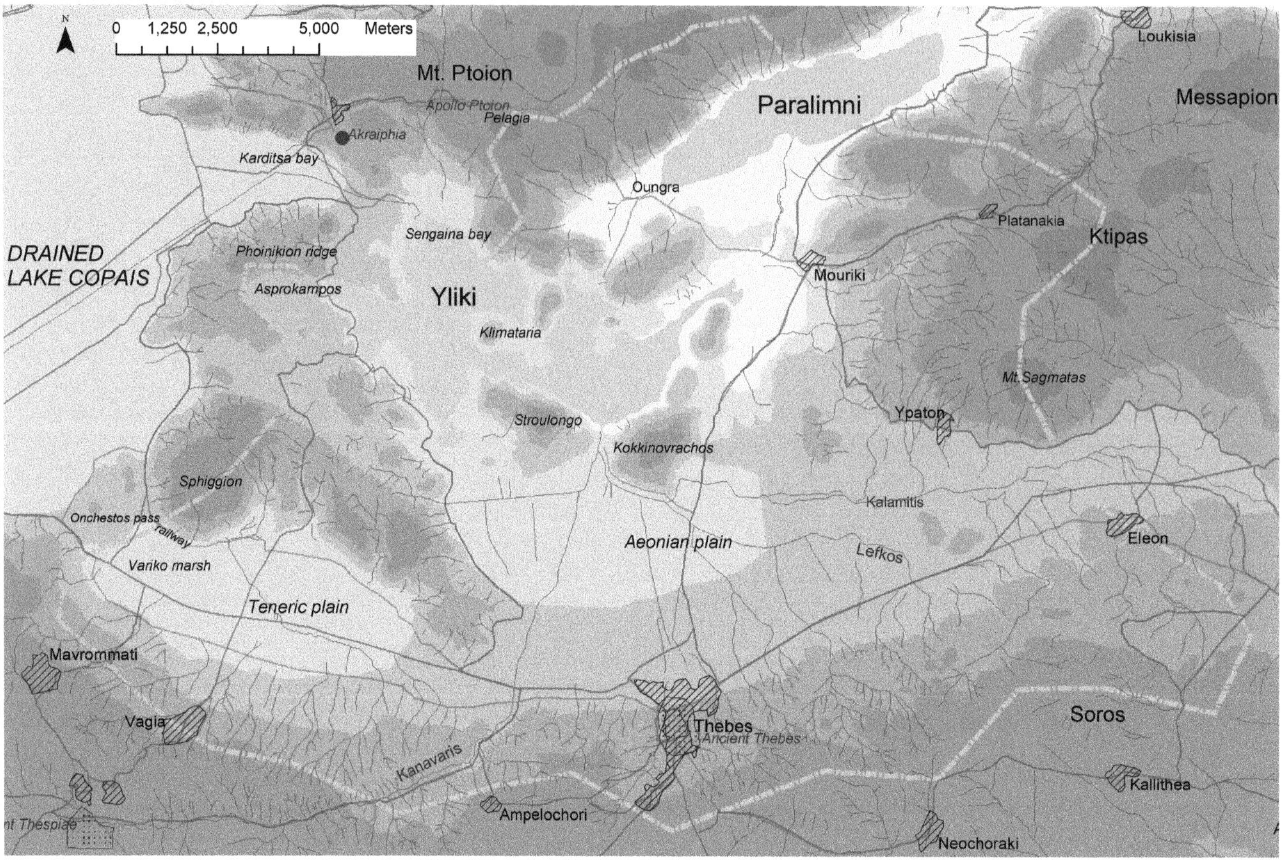

Fig.1. *Topographical setting of the chora of Thebes*

pass leads to the small plain of the hamlet of Yliki (formerly Oungra), at the W end of Paralimni; the E pass crosses the area before the modern village of Mouriki. As Philippson (1951: 495) notes, conglomerate and serpentine were revealed by the cut of the channel between the two lakes.

Boundaries

The borders of the area of Thebes are marked to the W by Mt.Phagas (Sphinx/Sphiggion), to the NW by the hills separating Yliki lake from the Copais basin, to the N by Mt.Ptoion rising above Paralimni lake, to the E the S ridge of Mt.Messapion/Ktipas with Mt.Sagmatas, and to the SE the Soros mountain range (Fossey 1988: 248).
Open boundaries remain the Paralimni lake itself (treated here as part of the Anthedon *chora* for its NE half) as well as the passage to the S of Mt.Sagmatas towards E Boeotia, and, primarily, the border to the S with the Parasopia area, which remains somewhat unclear. Conventionally, I have located the border line as crossing the elevated area of the Tertiary *tafel* running to the S of Thebes (see above). The Teneric plain also still remains an open question,[4] especially its W part, which could have belonged to Haliartos.[5] As stated recently (Aravantinos – Godart - Sacconi 2001: 355-358), during the Bronze Age (and in particular in the LH period), the whole of SE Boeotia, from Mt.Helicon to the W and extending across the Euboean Strait to Karystos and possibly beyond, belonged to the territories ruled by the palace at Thebes.

PHYSICAL LAND UNITS

The mountainous segment is almost absent, elevations higher than 600m asl being present only in the highest border areas on the Messapion and Ptoion chains.
Within plain and hilly segments, apart from the largest represented class, P1_P2 (Teneric and Aeonian plain, the bays around the lakes, the inner plain of Mouriki), the rest of the land zones are quite balanced. The plain morphology of the wider plain area is in contrast with the variegate character of the hilly landscape constituting the elevated area surrounding the plain.
Plateau features within the hilly landscape are quite well represented, and small plateaus constitute the main feature of the landscape of the elevated area of the Tertiary *tafel* running below Thebes to the E and the W (see table 1 and fig.2 in chapter II.1).

Hilly landscape	33.3%
Mountainous landscape	1.4%
Plain	65.3%

[4] Fossey (1988: 249) considers it as geographically not strictly belonging to the Thebais.
[5] Attempts have been also made to assign it to the Thespike *chora* (Fossey 1988: 249).

1	P1_P2	lacustrine basin, valley	44.4%
2	P3	gentle slope	6.6%
3	P4	foothill	14.2%
4	H1	plateau	12.3 %
5	H2	gentle slope	1.5%
6	H3	moderate slope	5%
7	H4	severe slope	7.1%
8	H5	very severe slope	7.5%
9	M1	plateau	0.5%
10	M2	plateau/gentle slope	0.3%
11	M3	moderate slope	0.5%
12	M4	very severe slope	0.2%

Table 1. *Percentage of the different physiographical classes present in the Levadeia area (P=plain; H=hill; M=mountain).*

RESOURCES

Generally speaking, the wider area of the *chora* of Thebes is mainly comprised of class F (fertile) areas, being constituted by the Teneric and Aeonian plains, the bays around the lakes, the inner plain of Mouriki, as well as most of the elevated area of the Tertiary *tafel* running behind Thebes.[6] Thebes took away (426 BC) most of the upland plateau to its S towards Plataea, so its wealth derived from its lowland N plain of pre-Holocene colluvia/alluvia and its southern uplands of Neogen formation, in addition to the abundant local water.
Elevated limestone areas represent the class U (unfertile), not suitable for agriculture but often used for grazing. Some areas can be considered as class LF (low fertile) in the N half of the *chora*, and are mainly constituted by some limestone and sandstone formations which are not available for agriculture in the more elevated and steep areas (for instance on the Sagmatas/Ktipas ridge to the NE)[7] – see fig.6.
Apart from grazing activities in some rocky or elevated areas, the area of the lakes would also have provided a valuable resource for fishing.

THE ARCHAEOLOGICAL RECORD

1	**THEBES**	*Components T_1 to T_14*
2 – 7	**LH burials**	*Components T_15* (2), *T_16* (3), *T_17* (4), *T_35* (5), *T_45* (6), *T_46* (7)
5	**Ismenion**	*Components T_30 to T_34 and T_36 and T_37*
8 – 14	**Historical period burials**	*Components T_18* (8), *T_142* (9), *T_143* (10), *T_28 and T_29* (11), *T_36* (LR tombs at Ismenion - 5), *T_24 to T_27 and T_47* (12), *T_141* (13), *T_20 to T_23* at Pyri (14)
15	**Pyri**	*Component T_19*
16	**Kapanitsa**	*Component T_139*
17	**Thebes Railway**	*Components T_153 to T_154*
18	**Tachi**	*Components T_38 to T_43; T_148 and T_149*
19	**Thebes W**	*Component T_44*
20	**Kabeirion**	*Components T_49 to T_54*
21	**Kabeirion - Agios Nikolaos**	*Component T_48*
22	**Vagia-Provatari**	*Components T_144 to T_147*
23 – 25	**Vagia-Mavrokampos, Vagia-Kritsola and Vagia-Darani**	*Components T_55 (23), T_56 (24), T_57 (25)*
26	**Toumpi**	*Component T_58*
27/28	**Lithares**	*Components T_60 to T_68* (27), *and T_69* (28)
29/30	**Lithares Stroulongo**	*Components T_70 to T_72* (29) *and T_151* (30)
31	**Vouliagma**	*Component T_59*
32/33	**Mouriki**	*Components T_75 to T_76* (32) *and T_78* (33)
34	**Yliki E peninsula**	*Components T_73 to T_74*
35/36	**Klimataria**	*Components T_109, T_110 and T_140* (35)*; Components T_111* (36)
37	**Klimataria N**	*Components T_112*
38	**Klimataria W**	*Components T_113 and T_114*
39/40	**Yliki Sengaina**	*Components T_104 and 105* (39), *T_106* (40)
41/42	**Yliki Boubali**	*Components T_107* (41) *and T_108* (42)
43	**Oungra Chelonokastro**	*Components T_85 and T_86*
44/45	**Paralimni Oungra (lake shore site)**	*Components T_87 to T_91* (44)*; T_92 to T_95* (45)
46	**Paralimni Oungra N**	*Components T_96 to T_98*
47	**Paralimni Oungra area**	*Component T_99*
48	**Paralimni S hills**	*Components T_100 to T_102*
49/50	**Paralimni Yliki**	*Components T_116* (49) *and T_117* (50)
51	**Mouriki – Kamelovrisi**	*Components T_103 and T_152*
52	**Platanaki**	*Components T_80 to T_84* and *T_79*
53	**Sambali**	*Component T_115*
54-56	**Ypaton**	*Components T_118* (54), *T_129* (55) and *T_119 to T_129* (56)
57	**Moni Sagmata**	*Component T_130*
58	**Mesovouni**	*Components T_131 to T_135*
59	**Stroma**	*Components T_155 and T_136 to T_138*

Table 2. *List of archaeological components and activity loci mapped in fig.2.*

[6] Fertile talus and fertile marls are to be found on the S edge of the Teneric and Aeonian plains respectively.
[7] For the classes see chapter I.2.1.

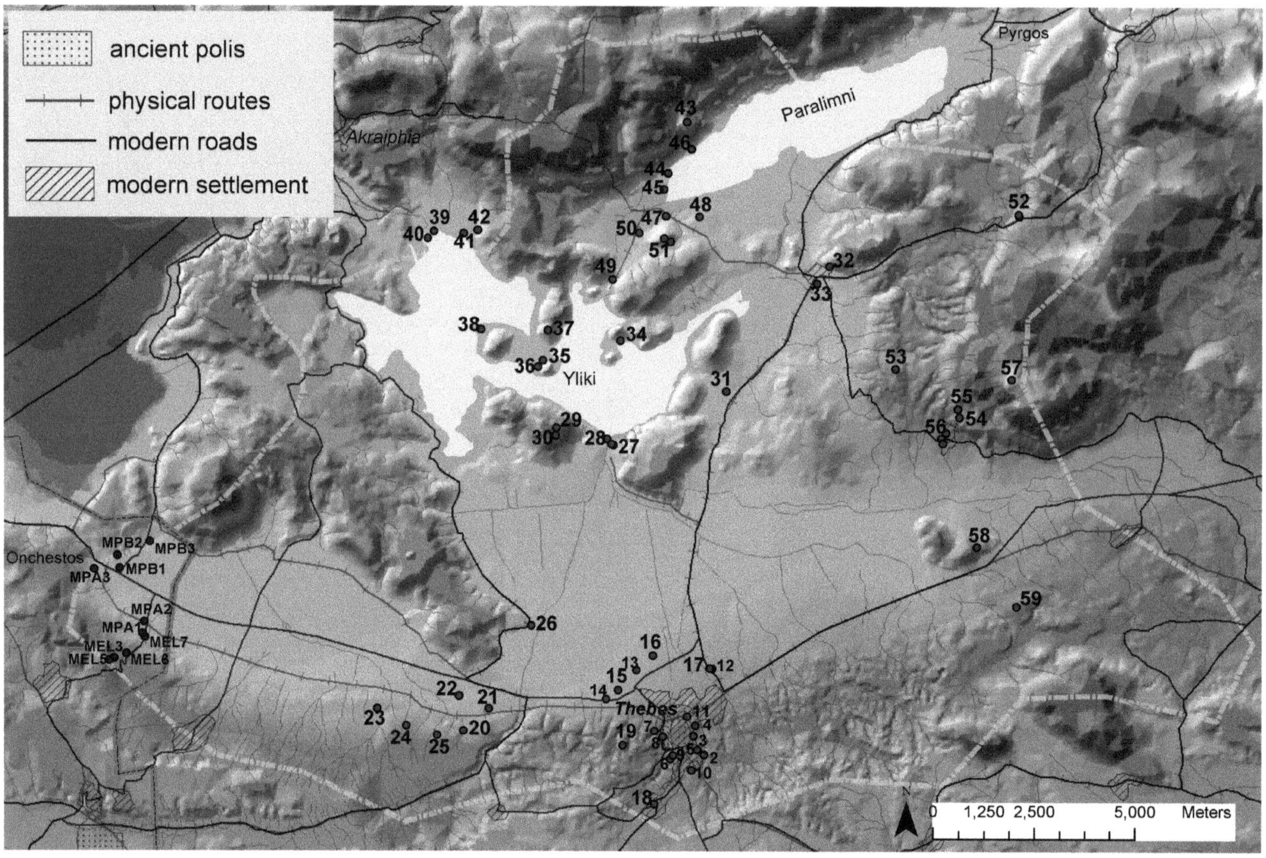

Fig.2. *Archaeological map of Thebais chora. In the SW corner of the map the area intensively and systematically surveyed is marked* (The extension of the lower town of Thebes is hard to determine with certainty and may vary considerably throughout the periods. What is suggested in the map is just a possibility to give an idea to the reader of the city's considerable extention. Within it, the Kadmeia is marked).

Rescue excavation constitutes one of the major providers of archaeological discoveries in the area (graph in fig.3), and this is due to various factors: the extended rescue excavation work in the modern town of Thebes, as both the prehistoric and the Greco-Roman settlement lie below it; rescue excavation in the vicinity of the major town sites, or in the surrounding areas, for roads or building activities, occurring due to the vicinity of Thebes. The set up of the natural gas pipeline (*Agogos Physikou Aeriou*) in the last decade, running from NW Boeotia and crossing the area of Thebes, constituted an important opportunity for the discovery of significant archaeological remains (such as, for instance, the Vagia-Provatari cemetery – *components T_144 to T_147*).

Within the wide *chora*, the situation in the area of the lakes is quite different, and gives us a different picture with regards to archaeological discoveries, mainly due to personal research interest and accidental discoveries, during the lowering of the water level of the lakes in the late 1980s and occasionally earlier (1965-66 for instance), which uncovered some previously unknown sites and clarified the extension of some settlement sites bordering the lakes (Yliki E peninsula, Klimataria, Yliki Sengaina, Yliki Boubali, Paralimni Oungra N). Generally speaking, settlement sites are quite well known in the area of Thebes, compared to other *chorai*.

We can note that 'personal or group interest' discoveries constitute the majority of the entries, mainly represented by the discoveries around the lakes or by research at specific well known sites (such as Lithares or Pyri by Thebes).

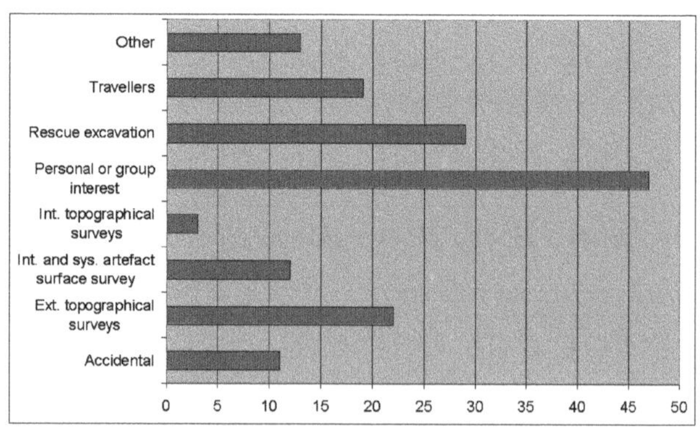

Fig.3. *Graph illustrating the proportion of components discovered within different research frameworks.*

A Systematic Intensive Surface Survey carried out in a small area in the *chora* of Thebes (area surveyed mapped in fig.2) in the 1980s by J.L. Bintliff A.M.Snodgrass and their team discovered sites listed in appendix I.12 (table 1 – *Survey Sites*) and mapped in fig.2.

The inclusion in the graph of the results of the intensive and systematic surface survey changes the picture only slightly, as only a small part of the *chora* has been investigated by intensive survey work.

In fig.4 we can see the relationship between known archaeological sites and the distance from the modern road network. Especially for the area of the lake, there is no connection between the two variables, the archaeological knowledge of that area being mainly due to specific archaeological questions and interest. This, in addition to rescue excavation activities in the area, which, as seen earlier, theoretically provide less biased discoveries, would tend to make the archaeological record less biased, at least in part.

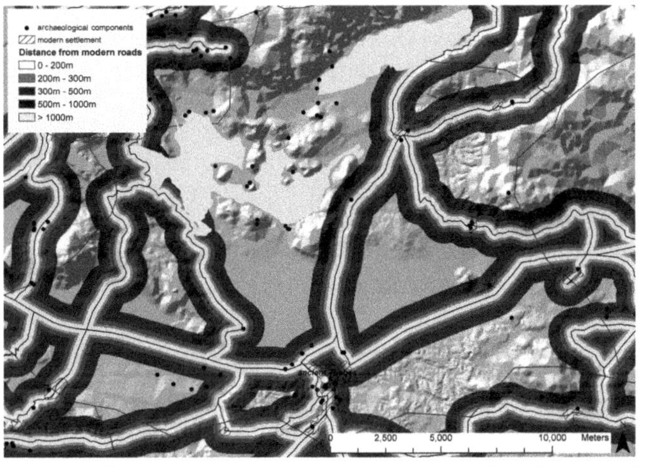

Fig.4. *Relationship between components and modern road network*

The ratio of known Prehistoric to Greco-Roman components is 48 to 101 (1:2.1), while among the historical periods, 53% are dated Archaic to Hellenistic, 21% Roman-Late Roman, and 13% are attributed to the general Greco-Roman period. As noted above, particularly for the case of Thebes, Bronze Age periods are certainly the most examined and investigated, due to the personal research interests of scholars and the availability of a rich archaeological record (Thebes and Lithares especially, but also the area of the lakes, rich in Prehistoric remains). As for the relative proportions of the historical periods, the picture of this area corresponds to that offered by the majority of *chorai* in the wider Boeotia region (see Thespiae *chora*, for instance).

ANALYSIS OF THE *CHORA* LANDSCAPE

PREHISTORIC PERIOD

As noted earlier, the prehistoric period (especially the Bronze Age) is quite well known for the *chora* of Thebes, due to the personal specific research interests of scholars as well as the availability of a rich prehistoric record.

Besides the important Neolithic settlement site at Toumpi, at the W end of the Aeonian plain (*component T_58*), few Neolithic sites are known in the area: on the Kadmeia (Thebes – *component T_1*); in the area of the modern centre of Thebes (*components T_19* and *T_153*, respectively to the NW –Pyri- and N -by the railway- of the Kadmeia); activity foci have also been reported by the Kabeirion sanctuary (*component T_49*), at Ypaton Tourleza (*component T_119*) and on Stroma hill (*component T_155*) – see fig.5 in chapter II.3.11.

For the EH period, the settlement by the railway line N of Thebes (*component T_154*) belongs to the series of settlements that exploited the natural resources of the wider area around Thebes before the foundation of the EH settlement on the Kadmeia (Thebes – *components T_2 and T_13*), since the Neolithic period (see above). The wider region was also occupied in the EH period, as apparently attested by the probable EH activity focus at Platanaki (*component T_80*) and the activity focus at Mesovouni (*component T_131*), as well as the settlement recognized at Ypaton Tourleza (with associated burials – *components T_120 and T_121*), together with the EH occupation in the area of the lakes (see below) – see fig.6 in chapter II.3.11.

Apart from the large extended MH settlement on the Kadmeia (Thebes - *components T_3 and T_14*), the picture from the MH period for the region is quite poor, being limited to the activity focus at Ypaton Tourleza (*component T_122*), and can be enriched adding the MH sites and activity foci known in the area of the lakes, to which can be added the probable activity focus at Platanaki (*component T_81*) – see fig.7 in chapter II.3.11.

The LH landscape of the region is mainly known through burials, in particular the chamber tombs (*components T_15 to T_17, T_35, T_45, T_46, T_149*) situated on the hills surrounding the main citadel on the Kadmeia (*component T_4*), with the addition of the tombs at Vouliagma, as usual facing the settlement site (at Lithares – *components T_63, T_68 and T_59*). Other LH activity foci are known in the region, such as *component T_123* at Ypaton Tourleza and *component T_136* on Stroma hill– see fig.8 in chapter II.3.11.

The prehistoric landscape of the *chora* is enriched by the archaeological record available for the area of the lakes, which offers a good example of settlement strategy in a peculiar landscape (see above).

In the area of Yliki lake, Neolithic is known at Lithares (*component T_60*)[8] and just as a flint concentration on the E peninsula (*component T_73*), while in the area of Paralimni lake Neolithic is reported in the NE bay (Paralimni Pyrgos – *component AN_26*[9]) and only probably in the SW bay (Paralimni Oungra – *component T_87*) – see fig.5 in chapter II.3.11.

The well known settlement site of Lithares marks the landscape of Yliki lake in the EH period (*component T_61* with associated burials *component T_66*), along with the activity foci attested on the nearby Stroulongo hill (*component T_70*), on the Klimataria peninsula (*component T_113*) and in the NW Sengaina bay (*component T_104*). On the shore of Paralimni, EH is

[8] The prehistoric village of Lithares marks the edge of the plateau where the Theban Plain descends into Yliki lake.

[9] The Paralimni - Pyrgos site at the E end of the lake is included in the Anthedon *chora* in the present work, while the Paralimni - Oungra bay at the SW end of Lake is included here in the Thebais *chora*.

reported in the SW Oungra bay (*component T_88*) as well as in the Paralimni NE bay (Paralimni Pyrgos – *component AN_27*- along with a burial place –*component AN_38*) – see fig.6 in chapter II.3.11.

By the Yliki lake, MH is known at Lithares (*component T_62* with burials *T_67*) and at the very end of the Klimataria peninsula (*component T_114*), probably with a settlement status, while on the shore of Paralimni MH is reported in the SW Oungra bay together with a burial place to the N (*components T_89 and T_96*, see also *component T_152*), as well as in the Paralimni NE bay (Paralimni Pyrgos – *component AN_28*) – see fig.7 in chapter II.3.11.

LH is attested at Lithares (*component T_63* with burials *T_68*), as well as on the Stroulongo hill (*component T_71*), while LH chamber tombs are known at Vouliagma (*component T_59*). As for the area of Paralimni lake, a LH fort is known at Oungra-Chelonokastro – *component T_85*- along with material on the Paralimni shore and burials –*components T_90 and T_97,* as well as *component T_103*, while a settlement site is reported in the NE bay (Paralimni Pyrgos – *component AN_29*- along with a burial place –*component AN_37*) – see fig.8 in chapter II.3.11.

GRECO-ROMAN ANTIQUITY[10]

Town level

Thebes was the major city of Boeotia for most of its history (see appendix I.12). It also lay in a central position geographically within the region of Boeotia. As Fossey remarks (1988: 199), and as is visible in the physical route map (fig.1 in chapter II.1), routes converge in the area, leading to Anthedon and the coast, to Athens and to Chalkis, joining Attica to Euboea, to the ports on the S coast and through the Kithaeron route to the Peloponnese, to the N of Greece round Copais and joining N and S Greece. The city had much land available and, as many Boeotian cities (such as Levadeia, Koroneia), lay between mixed land types.

Thebes lies at the edge of a line of hills running E-W and forming central Boeotia (Tertiary *tafel* of Thebes). The exact position chosen by the settlement along this line is mainly due to the presence of rich water springs. The site was always the main centre of the area, except during the Dark Ages when, after the fall of the Mycenaean palace, the reduced population was probably absorbed by rural communities in the surroundings.[11] This is testified by the apparent absence of archaeological remains on the *Kadmeia* from the 11th to 9th C BC, with the exception of a burial area.

Village level

The city of Thebes was highly nucleated, thus not leaving much space for other nucleated settlements, unlike the city of Athens, for instance, spread all over the region of Attica through dispersed *demoi* (see Bintliff 1997b). Nevertheless, the *polis* probably had some small satellite settlements very close to the city (Tachi for instance[12]). The 350ha suggested by Bintliff (1997b) as a probable extension of the city of Thebes in the Classical period would probably incorporate the suburbs.

In the wider Theban region, some 2nd rank small nucleated settlements are represented in the available archaeological record. At Vagia-Provatari, the distance of the Archaic/Classical cemetery (*components T_144 and T_145*) from the ancient town of Thebes, suggests that it probably belonged to another 2nd rank settlement in the vicinity (quite flourishing according to AD49 1994: 284-286), which in the Roman period would have been substituted by large rural sites (*components T_146 and T_147*). Ancient sources mention several towns which would have been in the *chora* of Thebes, such as *Peteon* (referred to by Strabo IX 410 - see E. Kirsten, RE 1937, s.v. *Peteon*: 1128), *Teumessos* (see Fossey, Topography 1988: 213-7 and Wallace 1979: 96-8), and *Glisas* (see Fossey 1988: 221-3). Attempts have been made to identify those towns with known archaeological evidence which could indicate nucleated settlements in the Greco-Roman period: a probable settlement site at Ypaton (*components T_125 and T_126*, only guessed from burials, *components T_124, T_127 and T_129*, and fortified hilltop *component T_118 - Glisas*?), a possible settlement at Platanaki (component *T_83 - Peteon*?), settlement evidence at Mesovouni (Classical and Late Roman - *components T_133 to T_135 – Teumessos*?) and Stroma (Classical/Hellenistic - *components T_137 and T_138 - Teumessos*?) - see appendix I.12. For settlement in the area of the lakes, see below.

Rural segment

The picture of the rural landscape of the Greco-Roman period available for the *chora* of Thebes comprises little evidence which could be interpreted as rural sites/activities. For the Classical period, a farmstead W of Thebes (*component T_44*) and three sites with uncertain function at Vagia-Mavrokampos (*components T_55 to T_57*[13]*);* for the Hellenistic period, the evidence at Lithares (*component T_64*) and Lithares Stroulongo (*component T_72*) could be from rural sites; for the Roman period, two *villae rusticae* found at Vagia-Provatari - *components T_146 and T_147*. The panorama is slightly enriched by the sites discovered by the Boeotia survey project (which covered a small area of the *chora* - see above), mainly represented by farms of Classical/Hellenistic date (see map in fig.2). We have evidence of rural settlement in the area of the lakes as

[10] Period maps are included in chapter II.4, figs.18-20-22-24-26-28.

[11] According to Papadaki (2000), the population would have moved to the lower surroundings, and probably to the S. This could explain the mention in the Catalogue of Ships (Iliad B505) of Ypothebai and not of Thebes itself (see among others Hope Simpson-Lazenby 1970: 31ff). Symeonoglou (1985: 60-63, 66, 89-91) suggests that Ypothebai must have been the N part of the Kadmeia, named Gourna.

[12] *Components T_38 to T_43 and T_148*. The area of Tachi today is incorporated into the urban area of Thebes itself.

[13] Probably to be interpreted as activity focuses probably to be linked with the rural settlement system in the area lining S the Teneric plain.

well, mainly concentrated in small hamlets exploiting the bays (see below).

Burial areas

The picture of burials in the *chora* of Thebes shows city cemeteries (such as at Pyri or by the railway) or burial areas somehow linked to the Thebes city site, ranging from Geometric to Late Roman times, well represented in the available archaeological record.[14] There is evidence of a monumental Macedonian-type tomb of the Hellenistic period W of Thebes (*component T_139*). Mazonaki and Faraklas (AE 1967: 15-19) suggest that the Classical/Hellenistic burial areas of the city would be in the Eastern and central part of the Theban plain (Kapanitsa), while the Archaic cemeteries would be mainly in the W part.

The archaeological record also includes burial evidence that could correspond to settlement sites, such as at Tachi (*components T_38* to *T_43* and *T_148*) and Vagia-Provatari (*components T_144 and T_145*).

Cult places/Religious areas

There have been several attempts to locate the extra-urban sanctuaries of Thebes mentioned especially by Pausanias. Some of them are clearly recognizable, such as the extramural sanctuary of Apollo Ismenios (*components T_30* to *T_33*) or the Kabeirion (strictly linked to the city of Thebes, from the Geometric to the Roman period - *components T_50 to T_54*). Some others are less easily identifiable, such as the sanctuary of Herakles Hippodetes (*component T_55*) located at Morokampos, or the sanctuary of Demeter Cabiria, located to the N of the Kabeirion,[15] or the sanctuary of the healing hero Amphiaraos - Paus IX 8.3, located by modern Tachi by Symeonoglou 1985.

The landscape of the NE edge of the Theban plain is marked by the presence of Mt. Sagmatas/Ypaton (746m asl), which juts into the plain of Thebes from Messapion chain and stops abruptly, creating a passage from the area of Thebes towards Eastern Boeotia (see chapter II.3.14). The limestone mountain, named Ypaton in ancient times,[16] indeed dominates the landscape, so that its ancient name, meaning 'the highest (mountain)', would make sense (Philippson 1951: 497). On its top, the existence of a temple of Zeus *Ypatos*[17] has been suggested,[18] on the site of the Sagmata monastery, where a Byzantine church is recorded. The epithet and the character of the divinity worshipped there could be linked to the character of the mountain, which would seem to be visible from many different areas of Boeotia and would be marked by distinctive cultural values, i.e. by both a high sacral and at the same time a socio-political meaning, as suggested by Schachter in his monograph on the cults of Boeotia (Schachter 1994).

The Onchestos sanctuary, listed in the *chora* of Haliartos (*components HA_28 to HA_32*), to which it geographically belongs, can in fact be considered as belonging to the Theban landscape of cults for historical and political reasons. Being situated on the Onchestos pass, between the Teneric plain and the Copais area, the federal sanctuary dedicated to Poseidon constitutes a landmark within the regional landscape, giving sensitive value to the distinction between NW and SE Boeotia for geographical and historical reasons (see chapter II.4).

Forts and fortifications

Thebes controlled the 4th-C system of forts and fortifications all over Boeotia, constructed to control the wider Boeotia area under the control of Thebes by means of a system of inter-communications (see final chapter). In the *chora* of Thebes, Sambali (*component T_115*) is an example, while the Classical/Hellenistic fortification above Ypaton should probably be linked with the settlement below (*component T_118*). A probably earlier –Archaic?- fort is known in the area of the lakes, at Oungra-Chelonokastro, at the SW end of Paralimni lake (*component T_86*), as well as the fortified acropolis at Paralimni Pyrgos (*components AN_32 and AN_33*), showing the importance of this pass from Anthedon to Copais and Thebes, which consequently gives strategic meaning to the area of Paralimni and in general to the area of the lakes.

LONG TERM SETTLEMENT TRENDS IN THE CHORA *LANDSCAPE*

The Thebais area is shared today by a number of village communities which occupy the natural settlement chambers around the Teneric and Aeonian plains, and the bays and inner plains around the lakes.

The W part of the plain today forms the territory of the modern village of Vagia (registered also in the Ottoman period in 1506-1642), situated on the hills S of the plain. In the Ottoman period, however, this area was divided among several communities occupying small settlement chambers across the elevated area of the Tertiary *tafel* south of the plain (deserted village of Darani and possibly also Porovanka). In the area between Thebes and Vagia several places have perhaps disappeared and remain identified. On the northern side of the plain lies the deserted village of Mazaraki in a sheltered valley immediately east of Sphiggion Mountain (recorded 1466-1646). For antiquity, archaeological evidence seems to suggest the presence of a second-rank settlement on the hills S of the Teneric plain close to Vagia (Vagia-Provatari) occupied in the Archaic/Classical period, showing that these settlement chambers were also inhabited in antiquity. As is clear from the map resulting from a cost-distance analysis based on first and second rank Greco-Roman settlement (fig.5),[19] there is enough room for at least one settlement chamber between Thebes

[14] See appendix I.12: *components T_18; T_24 to T_29 and T_47*; *T_20 to T_23* at Pyri; *T_36*; *T_141*; *T_142 and T_143.*

[15] On the map drown by Judeich AM 1988: 83

[16] The mountain is known also by the name of Mt. Sagmatas, and the same name is taken by a monastery –Moni Sagmata-which lies atop it.

[17] i.e. Zeus the highest, mentioned by Pausanias.

[18] There is no real archaeological evidence for it, though it could be buried beneath the monastery - see *component T_130*.

[19] See chapter II.3.1 – LONG TERM SETTLEMENT TRENDS.

and the W edge of the Teneric plain, exploited by rural farms/small nucleations (*components T_55 to T_57; T_144 and T_145*) or cult places (Kabeirion) and villas in the Roman period (*components T_146 and T_147*) – see fig.6.

The large plain of Thebes is occupied by the city of Thebes, which has always been a large administrative centre, continuing from late antiquity through the Medieval and Ottoman periods. The status of Thebes as a large city is reflected in the absence of secondary villages in its territory, with the exception of two possible Ottoman villages at the northern and southern edges of its territory. For antiquity, however, we could postulate a different picture. In the area around the city of Thebes most of the landscape must have been farmed by citizens resident in the city. Yet despite this, the present archaeological evidence shows the existence of additional rural nucleations close to the city (e.g. Tachi). Evidence of settlements can be found at Tachi (today a suburb of the town of Thebes), as well as at Mesovouni and Stroma to the E (see above VILLAGE LEVEL and appendix I.12).[20] Additional settlement chambers exist in the upland plateau S of the city towards Plataea, which belonged to Thebes in some periods of antiquity (see chapter II.3.11). This conforms to the picture from the Boeotia survey at Thespiae, where a predominance of medium-sized farms and hamlets was recorded in the vicinity of the urban landscape of Thespiae and Askra, representing either hamlets composed of several farming families, or the centres of wealthy estates related to the issue of oligarchic class in the countryside (chapter II.3.9 and Bintliff-Howard-Snodgrass 2007: 134-35, 150). Yet in the *chora* of Thespiae, imbalances were observed in the outer *chora*, which produced slim evidence for rural settlement. In the case of Thebais a number of rural nucleations are recorded, associated with the settlement chambers of modern communities (see fig.6). Today the eastern part of the plain is occupied by the modern village of Ypaton (formerly Syrtsi, recorded in the Ottoman defters in 1466-1642 and in 1688), a second-rank settlement also in the Greco-Roman period (see above VILLAGE LEVEL and appendix I.12). W of it, settlement chambers are created around the bays and inner plains in the area of the lakes, forming village communities occupied in the modern and pre-modern period as well as in ancient times (see below).

Also taking into account the evidence for the ancient village by Vagia (see above), we could infer a rural landscape of hamlets or small villages, rather than one of small and medium Classical farms at a distance from the city. This is also strengthened by the results of the Boeotia survey which covered a small part of the W section of the Teneric plain, and found only limited evidence for rural settlement, in contrast to the dense rural landscape N of the city of Thespiae and in the valley of the Muses. Probably also thinly inhabited was the eastern edge of the Aeonian plain, indicating possible imbalances in the rural settlement on the extremities of the *chora*.

The precise location chosen for the settlement of Thebes along the line of hills running E-W across central Boeotia (see above) is mainly due to the presence of rich water springs, which would also condition the position of some of the other settlements in an area without strong physical constraints defining community areas (see also chapter II.3.11).

The settlement in the area of the lakes

The landscape of the Yliki and Paralimni areas constitutes a peculiar landscape within the larger *chora* of Thebes.

In the area of the lakes, some activity foci of historical periods are known, mainly to be related with agro-pastoral activities, such as at the edges of Yliki by Lithares *components T_64, T_65* and *T_72* and some burials of earlier periods (*component T_69 and T_151* – Archaic and Geometric) which could lead one to assume the presence of a settlement (or large rural site) in the area. Several small settlements or rural sites exploited the different bays around the Yliki lake and at the edges of Paralimni, such as: the quite extended settlement at Mouriki (*components T_75 to T_77*) with associated burials (*component T_78*) from Archaic to Hellenistic; the Roman settlement known on the Klimataria peninsula (*component T_110* with associated burials *T_111*, with traces of an earlier occupation in Hellenistic times *component T_109*); the large settlement site at the SW end of Paralimni lake (from the Protogeometric/Geometric to the Archaic/Classical period - Paralimni Oungra *components T_92 to T_95*- jointly with a burial place to the N –*components T_98 and T_99 to T_101, T_116 and T_117*); the settlement known at Paralimni Pyrgos, at the NE end of Paralimni lake (from the Geometric to the Hellenistic period - *components AN_30 to AN_33*[21]); the rural site on the easternmost peninsula jutting from the N into the Yliki lake (*component T_74* - H); the small settlement (hamlet) in the NW bay, at Sengaina (*components T_105 and T_106* - Classical to Hellenistic) and the similar one on the N shore, at Boubali locality, dated to the Classical period (*component T_107*). Roman period settlement is known only at Klimataria (*component T_110* –Roman settlement- and *T_140* –Late Roman activity focus-), and in the form of burial evidence at Boubali (*component T_108*), and only as possible at Paralimni Pyrgos (*components AN_34* –Roman- *and AN_35* –Late Roman).[22]

[20] A potential settlement chamber in the area of Mesovouni and Stroma (E side of the Aeonian plain) has been marked on the map in fig.5 and included in the cost-distance analysis performed on first and second rank settlement sites, although the archaeological evidence could also represent rural nucleations rather than proper village sites, if one considers the significant presence of the city of Thebes.

[21] The Paralimni - Pyrgos site at the E end of the lake is included in the Anthedon *chora* in the present work, while the Paralimni - Oungra bay at the SW end of Lake is included here in the Thebais *chora*.

[22] For problems of settlement in the Yliki Paralimni area see also Fossey 1988: 225ff and Philippson-Kirsten 1951: 495.

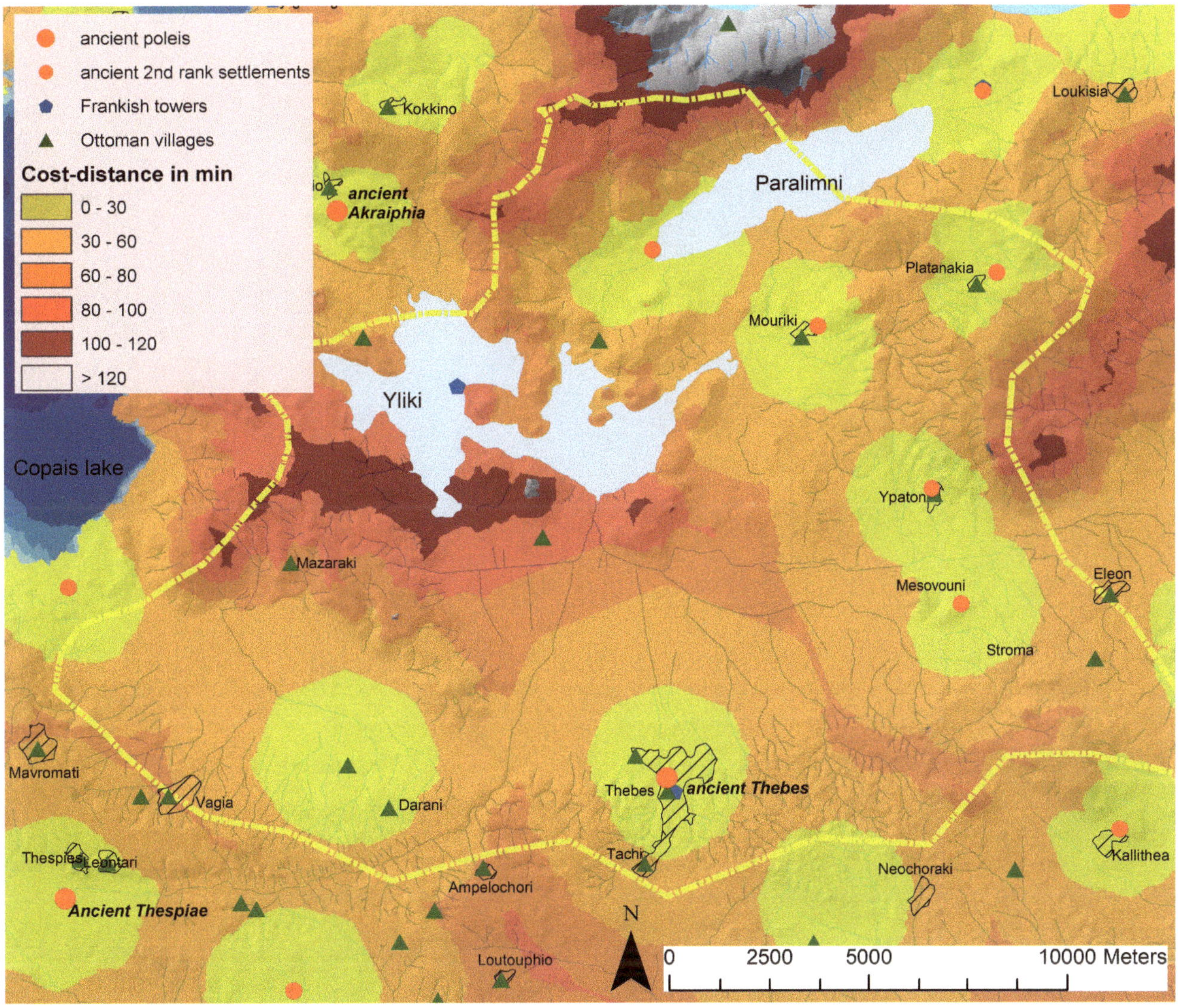

Fig.5. *Classified surface representing the cost-weighted distance (1/2 h walking and further ranges) from recognised 1st and 2nd rank ancient settlements (represented by larger and smaller dots). Areas without dots indicate potential settlement chambers. Ottoman villages as well as Frankish towers and the Frankish main centre of Thebes have also been added to the map to show their spatial relationship with the Greco-Roman settlement network and to appreciate potential settlement chambers.*

As seen earlier, on the basis of the archaeological record the settlement picture in the area of the lakes is quite clear for the Greco-Roman period regarding the settlement chambers which were in use. The edges of the lake were densely inhabited in almost every bay and peninsula. The existing evidence for antiquity would point possibly to the existence of an even denser ancient rural pattern than that represented by the modern and Ottoman and Medieval villages. An examination of the known settlement in the Ottoman period until the 19th century shows by the lake the secondary village of Yliki, formerly Oungra, recorded 1466-1688 (the village of Oungra/Yliki is more towards Yliki rather than Paralimni, in comparison with the Greco-Roman settlement, by the SW edge of Paralimni – see map in figs.5-6[23]), the deserted village of Sengaina on the NW edge of the lake,[24] and the modern village of Mouriki around an inner plain, recorded 1466-1646; E of Yliki Lake the secondary village of Platanakia is recorded in 1521-1688. All these locations were occupied in antiquity by settlement sites.[25]

[23] A fairly wide settlement chamber can therefore be located in this bridge-area between Yliki and Paralimni lake, where the settlement *focus* shifts through time. For antiquity, Fossey (1988: 239) suggests that the settlement was at Oungra/Yliki by the end of the 4th C BC, then moved to Klimataria. Fossey (1988: 243) also notes that the Hellenistic site found on the Yliki E peninsula (*component T_74*) may be part of the settlement organization in the area of the lakes which took place after the abandonment of the Oungra site. As noted above in the text, the Ottoman settlement is much closer to Yliki lake (see map), while in modern times (until the first half of the 20th century) at Paralimni Oungra was a family hamlet.

[24] A small hamlet existed in the Sengaina bay before it was swallowed up by the rise in water level of the Yliki lake resulting from the drainage of Copais (Philippson 1951: 495), while the small hamlet of Paralimni Oungra disappeared by the mid 20th century. In Philippson's days (Philippson 1951: 495) the area of the lake was already deserted, as it appears today.

[25] Poor archaeological evidence may attest the existence of a possible settlement site at Platanakia (see appendix I.12). The cost-distance allocation analysis performed on the basis of first

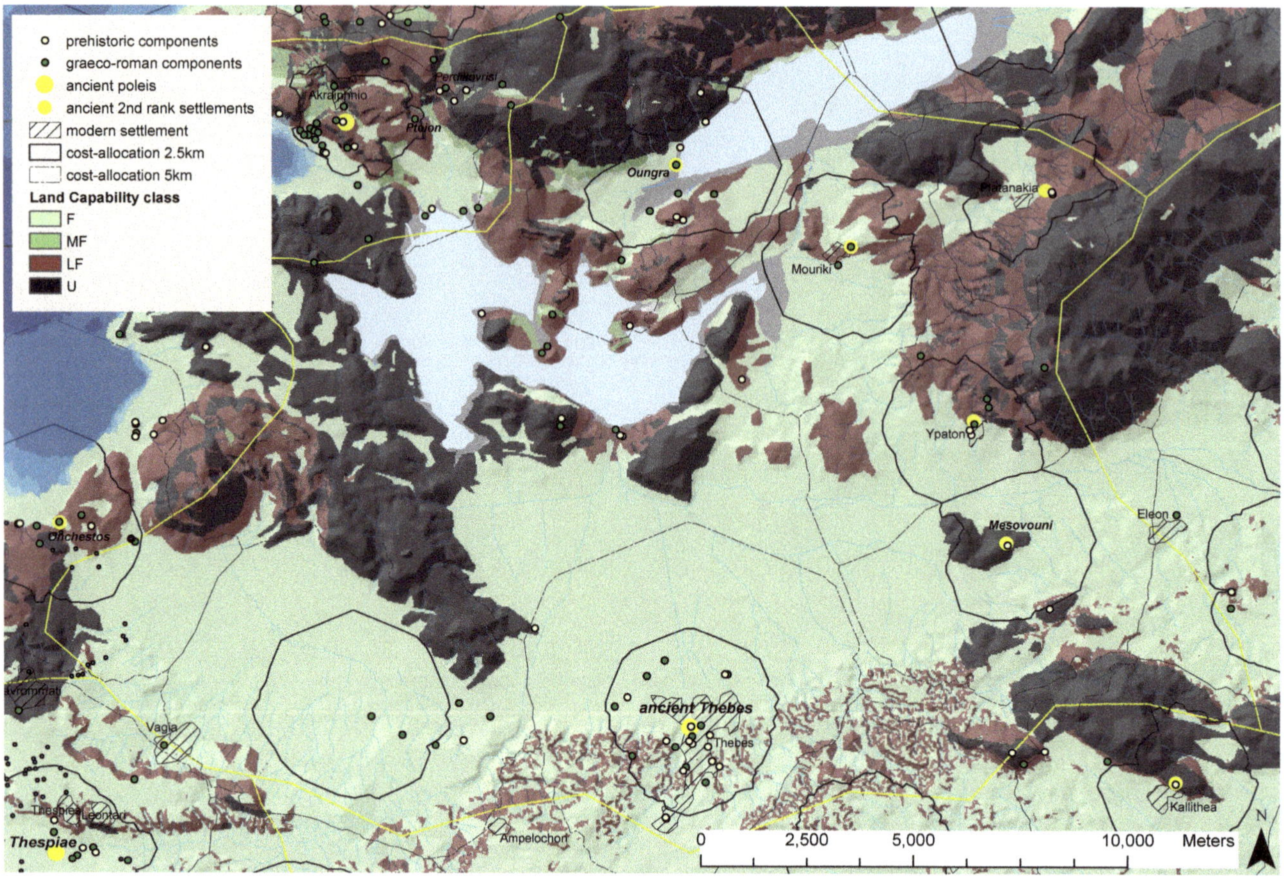

Fig.6. *Map showing the Greco-Roman settlement network, the polygons resulting from the cost-distance analysis (marking half an hour and one hour walking time distance) and dots representing the known archaeological components (same as in fig.2), with land capability information underlain.*

The known medieval (Frankish) settlement in the area is slightly different, related to the watchtowers located in strategic positions at the top of the Klimataria peninsula[26] and at Paralimni Pyrgos at the NE edge of the lake, on the route towards the Euboic sea. For both locations there is no evidence of settlement in the Ottoman and modern settlement pattern, while there is for Greco-Roman periods. The Greco-Roman occupation therefore shows a significant presence in habitation and economic activities around the lakes due to the distance from the main centre of Thebes but also to the peculiar character of the landscape, which allows for integrated and mixed economies but does not really allow for a larger settlement which can have either town or *proto-polis* status. The micro-landscapes at the edge of the lakes are potentially complete settlement chambers, i.e. they have within them all the necessary elements for settlement life: the water of the lake, upland/higher area, and the cultivable plain. Some of the settlement chambers available around the lakes disappeared as such after drainage of the Copais at the end of the 19th century.[27] We should, however, include them as dry fertile land in our reconstruction of past landscapes' behaviour. Each patch of available land was exploited,[28] and water and marsh based economies were performed. I would suggest that in the area of the lakes, constituting a very peculiar micro-environment, in antiquity the landscape was marked by mainly rural nucleations, such as hamlets or small villages, rather than real nucleated second rank settlements, and they would create a small-scale settlement system on their own.

and second rank Greco-Roman settlement (map in fig.5) would seem to allow for a small settlement gap in that area, already isolated by Bintliff 1994b fig.20, and corresponding to a small inner plain exploited by the modern and the Ottoman village located there (see in text).

[26] The Frankish bridge at Klimataria (on the E side of the lake) seems to run in a S direction to the S edge of the lake to allow direct travel across the Theban Plain southwards.

[27] The Yliki and Paralimni lakes are now considerably larger than they used to be, due to the drainage (of the 19th century) of the Copais lake. As noticed earlier, the majority of the known remains of settlement and activities (which were along the shores) lie underwater and were visible in dry periods of low water level (Fossey mentions the dry season of 1965-66 but there was also one in the late 1980s).

[28] Along with the larger cultivable areas, such as the bridge-zone between the lakes and the wider areas to the E (Mouriki and Platanaki), the Greco-Roman hamlets at Sengaina and Boubali, exploiting small bays at the edge of Yliki lake, were also linked to agricultural activities (AD 1981: 194), exploiting the fertile valley which was swallowed by water after the drainage of Copais and the subsequent increase of the water level of Yliki lake.

II.3.13

The Boeotian coastal area facing the Euboean sea N of Chalkis: Anthedonia and Skroponeri bay

TOPOGRAPHICAL SETTING

Geographically speaking, the Boeotian coastal area facing the Euboean sea N of Chalkis (the N Euboean Strait), including the Anthedon area proper and the Skroponeri bay (see also Fossey 1988: 251) can be seen as a whole. On the other hand, the two areas strictly speaking constitute two small topographic units and, if we are interested in the area under the control of a specific *polis*, then the *polis* of Anthedon probably did not always have direct control over the Skroponeri bay, though the bay, which seems to be the most isolated corner of ancient Boeotia, is reachable from Anthedon through a coastal pass.[1]

Therefore, when examining and describing some aspects, for instance the resources available to the *chora* of Anthedon, we will consider the areas as separated.

With regards to the Paralimni area, only the NE half of the lake is included in this *chora*, since it was probably under the control of Anthedon (or at least gravitating towards the area of Anthedon). The rest of the Paralimni and Yliki lake areas are included in the *chora* of Thebes. Therefore, the settlement history in the area of the lakes will be discussed in chapter II.3.12 - THE SETTLEMENT IN THE AREA OF THE LAKES.

Anthedon's territory is naturally defined by the sea and the mountains behind. The proper *Anthedonia* (Strabo IX 2.13) is, as Fossey describes it (1988: 251), a long, even, level coastal plain sloping gently between the Northern mountains (Ktipas and its foothills to the SE and S, Ptoion to the W) and the North Euboean Strait.

Between Ktipas and Ptoion, the low ridge of Rachi connects the coastal plain to the Paralimni lake (Fossey 1988: 251). The land which separates the Paralimni lake from the 2km distant coast is at an elevation of 87m

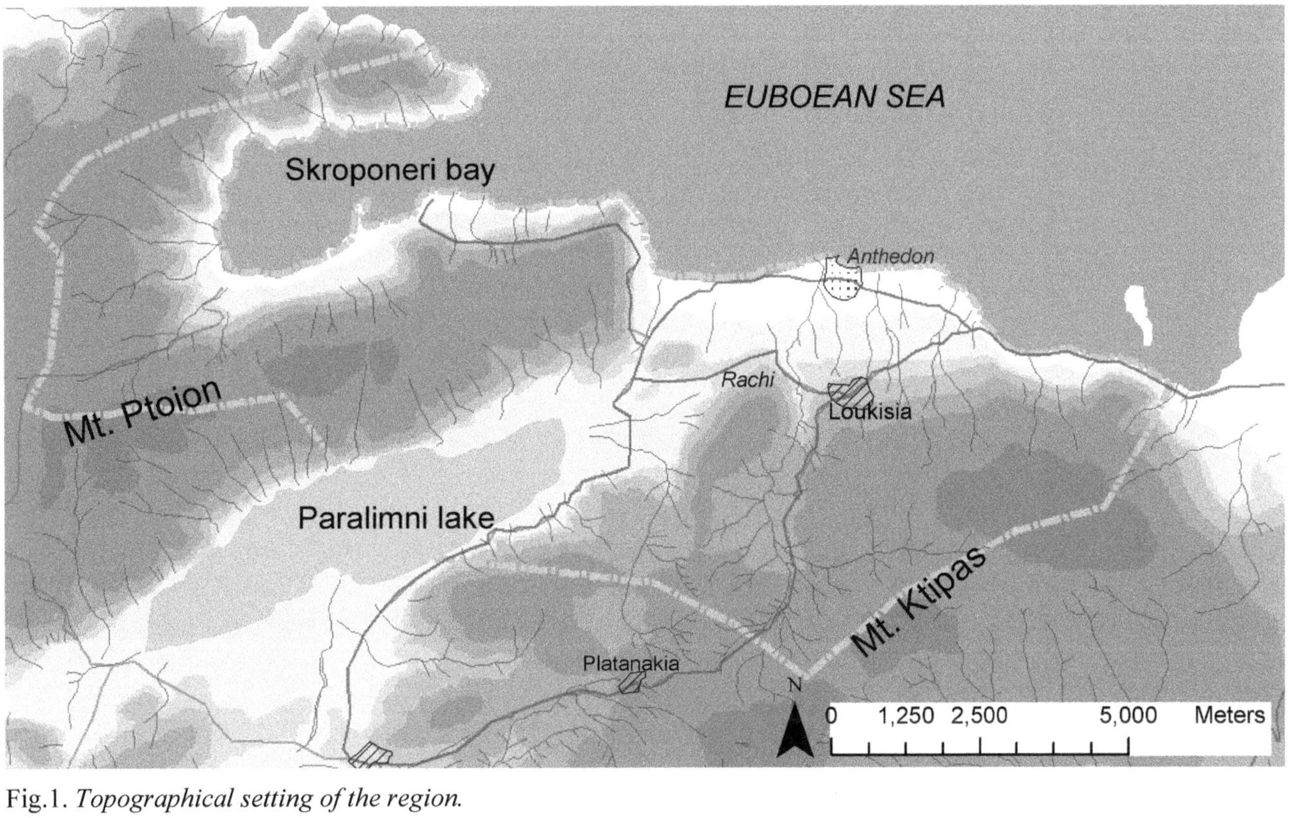

Fig.1. *Topographical setting of the region.*

[1] For a similar case of a wider geographical unit divided into smaller geographical and political units, see the Gulf of Corinth area.

(Philippson 1951: 496), while the shore is comprised of a long straight sandy beach. A tunnel discharges water from the lake into the sea when the lake's water level is high.

Boundaries
The borders of the region are marked in Fossey's map (1988: 250 fig. 31). The border of the *chora* probably passed through the middle of Paralimni lake. Mountain barriers are the Ktipas to the SE, and to the NW the mountainous peninsula which closes the Skroponeri bay to the N. Between the Skroponeri bay and the coastal plain of Anthedon is the NE spur of the Ptoion ridge, which closes the Anthedon area proper to the NW.

PHYSICAL LAND UNITS

All classes of physiographical position are present in the fairly small area, creating a landscape of variegated forms.
The mountainous segment is represented mainly by the highest peaks of Ptoion and Ktipas belonging to the area (see map in fig.1). Very few hillscape plateau features (H1) are present, and no mountain plateaus (M1). The landscape is mainly comprised of plain areas/valleys, foothills and some plateaus along the W slopes of the Ktipas ridge in the Anthedon area proper, while steep slopes mark the landscape of the Skroponeri bay (see fig.2 in chapter II.1).

Hilly landscape	48%
Mountainous landscape	7%
Plain	45%

1	P1_P2	lacustrine basin, valley	15.7%
2	P3	gentle slope	4.5%
3	P4	foothill	24.8%
4	H1	plateau	9.7%
5	H2	gentle slope	0.3%
6	H3	moderate slope	3.9%
7	H4	severe slope	9.5%
8	H5	very severe slope	24.5%
9	M1	plateau	1.1%
10	M2	plateau/gentle slope	0.92%
11	M3	moderate slope	2.3%
12	M4	very severe slope	2.8%

Table 1. *Percentage of the different physiographical classes present within the Anthedon area (P=plain; H=hill; M=mountain)*

RESOURCES

Anthedon was mainly a maritime city with little agricultural land, and the inhabitants lived from fishing, purple and murex, and boat-building (Philippson 1951: 496).
The plain of Anthedon is very fertile, being the result of alluvial deposits of the streams descending mainly from the Ktipas ridge (see fig.6). A good swathe of excellent Neogen land is available behind the site up to and beyond the modern village of Loukisia.
Heraclides (I 23-4) speaks about the economy of ancient Anthedon (based mainly on fishing and viticulture). For fishing (the fish of Anthedon were very well known in antiquity), see Athenaeus *Deipn.* VII 316 a. For viticulture, we know from Plutarch (*Quaest.Gr.* 19) that it had declined in Roman times (ancient sources reported by Fossey 1988: 263).[2] In the Late Roman period, the activities of the inhabitants probably became more and more focused on the harbour.
As noted by Fossey (1988: 251), and still today, mainly cereals, vegetables and fine citrus groves, as well as cotton fields, are in the area, with vineyards and grazing land on Mt.Ktipas, much as it probably was in antiquity also. Little fishing is done.
Just before Loukisia (following the road from Platanakia), one can note well-cultivated terraces on the talus and Tertiary formations immediately beyond the plain (included in class F=1 in the map in fig.6).

THE ARCHAEOLOGICAL RECORD

1	**ANTHEDON Mandraki – Kastri**	*Components AN_1 to AN_11*
2	**Anthedon Mandraki W**	*Component AN_13*
3	**Anthedon Mandraki SE**	*Component AN_12*
4	**Anthedon Paliambela**	*Component AN_40*
5	**Tourlo**	*Components AN_14 to AN_16 and AN_39*
6	**Skroponeri Peninsula**	*Components AN_17 to AN_23*
7	**Skroponeri Kastro**	*Components AN_24 to AN_25*

Table 2. *List of archaeological components and activity loci mapped in fig.2 – Anthedon and Skroponeri area.*

PARALIMNI AREA
Communication along the Paralimni lake was not easy. It is a lake of tectonic formation, and therefore with bordering slopes that are steep and rocky (see chapter II.1). There are openings only at the two ends, with very fertile zones formed by lacustrine deposits. The NE end can be interpreted as belonging to the area controlled by the *polis* of Anthedon (and this becomes valid especially when considering the difficulties in land communication between the two ends of the lake). On the other hand, the SW end has been included in the Thebais *chora* (chapter II.3.12), since it is much more easily approachable from there (and it was probably under the direct control of Thebes in the Greco-Roman period).

8	**Paralimni Pyrgos**	*Components AN_26 to AN_35*
9	**Paralimni Pyrgos E**	*Component AN_38*
10	**Paralimni Pyrgos NE**	*Components AN_36 and AN_37*

Table 3. *List of archaeological components and activity loci mapped in fig.2 – Paralimni area.*

[2] A decline in Roman times can be noted for Haliartos also, and in the case of Haliartos it is certainly due to a specific historical fact (the destruction of the city by the Romans in 171 BC).

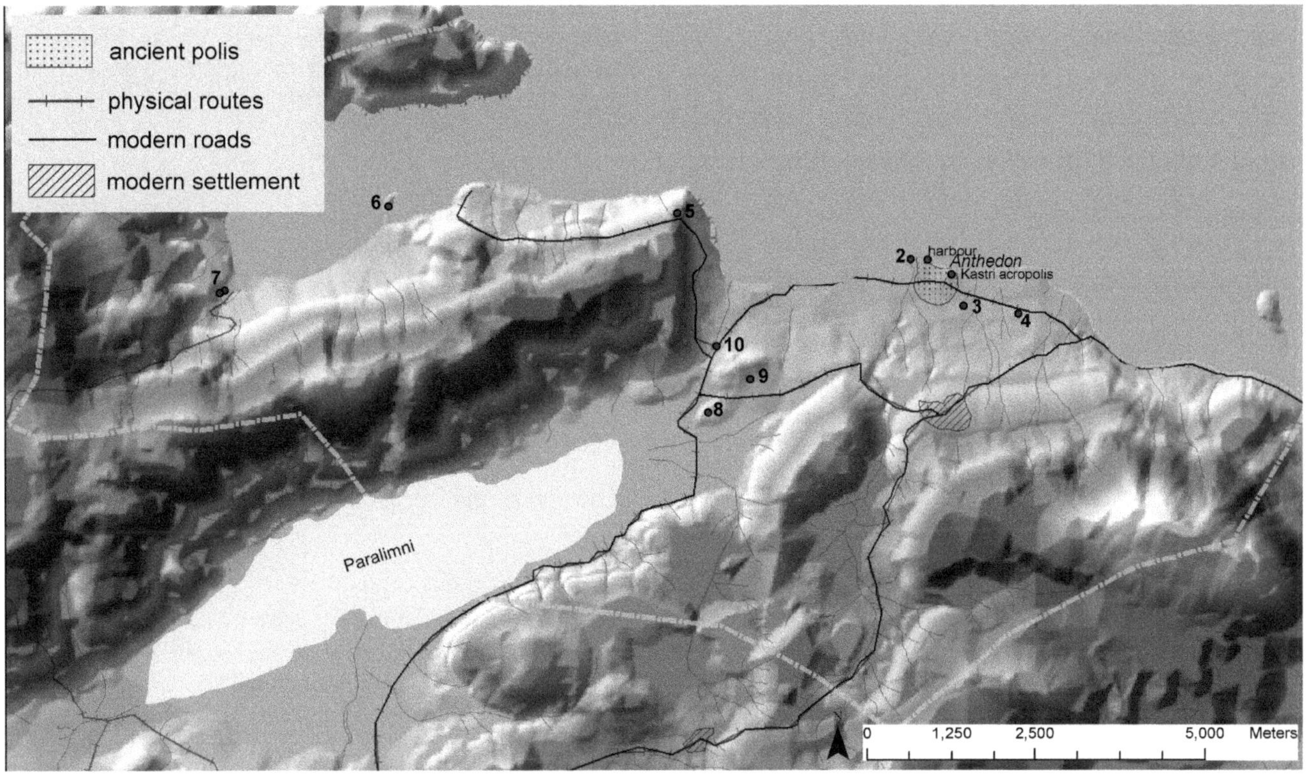

Fig.2. *Archaeological map of Anthedon chora.*

The only site known in the Anthedon area proper is, in fact, Anthedon itself, with its harbour, with the exception of a periurban sanctuary site and some Neolithic occupation by the city site. Other known sites from the wider area are at the Paralimni N edge, and further away in the Skroponeri bay, and a site along the coastal road that links Anthedon to Skroponeri (Tourlo) – see appendix I.13.

The fact that the archaeological record of the region is marked by research on the *polis* of Anthedon and its harbour is clear also from the graph (fig.3 – 'personal or group interest' value). Also, no rescue excavation has been carried out in the area, and this is mainly due to the marginality of the region itself, which means reduced road construction, no large scale or monitored building activities, etc.

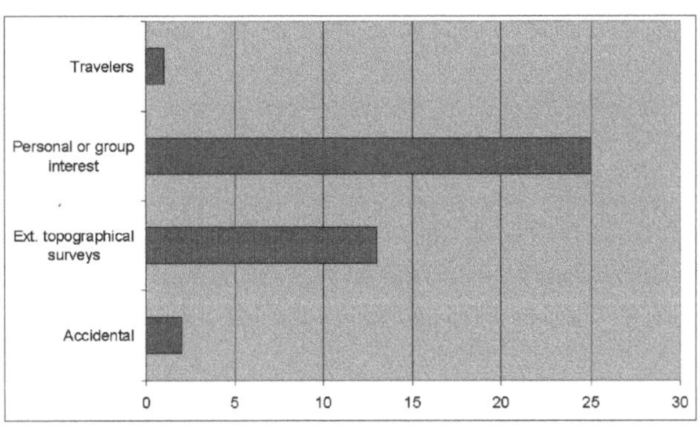

Fig.3. *Graph illustrating the proportion of components discovered within different research frameworks.*

In fig.4 we can see the relationship between known archaeological sites and the distance from the modern road network. Correspondence is indeed very strong, but it can be due in part to the lack of intensive and systematic surveys, and the consequent absence of known records related to the rural landscape, to the particular topographical situation of the area and to the pathways followed recurrently by past and present routes.

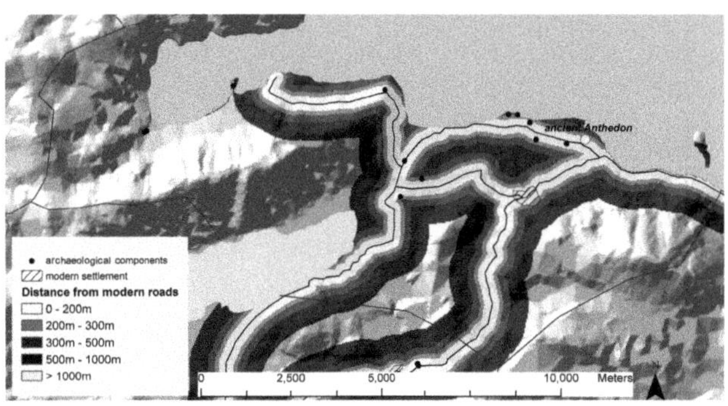

Fig.4. *Relationship between components and modern road network.*

The ratio of known Prehistoric to Greco-Roman components is 13 to 23 (1:1.7), while among the historical periods, 43% are dated Archaic to Hellenistic, 43% Roman-Late Roman, and 14% are attributed to the general Greco-Roman period.

A relatively large number of Prehistoric sites are known compared to historical ones, and this could be due mainly to research (predominantly topographical) focused on hilltop or prominent sites led by personal research interest (such as the search for Late Prehistoric hilltop settlements), but also, as pointed out earlier, to the relatively small number of sites known in total.

ANALYSIS OF THE *CHORA* LANDSCAPE

PREHISTORIC PERIOD

Prehistoric occupation is known for the site of ancient Anthedon itself. The only certain Neolithic activity focus known is on a hill outside the later city wall (*component AN_13*). Also at Tourlo (*components AN_14, AN_15 and AN_39*), traces of Prehistoric occupation are reported (probably since the Neolithic). The sites are both on low hills by the sea shore – see fig.5 in chapter II.3.11.

In the area at the N edge of Paralimni lake, a suitable landscape for Prehistoric occupation, Prehistoric activity foci are known at the Paralimni Pyrgos site and at a location to its E/NE, where traces of activity foci and burials have been found for the EH and LH periods[3]– see figs. 6 to 8 in chapter II.3.11.

GRECO-ROMAN ANTIQUITY[4]

Town level

Anthedon is the only town-level settlement present in the area. It was mainly a harbour city, which constituted a crucial passage towards Euboea, and appears to have flourished especially in Classical/Hellenistic periods, under the control of Thebes.

As mentioned earlier, the city might have known a period of decadence in Roman times, as historical sources (Strabo IX 405; Plutarchus *Quaest.Gr.* 19) seem to point out, as does the archaeological record, indirectly. A Roman grave ploughed up in a private plot (*component AN_41*) within the Kastri site could be an indication of a shrinking of the town site in the Roman period (see also below).

Village level

Second-rank settlements are not known for the area, though several attempts have been made to recognise settlement sites (*komai*) at the village of Loukisia, as well as at the Paralimni Pyrgos site (identified with ancient *Isos* by some scholars), and further in the Skroponeri bay (see appendix I.13). In fact, the town of Anthedon being mainly concerned with harbour activities, there might have been space for another nucleated settlement centre of agricultural activities. It is not necessary, however, to consider it a certainty, as the inhabitants of Anthedon, in addition to people living in farmsteads all over the landscape (a picture typical of other Boeotian areas, which can not be excluded here as long as no intensive research is done in the countryside), could have farmed the available area (see also below).

Rural segment

As noted earlier, the rural segment of the landscape is as yet unknown. Information on it could be provided only by an intensive and systematic programme of research throughout the countryside.

Burial areas

As far as the available archaeological record indicates, there is no evidence of burial areas from historical periods, though we might suppose burials linked at least to the known settlement sites. Exceptions are a few signs of a Roman cemetery area by the city site (*components AN_40 and AN_41*).

Cult places/Religious areas

Only an extramural sanctuary (*component AN_12*) is known for the area, in the immediate outskirts of Anthedon *polis*.

Forts and fortifications

Fossey (Teiresias suppl. ii (1979) 10) argues that the Skroponeri bay could have been a base – perhaps the main one - for the 4th C Boeotian fleet, and identifies as possible sites linked to this use the Skroponeri Kastro site – *component AN_24* (which Fossey mentions together with the peninsula site, *components AN_21 to AN_23*), as well as the evidence at Tourlo, interpreted as one of the 4th C watch posts protecting the Skroponeri naval base (Fossey 1988: 26).[5]

LONG TERM SETTLEMENT TRENDS IN THE CHORA *LANDSCAPE*

The main settlement in the area of Anthedon was always at the harbour site (Mandraki locality) and/or at the nearby acropolis ridge known as Kastri (Early Hellenistic to Late Roman times - *components AN_1 to AN_9*). The area constitutes a crucial passage towards Euboea. For the Neolithic period too (*component AN_13*), a settlement site has been recognised on a small hill by the shore to the W/NW of the later settlement location.

Today, the area as a whole contains only one village, Loukisia, 2km inland from the Mandraki/ancient Anthedon city site. Modern Loukisia is an Albanian village recorded in the Ottoman defters (1466-1646), with possible Byzantine finds also (Bintliff–Kiel in preparation). Being the main habitation focus, it occupies a central position in the *chora* territory. The ancient town of Anthedon has only a peripheral location, as it was geographically and economically orientated to the sea. Moreover, in antiquity the *chora* contained two settlement chambers, as the cost-distance analysis[6] – fig.5 - shows. The one gravitating on the main settlement was always the coastal plain.[7]

---

[3] Settlement in the area of the lakes is discussed in chapter II.3.12.

[4] Period maps are included in chapter II.4, figs.18-20-22-24-26-28.

[5] The *polis* of Anthedon was also fortified [AE559].

[6] See chapter II.3.1 – LONG TERM SETTLEMENT TRENDS.

[7] The importance of the coastal site of ancient Anthedon is reinforced by the use of the harbour in Middle Byzantine and Frankish times, as Port Lucaza (Bintliff – Kiel in preparation).

The N edge and part of the Paralimni lake was probably exploited by the second rank settlement which would possibly be in the Paralimni Pyrgos locality (see above – VILLAGE LEVEL), where the remains of a Frankish tower are also visible.[8] The area of Skroponeri could also be a potential settlement chamber, at least in some periods of history, as attested by the settlement gap in the cost-distance analysis as well as by the presence of a village in the Ottoman period (fig.5).

The harbour city flourished in Late Classical – Early Hellenistic times and probably knew great splendour as Thebes' naval base in the 4th C BC (after its importance as a harbour in the Homeric period, with a mention in the Catalogue of Ships). The city might have known a period of decadence in Roman times, as seems to be implied by historical sources (Strabo IX 405; Plutarch Quaest.Ge. 19), as well as by indirect archaeological evidence. A Roman grave ploughed up in a private plot (*component AN_41*) within the acropolis of Kastri could be an indication of a shrinking of the city site in the Roman period. Plutarch's reference to the reduction in wine production (for which the hills as well as the slopes of Mt.Ktipas are certainly suitable – see *Resources* above), as well as the apparent absence of Roman remains at the Paralimni Pyrgos site (if part of the Anthedon *chora*), might also be an indication of decay during the Roman period, which finds a parallel in other Boeotian cities studied intensively by the Boeotia survey project, especially Hyettos, Thespiae and Haliartos (for the latter, however, the decline is related to a precise historical event). In Late Roman times, the harbour and the town must have known another positive and flourishing phase, like the other harbour sites in the region (and in particular those towards the Gulf of Corinth – see chapter II.3.10).

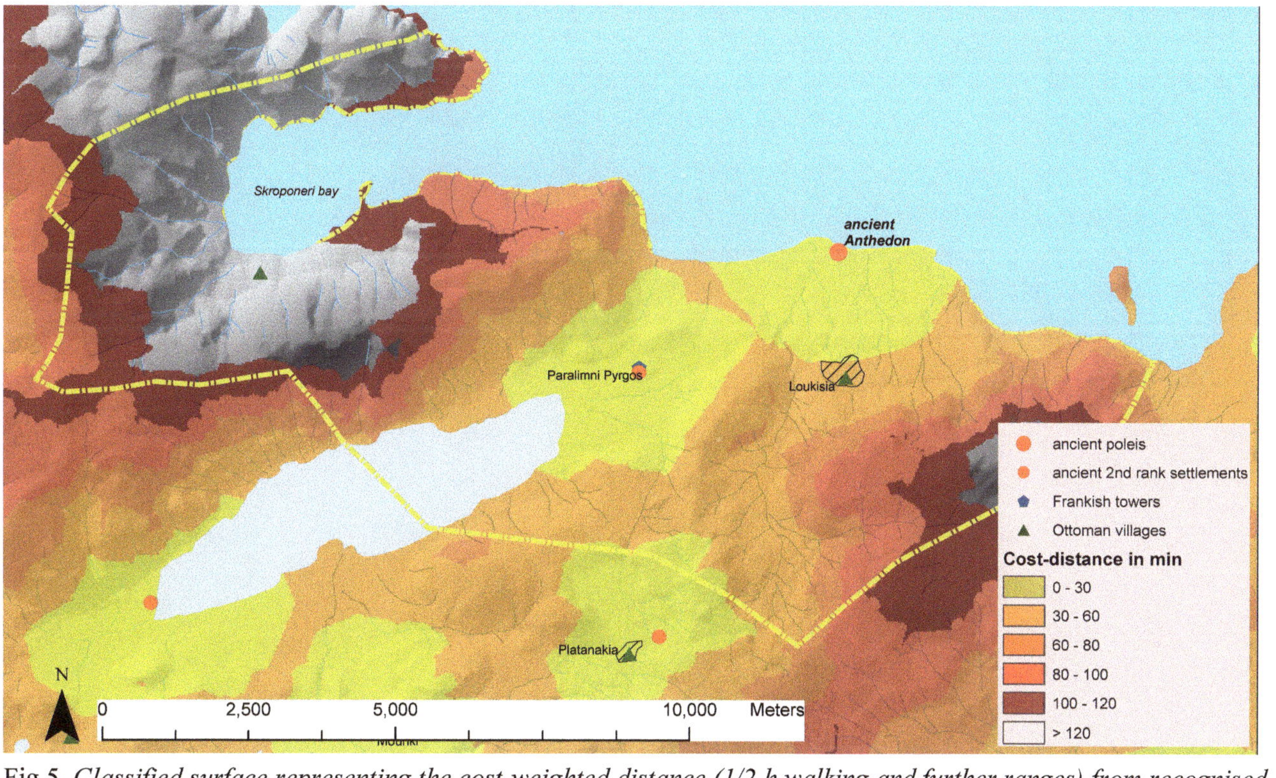

Fig.5. *Classified surface representing the cost-weighted distance (1/2 h walking and further ranges) from recognised 1st and 2nd rank ancient settlements (represented by larger and smaller dots). Areas without dots indicate potential settlement chambers. Ottoman villages and Frankish towers have also been added to the map to show their spatial relationship with the Greco-Roman settlement network and to appreciate potential settlement chambers.*

[8] In Frankish times a harbour was located by ancient Anthedon.

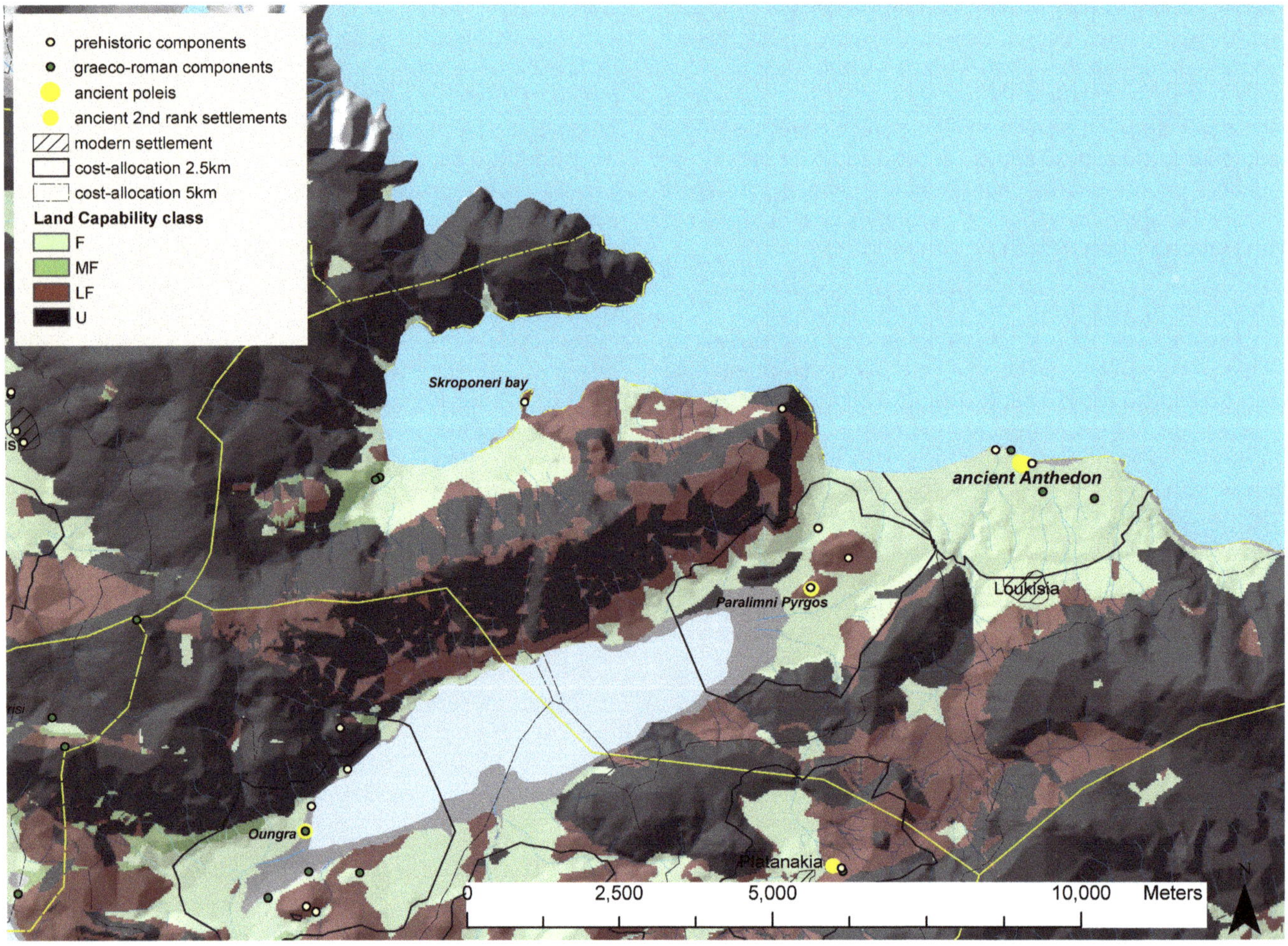

Fig.6. *Map showing the Greco-Roman settlement network, the polygons resulting from the cost-distance analysis (marking half an hour and one hour walking time distance) and dots representing the known archaeological components (same as in fig.2), with land capability information underlain.*

II.3.14

The Tanagra plain and Eastern Boeotia

TOPOGRAPHICAL SETTING

The area includes the plain to the N of ancient Tanagra, the low ridge of coastal hills to the E, the lower coastal plains beyond (Dilesi, Dramesi, and Vathy). Also included in the *chora* are the inland basin of Ritsona (which belonged to Tanagra at least in the Hellenistic-Roman period), the small coastal plain of Drosia (Chalia – which in some periods of history belonged to Tanagra), and the lower Asopos basin opposite ancient Tanagra. As M.H. Hansen (1996: 56, 104-5) pointed out, Herodotus and Thucydides normally refer to the *chora* of Tanagra as Τάναγρα,[1] while later sources use Ταναγραία,[2] Ταναγραική.[3]

The area is bordered to the N by the Messapion mountain chain. As Philippson (1951: 498) remarks, Mt.Sagmatas/Ktipas (SW section of Messapion), as a narrow line of limestone heights, separates the basin landscape with new deposits from the coastal area of Euripus,[4] and marks the low passage between the basin landscape of the Theban plain and the 'Tanagra' area (see below – PHYSICAL LAND UNITS).

An E spur of Mt.Ktipas, namely Galatsidesa in this NE section, reaches the Euboic sea with the promontory of Vessalas (the site of the ancient sanctuary of Artemis in Aulis), which divided the two gulfs of Mykro and Megalo Vathy,[5] while another spur to the NE abruptly reaches the steep coast of the Euboic sea E of ancient Anthedon (see chapter II.3.13),[6] marking a narrow pass where an important coastal road runs (Philippson 1951: 499). To the SE of it the rectangular peninsula of Chalia opens up, 7km long along the coast and 3 to 4km wide. Its SE edge is cut by the Euripus strait, which cuts the city of Chalkis into two parts.

To the S of the limestone ridges of Messapion/Ktipas and the gulfs, the landscape is marked by later alluvial deposits, along the Euboic sea (Philippson 1951: 499).

The Ritsona delta is a fertile coastal plain to the S of the Megalo Vathy bay, rich in wine, olives and cereals, where the village of Vathy is located (Philippson 1951: 516).

Small hamlets/villages, such as Pharos or Dramesi/Paralia Avlidos, exist also in the coastal plain to the S. This coastal plain is bordered to the S by the hill also known as *Kala Denta* (immediately to the S of Dramesi/Paralia Avlidos) along which runs a 'bad' road (as defined by Philippson 1951: 516) towards the modern village of Schimatari. In the fields were new pine trees in Philippson's time, while by the sea there were badland (eroded area) gorges. To the SE of Schimatari starts a stream that ends in a small delta by the modern village of Dilesi (identified with ancient Dilion – see appendix I.14).

A long and wide formation of tertiary hills, essentially marking the N watershed of the Asopos river valley, extends to the E from Schimatari with an elevation of 160m, runs parallel to the Dilesi stream mentioned above, and later descends to 80m by the outlet of the Asopos. These Tertiary hills, with mainly pine trees and bushes, continue to the coast, crossing the entire area between Schimatari and the sea.

In the interior, the small inner areas of Ritsona lie in a small basin surrounded by hills below the Ktipas ridge along its E side, marked by the passage of the Ritsona river which flows into the sea via the Ritsona delta (see above), while from Sphaides/Eleon to Dritsa/Harma there is a series of low limestone hills rising in the middle of the Tertiary formations marking the surrounding area (the 'corridor' – see appendix II –Eastern Boeotia; Philippson 1951: 512). In line with Dritsa/Harma following a NW-SE direction is the village of Bratsi/modern Tanagra, on the slopes of Soros, whose Eastern spur is the site of ancient Tanagra (see below), which overlooks the wider Tanagra plain.

Boundaries

The area controlled by the *polis* of Tanagra fluctuated in size over the centuries.

A recent work of A. Schachter (who investigated literature, inscriptions, coins, and archaeological remains, with the aim of defining the fluctuations of the border – Schachter 2004) showed that the Tanagrans were able to exploit land not normally identified as theirs, especially in southern parts of the *chora*.

Generally speaking, the topographically definable borders of the *chora* are: the Galatsideza ridge (N), the Ktipas

[1] Thucydides III 91.3-4

[2] Strabo IX 2.10; Pausanias IX 19.8, but also earlier Thucydides IV 76.4.

[3] Strabo IX 2.11

[4] Euripus is the ancient name of the strait dividing Boeotia from Euboea.

[5] In this area, the railway (and a modern road, not present in Philippson's time) runs along the coast, while the older Athens-Chalkis road runs through the inner part, on top of the low limestone hills (see fig.1 in chapter II.1).

[6] Here Ulrichs (1840) saw traces of an ancient road.

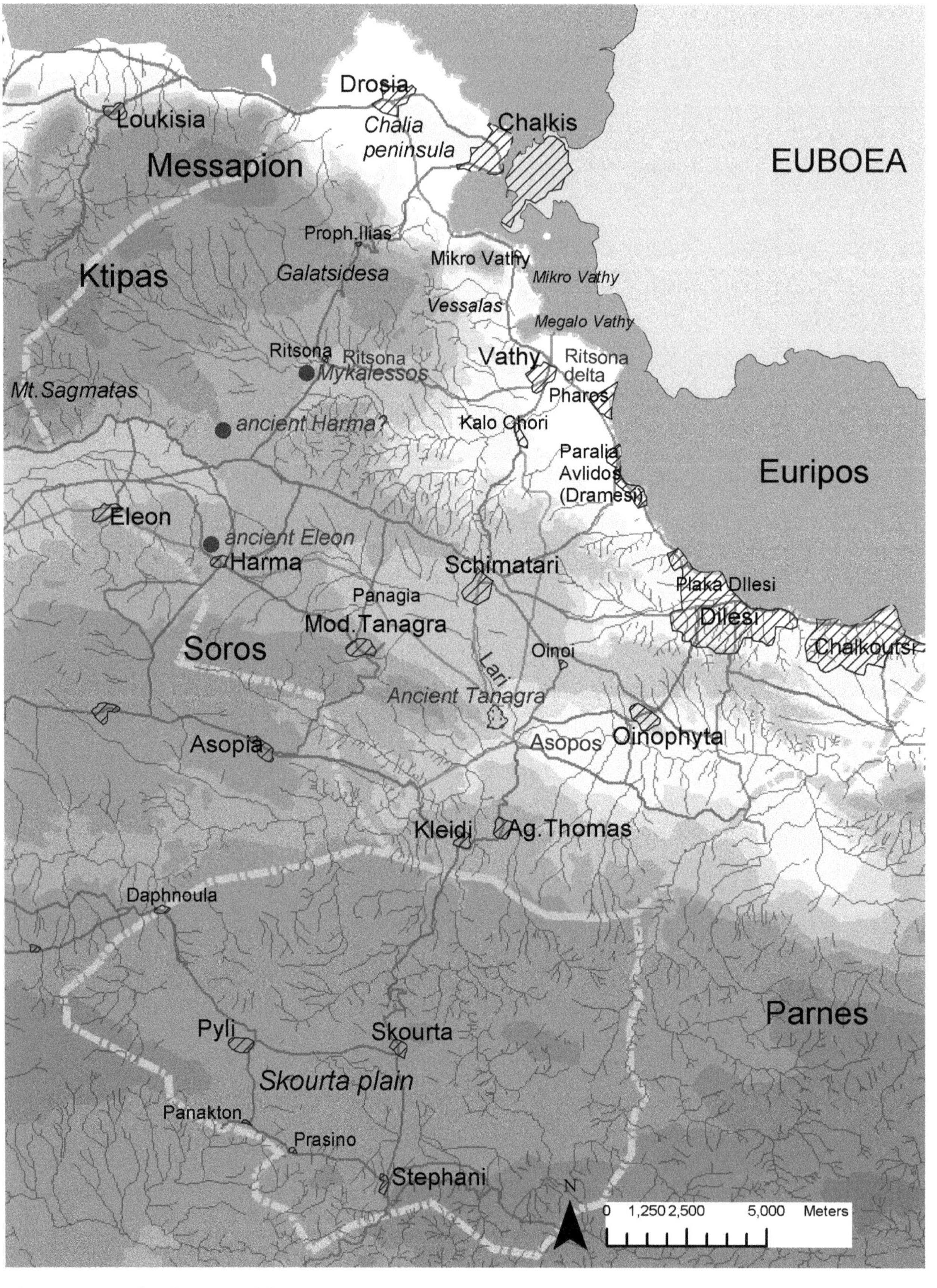

Fig.1. *Topographical setting of the region*

ridge (W), the Soros range and Parnes foothills (S), and the shores of the South Euboian gulf (NE) - Fossey 1988: 98-9.

Through the examination of the sources for the location of ancient Graia and Delion, Schachter's work points out that the boundaries on the E between Tanagra and Oropos (the *methoria* – the no-man's land between two *poleis*) would have been a fairly fluid area, around ten stades, that used to be somewhere between 1.5 and 2 km E of ancient Delion (Thuc. IV 9.4), probably W of the mouth of the Asopos (Plato, *Critias* 110d-e), and was probably not definitely fixed at a specific point (Schachter 2004: 52).

Considering the Northern boundary, the impression is that until some time in the 4th C BC the region N of Tanagra (as well as the control of the eastern seaboard) belonged to Thebes, but subsequently, and throughout the Hellenistic and Roman periods, to Tanagra (Schachter 2004: 54). Therefore, in the Classical period the boundary would have been lying S of the territories of Eleon, Harma, Pharai and Mykalessos (the so-called *tetrakomia* – see below), and Aulis, while during the 4th C the territory of Tanagra was extended to include at least Aulis if not the other towns in the region, which were certainly Tanagran in the Hellenistic and Roman periods (Schachter 2004: 59).

To the South, the Skourta plain – which was common pasture land shared by Boeotians, Athenians and even Corinthians (Munn & Zimmerman Munn 1990: 36-37) - can be regarded as *methoria*, no man's land between the territories of Attica and Boeotia (Eleusis and Tanagra), as pointed out by Schachter (2004: 56) following his examination of historical sources. Considering also the identification given by Munn (1988: 363-371) of the two watchtowers in the NE corner of the Skourta plain, at Tsoukrati and Limiko, as Boeotian, Schachter says that the territory of Tanagra extended into the Skourta plain to the W, and eastward along or into the northern flanks of mount Parnes, past Sphendaleis, and, clearly, up to the Oropia (in the 5th C too - Schachter 2004: 57). The border would have been at the NE corner of the Skourta plain.[7]

To the West, following Herodotus (IX 15.1) and Xenophon (*HG* V 4.49), Skolos[8] must have been the first place along the Asopos which was within Theban territory, so the border would have been not far to the E of it (Schachter 2004: 57), though the actual border area still remains vague.

PHYSICAL LAND UNITS

Plains constitute the majority of the *chora*, mainly marking the coastal area towards the E (see above and below *passim*). Mountainous landscape is very poorly represented, only at the NW edge, constituted by the Messapion ridge. Hilly landscape mainly marks the inner area running NS and is constituted (from N to S) by the lower slopes of Messapion, the low ridge geographically separating the Eastermost region from the Theban plain, and the low hilly morphologies of the same formation crossing the Thespiae and Parasopia & Plataea areas (Tertiary *tafel* of Thebes).

Hilly landscape	32.9%
Mountainous landscape	2.3%
Plain	64.8%

1	P1_P2	lacustrine basin, valley	37%
2	P3	gentle slope	10.8%
3	P4	foothill	17%
4	H1	plateau	12.8%
5	H2	gentle slope	3.3%
6	H3	moderate slope	6.3%
7	H4	severe slope	6.7%
8	H5	very severe slope	3.8%
9	M1	plateau	0.5%
10	M2	plateau/gentle slope	0.5%
11	M3	moderate slope	1%
12	M4	very severe slope	0.2%

Table 1. *Percentage of the different physiographical classes present in the area (P=plain; H=hill; M=mountain)*

As visible also from the map with land units (fig.2 in chapter II.1), the region does not have clearly defined geographical boundaries, presenting landscapes which recur several times in different locations, without therefore constituting real physical units. Landscapes composed of plains, for instance, constitute part of the coastal area towards the E, but also mark the landscape of the E part of the Theban plain probably belonging to ancient Tanagra. Foothills (P4) usually form a line running along the foot of the hills, but we can see areas with the same morphological characteristics (below 200m sl and slope $\geq$ 10%), though with much wider dispersion over the landscape, in the area E and NE of Tanagra and as far as the shoreline. Therefore, even the landscape of the coastal area is not even, with plain features (P1_P2) towards the sea and gentle slopes and foothills (P3 and P4) running along the edge (as we find commonly elsewhere, for instance in the valleys around Copais basin), but forms a variagate picture, formed by P3 and P4 features irregulary alternated with P1_P2 features, thus consituting a fairly irregular landscape, marked by low rolling hills.

RESOURCES

A map showing land potential for the Tanagrike area has not been realised, since the Geological base map for the area (Chalkis sheet) is not available; it has not yet been issued by IGME.

The landscape of this wide *chora* is variegated, and therefore so are the available resources. Also, some parts of the landscape had a changing history, which has altered land use and resource exploitation over time. For instance, one can consider the area between Dramesi and Ritsona (see below) for a long period as a zone of

[7] For the Skourta plain, see appendix I.14 – THE SKOURTA PLAIN SURVEY.

[8] For the location of Skolos see Schachter 1986: 133 and below in the text.

predators during the Middle Ages, when thick wild forest took the place of the productive cornfields and vineyards. Generally speaking, the wide *chora* is very fertile, both the coastal plain and the inner hills. The Tanagra plain can be considered one of the most fertile areas of Boeotia, unfortunately slowly but progressively occupied today by a widely expanding industrial area to the N of Athens. Observing it from the acropolis of ancient Tanagra, in the summer one notices that the predominant colour of the landscape is still yellow, as mostly cultivated as cereals (as it was in the past, most probably), compared to the Theban plain, currently mainly characterised by irrigation cultures.

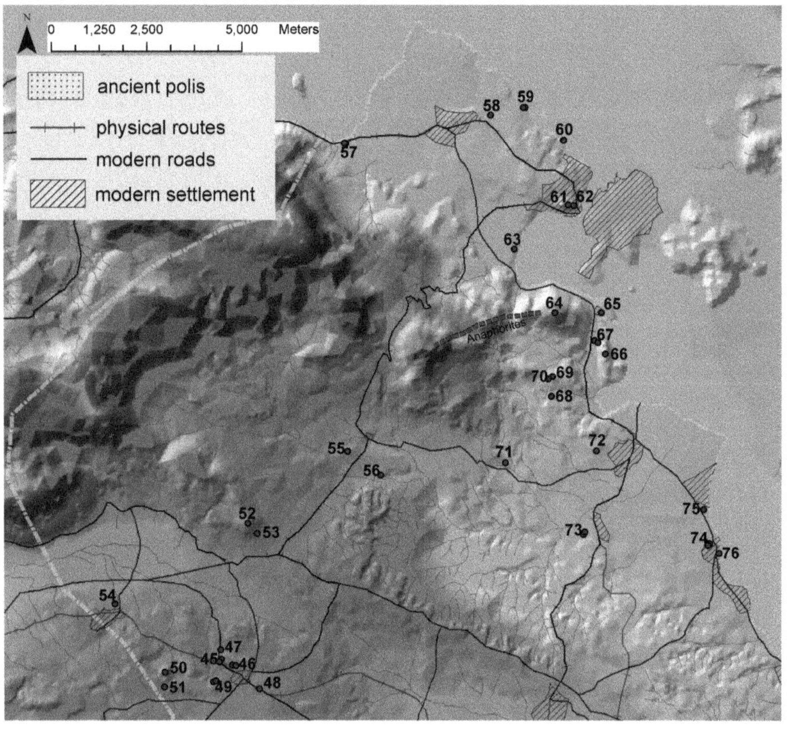

Fig.2. *Archaeological map of the Tanagrike chora (N part).*

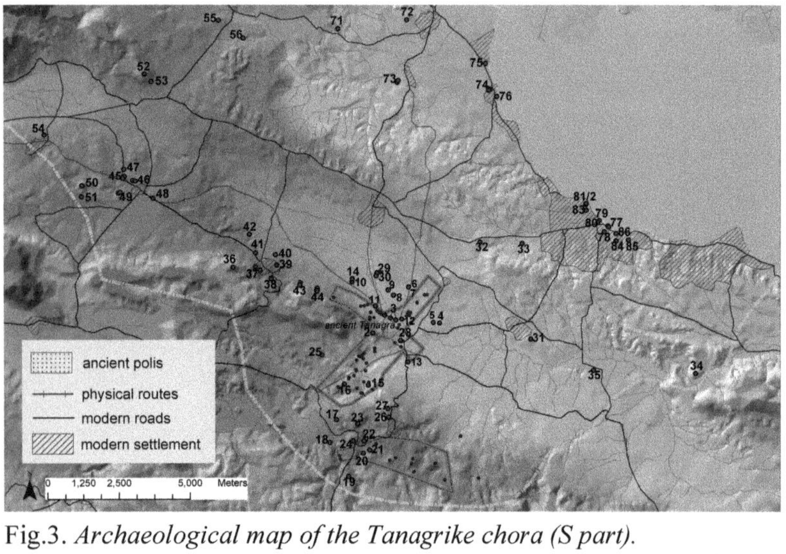

Fig.3. *Archaeological map of the Tanagrike chora (S part).*

A few areas can be considered as marked by peculiar characteristics with regards to land potential and land use. For instance, the Chalia peninsula, the Ritsona delta, and the Asopos lower valley.

The Chalia peninsula (see above) is consituted by limestone conglomerate and *poros* covered by red soil, cultivated mainly as vineyards and olive trees, especialy in the middle of the plain, where the village of Chalia lies (Philippson 1951: 499). In the SE part of the peninsula, by the Euripus, is a non-cultivated formation of rudist limestone, which also forms the Karababa hill (W part of modern Chalkis – see below). In Philippson's time, this part of the peninsula was covered only with maquis and frigana. The coast of the peninsula is steep with small flat sandy gulfs, while to the S of Chalkis the Euripos channel is a small plain crossed by the railways, with a salt-installation in Philippson's time.

The Ritsona delta is a fertile coastal plain to the S of the Megalo Vathy bay, formed by the deposits of the Ritsona stream[9] and rich in wine, olives and cereals (Philippson 1951: 516). The river no longer reaches the sea, and its former delta deposits consitute a promontory, with a lighthouse, just 500m from the Euboian coast.[10]

The lower Asopos valley, which crossed longitudinally the Southern part of the Tanagra *chora*, is marked by a fertile landscape of river terraces. The Asopos river runs through incised gorges from the Parasopia into the Tanagra area. From the site of ancient Tanagra (at the Grimada location – see below) it points to the E in a wide valley, marked in Philippson's time by pear and oak trees, used for grazing and husbandry activities (Philippson 1951: 519). Today it is mainly cultivated, with the lower terrace and the river bed recognisable through the presence of flourishing reeds. The N boundary of the lower Asopos valley is constituted by a line of low Tertiary hills, covered with bushes and pine trees in Philippson's time (see above).

[9] Several *kseropotamoi* join near the hamlet of Ritsona (180m asl), forming a wider stream (Ritsona) that runs towards E to the coast S of Aulis and empties into the plain of Vathy (Philippson 1951: 515). Along the N bank of the stream runs the road to Vathy.

[10] Together with the Lilados delta in Euboea the promontory creates a narrow pass of 500m (the 'Burtsi strait', the narrowest after Euripos).

THE ARCHAEOLOGICAL RECORD

1	**TANAGRA Grimada**	Components TA_1 to TA_8
2	**Grimada**	Component TA_220
3	**Extramural buildings by ancient Tanagra**	Components TA_224 to TA_226
4/5	**Kokkali**	Components TA_9 to TA_11 (4); TA_12, TA_13 and TA_198 (5)
6 – 13	**Other burial places**	Components TA_14 to TA_16 (6); TA_221 (7); TA_202 (8); TA_199 to TA_201 (9); TA_204 and TA_222 (10) ; TA_203 (11) ; TA_17 and TA_18 (12); TA_205 to TA_207 (13)
14	**Tanagra N**	Component TA_227
15	**Agios Konstantinos**	Components TA_19 to TA_29
16 – 26	**Kleidi burials**	Components TA_30 (16); TA_36 (17); TA_34 (18); TA_35 (19) ; TA_32 (20); TA_37 and TA_38 (21); TA_31 (22); TA_33 (23)
24/25	**Kleidi**	Component TA_208 (24) and TA_228 (25)
26/27	**Agios Thomas**	Components TA_40 (26); TA_41 and TA_42 (27)
28	**Lari**	Component TA_39
29/30	**Kesseri**	Components TA_43 and TA_44 (29) and TA_223 (30)
31	**Oinophyta**	Components TA_45 to TA_49
32/33	**Prophitis Ilias**	Components TA_50 to TA_54 (32) and TA_55 (33)
34	**Bathiza**	Components TA_195 to TA_197
35	**Asopos S**	Component TA_219
36 – 42	**Bratsi/Modern Tanagra**	Components TA_56 (36); TA_57, TA_68 to TA_70 (37), TA_62 (38) ; TA_65 to TA_67 (39); TA_64 (40); TA_63 (41); TA_58 to TA_61 (42)
43/44	**Gephyra/Ledeza and Dendron/Grava**	Components TA_71 and TA_72 (43); TA_73, TA_74, TA_209 and TA_210 (44)
45 – 48	**Dritsa/modern Harma – Pyrgos (ancient ELEON)**	Components TA_75 to TA_85 and TA_88 (45); TA_86 and TA_87 (46); T_89 and TA_90 (47) ; and TA_211 (48)
49	**Dritsa – Proph.Ilias**	Components TA_91 to TA_95
50/51	**Meg. Kastron**	Component TA_96 (50); TA_97 and TA_98 (51)
52/53	**Kastron Lykovouni (ancient HARMA?)**	Components TA_99 to TA_104 (52); TA_105 and TA_106 (53)
54	**Sphaides/modern Eleon**	Component TA_212
55	**Ritsona**	Components TA_107 to TA_114 and TA_213
56	**Ritsona Kamileza**	Components TA_115 to TA_118
57	**Lithosoros**	Components TA_119 to TA_125
58	**Drosia area**	Components TA_126 and TA_127
59	**Drosia – Ag.Minas**	Components TA_128 to TA_131
60	**Rodia**	Components TA_132 and TA_133
61/62	**Karababa**	Components TA_142 (61) and TA_143 (62)
63	**Glypha**	Components TA_134 to TA_141
64	**Mikro Vathy - Meg. Vouno**	Components TA_144; TA_146 and TA_147
65	**Mikro Vathy N**	Component TA_145
66	**Vathy Nisi**	Components TA_149 to TA_150
67	**Mikro Vathy (Artemis sanctuary)**	Components TA_151 to TA_155
68/69/70	**Vathy - Ag. Nikolaos**	Components TA_156 (68); TA_157 to TA_163 (69); TA_164 (70)
71	**Vathy - Proph.Ilias**	Component TA_148
72	**Vathy – Kouloureza**	Component TA_214
73	**Agios Panteleimon (ancient PHARAI?)**	Components TA_165 to TA_168
74	**Dramesi/Paralia Avlidos**	Components TA_169 to TA_176
75	**Pharos Avlidos**	Component TA_177 to TA_179
76	**Dramesi/Paralia Avlidos shore**	Component TA_180
77 – 86	**Dilesi**	Components TA_181 and TA_182 and TA_193 (77); TA_183 to TA_185 (78); TA_186 (79); TA_216 (80); TA_194 (81); TA_217 and TA_218 (82); TA_187 (83); TA_188 (84); TA_189, TA_190 and TA_215 (85); TA_191 and TA_192 (86)

Table 2. *List of archaeological components and activity loci mapped in figs.2-3.*

The area of ancient Tanagra and its surroundings is in general the best known within the wider *chora*, both for the presence of the ancient city and the strong indications given by the well known cemeteries (in the Mycenaean period with the *larnakes*, and in historical times with the 'Tanagrines' – see above). This might also explain the significant presence of personal or group interest within the panorama of archaeological discoveries in the area (see graph in fig.5).

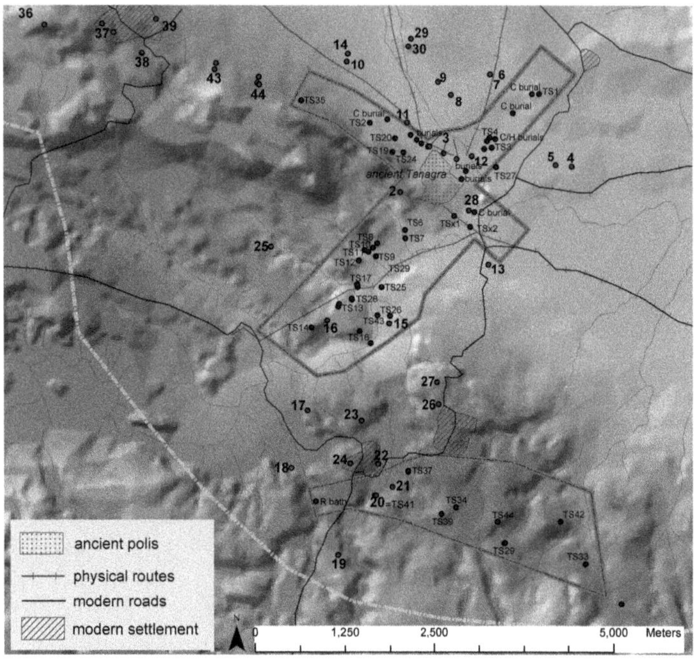

Fig.4. *Same as fig.3, but limited to the area area intensively and systematically surveyed, with discovered sites (listed in appendix I.14: table2) marked.*

Accidental discoveries also appear to a large degree, and are mainly represented by burials discovered through ploughing and robbing activities, and easily recognisable through the specific material culture associated. The quite high amount of 'other' is due to the discovery of evidence from other (minor) periods of occupation at major prehistoric or historical sites, such as traces of prehistoric activities found at large historical sites, such as Dritsa/modern Harma, or signs of Greco-Roman activities at well known prehistoric sites, such as Glypha. Rescue excavation also constitutes a section in the archaeological knowledge of the area, especially in recent years. See, for instance, the excavations carried out around the site of ancient Tanagra on the occasion of works for the natural gas pipe-line– 1997 (AD 52 (1997): 371ff), which uncovered burial areas and extramural buildings as well as rural sites (*components TA_222, and TA_224 to TA_227*), and parts of Greco-Roman roads leaving Tanagra [AE1862 and AE1863].

The high density of components around Tanagra is mainly due, as seen earlier, to research on cemeteries (prehistoric and Greco-Roman). An increase in the density of sites is recognisable also in the Bratsi/Modern Tanagra area (for the *larnakes* cemeteries mainly) as well as around Dritsa/Harma and Dilesi sites.

A Systematic Intensive Surface Survey was carried out in the area around ancient Tanagra (area surveyed mapped in fig.3) from 2000 to 2006 by J.L. Bintliff and a Leiden/Ljubljana team and discovered sites listed in appendix I.14 – THE TANAGRA SURVEY PROJECT - and mapped in fig.4. Even though limited to a relatively small area of the *chora*, surrounding ancient Tanagra, the data resulting from intensive and systematic surface surveys are quantitatively substantial in the picture of the available archaeological record for the wider area (see graph in fig.5). This is obviously due to the intensity of research in the small area around Tanagra in comparison with the poor data from other areas.

A new intensive survey has also been started in 2007 in the area of modern Arma and Eleon by the Eastern Boeotia Archaeological Project (EBAP). Also, a systematic surface survey has been carried out in the upland plain of Skourta, at the border between Boeotia and Attica, in the uplands to the S of the territory of Tanagra (Munn – Zimmerman-Munn 1987-9[11]). The research and results are briefly illustrated in appendix I.14 – THE SKOURTA PLAIN SURVEY. They have not been included in the graph in fig.5.

In fig.6 we can see the relationship between known archaeological sites and the distance from the modern road network. Correspondence, apart from hilltop sites which are obsiously off the roads, is striking indeed, though once again (see other *chorai* for similar cases) the older road to Thebes, as well as the coastal road, for instance, follows the same route as in antiquity, and the occupation might be linked to the ancient routes.

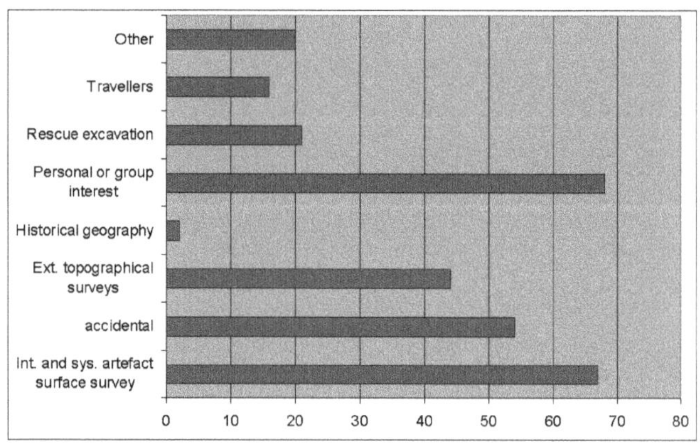

Fig.5. *Graph illustrating the proportion of components discovered within different research frameworks including Intensive and Systematic Artifact Surface Survey (Tanagra survey project).*

The ratio of known Prehistoric to Greco-Roman components is 74 to 154 (1:2.1), while among the historical periods, 58% are dated Archaic to Hellenistic, 22% Roman –Late Roman, and 20% are attributed to the general Greco-Roman period. The elevated presence of components of Prehistoric date is certainly due to the presence in the *chora* of the famous LH burials with larnakes (see above) which have led to research aiming to uncover Mycenaean cemeteries and sites and have drawn the interest of prehistorians.

[11] For further bibliography see appendix I.14.

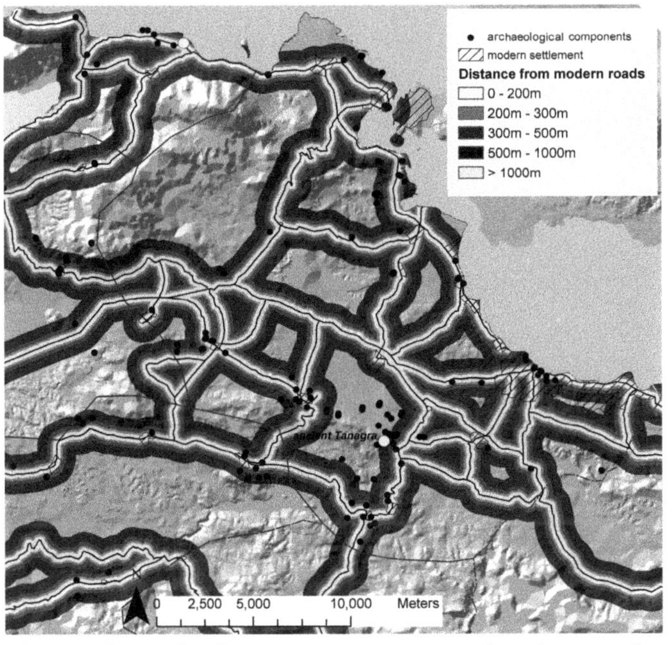

Fig.6. *Relationship between components and modern road network.*

ANALYSIS OF THE *CHORA* LANDSCAPE

PREHISTORIC PERIOD

The archaeological record available in Eastern Boeotia for Prehistoric periods is quite large.

For the Neolithic period, activity foci are known at the site of ancient Tanagra (*component TA_1*), at Ag. Konstantinos (*component TA_19*), at Proph. Ilias (*component T_50*), in the area of the Agios Thomas village (*component TA_42*), at Drytsa-Pyrgos (*component TA_75*), at Lithosoros (*component TA_119*), at Glypha (*component TA_134*), at Dramesi/Paralia Avlidos (*component TA_169*), and probably at Kokkali (*component TA_9*); and just a presence at Oinophyta (*component TA_45*). In an area close to the Asopos river, intensive survey work discovered an interesting extensive settlement pattern from the Neolithic period (see appendix I.14 – the Tanagra survey project) – see fig.5 in chapter II.3.11.

EH is attested at the site of ancient Tanagra (*component TA_2*), at Kokkali (*component TA_10*) and at Ag. Konstantinos (*component TA_20*), at Proph. Ilias (*component T_51*), at Bratsi by the Frankish tower (*component TA_58*), at Drytsa-Pyrgos (*component TA_76*) and on the Meg. Kastron hill (*component TA_96*), at Kastron Lykovouni (*component TA_99*), at Ritsona (*component TA_107*), at Vathy-Proph. Ilias (*component TA_148*), at Pharos Avlidos (*component TA_*178), at Bathiza (*component TA_195*), as well as probably in the area of the Agios Thomas village (*component TA_41*) and at Dritsa-Proph. Ilias (*component TA_95*). The EH presence at Oinophyta (*component TA_46*) is doubtful, while the burial site at Kesseri (*component TA_43*) is related to a probable settlement nearby. Important settlement sites are known at Lithosoros (*component TA_120* – EH I), on the contemporary Drosia-Ag. Minas promontory (*component TA_128* – EH I), and at Glypha (*component TA_135*) in the Drosia plain to the NE, as well as at Dramesi/Paralia Avlidos (*component TA_170*). EH presence is always situated on prominent sites, even if they are often quite low – see fig.6 in chapter II.3.11.

Generally speaking, MH Boeotia shows fewer sites than the EH period, and this is due to a preference for a nucleated rather than dispersed settlement pattern (see fig.5 in chapter II.3.11 for Tanagra, and also the *chora* of Thebes – for a different opinion see Fossey 1988 vol.II). MH nucleated settlement is known in elevated locations at Ag. Konstantinos (*component TA_21*) and at Drytsa-Pyrgos (*component TA_77*), and in prominent locations at Lithosoros (*component TA_121*) and Glypha (*component TA_136*), in the Drosia plain. The important settlement site at Glypha constitutes a parallel with the sites at Salganea and Dramesi along the Boeotian coast and with that at Magoulas and Amarinthos Euboea (AD 43 1988 (1993): 203).[12] A probable settlement site is known at Ritsona (*component TA_108*) and at the site of ancient Tanagra (*component TA_3*). Activity foci are reported at Bathiza (*component TA_196*) as well as at Proph. Ilias (*component TA_52*), while MH tombs 1.5km from the site (*component TA_55*) might be related to another settlement – see fig.7 in chapter II.3.11.

LH settlement is known on the Ag. Konstantinos hilltop (*component TA_22*), at Drytsa-Pyrgos (*component TA_78*[13]), at Kastron Lykovouni (*component TA_100*), and in lower locations at Proph. Ilias (*component T_53*), Ritsona (*component TA_109*) as well as Lithosoros (*component TA_122*) and Glypha (*component TA_137*). There is a little LH evidence from the Mikro Vathy (Aulis) area[14] and from the Dramesi area.[15] LH burials are well known in the area of Kleidi, Agios Thomas and Bratsi/modern Tanagra (*components TA_31 and TA_37*; *TA_40*; *TA_61 to TA_64*), most of which produced Mycenaean *larnakes*. Their presence could lead to the existence of settlement in the area, which has not been located yet but is certainly not on the site of Greco-Roman Tanagra (see appendix I.14: Tanagra Grimada and Kokkali sites). The LH burials at Gephyra/Ledeza and Dendron/Grava by Bratsi (*components TA_71 and TA_73*) are to be linked to corresponding small settlements next to them (*components TA_71 and TA_73*) – see fig.8 in chapter II.3.11.

---

[12] In general, the EH and MH settlement in NE Boeotia must be seen within the wider framework of Euboean settlement dynamics (among others Sampson AE1974: 5 and 14 and Sacket et al. 1966: 66 n.93). In addition to the sites listed here, a probable prehistoric focus is also known at Rodia (*component TA_132*).

[13] *Component TA_94* at Dritsa-Proph. Ilias could represent halo activities of the settlement at the Dritsa-Pyrgos site.

[14] LH burials (Mykro Vathy N – *component TA_145*), a possible LH fort (Vathy Nisi - *component TA_150*), LH potsherd scatters (*component TA_163*).

[15] LH activity focus and burials (Pharos Avlidos – *components TA_177 and TA_179*).

GRECO-ROMAN ANTIQUITY[16]

Town level

The ancient city of Tanagra is the main settlement of the extensive plain (300 sqkm) to the N of the ancient city that Strabo (IX.2.10) and Pausanias (IX.19.8) mention as *Tanagraia.* The city was located between two rivers, in a well defended spot and central to the Eastern part of Boeotia, of which it was the most important city. The Tanagra site shows occupation from Late Neolithic to Late Roman and even later periods, with a significant gap within mature Late Helladic.[17]

Before the addition of Aulis and the *Tetrakomia* (Harma, Eleon, Mykalessos, Pharai) during the Hellenistic period (Schachter 2004: 60), ancient Tanagra was located roughly in the middle of its *chora*, and it would have been a suitable meeting place for economic and social purposes.[18] After the annexing of NE Boeotia, in the area defined here as the wider *chora* of Tanagra, unlike other *chorai* of Boeotia, other town level sites were present (Aulis,[19] Harma,[20] Eleon,[21] Mykalessos,[22] Pharai[23]), with a certain autonomy throughout the Greco-Roman period, though probably never truly independent *poleis*. In the Classical period, four of these constituted the *Tetrakomia.*[24]

Village level

The wider region of Tanagra and Eastern Boeotia is quite peculiar compared to the other *chorai* of Boeotia. The Northern area is settled according to a system of towns which never reached the status of real *poleis* but worked as nucleated foci within the settlement chambers they occupied. Accordingly, in the Northern area there is no real room for 2nd rank settlements, with the exception of the areas towards the gulf of Euboea: the Chalia/Drosia plain, where attempts have been made to locate *Salganeus,*[25] *Chalia,*[26] *Hyria,*[27] the Mikro Vathy area[28] and the Dramesi/Paralia Avlidos area.[29]

On the other hand, the proper *chora* of Tanagra, which belonged to the city throughout its entire history, would allow for the existence of satellite settlement(s) of Tanagra. The sources mention *Stephon,*[30] *Schedia*[31] and *Gephyra,*[32] while the archaeological record allows for the existence of a settlement site S of Tanagra, just on the

[16] Period maps are included in chapter II.4, figs.18-20-22-24-26-28.

[17] See *components TA_1 to TA_8.* The great 4th C BC plan and rewalling at Tanagra attest to a large and wealthy city in Classical Greek times, and it is perhaps only with the arrival of Roman power in the Late Hellenistic (2nd to 1st centuries BC), when we might expect some radical change in its prosperity (considering the shrinkage of urban areas at cities such as Thespiae and Hyettos and the temporary or permanent cessation of occupation at the town of Haliartos and the village-town of Askra, plus drastic depopulation at rural sites). However, the examination of ceramic material collected at the site seems to indicate a fairly strong Flavian to Severan urban occupation (Poblome et al. 'The Late Hellenistic to Late Roman ceramic spectrum of Tanagra. Approaching the urban assemblage' in Bintliff et al. 2008). Later, there certainly seems to have been a large and busy city in the Late Roman period (ca. 400-600 AD), as the vast bulk of all surface finds from Tanagra belong to Late Roman forms.

[18] As Schachter points out (2004: 60), because of its location, Tanagra was open to a variety of influences from abroad. The Tanagraia, as events of the 5th C show, was particularly vulnerable to attack by land and sea (this is probably why Tanagra was loyal to Thebes for most of the Classical period). Another important area of access to Tanagra was the coast (see Tanagran participation in the foundation of Herakleia Pontike by the Megarians, ca. 560 BC – Herodotus IX 55).

[19] Aulis in Pausanias' time was a village, inhabited by potters, which belonged to Tanagra (Philippson 1951: 498). Most probably it should be located at the Vathy-Ag. Nikolaos site, as supposed by Blegen (1949).

[20] Attempts have been made to identify it with the Kastron Lykovouni site (but see discussion in appendix I.14 – *components TA_105 and TA_106*). The contemporary name *Harma*, following the 'Hellenisation' of village names, is attached to the village formerly called Dritsa (see next footnote).

[21] With occupation from Geometric to Late Roman (Dritsa-Pyrgos / modern Harma village *components TA_79 to TA_85*).

[22] Ancient Mykalessos has been located at Ritsona. The town site shows occupation probably from Geometric to Late Roman, with a gap in the Roman period, as regards surface material reported so far (*components TA_110 to TA_114 and TA_213*). No systematic research has been carried out at the site.

[23] The location of *Pharai* is still uncertain. Candidates are Agios Panteleimon (*components TA_165 and TA_166*), and less probably Bratsi/modern Tanagra village (*component TA_56*) and Asopia-Chlebotsari (see chapter II.1.11 and appendix I.11).

[24] According to Strabo, Pharai was a town belonging to the Tetrakomia.

[25] Karababa has been identified with historic Salganeus (though only Roman can be attested distinctively – *component TA_142*), while 'prehistoric Salganeus' has been located at Lithosoros (for discussion see Wallace 1969: 246ff.; Wallace 1979: 41; Fossey 1988: 80).

[26] For identification issues see Fossey 1988: 77-78.

[27] Fossey (1988: 76) locates *Hyria* (*Iliad* II 496) at the Glypha site, Frazer (1913) at the Vathy-Ag. Nikolaos site (which Blegen 1949 identifies as ancient *Aulis*), while attempts have been made to locate it at Dramesi / Paralia Avlidos, which has also been identified with Homeric *Graia* (Fossey 1970).

[28] In the Vathy area, on the low Vessalas ridge, some evidence mainly from earlier periods may indicate a village site (Vathy-Ag. Nikolaos - *components TA_158 to TA_162*) with associated burials (Vathy - Ag. Nikolaos *component TA_164*). According to Blegen (1949), *Aulis* was situated on the Vessala hill. Also located in the area is the sanctuary of Artemis (see *components TA_151 to TA_155*) and a settlement is known at the sanctuary site in the Late Roman period (*component TA_154*).

[29] On the Agios Panteleimon ridge are traces of a settlement (*components TA_165 and TA_166*) and burials (*components TA_167 and TA_168*).

[30] For *Stephon*, a possible location has been suggested at Bratsi/modern Tanagra (see appendix I.14). The site is mentioned by Plutarch (*QC* 37 [299C]) as first occupied by the founders of Tanagra, presumably a fortified place.

[31] Attempts have been made to locate it on the Ag. Konstantinos hilltop (Fossey 1988: 56 - see appendix I.14), but this location was probably a hilltop refuge site (occupation attested in the Late Roman period – *component TA_29*) rather than a rural village/*kome*. Some others locate it at Bratsi/modern Tanagra (for the question see Fossey 1988: 56).

[32] A village, to judge from the name –raft, or pontoon bridge-near the Asopos, perhaps on the N bank, perhaps occupied by the Gephyraioi: *Etymologicum Magnum* s.v. Γέφυρα.

basis of burials (Agios Thomas and Kleidi area[33]) and the fertile land available. There is very probably not enough evidence for a settlement at Oinophyta (*components TA_47 to TA_49*), presumably only the site of the battle that took place in 457 BC, won by the Athenians.[34] Also, the site at Bratsi could simply be a fort.

Delion, the port of Tanagra with a sanctuary of Apollo, has been located in the area of the modern village of Dilesi. Archaeological evidence seems to indicate that at least in Hellenistic, Roman and Late Roman times a settlement developed around the harbour (*components TA_191 and TA_192, TA_217 and TA_218*). Burials might be an indication for it (*components TA_185 and TA_186; TA_189 and TA_190; TA_215*).

A Late Roman settlement site is known at Ag.Konstantinos (*component TA_29*), probably with a defensive hilltop character.

Rural segment

The archaeological record includes some activity foci which could be linked to rural activities in the Archaic to Hellenistic periods: *components TA_59 and TA_60* (Archaic/Classical at Bratsi, by the Frankish tower), *component TA_44* (Classical/Hellenistic) which could correspond to *TA_223* near Agios Thomas and *component TA_209* (Archaic), as well as probably *component TA_39* (Classical) at Agios Thomas and a building S of Bratsi (*component TA_57*).

The Roman rural landscape is represented by *component TA_210* (Roman) at Dendron Grava and a Hellenistic or Roman rural site at Bathiza (*component TA_197*).

In the NE area, the rural picture is much less known[35]: a Classical rural site which could be represented by the evidence at Lithosoros (*component TA_125*), undetermined evidence (Geometric to Classical) on the Drosia-Ag. Minas promontory (*components TA_129 to TA_131*) as well as activity foci at Glypha (*components TA_138 to TA_141* - Geometric/Classical) and at Rodia (*component TA_133*).

We can distinguish the Late Roman rural sites, which are usually larger in extent: *component TA_227* at Bali and a probable rural activity represented by *component TA_98* at Meg.Kastron. In the NE area, Late Roman is reported at Glypha (*component TA_141*), at Vathy - Ag. Nikolaos (*component TA_156*), and at Mikro Vathy by the Artemis Sanctuary in the form of a small settlement with a bath (*component TA_154*).

Much of the rural landscape structure is known in the environs of the ancient city of Tanagra due to the intensive and systematic survey - Tanagra Survey Project - carried out in the area (see appendix I.14 – THE TANAGRA SURVEY PROJECT, *Greco-Roman Antiquity*). The picture of the rural landscape changes considerably from the Classical/Hellenistic to the Roman/Late Roman period, and changes are related to both the topographical setting and the size of the sites, as well as the distance from the city.

The small farms of Classical and Early Hellenistic date discovered by the survey are all at a distance from the city (from 1.8km and further), as the immediate outskirts of the city were farmed directly by inhabitants of the city itself.[36] The large number, the small size and the distance from the city of the small rural sites indicate a pattern of rural exploitation of the hinterland of a Classical city and possibly reflect agricultural plots. The position of the rural burials confirm the picture, as it reflects the position of family plots of people (see chapter II.3.3) either resident in the city, who were going from it to the fields, or of people living in the countryside in nearby farms, of which some were identified by the survey. The rural sites discovered represent both permanent habitation (TS1, TS25) and seasonal sites (TS11, TS12, TS16, TS17, and possibly TS7).[37] Family farms typical of the Classical period show a preference for a location on hill-slopes that naturally constitute suitable zones for olive trees and cereal cultivation,[38] as has also been noted elsewhere within Boeotia and Greece[39] (for details see appendix I.14 – THE TANAGRA SURVEY PROJECT, *Greco-Roman Antiquity* and fig.22 in chapter II.4). The population living in the rural sites in the Classical period represent only a small percentage (estimated around 20-25%) of the whole *polis* (*asty* + *chora*) population (Bintliff 1997b: 235). The general picture given by the density and distribution of material, which can also be observed in other survey projects in Greece, reflects the political importance that the countryside (*chora*) received in the framework of the city-state (*polis*) system, within which the rural segment was considered as an active and vital part.

In the Roman period, we observe changes in the exploitation of the rural landscape, and in particular in the Late Roman period, when traces of these socio-economic changes become particularly visible archaeologically. Those changes concern site size, distance from the town, as well as topographical location/characteristics, and testify to a different socio-economic system and transformations concerning land

[33] Most likely the settlement would have to be located somewhere between or beneath Kleidi or Agios Thomas, since cemeteries seem closer to the modern village area than to the Ag. Konstantinos site. Intensive survey work (Tanagra survey) between Agios Thomas and Kleidi did not locate any settlement site, and one could even suggest its presence below the modern village of Kleidi.

[34] Thucydides I 108, 2-3; Diodorus XI 81-83, 3; see appendix I.14 and Schachter 2004: 59 n.46.

[35] What is known is mainly just due to interest in Prehistoric sites, and usually the known Greco-Roman components are at the same location as the Prehistoric occupation.

[36] See elsewhere in Boeotia, at Thespiae for intance (chapter II.3.9 or bibliography on the matter).

[37] For a debate about permanently residential or seasonal farms in the Classical-Hellenistic period, see among others Foxhall 2004, Osborne 2001, Pettegrew 2001; Osborne 1992; Snodgrass 1990; Snodgrass 1987: 117-19; Osborne 1985b; Bintliff-Howard-Snodgrass 2007).

[38] To the S of ancient Tanagra, for instance, the small farms and groups of burials, which as suggested could be related to family plots, are located on the higher slopes of the Asopos valley, in the fertile hilly landscape of marls.

[39] See for instance *components T_55 to T_57* – appendix I.12 in the area of Vagia. For other areas of mainland Greece see for instance Cavanagh, W. – Mee C. - James P. 2005: 70 table 2.4 for Laconia.

property which characterise the period (mainly larger properties and specialisation in products). The picture resulting from the survey attests to the preference for lower locations within the valley, in contrast to the higher level on the slopes which constitutes the preferred location during Classical times (see above),[40] as well as a greater proximity to the city site.[41] Generally speaking, the mid and large size Roman rural sites (some of them fairly rich - TS2, TS3 and TS4) in the surroundings of the city reveal the important development of the countryside, while the noted preference for a lower location within valleys and basins[42] reflects the character of specialised and intensive cultivation which characterises the period, most probably grapes, as seems to be attested by the large quantity of amphora sherds[43] which were found at the sites as well as in the city (cf. Poblome et al. in Bintliff et al. 2008). We could therefore hypothesise a network of commercial villa estates aiming at a wider market than Tanagra itself. The main period of occupation of those sites in Late Roman times is in parallel with the peak of occupation of the Late Roman *polis* of Tanagra, in which one can note high activity from the 4th to the 6th C AD (for details see appendix I.14 – THE TANAGRA SURVEY PROJECT, *Greco-Roman Antiquity*).

Burial areas

Some burial areas in the vicinity of the Tanagra city site may be related to the city, such as the burials on Kokkali hill,[44] or in the immediate surroundings of Tanagra (*components TA_14 to TA_18; TA_199 to TA_207; TA_221 and TA_222*). On the other hand, some of those components could also represent cumulative cemeteries linked to rural settlements, along main roads (see Mesogeia), as *components TA_97, TA_212, TA_214, TA_180, TA_219* in the wider region might also do. Intensive and systematic artefact surface survey in the surroundings of Tanagra (see appendix I.14 – THE TANAGRA SURVEY PROJECT) discovered many burial areas which correspond both to city cemeteries and to burials connected with the rural occupation of the landscape (see RURAL SEGMENT above and the archaeological maps with survey sites fig.4[45]). In particular, none of the burial areas located in the S survey transect are on the natural roads from the city but they are rather located higher on the hill slopes and most probably linked to family plots and farms (see appendix I.14 – THE TANAGRA SURVEY PROJECT, *Greco-Roman Antiquity*).

The burials in the surroundings of Bratsi may also be linked to the rural occupation of the landscape (*components TA_65 to TA_67*) rather than to a settlement site (see above and appendix I.14). Moreover, burials that seem to create a circle around the village of Kleidi (*components TA_32 to TA_36; TA_38 and TA_228*) have been related to the Ag. Konstantinos site, but could be linked[46] to a settlement probably located further S, as yet undiscovered (see above, footnote 33).

The burial picture also indicates Greco-Roman burial places linked to town sites, such as at Dritsa-Pyrgos *components TA_86, TA_87, TA_89, TA_90 to TA_93 and TA_211* linked to ancient Eleon, and at Ritsona Kamileza *components TA_115 to TA_118* linked to ancient Mykalessos.

In the plain of Drosia, possible burial activities (*components TA_126 and TA_127 and component TA_143* at Karababa) constitute the sole archaeological evidence there for the Greco-Roman period, along with the Classical rural focus –*component TA_125* (mentioned above). Burials in Mikro Vathy (Aulis) area (*component TA_164*) and in the Dramesi area (*components TA_167 and TA_168*) could be linked to probable nucleated settlement exploiting the area.

Cult places/Religious areas

The sacred landscape of the region is mainly represented by the Classical to Roman Artemis Sanctuary at Mikro Vathy (*components TA_151 to TA_153*), an Archaic? sanctuary by Kleidi (*component TA_208*), and the Apollo sanctuary in the area of Dilesi, for which candidates are mainly *components TA_183 and TA_184* (on the Palaiochori ridge) and *component TA_187* (on a hill above church of Ag. Paraskevi). A possible cult place is at Agios Konstantinos (*components TA_23 and TA_24*).

Forts and fortifications

A Greco-Roman fort is known at Mikro Vathy-Meg. Vouno (*component TA_144*) and must be seen in connection with the Anaphorites fortified line, which probably marks the boundary between Boeotia and Chalkis when the latter had control of the plain of Chaleia below (see appendix I.14).

Bratsi (*component TA_56*) probably may also simply be a fort of the Hellenistic period rather than a fortified settlement, as well as the fortified hilltop of Kastron Lykovouni (*components TA_102 to TA_104*). A Greco-

[40] The location of the discovered sites (farms or villas) of the Roman/Late Roman period attests to the preference for these zones, in the S transect as well as in similar topographical settings/locations to the N and W of the city (fig.26 and fig.28 in chapter II.4).

[41] The same behaviour can be found at the outskirts of other Boeotian *poleis*, such as at Thespiae (see Thespike *chora* – chapter II.3.9, and discussion in chapter II.4).

[42] The preference for lower locations in this period can be noted also elsewhere in Boeotia according to the archaeological record available (see *components T_146 and T_147* in the area of Vagia, the villa sites in the lower Kephissos valley by ancient Chaironeia, as well as *component AK_53* by Akraiphia.

[43] Mainly large amphorae, locally made, normally reserved for long-distance trade.

[44] Those are the well known cemeteries which produced the terracotta figurines called *Tanagrines* (*components TA_12, TA_13, TA_198*).

[45] In particular, to the W of the city, by the city wall, for city cemeteries, and in the N and S survey transects for rural cemeteries.

[46] With the exception of the possible Late Roman burial site (*component TA_30*) which could be related most probably to the settlement at Ag. Konstantinos.

Roman fort is also reported at Vathy Nisi - *component TA_149*.

Other activities / unspecified activity areas

Ancient quarries are known in the area (*components TA_220 and TA_228*), as well as signs of activity interpreted as a pottery workshop S of Dilesi (*component TA_188*).

Harbour installations are known at Dilesi, where the ancient port of Tanagra (*Delion*) must be located (*components TA_181 and TA_182; TA_193; TA_216; TA_191 and TA_192*).[47]

Unspecified activity areas can be located in the surroundings of ancient Tanagra (*components TA_224 to TA_226*).

LONG TERM SETTLEMENT TRENDS IN THE CHORA *LANDSCAPE*

The ancient city of Tanagra is the main settlement in the basin of Tanagra, built at a strategic point. It controls two fertile plains, Laris and Asopos, as well as the hilly country of Tetartogene, mainly southwest of the city, where at heights of up to 200m are situated the present day villages of Ag. Thomas and Kleidi. Before the addition of Aulis and the *Tetrakomia* (Harma, Eleon, Mykalessos, Pharai) during the Hellenistic period, ancient Tanagra was located roughly in the middle of its *chora*. This area is divided today among four modern villages which go back to the Ottoman period. Modern Tanagra (formerly Bratsi, recorded 1506-1570) is located on the slopes of Soros, whose Eastern spur is the site of ancient Tanagra, overlooking the wider Tanagra plain. Modern Schimatari (recorded 1466-1642) is NE of the ancient city, controlling the plain north of the city but extending its territory to the sea to include the modern harbour of Plaka Dilesi, the harbour of ancient Tanagra. The hilly landscape and valleys south of the ancient city fall into the territory of two modern villages, Ag. Thomas (formerly Liatani recorded 1466-1688) and Kleidi (recorded 1506-1642). It is interesting to note that the borders of these modern village territories fall at the location of the ancient city, thus dividing the large plain which was controlled by the ancient city. Here, at the foot of the ancient city, the medieval site of Agios Thomas, also the site of a Frankish tower, could represent a possible continuation from the ancient city. In antiquity this large area, in which several settlement chambers are formed, would not have been exploited solely by the city.

In the immediate vicinity of the city, the systematic Tanagra survey located small farms of Classical and Early Hellenistic date that appear from a distance of 1700m from the city wall, once again showing that the immediate surrounding of the city were farmed by inhabitants of the city itself. With slight discrepancies in attributing patterns which we will not discuss here, this matches what has been claimed by Snodgrass (1987/9) concerning Classical farming around the Boeotian cities. Finds such as the loom-weights, lamps and cooking pots indicate more-or-less permanent habitation. The distribution of these sites and of the small burial components associated with them is indicative of the practices of land use. The small size and the distances from the city indicate a pattern of rural exploitation of the hinterland of a Classical city and possibly reflect agricultural plots. Especially in the southern transect surveyed (see map in fig.4, and figs. 22 and 24 in chapter II.4), characteristic is the existence of a site every 100 metres or so, conforming to the picture of small and medium sized Classical farms in the rural landscape observed in other parts of Boeotia but also in other Greek surveys. No nucleated settlements (hamlet level) were located in the immediate hinterland of the city, as attested around Thespiae and Thebes – see chapters II.3.9 and II.3.12, but only individual farmsteads.[48]

At greater distances from the city, the existence of second order villages is postulated by the ancient sources (see above VILLAGE LEVEL). The medieval settlement pattern also indicates the existence of several settlement chambers within the territories of modern villages or those of the Ottoman villages, which seem to reflect a settlement pattern of rural nucleations. In the area of the Ottoman village of Ginosati (predecessor of the modern village of Ag. Thomas, formerly Liatani), on the upland plateau and hills SE of the ancient city, the Tanagra survey located a series of Middle Byzantine villages about every 1500m, exploiting this agricultural landscape. The Tanagra survey located small farms of Classical/Early Hellenistic date in this area, as well as a larger rural estate of the Late Roman period.

The area N/NW of the ancient city was settled in antiquity according to a system of towns which operated as nucleated foci within the settlement chambers they occupied, with a certain autonomy throughout the Greco-Roman period though probably never truly independent *poleis* (Aulis, Harma, Eleon, Mykalessos, Pharai – see above). Some of these settlement chambers form foci of settlement in the modern and Ottoman periods. Modern Harma (formerly Dritsa and by the site of ancient Heleon – see above) is recorded from 1466 to 1688. The presence of a Frankish tower in the locality suggests Albanian recolonisation of an abandoned Greek village. Modern Eleon (formerly Sphaides), at its W, is also recorded in the Ottoman archives of 1466-1570). At the small modern hamlet of Ritsona, by the site of ancient Mykalessos, ruins of a deserted village could not be identified with any place in the Ottoman archive and could be of a later period.

The area towards the gulf of Euboea also presents a number of settlement chambers, occupied by the modern villages of Drosia (formerly Chalia, recorded from 1506 to 1688) and Vathy. The large number of secondary villages, especially in the territory of Vathy, most of which go back to the Ottoman period (Pharos, formerly Gierali, recorded from 1466 to 1642; Paralia Avlidos,

[47] The prosperity of the harbour in the Roman/Late Roman period fits well with its interpretation as the harbour of ancient Tanagra, quite prosperous in this period.

[48] See figs. 22 and 24 in chapter II.4, and appendix I.14 – THE TANAGRA SURVEY PROJECT, table 2.

formerly Dramesi, recorded from 1466 to 1688; Kalo Chorio, formerly Kribesi, possibly recorded from 1466 to 1642), and the deserted Ottoman villages of Kapandriti and Lavda, as well as the deserted village at Ritsona (see above), indicate the existence of several settlement chambers in this area. Indeed, for this area in antiquity attempts have been made to locate several second rank settlements, such as *Salganeus*, *Chalia*, *Hyria*, *Aulis*, as discussed under the archaeological entries listed in appendix I.14 (see also below), and at Ag. Panteleimon by Kalo Chorio evidence indicating an ancient settlement was found (see below and appendix I.14).

As seen above, the large *chora* of Tanagra constitutes a peculiar case within the Boeotian panorama. Hansen (1996: 105) also points out that the Aristotelian collection of *politeiai* seems to have included a *Constitution of Tanagra* (Plutarchus *Moralia* 299C) which reported a tradition that the territory of Tanagra had once been inhabited *κατὰ κώμας*. In terms of organisation of the landscape, we see a picture of decentralisation, where each *polis/kome* worked as an attraction factor. Several *kome*-type town centres mark the landscape of the Northern area, each dominating a settlement chamber: Mykalessos, Eleon, Harma, Aulis, with the addition of smaller settlement areas, such as in the area of Dramesi/Paralia Avlidos and in the Drosia/Chalia plain. On the other hand, the city of Tanagra dominates the Southern part of the *chora*, the proper *Tanagraia*. According to Schachter (2004: 60), before the addition of Aulis and the *Tetrakomia* (Harma, Eleon, Mykalessos, Pharai) during the Hellenistic period, the *chora* fell naturally into two sections, roughly equal in area, with the site of ancient Tanagra located roughly in the middle.[49] The city dominates a landscape of hills and plateaus, although its natural *pediada* (plain) seems to be that towards Attica.

The *kome*-type centres in the wider *chora* never became politically strong (see above) nor very large,[50] and this can be explained by considering the landscape character, marked by not very good soils, optimal for pine trees but not very good for much else. Even today the landscape is quite empty (the settlement at Ritsona, for instance, which has to be considered the successor of Mykalessos, lies abandoned, and in its place rose a hamlet of 3-4 houses around the deserted village), and settlement along the coast (Pharos, Paralia Avlidos, Dilesi), characterised by large extents irregularly occupied by pullulating houses, should not mislead us, as it is due mainly to summer houses and touristic activities by the coast (see fig.1). This seems to be confirmed by the recent past history: in the Ottoman period pirates controlled the area which was completely abandoned by the historical population, while in Early Modern times bandits took over the area, still unsafe for living. On the other hand, the city of Tanagra also never grew particularly large (the walled area is 31ha,[51] and in some periods it extended into the extramural area[52]), despite the large area of fertile low hills and plateaus it controlled, and its extended *chora* in some periods of history. This may be due to the fact that its territory was continously under threat from Thebes,[53] whose power was always greater throughout the Greco-Roman period. This tension both allowed the smaller communities to pursue their autonomy and stopped any centre from gaining leadership. Within the *chora* the balance was maintained by different factors, both historico-political and geographico-topographical. Between central Boeotia, dominated by Thebes, and Eastern Boeotia, there is only a band of hills running in an EW direction. Thebes lies at the edge of this line of hills, which, however, do not mark a natural division, and therefore the landscape is not characterised by a natural logic of space and areas of influence can only be set by arbitrary divisions. The settling of Thebes in that particular location somehow conditioned the position and size of the other settlement in the wider area of central and Eastern Boeotia, as pointed out earlier (chapters II.3.11 and II.3.12).

Several settlement chambers can be detected in the area, where living settlement can be noticed (even if usually shifted in location) from Late Prehistory until today. The examination of landscape data and the archaeological and archival record, as well as a cost-distance analysis[54] (fig.7) lead us to distinguish them as follows:

[49] This may explain at least in part the choice of Grimada as the urban centre of the *chora* (Schachter 2004).

[50] Generally speaking, the settlement chambers of the towns of Eastern Boeotia have never been well investigated (for instance, for Ritsona we know mainly only the rich cemeteries).

[51] The city size is comparable with Hyettos and Haliartos.

[52] As the Tanagra survey project is discovering (see Bintliff and Slapšak's preliminary report in Pharos 2007).

[53] The smaller communities in the plain towards Thebes (Mykalessos, Eleon, Harma) were threatened in particular, and for a long time they were under the influence of Thebes. At some point Tanagra even lost control of the port of Delion. This is why in this area the small *poleis* (joined in the *Tetrakomia*) managed to provide stronger resistance than in other regions (see the case of Askra in relation to Thespiae for instance, or the absence of town level settlement in the wider area of Thebes). Massive defensive walls, in many cases since the Archaic period, attest the effort of these small communities to demonstrate their autonomy and their pursuit of *polis* status in front of both Tanagra and Thebes. Mykalessos, for instance, retained limited autonomy at least as far as control over its small territory was concerned.

[54] See chapter II.3.1 – LONG TERM SETTLEMENT TRENDS.

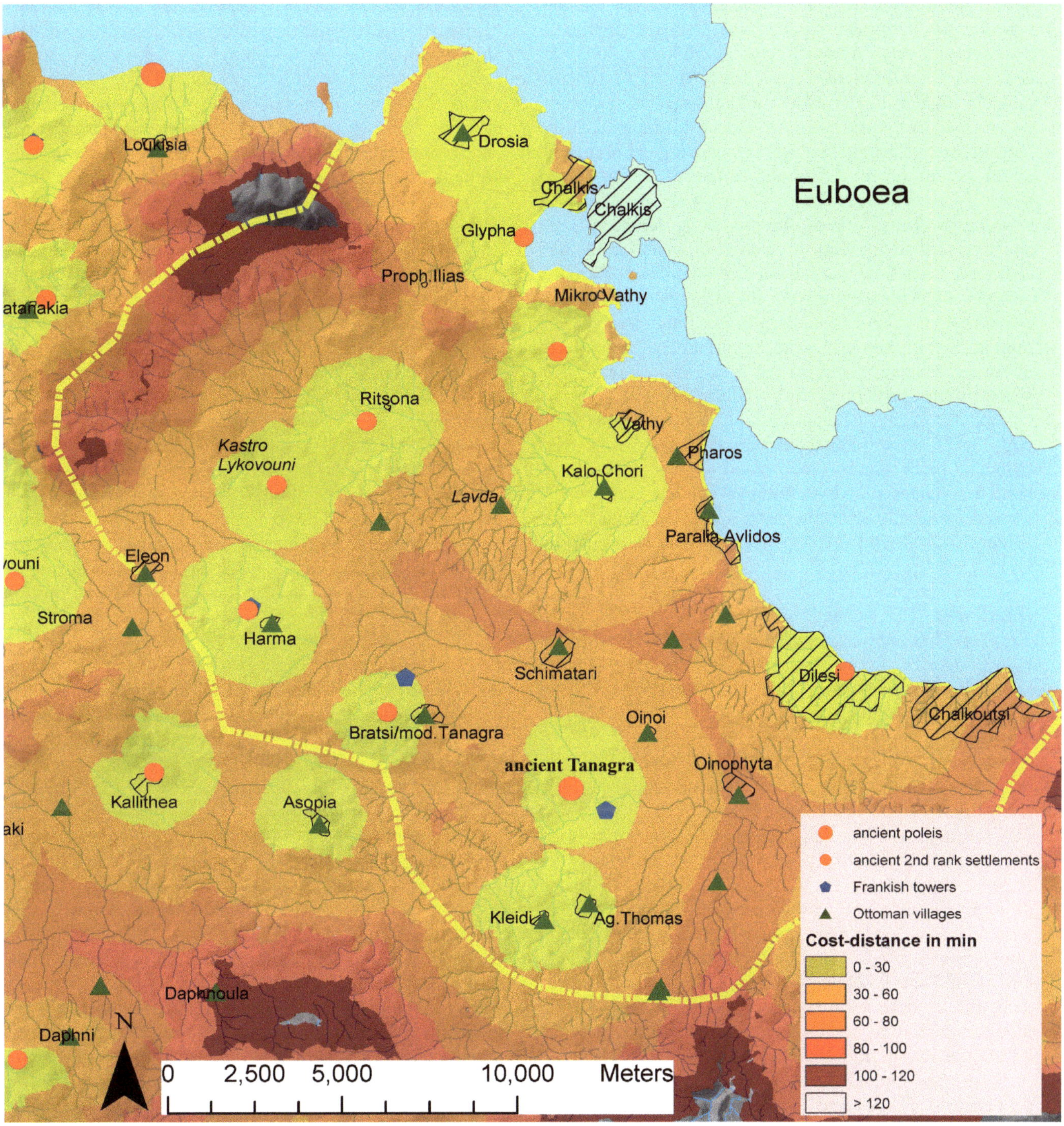

Fig.7. *Classified surface representing the cost-weighted distance (1/2 h walking and further ranges) from recognised 1st and 2nd rank ancient settlements (represented by larger and smaller dots). Areas without dots indicate potential settlement chambers. Ottoman villages and Frankish towers have also been added to the map to show their spatial relationship with the Greco-Roman settlement network and to appreciate potential settlement chambers.*

Tanagra area. The location of Grimada, site of the ancient *polis* of Tanagra, was occupied probably from the Neolithic, and certainly during the Bronze Age, while in the Late Bronze Age, settlement foci are elsewhere in the area (Bratsi, Agios Konstantinos-Agios Thomas).[55] The Grimada location is then occupied by a town level centre from the Geometric until the Late Roman period. In the Medieval period, the Grimada location is abandoned, in favour of locations some of which were already settlement foci in the late Prehistoric period (Bratsi, Agios Thomas, Kleidi - Ginosati in Medieval times, see Vionis 2006). For the Medieval (Frankish) period, there are remains of towers with associated settlements to the immediate E of ancient Tanagra (Ag.Thomas church) as well as to the NE of Bratsi village. In Ottoman times a hamlet was located on top of the earlier acropolis ridge

[55] The Grimada location was not the Mycenaean centre, which was, however, nearby, since we do have a lot of rich Mycenaean cemeteries in the environs.

within the walled area of the ancient city.[56] In the 15th century AD Ottoman records, the Tanagra district is covered with small Albanian hamlets, including the direct ancestor of the modern village, today renamed Tanagra after the ancient town nearby, but until recently retaining its original 'Arvanitic' name of Bratsi. In the area of the Bratsi/modernTanagra village, archaeological evidence could permit (even if without certainty) the existence there of an ancient settlement[57] as well as of a Medieval (Frankish) focus.[58] We observe both recurrent and shifting settlement choices within a wider settlement chamber, somehow natural in an open area like the *Tanagraia*, with fertile soil throughout, and without physical constraints. In modern times, Schimatari, in the middle of the plain and at an important road junction,[59] became the actual centre of the area, with several other second rank settlement foci (Bratsi/Modern Tanagra, Oinophyta, Agios Thomas, Kleidi and smaller Oinoi, with the Dilesi and Chalkoutsi coastal sites), some with Ottoman predecessors (see above). In the area of the modern villages of Kleidi and Ag.Thomas can also be located a second-rank settlement, according to the burial evidence all around the two village hills (see appendix I.14).[60]

Dilesi area. The ancient harbour of Delion (harbour of Tanagra) has been located in the area of modern Dilesi/Plaka Dilesi, in association with the Apollo sanctuary (see appendix I.14). A settlement grew up around these activities.

Eleon and Harma area. Ancient Eleon has been identified with the rich remains of an ancient town on a rise to the NW of the modern village of Harma. On the other hand, attempts have been made at identifying ancient Harma with the poor remains on the top of the Kastri Lykovouno ridge. In the map showing the results of a cost-distance analysis one can see that the latter spot would fill a settlement gap between ancient Eleon (modern Harma – occupied both in Frankish and Ottoman times) and Mykalessos (modern Ritsona). The land available to these town settlements was not very fertile, though agricultural exploitation can be carried out, as the traditional use of the landscape shows. This aspect, as well as their border location between the spheres of influence of Thebes and Tanagra, could help to explain the quick advantage that Tanagra took of them already in the Hellenistic period (see above).[61]

Ritsona area. Ancient Mykalessos lies in a remarkably distinctive micro-landscape, surrounded by more-or-less high mountains, which nowadays gives the impression of how the countryside must have looked like in Classical times. We know, however, that the landscape character changed many times, and the area was invaded by forest and wild vegetation after antiquity, being abandoned and at the mercy of pirates in the Middle Ages, and bandits in 17th and 18th century (see above).[62]
The site lies in an area characterised today by what may have been a typical 'Classical' landscape. Around the ancient town are fields in the plain, all the way to the mountains. At the end, where the higher mountains begin, they border the forest. On the slope, every now and then, there are vineyards and olive trees. On the other hand, despite the large basin available, the poor limestone and conglomerates lowlands which mark the area would never support anything larger than a *komopolis,* dependent on Tanagra since the Hellenistic period (see above).[63] In the post-Roman period we do not have any real centre after Mykalessos (Bronze Age site, and then early *polis*[64] absorbed by Tanagra). In the Ottoman archives we cannot find any corresponding record, while a deserted village is visible today by the tiny hamlet at the small road junction[65] in the Ritsona locality below the ancient acropolis.

Chalia/Drosia plain. There is little evidence for ancient occupation in the area, especially in earlier periods (Geometric to Classical), though mainly rural and burials without any identified settlement focus (see above RURAL SEGMENT, BURIAL AREAS). Several attempts have been made to locate the settlements of *Chalia, Hyria, Salganeus* in this area (see above and appendix I.14). In Prehistoric times the main settlement (*Salganeus*) was at Lithosoros, while in historical times it could have been at Karababa (see above VILLAGE LEVEL). Bintliff's Thiessen polygon analysis on Greco-Roman villages of Boeotia (Bintliff 1994b fig.20) includes three settlement foci in the area: Lithosoros, Chalia/Drosia and Glypha. These settlement chambers, in close proximity to each other, seem to be occupied in different periods, with shifting of the main settlement, or at least not always the three of them contemporaneously (also in the Ottoman period we have evidence for only one settlement, at Chalia/Drosia – see above). The archaeological evidence from Chalia/Drosia and Karababa is very poor (see appendix I.14), but might refer to the same settlement chamber, and we might suppose the existence of a settlement focus in

---

[56] A four-longhouse hamlet found during survey in 2001 (Farinetti – Sigalos 2002 and Bintliff et al. 2001).

[57] Attempts have been made at identifying it with ancient *Pharai* mentioned in ancient texts as a member of the *tetrakomia* (see appendix I.14 for discussion).

[58] A Frankish focus is attested by the presence of a tower.

[59] The Schimatari road-junction is in a strategic position for the market of Boeotia.

[60] See discussion under the site of Ag. Konstantinos (appendix I.14).

[61] The massive city walls by which the *town* of Eleon was surrounded, as still visible today, clearly indicate their twofold function: defensive, as is natural, but also symbolic, in order to visibly express a *polis* status which it never entirely had. This character would probably join Eleon with Mykalessos, Harma and Pharai, although the archaeological record is less rich in these other cases.

[62] In 1934, the wide basin of Ritsona was full of bushes (Philippson 1951: 515).

[63] Pausanias notes that the men of Tanagra use the land of the Ritsona basin (in addition to other places, i.e. ancient Harma area) for grazing (Pausanias I 34.2).

[64] Flourished in the mid 6th C BC (rich necropolis) and declined in the 5th C BC. Destroyed during the Peloponnesian war, but was still alive in the 4th C although under the control of Tanagra.

[65] Between the Thebes-Chalkis and Thebes-Vathy roads.

the area. Moreover, the evidence at Lithosoros indicates a settlement focus only in the Prehistoric period, while in historical times the site was probably only rural. In the cost-distance analysis (fig.7), three settlement areas were included, centered on the main known sites (Glypha, the area of Chalia/Drosia, and Karababa), simply to investigate a possible general settlement structure.

Mikro Vathy (Aulis) area. In the area between Mikro and Megalo Vathy[66] lay the sanctuary of Artemis, to which a small settlement could be associated. The Aulis settlement, however, which in Pausanias' time was a village inhabited by potters and belonging to Tanagra (Philippson 1951: 498), might be located in an inner location, at the Vathy-Ag.Nikolaos site. In the Late Roman period a small settlement/villa site took over the sanctuary. All these activities refer to the same settlement chamber, indicated in fig.7 as a result of a cost-distance analysis on Greco-Roman settlement.

Dramesi / Paralia Avlidos area. In Prehistoric times the settlement focus was at Dramesi/Paralia Avlidos (see appendix I.14). In Greco-Roman times a probable settlement is to be located at Ag. Panteleimon,[67] in association with burials (see appendix I.14), which would mean a shifting in the main settlement location from the coast to the interior. If the evidence represents a settlement site its location would fit a natural agricultural block suitable for sustaining a village. The settlement chamber would extend to the coast, allowing the settlement to exploit the coastal Vathy plain (see below - Ritsona delta), which in Prehistoric times would have had an important settlement at Dramesi/Paralia Avlidos,[68] while today the agricultural wealth supports four villages (Dramesi/Paralia Avlidos, Vathy, Gierali/Pharos, Kribesi/Kalo Chorio) as it did in the Ottoman past (Lavda, Dramesi, Gierali, Kribesi - see above and fig.7).

To the S of the *chora*, the upland Skourta plain constitutes a long-lived settlement chamber whose history must be linked with its border position between Boeotia and Attica (see above *–borders*, and appendix I.14).
In the wider region, possible settlement gaps can be noted in the area of Oinophyta (where, though, there is very probably not enough evidence for a settlement - see above VILLAGE LEVEL), as well as in the area to the NW of Kastri Lykovounou and Ritsona, where, however, the topographical characteristics (E fridges of Messapion ridge and land marked by low fertility) as well as the existence of the nearby settlement immediately to the E, would seem to preclude the existence of other settlement foci.

In conclusion, by looking at the settlement history and comparing the settlement situation with the LH, we notice some interesting elements arising after the general depopulation phenomena which had characterised the period immediately following the fall of the Mycenaean kingdoms. The general picture of the Geometric period is that of nucleated settlements, where we see the remnants of different disperse communities gathering together in certain centres, as already suggested by Fossey (1988). Also some new centres were created, as is the case for Tanagra, which gathered all the people from the neighbouring settlements (especially those along the Kerykion ridge, at the end of which Tanagra lies), and for Chalia-Agios Minas, along the NE shore. Moreover, the majority of the sites showing possible or very limited occupation during LH become settlements during the general recovery of the Geometric period (fig.8 in chapter II.3.11 and fig.18 in chapter II.4). In the Archaic period, after the recovery, the landscape fills up with settlements. The land is divided to create the apparently 'natural' mosaic of the territories of the emerging *poleis* (Bintliff 1994). Part of this pattern is comprised of the town level sites of Aulis, Harma, Eleon, Mykalessos, Pharai (see above) – fig.20 in chapter II.4. Moving to the Classical/Hellenistic period, the survey results have proved Tanagra to be at its flourishing peak (like the majority of Greek *poleis*). The great 4th C BC plan and rewalling attest to a large and wealthy city. As a consequence, the enlarged city territory overcomes the territories of the individual smaller towns, which retained a certain autonomy but never became truly independent *poleis* – figs. 22 and 24 in chapter II.4. In contrast to ancient writers' emphasis on the demographic and economic decline of Southern Greece by the era of the Early Roman Empire, indicated also in the shrinkage of urban areas observed in the Boetia survey, and in several other surveys in mainland Greece, the city of Tanagra shows signs of a fairly strong occupation in the Roman period as well.[69] This is in accordance with Strabo's (IX 2.25) consideration of Tanagra as a relatively well-to-do community. In the Late Roman period (ca. 400-600 AD) there seems to have certainly been a large and busy city, as the vast bulk of all surface finds from Tanagra belong to Late Roman forms (Bintliff and Poblome et al. in Bintliff et al. 2008). As for the port of Tanagra, ancient Delion, Strabo (IX 2.7 (403)) has it as a 'small town', while in the 2nd C, Pausanias (IX 20.1) seems to imply that it was just a sanctuary. The main peak of Delion, in fact, as of the other Eastern Boeotian harbour of Anthedon, is attested in the Hellenistic and Roman periods (AD 1981: 193). On the other hand, regarding the other town centres examined above in the wider area under the influence of Tanagra in Hellenistic times, Mykalessos and Harma are both known as having been deserted by Roman times. This would attest a decay of

[66] Megalo Vathy must have had importance in Prehistoric and Protohistoric period, as a harbour site.

[67] A possible interpretation of the site as a settlement and an identification with ancient *Pharai* is discussed in Fossey (1988: 97-8).

[68] The LH site has been identified with Homeric *Graia* (Fossey 1970: 3-22). On the other hand, Blegen (1949) felt this was ancient *Hyria*, disagreeing with Frazer's (1913) attribution of this to Meg. Vouno area (Vathy - Agios Nikolaos site, see above).

[69] The examination of ceramic material collected seems to indicate a fairly strong Flavian to Severan urban occupation (Poblome et al. in Bintliff et al. 2008).

the nucleated centres in Roman times, as the abandonment of the sanctuary at Aulis in Late Roman times and the establishment of a wealthy rural estate would also suggest - fig.26 in chapter II.4.

II.4

Exploring the socio-political and cultural landscapes of ancient Boeotia

The landscape approach applied in the present work, as discussed thoroughly in earlier chapters (chapter I.1 and chapter I.2.2), includes the revised concept of the settlement chamber and combines older landscape approaches, such as the German *Landeskunde* tradition, with more recent approaches, such as the Community area theory and the taskscape concept.

In the chapters concerning individual *chorai*, the approach has been applied in the discussion of and comments on available archaeological data and landscape history within the individual micro-regions. Our *chorai* can be considered micro-regions within the region of Boeotia, within which several community areas can be active (see chapter I.1), with one gravitating on the main settlement centre (the *polis*). According to the same approach, in this final chapter I will examine settlement trends in the wider Boeotia region, considering different levels of landscape occupation and activities.

ANCIENT BOEOTIAN SETTLEMENT

As seen earlier in this work (chapter I.1), while considering micro-regions, settlement chambers can be more or less defined, but they are certainly more easily distinguishable when clearly marked by physical constraints. By examining the hypothesised boundaries of the micro-regions/*chorai*, gravitating around a historical centre at the *polis* level, in the Boeotian regional panorama, one can easily see how some of them are more clearly defined through physical landscape characteristics (the valleys around the Copais basin, for instance, where the lake also played a role in defining boundaries – see chapters II.3.1, II.3.2 and II.3.4 and Farinetti 2008), while others are less easily definable (as in the case of the Parasopia area to the S/SE of the region). Within the micro-regions, we can localise micro-landscapes. Some of these can be considered to possess long term potential with regards to the presence of a nucleated settlement site, or an area of residential activities of some kind; others are not, and one cannot predict which character the site found there might have (see chapter I.1 and *chorai* chapters). Once again, the micro-landscapes are also more easily recognisable when there are marked physical constraints or environmental/physical characters (such upland plateaus, stream valleys, etc.).

As we saw in the individual *chorai* chapters, in some micro-landscapes a settlement can always be detected in the different chronological periods, even if a shift in location sometimes occurs. For instance, in the Late Helladic/Mycenaean period settlements were located mainly on prominent hills, and in the Early Helladic they were located on lower hills, while in the Greco-Roman period settlement/habitation areas became to a large extent nucleated into larger centres, but actually, even if the settlements and other activity *foci* moved within the area, people living in it exploited and used the same land, the same territory, and experienced the same landscape.

Therefore, one should examine archaeological settlement 'continuities' or 'shifts' on a particular micro-landscape rather than on a particular location (see chapter I.1 – THE 'SENSE' OF CONTINUITY). Through the comparison between Prehistoric, Greco-Roman, and Medieval and post-Medieval data (Frankish or Ottoman),[1] and the modern settlement network, we can investigate the complexities of matching archaeological settlement 'continuities' or 'shifts' with population and ethnic continuity. This has been done for potential micro-landscapes (and corresponding settlement chambers) within each single *chora* chapter.

In order to recognise settlement trends and individuate micro-landscape and potential settlement chambers through the examination of the modern settlement pattern, I have examined the distribution of the modern villages (GIS layer: *choria*) as well as the boundaries of modern administrative divisions (GIS layer: *koinotites*), in an attempt to progress from the actual situation (the 'known') to the archaeological reconstruction (the 'unknown').

In the map in fig.1 it can be noted that modern *koinotites* (i.e. areas administratively belonging to a village[2]) include different landscape zones, such as different geomorphological and agricultural zones, parts of plain areas and of valleys, of foothills and of inner areas of mountains. In exploring boundaries, it can be noted, for instance, how rivers play an active role as physical constraints: the Asopos river acts as a border line for the *koinotites* gravitating towards its valley, and so does the Melas (in another area of Boeotia, N Copais). Hillcrests

[1] For medieval and post-medieval period the information can come from highly informative historical sources (especially village tax registers for the Ottoman period).

[2] A *koinotita* might include one or more nucleated settlements, the main one of which gives the name to the *koinotita* itself.

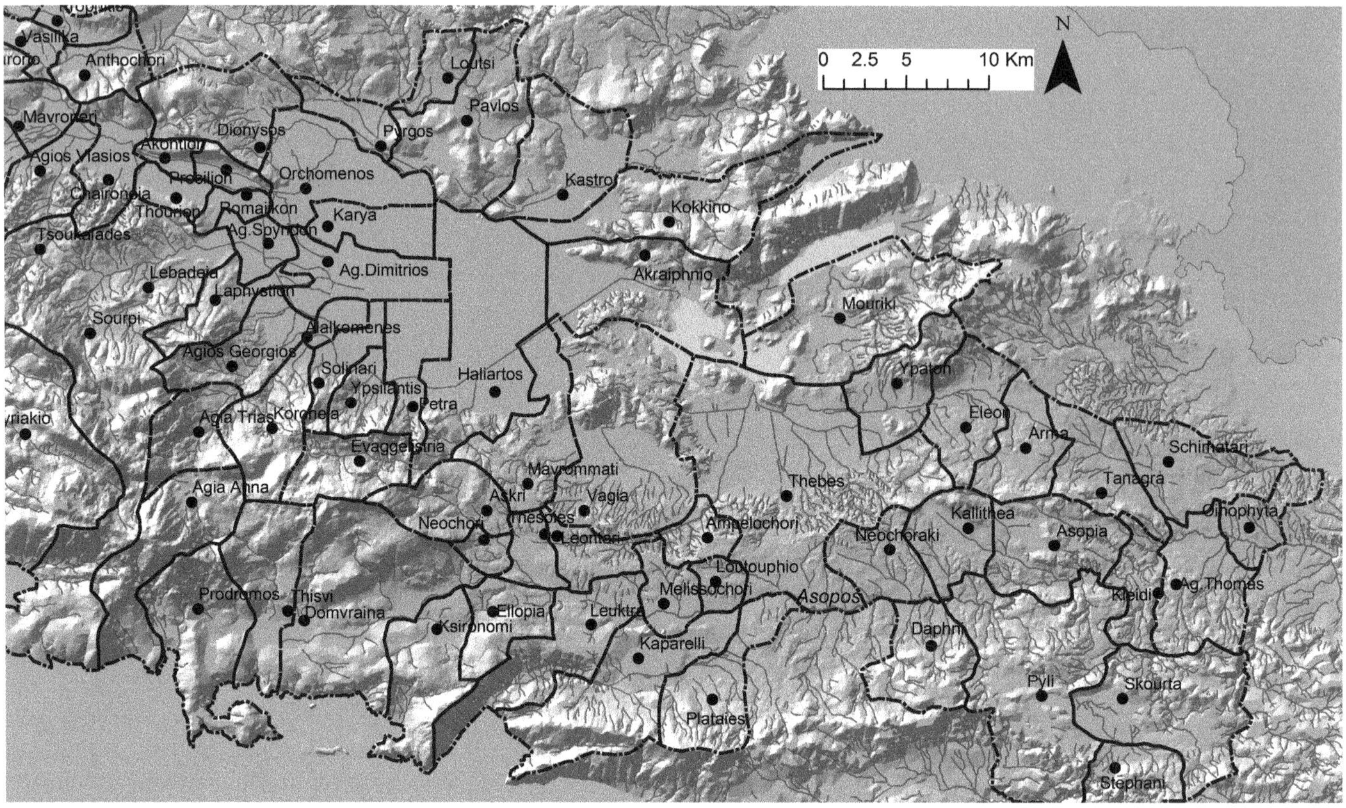

Fig.1. *Koinotites (administrative divisions) and choria (villages) of modern Boeotia, based on the 1960 map updated in 1983 (Hellenic Statistical Service).*

also play a role as boundary lines,[3] as they would probably have done in ancient times. The hypothesised ancient *chorai* border lines usually follow hillcrests – see fig.1 in chapter I.2.2). For the Copais basin, communities gravitating around it would today divide the farming land available after the drainage (according to administrative partitions), while in ancient times, when the lake existed and water covered most of the land, boundaries would have worked in a somewhat different way, and the lake itself would have constituted a physical boundary. The line of this boundary would have greatly fluctuated according to the level of the lake water and the land actually covered by water as well as marshy areas (for discussion on this see chapters on *chorai* around Copais and Farinetti 2008).

Using GIS, I have made an attempt to explore the ancient settlement network of the Boeotia region, especially as far as the relationships between them and their distribution in space is concerned. I have applied cost-surface analysis to different data sets available for Boeotia region, and will illustrate here some results, as well as some indications they may provide about Boeotian landscape issues.[4]

Different ranks of settlement were taken into account, often corresponding to different levels of information. One could note, for instance, how our degree of knowledge about ancient cities (*poleis*) is very high, while it is reduced considerably for towns and villages, and for some areas it can reach a very low level for the rural sites. The general picture of ancient Boeotia that we can obtain from site recognition carried out by means of archaeological, historical and extensive collection of data, has so far proved to be 'naturally' biased by a sort of either city-centred or historical place-oriented or archaeologists' interests-oriented approach. This has resulted in a lack of information concerning second rank settlements and rural sites, and the full range of occupation phases of well known sites.

I will examine here in detail the three main levels of occupation in the Greco-Roman period, focusing on the 1st and 2nd rank settlements, which are better known for the whole region and constitute the main subject of this work.

1st level – *poleis.*

As seen above, the *polis* level is the best known. Historical sources give us names and information on the *poleis* of ancient Boeotia, and the archaeological record concerning them is today quite rich and fairly precise

[3] An interesting analysis on watersheds (through GIS) has been carried out by A. Bevan on Prehistoric Kythera (Bevan 2002).

[4] For the methods employed in performing the analysis see chapter I.2.2. See also the application of the cost-weighted distance analysis in the long-term settlement analysis of each single *chora* (chapters II.3.1 to II.3.14).

with regards to the period of occupation, even when the city sites were not investigated by means of excavation. The location of the majority of the Boeotian *poleis* is known by means of historical, epigraphical and/or archaeological sources, and in general it can be noted that Boeotian cities were situated in a location where they could exploit their own fertile land, and usually between mixed land types.[5]

In fig.1 of chapter I.2.2 can be seen the *poleis* of Boeotia and the hypothesised extension of their territories, according to historical, epigraphical, and archaeological sources, as well as topographical criteria. The division into 'districts', which characterises the Boeotian political landscape, would seem to find its origin in Boeotian history itself and the reasons for its particular application. The first historical sources group the inhabitants of Boeotia mainly by large areas, corresponding to the districts where they settled. Boeotians, coming from the N, would have settled first in the W area of the Copais basin, then the Thespiae region, and only afterwards the Asopos area (Buck 1979: ch.5; Van Effenterre 1989).

As Buck (1979: 91) points out, most of Boeotia was parcelled out among various Boeotian communities by 950 at the latest; by the end of the 9th C BC several nucleated centres had been formed, often by synoecism of local villages at a convenient (and often Mycenaean) site, sometimes as continuations of Mycenaean centres (Thebes and Orchomenos, for instance). These first '*poleis*' included Orchomenos, Thebes, Thespiae, Koroneia, Haliartos, Akraiphia, Plataea and Tanagra. According to Buck (1979: 91), "*various 'communes' or districts composed of villages with no polis, though perhaps with a cult centre and place of refuge, still survived more or less independent units, including as it seems Chaeroneia, Lebadeia, Copae, Askra, the Tetrakomia, and Oropos*".

From 900 to 500 BC archaeological evidence (both for sites and cemeteries[6] - see map in figs.17 to 20) attests a general growth of population and prosperity in the whole region.[7] The development of the Boeotian concept of 'district', the basis of the later confederation(s), i.e. the grouping of various geographically close communities in order to pay taxes, and to cooperate in the supply and naming of representatives, as formalised during the 6th C BC (see below), would seem to find its root in an earlier stage, in particular in the first half of the 7th C, when some centres overcame (or tried to overcome) smaller or weaker neighbours. For instance, at Askra the population was expelled by Thespiae,[8] and Chaironeia was reduced to a '*perioecic*' condition (Roesch 1965), while Potnia was absorbed by Thebes, the larger neighbour city (Strabo IX 2.22). In the same time Thespiae, Tanagra[9] and Thebes enlarged their territories, and certainly the area separating the Copais basin (and maybe also the E part of the Copais itself, from Medeon to Ptoion) and as far as the hills which separate the Theban area proper from Eastern Boeotia (Tanagra area), was under the control of Thebes around 500 BC. On the other hand, W Boeotia was earlier controlled by Orchomenos, which, before the end of the 8th C BC attempted to gain control over large parts of Boeotia, starting from the smaller centres in neighbouring areas, before being stopped by Thebes (before the end of the 8th C). Thus, immediately before and during the period when Hesiod lived (about 700 BC), Orchomenos dominated Chaironeia[10] and Hyettos,[11] Olmones,[12] Hyria,[13] Lebadeia[14] and probably also Koroneia and Copai. Afterwards, probably after the defeat of Orchomenos by Thebes, Lebadeia and Koroneia became independent minor cities.[15] In the early periods of history, the nucleated centres present in the Boeotian 'districts' might be considered as *proto-poleis*, all of which could potentially become *poleis* (see Snodgrass 1980 and Bintliff 1994b). Some of them, however, would remain at the level of small towns/large villages (*komai*[16]), as they were incorporated by stronger centres, which would slowly become proper *poleis*, and would constitute the panorama of Boeotian cities taking an effective part in the 1st Boeotian League (figs.19-20).

---

[5] See for instance the case of Koroneia, Levadeia, Chaironeia, Thebes, Plataea and other Boeotian cities (see individual *chorai* chapters).

[6] More then 50 Geometric sites and more then 90 Archaic sites are archaeologically attested.

[7] Especially in central Boeotia, in the second half of the 6th C BC, as testified by the prosperity of cemeteries (the large and rich cemetery of Akraiphia – see chapter II.3.7) and of the Ptoion sanctuary (dedication of *kouroi* and tripods of high value), as well as of the Theban sanctuary of Kabiroi (rebuilding and restoration – see chapter II.3.12).

[8] The inhabitants of Askra were expelled by the Thespians and found refuge in Orchomenos, probably around 750/700 BC. Sources say that around 650/600 they settled Eutresis (Goldman 1931, quoting Plutarchus *Narr.amat.* 3. Only Fossey reports two Geometric sherds from the site [AE959] – *component TH_86*. In the late 6th C the attention of the Thespians became directed towards the gulf of Corinth and the harbours (Kreusis, Siphae, Thisbe and Chorseiai), but it was too late for Askra to become a real *polis*.

[9] Tanagra absorbed Graia and gained the control over Aulis. In 560-550 Tanagra participated with Megara in the foundation of *Eraclea Pontica*, on the Black Sea.

[10] See Roesch 1965.

[11] Hesiodus *frg.144*; Pausanias IX 36.6. Cf. Wilamowitz 1922: 19.

[12] Pausanias IX 24.3.

[13] See Hesiodus *frg.143*. As an example of *proto-poleis* which never became a real historical *poleis* see the case of Askra (in the *chora* of Thespiae), Chorseiai and Siphai (in the area of Boeotia towards the Gulf of Corinth), and the towns of the *Tetrakomia* in Eastern Boeotia within the political sphere of Tanagra.

[14] See Wilamowitz 1922: 18 f.

[15] Between 550 and 525 as reported by epigraphical evidence (inscription on a bronze helmet found at Olympia - SEG XI 1208; Oikonomos, AE 1925/26: 87-94; Jeffery 1961: 93-95, n° 11; many other scholars, however, accept the date of 447 BC for the inscription, suggested by Larsen 1960 (see Hennig 1974; Buck 1970: 225; Sordi 1968: 68) Jeffery dates it to the third quarter of 4th C BC, and this would testify that in that period Koroneia was at least relatively independent at the same time that Orchomenos was, despite its defeat by Thebes, still relatively strong.

[16] For the concept of *kome* see Hansen 1995.

Thus, the Boeotian idea of the 'district', already developed during the 7th C BC, became effective between the end of the 7th C and the beginning of the 6th C. Around 600 BC, Boeotia appears as divided into several districts, each one including several communities, some of which were about to absorb others. Thebes was culturally influential (see for instance the 'artistic' domination of Thebes visible in pottery styles[17]), politically important, and in progressive expansion for the control of territories, even though not yet the dominant city. Only in the following century (the 6th C) would history assign Thebes the task of unifying Boeotia or, rather, communities of Boeotians (considering the 'ethnic' character of the 1st League – see below).

According to some scholars, the Theban hegemony had already begun by the end of the 7th C BC.[18] According to others, only around the mid 6th C BC did Thebes enlarge its domination over smaller centres throughout the Teneric plain as far as Onchestos.[19] Most probably, it was around 525 BC (or in any case in the late 6th C) that the Confederation (the 1st League, the 'ethnic' confederation) originated,[20] when the full development of the society and institution of the Boeotian League occurred.

The Boeotian Confederation has been long considered an ethnic structure of traditional type, in which the sharing of dialect,[21] cults[22] and traditions (linked to a common origin from Boeotian ancestors – see above) plays a key role. According to this, the organism appears to be quite fluid, leaving to the member cities a fairly large degree of freedom,[23] even as far as the conducting of separate wars. The League was probably first of all a '*symmachia*', although implying other kinds of political ties as well (centralised structure, a sort of a 'state', with federal officers and common coinage), and was mainly an expression and instrument of Thebes' will (Ducat 1973: 70).

For the so-called 1st Boeotian League (504-479 BC) a total of seven Boeotarchs (representative members) was proposed: one each from Thebes, Tanagra, Thespiae, Koroneia, Haliartus, Akraiphia and the *Tetrakomia*. When Orchomenos, Levadeia, Copai and Anthedon joined, they also had one Beotarch each (Buck 1979: 110).

According to some scholars, the Boeotian Confederation loosened at the time of the Persian invasions, while around the mid 5th C are located two attempts to reconstitute it (Larsen 1968). In the so-called 2nd Boeotian League (existing from 447 BC), according to *Hell.Oxy.* (16 (11); text which clarifies the organisation of the Confederation in 395 BC, at the outbreak of the Corinthian War), there were eleven Boeotarchs. Of these, Thebes controlled four units (after having annexed Plataea, always seeking autonomy, and other small centres, probably in the Parasopia area); Orchomenos and Hyettos each provided two Boeotarchs[24]; Thespiae with Eutresis and Thisbe two units; while Tanagra represented one unit only. The other two units were each represented by three small centres that would send Boeotarchs in turn to the League council (Haliartos, Levadeia and Koroneia, and Akraiphia, Copai and Chaironeia, respectively).[25]

Examining the 'natural' history of the Greek *polis* in different historical periods (referring to Bintliff's 1994b article), in Boeotia we observe several processes concerning the relationship between communities on a territorial basis: the conflicts, takeovers and domination processes which characterise the region during the 8th C; the internal domination visible in the takeovers of smaller centres by larger settlements in the 7th C; the domination processes which lead to the swallowing up of the smaller settlements by the major centres all over the region in the 6th C; the phenomenon of the enlarged city territories and subordination which characterises the 4th C.

In the preceding paragraphs, I have tried to delineate general historical trends which worked within Boeotia at the *polis* level. In the following pages, I will discuss an attempt to explore the *polis* landscape of the region by means of GIS analysis, mainly the cost-distance analysis considered above, with special attention to the geographical 'districts' mentioned earlier as the key to the Boeotian League structure.

The result of the analysis can be seen in figs.2-3, where Thiessen/Voronoi tessellation (result of an allocation analysis employing a straight-line/Euclidean distance) is compared with the results of the tessellation of space based on cost-weighted distance from major *poleis*.[26]

---

[17] Coldstream 1968: 367ff; Desborough 1972: 202ff.

[18] Guillon 1943: 66; Sordi 1966: 21.

[19] Ducat 1973: 71; Schachter 1986, s.v. *Poseidon* (Onchestos).

[20] Its birth date "*rejoint ainsi en gros celle de la Ligue du Peloponnèse*" (Ducat 1973: 71).

[21] See Schachter 1966.

[22] See Schachter 1966: 13. See *Pamboiotia* for instance (Ziehen 1949, s.v. *Pamboiotia*, col.288).

[23] On the question of the autonomy of the Boeotian cities see Hansen 1996.

[24] In 447 Orchomenos would probably have controlled three units. Then it would have lost one when Chaironeia was taken from its control (Larsen 1960).

[25] For each Boeotarch the members were entitled to sixty councillors, and for each they provided 1000 hoplites and one hundred cavalry (*Hell. Oxy.* 19. 3-4). – from Schachter 2004: 46. For the organisation of the League see among others Larsen 1955 and 1968.

[26] For the methods employed in performing the analysis see chapter I.2.2.

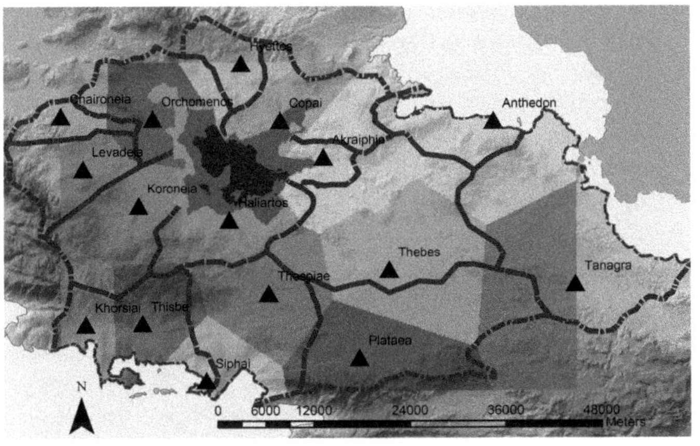

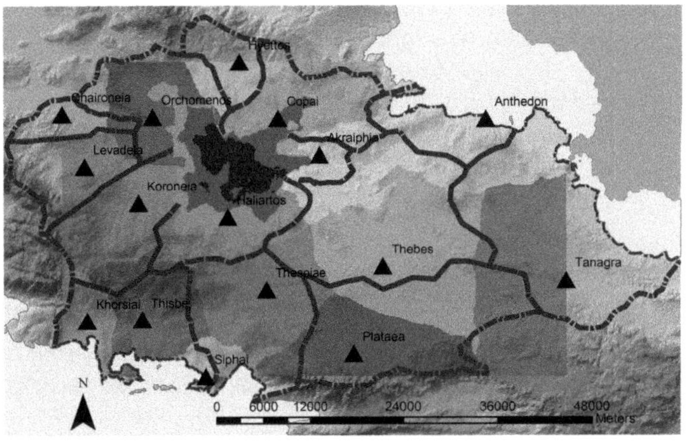

Fig. 2 (left) and fig.3 (right). *Thiessen/Voronoi tessellation (based on straight-line distance) compared with the results of the tessellation of space based on cost-weighted distance from major poleis.*

Some remarks can be made about the second map (cost-weighted allocation fig.3):

- The lakes work as boundaries in the area between Thebes, Akraiphia and Anthedon. Without the lake, the natural cost border line would probably behave somewhat differently.
- In the case of ancient Akraiphia, the vicinity to a major settlement area, such as that of Thebes, also plays a role in the definition of the border.
- The low hills separating Copais from the Theban plains act as borders for cultural/historical reasons, while they would not constitute a barrier as far as natural morphology is concerned, as the application of the cost-distance analysis shows. No physical constraints seem to operate here as border factors. The line of hills functions also as a division between W and E Boeotia.[27]
- As far as the borders between Thebes and the Plataea and the Parasopia area are concerned, the 'natural' territory of Thebes seems to have moved beyond the low hills of the Tertiary *tafel* (see chapter II.3.12) towards Plataea, including part of the Asopos basin/Parasopia. In this case the Asopos river would act as the natural border line, rather than the low hills to the N. The same is true for the boundaries of the modern *koinotites* (see above and fig.1). The modern *koinotita* of Thebes itself deserves a special examination: it today also reaches the Asopos river, as it would have probably done in the past (even if in the past it was larger, especially towards the N and W).
- The territory of Thebes is larger than the others, if we do not take into account the particular case of the wider Tanagra area discussed below. It is also 'naturally' larger and free of any clearly marked physical constraint. The principle of 'one's territory finishes where the territory of another begins' seems here to find its application, especially to the S towards Plataea and Parasopia. In fact, the strong influence of Thebes seems to have been a function since the beginning of the process of 'districts' formations.
- In contrast, the *polis* chambers in the Copais area are all well defined by physical constraints, and a comparison of the 'natural' areas (result of the cost-weighted allocation analysis) with the hypothesised historical boundaries of ancient *polis* territories (*chorai*) shows that the natural and political borders would correspond.
- The Tanagra area and the Easternmost area of Boeotia are characterised by an opening towards the Euboic sea as well as a dominant open landscape, allowing for an expansion along the N-S axis in certain periods of history and a fluctuating border in relation to the neighbouring major city of Thebes. Historical factors at a certain point of history (in particular in the Late Classical-Hellenistic period – see chapter II.3.14) influenced the settlement behaviour which would 'naturally' be represented by an area divided into small *polis* chambers. This could partially explain the tendency to autonomy which Mykalessos, Arma, Eleon and Pharai – the towns/*komai* of the *Tetrakomia* - always had (see chapter II.3.14). Conversely, since the 'incoming of the Boeotians' (as seen above, the last district they 'settled' was the lower Asopos valley and Tanagra area[28]) and in the 1st Boeotian League (Tanagra apparently sent only one Boeotarch, the unique representative of the Eastern part of Boeotia, otherwise unrepresented), Easternmost Boeotia seems to have always comprised a large single unit.[29]

---

[27] Topographical characteristics, rather than well-defined topographical limits, would mark this main distinction within the region. Also socio-political factors, as examined in detail later in the text (*2nd level*). Fossey (1978) speaks about the 'two Boeotias'. In the late Prehistoric and early Historical periods, the division would also work politically: the two main centres of those regions being Thebes in the SE area and Orchomenos in the NW area (see also Farinetti 2003 for the conflict between Thebes and Orchomenos).

[28] The northern part of the area in earlier periods was probably part of the Euboea region (as it was in Late Prehistory).

[29] In earlier periods it probably did not include the N part of the region, gravitating towards Euboea, while from the Late Classical-Early Hellenistic period the cities of the *Tetrakomia*

A comparison between the hypothesised extension of the *chorai* and the supposed extension of the corresponding city sites (according to the extensions suggested for the Boeotian *poleis* by Bintliff 1997 for the Classical period, and based mainly on the walled area) can give us an idea of the actual impact of some *poleis* on the territorial behaviour of the 'districts', which sometimes seems not to take into account the actual land available but to be dictated by other factors, examined in detail in the individual *chorai* chapters. Fig.4 is a histogram chart which compares *chorai* and *poleis* extensions (*polis* areas in sq.m. are exaggerated by a factor of 500 in order for the values to be graphically/visually comparable with the extensions in sq.m. of the *chorai*; *chorai* extensions are in ascending order). Generally speaking, one can note a certain regularity in the correspondence between *chora* areas and corresponding *polis* size. Some anomalous cases stand out though. As seen earlier, Thebes has available a large territory, but the oversized extension of the city (a supposed extension of 210ha compared to an average of ca. 20 to 50ha for the other *poleis*!) is mainly due to cultural and historical factors; on the contrary, Tanagra has (at least in some periods of history) an oversized territory compared to the actual size of the walled city. Thespiae, Koroneia, Orchomenos, Plataea would seem larger *poleis* compared to the actual extension of their *chora*. In the small-sized *poleis* regular trends seem to appear, with some exceptions, such as in the case of Haliartos and NE Copais, where the *chora* extension seems oversized compared to the *polis* area.

Clearly any comment must be taken with caution, as it is based on the comparison between two datasets (extension of *chorai* and of *poleis*) which cannot be proved in either case with certainty and are for the most part results of interpretation. Also, we should take into account that the micro-regions (or 'districts') would probably have been set up in very early periods, as seen above, while the population of the city sites, corresponding to the micro-regional centres, would have changed during time, often also considerably and within a short time span, as the archaeological record often attests. The area values employed in the graph are just indicative and probably of the Classical period.[30]

A final issue to be examined is the location of the major settlement (*polis*) within its *chora*.[31]

The distance from the geographical centre of the *chora* is greater in E Boeotia, as one can appreciate in fig.5, where the geometric centre[32] of the *chora* polygon is compared with the actual location of the *polis*.

In western Boeotia the *poleis* are very close to the geometric centre of their *chorai*. They deviate from the geometric centre of the *chora* only by a distance of between 600 and 3000m, since factors such as suitable topographical location (hilltops for the acropolis, dry land, strategic position), proximity to water and

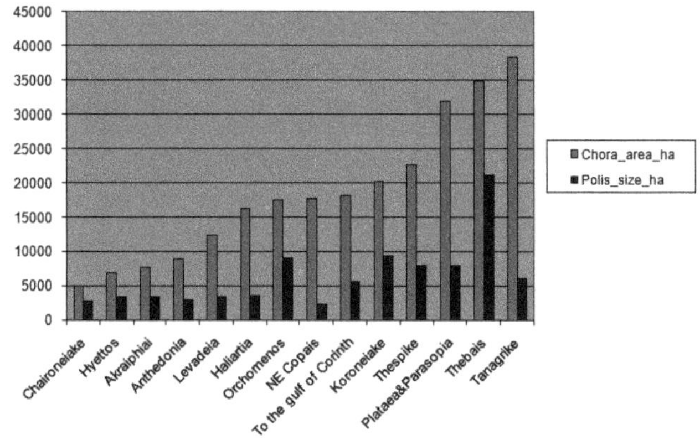

Fig.4. *Histogram chart which compares chorai and poleis extensions (polis areas in hectares, taken from Bintliff 1997, are exaggerated by a factor of 500 in order for the values to be graphically comparable with the extension in hectares of the chorai; chorai extensions are in ascending order).*

proximity to the contact point between two land-zones, play a role in the choice. This means that the cities, being in the centre of their *chora*, enjoy optimum exploitation of the whole *polis* territory. This central position is also related to the small size of the *chorai* of W Boeotia, as well as to the strong topographical constraints which define their boundaries.

On the other hand, in eastern Boeotia the deviation is larger (4000 to 7500m). The lack of strong topographical constraints corresponds to larger fluctuations of the boundaries over time and therefore to a non-central position of the main settlement. In some cases the *poleis* no longer stand at the centre of their *chora*, having incorporated the territories of former proto-*poleis*, e.g. Askra (in Thespiae *chora*), or of smaller *poleis*, e.g. Harma or Mykalessos (in the Tanagrike). In the case of Thebes, the historical settlement tradition (from Prehistoric times on the Kadmeia hill), as well as the position of the Kadmeia hill overlooking the Theban plains, create a natural centre of the area, whose boundaries, limited by few topographical constraints (mainly just the lakes to the N), fluctuate over time due to historical and political reasons. The strength of Thebes acts as a cultural constraint for the location of neighbouring *poleis* and the formation of their mutual borders. Plataea, for instance, does not lie in the middle of its *chora*, whose boundaries always fluctuated greatly, but in a suitable topographical location along the side of the Asopos plain opposite the location of Thebes.

(if not the Chalia peninsula also) were certainly part of the larger Tanagrike. Such a large unit with strong inner dynamics could even constitute a sort of 'third Boeotia', in addition to the main division into western and eastern Boeotia.

[30] Values are taken as they are from Bintliff's 1997 article.

[31] One can note how during the Greco-Roman period the major settlement tends to remain in the same location. Only in certain cases is there a shift, due to particular historical conditions or events, as in the case of Late Roman Orchomenos (shifted to Levadeia) or Haliartos (whose location shifted to Onchestos after the destruction of the *polis* by the Romans in 171 BC).

[32] i.e. centroid = geometric centre of a feature (polygon).

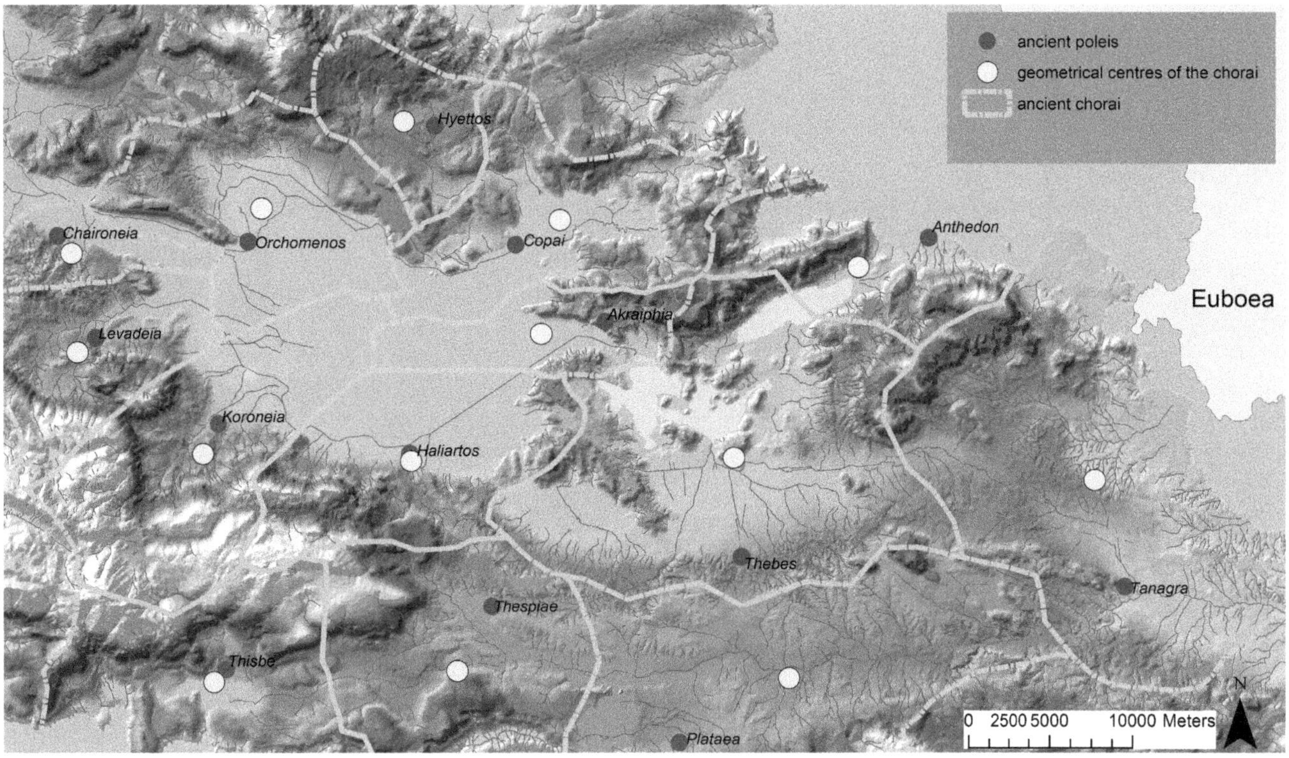

Fig.5. *Map showing the geometric centres of the chorai polygons compared to the location of the major Boeotian poleis.*

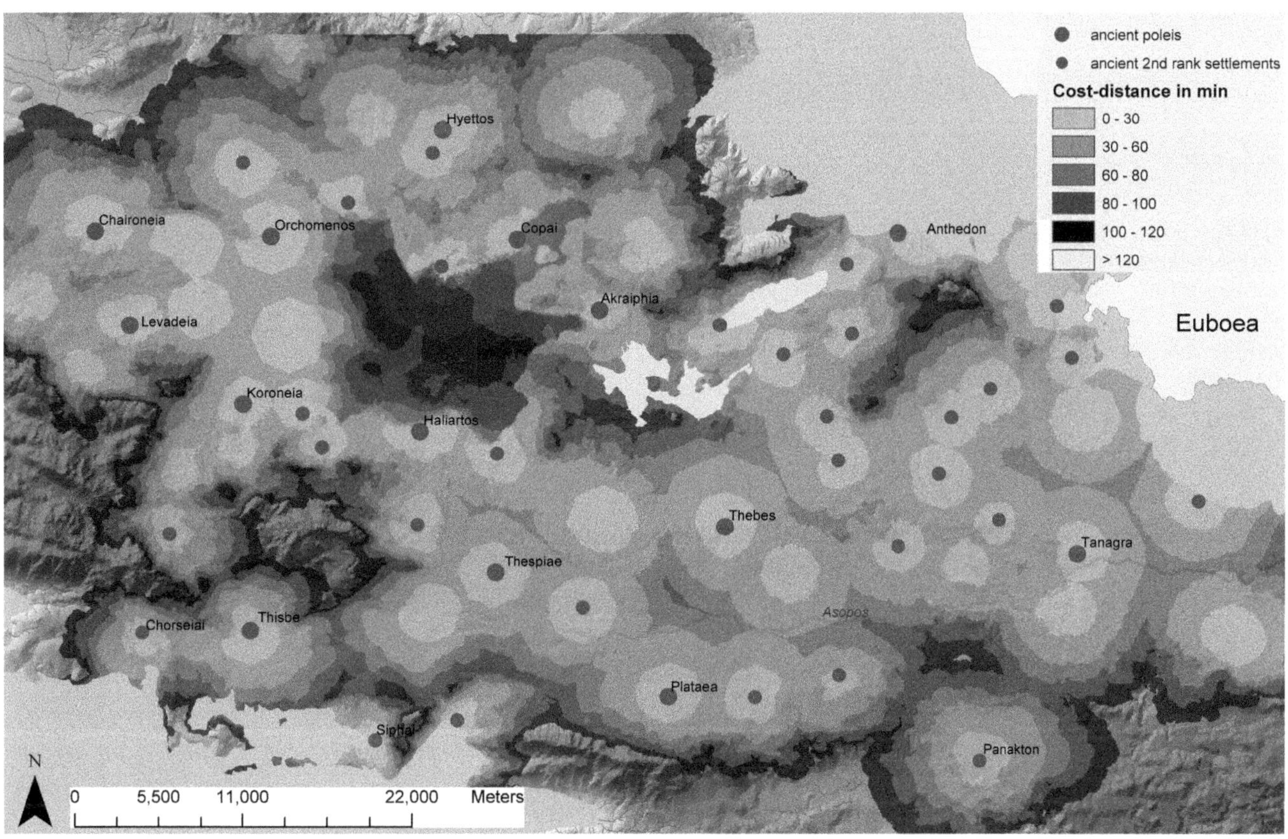

Fig.6. *Classified surface representing the cost-weighted distance from* 1st *(larger dots) and* 2nd *rank (smaller dots) nucleated centres. Areas without dots indicate potential settlement chambers (see chorai chapters for discussion).*

The location of the *poleis* within their *chora* also influences the location and distribution of the 2nd rank settlements.

2nd level – second rank settlements.

The available record for the second rank settlements, i.e. towns and large villages (*komai*), is certainly incomplete. We know this with certainty not only because of the existence of biases in archaeological research (indeed, villages do not have the attractive finds that a city does), but also because there are settlement names mentioned in the sources which have not yet found a suitable corresponding candidate, as in the case of small centres such as Pharai, Eteonos, Skolos, Schedia,[33] etc. For some of them, more than one identification has been suggested, as in the case of Aspledon, Tegyra, Schoinos and others (see *chorai* chapters and appendices). The result is that while cities have all been found and identified, there are still many villages to be found.

The GIS point layer used for the cost-surface analysis involving the 2nd rank settlements includes the main Boeotian *poleis* examined above as well as smaller nucleated settlements, including some that in an early (Geometric-Archaic) period could be considered *proto-poleis* (see above), and some that in the Classical-Hellenistic period are basic nucleated settlements, sometimes quite small, such as Eutresis or Teumessos, gravitating around the larger centres. In the archaeological record available, as seen in the *chorai* chapters, some settlement sites are only presumed on the basis of the presence of burials or a fortified acropolis, such as Pharai (located by some scholars at Bratsi/modern Tanagra), Ypaton, Erythrai, Hysiai, or the supposed villages by Kallithea, the Neochorachion/Soros area and Asopia (see *chorai* chapters – LONG-TERM SETTLEMENT TRENDS - for details). The layer on which the analyses are performed represents primarily a possible situation for the Classical period. A similar test was made by Bintliff (1997), in a study on the population of ancient Boeotia. He had plotted on the 1:200,000 modern administrative map of the province of Boeotia the known or suggested Classical village sites, together with the known *poleis*, and then used Thiessen Polygon analysis to suggest the most likely areas of the landscape exploited or serviced preferentially by each of these settlements. In general, one can note that the hill and plateau land of intermediate elevation made up of flysch and soft limestones are nearly always preferred by settlement in every period.

Here, I have applied the same cost-distance analysis, performed for the *polis* level, to the 1st and 2nd rank settlement network. The analysis produced results which are somewhat comparable with those of Bintliff 1997, although the purely geometric Thiessen polygons have been replaced by tessellations resulting from a cost allocation analysis (see above). The result gives us an idea of what could have been the large community areas gravitating towards cities, town and large villages (*poleis* and *komai*) in ancient Boeotia. Fig.6 maps the classified surface representing the cost-weighted distance from those nucleated centres. As stated above, 1st rank (*poleis*, centres of micro-regions) and 2nd rank nucleated centres are mapped together (1st rank *poleis* marked with a larger dot), as no ranks were differentiated in the analysis. In fact, the purpose was to delineate settlement trends over the landscape and try to understand settlement behaviour as well as territorial subdivisions, which could then explain rank differentiations.[34]

Poleis are therefore treated here with the same basic land module as villages. One should not forget, however, that all the larger cities, as also pointed out by Bintliff 1997, possessed an administrative territory enclosing the individual territories agriculturally exploited by their satellite settlements (villages and *komai*), as well as their own core exploitation zone utilised by the city and surrounding farms.

What stand out clearly from the map (fig.6) are the cases in which an actual boundary role is played by physical constraints, such as mountains (Helicon, Ptoion, etc.) or rivers (Asopos).

Mountains are often perceived as border land. Sometimes the highest mountainous areas are seen as no man's land between two communities (usually used for pastoral activities or for wood production), not really belonging to anybody and free to be exploited, and used as a border between communities. Mountains, and especially hillcrests (marking watersheds), present geographical characteristics which 'naturally' mark boundaries. They have been used through history, since ancient times, as administrative borders between communities (see, for antiquity, Daverio Rocchi 1988, and see also discussion on borders in the individual *chorai* chapters).

Within Boeotia, treaties concerning the definition of the borders between two *chorai* are epigraphically and historically known. Boundary stones have been found: AE1108 (boundary stone between Koroneia and Orchomenos - Fossey 1988: 501); AE1212 (boundary stone between Copai and Akraiphia[35]); AE1671 (boundary stone between Koroneia and Levadeia[36]). A treaty is reported[37] between Koroneia and Thespiae for the exploitation of the Koukoura/Agia Anna upland plateau, which also played an important role at a strategic level (see below and chapter II.3.1), as well as the

[33] There are 35 place names mentioned in historical sources that could be considered to be Boeotian village settlements, but of these Leuktra is probably only the locality of the famous battle rather than a village (see chapter II.3.9).

[34] Bintliff has argued (Bintliff 1994b) that this network of village/city territories arose during the Greek Dark Age and Archaic period, when there was little difference between the exploitative and administrative territories of all these settlements, and power was much more diffuse across the landscape (see also Bintliff 2000a).

[35] IG VII 279. See also SEG XXX 440 for a possible boundary stone once again between Copai and Akraiphia, datable to the 6th – 5th C BC. The text suggested by Roesch reads: [hόρος A]κραι[φιέον] [κ]αί Κοπ[αίον (*Teiresias* 10 [1980], AE/2).

[36] SEG XXIII 297; SEG XXXV 406.

[37] IG VII 1870.

Skourta plain.[38] The character of mountainous land as a border is marked also by the presence of forts and fortifications, which also gives a strategic role to mountainous zones and mountain passes. An example, for instance, is the area of Akraiphnia or Haliartos, where the role played by the mountainous area is very strong. These areas are situated in a central location within Boeotia, on the heights separating or overlooking the Copais basin to the NW and the Theban plains to the SE (see below as for their historical strategic importance).
As seen above, the borders are considered to be *eschatai* (the last land, the furthest land).[39] The borders are marked by strips of no-man's land, usually running along watersheds or around mountain peaks, although sometimes also in the lowlands (see for instance the borders between Boeotia and Attica to the SE of Dilesi, described in chapter II.3.14). Those areas/strips could be smaller or larger in extension. The fact that they are considered to be no-man's land does not automatically mean that they were not in use, even if in the majority of cases they were not permanently occupied. Such areas, especially when constituted by uplands, might have a high economic potential, for wood or pasturage land,[40] and were certainly exploited in periods of expansion, such as in Classical times. According to this perspective, it becomes clear that the borders not only mark the furthest areas of the communities (marking the boundary line between them, in ancient as in modern times), but also divide economic areas. It might be of interest to note that, in modern times, *eksokklisia* (small countryside chapels) on mountain peaks (often dedicated to Prophitis Ilias) often mark the borders between administrative divisions (Nixon-Price 2001: 416-7) and link the cult with agriculture or pasturage zones. This is also true in a way for ancient times, when cult places, besides their religious role, often had an additional twofold function of border markers and of signs of specific economic areas, to which the worshipped cult was often linked.
As already remarked, a constant tendency of the Greek *poleis* was to reach a stable border, individuated by reference features, creating a 'linear border' (Daverio Rocchi 1988: 25), or individuating 'no-man's land' as separating areas officially recognised. Therefore, to the border line was given a political importance as well as a sacral value.[41]

The map in fig.6 also clearly shows that the highest density of 2nd rank settlements can be observed in central and eastern Boeotia. Here we see large distances between the main *poleis* (Thebes, Plataea, Tanagra), and the intermediate areas are occupied by a ring of 2nd rank settlements around each of these *poleis* creating a 'village-based landscape' (see fig.6 and fig.8). On the other hand, in W Boeotia the density of *poleis* is higher and the number of 2nd rank villages smaller (fig.6). We usually observe only one 2nd rank settlement occupying the area between two *poleis*. Between Haliartos and Thespiae we find Askra, dependent on Thespiae and Onchestos, under the influence of Haliartos, creating a cross line between Haliartos and Thespiae. In the same way, between Haliartos and Koroneia we find the settlement of Alalkomenai, within the Koroneia *chora*. A low number of 2nd rank settlements are noted especially in the Copais area and towards the gulf of Corinth. This can be partially explained by the small distance between the *poleis* for topographical reasons in the Copais area, and by the mountainous character of the gulf area. When there is available land in the mountainous areas, then 2nd rank villages do appear, as for example in the case of the Koukoura/Agia Anna plateau (on Helicon, to the S of ancient Koroneia). Yet we should consider that a number of villages have not yet been identified, and they could correspond to the potential settlement chambers marked in the figure (fig.6).

As a next step, I applied cost-distance analysis considering distance ranges of 2.5km and 5km (½ h and 1h walking time respectively) from the settlement site (fig.7). The results are circles with 5km radius (calculated in time, 1h walking), as employed, in defining settlement catchment areas, by Renfrew-Wagstaff 1982 (Melos) and Bintliff (1977: 112) for Greece. On the other hand, Bintliff 1997c and 1999c (see above) calculates the 'standard village model' in circles of 2.5 km radius, based on a recurrent cross-cultural territory size associated with pre-industrial mixed-farming-based nucleated settlements of medium size (see above). The smaller radius around the nucleated centres is mapped in fig.8.

In both figures the empty plain area (with low rises) in the centre of the region to the W of Thebes is the most striking visually. This empty area without any known 1st or 2nd rank settlement sites, which is further than 1 hour's walk from the city of Thebes, as well as from the neighbouring centres, constitutes an area under the control of Thebes itself, and can be explained by cultural

[38] The Skourta upland plain was common pasture land shared by Boeotians, Athenians and even Corinthians (Munn - Zimmerman Munn 1990: 36-37; Schachter 2004: 56).

[39] For the concept of *eschatai* within the framework of the discussion of *polis* borders in the Classical period see, among others, Daverio Rocchi 1988.

[40] See examples from Tanagrike, Akraiphiai, as well as from the *chorai* marked by the presence of Mt.Helicon, such as Koronieake, Levadeia and Haliartos. For Boeotia, it can be pointed out that no intensive research was carried out in mountain lands, with the exception of mountain plateaus (e.g. Skourta plain – see appendix I.14), as field-by-field surveys devoted their attention mainly to ploughed lowland fields. On the other hand, as seen in the individual *chorai* chapters, some information on the occupation of the upland landscape can be inferred from the archaeological record also available from other sources, such as extensive topographical surveys or rescue excavation.

[41] The border line had a political value, as an entity juridically more apt to express a political boundary, perceived as the limit of the portion of land on which the city-state's power realises its control in a natural and stable way. The border must be intended in relation to what it includes, i.e. the appropriation of the concept of *polis* in its physical, social and political dimension (Daverio Rocchi 1988: 25-26).

factors, such as the importance of the centre of Thebes and its higher population potential, clearly in absence of any clearly marked physical constraint which could limit the expansion of the territory. In this situation, there is space for a settlement but the settlement is not there,[42] and if we consider that even Onchestos was probably nothing more than nucleated houses in the area immediately below the cult place, the territorial influence of Thebes would be even stronger.

It is striking that the limit of the 5km radius areas frequently coincide with the topographically corrected 'tessellation'. On the other hand, in certain areas (Eastern Boeotia especially) the 2.5km radius works better, showing a denser infill of the landscape in mature periods (in this case the illustration corresponds to the Boeotian settlement situation in the Classical period).

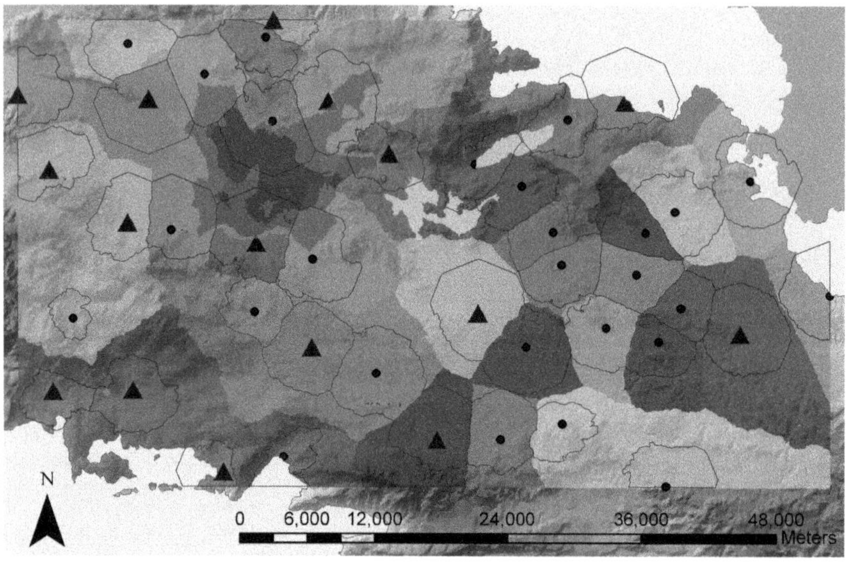

Fig.7. *5km radius area topographically corrected from poleis and 2nd rank settlements compared with the tessellation of space based on cost-weighted distance from the same points.*

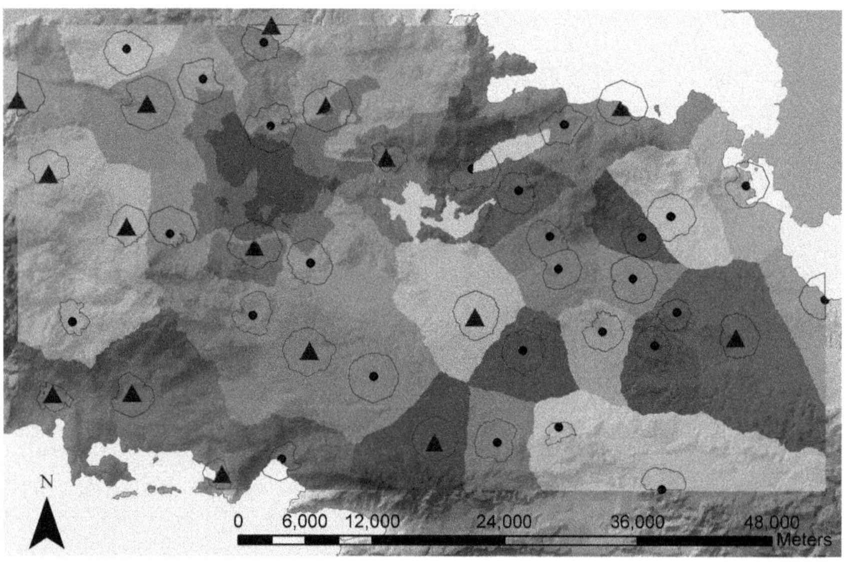

Fig.8. *2.5km radius area topographically corrected from poleis and 2nd rank settlements compared with the tessellation of space based on cost-weighted distance from the same points. In the picture we can appreciate the gaps and therefore the potential suitable places for other village settlements.*

Considering the graphs in fig. 9, it seems that the majority of the settlements is associated with a large proportion of fertile and mid-fertile land in their territory. Graph 1 shows the occurrence of F-MF (fertile – mid fertile), LF (low-fertile) and U (unsuitable for agriculture) land in the immediate territory of the 1st and 2nd rank settlements (mapped in fig.10), while graph 2 summarises and shows the occurrence of 1st and 2nd rank settlements presenting different proportions of F-MF land in their immediate territory.[43] As we can see, 42% of the settlement territories is characterized by fertile and mid-fertile land in a percentage of 80-100%. An additional 27% of the settlement territories presents fertile and mid-fertile land in a percentage of 60-80%. Therefore, the majority (69%) of immediate territories (½ hour walking time) of ancient 1st and 2nd rank settlements are characterised by the presence of more than 60% of F-MF (fertile-mid fertile) land, with a minimum percentage of 20% F-MF land in the case of the mountainous landscape of Chorseiai. In several cases of lower presence of F-MF land, LF land complements the areas suitable for agriculture. The analysis shows a definite preference for settlement locations which offer fertile territories in the immediate surroundings.

---

[42] Excluding an extensive settlement in the Roman imperial period on the Klimataria peninsula jutting into the Yliki lake – *component T_110.*

[43] The settlement areas of the Easternmost part of Boeotia are not included in the analysis as the geological map on which the land capability was based (see chapter I.2.1) is not available for the area.

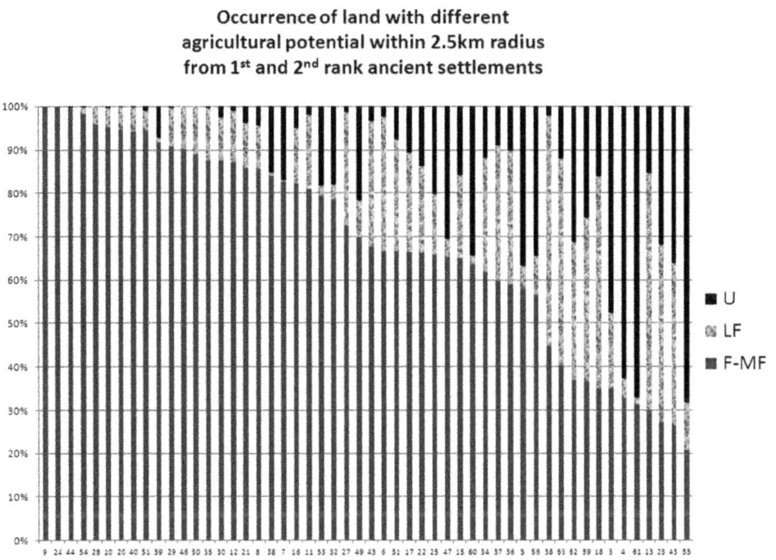

graph 1

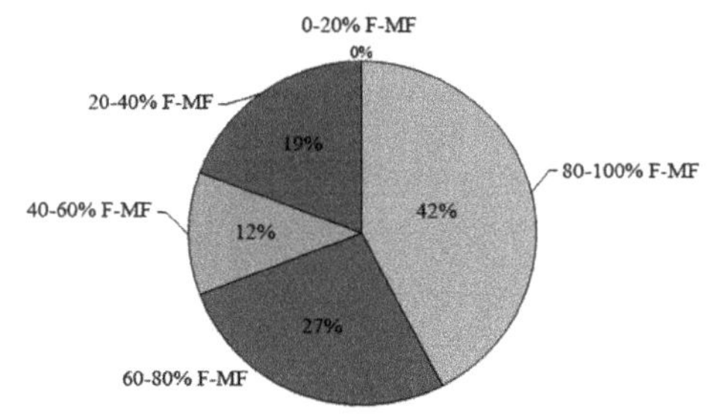

graph 2

Fig.9. Graph 1: *occurrence of F-MF (fertile – mid fertile) , LF (low-fertile) and U (unsuitable for agriculture) land in the immediate territory (1/2h walking time) of the 1st and 2nd rank settlements. The interested areas (numbered on the x axis of the graph) are marked in the map in fig.10.*
Graph 2: *occurrence of 1st and 2nd rank settlement territories with different percentages of F-MF land.*

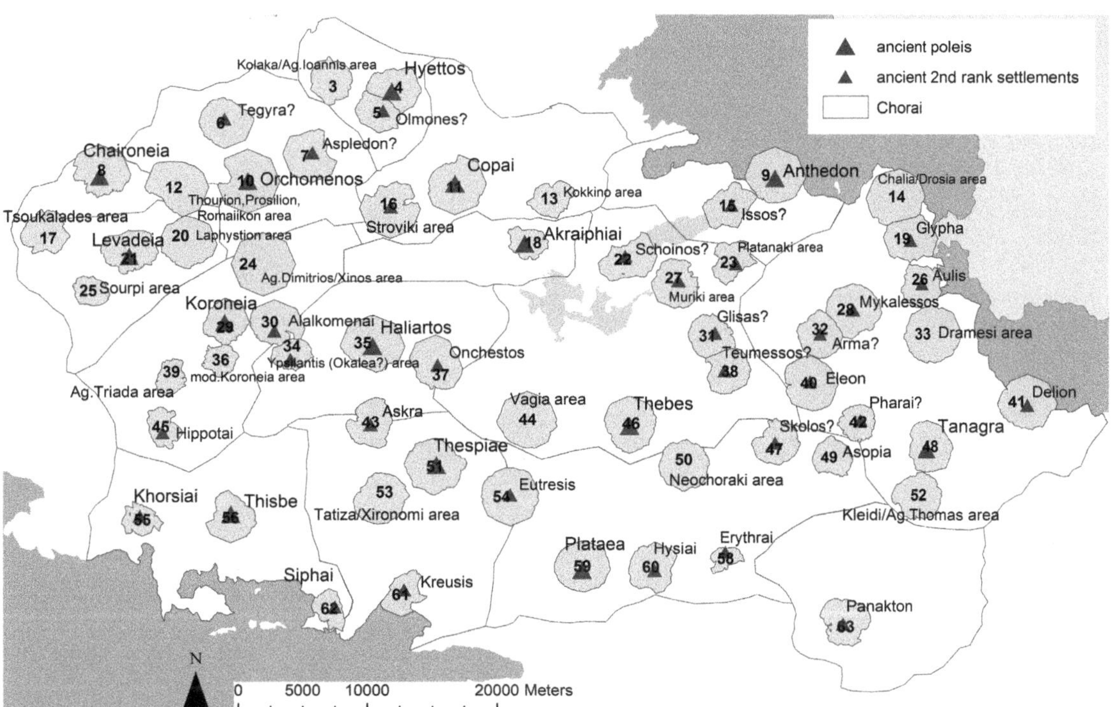

Fig.10. *Ancient 1st and 2nd rank settlement network with ½ hour walking time radius around them.*

Comparative analysis of the settlement network

The graph in fig.11a helps us to better compare the two main areas of the Boeotian landscape, marked by different geographical and topographical characteristics. The western part is dominated by the Copais and surrounding reliefs (Helicon up to the gulf of Corinth and the mountains to the E of Copais), while the eastern part is characterised by the presence of the two lakes Yliki and Paralimni, the central plains, the Tertiary *tafel* of Thebes, and the Eastern coastal plain towards the Euboic gulf. The two areas are comparable in size, measuring approximately 1,200 and 1,350 sq.km. respectively. The graph shows for each wide zone of Boeotia the percentage of settlement chambers associated with strong, medium, or low/absent topographical constraints (presence of mountain/ridges, lakes/marshes or permanent rivers), thus evaluating to what extent topography and/or cultural factors influence the presence of settlement chambers.

The graph in fig.11a shows a converse picture concerning the topographical constraints. In the landscape of W Boeotia, strong topographical constraints are actually defining settlement chambers. It seems also that marked topography leads to a greater fragmentation of the landscape between more *chorai* (9 *chorai* in the W compared to 5 *chorai* in the E). The more fragmented landscape in the W, as well as the larger number of defined *chorai* with relatively smaller size, is associated, as one would expect, with a smaller number of settlement chambers in the territory of each *chora* (three settlement chambers each, on average - see graph in fig.11b). When there is available land in the fragmented mountainous areas, where inner valleys open up, then more 2nd rank villages do appear, as for example in the case of Koroneia or Levadeia (with 6 and 4 settlement chambers respectively, mainly characterised by strong topographical constraints).

In eastern Boeotia the mild relief is associated with the existence of a smaller number of *chorai*. There is also, on average, a larger number of settlement chambers per *chora*, partially related to the larger size of the *chorai* in this Eastern part of Boeotia (see graph in fig.11b). The majority (60%) of the settlement chambers are associated with low topographical constraints (graph in fig.11a). The lack of strong physical constraints and the larger number of settlement chambers create a network of 2nd rank settlements associated with the main *poleis*, located quite regularly over the landscape. The only large gap, as noted earlier, is around Thebes (figs. 5-6-7). Low topographical constraints also allow for an open choice with regards to location of the main settlement (see above – *1st level*). In these open landscapes, mainly social and political factors, as well as the mutual distance factor, define settlement chambers exploiting the plains and the rolling hills of Eastern Boeotia.

The detection of settlement chambers over the landscape is possible not only with strong topography. Distinctive and recognisable landscape areas gravitating around settlement foci changing over time also exist in areas where the division of the landscape into settlement units is mainly due to cultural and social factors. Comparison of the settlement network in later periods and the pattern of known Prehistoric activity foci, examined in detail in each *chora* chapter, supports the existence of the landscape units defined as settlement chambers. In the Parasopia/Plataea area for example, the landscape is open without physical constraints but occupation is recurrent: all settlement chambers along the S side of the Asopos plain (and also Kallithea to the N) are occupied by a settlement in all periods. Also in the area of Thebes, settlement chambers are very weak, with low topographical constraints. Yet recurrent occupation in the same area and often on the same spot in all periods indicates the existence of settlement chambers (Hypaton, Platanaki and Mouriki).

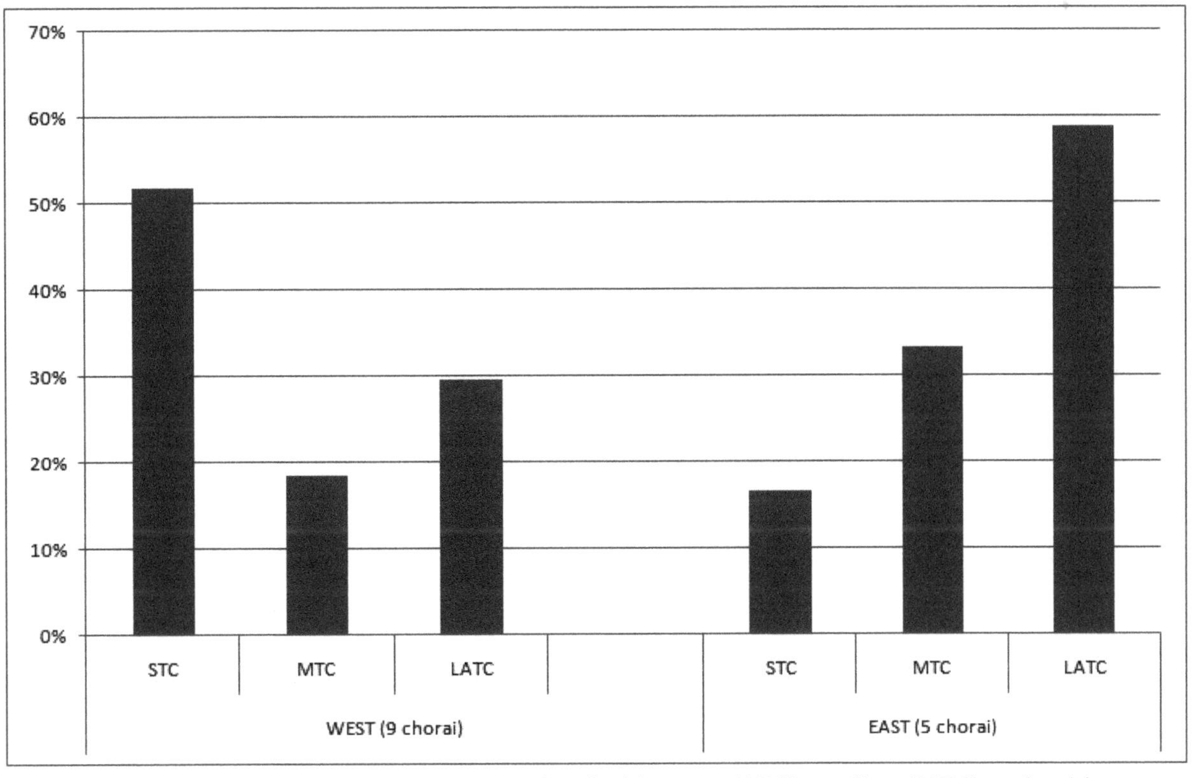

Fig.11a. Percentage of settlement chambers associated with strong (STC), medium (MTC), or low/absent topographical constraints (LATC) in W and E Boeotia.

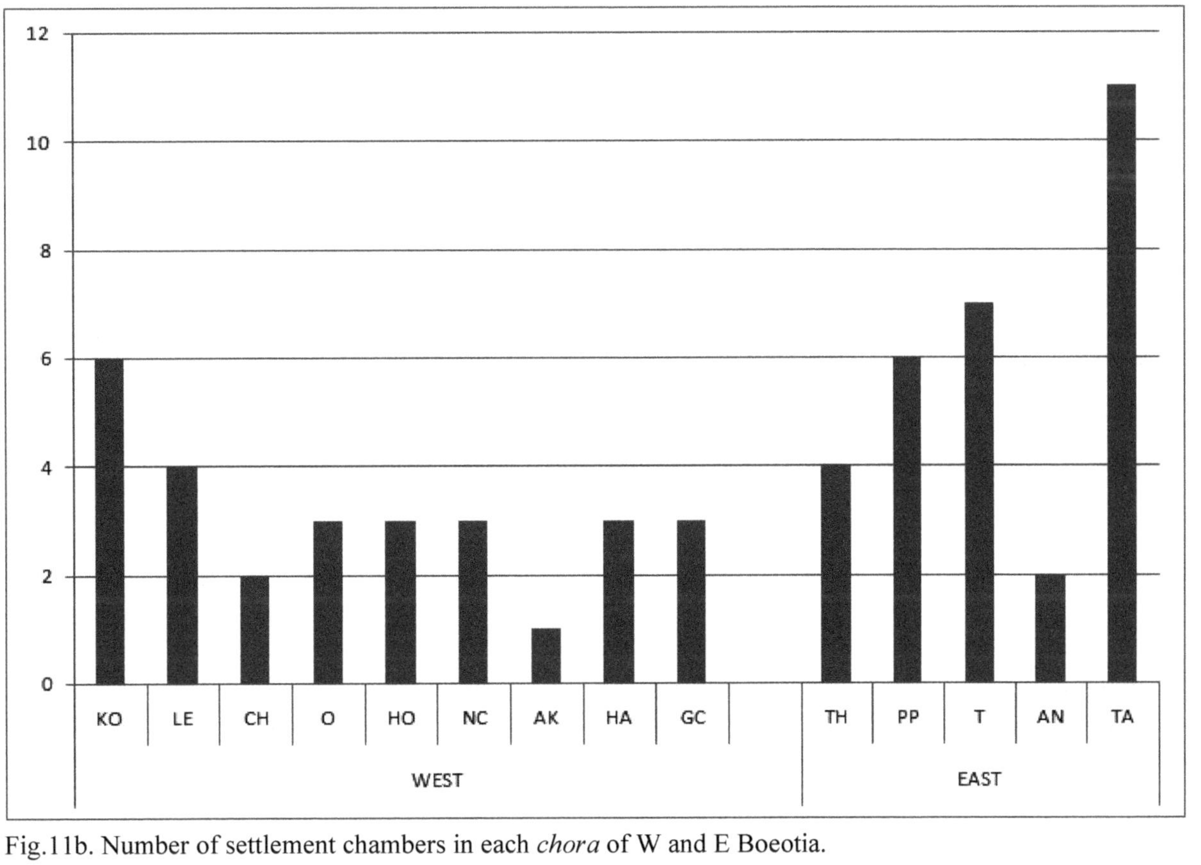

Fig.11b. Number of settlement chambers in each *chora* of W and E Boeotia.

Fig.11. *Topographical characteristics of the settlement chambers of W and E Boeotia.*

I will proceed next to a comparative analysis to investigate the issue of continuity over the Boeotian landscape, looking for the occurrence of settlement foci of different periods within a potential settlement chamber. The following periods are considered: Prehistoric, Greco-Roman, Frankish (represented in this analysis by the presence of known towers and castles), Ottoman and Modern. In this analysis, they are represented by the main settlements known for these periods, while for the Prehistoric period presence includes main activity foci recognised in the landscape (burials also), even without any particular evidence for a nucleated settlement. Within a settlement chamber one can point out some recurrent occurrences of the main periods, as shown in the graph in fig.12 and listed below. The list examines the continuity of occupation within the same settlement chamber but not necessarily on the same spot, although in many cases there is an actual overlapping of the main settlement site in the different periods in the same location, as examined in detail later.

ALL. Presence of a settlement focus in all of the main periods: ancient Koroneia (KO), Chaironeia (CH), Kastro (NC), Haliartos (HA), Askra (TH), Thisbi (GC), Plataea (PP), Thebes (T), ancient Tanagra, Bratsi/modern Tanagra and Harma (TA)
ALL but Fr. Presence of a settlement focus in all of the main periods except Frankish towers and castles: Orchomenos (O), Akraiphnio (AK), Eutresis/Melissochori (TH), Chorseiai (GC), Erythrai, Daphni and Kallithea (PP), Ypaton, Platanakia and Mouriki (T), Anthedon/Loukisia (AN)
ALL but Preh. Presence of a settlement focus in all of the main periods except Prehistoric: Levadeia (LE), Ypsilantis (HA)
ALL but Ott. Presence of a settlement focus in all of the main periods except Ottoman: Pyrgos (O)
Preh + Gr-Rom. Presence of Prehistoric activity and Greco-Roman settlement focus only: Siphai (GC), Oungra Paralimni (T), and probably Kastro Lykovouni and Glypha (TA)
Gr-Rom + Ott. Presence of Greco-Roman and Ottoman settlement focus only: Polygira (O), Onchestos (HA)

The majority of settlement chambers (22 of 31) show the presence of settlements of all the main periods, or of the four main periods with the exclusion of a Frankish tower. The lack of Frankish towers in certain settlement chambers is related to the fact that they are associated with larger territorial units, which correspond with the area of influence of Greco-Roman *poleis*, one often being present in each *chora*.[44] A larger number of Frankish towers, which could be associated with settlement chambers, is known from the wider Tanagrike and in the Thespiae area. The existence of four settlement chambers with Prehistoric and Greco-Roman finds only (see graph in fig.12) would mark a change in landscape strategy in later periods. Very weak settlement chambers (two), with a known settlement in only one period, are very rare, and they are present only in E Boeotia (see graph in fig.13), where open landscape does not allow an immediate recognition of separate potential settlement areas.
The graph in fig.13 compares W and E Boeotia in terms of multi-period settlement presence in the settlement chambers of the *chorai*. The fact that in Eastern Boeotia also, four or five periods are represented within each settlement chamber strengthens the hypothesis suggested earlier for the presence of settlement chambers in areas with a lack of strong topographical constraints also.
On the other hand, the presence of multi-period settlement foci in areas for which evidence for Greco-Roman 2nd rank settlements is lacking helps us to evaluate gaps in the landscape. Thus, in fourteen areas Greco-Roman settlement foci are hypothesised on the basis of the presence of settlements of other periods and of indications of Greco-Roman activity in the archaeological record (for example Ag.Triada and Mavrogeia/Ag.Dimitrios (KO), Kokkino (NC), Ellopia area (TH), Asopia (PP), where all the other main periods are represented with a settlement site).

The occurrence of settlements of different periods in the same area is related to the landscape history of the region. In the individual *chorai* chapters we noticed, for example, how in post-antiquity the main settlements of the region, corresponding to the former *poleis*, became in later periods smaller or larger villages (comparable to second rank settlements). Only Thebes and Levadeia remained as major centres throughout the Byzantine, Frankish and Ottoman periods. Socio-political factors and different strategies of land use might also affect the type and rank of habitation. Thus, for instance, the locations of second rank settlements of the Classical and Early Roman periods are taken up in Late Antiquity by large rural sites/villas. Yet we should also take into account the possible biases in the archaeological record, especially in the case of settlement chambers occupied by a *polis* settlement in the Greco-Roman period, where the presence of a settlement in all the main periods considered always occurs. In this case, intensive research on the historical *poleis* and their immediate surroundings often leads to a better understanding of the presence of other periods as well.[45]

[44] Bintliff 2000e suggests that in Boeotia, during the High Medieval/Frankish/Late Byzantine period, Frankish towers associated with settlements, and the two main Frankish castles of Levadeia and Thebes, were established on or near the long-established towns and villages of earlier periods, marking a continuous development from the landscape of Late Antiquity.

[45] For example the recent work at ancient Tanagra located the Prehistoric, Frankish and Ottoman presence in the immediate area of the ancient *polis*.

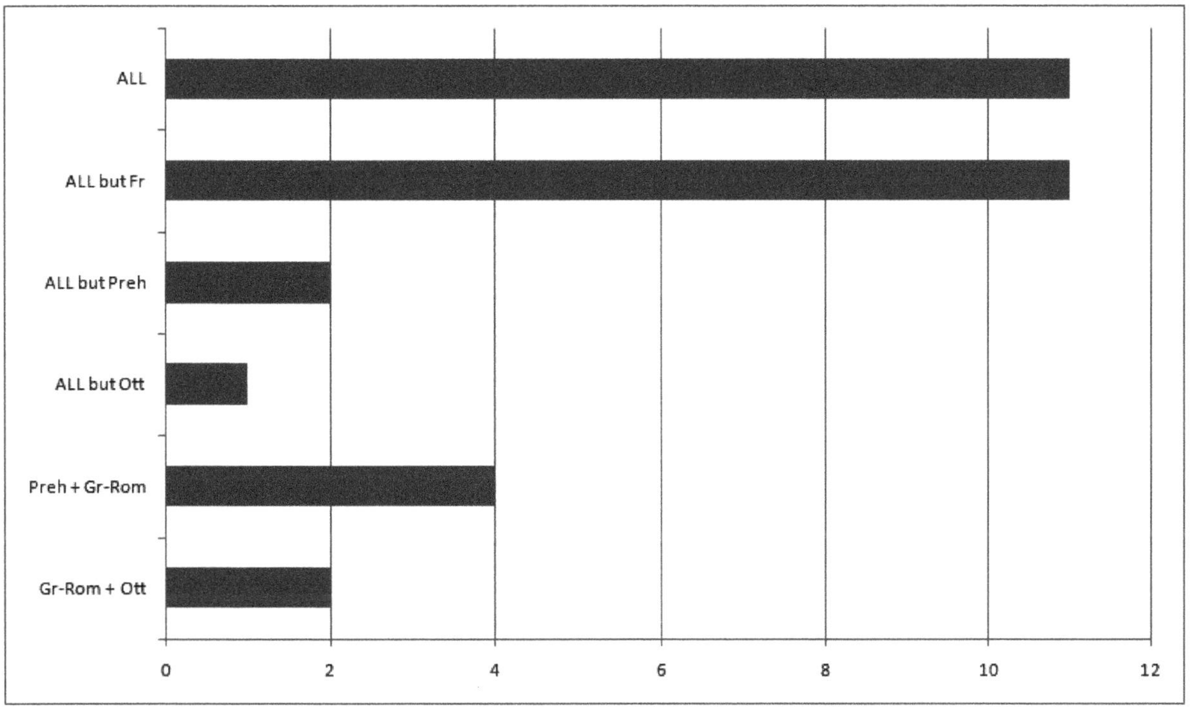

Fig.12. *Recurrent occurrences of settlement foci of the main periods considered (Prehistoric, Greco-Roman, Frankish, Ottoman, Modern) within recognised Boeotian settlement chambers.*

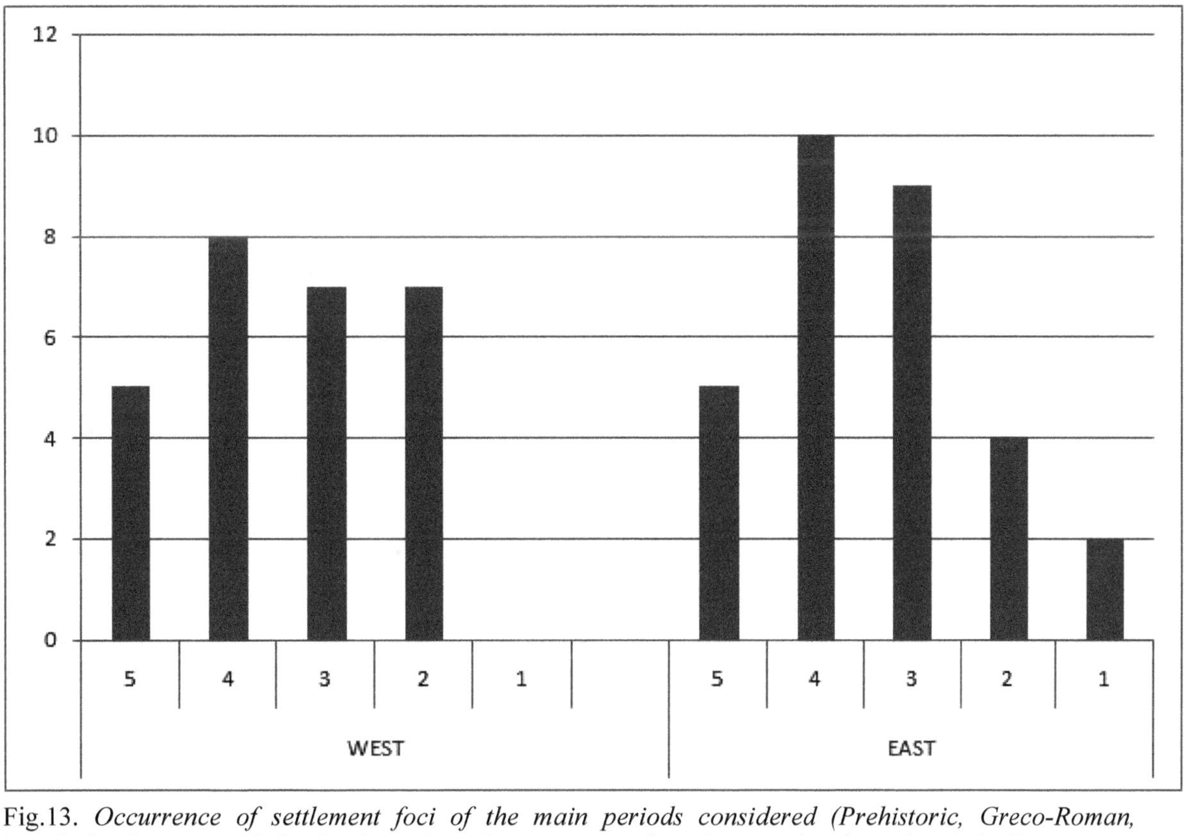

Fig.13. *Occurrence of settlement foci of the main periods considered (Prehistoric, Greco-Roman, Frankish, Ottoman, Modern) –1 to 5- within recognised settlement chambers (y axis). Comparison between W and E Boeotia.*

The locating of the main settlement on the same spot in different periods may mean, in terms of history of landscape, memory in addition to continuity. Apart from continuity within a certain settlement chamber, the preference for a precise location, creating recurrence, might indicate memory rather than any other specific reason of choice. Thus, in western Boeotia, for instance, which was less affected by population loss and subsequent Albanian colonisation in the Late Frankish period, villages registered in the Ottoman defters with Greek-speaking population form large nucleated villages close to the ancient Greco-Roman centre, keeping their traditional location and land use (Mediterranean polyculture). Another example, where there is memory but no ethnic continuity, is that of the newly established Albanian hamlets in the Late Frankish period, which in many cases occupy the location of deserted Byzantine villages. Remaining or ruined churches marking the positions of former Byzantine villages keep the memory in the landscape and mark locations which were resettled by the colonizers (see for example a series of medieval Byzantine villages in the Tanagrike studied by the recent Tanagra survey).

Examining the scale of movement of the main settlement across different periods, some remarks can be made. Some of the major Boeotian *poleis* lie below modern centres. This is the case of Thebes, Levadeia, Orchomenos, Chaironeia lower town, Copai/Kastro. In other cases, while the major (1st rank) settlement remained in the same location during the whole Greco-Roman period, in post-antiquity the corresponding settlement shifted its position. In the majority of cases the shift occurs within the same settlement chamber (for instance, Koroneia, Akraiphiai, Thespiae, Plataea, Anthedon) and the settlement often moved very close to the location of the Greco-Roman *polis*, frequently occupying the lower area immediately below the slopes of the acropolis (e.g. Haliartos, Tanagra, Chaironeia). In other cases there is a movement to the upland or inland in the Medieval and especially in the Ottoman period (e.g. Alalkomenai and Solinari (KO), Kokkino area (NC), Thespiae (TH), Plataea (PP), Anthedon and Loukisia (AN)). Comparing ancient and modern settlements, a shifting in location in the same topographical zone occurs in several cases: ancient Koroneia and Ag.Georgios (KO); Hyettos and Loutsi, Palaiokastron and Pavlo (HO); Akraiphiai and Akraiphnio (AK); ancient and modern Haliartos (HA); ancient Thisbe and modern Thisbi (GC); Ypaton (T); ancient Eleon and modern Harma (TA) – the main difference being that in the Greco-Roman period, naturally defended locations on 'natural' *acropoleis* were preferred.

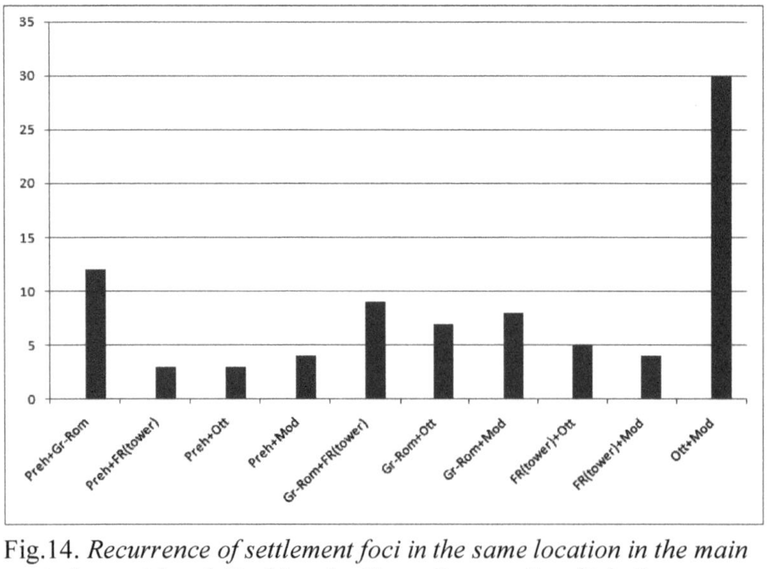

Fig.14. *Recurrence of settlement foci in the same location in the main periods considered (Prehistoric, Greco-Roman, Frankish, Ottoman, and Modern)*

One notices how settlements corresponding to ancient 2nd rank settlements move much more around the landscape in the long term, compared to settlements corresponding to ancient *poleis*, but often within the same settlement chamber, as far as the available archaeological record indicates. The average scale of movement of the main settlement focus within a settlement chamber is 25 minutes' walking distance, both in E and W Boeotia. The data set of which the mean was calculated included very low or null values, when the settlement occupies the same spot in the different periods. The minimum values are 0 min in the case of Orchomenos in W Boeotia and 0 min in five cases in E Boeotia. The maximum values are 60 min distance for the W (in the area of Kokkino; if we include the case of shifting of the main settlement from Haliartos to Onchestos in the Late Hellenistic/Early Roman period the distance reaches 65 min) and 56 min distance for the E (Oungra Paralimni). The fact that the maximum distance remains comparable in the E and W strengthens the idea that the presence of a settlement chamber is not affected by topographical reasons only (see above).

The choice of settling on the same spot recurs especially in certain periods, as shown in the graph in fig.14. Examining continuity of habitation at the same spot continuity can be seen in three broad succeeding periods (Prehistoric and Greco-Roman, Greco-Roman and Frankish, Ottoman and Modern) in several examples, indicating population continuity or same choice. Multi-period occupation (Prehistoric, Greco-Roman and Frankish) on the same spot is noted at very distinctive and prominent locations, as Pyrgos (O) and Pyrgos Paralimni (AN). Sometimes there are two preferred locations within the same settlement chamber chosen for the main occupation in different periods, such as in the case of ancient Koroneia and modern Ag.Georgios (KO), Copai/Kastro and Gla (NC), or Pyrgos and Magoula Pyrgos (O - where the latter is preferred only in Late Prehistory). As for non-contiguous periods, the few examples of habitation on the same spot may reflect similar choices in the landscape.

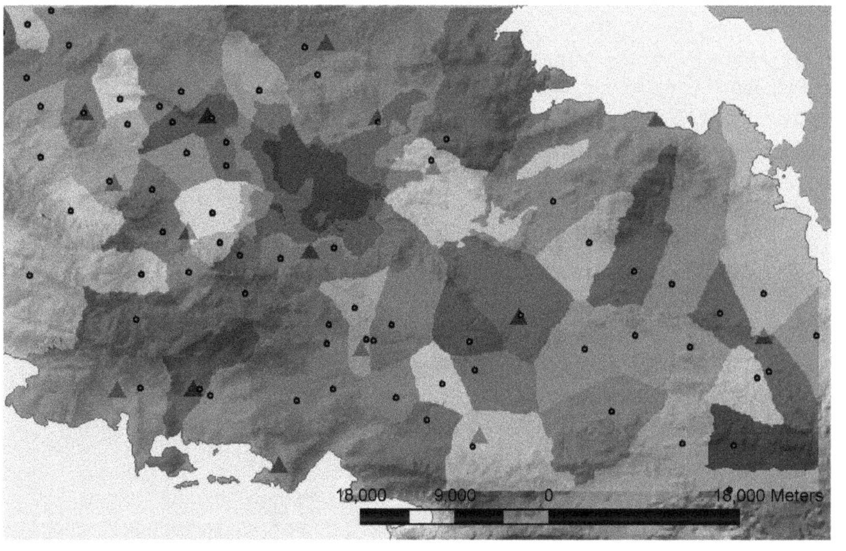

Fig.15. *Allocation based on cost-surface from the modern settlements (choria). Polis* triangles marking ancient *poleis* have been positioned in the map in order to make clearer the location of modern villages within the landscape and the comparison with the above maps. For village names see fig.4 in chapter I.2.2.

I also performed cost-distance analysis from the modern settlement network (GIS layer: *choria*). The results are shown in fig.15. In the map one can observe that the mountainous area to the W of Copais is distributed among numerous villages (dispersed pattern), compared to the villages in the plain. This is mainly for topographical reasons, as the 'tessellation', the result of the allocation analysis based on cost-surface, shows clearly. This does not seem to happen, though, in the Greco-Roman period, where around Copais we have several *polis* communities controlling the single valleys as well as the surrounding mountains (ridges jutting into the Copais and higher mountains behind). On the other hand, in the plain area to the E, the modern situation looks much more similar to the ancient one. There, in the absence of real/outstanding physical constraints, the communities divide the available land among them. Only Thebes, as seen above, and Thespiae, and Tanagra to a degree,[46] have enough strength (for historical cultural reasons) to control a larger territory. The settlement of Eutresis, though not so well investigated for historical periods of occupation (compared to the Prehistoric, which have attracted the interest of archaeologists at the site), seems to have played an important role with its location between Thespiae and Plataea (even though its available area would probably have been smaller, due to the presence of two 'important' neighbours to its E and W).

Furthermore, in the hilly landscape to the S of Thebes (between Thebes and Thespiae, and Thebes and Plataea), the modern landscape is marked by more settlements on the hills (Ambelochori, Vagia, Mavrommati, etc.) compared to the picture available for the Classical period. In considering the empty area in the centre of the region mentioned above and shown in the picture with *poleis* and 2nd rank settlements in the Classical period (fig.6), one can observe that in the modern landscape as well land is distributed among settlements which are positioned more or less in the same way as in ancient times, but two more settlements appear in the modern picture on the hills (Vagia and Ampelochori).

As stated by Bintliff (2000a), in Boeotia the top-level regional centres in the Greco-Roman period (as well as in Early Modern times) seem to present a spacing predicted by rural marketing theory, i.e. a day-return radius of 15km (comparable to Fossey's empirical estimate of 14km – Fossey 1988).[47] The *poleis* level examined above would seem to show a pattern of this sort. On the other hand, below this network is an infill of *komai*, villages, hamlets, farms and villas whose density rises and falls, period by period, according to economic (economic development or decline, impact of specific land use[48]) or social factors.

3rd level – rural sites.

The rural occupation of the landscape, probably the least well known segment of anthropic landscape, at least in the areas of Boeotia not covered by intensive surveys. I should point out here that the analysis of the Boeotian archaeological record carried out in this work has shown that at least for the region of Boeotia, from the picture provided only by extensive surveys and excavation data, the rural segment is certainly under-represented and the rural pattern difficult to understand, especially in the long-term. On the other hand, the information collected in smaller or larger windows through field-by-field survey helps in shedding light on general and local trends, and also allows one to understand to what extent the information from the 'extensive' record is lacking in order to produce a meaningful picture. Although we cannot assume total recovery of sites even with intensive and systematic surveys (Bintliff-Snodgrass 1985 assume for Boeotia a recovery of 50-60% of Classical sites), the high density of sites found in certain areas of Boeotia, such as the Valley of the Muses and the Thespiae SE

---

[46] See above in the text and chapter II.3.14.

[47] For the later periods of Prehistory a network of nucleated hamlet-village settlements at around 2-3 km radius of territory across the cultivable zones of the province has been hypothesised (Fossey 1988 and Bintliff 2000a). From EH onwards "*some of the villages may grow into regional central places and, in the most flourishing periods, many villages reach urban status*" (Bintliff 2000a).

[48] In this sense, both the advent of plough agriculture and the Secondary Products revolution in the Final Neolithic-Early Helladic period, as well as the introduction of large estates in the Roman-Late Roman period, would have had significant impact.

area, which probably indicates high rate of recovery, allows us to make some remarks on the rural pattern. The existence of gaps between the sites would point to the possible presence of undiscovered sites, at different degrees of density according to the period considered (Bintliff-Snodgrass 1985: table 5; Bintliff 1991c: 156ff. and fig.29). We will use as an example an area for which intensive and systematically collected surface data are available, such as the surroundings of ancient Thespiae as far as the *chora* of Haliartos and including the Valley of the Muses and the *kome* centre of Askra.[49]

In fig.16 one can see the rural sites known for the Classical – Early Hellenistic period in the area. Farms are located in the surroundings of the town level sites, responding to a model which sees them as starting at 1.3km from the city. The model (already suggested by Snodgrass 1987/9 as an urban-rural model for the Classical period, examining data from the Thespiae survey) seems to work in the case of Thespiae, Haliartos and Askra, as well as in another area where intensive survey work has been carried out recently, i.e. the surroundings of ancient Tanagra (where Classical-Early Hellenistic rural sites start at 1.8km from the city[50]). Farms appear to start at a 1.3km straight-line distance from the city, and are also present in the area outside the 1h walking radius from the nucleated settlement site. The rural sites discovered closer to the urban centres are all from the Geometric-Archaic period (see three sites S to Haliartos, one site S of Askra and one E of Thespiae – mapped in figs.19-20). Generally speaking, apart from the area immediately around the *poleis*, the rest of the land seems to be shared 'regularly' in the Classical period (ca. 300m radius each farm – 5min walking distance - see Snodgrass 1987/9).

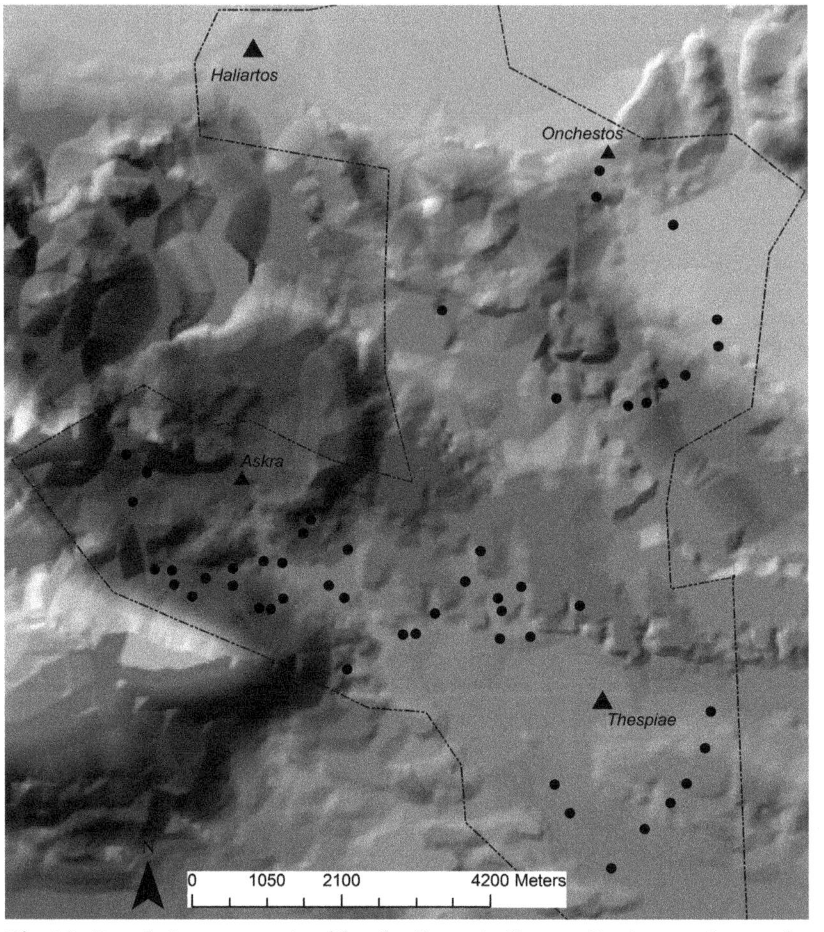

Fig.16. *Rural sites recognised by the Boeotia Survey Project as in use for the Classical – Early Hellenistic period in the area of Haliartos and Thespiae.*

Around Thespiae and Askra the pattern seems to respond perfectly to the urban-rural model and in general to the usual dispersed pattern recognisable also in other regions of Greece (Alcock 1993; Alcock 1999; Alcock 1993; Jameson – Runnels – van Andel 1994 for Argolid; Cavanagh et al 1996 and Cavanagh-Mee-James 2005 for Laconia). This is easily explainable as a system of small properties of land-owners and farmers exploiting the fertile area outside the area farmed by the town inhabitants themselves. In contrast, it is less easy to explain the pattern in the area N of Thespiae and E of Haliartos, namely the area where the Onchestos sanctuary lies (see chapter II.3.8). Here a settlement site would be possible by the Onchestos site also in the Classical-Hellenistic period (although better known is the settlement there in the Roman period, and in the Classical period it could have been only a small hamlet associated with the sanctuary), and the empty area to its S could be explained by the presence of unfertile rocky limestone hills. The other rural sites in the area around Onchestos are on the low hills of the Tertiary *tafel* and in the W part of the Teneric plain. Moreover, the emptiness around Haliartos might be explained by the relatively small area available which could be farmed directly by inhabitants of the city site. As seen above, the three rural sites known are all from the Geometric-Archaic period, when the city site was smaller.

The general filling of the landscape with rural settlement, which can be seen over all Boeotia in the Classical-Hellenistic period (see individual *chorai* chapters), shows

[49] The areas have been interested by the Boeotia Survey Project. Maps are accurate for the Thespiae S sector (published in Bintliff-Howard-Snodgrass 2007), while for other areas data are still at a preliminary stage.

[50] See also appendix I.14 – THE TANAGRA SURVEY PROJECT - for Tanagra. Also the absence of known rural sites in the small *chora* of ancient Akraiphia, if it is not a bias in research, could confirm the aforementioned Boeotian urban-rural model, being totally farmed by inhabitants of the city (see chapter II.3.7).

slight differences between the various *chorai*.[51] In the Thespiae area, we see the presence of medium-sized farms and hamlets rather than small farms (especially in the inner zones, where the urban centres of Thespiae and Askra are also located), and this seems to happen also in the environs of Thebes. On the other hand, finds in the area of Hyettos, Tanagra and Haliartos conform to the picture of relatively low rural density and a concentration of small to medium farms noticed also in other Greek surveys, in contrast to the picture of high rural density, including large Classical rural sites found in the survey sector S of the city of Thespiae.

Generally speaking, we can also say that farms in the Classical-Early Hellenistic period are usually not at network places (Bintliff-Howard-Snodgrass 2007). Their inhabitants choose the area they like, once living in the countryside is a matter of choice rather than a strict necessity.[52] Rural cemeteries (few of them are known from extensive survey, but some can be spotted by artefact surface surveys as they are easily recognisable due to their distinctive associated material culture[53]) are associated with farms but are usually not in the same location (Howard 2007), and not really associated with roads, while the city cemeteries were associated with the road network (see *chorai* chapters for examples).

In the Late Hellenistic - Early Roman period, rural farms are almost non-existent in the archaeological record, and in general the sites show a retraction of settlement intensity, in accordance with the major decline in overall activity in all Boeotian *chorai*.[54] For certain areas the decline is also marked by specific historical events, such as in the case of Haliartos, whose destruction by the Romans in 171 AD is followed by a spectacular collapse which also affects the rural landscape. For other areas, the decline seems less pronounced, as in the case of Tanagra.

In the Mid to Late Roman period, the landscape undergoes a radical transformation with the appearance of villa complexes associated with forms of dependent labour.

Around the late 4th C AD, a radical change occurs in many parts of Boeotia, with the population rising again and the appearance of Roman villa and hamlet sites (Bintliff 1991b fig.6; Bintliff-Howard-Snodgrass 2007 fig.9.15). It seems that some areas do not participate in this recovery, such as the *chora* of former Haliartos, which never recovered after the destruction.

The peak is observed in Late Roman times, when the total surface area of rural settlements has been measured as twice the Classical-Hellenistic component (in the Thespiae area), redirected into a new pattern of land use, associated with valley soils and marketable and transportable crops as a major product (Bintliff-Howard-Snodgrass 2007: 154-158). The phenomenon has been noted elsewhere in Greece, and seems to be recurrent in Greek mainland rural landscape (Alcock 1993; Jameson – Runnels – van Andel 1994 for Argolid[55]; Cavanagh et al 1996 and Cavanagh-Mee-James 2005 for Laconia). These new sites sometimes occupy the locations of previous rural sites, and sometimes choose new locations, usually somewhat prominent in the landscape, whether in open locations, or on lower terraces, within a river valley. The picture of the rural landscape in the Late Roman period differs considerably from the one of the Classical/Hellenistic period, and differences are related to both the topographical setting and the size of the sites, as well as to the distance from the city.[56] Along with the changing of the property system (Alcock 1993; Alcock 1999; Bintliff 1991b; Bintliff et al. 2008), the larger estates (run by elites with the subordinate work of local farmers and slaves) appear to be closer to the urban centre than were the rural sites of the previous periods. For this, see especially the case of Tanagra, with (Roman)-Late Roman villa sites close to the city, where in the Classical period there were urban cemeteries (see appendix I.14 – THE TANAGRA SURVEY PROJECT). Revealing is the case of the *polis* of Haliartos, destroyed in 171 AD and occupied apparently only by rural establishments in the Roman and in the Late Roman period. A large villa also occupies the site of the Onchestos sanctuary (see chapter II.3.8 for details). A general picture of known sites of the Roman and Late Roman periods can be found in figs.25-26 and figs.27-28 respectively.

[51] For general maps of Boeotia in the Classical-Early Hellenistic period see figs.21-22 and figs.23-24.

[52] Certainly, we should not forget that rural sites involved in analysis and evaluation of settlement pattern might not be contemporary, and some of them could be short-lived. A large debate is still going on as regards the question of residential/non-residential/temporary farms (see Foxhall 2004, Osborne 2001, Pettegrew 2001; Osborne 1992; Snodgrass 1990; Snodgrass 1987: 117-19; Osborne 1985b as well as the survey project publications).

[53] Namely very few freshly broken pieces of fine ware (black-glaze usually), resulting from recent ploughing.

[54] This was already noted by Alcock 1997, in addition to the Boeotia Survey Project (Bintliff 1991b for an overview on Roman Boeotia using survey data available up to that time).

[55] In the South Argolid survey they reach the same results analysing their survey data using Thiessen polygons (Jameson-Runnels-van Andel 1994, fig. 6.20).

[56] These locational changes are discussed in appendix I.14 - THE TANAGRA SURVEY PROJECT- and chapter II.3.14 – RURAL SEGMENT.

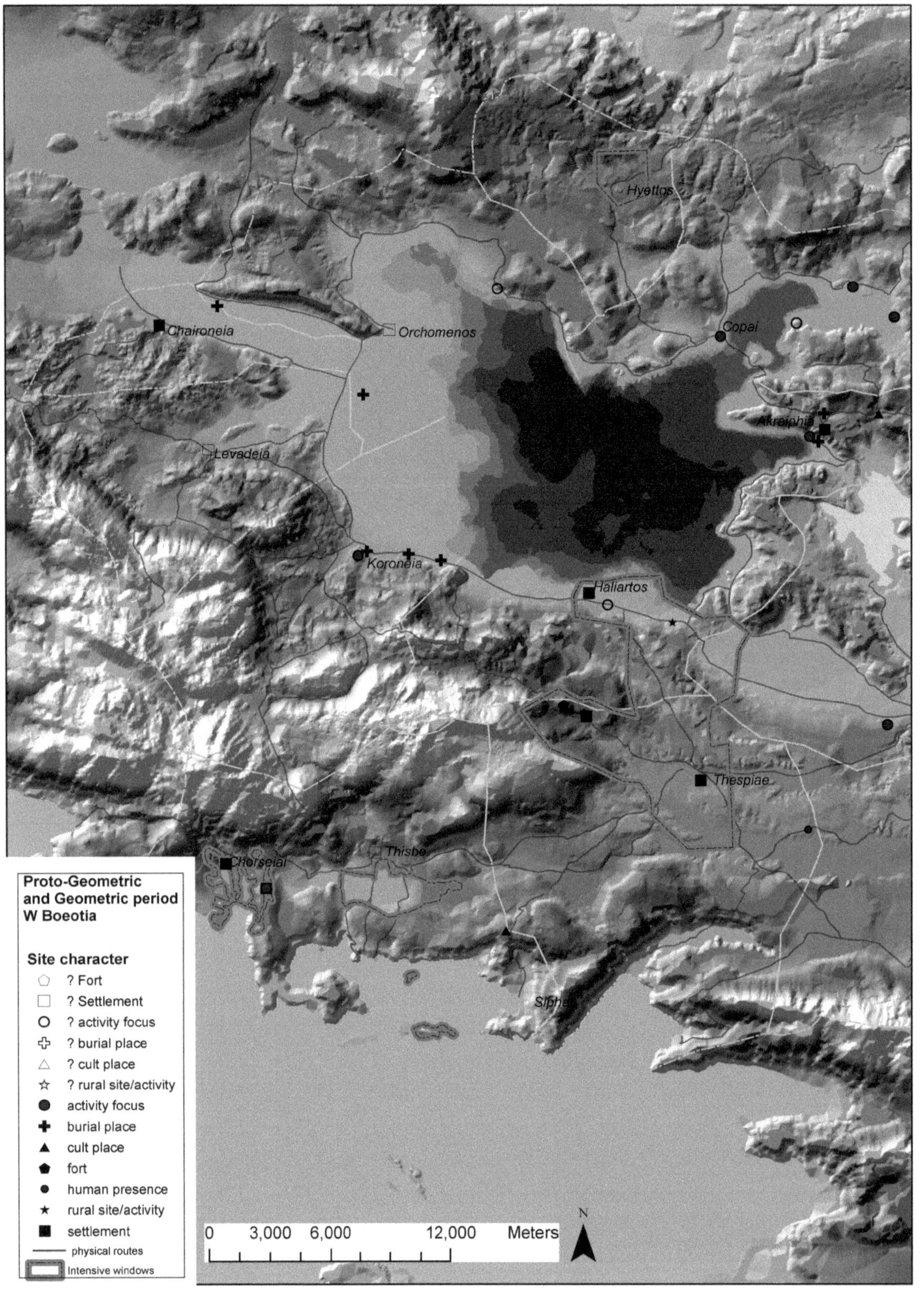

Fig.17. *PG and G map of W Boeotia*

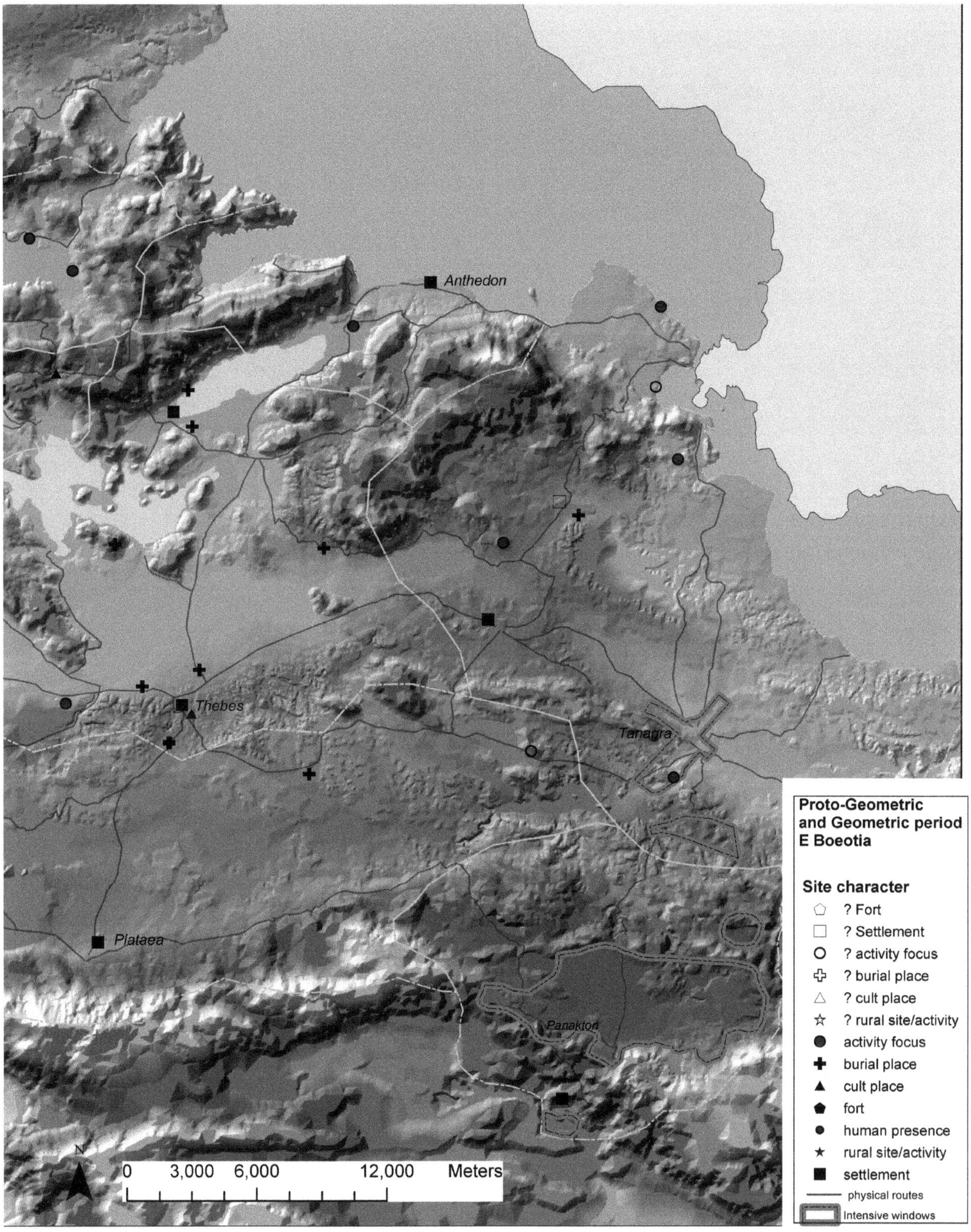

Fig.18. *PG and G map of E Boeotia*

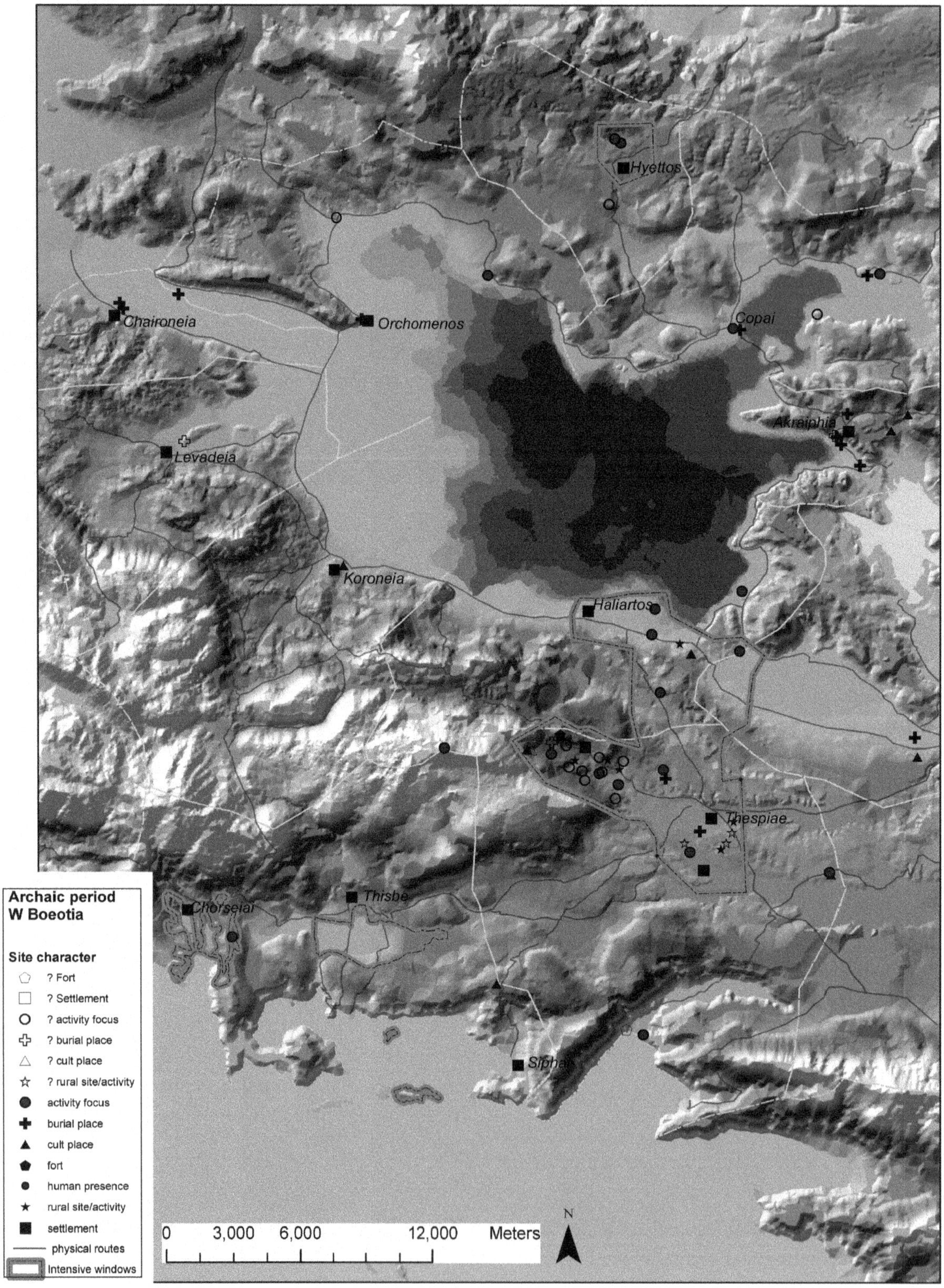

Fig.19. *A map of W Boeotia*

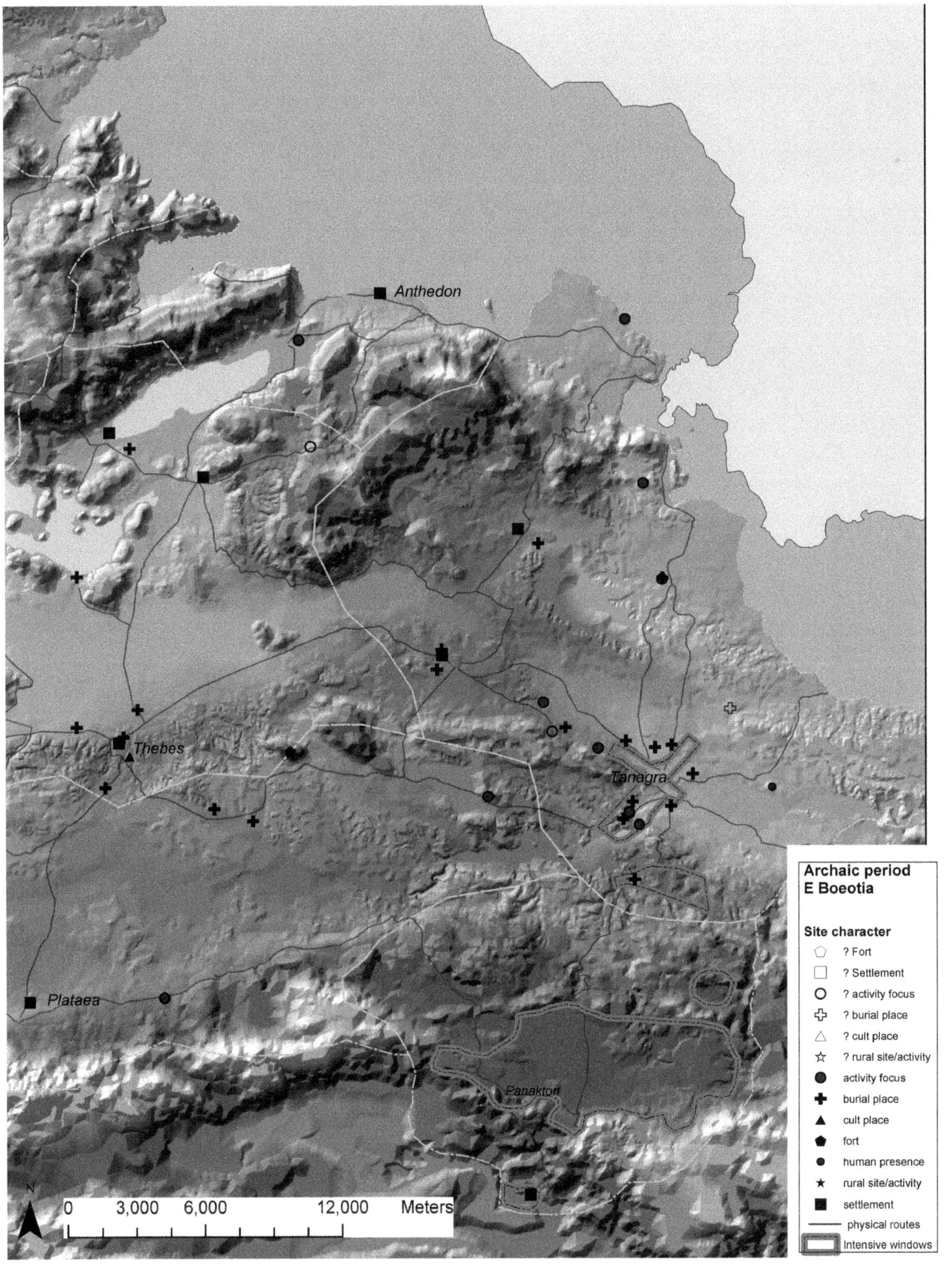

Fig.20. *A map of E Boeotia*

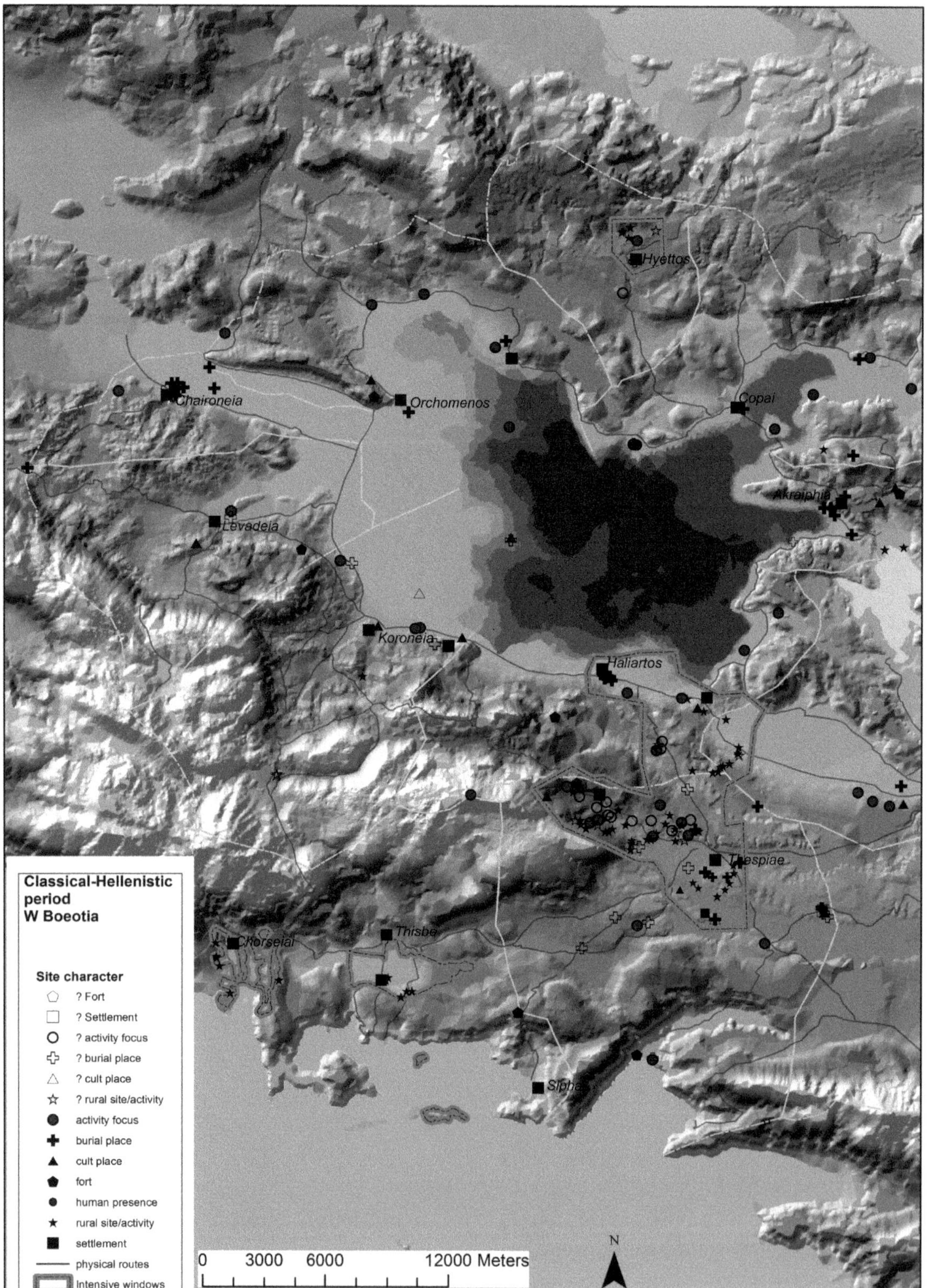

Fig.21. *C-H map of W Boeotia*

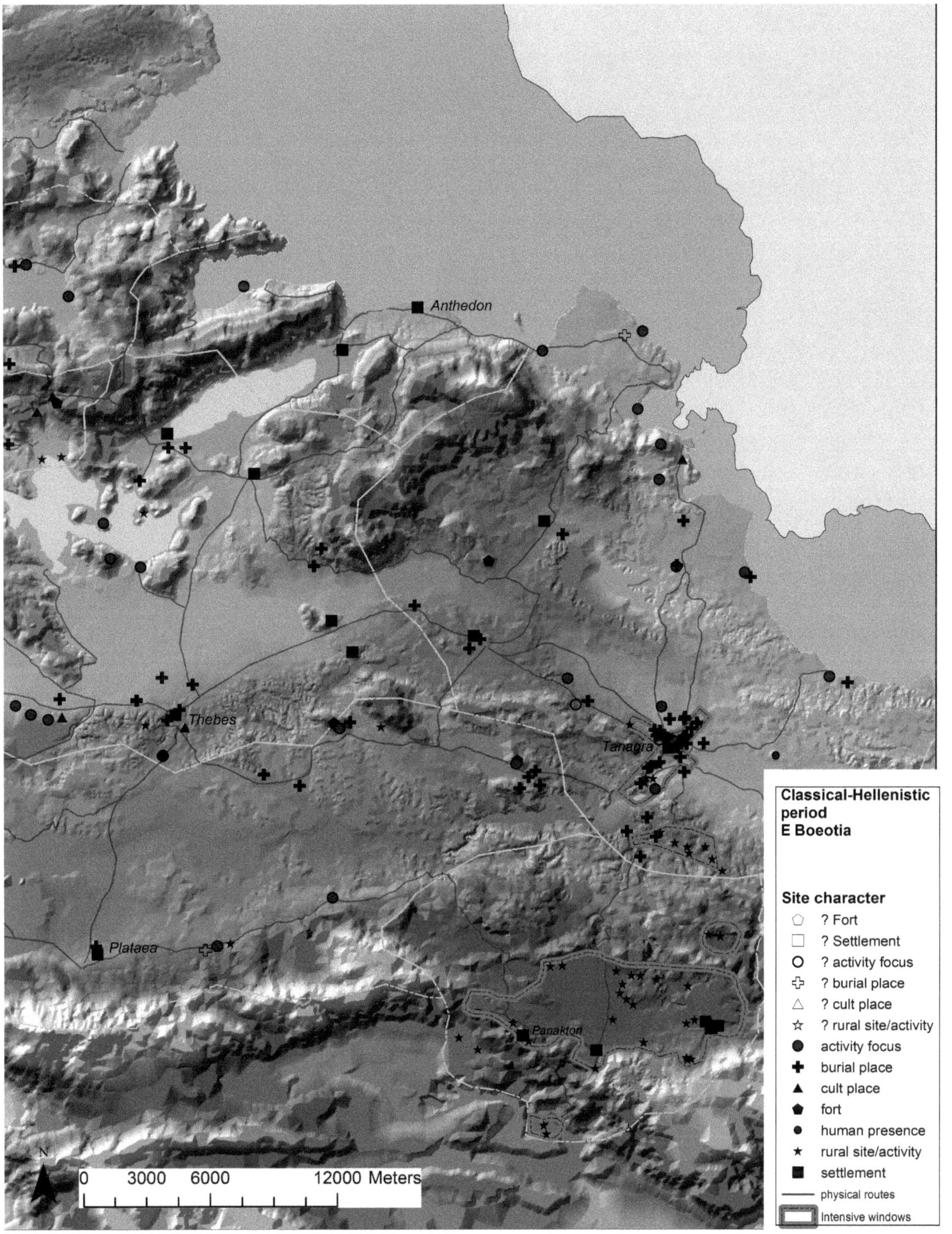

Fig.22. *C-H map of E Boeotia*

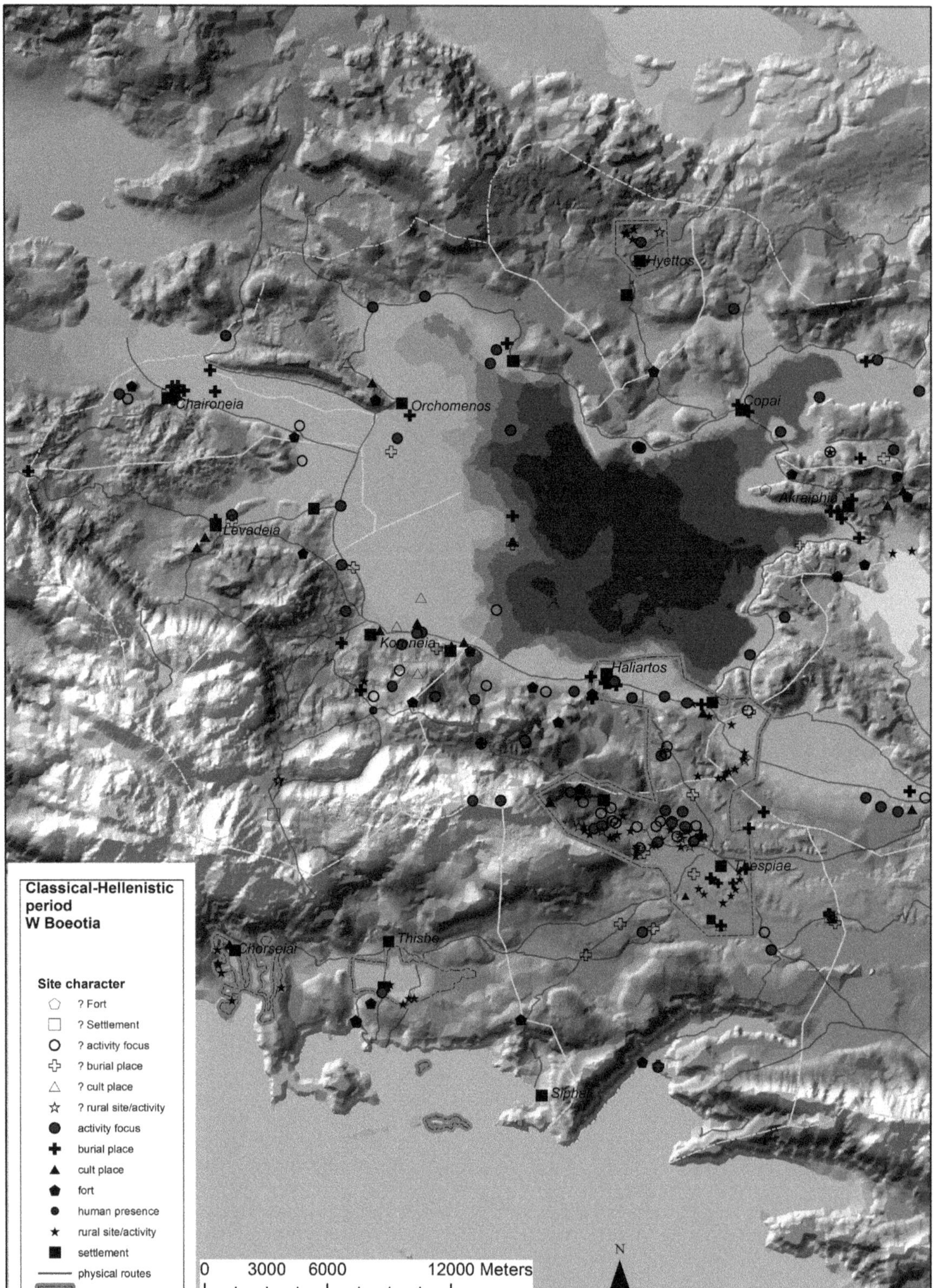

Fig.23. *C-H map of W Boeotia (with the addition of the components generally dated to the Greco-Roman period)*

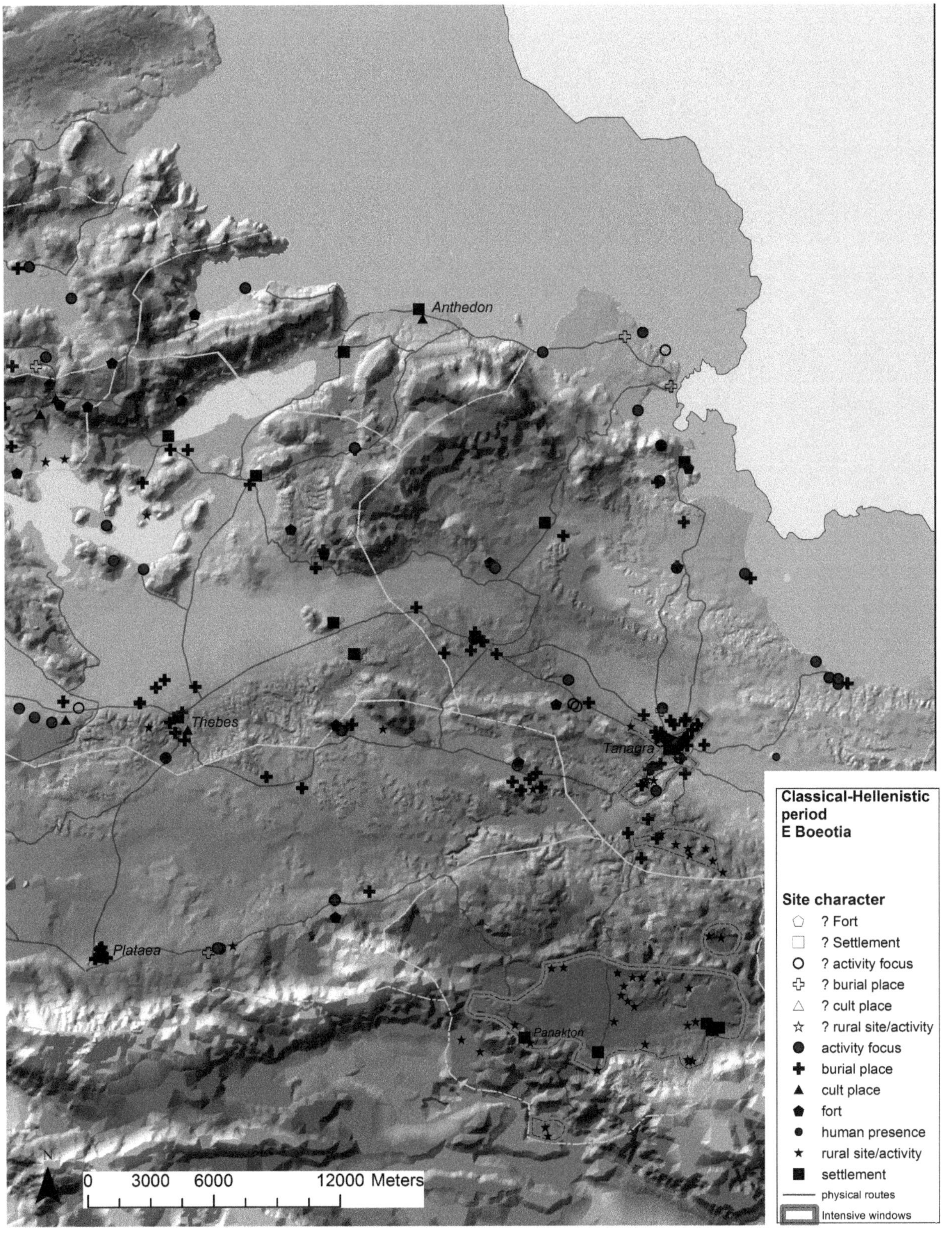

Fig.24. *C-H map of E Boeotia (with the addition of the components generally dated to the Greco-Roman period)*

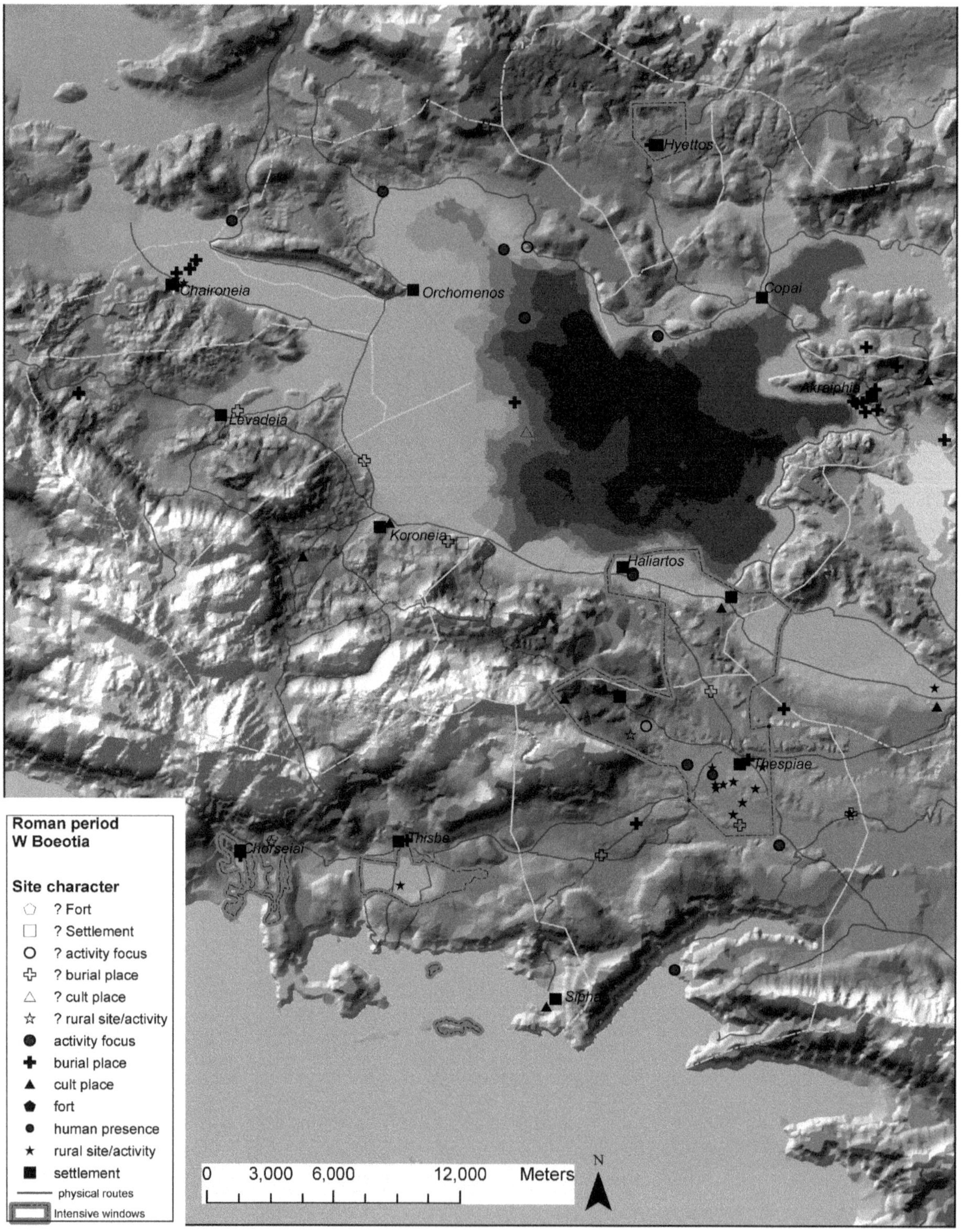

Fig.25. *R map of W Boeotia*

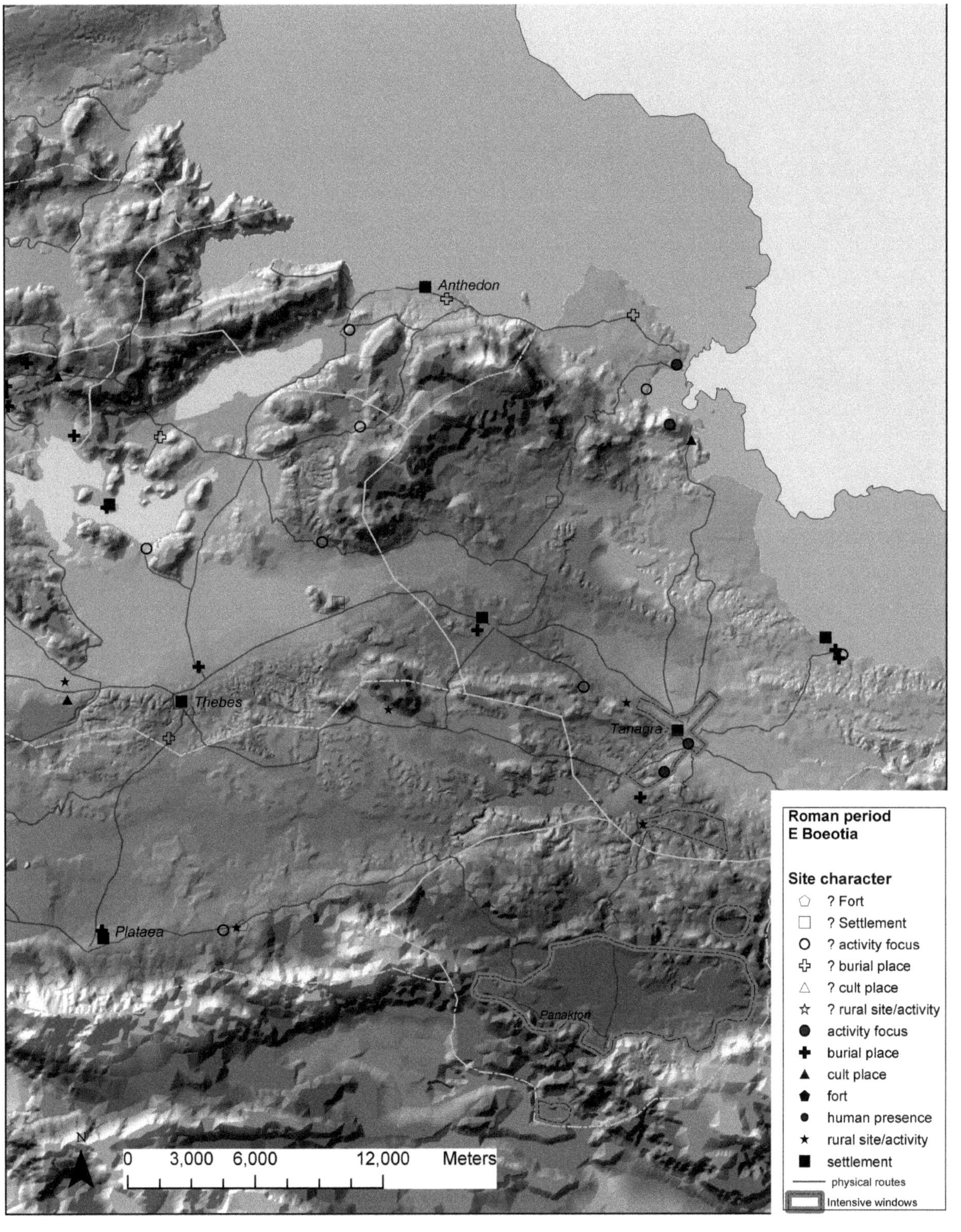

Fig.26. *R map of E Boeotia*

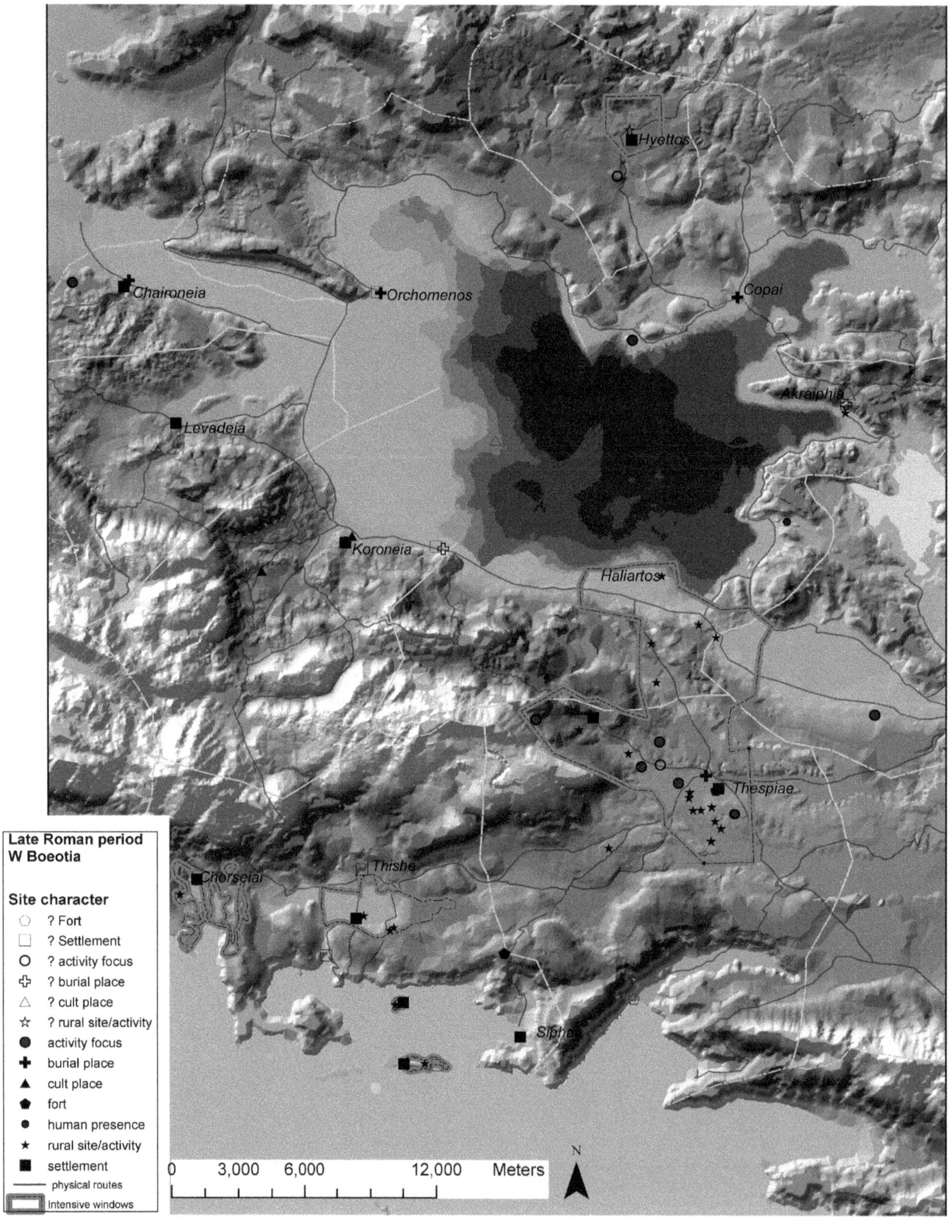

Fig.27. *LR map of W Boeotia*

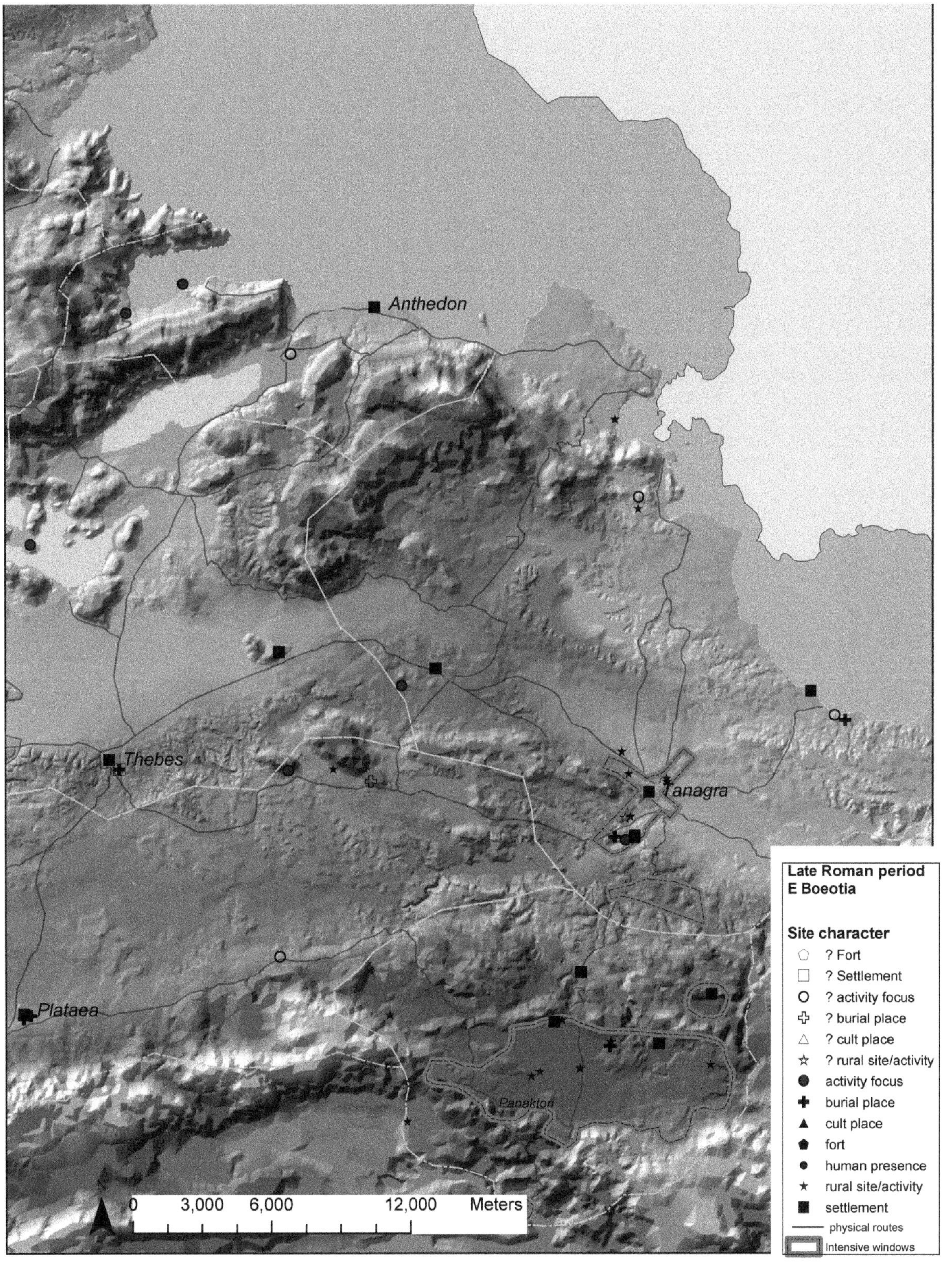

Fig.28. *LR map of E Boeotia*

THE LANDSCAPE OF THE BOEOTIAN CONFEDERATION

According to what has been pointed out earlier in this chapter, the district-structures which proved to be typical of the Boeotian landscape also constitute the territorial basis of the Boeotian Confederation. We could therefore focus on landscape structures/patterns which clearly marked the landscape and could be linked to historical dynamics put into effect by the Confederation.
On the basis of that which has been stated and discussed above, there are several diverse elements which can give an impression of the possible socio-political and cultural landscape of the Boeotian confederation. Some key themes have been chosen for analysis: the socio-political community chambers (examined above), as well as the fortified landscape of the confederation, and the federal sacred landscape (discussed in the following section).

The fortified landscape of the Confederation

Fortified settlements are often the subject of research within landscape archaeology, as well as forts, watch towers and fortifications considered in their mutual relationships and in relation to the surrounding environment and landscape (physical and cultural) features. The elements marking these landscapes are studied according to the topography and the environment in which they are situated, as well as according to their historical and socio-political values, in relation to the community which realises, uses and lives with them.
Fig.29 shows the Greco-Roman forts (including those both precisely and imprecisely located, as well as precisely or imprecisely dated[1]). As one can see, the majority of the Greco-Roman forts (mainly dating or at least used in the mid-late 4th C BC – Theban Hegemony – which especially in the S part of Boeotia have a typical style and masonry[2]) are located on the hills separating the 'two Boeotias' (i.e. the Copais area to the NW and the

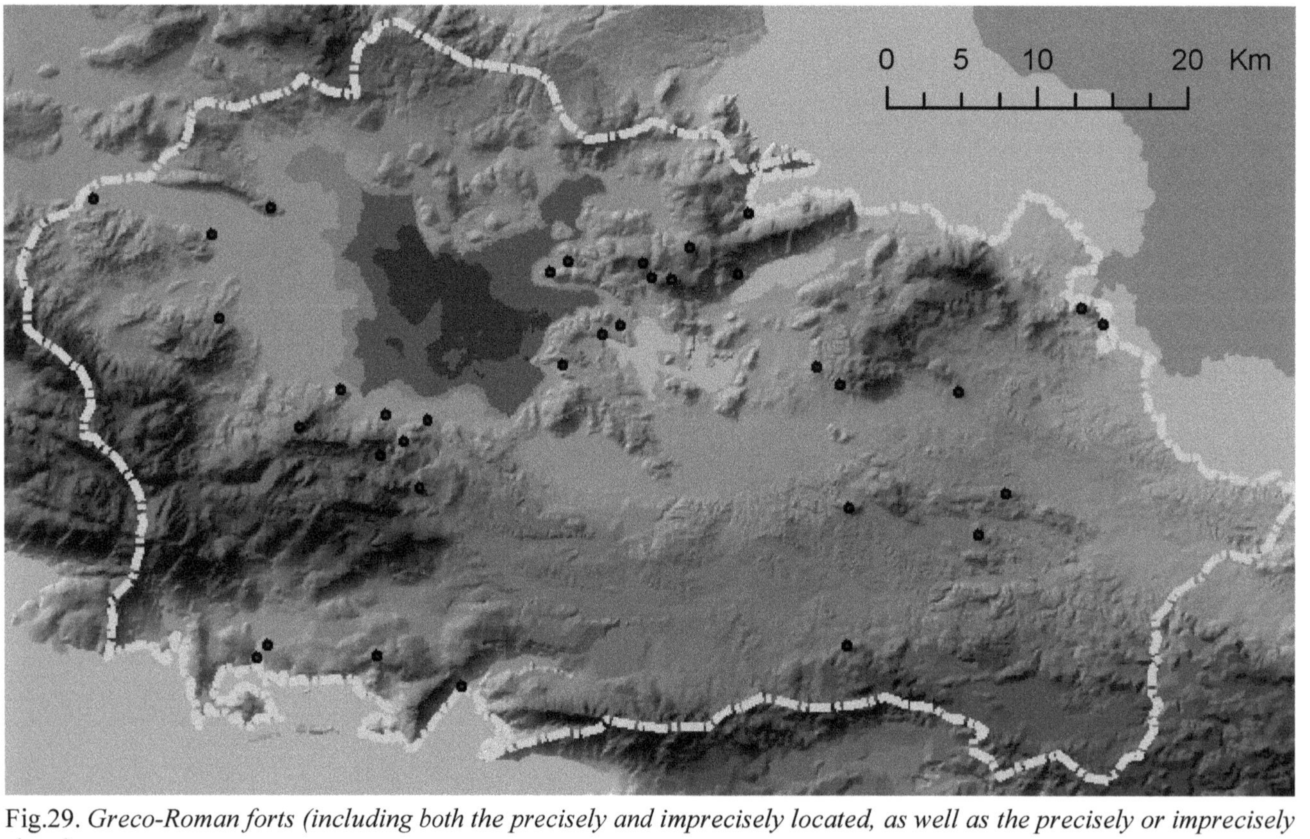

Fig.29. *Greco-Roman forts (including both the precisely and imprecisely located, as well as the precisely or imprecisely dated).*

[1] More certainly dated are the forts in the *chora* of Thespiae, carefully examined for masonry style by Fossey (1988: 142-3), and in general those in S Boeotia. The high presence of forts in the *chora* of Akraiphia is not dated with certainly (they are reported mainly by Lauffer, whose research was performed on the basis of Noack's work (AM 1894: 458-460), who thought the forts were Prehistoric – see chapter II.3.7).

[2] The typical S Boeotian style (mid-late 4th C) has been studied by Fossey 1992. On forts of mid-late 4th C see also, among others, remarks in Kallet-Marx 1989 and Camp II 1991 and Camp et al. 1992.

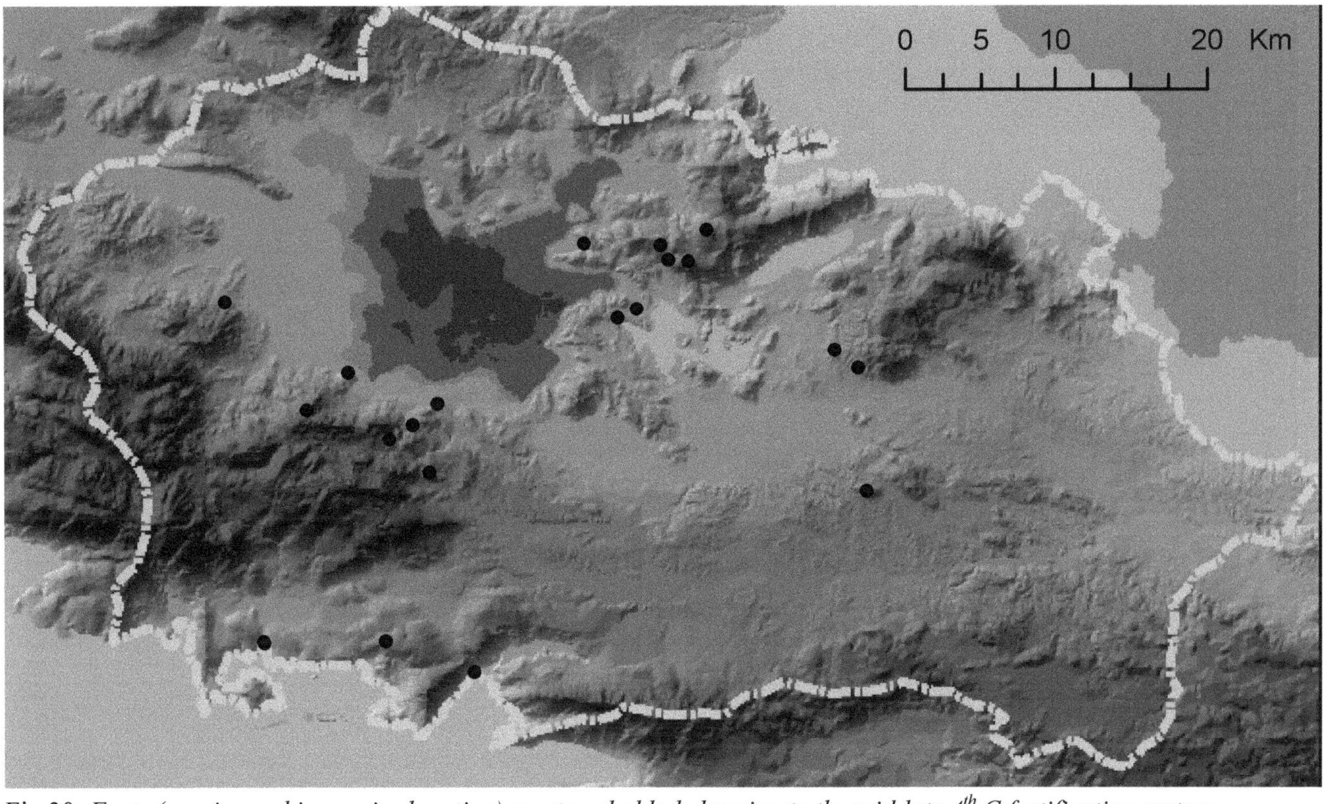

Fig.30. *Forts (precise and imprecise location) most probably belonging to the mid-late 4th C fortification system.*

Theban plains to the SE – see above).
Other forts have specific 'raisons d'être', such as those controlling bays or mountain passes (see for instance Ano Siphai – *component GC_13*). This is even clearer in fig.30, where the forts certainly or probably belonging to the mid-late 4th C BC – Theban Hegemony[3] – have been mapped. They are all concentrated in the central areas, with the exception of the forts overlooking and controlling the main harbours on the gulf of Corinth and those controlling the pass from the E plain of Boeotia to the Theban plain (the centrality of Thebes is evident from the layout of the forts that the city most probably controlled). The westernmost tower is located on Laphystion ridge (Camp II 1991).

Fig.31 displays the results of a viewshed analysis[4] carried out on the forts in certain positions and with most probability belonging to the mid-late 4th C fortification system. Only very general results can be obtained from the analysis and very general questions can be addressed, as the low resolution of the base DEM available and the scale of the map on which the positioning are based (see chapters 1.2.1 and 1.2.2) would not allow for further details and precision. In order not to bias the analysis further, the forts with uncertain or unknown position were removed from the dataset. In considering the entire region of Boeotia, half of it is probably controlled by these forts. The majority of the controlled area is constituted by the fertile lowland plains, the Copais, and the mountain passes, with the exception of the Asopos valley (Parasopia). The latter would have been controlled probably by some of the forts of uncertain or unknown position not included in this analysis. *Components PP_47* and *PP_56* (on the Soros and at Prophitis Ilias above the modern village of Asopia) would probably meet this requirement - once checked with viewshed analysis, it transpired that they controlled the Asopos valley, in the Parasopia and also towards Attica and the sea, especially *component PP_56*, though the date is not certainly 4th C. Worthy of remark is that all of Copais is controlled (always the hardest for Thebes to control – see above), as well as the crucial/strategic passage from Phokis through the Chaironeia basin (from which external invaders from the N always came throughout the whole of Boeotian history). Other strategic passes are also controlled, such as that from the S through the Koroneia basin. The forts of Classical-Hellenistic date are not there to control places but to relay enemy movements, since most of them are intervisible in chains.

[3] Theban hegemony over the whole of Boeotia, first of all, with the need for control over the other Boeotian *poleis*, which always sought autonomy (see Buck, Hansen, etc.).

[4] For the concept and methods of GIS viewshed analysis see among others van Leusen 2002: chapter 6 and Wheatley and Gillings 2002. Viewshed analysis has been performed here using the module/routine provided by ESRI ArcGIS 8.1.

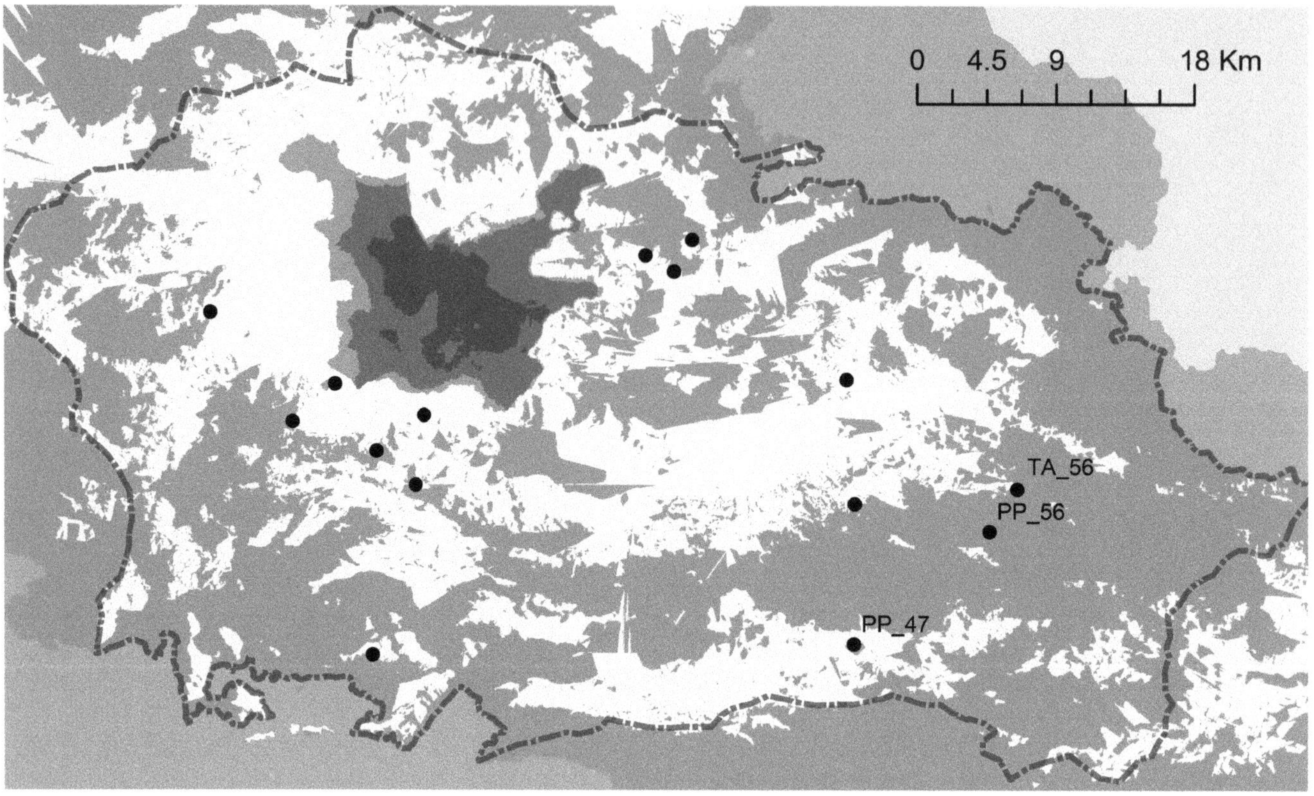

Fig.31. *Forts precisely located and with most probability belonging to the mid-late 4th C fortification system and viewshed analysis (visible areas are marked in light colour).*

The Tanagra area was apparently not controlled, or only partially, along the Asopos by *component PP_56* (4th C BC date uncertain), and in the N part of its territory by *component TA_56*. Probably the lack is partially due to the incompleteness of the archaeological record, but the examination of historical factors (not examined here because they fall outside the immediate purpose of the study) could provide a better explanation. Certainly, we do have the Anaphorites long wall in connection with the Megalovouno fort (*component TA_144*), and this would have constituted a historically defined border for Theban territory.[5]

The federal sacred landscape

In the Greco-Roman periods, archaeological evidence, and historical and epigraphical sources contribute to the understanding of features and relationships between elements of the past human 'sacred landscape' and the communities of which they are material expressions.
Within the Boeotian landscape, cult places are known which can be linked to landscape activity zones. For instance, cult places are known in mountainous areas suitable for pasturage, and they are dated to periods when pastoral activities flourish (examples in the Boeotian record could be the Agia Triada cave site, the Ptoion sanctuary). Cult places in mountainous or border areas, beside their cult/ritual role, often had an additional twofold function as border markers and signs of specific economic areas, to which the worshipped cult was often linked (see above). Cult places are also known in the plain and lowlands as possibly linked to agricultural activities or to the rural occupation of the landscape. Furthermore, cult places (or places of 'ritual'/sacred evidence) can be linked to the economic use and social relevance of marshy areas (see chapter II.3.1 and Farinetti 2008), e.g. the Classical-Hellenistic Herakles sanctuary by the source of the Melas – *component O_35* -,[6] the possible Classical-Hellenistic cult place site in the locality of Xinos attributed to Apollo – *component KO_69* -, the Agioi Taxiarchoi sanctuary of Roman date in the Pontsa locality – *components KO_31 and KO_32* -), or even to drainage works and activities. The Boeotian landscape is also marked by some physical landscape features which receive a sacred value/meaning in themselves, as in the case of the Sphiggion mountain, i.e. the mountain of the Sphinx, or the sacred landscape of Ypaton/Moni Sagmatas, between the Thebes and Tanagra areas.

When dealing with the landscape of the Boeotian Confederation, one could take a closer look at some cult places to which can be attached a particular meaning within the social landscape in the Classical period. In this

[5] As pointed out by Philippson (1951: 498), the road from Chalkis to Thebes would have run through the Anaphorites pass (400m asl).

[6] The links of Herakles to water and watery zones are strong. See Pappadakis AD 2 (1916): 217ff for a list of cults of Herakles in the Copais.

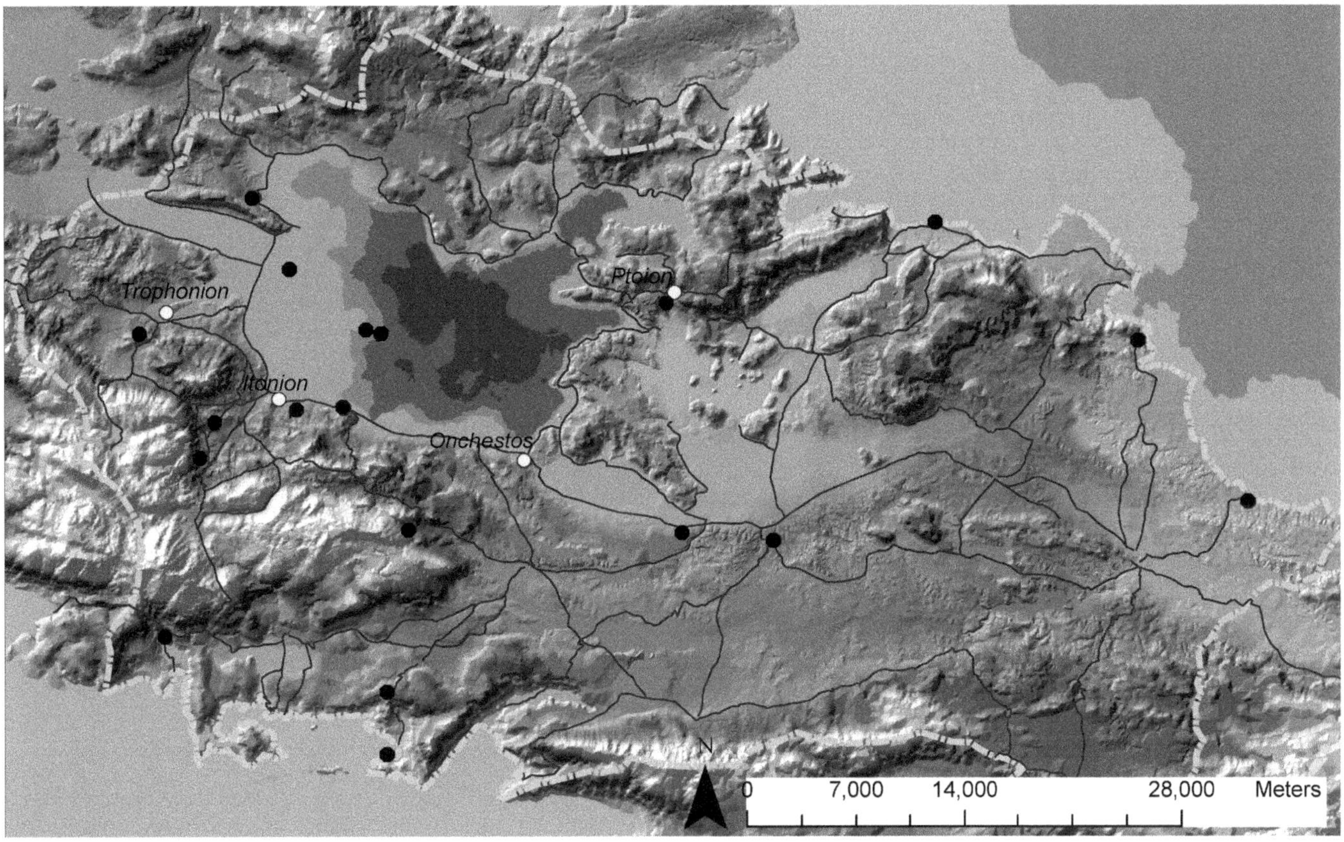

Fig.32. *Map showing Greco-Roman temples and sanctuaries and physical routes. Federal sanctuaries are marked in the light colour.*

way, we would find cult places marking the borders between communities, such as the temple of Zeus Laphystios (on the Laphystion ridge between Koroneia and Levadeia), the sanctuary of Poseidon at Onchestos marking a crucial passage between the Theban plains and the Copais area, or the sacred mountain of Ypaton between Thebes and Tanagra, as well as cult places which are linked to relationships between communities. The latter is the case for the federal sanctuaries, such as the Itonion by ancient Koroneia, dedicated to Athena Itonia, the Ptoion, dedicated to Apollo Ptoios on the Ptoion ridge, the sanctuary of Poseidon at Onchestos, and the prophetic shrine at Levadeia dedicated to the healing hero Trophonios. In addition, cult places are along and related to mountain passes, such as the Ptoion sanctuary, or along long-distance roads, which involve the issue of communication between communities).

In fig.32, we can see the cult places of the Greco-Roman period reported in the bibliographical record for Boeotia (comprehensive of federal, extramural and local sanctuaries) and their relationship with recognised physical routes. Cult places are located mostly along routes, either along roads exiting the city sites (extra-mural sanctuaries) or along long-distance communication routes, running both in the lowlands (for instance *component NC_32*, and *components KO_27/28* at Petra) and through mountain passes (for instance the sanctuary and *heroon* at Kastraki and Perdikovrisi on the Ptoion – *components AK_29 to AK_37*).

The federal sanctuaries are located along the main routes (see fig.32): at the pass between the Theban plain and Copais (Onchestos), along the main perilacustrine communication route joining Attica and Thebes to Levadeia-Orchomenos and Phokis (Itonion), along the mountain pass joining Copais with the area of Thebes (Ptoion), and along the route exiting Boeotia to the NW, at Levadeia (Trophonion). At the sanctuary of Athena Itonia (Itonion, see chapter II.3.1 and appendix I.1 for identification) the *Pamboiotia*[7] were celebrated, while at Onchestos, games, whose origin could be traced back to the Bronze Age, used to take place.[8] For the importance and socio-political meanings of the sanctuaries linked to the Confederation, see Schachter's book 1981-1994 and studies on the Confederation (Roesch 1965; Larsen 1968; Buck 1972 and 1979). For details of them see also the relevant archaeological components in the *chorai* chapters.

One could point out that even if non-federal cult places often lay along long-distance communication routes or along border lines as seen above, they were therefore anyhow linked to the relationship between communities, thus always involving the socio-political dynamics.

[7] Pausanias IX 34.1; Strabo IX 2.29. Ziehen 1949. The feast had an essentially military character (Roesch 1965: 107f.; Amit 1971: 57f.).

[8] Roesch 1965.

CONCLUDING REMARKS

One of the challenges of this work was to make the best use of the data sets available, setting up an environment within which heterogeneous, poor, incoherent and spatio-temporally incomplete data could be integrated in a meaningful way for the study of ancient landscape dynamics.

Different methods of enhancement have been used:

- Careful analysis and critical deconstruction and reconstruction of the archaeological record, as a way of monitoring metadata and constructing meaningful datasets, constituted by the enucleated components of the archaeological landscape.

- Integration with intensive survey results, which offer a more detailed picture of the anthropic use of the landscape through time. Integration primarily helped to provide a meaningful place in the landscape for known activity foci without a specific character, or for the few rural sites/activities discovered in more extensive research traditions. Sites previously floating in the landscape were now given a more meaningful interpretive context.

- Detailed analysis of aspects and levels of settlement and landscape use in the Greco-Roman period, and comparison with earlier or later periods of landscape occupation (Prehistoric and Ottoman-Modern) in order to detect similarities and differences, and enhance significances.

- The choice to analyse relatively small micro-landscapes, focusing primarily on the micro-region (each *polis chora*). This primarily helped to perform a more careful reference to the quality of collected data. It also helped the detection of micro-landscape trends within each micro-region, such as the relationship between the main *polis* and second rank settlements, the meaning of minor activity foci, as well as the evaluation of the use of diverse landscape zones.

- Application of theoretical approaches to the regional landscape, in particular the settlement chamber model, the community area theory and further implications, as a meaningful framework for the analysis of the available data sets and the long-term investigation of locational choices and physical and cultural characteristics of landscape zones and settlement chambers.

- Application of a land evaluation method for the definition of land capability. The results, together with the examination of other environmental variables, were used to evaluate the physical landscape characters of each *chora* landscape and therefore the agro-pastoral potential available for the *polis* territory, as well as to evaluate the character of less determined archaeological components (activity foci). This analytical framework also allowed qualitative evaluation of clearly defined and potential settlement chambers, and assisted in the process of their detection.

- Management of data within a GIS environment. The use of GIS helped the structured collection and management of the archaeological data, allowing for the integration of cultural and environmental landscape factors, acting as a useful tool for the analysis and assessment of micro-regional and regional landscape trends.

The theoretical approaches to the regional landscape applied in this work and illustrated in chapter I.1 proved to be useful for the understanding of the ancient landscapes of Boeotia. In particular, the concept of community area was central in the development of the settlement dynamic approach applied for the definition of settlement chambers over the Boeotian landscape, at the 1st and 2nd settlement level. It allowed to detect different areas with settlement potential in the long term, by analysing the presence and location of known settlements in different periods. In addition, the approach based on the concept of taskscape, which recognizes signs of tasks over the landscape so enhancing the community area theory which enucleates landscape zones, was applied in the interpretation of the archaeological record available. Within this framework, each single evidence was interpreted according to the character of a particular landscape zone, determined both by environmental characteristics and by the function, use and character of other archaeological landscape components. The analysis of the different activities in the landscape helped also in the detection of the wider settlement chambers.

Those analyses proved to be meaningful at the micro-regional level (see application in each individual *chora* study), provided the critical assessment of the archaeological record available. At a final stage, comparative analysis of the results obtained with the study at the micro-regional level was carried out and proved to be useful to detect settlement trends in different parts of Boeotia with specific environmental and cultural landscape characters (see chapter II.4 – COMPARATIVE ANALYSIS OF THE SETTLEMENT NETWORK), and partially enhanced also the comprehension of internal dynamics.

The methodology of assessing, evaluating and analysing the available data proved to be useful in the analysis of Boeotian landscape trends. Firstly, it helped in detecting micro-landscape dynamics and specific use of landscape areas, better evaluating the presence of known activity foci of more-or-less defined character within certain landscape zones.

Secondly, it proved useful in making the second rank settlements more visible, projecting windows in the landscape which would allow the second rank settlements to make their appearance. The second rank settlement level is otherwise somewhat neglected in regional studies in the area, as less clear and visible in comparison to first rank settlements (city centres/*poleis*) and to sites with a specific function (such as a cult place or necropolis).

Thirdly, a local level of analysis, resulting in landscape biographies of the various *chorai*, was integrated into a broad regional perspective. This allowed evaluation of historical patterns in terms of landscape settling in the region of Boeotia, detecting foci in the landscape around which settlement chambers would form, and exploring cultural choices over the regional landscape.

APPENDIX I

THE ARCHAEOLOGICAL RECORD

Abbreviations:

In the appendix the abbreviation *AE* followed by a number in square brackets, ex: [AE764], indicates a unit of archaeological evidence inserted into the database created for this work (see chapter I.2.2 for details). The units of archaeological evidence are listed in appendix III (available online). The abbreviations used for chronological periods are the same as in table 2 in chapter I.2.2. As for bibliographical references, the abbreviation AD stands for the journal *Archaiologikon Deltion,* AE for *Archaiologiki Ephimeris, Praktika* for *Praktika tis en Athinais Archaiologikis Etaireias*, CID for *Corpus des inscriptions de Delphes* - Paris 1977, while the other abbreviations used follow the American Journal of Archaeology abbreviation system as in the final bibliography. For ancient works, abbreviations follow the abbreviation system used in Liddell-Scott Greek-English lexicon.

Appendix I.1
Central Helicon: Koroneiake

THE ARCHAEOLOGICAL RECORD

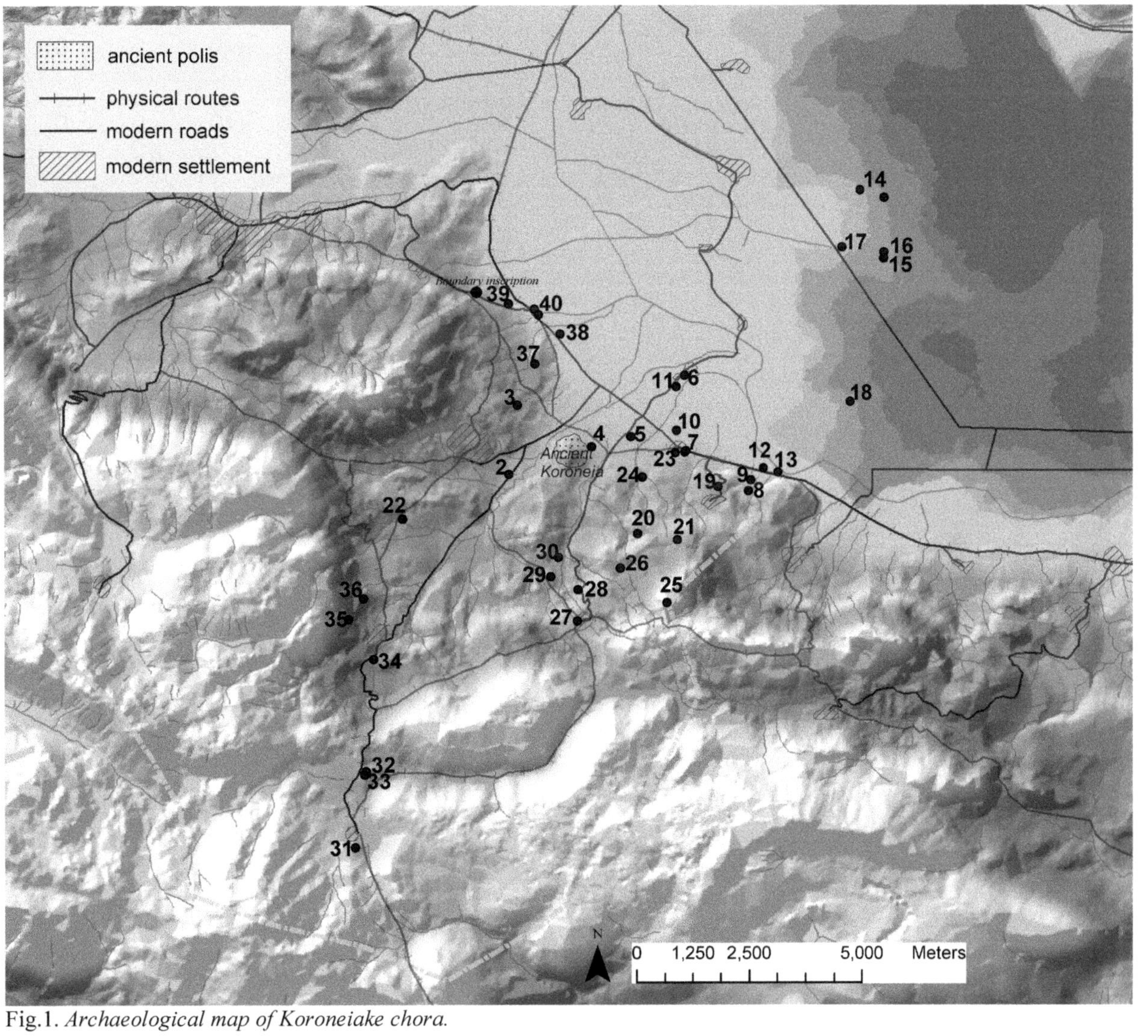

Fig.1. *Archaeological map of Koroneiake chora.*

1. PALAIA KORONEIA: *Components KO_1 to KO_10*

ID	Period	Type	Toponym	Discovery	Position
KO_1	C	settlement	Palaia Koroneia	Historical geography	certain
KO_2	N	? activity focus	Palaia Koroneia	Ext. topographical surveys	certain
KO_3	MH	? activity focus	Palaia Koroneia	Ext. topographical surveys	certain
KO_4	LH	? activity focus	Palaia Koroneia	Ext. topographical surveys	certain
KO_5	G	activity focus	Palaia Koroneia	Personal or group interest	certain
KO_6	A	settlement	Palaia Koroneia	Personal or group interest	certain
KO_7	H	settlement	Palaia Koroneia	Personal or group interest	certain
KO_8	R	settlement	Palaia Koroneia	Personal or group interest	certain
KO_9	LR	settlement	Palaia Koroneia	Personal or group interest	certain
KO_10	LR	burial place	Palaia Koroneia	Personal or group interest	certain

Components KO_1 to KO_10 refer to the ancient *polis* of Koroneia, including the settlement as well as burials and components within the *polis* area.

The ancient *polis*, situated at the NE end of the central ridge of Butsurati, is extended on two hills, one lower to the N, dominated today by the ruins of a Frankish tower, and another one higher to the S, the most prominent location of the ancient city. A fountain can be seen today below the ruins of the tower, and it most likely marks the place of a water source in antiquity also.

The two hills of the ancient *polis* of Koroneia are situated in the middle of two valleys, being the end part of a ridge which cuts the larger valley into two longitudinal sections (Butsurati). The Western one is wider, while the Eastern valley is narrower, and runs steeply down from the slopes of Helicon which constitute the Labethrion/Tilphousion ridge.

The city is mentioned in Strabo (IX 2.29) and Pausanias (IX 34.2). The site of Koroneia was already correctly identified by Leake 1835: 133-5.

Koroneia was a walled city when conquered by the Phokians in 353 (Hansen 1996: 91). There are remains of an undated polygonal circuit of walls around the acropolis, enclosing an area of 38ha (Fossey 1988: 325-6); another circuit of walls of an unknown date enclosing the lower town could still be seen in the 19th century (Frazer 1913: 70).

Small excavations carried out in the 1970s by the *ephoreia* (Spyropoulos) on the acropolis at different spots in the city area, recovered shallow archaeological strata. According to the excavator, the remains were disturbed by erosion and heavy cultivation, and although no prehistoric or Classical structures or deposits were found, structures datable to the LR - Early Christian period were (AD 1972: 317-318; AAA 1973: 385-392).[1] On the other hand, surface sherds (whose density on the ground is incredibly high, especially in some areas) indicate the presence of an urban settlement there since the G period (*components KO_1 and KO_5 to KO_8*), and excavations conducted by Pappadakis (BCH 1923: 521-2) on the acropolis E slope attest the presence of the R agora as well as structures referring to a LR settlement and cemetery (*components KO_9 and KO_10*). A theatre has been noticed for the city by travellers already [AE714] (Ross 1851 vol.I: 32, Gell 1819: 150, Frazer 1913: 170).

Surface sherds also seem to attest a possible presence of a Prehistoric activity focus at the site of the historical *polis*. Neo, MH and LH material are reported by prehistoric and multi-period gazetteers (HS&D 1979: G19; Fossey 1988: 326 and 1990d from his own fieldwork).

As Buck (1979: 58) states, in the Pherecydan tradition, several of Athamas' sons were eponymous heroes or founders of various Boeotian localities: Ptoos of Ptoon, Koronus of Koroneia; Onchestus of Onchestos, Orchomenus of Orchomenos, Sphingius of Mt. Sphinx. Most of his associations lie within the area that was probably under Orchomenian influence in the LH period.

Myth has it that Koroneia was one of the first cities conquered by the Boeotians, who came via Thermopyli and entered Boeotia through Chaironeia, and then conquered Koroneia and Orchomenos (Buck 1979: 80), and founded the sanctuary of Athena Itonia (see below). Archaeological data attest that most of Boeotia was parcelled out among various Boeotian communities, and by the end of the 9thC BC several *poleis* had been formed (Buck 1979: 91). Orchomenos, before the end of the 8thC, attempted to gain supremacy over all Boeotia, and it dominated Chaironeia, Hyettos, Olmones, Hyria, and Lebadeia and perhaps Koroneia and Copai before and during the time of Hesiod. The sanctuaries of Alalkomenai and of Itonian Athena lay fairly close to Orchomenos, and the Pamboeotia (taking place at the Itonion) would be easily under its control[2] (Buck 1979: 97). Lebadeia and Koroneia, probably after the defeat of Orchomenos by Thebes, broke loose and became independent if minor states. A fight between Orchomenos and Koroneia, with victory claimed by the former, occurred between 550 and 525 (dedication at Olimpia: Jeffery 1961; Buck 1979: 97).

The city was a member of the First Boeotian Federation (*Hell.Oxy.* XIX 3-4), the Second Federation (Diodorus Siculus XVI 35.3) and the Third Federation (CID II 74.50).

In 553/2 Koroneia was conquered by Onomarchos (Ephorus fr. 94a; Diodorus Siculus XVI 35.3) and became a Phokian stronghold (Demosthenes XIX 148; Diodorus Siculus XVI 58.1) until 346 when Philip had the city handed over to the Thebans (Demosthenes V 22, VI 13) who exposed it to an *andrapodismos* (Demosthenes XIX 112, 325). Shortly afterwards Koroneia must have been resettled, since in 337/6, one of the Boeotian *tamiai* in Delphi was a Koroneian citizen (CID II 74.50), as reported in Hansen 1996: 91.

Two important battles took place in front of ancient Koroneia in 447 and 394 BC.

Koroneia is reported to be one of the last Boeotian cities to fall into Roman hands together with Thisbe and Haliartos, in 171/0 BC ca. (Livy XLII 63.3). According to Livy (XLIII 4.11) the city was sacked and its population sold into slavery only to be later released by decree of the senate. Afterwards, Koroneia faded from the scene until the imperial period (Fossey 1990c), when the city showed vitality (see the Koroneia city epigraphical archive, mentioning also hydraulic work and activities - indication of a relative prosperity at Koroneia-, and then statues to the emperors[3] …). According to Alcock (1997) the type of Greek poleis in the Roman period were linked to the favour of the leading class in front of the emperor, and Koroneia

[1] Later Byzantine ceramics were found by the Frankish tower at the N foot of the acropolis hill (Lock 1986) and on the acropolis (Leiden-Ljubljana Koroneia city survey project 2005).

[2] Control of festivals was eagerly sought.

[3] An epigraphical archive, named the 'city archive', was found – see Fossey 1981-3. A statue of Hadrian was found in the water-reservoir discovered in the city site (BCH 1923: 521-522).

was probably in quite a favourable condition, as the archive seems to suggest.[4]

2. Ag. Athanasios: *Component KO_92*

ID	Period	Type	Toponym	Discovery	Position
KO_92	gr-rom	burial place	Ag.Athanasios	Personal or group interest	approximate

A small trial excavation by the church of Ag. Athanasios (probably to be located between the Ag. Georgios village and ancient Koroneia) brought to light some bases and some funerary *stelai* carrying names of ancient Koroneians [AE1673]. They are visible in the Museum of the church of Agia Triada in the village of Ag. Georgios (Pappadakis BCH 1923: 521-522).

3. Ag. Georgios - Xerokamara: *Component KO_87*

ID	Period	Type	Toponym	Discovery	Position
KO_87	gr-rom	activity focus	Ag.Georgios - Xerokamara	Int. topographical surveys	certain

Below the vulcanic E slope of Laphystion, 1.5km ca NE of Ag. Giorgios, a poor water source spills out, known as Xerokamara (Lauffer Kopais I: 119-120), that runs out today into a fountain made of ancient blocks, probably from ancient Koroneia [AE1165]. The slope goes down from Xerokamara another 100m ca and falls down into the 15m deep Phalaros valley. C sherds and tile fragments [AE 1166] are spread on the whole surface for an extension of 200-250m ca. from the fountain to the N (Lauffer Kopais I: sketch 129). The architectural fragments reused in the fountain might actually come from Palaia Koroneia and most probably are not associated with the site.[5]
Lauffer suggests that the place belonged probably to ancient Koroneia. It overlooked the road to the city and could have been either a fortification for the entrance into the Phalaros valley, as Lauffer seems to suggest, or a workshop. According to Lauffer, a settlement could not have existed in such a vicinity to the city (only 1km to the E on the right side of the Pontsa/Phalaros river). I visited the place, and today nearby it is a cave for building material and the surface in the whole area is very much disturbed. Actually the spread of the sherds over a quite extended area, according to Lauffer, might be due to post-depositional reasons. The place could be a workshop (which might have been linked to the quarry activities – see below *component KO_88* – Triapi) or a rural activity of some kind. It is probably too close to the city site to be interpreted as a farm. The corresponding component is left as 'activity focus' with a general attribution to the gr-rom period, since the C date given by Lauffer could not be proved in my visit. He certainly could have seen more in a less disturbed area at the time of his visit, even though he was not probably very familiar with historical Greek coarse ware (see below no.30 - Butsurati site).

4/5/6/7. ITONION: *Components KO_14 and KO_16 to KO_19* (4)*; KO_83* (5)*; KO_33 and KO_34* (6)*; KO_41* (7)

ID	Period	Type	Toponym	Discovery	Position
KO_14	A	cult place	Palaia Koroneia North	Personal or group interest	approximate
KO_16	C	cult place	Palaia Koroneia North	Personal or group interest	approximate
KO_17	H	cult place	Palaia Koroneia North	Personal or group interest	approximate
KO_18	R	cult place	Palaia Koroneia North	Personal or group interest	approximate
KO_19	LR	cult place	Palaia Koroneia North	Personal or group interest	approximate
KO_33	H	? cult place	Mamoura /Alalkomenai - Metamorphosis church	Ext. topographical surveys	approximate
KO_34	C	? cult place	Mamoura /Alalkomenai - Metamorphosis church	Ext. topographical surveys	approximate
KO_41	C	activity focus	Agoriani/Ag. Paraskevi	Personal or group interest	approximate
KO_83	gr-rom	? cult place	Thymari	Ext. topographical surveys	certain

Different hypotheses have been suggested for the identification of the ancient sanctuary of *Athena Itonia*,[6] mentioned by Strabo and Pausanias (Strabo IX 2.29; Pausanias IX 34.1) and known as a federal sanctuary of the Boeotian League:

A. **(4) Palaia Koroneia North** (*components KO_14 and KO_16 to KO_19*), according to the interpretation of Spyropoulos (mainly *Praktika* 1975: 392-414 and *Ergon* 1975: 12-26). At the NE foot of the ancient Koroneia hill, in the wider open space in front of it, Spyropoulos excavated some buildings, dated from the A period to the 4th C AD (A to R), whose remains were found deeply buried. Those structures appeared to the excavator as belonging to a sanctuary[7] and he wished to identify it with the Itonion. Papachatzis (1981: 216), among others, gives credit to this identification.

B. **(5) Thymari** (*component KO_83*), according to the interpretation of Pappadakis (AD 1916: 268) and Lauffer (Kopais I: 91ff). Lauffer locates it to the NW of Agoriani prehistoric mound, to the S of the new road to Kutumula/Koroneia, at the Thymari locality,[8] from where architectural fragments and surface walls are reported (Lauffer Kopais I: 91ff). The area is crossed by a palaeo-stream (see Philippson 1951: 42), probably the ancient Kyarios (see Lauffer's sketch in Kopais I: 91ff and chapter II.3.1 of this work for the identification of rivers in the area). To *component KO_83* the units of archaeological evidence AE1156,

[4] The so-called 'city-archive' was inscribed on the wall of a major building – probably in the Roman *stoa* by the Pontsa river (see below no.22). At least some inscribed pieces were reused in the monastery in the Pontsa locality.

[5] AE1165 (gr-rom reused) has not been assigned to any component.

[6] Schachter (1981: 122-7) from textual evidence demonstrates its continued existence from at least the 6^{th}C BC (and perhaps earlier) down to the 2^{nd}C AD. Though sources tell us that in 191 BC the Itonian was devastated by the Roman consul Acilius Glabrion (Acilio Glabrione) after the victory at the Thermopylae, Pausanias (IX 34.1) says that the general assembly of the Boeotian League still met there. At the sanctuary of Athena Itonia the *Pamboeotia* festival was celebrated annually (Buck 1979: 88-9 – Pausanias IX 34.1; Strabo IX 2.29; Polybius IV 3.5, IX 34.11). The *Pamboeotia* were supervised by a group of delegates chosen by and from the constituent Boeotian states. Certain evidence for the use of the Athena Itonia sanctuary as the Federal Centre comes only from the H and R times. The earliest datable document is the treaty recorded on IG IX^2 1.170 (Schachter 1981: 126).

[7] Concerning a tripod base in this sanctuary -reused- cf. Amandry BCH 1978: 565-9

[8] Between the Kyarios river and the new road to Kutumula/Koroneia, the area is crossed by a palaeo-stream (see also Philippson 1951: 42)

AE1157 and AE1159 have been assigned. The big limestone worked block found during rescue excavation at Petra Nikola [AE1158] could be part of the cult place, marked by the presence of architectural fragments and surface walls in the Thymari area, but this remains just an interpretation.

C. **(6) Mamoura/Alalkomenai** (*components KO_33 and KO_34*), according to the interpretation of Fossey (1988: 331) and Pritchett (1969: 85-9, pll. 57-64), who locate it in the Copais plain, in the area of modern Mamoura/Alalkomenai. Mamoura identification is very vague, as it is only based on epigraphical evidence [AE757] (Foucart BCH 1885: 427-433; Adler RE 1916 s.v. *Itonia*: 2375-6) and architectural pieces [AE755] (Pritchett 1969: 85-9, pll. 57-64 and AE1674 - AD 1915: 42 no.4) found at the church of the Metamorphosis, few metres NE of Mamoura/Alalkomenai village.

D. **(7) Agoriani/Ag.Paraskevi** (*component KO_41*). A short distance further away from the modern village, in the plain, by the Alalkomenai Railway station and S of the settlement of Ag. Paraskevi, nearby the prehistoric mound (*components KO_36 to KO_39*) on the road Thebes-Levadeia on top of the gentle slope, excavations (AD 1973: 272) found a building, containing a hoard of mid 5th C BC Boiotian coins, interpreted as a temple.[9]

Whatever the actual identification might be, there are four different cult places (or probably three, if one excludes Mamoura's vague evidence) to be added to the general picture of the cult places of the Koroneia *chora* (A: from A to R period; B: generally gr-rom period; C: C to H?; D: C and probably later).

Mamoura (recently renamed Alalkomenai) seems to be a medieval settlement (which reused ancient blocks for building construction) – Papachatzis 1981: 214 after Pappadakis AD 1916: 268. The settlement was at the border of the lake (fig.1 in chapter II.3.1).

For a possible identification of the Itonion, also see below *component KO_84* (Mamoura - Ag. Paraskevi), identified by some scholars with the sanctuary of Athena Tritonia. The component could be joined to *component KO_41.*

7. Agoriani/Ag.Paraskevi: *Components KO_40 and KO_41*

ID	Period	Type	Toponym	Discovery	Position
KO_40	G	burial place	Agoriani/Ag. Paraskevi	Rescue excavation	approximate
KO_41	C	activity focus	Agoriani/Ag. Paraskevi	Rescue excavation	approximate

On top of a gentle slope near the prehistoric mound (*components KO_36 to KO_39*), S of the settlement of Ag. Paraskevi, on the road Thebes-Levadeia, excavation works discovered remains of a LG cemetery (mainly cremation) - *component KO_40* (AD 1973: 272) - as well as a building of Classical date, containing a hoard of mid 5^{th}C BC Boeotian coins (*component KO_41* – AD 1973: 272; see above – ITONION).

8. ALALKOMENAI: *Components KO_21; KO_23 to KO_25*

ID	Period	Type	Toponym	Discovery	Position
KO_21	C	settlement	Solinari Vouno	Travellers	certain
KO_23	Preh	human presence	Solinari Vouno	Ext. topographical surveys	certain
KO_24	H	? Settlement	Solinari Vouno	Ext. topographical surveys	certain
KO_25	R	? Settlement	Solinari Vouno	Ext. topographical surveys	certain

Components KO_21, KO_24 and KO_25: at the entrance to, and in the S corner of, the small valley at the W foot of Mount Petra-Tilphousion, on a very small ridge (known as *Vouno*) at the very foot of the hills below Solinari (ca 1km E of the village). The ridge is formed into slight terraces. On one of these Knauss (Teiresias 1987) mentions a shrine dedicated to Konstantinos and Eleni. A sketch of the place is in Lauffer (Kopais I: 73, sketch 82).

Evidence has been found attesting the presence of a C, H, R site, probably a settlement site, at the entrance to the small valley. The identification of this place with the ancient *komai* of Alalkomenai is accepted by the majority of the scholars that visited or surveyed the area: Fossey 1988: 334-335, Knauss et al. Kopais 3: 42-68 (for other hypotheses see below).[10]

The settlement area was located first on the basis of surface material [AE737; AE738; AE739] and traces of fortification: incoherent ('polygonal' – Leake 1835: 135) walls [AE736], described and drawn by Lauffer (Kopais I: 73, sketch 82), and first noticed in the Vouno locality by Leake (1835: 135-6), by a small stream. The latter should be the Triton stream, according to Pausanias (IX 33.5) and Strabo (IX 2.36) accounts.

By the Triton river was also located the sanctuary of Athena Tritonia, known from the ancient sources and for which different hypotheses of identification have been suggested, varying according to the location of the site identified as the ancient *kome* of Alalkomenai, which the sanctuary was located next to (see below).

On the other hand, Lauffer (Kopais I: 73) suggests (following Pappadakis) a different location for the ancient *kome*, N of the small modern hamlet of Ag. Paraskevi, where other scholars place the Itonion sanctuary (see above - ITONION). They identify the cult place (*component KO_84*) as the sanctuary of Athena Tritonia/Alalkomenia (see below) and place ancient Alalkomenai nearby.

The ancient *kome* was called Alalkomenai, while its territory was called *Alalkomenia* (Hansen 1996: 79-80), and was very small (at least including the fortified mountain Tilphousion, according to ancient sources – Ephorus *FrGrHist* 70 fr. 153; Diodorus Siculus XIX 53.7).

The cult of Athena at the *Alalkomenion* (located at 300m ca N of Alalkomenai – Knauss et al Kopais 3: 50) was presumably an old one (Schachter 1981: 12), dating back to the A period. This fact might indicate that at least in the Archaic period Alalkomenai was a *polis* - Hansen 1995: 32-3- with the cult of its protecting god located in an extra-mural sanctuary (according to Stephanus Byzantius 68.18-9 the goddess had the epithet '*polias*').

Strabo (IX 2.36) retrospectively describes the site as a small *polis* (*polichni*) situated in a plain. According to Pausanias (IX 33.5) the settlement was just a *kome*, a '*polichni*' (just a shade of a polis) in decay at the time of his passage, 'low down at the foot of a mountain, not very high up'.

[9] Excavation work also discovered remains of a LG cemetery (mainly cremation) - *component KO_40* (AD 1973: 272).

[10] The ancient settlement of Alalkomenai is at the E edge of the basin and by the Petra promontory and Thilphoussa spring, where the old edge of the lake and today the Kephisos channel runs exactly at its foothill (Philippson 1951: 474; see fig.12 in chapter II.3.1).

Components KO_23: Prehistoric traces (obsidian chips) were noticed at the site by Fossey (1988: 332).

9/10. Sanctuary of ATHENA TRITONIA /ALALKOMENIA: *Components KO_63* (9) and *KO_84* (10)

ID	Period	Type	Toponym	Discovery	Position
KO_63	gr-rom	cult place	Solinari NE	Personal or group interest	certain
KO_84	gr-rom	cult place	Mamoura - Ag. Paraskevi	Ext. topographical surveys	certain

Component KO_63 and *component KO_84* could be characterized as gr-rom cult places. Attempts have been made to identify them with the sanctuary of Athena Tritonia-Alalkomenia.

9. *Component KO_63* (Solinari NE): a gr-rom cult place represented mainly by architectural fragments [AE1111] found to the N of ancient Alalkomenai, between the town site and the modern road, by the Triton stream. If we accept the suggestion of Knauss et al. (Kopais 3: 54 and map 2.18) this component has to be identified with the sanctuary of Athena Alalkomenia/Tritonia. Inscriptional evidence [AE1430] and architectural blocks reused in churches in the area [AE1431] could also be related to that cult place/sanctuary.

10. *Component KO_84* (Mamoura - Ag. Paraskevi): at the Livadi Dasou locality, between the Agoriani/Ag.Paraskevi hamlet and the railway station of modern Alalkomenai, between the road and the dam, ancient remains have been found at three spots, some excavated [AE754 - orthostats walls and foundations and AE756 - light foundations with column bases], others [AE1160] only seen on the surface and mentioned by Pappadakis (AD 1916). In addition, sherd concentrations have been found all over the wide area between the small hamlet of Ag. Paraskevi and the railway [AE1161 – recorded by Lauffer (Kopais I: 98)]. Those remains have been interpreted by Pappadakis (AD 1916) and Lauffer (Kopais I: 98) as the Athena Tritonia sanctuary [AE756 in particular]. They place ancient Alalkomenai nearby (300m ca) to the SE, as archaeologically represented probably by AE754 (excavated orthostats walls and foundations)[11] - see discussion above - ALALKOMENAI.
On the other hand, Knauss et al. (Kopais 3: 54 and map 2.18) do not agree and see *component KO_63* as the most suitable for the sanctuary and *components KO_21 and KO_24/25* for ancient Alalkomenai. Fossey does not seems to agree either.[12]
I think that the evidence found at Mamoura - Ag. Paraskevi could lead to the interpretation of the area as a cult place area (*component KO_84*), probably a large sanctuary place, which could even suit the identification with the large pan-Boeotian *Itonion*,[13] rather than as a town/*kome* and the nearby local sanctuary. It does not look like a proper location for a town - too far down in the plain. Therefore, I included the evidence AE754 (excavated orthostats walls and foundations) in the cult place *component KO_84* only, instead of also creating another component as 'settlement'. Scatters of sherds and finding spots of architectural features and walls (some of them excavated) extend over a quite large area (700x700m ca, following Lauffer description), quite reasonable for a pan-Boeotian sanctuary area. The *component KO_84*, therefore, may even be joined to the *component KO_41* (see above), representing a large sanctuary (*Itonion*?). On the map they (*components KO_84 and KO_41*) are mapped separately (fig.1: nos. 7 and 10).

11. Mamoura/Alalkomenai: *Component KO_35*

ID	Period	Type	Toponym	Discovery	Position
KO_35	unknown	? burial place	Mamoura /Alalkomenai	accidental	imprecise

Inscriptional evidence [tombstones AE758] from the area of Mamoura indicates the presence of a cemetery area nearby (*component KO_35*). According to Fossey (1988: 332), the tombstones from the area of Mamoura may indicate that a small settlement existed by the sanctuary, as often happens in the case of religious centres. In my opinion, there is no reason to infer this, since known evidence could represent a cemetery and sanctuary on the road, linked either with Alalkomenai or Koroneia. Though one should keep in mind that epigraphical evidence is very movable, a burial area might have been related to the cult place (sanctuary) at a short distance (see discussion above on *Itonion*),[14] or it could have been a cemetery area along the road at the edge of the basin, linked either with Alalkomenai or Koroneia (most probably Koroneia - see location on map fig.1).

12. Solinari – Nat Road: *Components KO_93 and KO_94*

ID	Period	Type	Toponym	Discovery	Position
KO_93	G	burial place	Solinari - Nat Road	Rescue excavation	certain
KO_94	LR	? Settlement	Solinari - Nat Road	Rescue excavation	certain

Component KO_93: EG cemetery found by means of rescue excavations and systematically excavated (AD 1994: 284-286), between the Solinari village and the drained Copais, to the right side of the old National road Athens-Lamia.
Component KO_94: Rescue excavations for the positioning of a gas-pipe (AD 1994: 284-286) found structures [AE1777] belonging to a settlement of Byzantine date together with a cemetery of G date (see *component KO_93*). Date is imprecise (indicated as Byzantine generally), but Greek scholars predominantly consider as Early Byzantine remains from the LR – Early Christian period as Early Byzantine.

13. Solinari Agios Ioannis: *Components KO_27, KO_28* (Tilphousion?) and *KO_29*

ID	Period	Type	Toponym	Discovery	Position
KO_27	C	cult place	Solinari Agios Ioannis	Travellers	certain
KO_28	H	cult place	Solinari Agios Ioannis	Travellers	certain
KO_29	LR	? burial place	Solinari Agios Ioannis	Other	certain

[11] Pappadakis and Lauffer are somehow forced to the interpretation of some of these evidences as the ancient town of Alalkomenai, as they recognize some of the evidence as the sanctuary of Athena Alalkomenia, which is supposed to be close to the town.
[12] Fossey (1988) does not seem to mention the sanctuary, but he locates ancient Alalkomenai at Solinari.
[13] For other hypothesis of identification see above ITONION.

[14] Cemeteries and burial areas linked to religious centres are quite common in the Greco-Roman period. See also, in the area, tombstones found at the site of the cult place by the Pontsa river (*component KO_30 and KO_32*).

The spur of Petra –Tilphosion or Tilphousion in ancient times[15]- constitutes a rocky projection of the ridge on which Koutoumoula/Koroneia is (otherwise called Libethrion/Tilphousion), towards the Copais basin.
At the foot of Petra rocky spur, along the passage which bordered the Copais lake, by the church of Agios Ioannis (Theologos for Fossey, the Baptist for Spyropoulos), Leake (1835: 135-136) noticed traces of ancient structures under the ruined chapel (ruined on his days, then reconstructed – picture in Papachatzis 1981: 210) as well as ancient sherds. Excavations by Spyropoulos (AAA 1973: 381-385; AD 1973: 271) discovered structures that could belong to a temple of C and H date. The excavator suggests an identification with the shrine of Apollo Tilphosios mentioned by Strabo and Pausanias (the same identification is also given in Knauss et al. Kopais 3).
Strabo (IX 413) refers to the temple of Apollon Tilphosios jointly with a a *krini* (fountain), that comes up from under the Tilphousion *oros* (mountain). It was most likely a karstic source, as many others around the Copais basin, that spilled out fresh water. Pausanias (IX 33.6-7) mentioned the sanctuary after the *kome* of Alalkomenai, in a totally flat area, close to the *kome* and by a small stream, called Triton. The Triton can be identified with the small stream by the Solinari village (Papachatzis 1981: 214 n.6). According to Lauffer (Kopais I: 67), the sanctuary site has to be looked for in the area below the rocky spur, between the ridge itself and the road, where also in antiquity there was enough space for a road between the rock and the lake.
According to Philippson (1951: 451), earlier the Copais lake reached up to the Petra spur, while today the channelled Kephisos would free up the passage for the road and the railway. This certainly was not true in ancient times according to the model (see chapter II.3.1 – PERILACUSTRINE MARSHY LANDSCAPE) and according to the archaeological remains (C and H – *components KO_27 and KO_28*) found in front of the spring at the foot of the spur. Therefore, the road used to run presumably here also in antiquity.
Papachatzis (1981: 214 n.6) suggests locating here the settlement of *Athinai* (identified with Kalami-Lioma site by others), on the basis of the passage of Strabo (IX 406): 'along the Triton were the small *poleis* (*polichnes*) Athinai and Eleusis, that were flooded by the lake'.[16]
A fort is also known to be on top of the ridge (see appendix I.8 for details).

THILPOUSIA source – almost disappeared today, the water spring mentioned in ancient texts[17] was still visible at the beginning of the century and in 1938 Lauffer saw it (Kopais I: 38 and 67,[18] and sketch 68). No archaeological remains have been found at the site, apparently.[19]

Lauffer (Kopais I: 98ff) reports that Pappadakis (AD 1916: 257, n.2) noticed an inscription of a decree (*psephisma* - IG VII 2858) close to the Tilphousia spring, at the foot of the Petra mountain, 'by the Agoriani village by ancient Koroneia', which was then brought to Lebadeia. In his report Pappadakis notes how the inscription was found by the '*Skepasti Vrisi*', by ancient Alalkomenai (20 minutes from the hamlet of Agoriani/Ag.Paraskevi, as stated by the author) and not by the Tilphousia source (that is 3km from the Skepasti one), at the foot of the Petra/Tilphousion mountain, 50 *stadia* distance from Haliartos, as mentioned in Pausania.
The inscription is described in IG (VII 2858) as a 'tabula' that starts with the formula: *Thios. Eratonos archontos Boiotois, Amphot[tei]/ Ischinao elekse*, and it should be from the period of the Confederation. Additionally, in IG the inscription is reported as having been found by the Tilphousia spring. The misunderstanding is probably only apparent. Considering that Pappadakis identifies as ancient Alalkomenai the *component KO_84* (see above), the *psephisma* was not found at the foot of Petra/Tilphousion spur, but by the Skepasti Vrisi by the Agoriani village, which I could not locate exactly.

Component KO_29: by the church of Agios Ioannis (Theologos for Fossey, the Baptist for Spyropoulos), along the passage which bordered Copais lake, at the foot of Petra (ancient Tilphousion), a grave was found and described by the excavator (Spyropoulos AAA 1973: 81-385; AD 1971: 271) as of a 'very late date' [AE743]. Later on rescue excavations for the positioning of a gas-pipe (AD 1994: 284-286) found a cemetery of Byzantine date [AE1676] together with a settlement (see *component KO_94*) a little further on. Date is imprecise (indicated as Byzantine generally), but burial and settlement could be linked.

14/15. Ag.Dimitrios East: *Components KO_11 and KO_20* (14); *KO_70* (15)

ID	Period	Type	Toponym	Discovery	Position
KO_11	gr-rom	burial place	Ag.Dimitrios East	Personal or group interest	approximate
KO_20	R	burial place	Ag.Dimitrios East	Ext. topographical surveys	approximate
KO_70	H	? burial place	Xinos	Personal or group interest	approximate

Components KO_20 and KO_11: Fossey (1988: 326) lists under Koroneia an evidence for burials of a Roman date (AA 1937: 19). Checking the original bibliographical record, it turned out to be represented by different units of archaeological evidence:

- AE735: a necropolis of Imperial date, probably represented on the surface by a concentration of surface sherds (3 km E of Degle and the Ag. Dimitrios village).
- AE1134 (gr-rom epigraphical evidence): inscriptional evidence, found 7km (?) E of the Ag. Dimitrios village, close to the end of the Central Channel, where a grave tumulus [AE1135] is also known. The inscription is published in IG VII 2931: a tombstone (?) with the name Demetrios.
- AE1135: a grave tumulus (isolated tomb/tumulus) noticed 7km E of Ag. Dimitrios village, close to the end of the Central Channel, where the inscription IG VII 2931 [AE1134] was also found (probably in the

[15] Several variant spellings for the name – see Meyer *Der kleine Pauly* 5 (1975): 831. For reference on ancient sources see Fossey 1990f: 169-204 and Hansen 1996: 76.
[16] See Kalami – Lioma for *Athinai* (though not by the Triton) and Xinos for *Eleusis*. On the question of identification see Fossey 1990d.
[17] Pindarus Fr. 198b; Aristophanes Boeotus *FgrHist* 379 (*krini*); Ephorus *FgrHist* 70 and Theopompus *FgrHist* 115 (*oros*).
[18] The Tilphousia spring, Lauffer says, is to be recognized in the fresh water spring at the N foot of the Petra promontory.
[19] Below the ridge, by the Tilphousa spring, the tomb of Teiresias is mentioned in ancient texts (Strabo IX 411, for the spring see also Pindarus fr. 181). A sanctuary of Apollo is also reported in this locality by Strabo (see above in the text). In the *Hymn to Apollo* 244-276; 375-387 an oracle by the Tilphousia source is mentioned. Ross 1851 (vol. I: 31) saw here an ancient '*Teinanhäufungen*' (associated to the tomb of Teiresias). When Lauffer visited the place (1938), those remains were no more visible, due to the construction of the road Thebes-Lebadeia and the railway crossing the area. Between them only few meters were left, in which Lauffer saw what he recognized as the Tilphousia spring (Kopais I: sketch 68). Today, the road has been enlarged even more, and the water source was obliterated.

area of Xinos – see Knauss et al. KOPAIS 3: 17ff – if this tumulus corresponds to the one Kambanis (1892) saw – see below *component KO_11*)

Accordingly, two components were created: *component KO_20* (a R burial place [AE735]) and *component KO_11* (a gr-rom burial place [AE1134 + AE1135]).

Component KO_11 might represent the tumulus site seen by Kambanis in 1892 (a tumulus with a marble block), and it is therefore probably to be located in the area of Xinos – see Knauss et al. KOPAIS 3: 17ff. Knauss though seems to imply that the '*inselartigen*' which Kambanis noticed was actually a small mound formed by the debris of a settlement (a sort of *tell*, 95.5m high), with burned clay pieces (perhaps, as Knauss says, from the mud bricks that the habitations were made of, probably not unlike some houses in the present day village of Xinos nearby). Following this interpretation, probably what Kambanis saw was not a tumulus, and therefore we could not locate *component KO_11* by Xinos. The location of the site remains therefore imprecise.

Component KO_70: Another burial evidence is known from the area, and has been recorded as AE1120. It is constituted by a HL tombstone, found in the vicinity of Xinos and mentioned by Knauss et al. KOPAIS 3: 17ff.

All those evidences could also be clustered together at the end, probably forming a cemetery area crossed by the modern channel. The area in the past might have been crossed by a road (in the area just outside the lake - see map in fig.1) and populated with burials and cult places (see no.16 - XINOS).

16. Xinos: *Components KO_65 to KO_72*

ID	Period	Type	Toponym	Discovery	Position
KO_65	MH	settlement	Xinos	Personal or group interest	certain
KO_66	LH	settlement	Xinos	Personal or group interest	certain
KO_67	C	? cult place	Xinos	Personal or group interest	certain
KO_68	C	? burial place	Xinos	Personal or group interest	approximate
KO_69	H	cult place	Xinos	Personal or group interest	certain
KO_70	H	? burial place	Xinos	Personal or group interest	approximate
KO_71	R	? cult place	Xinos	Personal or group interest	certain
KO_72	LR	? cult place	Xinos	Personal or group interest	certain

Components KO_65 to KO_72: In SW Copais, to the E of the former delta of the torrent Herkyna (and the overflow of the Kephisos) is the area of Xinos, probably a small former islet in the lake. Knauss et al. (Kopais 3: 17ff) found there, while following the 95 m. contour lines as the supposed regular extension of the Copais lake (see chapter II.3.1 – PERILACUSTRINE MARSHY LANDSCAPE), traces of a MH, LH and C activity focus (*components KO_65, KO_66 and KO_67*), marked by a concentration of surface sherds [AE1113 (MH), AE1114 (LH) and AE1115 (C)]. In the Prehistoric periods the evidence might represent a temporary or permanent occupation site, while for the historical periods Knauss et al. suggest the presence of a cult place. Also a burial area seems to be attested (*components KO_68 to KO_70*) for the gr-rom period. The assumption is based only on epigraphical evidence (Knauss et al. Kopais 3: 17ff [AE1121 and AE1120]), very movable, in apparent association with surface sherds [AE1115 for C period].[20] One could only suppose that the available evidence could represent a burial area close to the cult place (association quite common in the Greco-Roman world).

Components KO_67 and KO_69 and KO_71 and KO_72: the sherds (and black glazed tile) of the C period found at this place [AE1115], together with the inscription of Hellenistic times [AE1119 - dedication to Apollo Paiaion] and the presence of gr-rom architectural fragments [AE1118 - including a column drum], may be interpreted as traces of a cult place activity in C times (Knauss et al., KOPAIS 3: 17ff). Attempts to identify the site as a sanctuary to Apollo Paiaion have been made, considering the dedication to Apollo Paiaion found in the area [AE1119]. The Hellenistic dedication suggests the certain presence in the area of a cult place, and therefore the component for the H period is not accompanied by a question mark (*component KO_69*: H cult place). The presence of Roman *sigillata* ware and Late Roman potsherds (Knauss et al. Kopais 3: 17ff) can probably be linked to the continuation of the HL (and maybe earlier) cult place into the Roman and Late Roman periods. Even if inscriptional evidence is very much movable, its association with other remains could lead to the identification of the place as a cult area, at least in the H period. Considering only the known evidence for the C period [AE1115 and AE1118], the site could also have been a small rural site in C times, and it is only its most probable use as cult place in the following period (H and probably onwards) that led us to characterize the component for the C period as a possible cult place (rather than as a possible rural site/activity). On the other hand, in prehistoric times, the activity focus at the site has to be characterized as a form of dwelling associated with rural activities in the marshy area at the very border of the swampy lake (see chapter II.3.1 – PERILACUSTRINE MARSHY LANDSCAPE). An identification of the site with Homeric *Eleusis* has been attempted by Knauss et al.[21]

17. Xinos: *Components KO_64*

ID	Period	Type	Toponym	Discovery	Position
KO_64	R	infrastructure	Xinos	Personal or group interest	approximate

Component KO_64: to the W of Xinos site remains of a dike are reported, noticed already by Kambanis 1892 and then by Knauss et al. (Kopais 3 - map 2.2). Running in a SE direction, it crosses Xinos site.

18. Petra – Triton: *Components KO_73, KO_74 and KO_85*

ID	Period	Type	Toponym	Discovery	Position
KO_73	MH	? Settlement	Petra - Triton	Personal or group interest	approximate
KO_74	gr-rom	? activity focus	Petra - Triton	Personal or group interest	approximate
KO_85	LH	? activity focus	Petra - Triton	Personal or group interest	approximate

[20] Sherds from the Hellenistic period are usually not noticed by people who are not familiar with pottery from the gr-rom period (almost everything recognized as not prehistoric and as pre-Roman is usually mentioned as C), but could have easily been there.

[21] Along the Triton were the small *poleis* of Athenai and Eleusis, that were flooded by the lake (Strabo IX 406) (see Kalami – Lioma for Athenai and Papachatzis 1981: 214 n.6). As for the question of identification see Fossey 1990d.

In the area of the former delta of the joined streams Triton and Phalaros/Pontsa, in the locality of Petra, Knauss et al. (Kopais 3 1990: 62ff) noticed the presence of a large concentration of potsherds. Minyan (MH) sherds have been noticed, while the rest of the material could be generally dated as prehistoric, but authors do not specify. It was probably also an activity focus in the LH (Late Mycenaean) period. An inscribed tile fragment (R?) [AE1126] has also been found, at the foot of the stone bump noticed in the area [AE1125 – undatable building material], but it cannot attest anything more than a probable activity focus. Further research in the area could tell us about the presence of other material from historical periods, maybe among the potsherds of unspecified date.
An identification of the site with Homeric *Athenai* has been attempted by Knauss et al. (Kopais 3: 62ff).

19. Solinari: *Components KO_22 and KO_26*

ID	Period	Type	Toponym	Discovery	Position
KO_22	H	? burial place	Solinari	accidental	imprecise
KO_26	R	? burial place	Solinari	accidental	imprecise

Inscriptional evidence, which is very movable, indicates the presence of burials in the area of the Solinari village (one Hellenistic tombstone, IG VII 2976, and a Roman metric epitaph, *Chiron* 1976, 19 no.2).

20. Solinari - Agios Ioannis: *Component KO_82*

ID	Period	Type	Toponym	Discovery	Position
KO_82	gr-rom	? activity focus	Solinari - Agios Ioannis	Int. topographical surveys	approximate

According to Lauffer (Kopais I: 90), in the church of Ag.Ioannis, at Solinari, ancient architectural fragments were reused. According to Lauffer, they would be material from ancient Koroneia, and one could agree since the distance is relatively short (2.3 km straight-line). The church lies ca 500m to the W of the lower Ag. Paraskevi chapel (the one with the new church – see *component KO_30* Solinari - Agia Paraskevi), where the ruins of two ancient chapels are (see Lauffer Kopais I: 90, sketch 90). Ag.Ioannis is the Eastern one.

21/22. Sanctuary of Herakles Charops (*Charopeion*)

21. Solinari - Agia Paraskevi: *Component KO_30*

ID	Period	Type	Toponym	Discovery	Position
KO_30	gr-rom	? cult place	Solinari - Agia Paraskevi	Int. topographical surveys	certain

Epigraphical evidence reused [AE1101] in the old church of Agia Paraskevi, in the convent above the Solinari village. Though very movable, together with architectural fragments [AE1151] and surface walls (probably gr-rom) [AE1152] found in the vicinity, it could represent the evidence for a cult place in the area, although architectural fragments and reused material might come from Palaia Koroneia, as Lauffer (Kopais I: 88) notes. Evidence is not strong enough to be attributed a high degree of certainty and a component has been created as 'cult place'. A water source is nearby (Lauffer Kopais I: 88). Attempts have been made to identify it with the sanctuary of Herakles Charops (*Charopeion*), mentioned in ancient texts and never localized with certainty.

22. Pontsa - Agioi Taxiarchoi: *Components KO_31 and KO_32* (and *component KO_61)*

ID	Period	Type	Toponym	Discovery	Position
KO_31	R	cult place	Pontsa - Agioi Taxiarchoi	Ext. topographical surveys	certain
KO_32	LR	cult place	Pontsa - Agioi Taxiarchoi	Ext. topographical surveys	certain
KO_61	R	? Infrastructure	Pontsa - Agioi Taxiarchoi	Ext. topographical surveys	imprecise

By one of the sources of the Pontsa/Phalaros river, in the Pontsa locality, in the Ag. Georgios village area, close to a modern watermill, is a church dedicated to the Ag. Taxiarchoi (probably built on an earlier church of Middle Byzantine period). A Roman *stoa* [AE1103] along with a terrace wall that today sustains the road have been located and excavated by the local *ephoreia* (AD 1972: 316-7; AD 1973: 272; Fossey 1973: 452; Fossey 1974: 123). At the site an 'archive' of inscriptions has been found (Pappadakis BCH 1920: 388; BCH 1921: 522; Fossey 1982-3 and 1991: chapter 1), containing among others: funerary inscriptions, 20-30 in number, and an inscription mentioning the selling of certain terrains to the Sanctuary of Egyptian gods [AE1106].
An attempt of identification with the Sanctuary of Herakles Charops (*Charopeion*) has been made by Papachatzis (1981: 219). Though the vicinity of fresh water spilling out and running along the *stoa* can be in favour of a sanctuary of Herakles (a healing hero), the identification cannot be supported with certainty. Fresh water sources have been noticed also by the Solinari - Agia Paraskevi site (*component KO_30*), though probably not forming a stream as at the Pontsa – Ag. Taxiarchoi location. Regardless, even if of uncertain attribution, both components (*KO_30* and *KO_31/32*) can be assigned the character of cult places, more certainly the Pontsa - Agioi Taxiarchoi (probably a sanctuary if one takes into account the importance of the structures found at the site), and only probably the Solinari - Agia Paraskevi one.

Component KO_61: Among the inscriptions found to be reused in the church dedicated to Ag. Taxiarchoi (Pontsa locality), by the site of the Roman and Late Roman cult place (*components KO_31 and KO_32*), an inscription was found, mentioning hydraulic activities under the emperor Hadrian, which interested Copais and probably one or both of the two rivers which flow through Koroneia territory, the Kyarios and the Phalaros (Pausanias IX 34.5; see chapter II.3.1 – TOPOGRAPHICAL SETTING). The inscription was probably preserved in the sanctuary nearby (for the inscription see Fossey 1982-3 and 1991: chapter 1). The epigraphical evidence, referring to some work in the Hadrian period on the water of the Pontsa stream, in order to prevent impaludation and flooding, can give us information about the character of the landscape, which was marked by a surplus of water, at least in Roman Imperial times, which needed to be channelized and drained to get more land for agriculture (see chapter II.3.1 – PERILACUSTRINE MARSHY LANDSCAPE).

Kalami/Ag. Paraskevi: two water sources near the Prehistoric mound Kalami – Lioma (see below no.39), one 200m to the SE (Keramidi source) and another one 50m to the W, are reported. A third water source in the area, close to the Ag.Ioannis chapel, 400m to the SW of the prehistoric mound, spills out warm volcanic water (after Pausanias IX 34.5). Pappadakis (AD 1916: 258) has suggested the location here of the sanctuary of Herakles Charops (*Charopeion*). No archaeological remains would prove this identification.

------

23. Agoriani-Dekedes: *Components KO_36 to KO_39 and KO_12*

ID	Period	Type	Toponym	Discovery	Position
KO_36	N	activity focus	Agoriani - Dekedes	Personal or group interest	certain
KO_37	EH	activity focus	Agoriani - Dekedes	Personal or group interest	certain
KO_38	MH	settlement	Agoriani - Dekedes	Personal or group interest	certain
KO_39	LH	settlement	Agoriani - Dekedes	Personal or group interest	certain
KO_12	C	activity focus	Agoriani - Dekedes	Other	approximate

Low hill with a flat top, on the main road along the S edge of Copais, coming from Levadeia between the main turnings for Agios Georgios and for Solinari, to the S of the small hamlet of Ag.Paraskevi.

Visited because of its elevated position on the Copais basin, the hill shows signs of prehistoric presence (Neo to LH), marking a hilltop settlement site. It was possibly occupied at the beginning of Early Bronze Age and certainly occupied during Middle Bronze Age, it shows signs of strong occupation in LH III times and it was abandoned in Late Mycenaean times (*components KO_36 to KO_39*). The surface sherds go from Neolithic to Late Mycenaean: the main ceramic types of each period are represented (Lauffer 1940; Lauffer Kopais I: 101ff; Fossey 1988). Potsherds and other artefacts, including obsidian flints and stone axes [AE1105], as well as bones, seem to lie especially immediately below the terrace limit.
Knauss et al. (Kopais 3: 66) seem to imply that here should be the prehistoric centre of the area (the Homeric Koroneia[22]), which then moved to Palaia Koroneia in historical times (see chapter II.3.1 for discussion). On the other hand, Lauffer (Kopais I: 101ff) says the site can be identified with *Eleusis* (mentioned in Strabo IX 406).[23]

No evidence for any significant later occupation, 'later than Mycenaean' according to Fossey (1988: 335), is available for the site. On the other hand, Lauffer (Kopais I: 102) notes how on the flat top of the hill, where the prehistoric sherds were found in lower quantity in respect to the slopes, also isolated C sherds were found, in particular to the W, showing, he says, a later life of the place. Fossey (1988: 336) also found few isolated C black glaze sherds on the spot (*component KO_12*). I would suggest that the C traces of activity focus (few scattered sherds among prehistoric material) on the flat top of the hill can be a sign of the presence of a rural site there (though no tiles are specifically mentioned) or rather of a rural C cemetery (usually spotted by the presence of few black glazed sherds). The relationship between this evidence and the building excavated on the slope closer to the road and reported by Spyropoulos (AA 1973: 272) is not clear, since the exact position of these trenches, which uncovered a LG cemetery as well[24] (see no.7 - Agoriani/Ag.Paraskevi) is not clear. The unit of archaeological evidence AE753 (C surface sherds), for the time being is associated with a question mark to *component KO_41* (cult place, see above – ITONION), but forms also *component KO_12* on its own.

24. Solinari – Kaminia: *Components KO_86*

ID	Period	Type	Toponym	Discovery	Position
KO_86	gr-rom	activity focus	Solinari - Kaminia	Int. topographical surveys	certain

Between Trismagules and Agoriani runs a stream valley, called Kaminia because of the presence of an abandoned tile and brick workshop. To the NE of the Magoula mound (which is SE of Palaia Koroneia), immediately to the S of the confluence of the two small creeks, on a plateau, 700m W from Solinari, surface walls [AE1162], sherds and tiles [AE1163] are reported (Lauffer Kopais: 115ff).
Lauffer suggests that the evidence might correspond to a local sanctuary of Koroneia city (see Pausanias IX 34.7), dedicated to Koronos or Koronios. The R inscription IG VII 2873 [AE1164], found in Solinari village reused in a wall of Ag. Nikolaos chapel, 700 metres to the W of the site (Lauffer 1976: 41, n.49), mentions a temple dedication to Hero Koronios. According to Lauffer, the location of the place can speak for a possible sanctuary, similar to other situations in the Copais. In considering the position of the site and the evidence though, it could also be a rural site, and it would be worth a revisit. For the time being the component was recorded in the database as 'activity focus', generally dated to the gr-rom period.
Lauffer (Kopais I: 115ff) links this inscriptional evidence AE1164 to the remains found at Kaminia, 700m to the W (*component KO_86*). It corresponds very probably to a cult place in the area, but we cannot guess where it could have been. The AE therefore has not been assigned to a component, nor mapped.

25. Paliothiva: *Component KO_80*

ID	Period	Type	Toponym	Discovery	Position
KO_80	gr-rom	fort	Paliothiva	Travellers	certain

On the top of the PALIOTHIVA ridge (otherwise called the Libethrion/Tilphousion ridge) remains of a fortification (gr-rom fort) are visible, already seen by Ross 1851 vol. I: 31 [AE1149]. The fortified place on the plateau on the top of Paliothiva was apparently built in different times. A plan is in Lauffer (Kopais I: 82-6, sketch 94) and in Fossey 1990f (sketch with a plan in fig.2 p.173). Ancient tile fragments [AE1150] are reported on every side at the foot.
The site lay just above the saddle carrying one of the two routes leading out of the Zagara valley, as marked on the 1:100,000 Greek General Staff map (1943), as reported in Pritchett (1985: 155).[25] Probably it has to be interpreted as a post of observation rather than a refuge site (Pritchett 1985: 155), with a view

[22] Knauss et al. (Kopais 3: 66) point out that Koroneia hill for prehistoric occupation would be much higher compared to the other sites of Homeric centres in the Copais (240m compared to an average of 130/160m high). But also Kalami - Lioma can be seen as a potential centre for the Homeric Koroneia.
[23] For the question of identification see also Fossey 1990d.

[24] Probably to be positioned not on the slope of the prehistoric mound, but closer to the road.
[25] Pritchett lists the reference in travellers on Tilphousion-Palaiothiva: Ross 1851 vol. I: 31-32; Bursian 1862: 234; Bolte RE 1912 s.v. *Helikon*: 6; Schober RE s.v. *Tilphossion* (1936): 1045.

extending beyond Levadeia and Orchomenos to the W as well as to Haliartos and other sites to the E.

26. Tsoroko – Ag.Paraskevi: *Component KO_81*

ID	Period	Type	Toponym	Discovery	Position
KO_81	gr-rom	activity focus	Tsoroko - Ag.Paraskevi	Int. topographical surveys	certain

Tsoroko corresponds to the SW slope of Palaiothiva, where the upper chapel of Agia Paraskevi is (1.5km NE of the modern village of Kutumula/Koroneia). A fresh water source is nearby (Lauffer Kopais I: 89). Ancient foundation walls were found in the vicinity of the chapel (Lauffer Kopais I, sketch 100). A terrace wall was also noticed as forming a flat area, ca 100m below the chapel (sketch 101), quite clearly marked also on the 1:50,000 GYS map (forming a plateau).
The corresponding component has been created as a gr-rom activity focus, which could possibly be interpreted as a cult place on a terrace in a quite high position. Yet though the interpretation is much too vague and can be supported only by very movable evidence: an inscription [AE1153], built in the church nearby, refers to a temple of Artemis, together with walls sustaining the terrace [AE1154], which could have been realized with an aim different to agricultural purposes.

27. Kutumula/Koroneia - Ag.Taxiarchoi: *Component KO_75*

ID	Period	Type	Toponym	Discovery	Position
KO_75	gr-rom	human presence	Kutumula/Koroneia - Ag.Taxiarchoi	Int. topographical surveys	approximate

The finding of terracotta (a figurine) in the modern village of Kutumula/Koroneia near the chapel of Ag. Taxiarchoi, by which the Kephalovrisi source is located, can suggest human presence only for the gr-rom period, as general indication.

28. Kutumula/Koroneia - Ag.Giorgios: *Component KO_76*

ID	Period	Type	Toponym	Discovery	Position
KO_76	gr-rom	? activity focus	Kutumula /Koroneia - Ag.Giorgios	Int. topographical surveys	approximate

In front of the destroyed chapel of Ag. Georgios, to the NW of Kutumula/Koroneia village, a carved small pillar (*Profilpfeiler*) carrying an inscription lies on the ground [AE1139]. In the church is also built a base [AE1140] (Lauffer Kopais I 1986: 77-8). The evidence could correspond to a possible sign of activity focus (a cult place?) at the spot or in the vicinity. The evidence AE1138 (gr-rom reused) has been assigned to this component with a question mark, since we cannot say whether all the evidences representing this component (*KO_76*) belong to an archaeologically meaningful unit or not. As stated several times, epigraphical evidence and architectural fragments can be very movable (in this case the possibility is strengthened by the lack of building material). The component is marked with a question mark and has been positioned on the map at the spot of the Ag.Georgios chapel, though the evidence may probably belong to one of the already entered and mapped components.
Lauffer (Kopais I: 77-8) notes that the architectural material [AE1138] found reused in the other chapel of the village, could be from Palaio Koroneia, even though it is not impossible that an ancient settlement could have been even here, where the modern village of Kutumula/Koroneia is (see below – *component KO_77*- for the evidence of a cemetery area that would support this hypothesis). I would not agree, since there is apparently no place for another settlement (village) site there in gr-rom times (see chapter II.3.1), and the evidence, if not too weak for anything, could also refer to an extra urban cult place. See for a similar evidence *component KO_82* (Solinari – Ag.Ioannis), where though the architectural fragments reused could very probably have come from Koroneia, which the site is much closer to than *component KO_76*.

29. Koroneia – Kaminia: *Component KO_77*

ID	Period	Type	Toponym	Discovery	Position
KO_77	gr-rom	burial place	Koroneia - Kaminia	Int. topographical surveys	certain

In the inner Kakaris valley, 600m NW of Koroneia village, is a flat top hill, called Keramidopilion or Kaminia because of the presence of abandoned tile and brick workshops in the area (it is the inner SE hill of the Palaio Koroneia ridge - see sketch 85 in Lauffer Kopais I). There, ancient blocks of *poros* (porous limestone - travertine) have been found, as well as several bones and tiles from tile-roofed tombs (one of which is inscribed with *My* and *Omega* letters – Lauffer Kopais I: sketch 88). A component has been created as a burial place with a general gr-rom chronology. The site would be worth a revisit.
Lauffer says it could have been the cemetery place for the village supposed at Kutumula /Koroneia (see above *component KO_76*).

30. Butsurati: *Components KO_78 and KO_79*

ID	Period	Type	Toponym	Discovery	Position
KO_78	C	rural site/activity	Butsurati	Int. topographical surveys	approximate
KO_79	Preh	activity focus	Butsurati	Int. topographical surveys	approximate

On the SE slopes of the Megalo Butsurati ridge, the terrace-formation of the Kakaris valley, remains of ancient activities have been noticed by Lauffer during his intensive visit over the area (Lauffer Kopais I: 80).

Component KO_78: Few C black glaze sherds and a larger number of tile fragments are reported. Lauffer (Kopais I: 80) points out that the place would have belonged to Koroneia, because it is only 2km away. This is the natural conclusion in a logic of a survey oriented to the city / main settlement. For us, interested in the landscape as a whole, in its urban and rural forms, it is of interest that the place can be considered with certainty as a rural site/activity (farm) surrounding ancient Koroneia, exploiting the territory nearby (higher-up hill area, inner valley, good for slope agriculture and probably also suitable starting point for taking sheep for grazing to the upland plateau in summer seasons), rather than the fact that the place would have belonged to Koroneia, especially as it is not a self-standing settlement (see also no.3 Ag.Georgios - Xerokamara site).

Component KO_79: On the same site, a prehistoric activity focus was noticed by Lauffer. We cannot know (unless a detailed geomorphological mapping would be carried out) whether we are in the presence of an ancient river terrace

showing palaeo-soils from the prehistoric period,[26] but it could be, it seems, a very suitable location for a rural site (farm and *mandra* together) from which to take animals for grazing.
As Lauffer (Kopais I: 80) notes, very few prehistoric sites far from the Copais basin, like this one, are known. This is true, and corresponds to a bias in research. As we noticed many times, archaeologists have been looking especially for hilltop sites around the lake. Only Lauffer's intensive walk over the landscape of this *chora* allowed us to obtain knowledge of this small site in the inner Kakaris valley.

31. Koukoura/Agia Anna: *Component KO_13*

ID	Period	Type	Toponym	Discovery	Position
KO_13	gr-rom	? Settlement	Koukoura/ Agia Anna	Travellers	certain

Attempts have been made to locate ancient *Hippotai* (mentioned in Plutarch as a *kome* between Thisbe and Koroneia[27]) in the small basin of Koukoura/Agia Anna, in the hearth of Helicon (central ridge of Helicon: Palaiovouna), at the meeting point of the roads from Stiris to the valley of the Muses, and from Thisbe to Koroneia.[28] Today the village in the area is Agia Anna (formerly Koukoura). A settlement could have exploited this upland plateau also in antiquity, and it was probably located on the spot of the modern village or just above it (Roesch 1965; Fossey 1988: 339-40). The plateau played a central strategic role, and it would have been of interest to control it both for Thisbe and Koroneia. Ancient sources (Plutarchus *Moralia* 775A) give information about a treaty that split the territory of ancient Hippotai between Thisbe and Koroneia. The place also played an important role in the history of events (Burn 1949 and Buckler 1980: 55-58). Invasions and attacks from Phokis came from the pass at the locality of *Omorphi Lakka* (3rd Sacred War, and other attacks against Chorseiai and Koroneia). Today also through the pass runs a road (see map in fig.1) to the W of Agia Anna.

Only traces of fortification (in rough, dry stone masonry) have been noticed to the SW of the modern village (Fossey 1988: 339),[29] on a rugged and rocky hill. They had been reported already by Dodwell (1819: 259) as "*some ruins and large blocks of stone*", although he apparently did not visit the place (see Roesch 1965: 53 n.1).
According to Fossey (1988: 430-434; 1990e: 205-7), the place might have played the role of a refuge settlement during Dark Ages.[30] Fossey also notes: "*The plain is flat and could indeed have been conducive to horse-rearing (which might explain the ancient name)*".
Burn (1949: 317) describes the place thus: "*Helikon is formed by three parallel ridges. All except but the culminating one are cut through by the Phalaros valley and by that curious 'upper landing', above its head waters, at Koukoura.* [...] *It is a very curious place, this Koukoura basin.* [...] *It does not look like a glacier-scoop, and must once have been drained by the visible, as now by the underground, head-waters of the Phalaros. From it, east, south-east, west and north-west, radiate valleys which may once have held streams, but are now dry except immediately after rain, and all the better going on that account*".

32/33. Milia (32): *Components KO_49 and KO_50* and Milia S (33): *Components KO_95 and KO_96*

ID	Period	Type	Toponym	Discovery	Position
KO_49	C	? rural site/activity	Milia	Ext. topographical surveys	certain
KO_50	gr-rom	? rural site/activity	Milia	Ext. topographical surveys	certain
KO_95	C	? rural site/activity	Milia S	Ext. topographical surveys	certain
KO_96	gr-rom	? rural site/activity	Milia S	Ext. topographical surveys	certain

2km SW of Steveniko, to the immediate E of the road to Koukoura and near the edge of the plain of that village, on two adjacent small hillocks (Fossey 1988: 340). The name of the locality is reported as Milia. On both hillocks –Milia and Milia S (South) as recorded in the database– are several alignments of stones [AE768] and tile fragments [AE767], the remains of earlier buildings associated with undated material (tile fragments make us think of a rural site from the C period, but the hypothesis should be checked with further research). Two components have been created for each hill, one as possible rural site/activity of C period, and the other one as possible rural site/activity of gr-rom period. Only Fossey seems to report these evidences.

34. Agia Triada - plateau: *Component KO_59*

ID	Period	Type	Toponym	Discovery	Position
KO_59	EH	? Settlement	Agia Triada - plateau	Ext. topographical surveys	certain

On a plateau inside the village of Agia Triada, traces of an EH II-III occupation was noticed by the Speleologic Ephorate in 1995 (Ergon ΥΠΠΟ 2: 155; 3: 171). The settlement (?) site was probably linked with the so-called 'secondary product revolution', which left signs in the landscape (see no.35 - Agia Triada cave site and discussion in chapter II.3.1 – PREHISTORIC LANDSCAPE).

35. Agia Triada cave: *Components KO_51 to KO_58*

ID	Period	Type	Toponym	Discovery	Position
KO_51	N	activity focus	Agia Triada - cave	Other	approximate
KO_52	EH	activity focus	Agia Triada - cave	Other	approximate
KO_53	MH	activity focus	Agia Triada - cave	Other	approximate
KO_54	LH	activity focus	Agia Triada - cave	Other	approximate
KO_55	A	cult place	Agia Triada - cave	Ext. topographical surveys	approximate

[26] As in the Tanagra South transect case (see appendix I.14 – THE TANAGRA SURVEY). Probably we are not in the presence of an ancient river terrace showing palaeo-soils from the Prehistoric period, also considering the quite high up location in the inner river valley.
[27] In Plutarchus *Moralia* 775A Hippotai is mentioned as a *kome*, village destroyed by the Boeotians. Afterwards, the territory was divided between Koroneia and Thisbe, that continued to have a common border down into Roman times (IG VII 2870).
[28] Pritchett 1985: 151-156 ('Routes over Helicon from Thespiae towards Levadeia').
[29] Fossey (1988: 339) notes: "*Sherds and inner structure remains, if any, difficult to see because of the presence of a fairly thick scrub and tree cover*".
[30] A Dark-Age settlement of the type supposed for Melainai (Skourta plain), and of the type seen at Mali (see Fossey 1985a).

ID	Period	Type	Toponym	Discovery	Position
KO_56	C	cult place	Agia Triada - cave	Ext. topographical surveys	approximate
KO_57	H	cult place	Agia Triada - cave	Ext. topographical surveys	approximate
KO_58	R	? cult place	Agia Triada - cave	Ext. topographical surveys	approximate

A cave that opens up in the limestone rocky slopes just above the modern village of Agia Triada, is clearly visible from a church dedicated to the Vierge just below it. The place is known from literature as the 'Cave of the Nymphs of Koroneia (Agia Triada)'. Excavations have been carried out near the cave mouth (Ergon ΥΠΠΟ 2: 155; 3: 171; AReports 2000-2001: 56). Prehistoric levels have been uncovered (Minyan and Mycenaean sherds and burnt layers covering strata with EBA pottery, N –final Neolithic- sherds and few obsidian blades and animal bones). They probably correspond to a seasonal rather than permanent occupation. Good water sources are nearby but the location is quite far from agricultural land. The cave site has probably to be linked with the development of stock-raising (sheep and goats mainly) at the turn of N and EB period. The site very probably might have leaved together with the EH II-III site on a plateau within the village of Ag.Triada (*component KO_59*).[31] At the cave finds from the historical period are also reported, from the Archaic to the Late Hellenistic period, which probably refer to a cult place, identified as the cult place dedicated to the Leivethrides Nymphs mentioned by Pausanias and Strabo.

36. Agia Triada: *Components KO_60*

ID	Period	Type	Toponym	Discovery	Position
KO_60	EH	activity focus	Agia Triada	Personal or group interest	approximate

In December 1989 a few sherds and bones on the ground were noticed by Y. Suto (AAA 20: 66-74) N of the hilltop of the village, on the steep slope, at the edge of the road. Sherds are not described in totality, but one sherd is mentioned as belonging to a *Depas* shape ware (from Asia Minor), dated to the EH period. Suto says it cannot be from a grave as this kind of ware has never been found in the burial context; he comments that despite the presence of many bones fragments, it could be a settlement site.

37. Triapi: *Component KO_88*

ID	Period	Type	Toponym	Discovery	Position
KO_88	gr-rom	special activity	Triapi	Int. topographical surveys	approximate

Some ancient quarries, probably for building material, that must have belonged to Koroneia, are known (Lauffer Kopais I: 120) on the ridge called Triapi or Trapedi, that goes to the N of Xerokamara until the road and the N slope of the Laphystion. This ridge can be considered the limit of the Phalaros/Pontsa valley (Lauffer Kopais I: sketch 130).

[31] As for the FN/EH occupation see chapter II.3.1 – PREHISTORIC PERIOD.

38. Magoula Kalami: *Components KO_89 and KO_90*

ID	Period	Type	Toponym	Discovery	Position
KO_89	Preh	settlement	Magoula Kalami	Int. topographical surveys	approximate
KO_90	EH	? Settlement	Magoula Kalami	Int. topographical surveys	approximate

A lower, ca. 2m high and 10m long *magoula* lies 300m SE of Kalami below the Triapi ridge (see *component KO_88*) in the plain (Lauffer Kopais I: 120-1, sketch 131). Lauffer reports it as the site of an earlier settlement. He reports it as a 'very ancient' site, since he noticed few scattered prehistoric sherds (coarse ware) lying still on the top, while the majority of the material lies at the foot of the *magoula*. Lauffer notes how the soil rich in material that we see today in the plain constituted the upper part of the mound (tell), most of which would have been eroded down. Its top was probably flattened more and more by agricultural activities.
It could be a small EH site. Lauffer notes that the *magoula* does not present any similarity with the Helladic sites of the Copais, while it rather reminds one of the flat *magoulas* of Phokis plain (Agia Marina, Agia Paraskevi in Phokis, as well as the Magoula Balomenou site close to Chaironeia). A marshy area lies in front of it (see no.40 - Kalami/Rachi site and chapter II.3.1 - TOPOGRAPHICAL SETTING).
A general prehistoric component (*component KO_89*) has been created (allowing for the possible presence of other prehistoric period activities at the site) as well as a possible specific EH component with a question mark (*component KO_90*). The character of the site is certainly hard to define.

39. Kalami – Lioma: *Components KO_42 to KO_47*

ID	Period	Type	Toponym	Discovery	Position
KO_42	N	activity focus	Kalami - Lioma	Personal or group interest	certain
KO_43	EH	settlement	Kalami - Lioma	Personal or group interest	certain
KO_44	MH	settlement	Kalami - Lioma	Personal or group interest	certain
KO_45	LH	activity focus	Kalami - Lioma	Personal or group interest	certain
KO_46	C	? rural site/activity	Kalami - Lioma	Other	certain
KO_47	H	activity focus	Kalami - Lioma	Other	certain

The large mound at the NE foot of Mount Granitsa/Laphistion,[32] visited because of its elevated position over the Copais basin, shows signs of an important prehistoric site (Lauffer Kopais I: 122-3, sketch 136). Located 300m W of Kalami hamlet, this site is even larger in extension than the Agoriani - Dekedes site. The hill is ca 30m high and 50m long with some terraces on the N slope. Material is reported on the flat top and on the slope and lower terrace, and it is mainly prehistoric. The site has been interpreted as a large pre-Mycenaean hilltop settlement (*Stadtberg*), where only few LH materials were found [AE762]. Lauffer describes it as a typical settlement in front of a higher ridge (the Laphystion). Like the Agoriani – Dekedes site, the mound does not rise straight away

[32] Nice picture of Kalami prehistoric mound in Lauffer AA 1940: 182.

from the Copais basin bottom, but from an elevation of ca 10m above the Kephisos river level.
The main peak of occupation seems to be EH and MH (*components KO_43 and KO_44*), although quantitative differentiation cannot be made with the level of data available. Frequentation seems to start in the Neolithic period, when we apparently have signs of a focus of activity (*component KO_42*), without enough data available to suggest the degree of occupation. Signs of activity focus are also attested for the LH period (*component KO_45*), until LH IIIC according to Fossey (1988: 337).
Two fresh water sources are nearby the mound: one 200m to the SE (Keramidi source), another one 50m to the W. A third water source in the area, close to the Ag.Ioannis chapel, 400m to the SW of the prehistoric mound, spills out warm volcanic water (after Pausanias IX 34.5 from Pappadakis AD 1916: 258).[33] The combination of two factors, elevated position and proximity to water, would explain the development on the mound of a prehistoric settlement, since the Neolithic period (jointly with the presence of a permanent marshy area nearby. See chapter II.3.1 – TOPOGRAPHICAL SETTING).
For Lauffer (Kopais I: 122-3) the site does not seem to carry signs of later (post-Mycenaean) occupation, neither on the top nor on the slope or at the foot of the hill, unlike the other Mycenaean hilltops in the area. On the other hand, Fossey (1988) reports small quantities of gr-rom pottery and tile sherds (black glaze) visible on the surface, and Pappadakis (AD 1916: 258) has suggested that there might have been a small Classical shrine on the site. In the database a component (*component KO_46*) has been created as a possible rural site from the C period, because of the presence of black glazed sherds associated with tile fragments.[34] Also H surface material is reported from the site by Fossey (1988: 337).[35] We therefore could consider the later occupation as a C-HL one (probably a rural site).

Fossey would like to identify the site with Homeric *Mideia*, for which see appendix I.2 (Lebadeia). Lauffer (Kopais I) says it can be identified with *Athenai*, while others do not agree (Knauss for instance – see above).[36]

40. Kalami/Rachi: *Components KO_48, KO_62 and KO_91* (marshy area)

ID	Period	Type	Toponym	Discovery	Position
KO_48	H	? burial place	Kalami /Rachi	accidental	imprecise
KO_62	R	? burial place	Kalami /Rachi	Rescue excavation	approximate
KO_91	gr-rom	? Infrastructure	Kalami /Rachi	Rescue excavation	approximate

Component KO_48 and KO_62: Nearby the prehistoric site of Kalami – Lioma, in the area of the modern hamlet Kalami/Rachi, in a marshy area by the old national road Athens-Levadeia, a burial (a tile-roofed tomb) of a possible Roman date [AE1107] was found (*component KO_62* - AD 1972: 324). Also some inscriptional evidence from the H period, referring to burials, are known from the area (*component KO_48*) and could be linked to a burial area along the road exiting the city valley.

Component KO_91: Well remains [AE1672] to the W of the burial described above. They were not investigated as the area was marshy and a chronology is not available. The well (if gr-rom) could be linked to the gr-rom rural site (see above) on the Lioma mound or to a possible habitation area nearby the burial (*component KO_62*) close to which the well was found. The area here is always indicated as marshy – see Philippson's description (1951: 474) of the marshy area around Kalami and see fig.12 in chapter II.3.1.

[33] The presence of those fresh water springs has suggested the location here of the sanctuary of Herakles Charops (see above in text for discussion).
[34] Does Fossey mean that tile fragments were black glazed? Or potsherds only are black glazed and tiles are not?
[35] It could be the case as in many other multi-period sites: Fossey (or people expert in historical period material) recognizes different phases in potsherds while other visitors of the site (in this case, as in many other, prehistorians) do not (see chapter II.2).
[36] For the question of identification see Fossey 1990d.

Appendix I.2
Northern Helicon: Levadeia

THE ARCHAEOLOGICAL RECORD

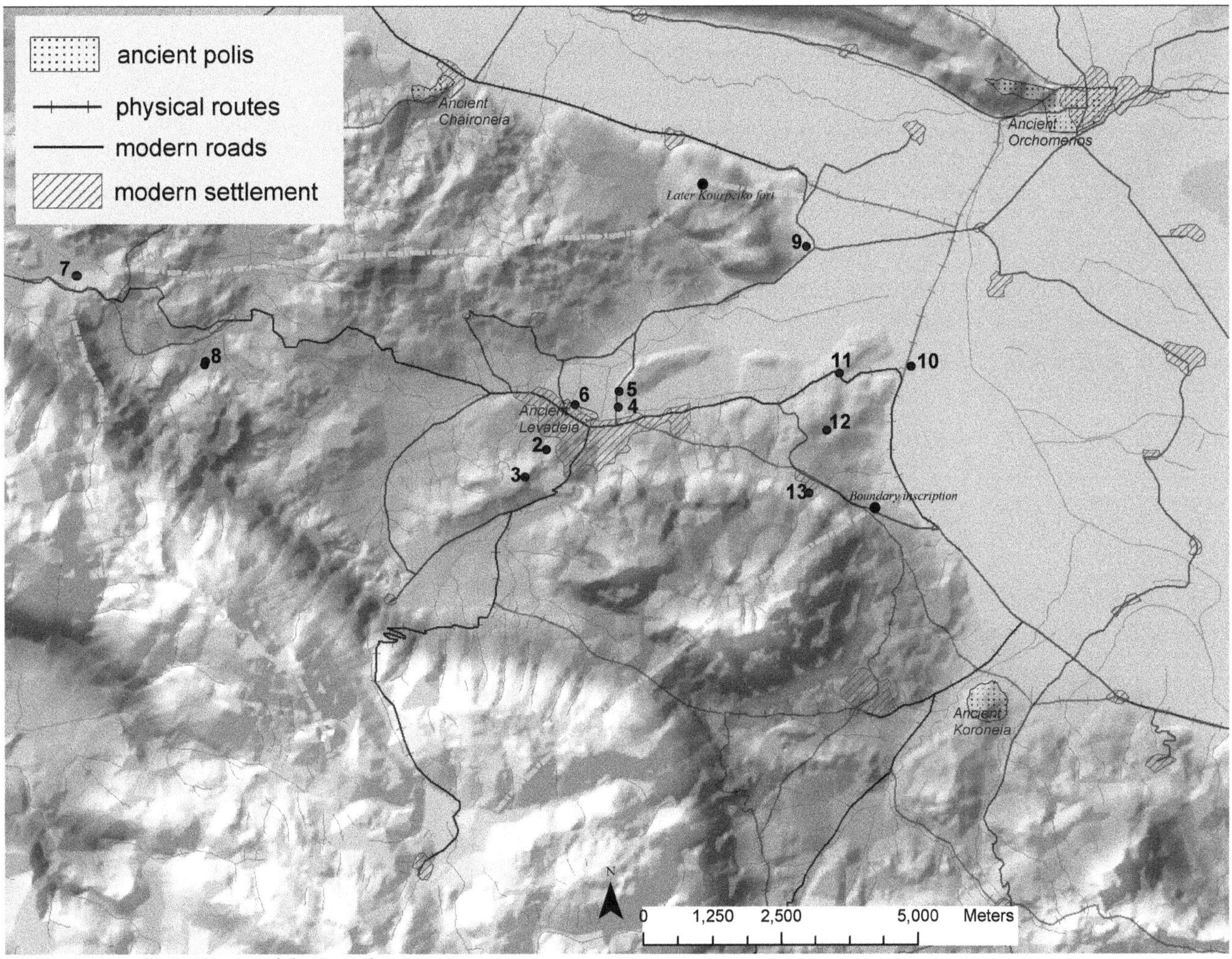

Fig.1. *Archaeological map of the Levadeia area*

1. LEVADEIA: *Components LE_1 to LE_6*

ID	Period	Type	Toponym	Discovery	Position
LE_1	G	? Settlement	Levadeia	Other	imprecise
LE_27	A	Settlement	Levadeia	Other	certain
LE_2	C	settlement	Levadeia	Historical geography	certain
LE_3	H	settlement	Levadeia	Historical geography	certain
LE_4	R	settlement	Levadeia	Historical geography	certain
LE_5	LR	settlement	Levadeia	Rescue excavation	certain
LE_6	LH	? human presence	Levadeia area	Other	imprecise

Components LE_1 to *LE_6* refer to the activity locus corresponding to the ancient *polis* of Levadeia.
The location of the modern town of Levadeia and the ancient homonymous *polis* more or less coincide. The town lies at the SW corner of its valley, on the Northern foothills of the Laphystion mountain, and faces the SW end of the Tryphiolithari ridge. To the S and W of the town the mountain sides rise very steeply.[1] To the N and the E the city site is surrounded by land suitable for agriculture.

While Leake (1835: 120-1) and Ulrichs (1840: 166-7) rightly considered that ancient Levadeia should be located in the low-lying position of the modern town (according to Pausanias description of Levadeia as 'low-lying' IX 39.1), other travellers (Ross 1851 vol.I: 38-9; Gell 1819: 149, 156-7; Frazer 1913: 197) suggested a much higher location on the Tryphiolithari ridge, where only few remains leading apparently to a burial area (see below – *components LE_11 and LE_12*) have been found.

Pausanias (IX 38, 39.1) says that originally the town was higher up on the mountains and was called Mideia (from Aspledon's

[1] Description of the ridge behind the city in Philippson 1951: 445

mother). Later on though, Levados came from Athens and people moved down the slope and the *polis* was given the name Levadeia after him. In contrast, Strabo (I 3.18) says that Mideia was a town bordering the Copais lake, and was swallowed up by the lake together with ancient Arne. Papachatzis (1981: 50) suggests that that people of Levadeia could have given to their *polis* the name Mideia because they wanted to show it as mentioned in Homeric Ilias (B 507) and therefore as one of the more ancient *poleis* of Boeotia.
This was not the case, and Levadeia does not seem to be occupied in earlier times, or at least must have been very small. As for LH only a piece[2] is reported as coming from the area, without exact provenance. Fossey (1990d: 57) correctly notes how no prehistoric material can with certainty be attributed to the site, while on the basis of that single sherd Hope Simpson (1965: 140 – n.412) seems to base the existence of a Mycenaean settlement.[3] The earliest possible archaeological evidence for the occupation of the *polis* itself dates to the 8th C BC.[4] Sherds, inscriptions and even some building remains attest occupation in the succeeding C, H and R periods.[5] The city lived until the Late Roman period. Although not very substantial in Hellenistic times (its muster-lists show a population smaller than Akraiphia, Thespiae and certainly Orchomenos) it is presented by Pausanias (IX 39.1 to 40.2) as among the most prosperous cities of Greece by his time.

Due to the fact that the ancient city lies beneath the modern one, the earliest periods of occupation, as well as the location of the ancient city centre and its relationship with its sanctuaries are not easy to clarify. On the other hand, rescue excavations beneath the modern town suggest that ancient Levadeia was probably built in the lower area of the modern town.[6] On the basis of the analysis of the known archaeological evidence mapped on the 1:5,000 GYS map, Turner (PhD dissertation 1995 and summary in AJA 97 1993: 345) identified three main nuclei within the intra-urban plan: i) funerary monuments on the hill of Tryphiolithari (NE of town); ii) religious centre on the Western bank of the Herkyna river (monuments related to the oracle of Trophonios and the federal sanctuary of Zeus Basileus); iii) civic monuments in the NE quadrant of the modern town, between the funerary and religious areas.

Pausanias' text (IX 39.4) refers to many temples and sanctuaries belonging to the ancient city of Levadeia. This is why many researchers have attempted to identify these cult places with known archaeological remains and inscriptional evidence. Most of them would have been inside the city, others in the immediate outskirts, especially on the mountains above the city site. It is not the purpose of our work to deal with all these identification problems and discussions. An update account on the matter can be found in Turner, dissertation 1995. We will deal in the following paragraphs with a select number of cases related to the extra-urban sanctuaries.

2. Levadeia – Herkyna springs: *Component LE_10*

ID	Period	Type	Toponym	Discovery	Position
LE_10	gr-rom	cult place	Herkyna springs	Travellers	approximate

The famous sanctuary of Trophonios, the feature for which Levadeia is mainly known, is still poorly known archaeologically. Traditionally, its location was thought to be on the W bank of the Herkina river, by the fresh water complex karstic springs, opposite the ancient *polis*. Pausanias describes the cult place as an *alsos* (Paus IX 38, 39.2), a 'sacred place with trees', a sort of sacred wood.[7] The area by the Kria spring (as it is called today), independently from its relationship with the oracle of Trophonios, is full with ancient building material and architectural fragments also reused in later constructions (especially in the biggest tower of the medieval fortress of Levadeia – see pictures in Papachatzis 1981: 50). Furthermore, a statue (woman without head, dated to the 4th C BC) has been found during rescue excavations in the area (AD 1994: 286-287).
On the other hand, scholars more recently suggested a higher up location for the sanctuary by the river but on the slopes of the Proph.Ilias hill (see below).
By the Herkyna/Krya springs a certain cult place was located - a cave sanctuary probably dedicated to Pan and the Nymphs (*component LE_10*),[8] testifying once again to the probable religious vocation of the area. At the entrance of the cave, on a rock, an inscription is engraved, datable to the second half of 2ndC BC. Usually epigraphical evidence is a very movable indicator of the presence of a site, but in this case, since the inscription is carved directly on the rock at the entrance of the cave, there is no doubt.

3. Levadeia - Proph.Ilias: *Components LE_7 to LE_9*

ID	Period	Type	Toponym	Discovery	Position
LE_7	C	? cult place	Levadeia-Proph.Ilias	Rescue excavation	certain
LE_8	H	cult place	Levadeia-Proph.Ilias	Rescue excavation	certain
LE_9	LR	? cult place	Levadeia-Proph.Ilias SW	Rescue excavation	certain

Components LE_7 and LE_8: The remains of the temple of Zeus Basileus on the Proph. Ilias hill were accidentally found during wartime activities and quarrying for building stones. Pausanias (IX 39.4) gives mention of the temple on the hill where the church of Prophitis Ilias is now. Pharaklas first excavated at the site (Pharaklas AD 1967: 244-5; AAA 1969: 228-232). The visible remains belong to the 3rd century BC, but among the debris were found also *poros* (porous limestone -

[2] A LH III *rython* [AE787] in the Athens National Museum, registered as coming from Levadeia (*component LE_6*).
[3] See also Syriopoulos 1968: 118, YE59.
[4] Slight is the evidence for the occupation of the town in the Geometric period (Turner 1995). Scarce also are the archaeological finds from the Archaic period: a sculpted grave stele dated to the Archaic period and some epichoric stelae which may also date to this period - see *component LE_1*.
[5] C layers are less clearly recognisable compared to the H and R ones: BCH 1940/1: 37 n.2 (a Hellenistic building); AE 1953/4: 235-236; BCH 1953: 219 (a Roman aqueduct); AD 1965: 241-2 (a 4th C BC building); AD 1983: 135 (R structures); AD 1995: 301. For the R period see Fossey 1990c. On coins of the R period (Augustus) olive branches are illustrated (Papachatzis 1981: 250 fig. 303).
[6] The ancient city Eastern end is marked probably by the modern Chrisikou property, 100m NE to the plot *Tolia brothers* excavated for the new flat block [AE1181, AE1174, AE1110] (Piteros 1995: 481-507).
[7] Papachatzis 1981: 245: The *Trophonion* is a '*manteion*', together with the Boeotian Amphiaraion of Thebes and the sanctuary of Apollon on the Ptoion. In Plutarch's time (around 100 AD), among the manteia the Amphiaraion and the Ptoion were not existing anymore, while the Trophonion was (Plutarch *Pelop.* 16). The *Trophonion* is mentioned by Herodotus as well (1.46-47; 8.134). Around 175 AD, when Pausanias visited the sanctuary, he found it in its pick of prosperity. Hadrian had visited the place a few years earlier and probably had given munificence. In the 3rdC the *manteion* is still in life (even after the anti idolatric laws of Theodosios II).
[8] IG VII 3094 - dedication to a woman and a man (Schachter 1986: 187). See also Mitropoulou 1995.

travertine) fragments (some carved with floral reliefs, and two pieces from a geison in the Doric order) indicating the existence of an earlier, smaller Doric building (Pharaklas AAA 1969: 228-232). More recent excavations in the area, carried out by the local *ephoreia*, brought to light the whole large Hellenistic temple building (60x23m), whose construction was probably never completed (AD 1997: 392). In the construction of the sanctuary, probably to be interpreted as a federal sanctuary of the Boeotian confederation as Turner suggests (on the basis of historical and epigraphical evidence - see Turner's dissertation 1995 and also briefly the summary of her work in AJA 1992: 365 and 1993: 345), black granite from the Helicon was used among other material, and precisely from the area of the modern village of Surpi/Analypsi.

Component LE_9: SW of the temple of Zeus Basileus an underground circular structure (LR[9]) was found, interpreted by Pharaklas (AAA 1969, 228-232) as the *manteion* of Trophonios, which would have been at this location and not at the Herkyna sources, as formerly believed. Turner (1995) also seems to agree in locating the *manteion* of Trophonios up on the Proph. Ilias location, identifying it with the walls visible on the surface next to the church of Ag. Ioannis Prodromos, just down the slope much closer to the Herkyna river bank (to the left side of the river). In any case, the religious function of the structure can be most probably generally accepted, and I characterised the activity focus as a LR cult place.

Pausanias (IX 39.4) mentions more than one temple in the vicinity of the oracle.

4. Levadeia NE: *Components LE_14 and LE_15*

ID	Period	Type	Toponym	Discovery	Position
LE_14	H	? burial place	Levadeia NE	Rescue excavation	approximate
LE_15	R	? burial place	Levadeia NE	Rescue excavation	approximate

One tomb excavated (AD 1967: 243-4 - H or R tomb) to the NE of the town. It cannot be located precisely. One could infer from the evidence (along with burial evidence found in the vicinity of the Ag. Vlassios church - see *components LE_11 and LE_12*), as Fossey does (1988: 346) as well as Turner, that certainly one tomb area lies to the NE of the modern town, to the SW slopes and foothills of Tryphiolithari.

Tombstones have also been brought to local authorities from field owners from areas around Levadeia (Kyriaki Levadeias and Distomo Levadeias - AD 1972: 323) [AE1873].

5. Levadeia – Ag.Vlassios : *Components LE_11, LE_12 and LE_13*

ID	Period	Type	Toponym	Discovery	Position
LE_11	A	? burial place	Levadeia-Ag.Vlasios	Int. topographical surveys	approximate
LE_12	C	? burial place	Levadeia-Ag.Vlasios	Int. topographical surveys	certain
LE_13	H	activity focus	Levadeia-Ag.Vlasios	Other	certain

Tombstones found built in and in the vicinity of the Ag. Vlassios church jointly with sherd concentrations, marble blocks, and grave goods reported by local people [AE779; AE1247; AE1248; AE1175; AE778; AE785] testify to the presence of a burial area.

Near the church of Ag. Vlasios NE of Levadeia, on the slope of the central small ridge which crosses the Levadeia plain, archaeological evidence is reported by Lauffer and Fossey. The discovery of the site was done in the framework of an intensive topographical survey (Lauffer Kopais I: 145), but the site is known also to Fossey (1988: 346) through information given by local people. Lauffer (Kopais I: 145) attempts to identify the site with one of the extra-mural temples mentioned by Pausanias as belonging to ancient Levadeia (IX 39.4), while I would follow the interpretation given by Fossey (1988: 346), who reads the evidence as part of a possible large cemetery of the Levadeia town in the C period (and probably also in the A period, if we consider the archaic inscribed tombstone built in the church as coming from the immediate vicinity). Also the inhabitants of the small hamlet around the church, as reported by Fossey, refer to the finding of tombs in the area.

For the H period, only a terracotta figurine [AE786] found in the area can be surely datable (Fossey 1988: 346), but it is probable that the burial use of the area would have continued in the H period as well (the component type was left as 'activity focus' for the Hellenistic period, and it could be interpreted as a burial place because of supposed continuity of use).

6. Levadeia N: *Component LE_24*

ID	Period	Type	Toponym	Discovery	Position
LE_24	gr-rom	burial place	Levadeia N	Rescue excavation	certain

A necropolis was excavated at the modern hospital of Levadeia (N of town) by means of rescue excavation (AD 1995: 301).

7. Karakolithos: *Components LE_16 to LE_18*

ID	Period	Type	Toponym	Discovery	Position
LE_16	C	burial place	Karakolithos	Rescue excavation	certain
LE_17	H	burial place	Karakolithos	Rescue excavation	certain
LE_18	H	? Fort	Karakolithos	Personal or group interest	certain

Component LE_18: On Karakolithos, remains of a H (?) fortification are known (Bursian 1862: 183; Lolling 1989: 608; Fossey 1986: 59-61).

Components LE_16 and LE_17: Few metres W of the Karakolithos fort, traces of a C and H cemetery have been found (AD 1978: 149-153).

According to Dasios (1995: 245-255), these remains would testify to the presence of an upland settlement in the area (along with Tsoukalades evidence). For the time being its position could not be found– see also below no.8.

8. Levadeia-Delphi National Road – Tsoukalades: *Components LE_19 and LE_25*

ID	Period	Type	Toponym	Discovery	Position
LE_19	R	burial place	Levadeia-Delphi National Road	Rescue excavation	certain
LE_25	R	burial place	Levadeia-Delphi National Road	Rescue excavation	certain

[9] This dating rests on two sherds which from their description seem to be late Roman ripple ware of which the date may be much later than the mid 3rd century AD proposed by the excavator. Unfortunately the report is not sufficiently detailed to permit reassessment of the finds (Pharaklas AD 1967: 244-5; Pharaklas AAA 1969: 228-232; Fossey 1988: 345).

On the S side (left-hand) of the national road between Levadeia-Delphi (between the tunnel and the junction to the Tsoukalades village), isolated tombs of the R period (tile-roofed tombs) have been found during rescue excavations (AD 1982: 170). No grave-goods and no surface sherds have been noticed. Tombs have been dated according to the tile shape. Later on, on the N side (right-hand) of the road, other tombs datable to the same period have been found (AD 1995: 304), without grave goods but with a bronze earring.
Those evidences might either represent a larger burial area (which could have been linked to a village settlement existing there) or small rural cemeteries along the route to the Parnassos. The position of the evidence is not precisely given (nor the exact distance between the two discoveries) and therefore I preferred to create two separate components. Regardless, they testify to the presence of settlement activities in this upland area, as do *components LE_16 and LE_17* (see above no.7),[10] probably linked to upland pasturage and the exploitation of the small alluvial valley.

9. Thourion-Ag.Ioannis: *Component LE_23*

ID	Period	Type	Toponym	Discovery	Position
LE_23	gr-rom	? activity focus	Thourion-Ag.Ioannis	Int. topographical surveys	certain

Below the E slopes of the Thourion mountain, on a gentle slope rising up from the Copais, by the church of Ag.Ioannis, medieval and ancient sherds have been noticed by Lauffer on the church terrace (Lauffer Kopais I: 146). The undetermined evidence reported from there, though, does not allow us for any interpretation. I can only note that the place at Ag.Ioannis (Thourion East), on the edge of the promontory jutting into the Copais basin, was overlooking the meeting place of the roads coming from Orchomenos, Chaironeia, Levadeia and Koroneia (fig.1), probably also in ancient time as it is today, and it would mark the NE corner of the *chora*.

10. Mavrogeia: *Components LE_20*

ID	Period	Type	Toponym	Discovery	Position
LE_20	gr-rom	activity focus	Mavrogeia	Int. topographical surveys	approximate

In terms of archaeological evidence found there, *component LE_20* carries signs of permanent occupation (a large pithos fragment, large painted vase fragment, tile fragments, as well as large stone blocks visible in the fields), whose character cannot be easily identified. In terms of landscape, its position overlooking the passage of the peri-Copais road led Lauffer (Kopais I: 139) to the interpretation of the site as a fort, controlling the passage of the peri-Copais road, that would pass from this location connecting Orchomenos and Levadeia with Koroneia and Haliartos.[11]
On the other hand, landscape considerations would lead me to interpret the component as a large Greco-Roman rural site by the road, exploiting the fertile area at the edge of the Copais basin. The date remains undetermined unfortunately, and the site would need a revisit to clarify it.

11. Neraidogorna-Ag.Taxiarchoi: *Component LE_21*

ID	Period	Type	Toponym	Discovery	Position
LE_21	gr-rom	settlement	Neraidogorna-Ag.Taxiarchoi	Int. topographical surveys	certain

At the entrance of the Levadeia valley, W of Ag.Taxiarchoi church, below and on the slope of Moulki hill, as well as on the Kazakines slope, Lauffer (Kopais I: 140-1, sketch 148) mentions the presence of a quantity of sherds and tile fragments on the slopes on both sides of the road near the church dedicated to the Ag. Taxiarchoi [AE1244, AE1245], at the entrance of the Levadeia valley. Lauffer interprets the evidence as signs of a large settlement site. I would agree with this hypothesis and I therefore created a 'settlement' component, probably with a hamlet or village status. On the other hand, if the evidence would turn out as constituted by material dating back to the R period, the component could also be interpreted as a large rural (villa) site. Its low position could support this interpretation, in considering the usual location of large villa sites in Boeotia, according to the results of the Boeotia Survey Project (see appendix I.14 – THE TANAGRA SURVEY). The site lies today across the road and in the past would have been probably along the main road from Thebes at its turning towards Levadeia.
As far as the other two units of archaeological evidence from the area are concerned (architectural fragments and blocks from nearby locations reused in the church [AE1242, AE1243]), they could have easily come from Levadeia.
Other scholars identify the remains visible in and by the church dedicated to the Ag. Taxiarchoi with the temple of Zeus Laphystion,[12] within the tradition of the quest for places mentioned in ancient sources.

12. Laphystion-Kazakines: *Component LE_22*

ID	Period	Type	Toponym	Discovery	Position
LE_22	Preh	human presence	Laphystion-Kazakines	Int. topographical surveys	approximate

To the N of Granitsa/Laphystion village, on the Kazakines slopes, above *component LE_21*, Lauffer (Kopais I: 141, sketch 148) noticed obsidian flints. He comments how the presence of a prehistoric site is notable here, in such an inner location compared to other sites around Copais. Few obsidian flints solely might indicate a 'human presence' rather that an actual site. The archaeological evidence [AE1246] would need a further look to be assigned a more specific character.

13. Laphystion: *Component LE_26*

ID	Period	Type	Toponym	Discovery	Position
LE_26	C	fort	Laphystion	Travellers	certain

To the S of Granitsa/Laphystion village, on a rocky eminence 1600m ca to the W of the church of Agios Ioannis, remains of a tower are reported by Dodwell 1819: 245, Leake 1835: 133, and Gell 1819: 140, and was drawn by Simone Pomardi for Dodwell (Camp II 1991: fig.1). Today the tower has disappeared and on the spot is a private house (Ath. Skorliakos). The site is largely described by Camp (Camp II 1991: fig.1 and fig.4). The square shape tower is probably to be dated to the 2nd or 3rd quarter of

[10] An identification with ancient *Trachinos* has been suggested for the area of the modern Tsoukalades village.
[11] Lauffer (Kopais I: 139) suggests it could be the Metachoion fort mentioned by Stephanus Byzantius (448.19), which would lay between Orchomenos and Koroneia.
[12] On the other hand, Lauffer identifies the temple of Zeus Laphystion (which Pausanias IX 34.5 says it was 20 stadia from Koroneia) with the Kardavitsa height, 700m high W of the Profitis Ilias peak. Apparently he did not find any actual visible remains there, and therefore there is no associated entry in the database.

the 4thC BC and presumably belonged to the system of watchtowers built by the Boeotian confederacy under the Theban domination (together with the towers at Evangelistria, Askra and Mavrovouni, for instance[13]), for whom it would constitute the Westernmost tower. The actual function of the tower must have been though to guard both the principal road through central Greece[14] as well as the passage between Koroneia and Levadeia *chora*. The tower has good sight-lines with both the fort at Petra/Tilphousion (*component HA_74* – no.42 in appendix I.8) and that at Kariopoula (*component HA_61* – no.29 in appendix I.8), and Northward to Orchomenos, but none with Levadeia and Koroneia fortified towns.

[13] See correspondent entries in appendix I.7, I.8, I.9, chapter II.4 – FORTIFIED LANDSCAPE– and Kallet-Marx 1989: 301-311.

[14] It is known that the ancient route closely followed the line of the road that brought early travellers to the tower site (Pausanias IX 33-34 and the Homeric *Hymn to Demeter*: lines 224-76).

Appendix I.3
Chaironeia valley: Chaironeiake

THE ARCHAEOLOGICAL RECORD

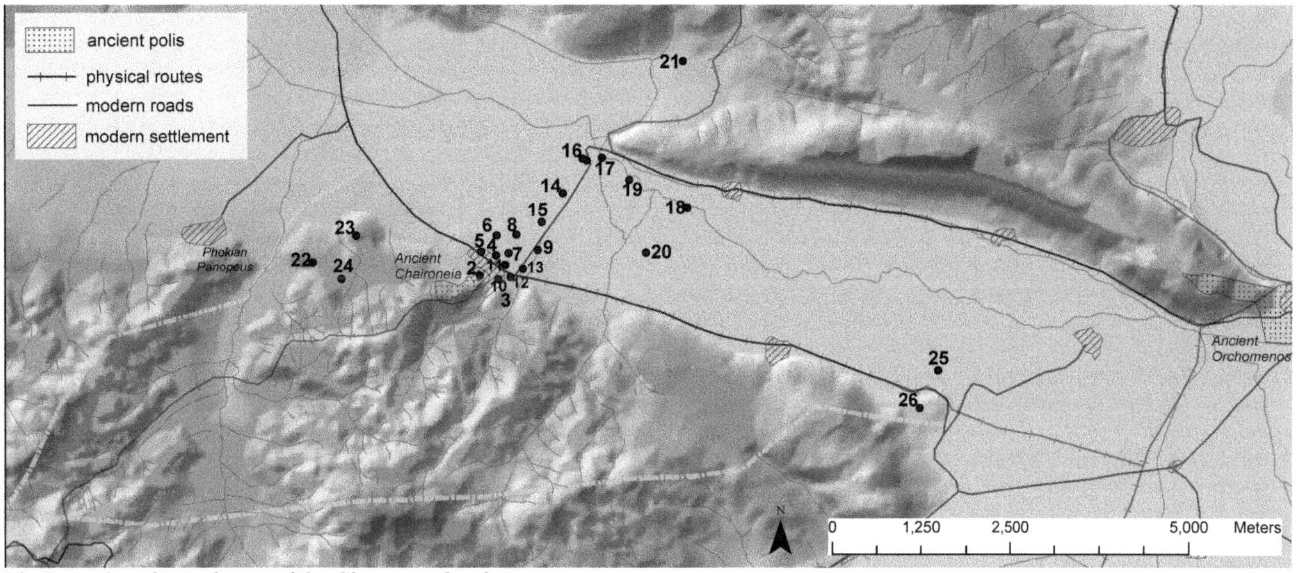

Fig.1. *Archaeological map of the Chaironeiake chora.*

1. CHAIRONEIA: *Components CH_1 to CH_9 and CH_38*

ID	Period	Type	Toponym	Discovery	Position
CH_1	A	settlement	Chaironeia	Other	certain
CH_2	C	settlement	Chaironeia	Travellers	certain
CH_3	R	settlement	Chaironeia	Travellers	certain
CH_4	H	settlement	Chaironeia	Travellers	certain
CH_5	LR	settlement	Chaironeia	Travellers	certain
CH_6	G	settlement	Chaironeia	Ext. topographical surveys	certain
CH_7	N	activity focus	Chaironeia	Ext. topographical surveys	certain
CH_8	EH	activity focus	Chaironeia	Ext. topographical surveys	certain
CH_9	LH	activity focus	Chaironeia	Ext. topographical surveys	certain
CH_38	gr-rom	cult place	Chaironeia	Personal or group interest	certain

The acropolis of the ancient *polis* of Chaironeia lies on a rocky steep spur above the village of Kapraina/Chaironeia.[1] The lower settlement is situated in the area of the modern village. It was located at the entrance of the narrow valley of the Kapraina stream, along the S side of the Chaironeia basin, and it was built on both sides of the stream. A possible extension of the lower settlement is to the NE of the walled acropolis (see Fossey 1988: 378 and fig.50), and would lie today underneath a part of the modern village area.[2]

Components CH_1 to CH_6 (and probably also *CH_38*) refer to the *polis* settlement of ancient Chaironeia, while *components CH_7 to CH_9* refer to the earlier prehistoric occupation within the later *polis* site.

Components CH_7 to CH_9: At the site was found some prehistoric material, which probably represents an activity focus (if not a settlement site). Only Fossey (Fossey 1990d, but not in 1988) and HS&D (1979: G46) report the presence of Neolithic material from the city site, on the historical acropolis and on its N slope (*component CH_7*). They alone also report the presence of Early Helladic material (including EH III) from the city site (*component CH_8* - Fossey 1988: 375 and HS&D 1979: G46). Components (*CH_7 and CH_8*) have been left as 'activity focuses' since evidence [AE811 and AE803] is too scarce to attest with certainty the presence of a habitation/settlement site. Traces of a habitation site in the Middle Helladic period[3] are not known from the site, while a large Matt painted pithos in Chaironeia museum is recorded as coming from an unpublished tomb at or near the village (Fossey 1988: 380) – see below (*component CH_10*). As for the Late Helladic period, traces of an activity focus have been found within the W circuit of the acropolis walling and on the slope to the N (Fossey 1988: 375 and HS&D 1979: G46). According to the analysis Fossey made of the city walling, the earliest phase (the so-called W circuit) consists of a large circuit in Cyclopean masonry [AE792] encircling the whole mountain top (Fossey 1988: 378 plan

[1] Papachatzis 1981: 260 pl. 314; general air view in Schoder 1974: 94-5; Fossey 1988: 375.

[2] The area of the ancient lower town is traced approximately in fig.1, but it was probably wider in the C period, when the burials appear at a further distance from the city compared to later (H/R) ones (see chapter II.3.3 – BURIAL AREAS).

[3] Prehistoric Gazetteers do not report MH material from city site.

fig.50[4] and Sotiriadis AM 1903: 324 pl.4).[5] Also LH burial places are known from the area (*component CH_11*).
According to Fossey, the settlement in the LH period must have extended to the top (later the ancient acropolis) and the N slope.[6] According to Fossey, thus, the site might have been an important fortified site in the LH landscape. We must take into account, though, that HS&D 1979 only report unspecified LH from the site (probably taken for granted from Hope Simpson 1965), and they do not mention a LH fortification (apparently noted only by Fossey). On the other hand, Fossey made a detailed study of the fortification and site (1990b). The different levels of research might be due to the incongruence in information, and the character of the LH *component CH_9* has been left as a generic 'activity focus'.

Components CH_1 to CH_6: As for the settlement site in historical periods, there is plenty of evidence both as for the fortified acropolis area as well as for the lower settlement (to be located probably N and E of the acropolis).[7] In the ancient *polis* and modern village area material from the G and A periods were noticed, as well as material and structural remains from the C, H, R, LR periods (some of which were systematically excavated[8]), among which most impressive is the theatre. Also G and A material probably correspond to a settlement site (*components CH_6 and CH_1*) already in those early periods (as for the A period, the existence of a fortification around the acropolis can also probably be assumed [AE794]). In the gr-rom period the acropolis was fortified [AE788]; what is mainly visible today is an ashlar circuit on the upper East slope of the mountain (probably of 4th C BC date).

Chaironeia is situated at the opening from Phokis to Boeotia, along the overland route from the N to Attica and Southern Greece. The plain in front of Chaironeia offered the battlefield for two events known from history: 338 BC (Macedonians VS Greece) and 86 BC (Sylla VS Mithridate). Pritchett (1958: 309) reports how C sherds were found below the modern village (see above), while in the Lykouresi basin[9] (to the S of the Chaironeia settlement site) a large quantity of surface sherds could be seen on the surface, but none appear to be C (they are unglazed tiles and pottery of a deep red colour, which could be Roman). On the basis of this, Pritchett concludes that the area was empty of habitation or other activities in the C period and therefore could have been where the Chaironeia battle took place. On the other hand, no-systematic personal visits of members of the Boeotia project (J.L.Bintliff pers. comm.) report several Greco-Roman pottery scatters (including C finds), likely representing small rural sites.

Component CH_38: Leake (115) had already suggested that the church of Panagia by the Turkish fountain near to the theatre of ancient Chaironeia would stand on the site of an ancient temple, either that of Artemis Eilethyia or that of Serapis, as he would suppose. In fact, by the church of Panagia, a part of the sanctuary probably dedicated to Asklepios - a deity frequently mentioned in inscriptions from the village - may have been found during an excavation, though few details are given, in BCH 1952: 224 and are reported by Fossey (1988: 379). An inscribed 'trone' [AE796] was found during a small excavation (together with a small statue of a young man). The inscription seems to mention Asklepios, and according to the excavator (Chr. Christou) is probably to be referred to the consecration of an area to the deity. The evidence would represent probably a cult place within the *polis*, but it could also have been an extramural sanctuary at the immediate outskirt of the city. As one cannot confirm it with certainty a different component has been created.

2/3. CHAIRONEIA area: *Components CH_10* (2) and *CH_11* (3)

ID	Period	Type	Toponym	Discovery	Position
CH_10	MH	? burial place	Chaironeia	Other	imprecise
CH_11	LH	burial place	Chaironeia E	Other	approximate

A large matt painted pithos in the Chaironeia museum is recorded as coming from an unpublished tomb at or near the village (Fossey 1988: 380). A component (*CH_10*) has been created as a possible MH burial place in the area. LH burial places are also known from the area (*component CH_11*).[10]

4. Chaironeia – dioikitirio: *Components CH_26 and CH_45*

ID	Period	Type	Toponym	Discovery	Position
CH_26	R	rural site/activity	Chaironeia – dioikitirio	Rescue excavation	certain
CH_45	LR	burial place	Chaironeia – dioikitirio	Rescue excavation	certain

Along the main road of the Chaironeia village, N of the road, a 'suburban villa' was found and excavated during public work in 2002 on the *oikopedo dioikitiriou* (information from poster exhibition at Thebes museum - 2002).
From the mid 3rdC AD the site seems to be occupied by burial areas with graves. This could be associated with labour force for the villa and a probable different organization of functional spaces within the site, rather than to an abandonment of the villa site and the decay of the property system of large estates, as suggested by the panel of the exhibition at the Museum of Thebes (2002).[11]

---

[4] See also Fossey Papers 1990b: 100ff.

[5] Fossey (1988: 378) adds: "*From the NW corner of this circuit another Cyclopaean wall runs Northwards down the mountain slope towards the plain below. This wall may suggest that, at the time of the Cyclopaean circuit, the lower settlement lay at the North foot of the acropolis*".

[6] Fossey mentions prehistoric sherds outside the W circuit to the N (W of the historical polis). One piece of obsidian lame was also found within the circuit by Fossey (1990b: 100ff).

[7] Fossey 1988: 380 reports on the findings in the modern village area which confirmed the location of the lower settlement: in the rock face at the NE foot is carved the small theatre (description by Dilke BSA 1950: 35-7. Photographs in Papachatzis 1981: 260-2, pll. 315 and 317); many ancient blocks in its vicinity [AE795] especially in the nearby fountain; cf. Frazer 1913: 205-208. For the relative positions of the village, theatre and acropolis see Fossey 1988 fig.50. Classical potsherds and tile fragments [AE1185] have been found in the area of the modern village (Pritchett 1958: 309 saw material from excavation at the Panagia church). A Roman colonnaded building [AE1183] at the W edge of the village (BCH 1925: 456), and two Roman buildings with hypocaust [AE1184] (Kahrstedt 1954: 109; AD 1967: 243) are also known. Recent research (mainly rescue excavations during building activities in the modern village) has added to the picture new data, especially from the R and LR periods [AE798; AE1691; AE1679].

[8] AE798; AE1691; AE1679.

[9] It can be spoken of as the Likouresi basin, since it is not a stream valley as most land is well above the stream level.

[10] Two LH chamber tombs [AE1295] and signs of probable others on the E slope [AE800] (HS&D 1979: G46 from AD 1969: 179). It is said that the tombs were noticed for the first time, therefore probably AE1295 is not the same evidence as AE800 (the chamber tomb mentioned in Syriopoulos 1968 and Hope Simpson 1965). There is no location given for AE800, and it has been decided to list it under the same component entry (*component CH_11*).

[11] The villa / large estate system seems to begin in Boeotia usually in Mid Roman times (200-400 AD) and reach their maximum size and extent in the Late Roman period (400-600 AD).

5. Chaironeia N: ***Component CH_12***

ID	Period	Type	Toponym	Discovery	Position
CH_12	H	? burial place	Chaironeia N	Rescue excavation	approximate

To the N of the Kapraina/Chaironeia village, by the road towards the W, two tombs were discovered, made of reused material, with *lekythos* and glass finds of Hellenistic (?) date (BCH 1923: 522). Chronological attribution is not sure, and the component is recorded and mapped as probable.

6. Chaironeia – Plateia Scala and Vourlies: ***Components CH_46 to CH_49***

ID	Period	Type	Toponym	Discovery	Position
CH_46	A	burial place	Chaironeia – Plateia Scala and Vourlies	Rescue excavation	certain
CH_47	C	burial place	Chaironeia – Plateia Scala and Vourlies	Rescue excavation	certain
CH_48	H	burial place	Chaironeia – Plateia Scala and Vourlies	Rescue excavation	certain
CH_49	R	burial place	Chaironeia – Plateia Scala and Vourlies	Rescue excavation	certain

In the locality of Plateia Scala and Vourlies during rescue excavation in 2001 a cemetery was found dated to the A, C, H and R periods, up to mid 2nd C AD (information from poster exhibition in Thebes museum - 2002). It may be considered as the Chaironeia city cemetery, due to the extension and type of burials.
During the installation of the gas pipe-line in 1993 by Chaironeia (I. Galanou property at the 106th km of the pipe-line), an inscribed stele was brought to light [AE1685]. Excavation followed and revealed a small Christian building (AD 1993: 180). No component has been created for the AE1685 entry, because it is the only evidence available and has to be considered too movable. As for the excavated evidence defined as 'Christian', it is not specified whether it can be considered as LR, and therefore a component has not been created but only a mention is given here in the text.

7/8. Chaironeia - Rema: ***Components CH_34 and CH_35*** **(7);** ***CH_36*** **(8)**

ID	Period	Type	Toponym	Discovery	Position
CH_34	A	burial place	Chaironeia - Rema	accidental	certain
CH_35	C	burial place	Chaironeia - Rema	accidental	certain
CH_36	C	burial place	Chaironeia - Rema	accidental	certain

Along the bank and the river bed of the man-made stream that runs through Chaironeia in the NS direction (Ksirorema or Vroulia), 6 tile-roofed and cist tombs in two groups have been excavated, 180-200m -*components CH_34 and CH_35*- and 610-630m -*component CH_36*- NE of the central road of the village (ca 600m from the ancient theatre). The tombs date from the 6th to the 3rdC BC (AD 1994: 287-289). These pieces of evidence can help to interpret the areas as rural burial areas associated with isolated farms, as do the groups of cist tombs recorded as *component CH_50* (see below no.9).

9. Chaironeia - Kovras/Agortses: ***Component CH_50***

ID	Period	Type	Toponym	Discovery	Position
CH_50	H	burial place	Chaironeia - Kovras/Agortses	Rescue excavation	certain

A group of cist-tombs, dated to the H period (from 3rdC BC to 1st half of 2nd BC) related to a farm, were found in the area of the Chaironeia - Kovras/Agortses locality (information from a poster exhibition in Thebes museum - 2002).

10/11. Chaironeia museum: ***Components*** **(10)** ***CH_15, CH_16 and*** **(11)** ***CH_37***

ID	Period	Type	Toponym	Discovery	Position
CH_15	R	burial place	Chaironeia museum	Rescue excavation	certain
CH_16	H	? burial place	Chaironeia museum	Rescue excavation	certain
CH_37	C	burial place	Chaironeia museum	accidental	approximate

Components CH_15 and CH_16: Discovered accidentally and excavated along the road behind Chaironeia museum (AD 1968: 223) was a cist-tomb carved into the rock, with glass finds inside: R (or H?). The evidence lies outside the supposed area of the lower town, but no attempts have yet been made to precisely locate the edge of the settlement in the different periods and it cannot be stated with certainty whether the burial would represent an extramural necropolis linked to the city or a rural cemetery.
Component CH_37: 100m from the museum, on the central road of Chaironeia (Atalali plot) a group of vases, figurines and objects, were found and delivered to the local *ephoreia* (AD 1994: 287-289). They can be dated to the 5th C BC, and probably are from cremation burials (*taphiki pira*).

12. Chaironeia East: ***Component CH_19***

ID	Period	Type	Toponym	Discovery	Position
CH_19	H	burial place	Chaironeia East	Travellers	certain

Component CH_19: just to the E of the modern village, next to the museum courtyard, the restored lion monument is visible marking the *Polyandrion* of the 254 Thebans killed in the battle of 338 BC (Papachatzis 1981: 261 pl. 316; aerial view in Schoder 1974: 96).

13/14. Chaironeia North and Chaironeia bridge: (13) ***Component CH_14;*** **(14)** ***components CH_13 and CH_17***

ID	Period	Type	Toponym	Discovery	Position
CH_13	R	rural site/activity	Chaironeia North/Magoula Balomenou	Travellers	precise
CH_14	R	rural site/activity	Chaironeia bridge	Rescue excavation	certain
CH_17	R	burial place	Chaironeia bridge	Personal or group interest	approximate

Component CH_13: Roman structures have been excavated[12] (Praktika 1902: 55; AJA 1904: 357) N of Chaironeia, between the city and the Kephisos, 300m SW of the modern bridge. Those structures, by the Magoula Balomenou site (see below no.16-17), might correspond to one of the two Roman brick buildings whose surface remains were seen by Dodwell 1819: 224 (see Kahrstedt 1954: 109, who interpreted them as villa sites). Fossey (1988: 379 and 1990c: 58) reports the evidence which constitute *component CH_13*, referring to it as a 'Roman farmstead'. The other brick building seen by the traveller might be *component CH_14*, excavated later and described here below.

Component CH_14: after the bridge at the entrance to the town, right to the road coming from Levadeia, rescue excavations – OTE telephone company– discovered part of an hypocaust and sherds from the R period (AD 1967: 243). The component might correspond to the second of the two Roman brick buildings seen by Dodwell 1819: 224 (see Kahrstedt 1954: 109) and interpreted as villa sites. The other excavated one might be *component CH_13* (see above).[13]

We cannot be sure whether those two Roman buildings, excavated later, correspond to the two brick buildings seen by Dodwell, which are probably the same as the ones mentioned also in Kahrstedt 1954: 109. As the identification is not certain, one could even hypothesize the existence of a third (or a third and a fourth) brick building seen by the traveller and not visible anymore.[14]

Component CH_17: a cemetery of R date has been found during excavation next to the R building with bath - *component CH_13* (Praktika 1902: 55; AJA 1904: 357).

15. Varka: *Component CH_18*

ID	Period	Type	Toponym	Discovery	Position
CH_18	R	burial place	Varka	accidental	approximate

About 3km NE of Chaironeia village, in the Varka locality, tomb remains from the 2[nd]C BC are mentioned in an enclosure (AD 1973-4: 442-3). Accidentally two tombstones where found while ploughing and then a small excavation followed.

16/17. Magoula Balomenou: *Components CH_21 to CH_24 and CH_13* (16); *component CH_25* (17)

ID	Period	Type	Toponym	Discovery	Position
CH_21	N	settlement	Magoula Balomenou	Personal or group interest	precise
CH_22	EH	? activity focus	Magoula Balomenou	Personal or group interest	precise
CH_23	MH	settlement	Magoula Balomenou	Rescue excavation	certain
CH_24	LH	? activity focus	Magoula Balomenou	Personal or group interest	precise
CH_25	MH	burial place	Magoula Balomenou	Rescue excavation	certain
CH_13	R	rural site/activity	Chaironeia North/Magoula Balomenou	Travellers	precise

A low mound can be seen NE of Chaironeia, N of the railway line and W of the bench of the Kephisos river, 300m SW of the modern bridge, in the locality of Gilophos (as mentioned in Syriopoulos 1968: 31). The mound is a tell, created by the superimposition of the occupation layers (mainly Neolithic period – see below). The dimensions of the *toumba*, as reported by Tzavella-Evjen, are N-S 155m and E-W 140m, measured through core-sampling (0.025m diameter).
Sotiriadis discovered the site at the very beginning of the 20th century, while searching for the Macedonian tumulus for the 338 BC battle (PAE 1902: 55ff).[15] When he got to the site, the surface was cultivated and a piece of wall was visible.[16] During excavation, the upper layers found presented R (or probably LR) structures (recorded as *component CH_13* – see above no.13-14) and graves (*component CH_17* – see above no.13-14).[17]

Components CH_21 and CH_22: According to the account of Sotiriadis, prehistoric layers were found disturbed by later activities. The date of prehistoric material is not specified by Sotiriadis (who mentions generally prehistoric levels),[18] but it was examined later by Tzavella-Evjen. Sotiriadis notes how all the periods (from NEO to R?) were found in the excavation, while Tzavella-Evjen reports only EH, MH and C material[19] from the museum, where the excavation material was brought. This material suggests the presence of activities in those periods, though the relative layers were disturbed by later occupation (as stated by Sotiriadis), and probably partially by Sotiriadis' excavation as well. Tzavella-Evjen could not retrieve strata from that period's digging activities at the site years later, when her excavation revealed a Neolithic occupation at the site.[20]
Three phases of Middle Neolithic occupation (5,000-4,000 B.C.) were noticed: the earliest at 4.30m depth, the middle at 3.30-3.90m, the more recent at 0.80-3.30m (the latter with 3 sub-phases). As stated by Tzavella-Evjen, some pottery found by Sotiriadis can be dated to the Late Neolithic (after 4,000 BC), probably from the upper layer of the *toumba*, which have been lost, as seen earlier.

Component CH_24: On the *toumba* site, weak evidence for activities in the LH period is present. Only 5 LHIIIB sherds in the Chaironeia museum, reported as from the Magoula site and mentioned in the prehistoric gazetteers, might be evidence of a burial place.

[12] Foundation walls, two cisterns, one probably belonging to a bath, a paved floor (along with lots of potsherds from the R period). The excavated structures [AE1184 - constituting *component CH_13*] are probably marked in the map in AD 1996: 285 (fig.14), although the absence of a map key does not allow for a certain identification.
[13] AE1678 (*component CH_14*) is probably mapped in AD 1996: 285 (fig.14), without map key.
[14] For comments on R rural sites in the area see chapter II.3.3 - RURAL SEGMENT.

[15] See Fossey 1988: 379, as well as Syriopoulos 1968: 30-1, N22 and Pritchett 1958: 308 n.19 for a summary of previous research.
[16] The wall had probably been seen already by Dodwell (see *component CH_13*).
[17] The knowledge of the later (R) activity focus on the site, in this case, is not due to the interest in the place because of the presence of a major site in the prehistoric period, as it happens in many other cases, especially around Copais – see Kalami-Lioma (Koroneia area) and others. It is rather due to a visible surface wall (probably already seen by Dodwell), and secondly to its characteristic mound/tell shape for which it was visited later by Sotiriadis (personal interest).
[18] Sotiriadis, Praktika 1902: 55ff; 1904: 40ff; 1907: 109; 1909: 123ff; AE 1908: 64ff; AM 1905: 120-129; REG 1912: 263-270.
[19] C material (if any) has probably to be considered off-site.
[20] Tzavella-Evjen Praktika 1986: 55ff; Ergon 1986: 90ff; AJA 91 1987: 320; A' *diethnes synedrio Boiotikon meleton* 1988: 37-41; B' *diethnes synedrio Boiotikon meleton* 1995: 63-67.

Components CH_23 and CH_25: Tzavella-Evjen in 1989 extended the excavation (rescue excavation – due to the opening by local farmers of a drainage ditch, where architectural fragments were found) and discovered a large settlement of the MH period (*component CH_23*),[21] extending to the S and E side of the Neolithic *toumba* towards the Kephisos river. The trench contained Neolithic sherds, lithic tools and a bone idol of human shape. The remains of walls dated back to the MH period (ca. 2000-1700 BC). MH pottery was found, mainly coarse ware but also a few pieces of Minyan ware of fine quality.[22] MH burials were also discovered (*component CH_25*).

The Kephisos river course has shifted during time, and it certainly would have been in a different location in the MH period (though it cannot be said exactly where). This would be proved by the appearance of archaeological remains inside the river (Praktika 1989: 158, n.17 – MH tombs *component CH_25*) and therefore the limit of the MH settlement would not be marked by the river meander.
During the excavation Sotiriadis found layers of 'empty soil'. He thought they might be the result of the practice Neolithic people had to cover the tombs, while later on he interpreted them as alluvial deposits from the river, until he reached the point to say that the whole *toumba* was a '*heroon*' dedicated to the Kephisos god. Tzavella-Evjen (1988: 37-41) does not agree with either of the two interpretations. She points out that the river was not running in the present day location, and that the form of the deposits presents a lacustrine character (*limnaias proelefsis*). She therefore suggests that probably the marshy lake would have reached an area very nearby the *toumba/magoula* and had an effect on the whole area. We would exclude the presence of the Copais lake or marshes in this area,[23] although it cannot be excluded that this effect might be due to the presence of another former marshy area nearby, which could be read on the map (GYS 1:50,000 map) in the toponym *Loutsa* (see below no.21). The probable presence of the marshy area to the immediate N of the Kephisos river, and opposite the Magoula Balomenou site, might constitute a critical element which would have given a special character to the surrounding landscape, taking into consideration also the presence of a probable settlement or large rural activity to the other side (Agioi Apostoloi site – see below no.21).

Joining all the information examined above, it can be concluded that:
- at the *toumba* site there is Neolithic occupation by the river forming the tell/mound (*magoula/toumba*).
- On the *toumba* site, EH material is also present (found in the Sotiriadis excavation), suggesting a probable focus of activity in the EH period.
- Later on, a MH settlement was located just below the *magoula* (to its S and E), extending until the river. Probably the river shifted its course through times, since MH tombs were found in its present river bed. A burial area was therefore located between the settlement and the river.
- On the *toumba* site, we probably have low-level evidence of activities for the LH period, but nothing more than weak signs. The five sherds in the Chaironeia museum dated to the LHIIIB period reported as from the Magoula site (decorated or distinctive sherds, I suppose), might be evidence of a burial place nearby.
- Yet on the *toumba* site, we do have evidence for occupation in the R (not better specified – *component CH_13*) period, where a rural site was probably located.

18/19. Akontion – Kephisos: (18) *Components CH_40, CH_41 and* (19) *component CH_42*

ID	Period	Type	Toponym	Discovery	Position
CH_40	G	burial place	Akontion - Kephisos	Rescue excavation	precise
CH_41	A	burial place	Akontion - Kephisos	Rescue excavation	precise
CH_42	C	burial place	Akontion - Kephisos	Rescue excavation	precise

During rescue excavations in occasion of the public work for the widening of the Kephisos river bed, four groups of graves belonging to an ancient cemetery were excavated (AD 1993: 183-7), for 2km along the river in the area of the Akontion village and the bridge of Chaironeia-Akontion. Excavated burials locations are marked in AD 1996: 284 (fig.14). Below the graves are thick alluvial deposits. Among the graves, two groups are dated to the LG-A period,[24] while the other two groups (five tombs only in total) are dated to the C period.[25]

20. Chaironeia plain: *Component CH_20*

ID	Period	Type	Toponym	Discovery	Position
CH_20	H	burial place	Chaironeia plain	Travellers	approximate

Component CH_20: Out in the plain, ca. 2.5km to the E of the Magoula Balomenou site, by the Kephisos river, 3km E of Chaironeia and, as Sotiriadis (Praktika 1902: 58) notes, closer to the Bramaga village, is the burial mound for the Macedonians killed in the 338 BC battle[26] (Cooley, AJA 1904: 81-2 and Sotiriadis, Praktika 1903[27]).

21. Agioi Apostoloi - Agios Georgios: *Components CH_31 to CH_33*

ID	Period	Type	Toponym	Discovery	Position
CH_31	C	activity focus	Agioi Apostoloi - Agios Georgios	Personal or group interest	approximate
CH_32	H	activity focus	Agioi Apostoloi - Agios Georgios	Personal or group interest	approximate
CH_33	R	activity focus	Agioi Apostoloi - Agios Georgios	Personal or group interest	approximate

In a small valley bordered by the Akontion (Dourdouvana), the Idilion (Vetriza) and the Iphanteion (Paliovounas), in the area around the deserted village of Agioi Apostoloi, especially around and to the W of the church of Agios Georgios and towards the plain (*kampo*), an extended scatter of sherds is reported by Dasios (1995a). He suggests the presence of an extended habitation (*ektetamenis katoikisis*) of Roman and

[21] Ergon 1989: 49ff; Praktika 1989: 146ff; AJA 1991: 306

[22] A cist tomb was found in the S trench, consisting in a rideposition inside a MH wall foundation (ritual?). Tzavella-Evjen has written a lot about this discovery and its ritual implications, whose examination would lead us beyond the purpose of the present work.

[23] On the shifting of the Kephisos river bed and related issues see chapter II.3.3 – LANDSCAPE IN THE LONG TERM. On the modelled extension of the Copais lake see chapter II.3.1 and Farinetti 2008.

[24] The LG-A tombs are contemporary to the cemeteries of Akraiphnio, Ritsona and Pyri Thivon.

[25] Pottery from the cemetery was presented by Sarandopoulou 2000.

[26] Plutarch *Alex*. 9.3.

[27] Two mounds were dug by Sotiriadis. One is Magoula Balomenou site and the other one is the Macedonian tumulus of the 338 BC battle (Sotiriadis, Praktika 1902: 53-59 and Pritchett, AJA 1958: 307ff).

earlier (C-HL) periods. The scattered sherds are mentioned apparently by Dasios only, who notes how Plutarch (*Syllas* 17.5) mentions *Assioi* in this area.
The area of Agioi Apostoloi-Agios Georgios may belong either to Phokis (Panopeus) or to Boeotia (either to Chaironeia or Orchomenos), as the borders in the Kephisos valley are not clear. Dasios suggests an attribution to Phokian Panopeus (see chapter II.3.3 – BOUNDARIES). Dasios (1995b) states that certainly to Panopeus belonged the whole Idilion mountain and probably at least a part of the small valley, to get the NW hillsides of the Yphantion (Paliovouna) where a (Phokian) fort controlling the border was (at Mavrokastro), whose the remains can still be seen.[28]

In the vicinity is an area called *Loutsa*, to the orographic left of Kephisos, with traces of canalizations visible on the map (GYS 1:50,000). Loutsa is an arvanitic word for 'wet'. It was thus probably a marshy area. To the S of this 'wet' area is also the Magoula Balomenou site (see above no.16-17).

22. Agios Charalambos: *Components CH_27 to CH_29*

ID	Period	Type	Toponym	Discovery	Position
CH_27	H	activity focus	Agios Charalambos	Ext. topographical surveys	certain
CH_28	LR	activity focus	Agios Charalambos	Ext. topographical surveys	certain
CH_29	LH	? activity focus	Agios Charalambos	Ext. topographical surveys	certain

In a small valley between the Karvouni ridge to the S and an outlier of the same ridge to the N, facing westwards Agios Vlasios (ancient Panopeus), traces of ancient activities are reported on the N and S slope and in the valley bottom (Pritchett AJA 1958: 307-8; Fossey 1988: 383). In the valley is reported a spring by the newly constructed church of Agios Ioannis (Fossey 1988: 383), while the church of Agios Charalambos lies in the W part of the valley.

Components CH_27 and CH_28: The interpretation suggested by Fossey for the site as a sanctuary[29] is mainly due to the Schachter's and Pritchett's accounts for the reused blocks in the ancient Byzantine church (the old church of Agios Georgios probably), and it is given value by the fact that Fossey himself during his visit noticed on the site surface walls on the slopes, as well as H and LR potsherds.[30] On the other hand, both the LR and the H evidence could also represent a different kind of activity focus, such as a rural site, and this would probably be the case at least for the H period (see also discussion below – Isoma site).

Component CH_29: Fossey noticed at the site LH surface sherds (Fossey 1988: 383), but we cannot be certain of what they might represent. The site would deserve a revisit.

[28] The fort is mentioned also by Philippson 1951: 426 - map p.747, and it was mentioned by Diodorus XVI 58.4. The fort is not included into the database of this work because it was considered as being Phokian.
[29] Fossey (1988: 384) also attempts a possible identification with the sanctuary of Apollon Thourios (see though the recent discovery at Isoma – no.23-24).
[30] The ancient blocks could also be coming from the ancient city (Panopeus). On the other hand, the situation may be similar to the one at the Pontsa site (in Koroneia area – see appendix I.1), where architectural fragments from a Late Roman sanctuary at the site are employed in the construction of a Byzantine church, which was usually then surrounded by a small hamlet.

23/24. Isoma: *Components* (23) *CH_43 and* (24) *CH_44*

ID	Period	Type	Toponym	Discovery	Position
CH_43	gr-rom	fort	Isoma	Ext. topographical surveys	certain
CH_44	gr-rom	? activity focus	Isoma SW	Ext. topographical surveys	certain

Component CH_43: On the hill Isoma, a roughly conical rise jutting northwestward into the Kephisos plain, to the W of Chaironeia, a fortification wall has been noticed by Camp et al. (AJA 1992: 443-455, plan of the fort in fig.12). They suggest an identification with the Mont Thourion of Plutarch (*Sylla* 17.6-8). It is a rubble wall, today poorly preserved. As the authors say, there is no indication of date, but the trophy, commemorating the victory of Sylla over Mithridate VI at Chaironeia in 86 BC [AE1693], that they found on the same rise, could be connected with the building of the fort.

Component CH_44: In the vicinity, Camp et al. 1992 suggest the presence of the remains of the temple of Apollon Thourios, mentioned by Plutarch (*Sylla* 17.6-8), whose exact location is unknown (Schachter BICS 1967: 7 – see above Agios Charalambos site). The site, apparently never noticed earlier, is recorded in the database as a possible cult place. At the confluence of several streams flowing roughly Northward off the ridge of Agios Dimitrios towards Isoma hill, they noticed a large terraced area, located ca. 500m S-SW of Isoma hill, on the W bank of the stream, and occupied today by the shrine of the Metamorphosis. Within this area, remains of late walls are visible, with numerous early blocks and architectural fragments built in. Opposite, on the E bank of the stream, they found archaeological fragments in some piles of rubble derived from collapsed huts. In this area Byzantine pottery in large quantity can be noticed, where surface remains of a settlement grew up around a church.[31] The terraced area, they say, could be what remains of a platform for a temple (C or H date seems likely considering the architectural pieces found). One should keep in mind, though, that the architectural pieces (reused in both locations) might have been taken from a place in the vicinity.

According to Plutarch (*Sylla* 17.6-8), at the foot of the Thourion mountain the river Morios and the temple of Apollo Thourios were located. As mentioned by Camp et al. 1992, Leake had identified Morios as the Vathyrema and suggested that a Turkish village he called Mera, located near the banks of that stream (by Ag.Charalambos church), preserved the ancient name Morios. Local farmers confirm that a Turkish village named Megara once existed near the church.
By the church of Agios Charalambos, Fossey detected some remains (see above), which though Camp et al., after a visit to the place, do not consider substantial. They note that visible surface sherds were mainly medieval, and consequently the surface walls Fossey saw were probably later as well, linked to the settlement site by the church. On the other hand, Pritchett reports ancient blocks built into a ruined chapel near the shrine of Agios Vasileios just to the NW of Orthopagus, though he does not specifically associate them with the Apollo temple. The two sites – Fossey's and Pritchett's ones– are nearly 1km apart, on opposite sides of Hammond's hill (as called in Camp et al.'s work), which corresponds probably to the hill above Ag.Charalambos.

[31] The Byzantine church and settlement could have developed on the site of the ancient temple, marking the continuity of a sacred vocation, as often happens in the area (see Pontsa site -Koroneia *chora*, appendix I.1- and Agios Charalambos site above).

25. Thourion - Koumoulia: *Component CH_39*

ID	Period	Type	Toponym	Discovery	Position
CH_39	gr-rom	? activity focus	Thourion - Koumoulia	Rescue excavation	certain

At Koumoulia locality, before the Thourion village, below the Frankish tower, rescue excavations during the installation of the gas pipeline (in 1993 – at the 104th km) brought to light pottery and badly constructed walls (AD 1993: 180). No exact chronology is given.

26. Thourion East: *Component CH_30*

ID	Period	Type	Toponym	Discovery	Position
CH_30	gr-rom	fort	Thourion East	Int. topographical surveys	certain

On a round shaped rise between the Frankish tower location and the Kephisos valley, on a flysch area on the NE descending slope of Thourion, Lauffer (Kopais I: 147, sketch 154) reports traces of an ancient tower, whose square perimeter is visible on the ground.

Appendix I.4
The Copais area: Orchomenos

THE ARCHAEOLOGICAL RECORD

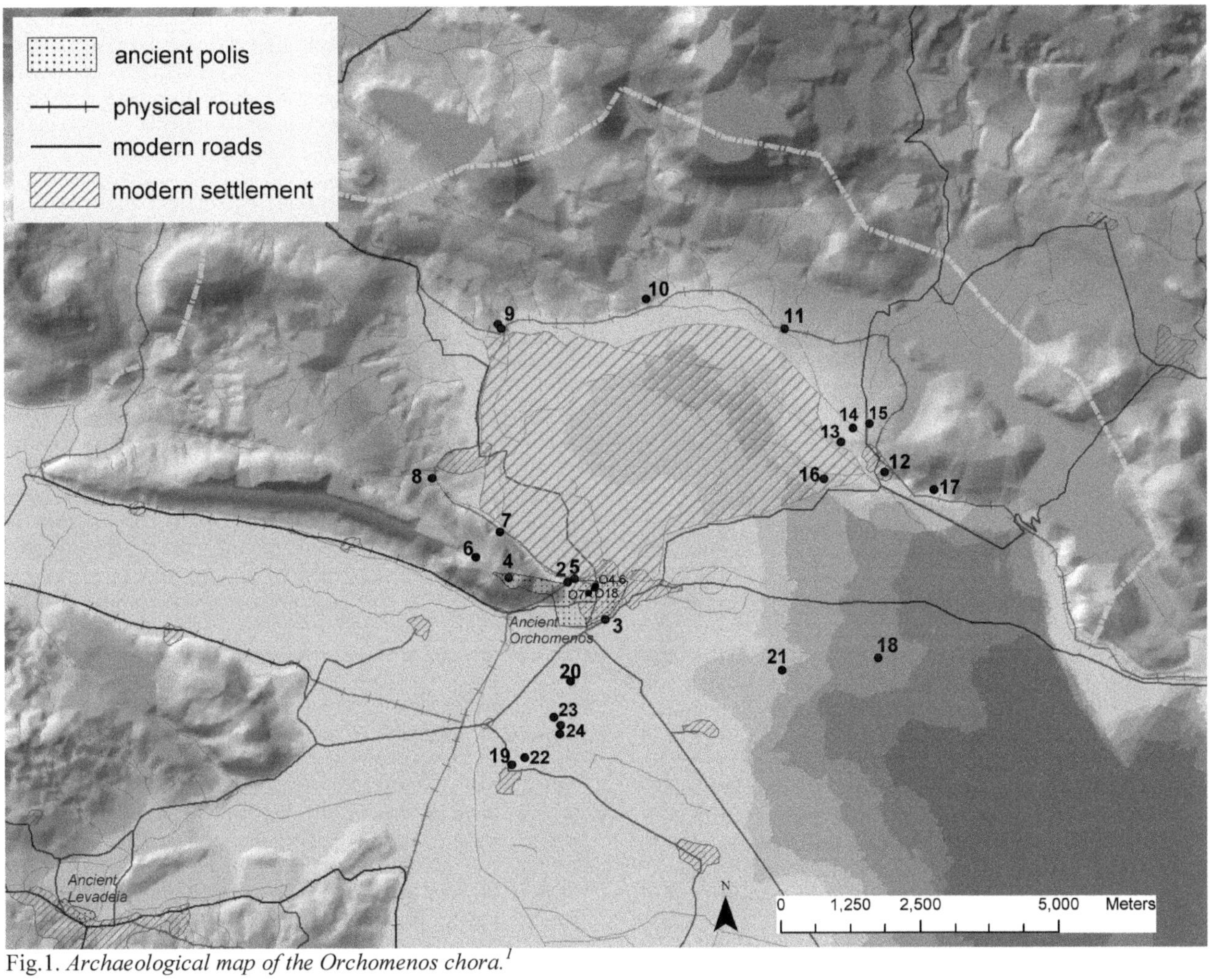

Fig.1. *Archaeological map of the Orchomenos chora.*[1]

[1] Prehistoric and LR burials within the city area (*components O_4, O_6, O_7* and *O_18*), are marked on the map.

1. ORCHOMENOS[2]: *Components O_1 to O_11; O_13 and O_14; O_16 to O_19*

ID	Period	Type	Toponym	Discovery	Position
O_1	N	settlement	Orchomenos	Other	certain
O_2	EH	settlement	Orchomenos	Personal or group interest	certain
O_3	MH	settlement	Orchomenos	Personal or group interest	certain
O_4	MH	burial place	Orchomenos	Personal or group interest	certain
O_5	LH	settlement	Orchomenos	Personal or group interest	certain
O_6	LH	burial place	Orchomenos	Personal or group interest	certain
O_7	LH	burial place	Orchomenos	Travellers	certain
O_8	SubM	? Settlement	Orchomenos	Other	approximate
O_9	PG	? Settlement	Orchomenos	Other	approximate
O_10	G	? Settlement	Orchomenos	Other	approximate
O_11	A	settlement	Orchomenos	Personal or group interest	certain
O_13	C	settlement	Orchomenos	Historical geography	certain
O_14	H	settlement	Orchomenos	Personal or group interest	certain
O_16	H	fort	Orchomenos	Travellers	precise
O_17	R	settlement	Orchomenos	Historical geography	certain
O_18	LR	burial place	Orchomenos	Rescue excavation	certain
O_19	LR	? Settlement	Orchomenos	Other	approximate

The ancient settlement site of Orchomenos lay at the foot of the E end of the Akontion ridge, at the NW edge of the Copais basin. From the top of Akontion, above Orchomenos, one can have the best look at the entire Copais basin.
As Philippson notes (1951: 471), as one comes from Levadeia through the fertile plain of Kephisos he sees the soft limestone rock of Akontion, with a small rock pick. The Akontion is the '*polis* ridge' of Orchomenos, called in modern times Dourdouvana and currently Akontion again. It is named after its characteristic dart shape. The toponym has to be included in the series of orographic names (mainly related to ridges and mountains whose slope or foothills host a settlement or city site) derived from the military equipment, as for instance Aspis at Argo and Kerykion at Tanagra, yet in Boeotia.
The slopes of Akontion suddenly end with a peak (369m asl) carrying the ancient wall (see below), and then towards the E they drop with a regular slope (ca 12 degrees of slope) and end with a small promontory-like feature that sinks down into the marshy area (Philippson 1951: 471). The Bronze Age and Greco-Roman settlement lay on the lower spur and extended into the plain.
To the E end of the Akontion spur at the edge of the plain, lies the monastery of Skripou (built in 874),[3] probably on the place or near a former sanctuary of the *Charites*. In the vicinity, on the abrupt N foot of the Orchomenos promontory, springs out the Akidalia spring (one of the sources of the Melas river). The medieval village of Skripou spreads SE from the end of the Akontion in the plain. Next to it, on both sides of the Kephisos river, which runs along the S foot of the Akontion, lies the large village of Petromagoula. Currently the river bed there is dry, while in previous times it was not. In Philippson's times (1951) the two villages (Skripou and Petromagoula) together had the largest population in Copais. In present times the two villages form modern Orchomenos.

Ancient Orchomenos was located at a very nice spot, on a rock spur defended by the river, in the middle of a small area. The *polis*, with Tsamali bay to the N and the E and the long and narrow ridge of Akontion to the NW and the Kephisos valley to the W, was in a strategic position to control incoming routes to Boeotia from Lokris Opuntian (through the area of ancient Hyettos and the mountains behind Topolia/Kastro, as well as through the opening between Prophitis Ilias and Chlomon mountain to the N of Orchomenos) and from Phokis through the Kephisos valley (fig.1 in chapter II.3.4).
The settlement site at Orchomenos has a long documented story that extended since the Bronze Age until today, and is mentioned already in Homer. The conflict with Thebes within Boeotia marks its history throughout all periods, with a crucial importance for city power and life especially in Mycenaean and Archaic times.[4] Thebans destroyed it in 364 but then Philip the Fifth and Alexander the Great fortified it and gave it a new life during the Hellenistic period.

Component O_1: Excavated layers [AE429], from old and recent excavations, since the end of the 19th century (de Ridder 1895), attested the presence at the site of a Neolithic settlement. From the Neolithic period sherds were found but apparently no structures were noticed. Kunze (Orchomenos II) published in 1931 the Neolithic pottery known from the site already reported by Bulle (1908). Caskey (1960: 159ff) comments on the varieties and age of the Neolithic at Orchomenos. Several excavations have been carried out later on below the modern settlement site (mainly by Sotiriadis and more recently by the local ephorate) and found Neolithic among and below levels from later periods. As stated by HS&D (1979: G1), excavation data and visits at the place lead to conclude that the lower part of the Eastern spur of Dourdouvana/Akontion was the centre of the prehistoric site, and is in that area where Neolithic levels were also found. Fossey (1988: 354) describes the lower area, occupied by the prehistoric settlement, as 'not rocky'. I would rather say, as HS&D (1979: G1) say, that the prehistoric settlement lay on the extreme spur of the Akontion ridge (resembling under this respect other contemporary settlements in the Copais area – Kalami-Lioma, Agoriani-Dekedes, etc.). The expression 'not rocky' used by Fossey probably stands for the area not strictly on the bare rock, but covered by the soil eroded away of the calcareous steep slopes of Akontion, as it constitutes the lower area of the Akontion hill-side.

Component O_2: Excavated layers [AE430] also testify the existence of an Early Helladic settlement site in the lower part of the eastern spur of Akontion. Structures of notable entity from the EH period were found at the site (Caskey 1960: 159ff), since de Ridder excavation (1895). Kunze (Orchomenos III) published in 1934 the Early Helladic pottery known from the site already reported by Bulle (1908). Sotiriadis' digs and more recently excavation work carried out by the local *ephoreia* brought to light EH layers, pottery and structural remains. EH

[2] Bibliography on the city site until the mid 1980s has been collected by Fossey 1988: 352-3. More recent references are referred to here, under the individual components.

[3] The surface level around the church is much higher than when it was built, as noted by Philippson (1951: 472).

[4] Orchomenos, before the end of the 8th C, attempted to gain supremacy over all Boeotia, and dominated Chaironeia, Hyettos, Olmones, Hyria, and Lebadeia and perhaps Koroneia and Copai before and during the time of Hesiod (Buck 1979: 97). See also Farinetti 2003 for Orchomenos and the Archaic period.

layers were noticed also by Fittschen[5] at the Asklepion, higher up on the E end slopes of the Akontion[6] - a fact that would prove –against Bulle– that the area (on the slope of Akontion, higher up compared to the lower settlement site) was settled from the EH period at the latest.

Component O_3: a Middle Helladic settlement is also testified by the presence of excavated layers and structures [AE431], since earlier excavations of de Ridder (1895) up to Sotiriadis (AM 1905: 130). Also recent excavations by the local *ephoreia* interested MH layers. The very recent Sarri 2010 constitutes a complete study on MH pottery from early excavations and in general on MH Orchomenos.

Component O_5: remains of a Late Helladic settlement are known from the site and have been excavated (Sotiriadis' excavation – AM 1905: 130 - and more recent excavations by the local ephorate - Spyropoulos AAA 1974: 324). The widely known structure from the Mycenaean period is the Treasury of Minyas (*component O_7*), while the existence of a 'palace' in the area occupied later by the Skripou church was inferred from finds of wall frescoes among structures[7] of that period (Bulle 1908: 71-85).[8] As Fossey (1988: 356) notes, it is to the density of Byzantine settlement that Bulle attributes (1895: 12) the disappearance of much of the Mycenaean and later remains. HS&D (1979: G1) notes how LH layers have been eroded on the slopes of Akontion, while elsewhere ('on acropolis', they say, probably meaning the lower area though) LH levels and fresco fragments are preserved (over earlier levels).

Components O_4, O_6 and O_7: a MH (*component O_4*) and LH I burial area (*component O_6*) has been excavated [AE437 and AE438] in the lower E spur of Akontion, North of the Treasury of Minyas (monumental tomb of Mycenaean period – *component O_7*), W of Skripou church along the road, in the Melissomandri locality (Spyropoulos AD 1971: 218-9; AD 1972: 312; AAA 1974: 324).[9] Other LH burial places are known from the vicinity: see below *component O_22* (a tumulus of LH III date apparently) and *component O_24* (an extensive chamber tomb cemetery).

Component O_8: some isolated finds [AE443] are reported (HS&D 1979) from the Sub Mycenaean period.

Component O_9 and O_10: rescue excavations (AD 1993: 183-187) found (apart from remains of buildings of an R date [AE435] decorated geometric potsherds in a deeper sample section [AE1696] to the S of the LH so-called 'palace', and some PG and G isolated finds are reported from the site [AE1133 and AE444]. They could be related to a habitation or burial area. Other burials are also known from the vicinity, which may lead to the location of a settlement site in the geometric period at the Orchomenos site.

Components O_11, O_13, O_14, O_16 and O_17: the existence of a settlement site is well documented for Archaic,[10] Classical, Hellenistic[11] and Roman[12] periods. The settlement was probably defended since the A period.[13] The Hellenistic fortification [AE1194] enclosing the E spur of Akontion up to its E summit (502m asl) is quite impressive and well known[14]; hence it has been created a separate component (*component O_16*) as the H fort standing alone.

We cannot know whether the surface walls [AE436] immediately S of Skripou/Orchomenos village mentioned by Lauffer (1974) are to be linked to the R *polis* settlement of Orchomenos or to a rural activity (villa?). Due to the uncertainty of the information, they have been added to the *component O_17* only as a possibility. The evidence AE436 could although constitute a separate component as a rural site/activity.

Component O_19: Fortification might have still been in use during the LR period, though there is no actual evidence of a settlement in that period, but some early Christian burials [AE442, *component O_18*] could lead to the hypothesis of a continuity of habitation in that period.

Component O_18: in 1954 LR graves (Early Christian [AE442] - BCH 1956: 295) have been discovered during the construction of a road between the church and the 'archaeological site' (treasury of Minyas?).

2. Orchomenos: *Component O_12*

ID	Period	Type	Toponym	Discovery	Position
O_12	A	burial place	Orchomenos	Other	certain

Archaic burials are known, since de Ridder (1985: 137-224), in and around the so-called *Asklepieion.*

3. Orchomenos: *Component O_15*

ID	Period	Type	Toponym	Discovery	Position
O_15	H	burial place	Orchomenos	accidental	approximate

A small group of Hellenistic graves have been excavated immediately to the SE of Orchomenos, in the Mesambelia locality (AD 1972: 324).

[5] The recent work of the DAI equipe lead by K.Fittschen (AReports 1997-1998; 1998-1999; BCH 1999: 730), aiming to document the monuments of historic times and update the plan of Lauffer (RE Suppl. XIV 1974: 294 fig.1), reexamined (with drawings, plans, study) the fortification wall on the Akontion, and several other buildings. Among them especially: the so-called Asklepios temple (excavated by the French), the so-called Archaic temple excavated by Bulle, the rocky terrace with the sanctuary of Petakas outside the ancient city (also examined by the French, above the most westerly sources of the river Melas, connected by the French with the sanctuary of Herakles).

[6] Amongst them an EH clay seal (Fittschen, AReports 1997-1998; 1998-1999).

[7] Three 'megaron' units, apparently part of the 'palace' (Spyropoulos AAA 1974, 316-323; cf. BCH 1975 pll. 111-3 and AD 1972: 258-263).

[8] There is a long debate on the question of the existence of a Mycenaean 'palace' at Orchomenos, and on its possible identification with the complex where fresco fragments were found, by the Skripou church. This certainly would be of interest to investigate the socio-political balances in the area, but for the purposes of the present work it is probably enough knowing the presence of an important settlement in the area (as the Treasury of Minyas seems to attest).

[9] Mycenaean structures (the so-called 'palace') did not disturb the earlier cemetery. The MH – LH I burial area extended until the area to the N of the Treasury of Minyas.

[10] The excavated structures for the Archaic period belong to cult places mainly [AE433]. They attest though, together with other few remains, the presence of a settlement in the A period at the site, of which they would represent intrasite temples and cult places.

[11] The theatre is the most impressive remain from the period.

[12] In addition to public buildings (theatre, *stoa*) rescue excavations found remains of buildings of the R date of an unspecified function S of the church and SE of the ancient theater [AE435 and AE1197].

[13] Archaic defense wall (circuit) on the E slope of the Akontion reported by Spyropoulos [AE271].

[14] On the H fortification focused the work of Fittschen's team – see footnote n.5.

4. Orchomenos: *Component O_16*

ID	Period	Type	Toponym	Discovery	Position
O_16	H	fort	Orchomenos	Travellers	precise

The Hellenistic fortification [AE1194] enclosing the E spur of Akontion up to its E summit (502m asl).

5. Orchomenos neighborhood: *Component O_24*

ID	Period	Type	Toponym	Discovery	Position
O_24	LH	burial place	Orchomenos neighborhood	Personal or group interest	imprecise

In the neighborhood of Orchomenos an extensive chamber tombs cemetery is reported [AE1198]. HS&D 1979 do not give any more details of a precise location, but this evidence is certainly not at the same location as AE439 – *component O_22* (see below no.20).

6. Orchomenos quarry: *Component O_21*

ID	Period	Type	Toponym	Discovery	Position
O_21	gr-rom	special activity	Orchomenos quarry	Personal or group interest	approximate

Ca 300m further W of Orchomenos-*Kastro* Fittschen (AReports 1998-99: 56) reports the quarries [AE1297], from which the building material for the fortification was carved out.

7. Melas source: *Components O_35 and O_36*

ID	Period	Type	Toponym	Discovery	Position
O_35	C	cult place	Melas Source	Personal or group interest	certain
O_36	H	cult place	Melas Source	Personal or group interest	certain

In correspondence to the westernmost spring/source of the Melas river, along the eastern side of the Akontion ridge, fine pottery, figurines and metal votive objects [AE1256], along with visible surface walls [AE1255] as well as structures carved in the rock [AE1254], mark the presence of a cult place in the C-H period. The evidence was known to de Ridder (BCH 1895: 150ff.).[15] Recent work on the area has been carried out by Fittschen (AReports 1998/9: 56 and fig.64; BCH 1999: 730).
The site has been interpreted as a cult place (probably a temple) dedicated to Herakles, mentioned by Pausanias (IX 38.7).[16] The link that traditionally relates the figure of Herakles to the water element would suit the case, being the site by the point where the river Melas leaves the mountain spur.

8. Tsamali: *Component O_37*

ID	Period	Type	Toponym	Discovery	Position
O_37	C	? cult place	Tsamali	Personal or group interest	approximate

On the slope above Tsamali/Dionysos village, along the W edge of the so called Tsamali bay (the NW bay of Copais), Bulle excavated the remains [AE603] of a small sanctuary of Classical date (the results of the trial excavation in Bulle 1908: 116 are summarized in Lauffer 1974: 325 and Kopais I: 149).[17] Only 5th and 4thC BC fine, decorated pottery was found during excavation, which might lead to the interpretation of the site as a cult place (the associated component has been recorded as 'cult place?'). Bulle suggested the identification with a sanctuary dedicated to Dionysos, after whom, as reported by Fossey (1988: 363-4), the village would have recently been renamed Dionysos.

Lauffer (Kopais I: 151) suggests the presence in the past of a route leaving Orchomenos, passing by the two cult places (Melas source and Tsamali) and leading towards Polygira site and then all around the bay (see fig.1).

9. Polygira: *Components O_28 to O_34*

ID	Period	Type	Toponym	Discovery	Position
O_28	N	? activity focus	Polygira	Personal or group interest	certain
O_29	EH	activity focus	Polygira	Personal or group interest	certain
O_30	MH	activity focus	Polygira	Personal or group interest	certain
O_31	LH	activity focus	Polygira	Personal or group interest	certain
O_32	A	? activity focus	Polygira	Ext. topographical surveys	certain
O_33	C	activity focus	Polygira	Personal or group interest	certain
O_34	R	activity focus	Polygira	Personal or group interest	certain

As reported by Fossey (1988: 360), at the furthest corner of the Copais basin's NW bay, about 2km N of the village of Tsamali, is the source of the river Polygira which flows around the N side of the bay until it joins the Melas river about 1km W of Pyrgos. For much of its course the Polygira river flows through marshy ground, but around its source the land is better drained. The source itself is a copious spring or double spring (called *Kephalovrisi* or *Kephalari*) at the end of a long stony ridge running out SE into the bay. On the ridge remains of ancient occupation were found. The site is known since Bulle, who excavated test trenches at the site (1908: 116ff), and it was inhabited in recent times by an outpost of the monastery at Skripou.

Syriopoulos (1968: Neo20) lists Neolithic material from the site, reported also in HS&D 1979: G2 and in later works (that quoted them), but no mention of Neolithic is in Bulle's report on the excavation nor in Hope Simpson 1965. EH (EH I) is reported by Hope Simpson (1965), Syriopoulos (1968: EH29) and HS&D (1979: G2). MH and LH surface sherds have been tested by excavation by Bulle (1908: 116-119), who uncovered MH and LH potsherds from excavated strata.
Generally speaking, the site was occupied in prehistoric times (as a settlement site probably). It was probably fortified, according to what is reported by HS&D 1979: G2, who mention Tegyra as a "*hilltop surrounded by a circuit wall ca 2.5m*

[15] Lauffer (Kopais I: 148) says it was probably here that Lolling saw remains of a polygonal wall, though in the database the evidence has been entered as seen at Polygira site (see below).

[16] The sanctuary of Petakas on a rocky terrace outside the ancient city, above the most westerly sources of the river Melas, was connected (probably correctly) by de Ridder with the sanctuary of Herakles. The setting for six tripods preserved in the rock still gives an idea of the former importance of the sanctuary (Fittschen, AReports 1998/9: 56 and fig.64; BCH 1999: 730).

[17] In the former swampy area of Tsamali, from the W a schist structure gets into the plain that borders the marsh to the W. On it is the Tsamali village - 107m slm (Philippson 1951: 476). Philippson also notes the archaeological remains.

thick". Fossey (1988) does not mention any fortification, neither does Lauffer,[18] while Knauss et al. (Kopais 3) note how the fortified site controlled the entrance to the mountain pass towards Abai and Hyampolis, and then Opus in the Euboic gulf (through the path followed by Leake 1835: 161-4).[19] A misleading probably leads to confusion between the Polygira and the Avrokastro sites (see below). The same misunderstanding could have also led to the mention of Neolithic material from the site (mentioned in Syriopoulos 1968 and in HS&D 1979 but not in Bulle 1908 nor in Hope Simpson 1965), which cannot therefore be considered as reliable.

Fossey (1988: 362-3) identifies the site above Polygira source as the earlier settlement site of ancient Aspledon, whose C and H fortified settlement would have moved to the Avrokastro site. On the other hand, Lauffer (Kopais I: 162), and Knauss et al. (Kopais 3) later, identified the evidence above the Polygira source as ancient Tegyra. In the remains in the area below the Polygira site, they see the sanctuary of Apollo Tegyraios, close to the Polygira river source.[20]
Gazetteers do report prehistoric evidence but do not mention later material from the site. Archaic surface material from the site is mentioned apparently only by Fossey (1988: 360), while C (black glaze) and R (including *terra sigillata*) surface material were reported already by Bulle (1908: 116ff - lots of surface sherds at the W foothills and C sherds on top of the terrace above the surface walls).
Bulle (1908: 116ff) already, and then Lauffer (Kopais I: 162) and Fossey (1988) do report surface walls E and N of the spring. Lauffer had located the temple of Apollo Tegyraios by the spring at the Polygira site (AA 1940; AD 1971),[21] but later (Kopais I: 152) he only referred to structures by the spring without mentioning the sanctuary anymore. Lauffer himself says that probably the traces of buildings are too well preserved to be ancient (also Bulle would suggest that the surface walls visible belong to recent periods), and we know that the site, as mentioned earlier, was inhabited in recent times, by an outpost of the monastery at Skripou.[22] On the other hand, Fossey (1988: 361) suggests that the traces of walling may mark the probable location of a lower settlement (but this is to support his hypothesis of identification of the site with ancient Aspledon).

If Fossey's (1988) interpretation (and identification with ancient Aspledon) had been taken into account, the component for the historical period would have been a settlement (small scale settlement); unlikely, if Lauffer's (1940 and 1971, but not Kopais I) and Knauss et al.'s (Kopais 3) interpretation (and identification with the sanctuary of Apollo Tegyraios) was taken into account, the component would have become a cult place. I preferred to record the evidence as an indistinct C and R activity focus.

If we identify the site with ancient Tegyra, probably a settlement ceased to exist there after earlier historical periods and the walling we see are mainly later (Byzantine-Medieval). The few other remains, probably C, may mark either a cult place or something rather smaller, but not a proper settlement site in that period (see also discussion in chapter II.3.4 – LONG TERM SETTLEMENT TRENDS). What Fossey reports as C and R pottery are mainly fine ware sherds (black glaze and *terra sigillata*). This could actually lead to the interpretation of the site as a cult place. We should take into account though that he may report only clearly datable wares, in accordance to the fact that he says '*including* terra sigillata' for the R material.

10. Avrokastro–Mavromandili: *Component O_44 to O_46*

ID	Period	Type	Toponym	Discovery	Position
O_44	LH	? Fort	Avrokastro-Mavromandili	Travellers	certain
O_45	C	activity focus	Avrokastro-Mavromandili	Travellers	certain
O_46	H	activity focus	Avrokastro-Mavromandili	Ext. topographical surveys	certain

A spur on the N side of the Tsamali gulf (2km W of Polygira site), marked as Mavromandili on the 1:50,000 GYS map.
The position of the site is misled because of a wrong positioning made by Curtius 1892 (map in Tafel VIII) who located it at the NW corner of the Tsamali bay.[23] What Lolling had seen (1889: 193, in the locality he calls Avrokastro) and Bulle had visited (1908: 119), has to be located at the spur marked on the 1:50,000 GYS map as Mavromandili.
The place is already known by travellers (Lolling 1889: 193 and Frazer 1913: 195) that noticed traces, today almost invisible, of an ancient circuit which could be dated either LH (cyclopean walling?) or gr-rom. Lauffer (Kopais I: 154) visited the place and reported traces of cyclopean walling, while Bulle (1908: 119) could not see the walling but noticed C sherds at the foothill, and Fossey (1988: 361) saw traces of walling on the ground and sherds (C and H [AE600 and AE601]). We cannot know if there is a strict connection between fortification and potsherds (C and H fortified site?). C sherds were seen by Bulle (1908: 119) and Fossey, and H sherds apparently only by Fossey,[24] who also noticed the presence of ancient building blocks in the fields immediately S and SE of the hill. Evidence is probably too scarce to recognize the character of the activity focus (*components O_45 and O_46*). A LH component has been created also, with a probable character of fort – *component O_44*. We could consider though the option of the existence, if the circuit was Greco-Roman, of a fort at this hilltop site (either readapted from LH -if LH evidence are reliable- or build up *ex novo* in historical times).
The site does not seem to be mentioned (nor mapped) in Knauss et al. Kopais 3, while Fossey interprets it as a C and H fortified settlement (1988: 361-363 – Avrokastro site) and identifies it with ancient Aspledon, extension of the Polygira site occupied in an earlier period. According to Fossey, the Avrokastro site would have been abandoned by Roman times because of shortage in water (but it seems to be a fresh water source nearby!), while R material found at the Polygira site would suggest the continuity of this other settlement site. The

[18] Lauffer (Kopais I: 152) reports of a long connecting wall between the two small hilltops [AE1262] that reminds the prehistoric dike of Spitia Katavothra in the NW bay of Copais (see appendix I.6). It could be a terrace wall, or an infrastructure, or a sort of fortification (Bulle 1908: 117 has it as ancient dike). The evidence could be seen as a prehistoric fortification wall, but it would not be the fortification mentioned by HS&D 1979: G2, as they mention a hilltop surrounded by a circuit wall.
[19] On the communication routes in the area see Kopanias 2008: fig.5.
[20] For the Apollo Tegyraios site see Plutarch *Pelop.* 16.5f, reported in Lauffer Kopais I: 162.
[21] Many of the striking apsidal houses remaining here near Orchomenos proved to be ancient. Ancient Tegyra was here and not at Pyrgos to the East (Lauffer 1940: 185). "*Here below the early settlement we find the two springs linked with the foundation of the Apollo Sanctuary, exact to Plutarch's description (Pelop.16)*" (Lauffer 1940: 187). The same identification is reported by Knauss et al. Kopais 3.
[22] Bulle 1908: 118 and plan fig. 30.

[23] Fossey (1988: 361) locates it on the tall hill with smooth steep sides (a spur from Milovouni) 1km NW of the deserted village of *Erimakalyvia*, in correspondence to the Curtius' wrong positioning rather than to the actual location visited by Lolling and Bulle.
[24] See elsewhere (*passim* and chapter II.2) for Fossey gazetteer reporting H material because of his better knowledge in recognizing fine wares especially from this period among other material.

hypothesis is based mainly on the lack of H pottery at the Polygira's site (consideration based on extensive and not systematic surface research and on the recognition of fine wares mainly) and on the fact that there were no other place-names mentioned by the sources to be attributed to one of the two sites. See discussion in chapter II.3.4 – LONG TERM SETTLEMENT TRENDS.

11. Erima Kalyvia: *Component O_47*

ID	Period	Type	Toponym	Discovery	Position
O_47	Preh	burial place	Erima Kalyvia	Int. topographical surveys	certain

30m ca W of the chapel of Erima Kalyvia, Lauffer (Kopais I: 156) mentions a cist grave, probably of a prehistoric date.
As Fossey notes (1988: 361) Erima Kalyvia was probably an offshoot of the village of Kolaka on Mt. Khlomon to the N, and the people of that village cultivate all the land in this area. A settlement might have existed in this area also in the prehistoric period, and even later, since the area could constitute a suitable settlement chamber (see chapter II.3.4 – LONG TERM SETTLEMENT TRENDS for a discussion on this).

12. Pyrgos: *Components O_48 to O_56*

ID	Period	Type	Toponym	Discovery	Position
O_48	N	? activity focus	Pyrgos	Personal or group interest	certain
O_49	EH	? activity focus	Pyrgos	Personal or group interest	certain
O_50	MH	activity focus	Pyrgos	Personal or group interest	certain
O_51	LH	activity focus	Pyrgos	Travellers	certain
O_52	G	? activity focus	Pyrgos	Ext. topographical surveys	certain
O_53	A	activity focus	Pyrgos	Ext. topographical surveys	certain
O_54	C	settlement	Pyrgos	Travellers	certain
O_55	H	? Settlement	Pyrgos	Ext. topographical surveys	certain
O_56	R	? activity focus	Pyrgos	Ext. topographical surveys	certain

A steep hill to the immediate SE of modern Pyrgos village, on which stand the remains of a Frankish tower. The place is mentioned by different travellers (Fiehn, RE 1903: 124 s.v. *Tegyra*; Leake 1835: 159-160; Ulrichs 1840: 196; Frazer 1913: 195), by Schliemann 1881 also, and it has been investigated by Bulle (1908: 119-120). Afterwards, material from the site is mentioned by different gazetteers and topographical works: Wace - Thompson 1912: 196; Ålin 1962: 122; Hope Simpson 1965: 115 no. 399; Syriopoulos 1968: Neo21 – EH20 – MH31 – LH49; HS&D 1979: G4; Syriopoulos 1983-84: Neo245 - LH 205 - MH 151 - LH II33 - LH IIIA 75 - LH IIIIB 108; Buck 1979: 7-8 ; Wallace 1979: 163-4.

Although the place had been visited by many travellers, Bulle (1908: 119ff) first mentioned the prehistoric surface evidence (N, EH and MH).
As for the Neolithic period, no Neolithic material is apparently reported by any gazetteer. Bulle's account (1908: 119ff) on Neolithic from the site is, as noted by Fossey (1988: 367ff), "*much taken up with the description of a steatopygous statuette found in the vicinity; this object, whose non-Kykladic nature was pointed out by Wace and Thompson (*1912: 197), *is perhaps better associated with the nearby Neolithic site of Magoula*" (*component O_62*).

As for the EH period, HS&D 1979 report an uncertain presence at the site, while Fossey has EH I and II present. Certain EH occupation was found at the *magoula* site nearby – see below *component O_63*. As for the MH and the LH period (III A-C), HS&D 1979 report surface sherds of a high quality. Remains of a LH fortification [AE625] in Cyclopean walling style are also reported (Hope Simpson 1965 and HS&D 1979: G4), extending over an area of ca 250x150m (Knauss et al. Kopais 3: 68ff), probably part of the Mycenaean fortress system controlling Copais.[25] Due to its position along the Minyan great channel (see fig.11 in chapter II.3.1 – drainage work in Copais), the site in the Mycenaean period could have played a crucial role in the control of the drainage system of Copais in the NW area.
The historical phases at the site are represented as the following: G pottery on the site is reported apparently only by Fossey (1988: 367), while HS&D 1979 (G4) have A and C surface material. An earlier occupation at the site (either G or A) seems attested either way. Historical period (C-H probably) may belong the surface walls [AE606] reported by Bulle (1908: 119ff) jointly with C sherds, as well as traces of a gr-rom fortification [AE605]. In addition, a number of ancient blocks were built into the later tower [AE1209]. H surface material is listed in Fossey's gazetteer, while a later (R) occupation is probably to be excluded.

Some scholars (Fossey 1988: 369-370, and Schliemann 1882: 160 already[26]) identify the Pyrgos hilltop site with ancient Tegyra, while Lauffer (Kopais I: 156) and Knauss et al. (Kopais 3: 68ff) identify it with the acropolis of ancient Aspledon.[27]

13. Pyrgos *magoula*: *Components O_62 to O_66*

ID	Period	Type	Toponym	Discovery	Position
O_62	N	activity focus	Pyrgos Magoula	Personal or group interest	certain
O_63	EH	activity focus	Pyrgos Magoula	Personal or group interest	certain
O_64	MH	activity focus	Pyrgos Magoula	Personal or group interest	certain
O_65	LH	? activity focus	Pyrgos Magoula	Personal or group interest	certain
O_66	C	activity focus	Pyrgos Magoula	Personal or group interest	certain

The *magoula* is a large limestone rocky hill, rising out of the Copais plain about 1km NW of the Pyrgos hill (Fossey 1988: 368, from Albanian *Magule* = hill, mound).

[25] Fossey during his visit was not able to locate the remains of this LH fortification (Fossey 1988: 368).
[26] Schliemann (1882: 160f) identified Xeropyrgo (Pyrgos) with Tegyra, since Plutarch (*Pelop.*) mentioned it as standing not far from Orchomenos, above the marshes of the Melas, and the two cities were joined by a road, which led through a pass formed by these marshes (see chapter II.1.4 and appendix I.4 *passim* for a discussion on this issues of identification and routes). On his way from Orchomenos to Pyrgos, he travelled along a road for the most part across the rocks, very bad and full of stones, for 16 miles (5 hours on horseback).
[27] As noted by Knauss et al. (Kopais 3: 68ff), Orchomenos and Aspledon are mentioned together in the Homeric *Catalogue of Ships* (Homer *Il.* 2.511). Strabo (IX 2.41) says that Aspledon is 20 stadia (= 4km ca) from Orchomenos and the Melas river runs between the two. Strabo's words can tell us that Aspledon site would be the second one, by the Pyrgos village, on the hilltop where currently one can also see the remains of a Frankish tower. The late Mycenaean fort controlled the entrance to the mountain region of Hyettos-Olmones (Knauss et al. Kopais 1: 44). This place is widely recognized today as the acropolis of Aspledon, despite the fact that on the GYS 1: 50,000 map (1971) ancient Aspledon is located by the sources of the Polygira site which is instead recognized today as Tegyra, with the temple of Apollo Tegyraios close to it (see *components O_32 to O_34*).

The *magoula* is ca 300m long, ca 200m wide and 25m high (Knauss et al. Kopais 3: 68ff). On all sides, except the S, the hill rises out of the plain in sheer cliffs (Bulle 1908: 115-124, pl. 34). To the SE and W are fresh water springs which create a certain marsh-like quality around the site even today (Fossey 1988: 368). In the past the whole bay was occupied by marshes (see chapter II.3.4 – TOPOGRAPHICAL SETTING).
Bulle visited the site (1903) and investigated it first.[28] Lauffer then visited the place and described the visible remains (Lauffer Kopais I: 162-171). Potsherds of different periods as well as architectural remains were found at the site.[29]
I have chosen to characterize the components as activity focus for the N, EH (EH II as reported in HS&D 1979: G5), MH periods and as possible activity focus for the LH period (LH I/II? and LH III A-B are reported in HS&D 1979 G5, LH I/II is reported in Syriopoulos 1983-4). This is despite Lauffer's remarks that N and EH are better and larger represented than MH and LH and despite the fact that Bulle's dig found N, EH and MH but not LH. Probably either Lauffer's remarks and Bulle's excavation results can be interpreted as a sign of a different level of occupation in later prehistory, or the disparity is due to post-depositional reasons (which washed away the top most layer).
Lauffer (Kopais I: 163-4) remarks that only two other sites in the Copais are similar to this in shape and height: Nisi by Stroviki (*component NC_1*) and Gla (*component NC_70*).[30]

According to Lauffer (Kopais I:170), the presence of fresh water springs at the foot of the hill would allow for the possibility of a settlement here which would have been in a close relationship with the settlement at the Pyrgos site. We do not have any certain information about the *durée* of settlement there and neither on the precise settlement site on the *magoula*, but we could hypothesize that the site would be abandoned at the beginning of the Bronze Age and reoccupied in the 5th C BC, as we will see below, probably with a different occupation character.
Prehistoric material constitutes the majority of material,[31] but also some Classical potsherds have been found (the corresponding *component O_66* has been recorded as C activity focus). Moreover, Bulle's investigation revealed, at the W end of the hill, terrace walls of laid trapezoidal masonry (Bulle 1908: 121-124, plan 35-6[32]; see also Lauffer Kopais I for a sketch –171– of the remains on the *magoula*). The structural remains visible at the site [AE1273] have been interpreted by Bulle as a sanctuary site of historical age, while Lauffer does not agree and would rather interpret them as a fort used in later prehistory and part of the system of fortification in N and NE Copais (even though the masonry is described as 'trapezoidal', this would explain the scarcity of material from the LH period).[33] For the time being, until new research on the site would give us any further clue for the interpretation, those structural remains are entered as unknown surface walls [AE1273] and associated with a question mark to the LH and the C component (*components O_65 and O_66*). As for the remains dated to the historical period (suggested as C but could also be A or H, as there is an apparent lack of directly associated material) evidence could even represent a rural activity site. The wall of trapezoidal masonry could even be a terrace wall or similar, if the structure is not associated with the earlier –late prehistoric– occupation. On the other hand, the surface wall remains and the C material probably to be associated to them (although material from the historical period has been reported by Lauffer from the flat area and not in direct association with the structures) might really represent a cult place. Erosion processes, which probably uncovered the remains by eroding the soil away, jointly with heavy cultivation activities, which might have destroyed further remains, as Bulle had pointed out, must have affected the archaeological record. As Lauffer (Kopais I: 168) notes, soil is reduced to a minimum on the top.
An interesting feature from the site is constituted by series of one or many holes on top of the rocks [AE1271] in the W part of the *magoula*, and especially above the spring, whose meaning is still to be clarified (Lauffer Kopais I: 168).

14. Pyrgos NW: *Component O_57*

ID	Period	Type	Toponym	Discovery	Position
O_57	LH	burial place	Pyrgos NW	Personal or group interest	approximate

Probably the two entries, AE608 (the far side of the Pyrgos village) and AE1210 (in the torrent bed NW of Pyrgos), refer to the same archaeological evidence, representing a LH burial area in an imprecise location to the NW far side of Pyrgos, probably in the torrent bed of a stream.[34]

15. Pyrgos-Trassa: *Component O_58*

ID	Period	Type	Toponym	Discovery	Position
O_58	C	burial place	Pyrgos-Trassa	Int. topographical surveys	certain

In a flat, marshy zone to the W of the Pyrgos village, architectural fragments [AE1270] (including sarcophagus fragment), sherds (including black glaze) and tiles [AE1269], as well as bone fragments and surface walls [AE1267 and AE1268] extend over a large area, on both sides of the road by the Ag.Ioannis stream. The evidence was noticed by Lauffer (Kopais I: 159).

16. Pyrgos-Melas dam: *Components O_59 to O_61*

ID	Period	Type	Toponym	Discovery	Position
O_59	gr-rom	activity focus	Pyrgos-	Int.	certain

[28] Account of the excavation in Bulle 1908: 115-124.
[29] See HS&D 1979 (G1-G5) for dating of the material and the site (from previous accounts).
[30] See discussion in chapter II.3.4 - PREHISTORIC PERIOD – and appendix I.6 for the similar sites.
[31] Mainly N and EH and comparatively less MH and LH material (see above in text).
[32] As in Fossey 1988 - plan p. 371: "*inside the trapezoidal masonry walls are the remains of a large rectangular structure which Bulle suggested might be an aperipteral temple (5th C BC) of which the* pronaos *at the E end has completely perished. Bulle suggested that the absence of any other structural remains on the hill is probably due to intensive modern agricultural work on the flat area on the top of the* magoula".
[33] Bulle (1908) identified the *magoula* with the mount known from Plutarch (*Pelop.* 16) as 'Delos', which was where the Melas marshes finish. Lauffer (Kopais I: 162-171) notes how Bulle thought the remains of the 'Classical' building they found could be the remains of the Apollo sanctuary which Plutarch mentions (*Pelop.* 16.5), but he disagrees (162 and 170), and recognizes the Delos mount with the Aspledon acropolis, which also marks the landscape. See Lauffer's sketch 171 (Kopais I) for a plan of remains. See also Knauss et al. Kopais 3: 68ff.
[34] AE608 has been assigned with a question mark to the *component O_57*. The evidence is constituted by Mycenaean tombs found by Bulle on the far side of the village and associated - probably correctly- with the Pyrgos site by Hope Simpson 1965 (Fossey 1988: 368). AE1210 (Syriopoulos 1968) is constituted by Mycenaean tombs found in the torrent bed NW of Pyrgos.

ID	Period	Type	Toponym	Discovery	Position
			Melas dam	topographical surveys	
O_60	R	activity focus	Pyrgos-Melas dam	Int. topographical surveys	certain
O_61	MH	activity focus	Pyrgos-Melas dam	Personal or group interest	certain

A wide area ca 1km SW of the Pyrgos village, to the S side of the Melas dam, and 700m ca to the S of Pyrgos Magoula, that carries traces of an ancient settlement, as noticed by Lauffer (Kopais I: 161ff), has been surveyed by the German team working on Copais (Knauss et al. Kopais 3).

Component O_61: MH (Minyans) sherds are reported in the area by Knauss et al. (Kopais 3: 68ff),[35] in particular in a field cultivating cotton to the S of the dam, where the Polygira's outlet into the Melas river was before the drainage (see Knauss et al. Kopais 3: map 2.26). MH material, in such a low location, could represent a burial area, which could probably be referred to as the MH settlement at the Magoula Pyrgos site.

Component O_59 and O_60: Group of remains (architectural fragments and potsherds [AE1232, AE1233, AE1235, AE1236]) over an extended area, at the ancient outlet of the Melas river into the lake, just above the 95m contour lines (limit of the lake), like the other sites found by Knauss et al. (see chapter II.3.1 and Farinetti 2008). The site is by the Melas (dam) nowadays, but the ancient course of the river was somewhere else (see Knauss et al. Kopais 3 map 2.26). The components created from the archaeological evidence available were two: a gr-rom activity focus, and a R activity focus, the latter (*component O_60*) characterized by a more specific chronological attribution since the sherds seen at the site are reported as R (only R by Lauffer, from MH to R by Knauss et al., but with the majority characterised as R, *terra sigillata* included). We cannot exclude though, without a detailed study of the surface remains, the presence of an activity focus in other periods of the gr-rom range (*component O_59*).[36]
Interpreted as a Pyrgos site lower-town (Aspledon lower?)[37] by Knauss et al. (Kopais 3: 68ff),[38] the remains were noticed first by Lauffer (Kopais I: 161). Knauss et al. note how the site is located at the border of the lake, and gets fresh water from the Melas.
The sherds extended until the former outlet of the Poligyra river into the Melas, where, along with blocks (already seen by Lauffer) and a large quantity of building material (Knauss et al. Kopais 3: 68ff and also Kopais 1: 46ff), MH sherds were also noticed (see above, *component O_61*).

ASPLEDON: identification issues

What follows is a summary of the questions, still open, concerning the identification of ancient Aspledon.

Lauffer (Kopais I: 151) reports different hypothesis for the identification of ancient Aspledon, that have been discussed *passim* above and are summed up here:

- Leake (1835: 162) searched for the city in the area of Tsamali, but Strabo (IX 415) mentions Aspledon as at a distance of 20 *stadia* from Orchomenos, and Pausanias (IX 38.9) says that the town was abandoned because of lack of water. Tsamali is only at 1.5km from Orchomenos and it is close to the Melas sources. Moreover, Lauffer notes how Strabo (IX 415) describes Aspledon as *eudeleios*, i.e. 'visible from a distance', and this would not suit Tsamali site, at least the cult place, as it is on a flat slope looking E. Furthermore, archaeological evidence there seems too scarce to represent a settlement site, at least from the present knowledge .
- Another candidate for Aspledon is the Polygira site. The place described by Bulle (1908: 116ff) is not a specific site, but rather the whole angle of the NW basin, between two small hilltops. See above under the site, for details on the archaeological evidence from the area. Curtius 1892 identified the place with ancient Aspledon. Lauffer notes that here prehistoric material was found by Bulle's excavations, while at Tsamali no prehistoric material was found. The vicinity of the Kephalari spring can also lead to the interpretation of the site as ancient Tegyra rather than Aspledon.[39]
- Lolling 1889: 193, Forchammer 1837: 177, and Bulle 1908: 119 identified Aspledon with another hilltop, called Avrokastro-Mavromandili, as does Fossey 1988: 360. There are also visible tracts of cyclopean walling going down the slope. Lauffer would not agree, but he hypothesizes here the existence of the Homeric Aspledon (with the Cyclopaean wall and cist grave as evidence for it), before it moved to the Pyrgos location.
- Dodwell 1819: 233, identifies with ancient Aspledon the site by the modern village of Pyrgos, and Lauffer agrees. On the other hand, Lolling 1889: 190ff. would identify this site with ancient Tegyra, as well as Fossey 1988: 367f later. Materials from prehistoric to Classical periods have been found there. On the top of the hill Lauffer locates the acropolis of the earlier town, while in the C period the main part of the town would have developed in the plain in front. In the Trassa locality (swamp area) Lauffer noticed traces of surface walls and in the plain black glazed potsherds, tiles and bones - probably an indicator of the presence of a vast necropolis area. The settlement area should be probably located in the area S of the Magoula Pyrgos site (see Knauss et al.'s lower Aspledon). Aspledon was a quite large town in the Copais, bigger than the modern village of Pyrgos (Lauffer Kopais I: 160), even though it is not reported by Strabo IX 410 (it had probably disappeared by the Roman period). Lauffer (Kopais I: 161) reports how Spartans from Lokris, leading to Orchomenos, entered the Copais through the Ag. Ioannis valley, which ends at the Trassa area.[40] Here it would have been the road crossing the Copais.

[35] MH material not noticed by Lauffer (who mentioned only R – Kopais I: 161).
[36] Knauss et al. (Kopais 3: 68ff) report lots of sherds ranging from MH to R in the cotton fields to the S of the former outlet of the Poligyra river into the Melas.
[37] See discussion on identification of ancient Aspledon below in text.
[38] I have got the impression that they attempt the identification of this site as the lower town of Aspledon mainly because they could not find another place name mentioned by ancient sources to identify with those extensive remains found.
[39] The ancient Aspledon could not have been by the large *Kephalari* spring because Pausanias says it got deserted because of water shortage. Aspledon should have been by the Mavromandili area, more to the W (Philippson 1951: 476).
[40] Spartans –Lauffer says- would have got lost in the swamps..

Furthermore, Fossey (1988: 360) believes that the Poligyra site would have to be identified with ancient Aspledon (as traditionally intended; the name Aspledon is marked on the GYS topographical map 1:50,000 – 1971), that moved then to the Avrokastro site in C and HL times. See chapter II.3.4 – LONG TERM SETTLEMENT TRENDS for a discussion on this.

------

17. Mavrospilia: *Component O_67*

ID	Period	Type	Toponym	Discovery	Position
O_67	Preh	activity focus	Mavrospilia	Ext. topographical surveys	approximate

In AD 1972: 315 and AAA 1973: 202 the excavation of a cave site is mentioned by Spyropoulos as a trial dig with few finds (few sherds and one or two tools [AE604]), just slightly to the E of Pyrgos, in the S mountainous lump which surrounds the Copais. Chronology of the finds is not specified (the component is dated generally as prehistoric), but it could probably be Neolithic. The cave was investigated by the Italian Speleological Group in the 1980s (Bertolani-Rossi 1983).

18. Old Orchomenos: *Components O_23, O_38 to O_43*

ID	Period	Type	Toponym	Discovery	Position
O_23	gr-rom	activity focus	Old Orchomenos	Personal or group interest	
O_38	N	? activity focus	Old Orchomenos	Personal or group interest	certain
O_39	EH	activity focus	Old Orchomenos	Personal or group interest	certain
O_40	MH	settlement	Old Orchomenos	Personal or group interest	certain
O_41	LH	activity focus	Old Orchomenos	Personal or group interest	certain
O_42	C	activity focus	Old Orchomenos	Personal or group interest	certain
O_43	R	activity focus	Old Orchomenos	Personal or group interest	certain

5km SE of Orchomenos, along the 95m contour line, 5km N of the Xinos site (*component KO_64 to KO_72*) is a place, a little bit higher than the rest of the plain, jutting out like a peninsula over the area that was formerly occupied by the marshy lake (Knauss et al. Kopais 3: 32ff and map 2.8). According to Knauss, probably in ancient times the outlet of an older, N arm of Kephisos was located into the lake at this place (difficult to say).[41]

The site has been recognized as a MH habitation site (concentration of potsherds, among which oil lamps were found) – *component O_40*. The settlement was living in parallel with the MH settlement at Orchomenos (at the foot of Akontion)[42] and the one at Magoula Balomenou site – *component CH_23* – as well as the one at Pyrgos and Pyrgos Magoula – see chapter II.3.4 – PREHISTORIC PERIOD.

Signs of activities in the N(?) [AE1229] and EH period [AE1228][43] have also been noticed (*components O_38 and O_39*), as well as in the LH period ([AE1223] *component O_41*).

The authors suggest the identification of the place with old-Orchomenos, abandoned because it was 'swallowed up' by the lake, as reported by Strabo (IX 2.18 - together with Eleusis and Athenai - and IX 2.42). The inhabitants, according to the ancient author, protected themselves by moving to a higher position at Orchomenos (located at an elevation from 95m to 100m asl). The place is 5km from the foot of the Akontion ridge and lies in a polder (Knauss et al. Kopais 1: 225). As Knauss et al. note, the inhabitants of the site could get fresh water from the Kephisos or from the sources of the Melas, and the place could have had an agricultural vocation, being close to the *Hyppia* plain.[44]

Components O_42, O_43 and O_23: material from historical periods is also known from the site. We cannot totally exclude the possibility that off-site material from later periods turned out to be interpreted as a site once found in a prehistoric site context. A closer and more intensive look at the material on the surface could help in the interpretation. For the time being, we can hypothesize the presence of a probable rural activity focus at the place. Knauss et al. (Kopais 3) suggest the presence of a cult place at the site, as at the Xinos site, identified by the authors as ancient *Eleusis*,[45] and later as a cult place in gr-rom period (see chapter II.3.1 - *components KO_65 to KO_72*). On the other hand, at the Xinos site, epigraphical evidence and architectural fragments could lead to that interpretation, while in this case the evidence is less indicative, though architectural fragments and building material have been noticed [AE1230]. Thus, even if I left the component as a general activity focus, I would prefer to interpret the component of the gr-rom period as a possible rural site (a villa in the R period?), by the marshy area, for lake-related activities (the created components are C and R ones according to the chronological indication of the surveyors, though we cannot exclude the presence of intermediate, earlier or later periods, indicated as gr-rom in *component O_23*). Knauss et al. themselves note how the place could have been used to continue their agricultural activities, after the supposed abandonment of the habitation site, since its location in the middle of the basin was closer to the fields, even if they were living in a more secure place at the foot of Akontion.

19. Vranesi tumulus: *Components O_26 and O_27*

ID	Period	Type	Toponym	Discovery	Position
O_26	PG	burial place	Vranesi tumulus	Travellers	approximate
O_27	G	burial place	Vranesi tumulus	Travellers	approximate

A *magoula* with a conical shape S of Orchomenos by the village of Vranesi/Ag.Spyridon, about 5-6km SW of Skripou.

The site, already noticed by Ross (1851: 39), was excavated by Sotiriadis at the beginning of the 20th century (Praktika 1904: 39-40; Praktika 1907: 109; AM 1905: 132-3). He was probably looking for a LH site when he started the excavation. The *magoula* looked like one of the many *magoulas* in N Copais,

[41] The site is called Old Orchomenos by Knauss et al. According to our reconstruction of the lake fluctuations, the place is apparently a little bit further back compared to the reconstructed 95m contour (see chapter II.3.1 – PERILACUSTRINE MARSHY LANDSCAPE). On the other hand, I located the site on the basis of the map published in Knauss et al. Kopais 3, map 2.8. I could not visit the place, but probably a more accurate GPS positioning is needed.

[42] and the one at Magoula Balomenou site – component CH_23 as well as the one at Pyrgos and Pyrgos Magoula – see chapter II.3.4 - PREHISTORIC PERIOD.

[43] Dated to the EH period are two fragments of red ware dishes with a X inscribed – correlated to a type of ware in Kunze 1934. Knauss et al. Kopais 3 suggest a former cult of the Charites.

[44] The *Hyppia* plain is known, in association with Orchomenos, by ancient sources (Theophrastus *HP* IV 11.8) as the largest and nicest plain of Boeotia (Plutarch *Sylla* 20), where the infamous Minyan cavalry was trained.

[45] In parallel with the identification of this site as 'old-Orchomenos'.

usually in the LH settlements or tumuli. Although this tumulus lay among several small rises on the plain which (from surface sherds, according to Sotiriadis) should correspond to H or R burials (Praktika 1904: 39), see *component O_20* (Vranesi). In actual fact, the Vranesi *magoula* turned out to be a funerary tumulus, containing several PG and G (MG) tombs.[46]
Desborough (1972: 202) suggests that the site may be associated with Orchomenos, while Fossey (1990d) is not convinced and believes that 6km is too long a distance when there are contemporary cemeteries nearby (Coldstream 1968: 198-9, 411). One could hypothesize the existence of another geometric settlement closer to the burial area, abandoned during the 8thC BC (material from the graves: PG-MG), since movement of settlements in the SW Copais during the Dark Ages seems to be not infrequent. On the other hand, the material noticed by Sotiriadis spread in the area (Hell –C? – and R fine wares), seems to testify the presence of burials in the area in later periods also, and those were probably linked to Orchomenos.

20. Orchomenos tumulus: *Component O_22*

ID	Period	Type	Toponym	Discovery	Position
O_22	LH	burial place	Orchomenos tumulus	Personal or group interest	approximate

The site is a tumulus that looks like a *magoula*. It is similar to the *magoula* by Vranesi (*components O_26 and O_27*), but much closer to the Orchomenos village (to its S). Sotiriadis (Praktika 1903: 41; AM 1905: 129-132) excavated it and identified it as a LH tumulus, with uncertain occupations from other periods also.[47]

Sotiriadis noticed a lot of these *magoulas* in the area (N Copais), and investigated some of them (the more elevated probably). He interpreted all of them as signs for the presence of burial places. On the other hand, Lauffer (Kopais I: 139) notes how the Copais area in general is full with these *magoulas*, typical of the Phokian landscape, that mark, in the majority of cases, prehistoric sites (habitation sites or burial places/tumuli, as in this case). Lauffer himself mentions the analogy with other tumuli in the area seen by Sotiriadis (AM 1905: 129).[48]

Sotiriadis (Praktika 1904: 37) during excavation noted that the deposits of Kephisos were 3m high at the site (and therefore, as he notes, the tumulus was once standing out of the plain by 8m, while at the time of the excavation –beginning of the 20th century- it was only 5m high). He also notes that during the excavation of the tumulus underground water appeared at a certain depth (3m ca), during the summertime, characterised by strong water withdrawal. But the original surface of it lies, as said above, 3m deeper than the surface today. The underground water would therefore be found at a deeper level only around September-October, when the heavy irrigation of the fields stops (and with it the heavy water withdrawal stops) and the added pressure raises the level of the aquifer.

[46] Without any prehistoric nor LH material. G material examined by Andreiomenou AE 1985: 57-84.
[47] There are doubts on full stratigraphy, as noted by Syriopoulos 1968.
[48] I do not think one of the two *magoulas* (*components O_69 and O_70*) noticed by Lauffer (Kopais I: 139) refers to this tumulus. It is likely that they both refer to the other 'bumps' noticed already by Sotiriadis (Praktika 1903: 41).

21. Orchomenos Magoula SE: *Component O_25*

ID	Period	Type	Toponym	Discovery	Position
O_25	unknown	activity focus	Orchomenos Magoula SE	Personal or group interest	certain

A *magoula* (with ancient sherds) has been noticed by Knauss et al. and it is noted on plan 2.8[49] (Kopais 3). The locality is reported as Magoula on the 1:5,000 GYS map, SE of the Orchomenos village.
Can the site be the Orchomenos tumulus site mentioned above (*component O_22*)? Or rather, since the sherds are reported generically as 'ancient', could it be one of the low 'bumps' noticed by Sotiriadis, characterized by scatters of H and R sherds? Probably the *magoula* in question is very much off the range area of the other *magoula*, being much further E (see GYS map 1:5,000 and 1:50,000).

22/23/24. Vranesi: *Components O_68* (22), *O_20* (23) *and O_69 to O_70* (24)

ID	Period	Type	Toponym	Discovery	Position
O_20	gr-rom	? burial place	Vranesi	Other	approximate
O_68	unknown	? activity focus	Vranesi	Int. topographical surveys	approximate
O_69	unknown	? activity focus	Vranesi	Int. topographical surveys	approximate
O_70	unknown	? activity focus	Vranesi	Int. topographical surveys	approximate

Component O_68: The evidence AE1697 is described by Lauffer (Kopais I: 138) as a flat *magoula* 300m ca NE of the Vranesi/Ag.Spyridon village and 3.5km ca. from Orchomenos, similar to the Phokian *magoulas*, which mark typical prehistoric settlement sites. The *magoula* constituting AE1697 is 2m high, 15m wide and 40m long (sketch 145), but no sherds were found on the surface (Lauffer Kopais I: 138).[50]
Component O_20: AE1697 could be, as well as AE1239, one of the low tumuli (interpreted as necropolis) seen by Sotiriadis (Praktika 1904: 39).

Components O_69 and O_70: These two entries represent two tumuli (5m and 3m high - diameter 10m and 5m) noticed by Lauffer (Kopais I: 139). Probably neither of these is to be identified with the Orchomenos tumulus excavated by Sotiriadis (*component O_22*). Probably they both refer to the other 'bumps' noticed already by Sotiriadis (Praktika 1903: 41) – see *component O_20.* Lauffer himself mentions the analogy with other tumuli in the area (Sotiriadis AM 1905: 129).

[49] Apparently there is no mention of it in the text.
[50] The evidence AE1697 is recorded as 'unknown' because it represents just a topographical feature only, without further research.

Appendix I.5
The Northern mountains of the Copaïs: HYETTIA

THE ARCHAEOLOGICAL RECORD

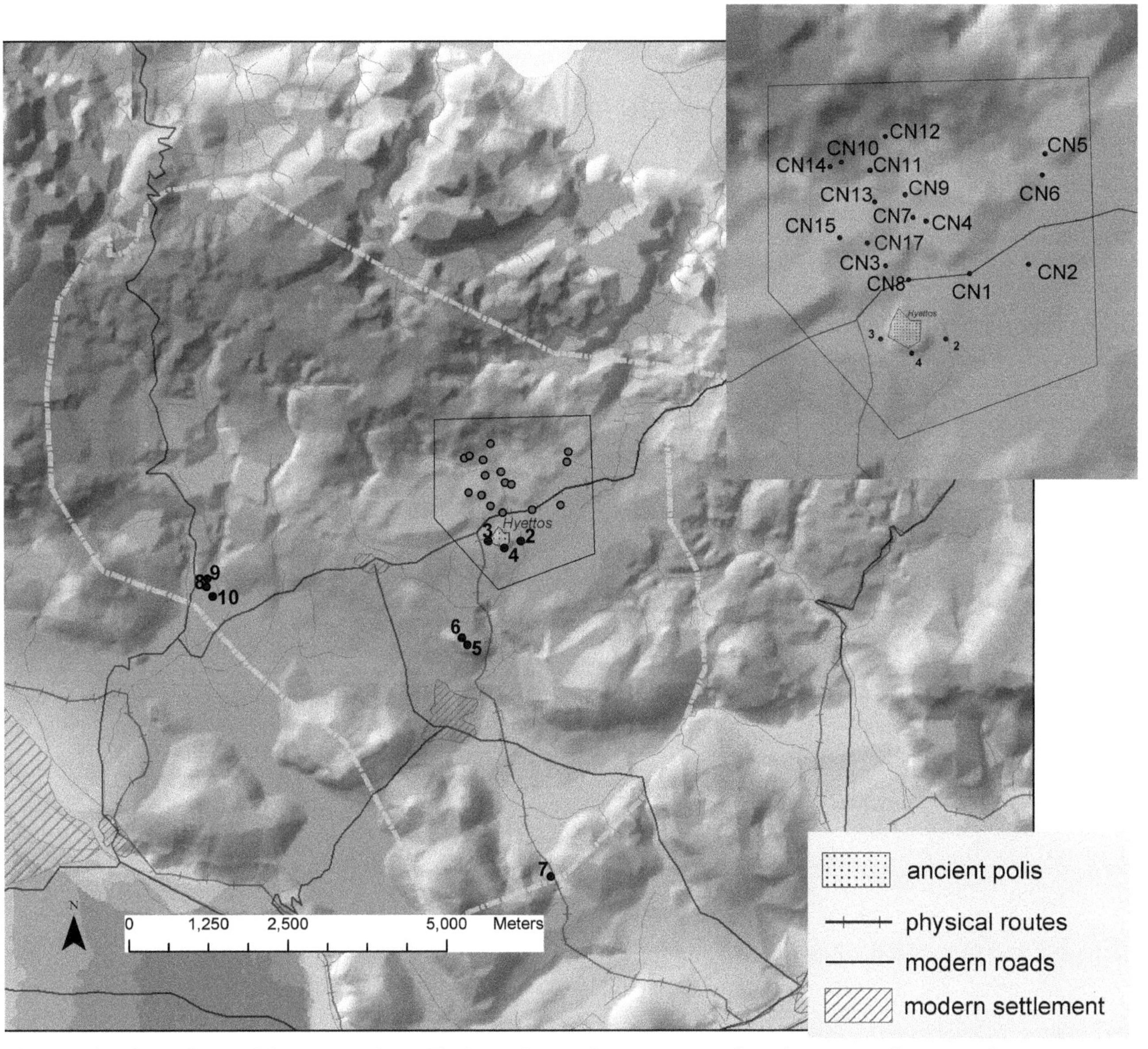

Fig.1. *Archaeological map of the Hyettos chora. The box indicates the area intensively and systematically surveyed.*

1. HYETTOS: *Components HO_1 to HO_6*

ID	Period	Type	Toponym	Discovery	Position
HO_1	A	settlement	Hyettos	Travellers	certain
HO_2	C	settlement	Hyettos	Travellers	certain
HO_3	H	settlement	Hyettos	Travellers	certain
HO_4	R	settlement	Hyettos	Travellers	certain
HO_5	LR	settlement	Hyettos	Travellers	certain
HO_6	Preh	activity focus	Hyettos	Int. topographical surveys	certain

The acropolis of ancient Hyettos is located on a steep-sided hill surmounted by the small chapel of Agios Athanasios, 1km S of the *metochion* of Agios Georgios, NE of Loutsi village (Fossey 1988: 293).
The site was identified because of two inscriptions mentioning the name (Papachatzis 1981: 164). The location of the ancient city, though, is known since Girard 1881 and Oldfather 1916. A detailed study of the city site was also done by Étienne - Knoepfler 1976, on excavations at the city site. Concentrations

of surface sherds and visible surface walls are visible at the site. Intensive survey works in the urban area (Bintliff 1992d)[1] showed the location of the lower city along the E side of the acropolis and to the S of it, as previously supposed (Fossey 1988: 293; Lauffer Kopais I: 182).

No prehistoric material at Hyettos city site is mentioned in prehistoric gazetteers, though Fossey (1988: 293) noticed obsidian flints at the acropolis foothills[2] and intensive surveys on the ancient acropolis showed a prehistoric focus on the N part of the hilltop.
A fortification wall[3] is clearly recognisable, and occupation is attested from the G-A to the LR period. By means of intensive artefact surface survey a large settlement has been recognised in the A and C period (with a flourishing in the Late Classical-Early Hellenistic period), progressively reducing its extension in Late Hellenistic (apparent concentration in the NE area), Roman and Late Roman (on top of the acropolis only) periods, showing a trend typical to several Boeotian city sites, with the exception of Thespiae and Tanagra especially (see dedicated sections).
Foundation myths link Hyettos with Orchomenos (Pausanias IX 36.7 and Hesiod *Great Eoiae*).

Burial places have been noticed (either accidentally or through rescue excavations or through extensive surveys in the area) to the W, S and E of the acropolis and are listed here below. Intensive systematic artefact surface survey (Boeotia survey project in the early 1990s – Bintliff 1992d) discovered other focuses of burial activities probably related to the city site (see table 1 -*Survey Sites*- below).

2. Hyettos East: *Component HO_7*

ID	Period	Type	Toponym	Discovery	Position
HO_7	gr-rom	burial place	Hyettos East	Ext. topographical surveys	approximate

Fossey (1988: 293) mentions tombs noticed in different torrent banks to the E of Hyettos, outside the settlement area.

3. Hyettos West: *Component HO_8*

ID	Period	Type	Toponym	Discovery	Position
HO_8	R	burial place	Hyettos West	Rescue excavation	certain

A rock-cut tomb of a Roman date has been excavated on a hill 200m W of the acropolis of ancient Hyettos, during a rescue excavation due to metal mining activities (AD 1915: 42). The tomb (recorded into the database as isolated tomb [AE975]) probably belongs to the ancient cemetery of the *polis* (see *component HO_7*).

4. Hyettos South: *Component HO_9*

ID	Period	Type	Toponym	Discovery	Position
HO_9	H	? burial place	Hyettos South	Other	approximate

A Hellenistic tombstone (BSA 1974: 121 no. 6) ploughed up to the S of the acropolis, might be evidence of a burial area, though it is very movable evidence and might come from the necropolis areas to the E and W of the acropolis.

5. Pavlon-Palaiokastron: *Components HO_10 to HO_13* (Olmones?)

ID	Period	Type	Toponym	Discovery	Position
HO_10	gr-rom	settlement	Pavlon-Palaiokastron	Travellers	certain
HO_11	A	? activity focus	Pavlon-Palaiokastron	Ext. topographical surveys	certain
HO_12	C	? activity focus	Pavlon-Palaiokastron	Ext. topographical surveys	certain
HO_13	LR	? activity focus	Pavlon-Palaiokastron	Ext. topographical surveys	certain

Identified by some as ancient Olmones, the fortified site lies on the Palaiokastron hill, 1700m ca SW of the Hyettos city site.[4] The place is called 'Kastro' or 'Palaiokastro', as reported in the 1:50,000 GYS map.

The location of a focus of activity (probably a settlement site, if not merely a refuge) seems to be certain in the gr-rom period, if one considers the concentration of sherds on top and on the slopes as well as visible surface walls and building remains [AE983; AE1281; AE1283; AE1338]. Nevertheless, there are no precise indications in the bibliography for the possible periods of occupation of the site; only little information is available. Fossey visited the site and reports (1988: 297) that he could not date almost any sherd. He recognized one sherd of late Roman ripple ware and a pyramidal loom weight of classical date. Yet as pointed out by Fossey, one of the possible buildings visible on the surface [AE984] seems to be, from its plan and alignment, an aperipteral temple.[5] Traces of ancient fortification [AE982] were marked already on Frazer's map (1913: facing p.110), and noted also by Lolling 1887: 128.

The site has been identified with ancient *Olmones* (Fossey 1988, Lauffer Kopais I: 176 and Lolling 1889: 180-88 already), but Étienne - Knoepfler 1976[6] disagree with the identification and prefer to consider it as a refuge site - see Fossey 1988: 297 and Fossey 1990e: 208.[7] In the database for this work, the corresponding gr-rom component has been listed as 'settlement', even though one cannot be certain, without further archaeological research, whether it was only a refuge settlement in use for short periods, or it was indeed ancient Olmones, reported by sources as a proper settlement site, or rather, most probably, a small settlement/hamlet dependent on Hyettos.

[1] The Hyettos survey constituted a survey project on an ancient urban site lead by J.L. Bintliff, following Thespies and Haliartos survey, and preceding the Tanagra survey (see chapter II.2).
[2] At the SE foothills of the Acropolis hill, is situated an abundant fresh water source, currently tapped by the village of Pavlon (Fossey 1988: 293). The same water source alimented the river running nearby (see sketch in Lauffer Kopais I: 175).
[3] See plan in Étienne - Knoepfler 1976, and Papachatzis 1981: 165.

[4] Fossey 1988: 296-7 fig. 39 (plan of visible surface walls and fortification in small stones); Lauffer Kopais I: 182 - sketch 187.
[5] The archaeological evidence AE984 is constituted by surface walls possibly to be attributed to the A period, as the aperipteral temple constitutes a group of early Boeotian temples. A preliminary account was given before the Archaeological Institute of America, December 1970 in New York; cf. AJA 1971: 201).
[6] Étienne - Knoepfler 1976 prefer to locate ancient Olmones at the Stroviki site, with Archaic and Hellenistic finds along with the remains of the prehistoric settlement (see appendix I.6 – *component NC_7 to NC_11*).
[7] Problems in the identification are related to the distances in *stadia* given by Pausanias. Pausanias (IX 24.3) says that Olmones was situated only about 12 *stadia* from Kopai, and comments that there was nothing worth looking at at Olmones - see also Papachatzis 1981.

6. Pavlon-Palaiokastron: *Component HO_14*

ID	Period	Type	Toponym	Discovery	Position
HO_14	gr-rom	special activity	Pavlon-Palaiokastron	Personal or group interest	certain

As still visible on the spot, and as already reported by Étienne – Knoepfler 1976 and later by Fossey (1988: 296), while the N slopes of the Palaiokastron hill are pierced at several points by modern iron workings, traces of ancient mining have also been noticed (Davies 1935) – *component HO_14*. The visible remains (or only some of them), considering the scarcity of diagnostic surface material reported (probably coarse ware), could even be related to mine activities there and/or in the vicinity.

7. Megalovouna-Kiapha: *Component HO_17*

ID	Period	Type	Toponym	Discovery	Position
HO_17	gr-rom	fort	Megalovouna-Kiapha	Int. topographical surveys	certain

In correspondence with Kiapha (Megalovouni) mountain pass, 3km S of Pavlon village, surface walls and architectural fragments [AE1284 - AE1285], as well as a lot of surface sherds near the structure remains [AE1286] are reported by Lauffer (Kopais I: 182-3, sketches 190-192). Lauffer (Kopais I: 184) interprets the evidence as a fort checking the mountain pass, controlling the road connecting the upland with Copais.[8] Traces of an ancient road [AE1280] at the mountain pass have been noticed below it (Lauffer Kopais I: 182).

8/9/10. Kolaka-Agios Ioannis: *Components HO_15* (8) *and HO_16* (9) *and HO_18* (10)

ID	Period	Type	Toponym	Discovery	Position
HO_15	unknown	activity focus	Kolaka-Agios Ioannis	Int. topographical surveys	certain
HO_16	LH	burial place	Kolaka-Agios Ioannis	accidental	approximate
HO_18	LH	burial place	Kolaka-Agios Ioannis	accidental	approximate

A hilly landscape characterized the area of the church of Agios Ioannis, located ca 4km NE of Pyrgos village and 1km W of Loutsi. The area is known as Kolaka, with the name of the village situated 5km to the N-NE. In the area, above the torrent Platania, 18 LH tombs have been found robbed (*component HO_16* - AD 1968: 223-4), as well as traces of an ancient activity focus (Lauffer Kopais I: 171-2, reported also by Knauss et al. Kopais 3: 68 – *component HO_15*). Lauffer mentions only the historical focus,[9] without any reference to the LH graves found in the same area. Knauss et al. (Kopais 3: 68ff) point out that the LH chamber tombs by Ag. Ioannis may suggest the existence of a settlement nearby. A group of 17 robbed chamber tombs (*component HO_18*) were found on a hill to the SE of *component HO_16* (AD 1968: 223-4).

The historical activity focus (*component HO_15*), cannot be dated (recorded as unknown), since the reference to 'glaze sherds' is much too vague (C/H? – Med?).

What follows is a preliminary version of the list of sites discovered and recorded during the Systematic Intensive Surface Survey carried out in the area around Hyettos (area surveyed mapped in fig.1) in the years 1989-1991 by J.L. Bintliff and his équipe (mainly from the University of Durham – UK – Bintliff 1992d).

Table 1. *List of intensive survey sites discovered during intensive and systematic survey work (Bintliff 1992d) – mapped in fig.1.*

SurveySITE code	CHRONOLOGY	CHARACTER
CN1	A	rural site
CN1	C-Ehl	rural site
CN1	G	rural site
CN1	LH	?activity focus
CN10	C	rural site
CN10	C-Ehl	rural site
CN10	Lhl-ER	rural site
CN11	A	activity focus
CN11	C	rural site
CN11	C-Ehl	rural site
CN12	C-Ehl	rural site
CN13	C	rural site
CN13	C-Ehl	rural site
CN14	H	rural site
CN14	Lhl-ER	rural site
CN14	LR	rural site
CN14	MR	rural site
CN14	R	rural site
CN15	*Med*	*rural site*
CN3	EH	settlement
CN3	Preh	?burial place
CN4	C-Ehl	activity focus
CN5	C	?rural site
CN5	C-Ehl	?rural site
CN5	LR	rural site
CN6	A	rural site
CN6	C	rural site
CN6	C-Ehl	rural site
CN6	Lhl-ER	rural site
CN6	R	rural site
CN7	C-Ehl	rural site
CN7	Lhl-ER	rural site
CN7	LR	rural site
CN7	MR	rural site
CN8	LR	?rural site
CN9	A	activity focus
CN9	C	?activity focus
CN9	G	activity focus

[8] The fort at Phtelio (Vristika locality) probably had a similar use –see appendix I.7. These kinds of forts were probably part of the 4thC Boeotian system of fortifications (see chapter II.3.5 – FORTS AND FORTIFICATIONS)

[9] Lauffer (Kopais I: 172) reports of glaze sherds and tile fragments [AE1275], along with surface walls and house foundations [AE1274].

Appendix I.6
The Copais area: Copai and the North-East bay

THE ARCHAEOLOGICAL RECORD

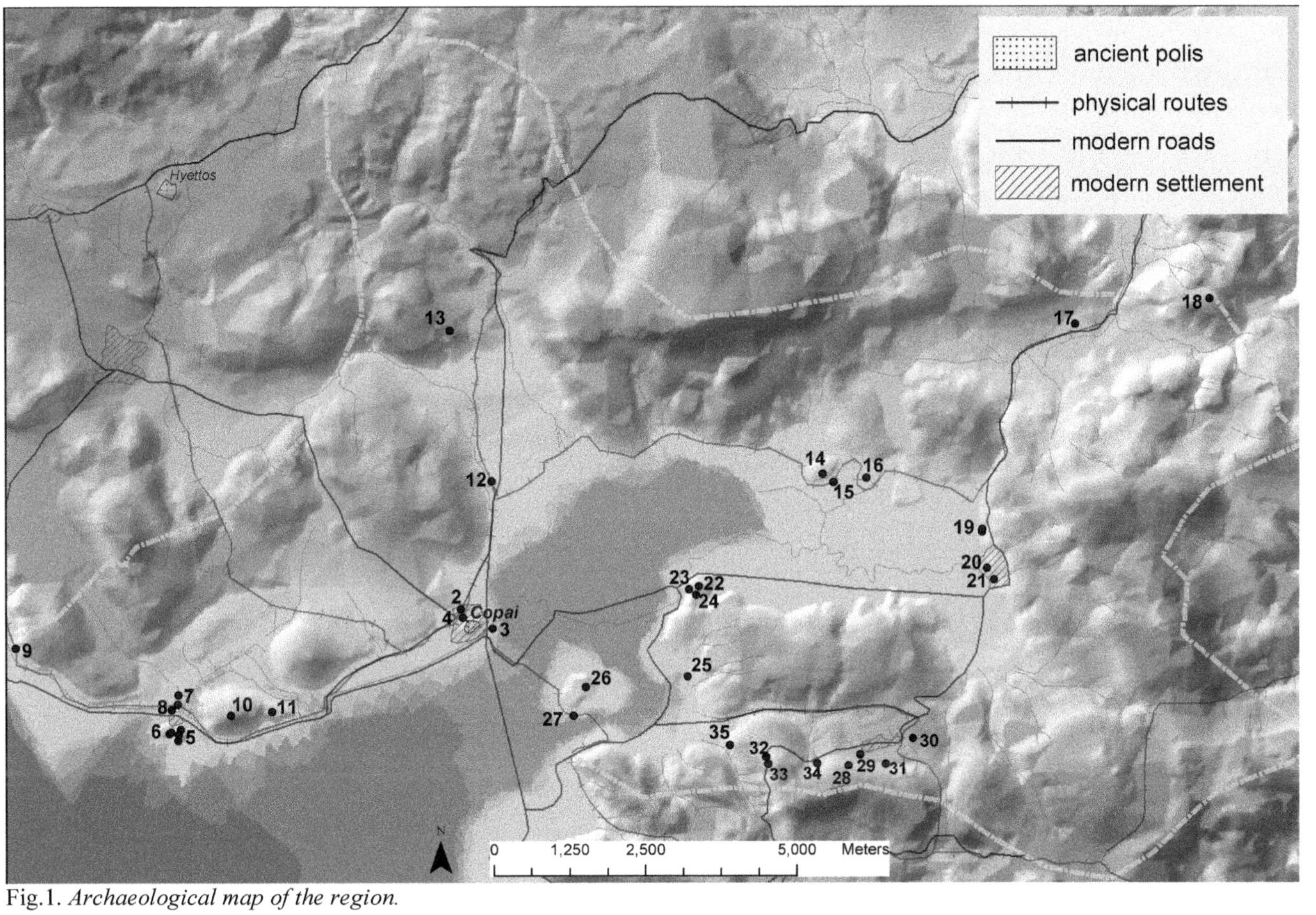

Fig.1. *Archaeological map of the region.*

1. Topolia/Kastro: *Components NC_15 to NC_24* (ancient *Copai*)

ID	Period	Type	Toponym	Discovery	Position
NC_15	N	activity focus	Topolia/Kastro	Personal or group interest	certain
NC_16	EH	activity focus	Topolia/Kastro	Personal or group interest	certain
NC_17	MH	activity focus	Topolia/Kastro	Personal or group interest	certain
NC_18	LH	activity focus	Topolia/Kastro	Travellers	certain
NC_19	PG	activity focus	Topolia/Kastro	Personal or group interest	certain
NC_20	G	activity focus	Topolia/Kastro	Personal or group interest	certain
NC_21	A	activity focus	Topolia/Kastro	Travellers	certain
NC_22	C	settlement	Topolia/Kastro	Ext. topographical surveys	certain
NC_23	H	settlement	Topolia/Kastro	Ext. topographical surveys	certain
NC_24	R	settlement	Topolia/Kastro	Ext. topographical surveys	certain

At the location of the modern village of Topolia/Kastro, on a steep-sided hill to the W of the New National Road, traces of ancient occupation have been recorded. The place was known as an ancient site (Copai) already to travellers (Leake 1835: 306; Forchhammer 1857: 179; Bursian 1862: 212; Vischer 1875: 580; Lolling 1889: 189; Frazer 1913: 131), who noticed surface structural remains, today mostly disappeared, since the modern village occupies exactly the same position of the ancient site. Schliemann (1882: 162) also knew the site.[1]

The place has an 'island character', joined to the 'mainland' to its NW side by a narrow isthmus (similar to the case of Prophitis Ilias W of Sphingion in Haliartos' *chora* - Lauffer

[1] Strabo (IX 411) tells us that Lake Copais was named after Copai. Schliemann (1882: 162) reports: in ancient times there was no common name for the lake, but it was called after the name of each city situated on its banks; thus *Copais* from Copai, *Haliartus* from Haliartos, and it has other names from other cities; in later times the predominating name, Copais, was applied to the whole lake, for the region of Copai forms the greatest gulf. The people of Topolia/Kastro procure from the lake numerous eels, which were already in antiquity renowned for their bulk and fatness, and which Pausanias commends. Schliemann (1882: 160) visited ancient Copai, on the island of Gla [*sic*], close to the N shore of the Lake of Copais, joined to the mainland by a narrow isthmus. Obviously Schliemann visited Topolia/Kastro and not Gla.

Kopais I: 212) which probably most of the time was submerged underwater during periods of high level of lake water. Pausanias (IX 24.1) tells us that the way to the town of Copai is from Akraiphia, and Strabo (IX 2.18) relates that Copai had once been very nearly 'swollen up' by the lake. In Leake's time (Leake 1835: 307) Topolia location was an 'island' surrounded by water during the great part of the year. The fields to the E of the village were flooded when visited by Forchhammer (1857: 179) in the year 1834.

Prehistoric evidence is reported in various gazetteers, but we do not have any idea of the intensity of occupation in the prehistoric period at the site, nor if the site was actually a settlement/habitation site in all prehistoric periods.
Neolithic surface sherds are reported from the site, as well as the EH (E slope especially – Hope Simpson 1965: 116) and MH material. Lauffer (Kopais I: 214) reports MH sherds to the SW of the village, by a fresh water spring (one of the few fresh water sources known in the N edge of the Copais, as Lauffer remarks). LH surface sherds were also found,[2] probably to be connected with traces of polygonal walling, already noticed by Frazer (1913: 132). Prehistoric evidence is reported in various gazetteers.

The place has been identified as ancient Copai. The Copai settlement belonged to the series of Greco-Roman settlements around Copais, and it is mentioned also in historical texts. From Copai, for instance, troops were sent to the Delion battle, in 424 BC (Thucidides IV 93.4), together with the ones from Haliartos and Koroneia. With Medeon, Koroneia and Haliartos, Copai is also mentioned in the Homeric Catalogue of Ships (*Il.* II 502). The town existed in Pausanias' time (IX 24.1).
PG material at the site was noticed by Fossey (1988: 277) and Desborough (1952: 195) discusses it, while G and A material is reported by Fossey (1988: 277), and mentioned in HS&D 1979 (see also AD 1967: 243).[3] C and H material is also reported. R sherds were noticed by Fossey (1988: 277) only, during his own fieldwork in the area. AD 1967: 243 mentions G to Byzantine material probably from a burial area by the church of Ag. Paraskevi (see *component NC_25*), without mentioning specifically R material though. As for the LR period, only burial evidence seems to be available (see below *components NC_25 and NC_29*).

2/3/4. Topolia/Kastro NW (2): *Component NC_25*, Topolia/Kastro East (3): *Components NC_26 to NC_28* and Topolia/Kastro village (4): *Component NC_29*

ID	Period	Type	Toponym	Discovery	Position
NC_25	gr-rom	burial place	Topolia/Kastro NW	accidental	certain
NC_26	A	burial place	Topolia/Kastro East	Rescue excavation	approximate
NC_27	C	burial place	Topolia/Kastro East	Rescue excavation	approximate
NC_28	H	burial place	Topolia/Kastro East	Rescue excavation	approximate
NC_29	LR	burial place	Topolia/Kastro village	Rescue excavation	approximate

The gr-rom necropolis (*component NC_25*), probably related to the town, was found at the mainland foothill in front of the Kastro 'island', not far from the chapel of Ag. Paraskevi. Illegal excavations [AE463], surface material [AE1713 – from G to Byzantine – AD 1967: 243], epigraphical evidence [AE1211] and reused material [AE1712] would attest it. Another cemetery area (from A to H – *components NC_26, NC_27 and NC_28*) was found through rescue excavations nearby Topolia/Kastro site, in the direction of Gla (AD 1964: 198-9).[4] Near the village, a Late Roman tomb was also excavated (AD 1970: 227-8), with two inscribed gravestones (3rd and 4th C AD) – *component NC_29.*

5/6. Stroviki-Nisi: E hill (5) *Components NC_1 to NC_4 and NC_7 to NC_10*, *NC_5* (South), *NC_6* (North)*;* W hill (6) *NC_11 and NC_87*

ID	Period	Type	Toponym	Discovery	Position
NC_1	N	settlement	Stroviki-Nisi	Int. topographical surveys	certain
NC_2	EH	activity focus	Stroviki-Nisi	Int. topographical surveys	certain
NC_3	MH	settlement	Stroviki-Nisi	Int. topographical surveys	certain
NC_4	LH	settlement	Stroviki-Nisi	Int. topographical surveys	certain
NC_5	unknown	burial place	Stroviki-Nisi	Other	certain
NC_6	MH	? burial place	Stroviki-Nisi	accidental	certain
NC_7	C	activity focus	Stroviki-Nisi	Int. topographical surveys	certain
NC_8	H	activity focus	Stroviki-Nisi	Int. topographical surveys	certain
NC_9	R	activity focus	Stroviki-Nisi	Ext. topographical surveys	certain
NC_10	LR	activity focus	Stroviki-Nisi	Ext. topographical surveys	certain
NC_11	C	activity focus	Stroviki-Nisi	Int. topographical surveys	certain
NC_87	C	burial place	Stroviki-Nisi	Personal or group interest	approximate

Above the modern village of Stroviki is a low rounded twofold hill called Nisi/Nisaki, crossed and cut by a cleft running from N to S. The name is due to the fact that, before the drainage of the Copais, the water of the Melas river was running between the '*nisi*' (island) and the small height where a chapel of Ag.Giorgios is, and therefore there was water in between (Lauffer Kopais I: 185) – see digital reconstruction of the lake fluctuations (chapter II.3.1 and Farinetti 2008). A canal built in the occasion of the Bronze Age drainage of the lake used to run immediately behind the Stroviki hill, in order to collect the extra water (Knauss Kopais 2: 180). Lauffer (Kopais I: 185) says there is no fresh water spring nearby.

On the E half (the higher –ca 30m- larger and less rocky top), there are clear signs of settlement from Neolithic to LH times. Occupation at the site was already known to Ulrichs 1840: 197 and Lolling 1889: 191. Lauffer (Kopais I: 185ff) describes a

[2] LH III material is reported by HopeSimpson 1965: 116. LH I and LH IIIA-B material is reported by Fossey 1988 [AE472].
[3] G material was found by the church of Ag. Paraskevi, probably to be associated with a burial area there (*component NC_25*).

[4] This cemetery area was discovered during rescue excavation in occasion of works of the *Organismos Copaidos* levelling of the area. 46 graves excavated, A-C-H periods attested. They only represent a small part of a larger cemetery.

quantity of fine wares easily recognizable from all periods. For the MH period he also describes imported ware known from the area only in the other two big settlements of Kalami-Lioma and Agoriani-Dekedes (see appendix I.1).
Surface signs, marking habitation sites, are mainly on the E side of the two hills SE of Stroviki village. Surface sherds are known from the Neolithic period (AA 1940: 186), from the EH period (AA 1940: 186; AD 1969: 179; AD 1971: 241), from the MH, in apparent association with cist graves[5] - *component NC_6* (AD 1969: 179; AD 1971: 241), and from the LH (AD 1971: 241).[6] All those periods are mentioned in the prehistoric gazetteers. HS&D 1979: G6 mention sherds spread over an area ca. 200m N-S by 150m E-W. Structures are reported in AD 1969: 179 [AE642, AE643, AE644].[7] Obsidian flints and animal bones are also reported (Lauffer Kopais I: 186).

Lauffer compares the site with the other two 'island' prehistoric sites in N Copais: Gla and Pyrgos *magoula*. Stroviki-Nisi and Gla know a strong pre-Mycenaean occupation, while Pyrgos *magoula* does not.

A section seen and drawn by Lauffer (Kopais I: 192) on the place shows a clear break of occupation after the LH period. Afterwards, there are clear signs of the C occupation (*component NC_7*), though from the information we have we cannot infer the character. C sherds on alluvial terrain in the profile examined represent probably, says Lauffer (Kopais I: 193), an indication of an increase in the lake level in intermediate periods, and therefore a lack of occupation in the area. On the other hand, this can be simply due to different choices in settlement over the landscape. The signs of occupation for the C period might represent a rural site on the higher easterly hill (with cemetery site on the western smaller hill – see below *component NC_11*). The evidence is obviously known because of the heavy presence of prehistoric signs of occupation, representing a large prehistoric settlement, but this does not mean that in C times the settlement site had to be at a village status as well just because of the large prehistoric settlement there. Activities within the landscape change over time (see below also for a possible location of a sanctuary here).

Fortification or terracing walls are visible [AE1339]. Lauffer (Kopais I: 195-6) considers them prehistoric, and apparently so imply also HS&D 1979: G6. The structural remains are especially on the N side, and down the N slope prehistoric sherds have been found in larger number.
Surface sherds from the H period are also reported (Fossey 1988: 282 and HS&D 1979: G6) as well as R and LR (Fossey 1988: 282 only). Simply considering the information that inscribed sherds[8] were found in one of the buildings visible on the surface has got a votive character, one could attribute the building remains and surface sherds (or the whole system of terracing walls and building remains) either to a cult place (sanctuary) of a historical period, or to a fort containing a religious building (or a votive activity focus), if we consider jointly also Fossey's attribution of the fortification walls to historical periods. Obviously this would not exclude a possible prehistoric fortification at the place, since it could also be the case of a prehistoric fortification reused as terracing for a sanctuary or as a fort in the gr-rom period. But all these are simply hypotheses, and a more detailed research on the site would be needed.

Components NC_11 and NC_87: Few C sherds are reported from the E slope of the lower W peak of Stroviki-Nisi (Lauffer Kopais I: 186) Signs of occupation here were also noticed generically by Fossey 1988: 282. They could indicate a burial place. C graves from the site are also mentioned in AD 1969: 179 (*component NC_87*). *Component NC_87* and *component NC_11* might belong to the same activity focus (burial character).

Lauffer attempts an identification of the prehistoric Stroviki with the Homeric *Mideia* quoted by Strabo (I 59.9, 413). In addition, some attempts have been made to identify the place (if interpreted as a settlement site) with ancient *Olmones* (Étienne-Knoepfler 1976).[9]

7. Stroviki-Ag.Georgios: *Components NC_88 and NC_89*

ID	Period	Type	Toponym	Discovery	Position
NC_88	unknown	? Fort	Stroviki-Ag.Georgios	Personal or group interest	certain
NC_89	Preh	? activity focus	Stroviki-Ag.Georgios	Personal or group interest	certain

To the SE of the rocky hill (on top of which is the small church of Ag. Georgios) NW of the Stroviki village, worked blocks probably from a polygonal ancient wall (AD 1969: 179) – *component NC_88*, while to the N of the rocky hill, remains of buildings, probably Prehistoric, have been noticed (AD 1969: 179) – *component NC_89*.

8. Stroviki-Ag.Georgios: *Component NC_90*

ID	Period	Type	Toponym	Discovery	Position
NC_90	unknown	burial place	Stroviki-Ag.Georgios	Personal or group interest	approximate

Between Stroviki-Ag.Georgios and the Stroviki village, W of the meteorological station, two graves made of stones, undisturbed by robbers (AD 1969: 179).

9. Baroutospilia: *Component NC_14*

ID	Period	Type	Toponym	Discovery	Position
NC_14	N	activity focus	Baroutospilia	Personal or group interest	approximate

Some distance W of Stroviki, a cave site was excavated, probably with signs of Neolithic occupation (AAA 1973: 202).

10/11. Tourlogianni (10): *Component NC_12* and Tourlogianni-Bazaraki (11): *Component NC_13*

ID	Period	Type	Toponym	Discovery	Position
NC_12	LH	fort	Tourlogianni	Travellers	certain
NC_13	LH	? Fort	Tourlogianni-Bazaraki	Travellers	certain

[5] On the N side of the hill foundations of two circuit walls and MH cist tombs were noticed (Syriopoulos 1968 [AE631]); W of meteorological station, two graves made of stones, not disturbed by robbers (one cist grave opened mentioned in AD 1969: 179 [AE1706]).

[6] Lauffer (Kopais I: 186ff) duplicates most of Fossey's material, except that he records LH I-III; Fossey found no Mycenaean material earlier than LH III, as is frequently the case in Boeotia.

[7] Easterly hill, immediately SE of village. Remains of prehistoric buildings interpreted as a prehistoric cult place (mainly because of the presence of some cup-form carvings in the rock - AD 1969 B: 179).

[8] AD 1973-4: 454. A 6th C inscribed vase and inscribed sherds were found at Stroviki (on top of the hill), which could lead to the location of a probable sanctuary NW of the plateau (5th C inscribed sherds reported also in AD 1971: 241).

[9] See Pavlon-Palaiokastron site in Hyettos *chora* for another possible location of the historical site (appendix I.5 - see Fossey 1988: 282).

On a conical peak NE of the modern village of Stroviki (known as Tourlogianni), Lolling (1889: 190) had already noticed wall foundations, and interpreted them as the remains of a prehistoric fortification. According to Fossey (1980: 162), Tourlogianni is part of the system of LH fortification (along with Stroviki-Nisi), while HS&D 1979: G6 believe that the fortification walls visible on top of the hill are historical (C or H in considering the sherds) as well as the remains of polygonal walls in the plain to the SE of Stroviki [AE1289[10]]. Two are the fortified hilltops on top of the Turlogianni hill, recorded as Tourlogianni – *component NC_12* – and Tourlogianni-Bazaraki –*component NC_13*.
Forchhammer (1857: 178) described Tourlogianni as an island, probably because the water of the lake was penetrating all around it when the water level was high (as in Stroviki-Nisi case - see digital model of the lake fluctuation in fig.1, described in chapter II.3.1). Lauffer (Kopais I: 207) believes that the fort on top of Tourlogianni-Bazaraki was LH/Mycenaean, intended to watch over the late prehistoric hydraulic works in the area between the foothill of Kymitis hill and Tourlogianni-Bazaraki hill (protecting wall [AE1291] and bridge [AE1292] remains - Lauffer Kopais I: 207 and sketch 216). Even if this is to be proved, the fort was overlooking a passage from Lokris to Copais – see also Kophia pass in Hyettos *chora* (appendix I.5: *component HO_17*).

12. Magoula Kavkala: *Components NC_30 and NC_31*

ID	Period	Type	Toponym	Discovery	Position
NC_30	N	activity focus	Magoula Kavkala	Int. topographical surveys	certain
NC_31	EH	activity focus	Magoula Kavkala	Int. topographical surveys	certain

2.2km N of Topolia-Kastro, ca 150m W of the motorway, on a small mound (*magoula*), is a Neolithic and EH activity focus, with a probable settlement/habitation character. Lauffer (AD 1971: 241-2) reports N and EH surface potsherds (though he does not mention it anymore in Kopais I), and so do French, HS&D 1979, Syriopoulos 1983-84 and Fossey 1988 (probably from Lauffer).

13. Alogopatisia-Kaphkala: *Component NC_32*

ID	Period	Type	Toponym	Discovery	Position
NC_32	gr-rom	activity focus	Alogopatisia-Kaphkala	Int. topographical surveys	certain

Along the path from Topolia to Martino, ca 5km N of Topolia, on the SE slope of Mavrovouni, in the Alogopatisia locality, surface walls, as well as signs of terrace walling, have been noticed by Lauffer (Kopais I: 247-8, sketch 246). The Alogopatisia locality lies in an alluvial valley, probably used for pasturage, in particular for horses, as the name states.
Being along the road from Copais to Eastern Lokris,[11] according to Lauffer the evidence can be interpreted as a cult place. There are many similarities with another cult place site, to the S border of Copais, by the Petra mountain (Koroneiake *chora* – appendix I.1. – *components KO_27 and KO_28*). In both cases, the cult place is exactly at a passage of a long-way road.

14/15. Chantza (14) and Chantza nearby (15): *Components NC_33 to NC_37*

ID	Period	Type	Toponym	Discovery	Position
NC_33	MH	activity focus	Chantza	Int. topographical surveys	certain
NC_34	LH	activity focus	Chantza	Travellers	certain
NC_35	Preh	? burial place	Chantza	Personal or group interest	certain
NC_36	A	burial place	Chantza nearby	accidental	approximate
NC_37	C	burial place	Chantza nearby	accidental	approximate

The western promontory of three peninsulas that jut out into the basin in far North Copais, W to Agios Ioannis promontory (*components NC_38 to NC_45*), is named Chantza. Fortification remains on top of it were first noticed by Noack (1894: 442f) that located here a prehistoric fortified settlement (Lauffer Kopais I: 242). Because of the exiguity of remains found at the site, as well as because the dating of wall remains was uncertain, the site did not get mentioned in the gazetteers of Mycenaean sites by Fimmen 1912, Wace-Thompson 1912 and Karo in RE Suppl. 6: 609. LH pottery was then found at the site and Noack's hypothesis was confirmed. Surface walls can be seen in olive grove fields, most of them almost destroyed. Typical Minyan ware was also found through new investigations (Lauffer Kopais I: 242) and the site was probably also a MH settlement site.[12]
The relationship between the site at Agios Ioannis at Spitia Katavothra and the one at Chantza is unclear. A dam is between the two peninsulas. This can lead to the hypothesis, according to Lauffer (Kopais I: 242-3), that Chantza would belong to the Agios Ioannis site, and they together would form a '*Art Doppelburg*', resembling this way the settlement situation at Megali Katavothra-Spilia Tsoutso sites, on two different promontories. Fossey also lists the Agios Ioannis and Chantza sites together (Fossey 1988: 287), as he does for the Megali Katavothra and Spilia Tsoutso sites (Fossey 1988: 286).
On the other hand, the site could also be a fort site, with no proper permanent habitation, linked to the Agios Ioannis site, which would be the proper settlement site. Fossey elsewhere (Fossey 1980: 159) listed Chantza as a fortified settlement, but then he thought that the situation would correspond rather to the one described in Fossey 1988, with Chantza as a fort related to the Agios Ioannis settlement, since Chantza's rocky interior probably could not have supported a proper settlement site. He therefore links the prehistoric tombs (MH and LH) found at the site with the Agios Ioannis prehistoric settlement site.
Components NC_36 and NC_37: Historical period burial areas were found in the vicinity of Chantza, mainly A and C tombs (including a couple of small tumuli). According to Fossey, the historical period burials would also relate to the Agios Ioannis historical period site (Fossey 1988: 287).

[10] Lauffer (Kopais I: 198, sketch 210) reports on a 'cyclopean' wall 2,5m thick seen by Kenny 1935: 210f, interpreted as a dam. It runs in the middle of the lake from the E side of Stroviki-Nisi towards the centre of the lake and then it turns towards the E and runs parallel in front of Tourlogianni hill.

[11] Traces of an ancient road can be found between Topolia and Martino (Lauffer Kopais I: 247).

[12] The presence of MH sherds is reported as uncertain in various gazetteers, except from Lauffer (Kopais I: 242) who noticed typical Minyan ware. Absidal buildings are also known from the site (Lauffer AA 1940: 185).

16. Agios Ioannis-Spitia Katavothra: *Components NC_38 to NC_47*

ID	Period	Type	Toponym	Discovery	Position
NC_38	N	? activity focus	Agios Ioannis-Spitia Katavothra	Personal or group interest	certain
NC_39	EH	? activity focus	Agios Ioannis-Spitia Katavothra	Personal or group interest	certain
NC_40	MH	settlement	Agios Ioannis-Spitia Katavothra	Personal or group interest	certain
NC_41	LH	settlement	Agios Ioannis-Spitia Katavothra	Travellers	certain
NC_42	PG	? activity focus	Agios Ioannis-Spitia Katavothra	Other	certain
NC_43	G	activity focus	Agios Ioannis-Spitia Katavothra	Other	certain
NC_44	A	activity focus	Agios Ioannis-Spitia Katavothra	Ext. topographical surveys	certain
NC_45	C	activity focus	Agios Ioannis-Spitia Katavothra	Other	certain
NC_46	MH	burial place	Agios Ioannis-Spitia Katavothra	accidental	certain
NC_47	LH	? burial place	Agios Ioannis-Spitia Katavothra	Other	certain

The promontory named after the chapel of Agios Ioannis on its top is the central promontory of three peninsulas that jut out into the basin in far NE Copais. On top of it lies a large prehistoric settlement site and a LH fortified settlement. The site was first noticed by Noack (1894: 444-8 and Taf. 13), which also describes the prehistoric fortification at the site (Lauffer Kopais I: 240, sketch 242; Fossey 1988 plan fig. 37; Knauss Kopais 2: 222 and figs.6.24 and 6.27). Heavy cultivation and robbery at the site led to the disappearance of most of the surface evidence seen by Noack.

Neolithic material is reported only by Lauffer, and only in AD 1971: 242. EH at the site is reported by Lauffer (few sherds EH III? - AD 1971 and Kopais I: 239), and Fossey (1988: 287), and as not sure by Hope Simpson 1965 and HS&D 1979. According to Lauffer (Kopais I: 239-40) EH occupation is not sure but probable, considering the continuity in later periods and the few sherds recognizable on the surface. Lauffer also remarks how signs of actual occupation differ from MH to Byzantine times - heavier for the prehistoric period, especially LH, and lighter for historical periods.

As for the MH period, there is evidence representing a settlement with graves in between,[13] while in the LH period at the site is a fortified settlement,[14] probably fairly developed (IIB-IIIB). Fossey (1988: 287) notes that there is abundant evidence that Agios Ioannis was a settlement before and much after the erection of the Mycenaean fortifications (unlike the Chantza site – see *component NC_34*). Moreover, Lauffer (Kopais I: 240-1) points out that the importance of the site cannot be understood without considering the drainage works undertaken at the foot of the hill to canalize the Melas into the Spitia Katavothra.[15] He remarks that at the Chantza site as well (according to this hypothesis also linked with the drainage system) the occupation would have started in pre-Mycenaean periods (see Chantza site – *component NC_33*). Fossey though does not agree with the importance assigned by Lauffer to Chanza in the MH period, neither with the existence of an actual settlement site at Chantza (see no.14 - Chantza site).

As for historical periods, signs of occupation are slighter, and are listed in prehistoric gazetteers, as they have been noticed by people interested mainly in the prehistoric occupation at the site. The historical activity focus there could have been a hamlet site, with recognisable traces of occupation from PG to C period (fertile land on the flat hill top). Only potsherds are known, but heavy cultivation at the site (and consequent erosion) may have destroyed structural surface remains, as both Fossey and Lauffer note, through which the character of the site might have been better understood. In other similar places, occupied in prehistoric periods, such as on Copais islets (Pyrgos *magoula* or Stroviki-Nisi), surface signs lead to the interpretation of the evidence as cult places in historical periods. A detailed study of pottery remains would give us a better idea of what the available evidence might represent. Anyhow, if we consider the A and C cemetery area found in the vicinity of Chantza, we should probably at least hypothesize the existence of a habitation area on the Agios Ioannis peninsula. It could also have been a small hamlet linked to a cult place, but most probably it was just a long-life (G to C) small group of houses.
As Lauffer notes, the peri-lacustris road would have gone along the foothills of the promontory.

17/18. Souvliki and Bazaraki: Souvliki *Component NC_80* (17) *and* Bazaraki *NC_81* (18)

ID	Period	Type	Toponym	Discovery	Position
NC_80	Preh	human presence	Souvliki	Int. topographical surveys	approximate
NC_81	LH	? Fort	Bazaraki	Int. topographical surveys	certain

Along the road from Copais to Larymna, to the E of it, lies the Bazaraki hill.[16] The top of the Bazaraki hill would probably have been the location of a LH fort. In the Souvliki locality, on a fertile plateau to the SE of the hill, Lauffer (Kopais I: 243) noticed obsidian scatters (according to him probably early prehistoric - N/EH[17]). In the past also the upland way to Larymna should have crossed this area (Lauffer Kopais I: 243).[18]

[13] Unlike the Spilia Tsoutso site, for this place Lauffer (Kopais I: 240) hypothesizes the existence of a settlement site together with the cist-grave burial place. This is mainly due to the presence of a cist-grave cemetery disturbed by robber-trenches on the flat top of the promontory, along with MH sherds on the surface here, unlike at the Spilia Tsoutso site (where no sherds have been found apparently).

[14] For discussion about the stratigraphical relationship of cist tombs with prehistoric fortification see Lauffer (Kopais I: 239). He remarks that the case of Mycenaean fortification on top of MH graves is reported also for other sites, taking Mycenae as an example. Some of the graves could also be LH (*component NC_47*). LH IIIB chamber tombs (?) are mentioned in Syriopoulos 1983-84 (LH IIIB 114).

[15] Maintaining this belief, Lauffer concludes that drainage works were probably already undertaken in the MH period. On the drainage work see Knauss (Kopais 2: 219ff and fig.6.24).

[16] In the area would also be the location of Pano-Larymna, the deserted medieval/ottoman village corresponding to modern Larymna by the sea.

[17] This interpretation is probably mainly due to the attempt to suggest a shifting of settlement from a lower location to an upper one, on top of Bazaraki hill, where the LH fort would have been – *component NC_81*.

[18] Lauffer suggests the hypothesis of a shifting of settlement in time between the two places (from earlier to later Bronze Age – from the area SE of the hill and the Bazaraki hill itself), as a situation similar to the shifting between the focuses in the depression to the S of Megali Katavothra site and the Megali Katavothra site itself (Lauffer Kopais I:

19. Spilia Tsoutso: *Components NC_48 to NC_50*

ID	Period	Type	Toponym	Discovery	Position
NC_48	N	settlement	Spilia Tsoutso	Personal or group interest	approximate
NC_49	EH	? activity focus	Spilia Tsoutso	Personal or group interest	approximate
NC_50	MH	burial place	Spilia Tsoutso - above	Personal or group interest	approximate

To the N of Megali Katavothra is a cave, which is in fact a *katavothra* (the opening of a subterranean natural channel), called Spilia Tsoutso, by which traces of prehistoric (Neolithic) activities were found (HS&D 1979: G13 and other prehistoric gazetteers; Fossey 1988: 286[19] and Lauffer Kopais I: 233).
Only Fossey (1988: 286) seems to report EH material (EH II) from the site.
Lauffer (Kopais I: 233) reports MH cist graves on the flat surface above the cave[20] (on the spur under which the *katavothra* Spilia Tsoutso is) and other similar tombs on the S slopes by the Ag. Ioannis chapel (MH cist graves described in Lauffer Kopais I: 234-5, sketch 237) – *component NC_50*. Here, on the S slope, he also noticed remains of a round shape structure (sketch 237), probably to be associated with the graves (a similar structure has been noticed also at Pyrgos-Chuntiklissa – see no.24).
In ancient times, the road should have been going along this upland area unlike the modern road that goes by the foothills in the basin (since the lake was not drained). Traces of old paths are still visible (Lauffer Kopais I: 233), and also traces of a more ancient upland path have been noticed in this necropolis area (Lauffer Kopais I: sketch 241).
Lauffer says that no sherds have been found in association, and surface walls neither. Therefore it was probably not a settlement site[21] but only the large necropolis associated to the large settlement site at Megali Katavothra. It would be a situation similar to others in the Copais NE area (see Topolia-Kastro, Medeon, etc.). Lauffer hypothesizes also that the small gulf between the two sites (Megali Katavothra and Spilia Tsoutso) would have been protected by a dam (like the one seen by Noack 1894: 443 at Spitia Katavothra – see above).

20. Agios Ioannis-Megali Katavothra: *Components NC_51 to NC_56*

ID	Period	Type	Toponym	Discovery	Position
NC_51	N	activity focus	Agios Ioannis-Megali Katavothra	Int. topographical surveys	certain
NC_52	EH	activity focus	Agios Ioannis-Megali Katavothra	Int. topographical surveys	certain
NC_53	MH	settlement	Agios Ioannis-Megali Katavothra	Int. topographical surveys	certain
NC_54	LH	settlement	Agios Ioannis-Megali Katavothra	Personal or group interest	certain
NC_55	G	activity focus	Agios Ioannis-Megali Katavothra	Other	certain
NC_56	C	activity focus	Agios Ioannis-Megali Katavothra	Other	certain

Above the main swallow hole (*Megali Katavothra*) on the E side of the Copais NE Bay, is a site which apparently presents a continuity of occupation from Neolithic to the Byzantine period (Lauffer, Gnomon 1952: 483). As Lauffer notes (Kopais I: 232), despite the fact that most of the plateau is today built, many open fields were still visible at the time of his visit.
Only Lauffer (Kopais I: 231) seems to report Neolithic (*rotpoliert Ware*) – *component NC_51*– and EH (few sherds of Urfirnis ware) at the site, which he visited personally. MH surface material is reported from the site by Fossey (1988: 286 - Grey Minyan and Matt painted) and Lauffer (Kopais I: 231 - numerous MH sherds). No mention of MH is made by Hope Simpson 1965 or other prehistoric gazetteers, while the site is reported as a small LH settlement. No structural remains are visible, and apparently the site was not fortified.[22]

Historical occupation at the site is known because of the interest in the prehistoric site, and in general in the prehistoric occupation in the area.
G material is reported only by Fossey (1988: 286),[23] while C material reported for the site by different gazetteers and topographical works might be interpreted as a rural activity focus or a farmstead.

21. Megali Katavothra SE: *Components NC_57 to NC_58*

ID	Period	Type	Toponym	Discovery	Position
NC_57	N	activity focus	Megali Katavothra SE	Int. topographical surveys	certain
NC_58	EH	activity focus	Megali Katavothra SE	Int. topographical surveys	certain

In a depression/sattle to the SE of Megali Katavothra, 200m SE of the chapel of Ag. Ioannis by the nickel mines (*Metalleia* locality, as marked on the 1:50,000 GYS map, by Neo Kokkino), Late Neolithic and EH potsherds have been noticed (Lauffer Kopais I: 227-9, sketches 217 and 233).[24] The depression which characterizes this place is in fact of karstic formation, and forms a terrace area (Philippson 1951: 487).

22/23/24. Pyrgos-Agia Marina: *Components NC_59 to NC_66* (22), *NC_67* (23) and Pyrgos-Agia Marina-Chuntiklissa: *Component NC_68* (24)

ID	Period	Type	Toponym	Discovery	Position
NC_59	N	? activity focus	Pyrgos-Agia Marina	Ext. topographical surveys	certain

243-4). Despite being a plausible idea, a few obsidian scatters are too scarce as evidence to lead to such a conclusion.

[19] Fossey 1988: 286 lists Megali Katavothra and Spilia Tsoutso sites together.

[20] The first mention of the evidence is in JHS 1944: 23.

[21] Unlike Agios Ioannis-Spitia Katavothra site (no.16), where MH burials were found in association with a settlement site (whose presence would be testified, according to Lauffer, by surface sherds, lacking in this case).

[22] Lauffer (Kopais I: 232) notes the absence of fortification as a peculiar thing, unlikely in all known Prehistoric sites in the area. He suggests that it could be due to a catastrophic event, using, in my opinion, an explanation too extreme, which might simply be explained with a different character of the site. We should also take into account that the area is built over.

[23] Fossey obtained information on the site from H.W.Catling, who informed him of the results of his visit to the site in the 1960s. They probably found fine ware from the G period, as they could easily spot it on the ground.

[24] This location has nothing to do with the Ag.Ioannis chapel near Megali Katavothra, 4km ca to the NW.

ID	Period	Type	Toponym	Discovery	Position
NC_60	EH	? activity focus	Pyrgos-Agia Marina	Ext. topographical surveys	certain
NC_61	MH	? Settlement	Pyrgos-Agia Marina	Personal or group interest	certain
NC_62	LH	settlement	Pyrgos-Agia Marina	Travellers	certain
NC_63	G	? activity focus	Pyrgos-Agia Marina	Ext. topographical surveys	certain
NC_64	A	? activity focus	Pyrgos-Agia Marina	Ext. topographical surveys	certain
NC_65	C	activity focus	Pyrgos-Agia Marina	Ext. topographical surveys	certain
NC_66	R	? human presence	Pyrgos-Agia Marina	Ext. topographical surveys	certain
NC_67	MH	? burial place	Pyrgos-Agia Marina	Other	certain
NC_68	MH	burial place	Pyrgos-Agia Marina-Chuntiklissa	Int. topographical surveys	certain

The steep-sided promontory of Pyrgos Agia Marina lies at the entrance to the Copais NE bay, on the opposite side of Topolia-Kastro, ca 1km NE from the church of Agia Marina, after which it is named.
At the NW foot of the hill is located the important junction between two Mycenaean drainage channels (Fossey 1988: 283, Knauss et al. Kopais 3).
The location of a prehistoric fortified settlement at the site was first reported by Noack (1894: 445ff). Several periods are represented: N and EH (probably[25]), MH (mentioned in different prehistoric gazetteers), LH (mentioned in prehistoric gazetteers as a fortified settlement).

A MH extended necropolis area has been noticed in a saddle to the S, linking the Pyrgos hilltop with the Chuntiklissa area (Lauffer Kopais I: 217 – *component NC_68*), with cist tombs similar to the ones found at Spitia Tsoutso and Spitia Katavothra. In the three cases tombs had been disturbed by robber trenches.
Also some other cist graves (*component NC_67*) were found in the W area of the hilltop (Fossey 1988: 283), inside the remains of a gateway of the later fortification.[26]
Lauffer (Kopais I: 218) points out that the presence of such a high number of tombs from the MH period (compared to the exiguity of surface remains from the MH period at the site) may lead one to consider the possibility of a large necropolis to be associated with another big settlement in the period in the area. Lauffer would suggest the settlement of Gla, taking into account the visible signs of a dam running from Gla towards this area as well as the similarity with other cases, such as the necropolis in front of the settlement site on the mainland found both at Copai (see above and Medeon site – appendix I.8).
We do not know whether before the LH fortification existed a MH fortification (and fortified settlement). The presence of burials that could be MH (*components NC_67 and NC_68*), both outside and inside the fortification walls, seems to lead to a negative answer, as well as the presence of such a large necropolis in no correspondence with the apparent exiguity of MH surface material, as we saw earlier. Uncertainty remains (Lauffer Kopais I: 218).

Only Fossey (1988: 283) and earlier Ålin (1962: 123) seem to report surface material from historical periods. Fossey lists G, A, and C periods, represented probably by fine wares, easily distinguishable also for periods such as G and A (he mentions black glaze). A generic C material is reported in Ålin, who is clearly interested in late prehistoric periods and not an expert knower of gr-rom wares. R material is reported by Fossey with a question mark, and would probably represent off-site material and not a focus. Otherwise, in historical periods Fossey would see at the place a small settlement. Available information is probably too scarce to allow us to give an answer and assign a character to the components.

25. Agia Marina SE: *Component NC_69*

ID	Period	Type	Toponym	Discovery	Position
NC_69	Preh	? Fort	Agia Marina SE	Int. topographical surveys	certain

On the W slope of the Nisi ridge, SE of the Agia Marina church, remains of a fortification, probably prehistoric, were noticed apparently only by Lauffer (Kopais I: 219, sketches 220 and 224). They could have been in connection with a dam [AE1304].

26. Gla: *Components NC_70 to NC_74 and NC_91*

ID	Period	Type	Toponym	Discovery	Position
NC_70	N	settlement	Gla	Other	certain
NC_71	MH	? activity focus	Gla	Other	certain
NC_72	LH	settlement	Gla	Travellers	certain
NC_73	C	activity focus	Gla	Other	certain
NC_74	H	activity focus	Gla	Other	certain
NC_91	LH	? Settlement	Gla W	Personal or group interest	certain

Gla is named the well known islet in the middle of the entrance to the NE Copais bay, SE of Topolia/Kastro, 0.5km from the Phtelia promontory and from the small gulf of the E edge, below Kokkino. The dark limestone rocky islet is 750m long in the direction W-NW/E-SE. From afar it looks like a levelled hill with vertical sides on which open up *katavothrai* and caves. The NW side of the islet is wider and its NE part, where the focus of the archaeological remains are, is the highest. A lower edge of the islet extended towards E, in the direction of Kokkino. Gla controlled the entrance towards the Topolia bay and the *katavotrai*, the road towards the edge and the joining point of the two main channels of prehistoric drainage; the S main drainage channel used to run below the E edge of the walled area (see map in fig.11 in chapter II.3.1).

The prehistoric fortified settlement was already known to Schliemann (1881) and travellers.[27] It was then excavated by de Ridder (twenty two days of excavations in 1893) and afterwards by Iakovidis, who carried out research on the site for a long period. The results of this research have been published in many monographies and articles, and recently (2003) in a book entitled: *Gla and the Kopais in the 13th century B.C.*
Neolithic (sherds noticed during excavations at different locations on the plateau – Fossey 1988: 288 and Iakovidis 2003, who summarize previous research) and MH (Fossey 1988: 288

[25] Only Fossey (1988: 283) mentions Neo, as well as R.C.S. Felsch, who visited the site in 1978 (personal comment given to J.L. Bintliff). As for EH, Fossey only (1988: 283) reports it, while no mention of EH at the site can be found in prehistoric gazetteers.

[26] As for MH tombs below LH fortification see also the case of *component NC_6* (Stroviki).

[27] Dodwell 1819; Leake 1835; Forchhammer 1837; Ulrichs 1840; Ross 1851; Vischer 1875; Noack 1894.

from Wace - Thompson 1912: 193) have also been reported from the site,[28] whose main focus though is LH (Mycenaean acropolis). As Iakovidis (2003) states, the rock was settled during the Middle and Late Neolithic period, while in Mycenaean times a strong Cyclopic fortification[29] [AE518] enclosed the plateau, with several structures in it (see Iakovidis 2003: plan 1).
Iakovidis (2003) interpretes the LH site as a vast fortress, built as part of the drainage work controlled by Orchomenos, aiming at the exploitation of the Copais plain, and belonging to the Orchomenos Mycenaean palace. The earlier Mycenaean sherds at the site are late LH IIIA2 or beginning LH IIIB1, while destruction layers contain nothing later than advanced LH IIIB2 examples. Gla shows all signs of a violent destruction by human agency.

As in many other major prehistoric sites in the Copais area, evidence of an occupation in historical periods has been noticed. The occasion for the discovery is, as usual, due to the intensity of research in order to get a better knowledge of the prehistoric occupation. Surface material from historical periods (C and H) are reported by HS&D 1979: G9. Iakovidis 2003 reports on Hellenistic material noticed during excavation in the area of the LH buildings and at the gates.[30]

Component NC_91: Below the Gla's rocky hill, and to its W, a polder (i.e. an area drained at the time of the late Prehistoric drainage of the Copais) has been detected by the team led by the idraulic engineer J. Knauss working in the area (Kopais 2). In the polder area, Knauss' équipe recognised traces[31] of a possible Mycenaean-Minyan settlement (Kopais 2: 216 and fig. 6.21). They attempt to identify it with the Homeric *Arne*, mentioned in the Catalogue of Ships.

27. Gla-Talantouseza: *Component NC_75*

ID	Period	Type	Toponym	Discovery	Position
NC_75	MH	burial place	Gla-Talantouseza	Other	approximate

In the Talantouseza locality, to the S of the Mycenaean acropolis, MH cist tombs are reported, as looted in the past (AD 1970: 233).
Fossey remarks that their location is too imprecisely given to indicate which settlement they belonged to. In fact, since Gla was probably not occupied in the MH period (see above), those tombs would belong to another site. Also the MH tombs at the Pyrgos-Agia Marina site create a small problem in attribution, apparently. See discussion above – no.22/23/24.

28/29/30/31. Kokkino near (28): *Component NC_76* – Kokkino W (29): *Component NC_77* - Kokkino NE (30): *Component NC_78* - Kokkino S (31): *Component NC_79*

ID	Period	Type	Toponym	Discovery	Position
NC_76	LH	activity focus	Kokkino near	Personal or group interest	imprecise
NC_77	Preh	activity focus	Kokkino W	Int. topographical surveys	certain
NC_78	gr-rom	activity focus	Kokkino NE	Int. topographical surveys	certain
NC_79	gr-rom	? burial place	Kokkino S	Int. topographical surveys	certain

Component NC_76: near the village of Kokkino (location no better specified) LH pots were noticed (Hope Simpson 1965: 120-121, quoted by Syriopoulos 1968 and HS&D 1979 – LH?).
Component NC_77: obsidian flints on a 30m large terrace at the W end of Kokkino village, crossed by the road (Lauffer Kopais I: 225-6).
Component NC_78: on a terrace near the village cemetery of Kokkino, Lauffer (Kopais I: 226) noticed surface walls and bumps of stones.
Component NC_79: to the E of the fresh water spring to the S of the Kokkino village an incineration burial place was noticed by Lauffer (Kopais I: 226). It could be gr-rom in date.

The area of Kokkino could constitute a settlement chamber, even today rich in wine yard, and including the valley running EW down to the Copais basin. *Component NC_78*, represented by AE1312 and AE1313, along with AE1315 (reused), near the village cemetery of Kokkino, would do for the settlement site of the area probably, as Lauffer says. There are water sources nearby (marked on IGME 1:50,000 geological map). On the other hand, those few pieces of evidence (and in no apparent association with pottery) are the only material signs of a possible settlement in historical times here, with the exception of a burial place (incineration) seen by Lauffer by a perennial water source – *component NC_79*. Lauffer notes (Kopais I: 227) that architectural fragments reused in the two chapels E of the Kokkino village [AE1315] are commonly thought to come from the Apollo sanctuary. He thinks though that the evidence could be a sign of a different cult place much closer and related to the settlement (?) site he hypothesizes to the N side of Kokkino (see above *component NC_78*).
The area is also well linked as far as communication routes are concerned. To the S, through the pass leading to the Apollo temple (Ptoion) runs a way towards Yliki lake and Thebes, while to the N the area is linked with the Copais basin. After the Byzantine period the settlement would have moved to the N, closer to the Megali Katavothra, by Neo Kokkino, W of the '*Metalleia*' locality, marked on the 1:50,000 GYS map. A deserted village was seen by Lolling (1889: 162) there.

32. Megalovouno NE: *Component NC_84*

ID	Period	Type	Toponym	Discovery	Position
NC_84	gr-rom	? special activity	Megalovouno NE	Int. topographical surveys	certain

Component NC_84: 400m ca from the place where the road from Akraiphnio to Kokkino bends to the E, right side of the road to the N of a R burial area (*component NC_85*), Lauffer noticed a quarry, which could be ancient (Kopais I: 223, sketch 229).

33/34. Megalovouno NE: *Components NC_85* (33) *and NC_86* (34)

ID	Period	Type	Toponym	Discovery	Position
NC_85	R	burial place	Megalovouno NE	Int. topographical surveys	certain
NC_86	C	burial place	Megalovouno NE	Int. topographical surveys	certain

[28] Wace - Thompson (1912: 193) mention some Minyan ware, but do not specify their source of information. Iakovidis (2003) shows perplexity, and allows for the possibility that those were LH sherds burnt grey. The existence of a MH settlement at Gla is implied by Lauffer (Kopais I), while discussing the Pyrgos-Agia Marina MH burial evidence (see component Pyrgos-Agia Marina).
[29] Gla is the Albanian equivalent of the Greek name '*Kastro*' (Iakovidis 2003).
[30] A few medieval sherds and two fragments of a LG vase were also found (Iakovidis 2003).
[31] Those settlement traces were not better specified in the source.

Component NC_85: by *component NC_84*, two glaze tile covered graves, with glaze ceramic and lamps inside and on the surface were noticed near the road by Lauffer. A third tomb was also noticed 200m to the N (Lauffer Kopais I: 223, sketch 229).
Component NC_86: above the road Akraiphnio-Kokkino, E of Kokkino, where the road crosses a stream, on a terrace, Lauffer (Kopais I: 223-4, sketch 230) noticed tile fragments, bones and C pots.

Those traces of necropolis found to the NW and N foot of the Megalovouno ridge could refer to burial areas connected to the ancient town of Akraiphia, but could also be interpreted as rural burials, still in the *chora* of Copai, or linked to a village settlement there (as suggested by Lauffer Kopais I: 225 – see chapter II.3.6 LONG TERM SETTLEMENT TRENDS). Therefore, for the time being the Megalovouno components have been entered under the NC *chora*. The definition of the exact border between ancient Akraiphia and Copai in this upland area is still problematic.

The area of Kokkino-Megalovouno was strategic as it would mark the border between Gla, Kokkino and Karditsa (says Lauffer Kopais I: 257) and was the cross point for various roads leading to Akraiphia, Gla and Copais. Even if Gla does not play a central role in most periods of history (unlike what Lauffer seems to believe and too often implies), its wider area is indeed at a crucial location. The modern road between Akraiphnio-Kokkino follows almost the same path as the ancient one (Lauffer Kopais I: 222, sketches 250 and 253), and crossed another path leading to Gla, seen by Noack (1894: 462).[32]

35. Megalovouno-Blaitesa: *Components NC_82* and *NC_83*

ID	Period	Type	Toponym	Discovery	Position
NC_82	C	rural site/activity	Megalovouno-Blaitesa	Travellers	certain
NC_83	gr-rom	? activity focus	Megalovouno-Blaitesa	Travellers	certain

On the lower spur of the W slope of Megalovouno, called *Blaitesa* (from the Albanian word for beehives), 2.5km ca NW from Karditsa. On a flat terrace (200m long) above the 100m contour line, remains are visible 30m ca to the orographic right of a stream (Lauffer Kopais I: 252).
Surface walls (built with irregular big stones) and building material [AE1345 and AE1346] can be seen at the site, as well as C sherds [AE1347] among the surface walls, which lead to a dating of the site to the C period (Lauffer Kopais I: 256-7, sketch 251). Structural remains had already been noticed by Noack (1894: 462), who interpreted the remains as a LH site. We cannot exclude the possibility that the period of occupation of the site could be other than C, but further surface research would be needed in order to get a better picture of the date of surface sherds. For the time being the component has been recorded as C. We cannot be sure either that the walls are pertinent to the site, and not built later with material reused. The site could be interpreted as rural in character, though we do not have any clue regarding the extension of the site, nor of its status (farmstead, hamlet?). A second component has been created as gr-rom –with question mark– taking into account the fact that this site could have other periods of occupation.

The burial place on the road between Akraiphnio-Kokkino just below this settlement site (*component NC_85* – see above) could be linked to this component.

[32] Noack 1894: 460f (fig.17) describes a path that would join Karditsa with Gla, and then Copais, crossing the Souvliki area, and another path which from Karditsa would pass between Phtelio and Mytikas and then would reach the Copais (Lauffer Kopais I: 259, sketch 253).

Appendix I.7
The Eastern Copais area: Akraiphiai

THE ARCHAEOLOGICAL RECORD

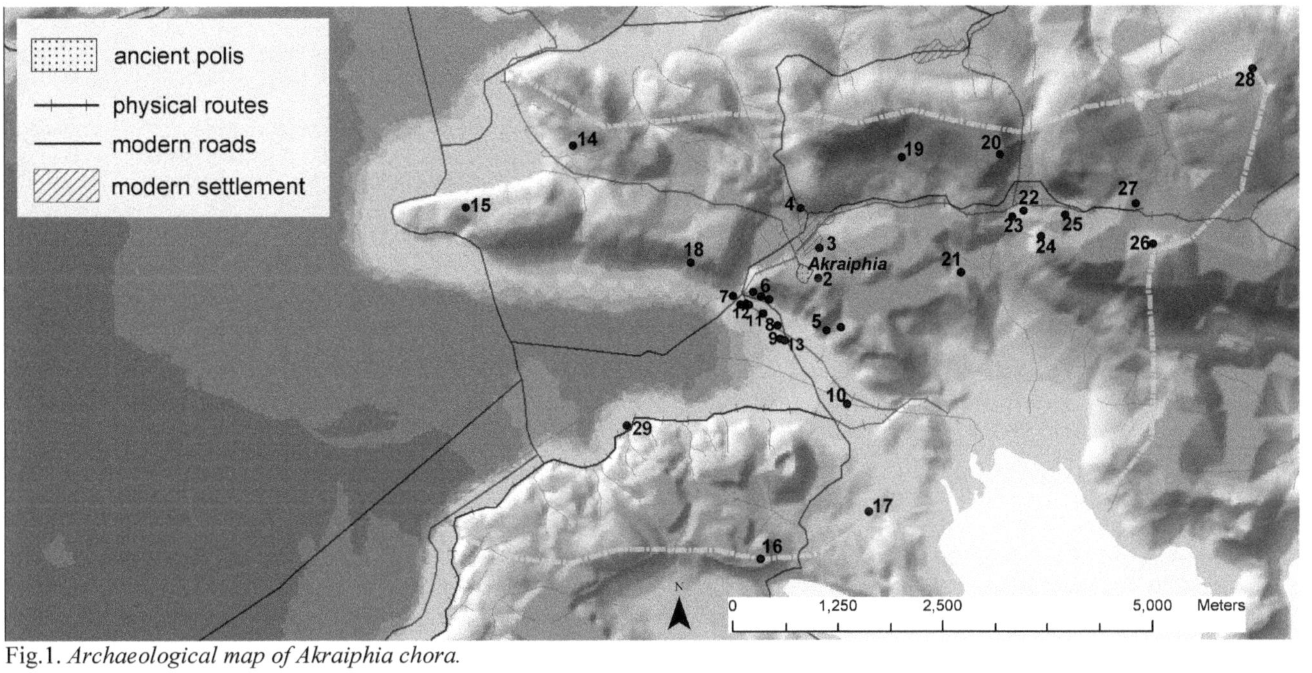

Fig.1. *Archaeological map of Akraiphia chora.*

1. Akraiphnion Skopià: *Components AK_1 to AK_6*

ID	Period	Type	Toponym	Discovery	Position
AK_1	G	settlement	Akraiphnion Skopià	Ext. topographical surveys	certain
AK_2	A	settlement	Akraiphnion Skopià	Travellers	certain
AK_3	C	settlement	Akraiphnion Skopià	Travellers	certain
AK_4	H	settlement	Akraiphnion Skopià	Travellers	certain
AK_5	R	settlement	Akraiphnion Skopià	Travellers	certain
AK_6	LR	? burial place	Akraiphnion Skopià	Personal or group interest	certain

To the S of the modern village of Karditsa/Akraiphnion, on a large, rocky hill (named Skopià) lie the remains of the ancient *polis* of Akraiphia.[1] The hill is overlooking a deep eastern bay of Copais (Stenà/Karditsa bay) which is separated from the Yliki lake only by a low ridge, ca 1.5km to the SE of the site of ancient Akraiphia.

The site was already known to travellers (see for instance Ulrichs 1840: 242ff and Frazer 1913: 97-99). Guillon and Feyel (BCH 1936: 461) carried out excavations at the site, uncovering the complete perimeter of the circuit wall[2] and locating the *agora*, as well as public buildings and many inscriptions.[3] Immediately afterwards, Chorapas' excavations were also published in JHS 1936: 19 and AA 1936: 123.

Fossey (1988: 269) reports G material from the site, though it is not clear whether he noticed it (or he found it reported) as from the city site or only from cemeteries. In any case, the presence of rich G (LG) cemeteries, attested trough excavations, starting from Chorapas' excavations and then with the many Andreiomenou's ones as well as more recent rescue excavations (see below for references), makes the existence of a G settlement at the site fairly sure.

Settlement components were created for G, A, C, H and R periods. Probably in the G period the settlement had a *protopolis* status, although Akraiphia seems to be much more developed in earlier historical periods compared to other Boeotian *poleis*.[4] At the city site it seems that there are no traces of occupation in the Neolithic or Bronze Age periods, with the exception of traces of a prehistoric fortification and the presence of MH sherds found below the acropolis hill (see below *component AK_60 and AK_28* – Lauffer 1940 and Knauss Kopais 2: 247ff).

Component AK_6: A cemetery area found in 1940 inside the W part of the acropolis circuit by Threpsiadis (AE 1973: 83-5 no.

[1] Position shown on Guillon's map (Guillon 1943) and in Papachatzis 1981: 146-150.

[2] The circuit wall dates back to the Late Classical-Early Hellenistic period (Garlan 1974), and is probably to be linked to the independency of the *polis* after the defeat of Thebes (Guillon 1948: 105).

[3] In the *agora* a large number of inscriptions were found, as well as a large cistern or monumental fountain close to which was found an altar with an inscribed base, dedicated to Zeus Soter (BCH 1936: 461).

[4] For a history of the *polis* see Hansen 1996: 78-9.

8; Fossey 1988: plan fig.35), indicated as of a 'later' date, may correspond to a shrinking of the city site in LR periods, as occurred in other Boeotian poleis as well (see Hyettos, for instance, or Thespiae, investigated through intensive artefact surface survey).
Information on the *polis* of Akraiphia in historical periods are indirectly given by the extended excavation carried out at the many cemeteries connected with the *polis* (see below: *components AK_7 to AK_18*), largely studied by Andreiomenou (monography in 1980 and then several articles and reports). The city life was also in connection with the presence of the Ptoion sanctuary above it, located within the *polis* territory (see below no.22). Major work at the site and the wider area was conducted by the French School, starting with the book of P. Guillon (1943), and then published in archaeological reports, recently by Ch. Müller mainly.

2. Skopià: *Components AK_28 and AK_60*

ID	Period	Type	Toponym	Discovery	Position
AK_28	MH	activity focus	Skopià	Personal or group interest	certain
AK_60	LH	? Fort	Akraiphnion Skopià	Personal or group interest	approximate

Knauss reports in Teiresias 1987 and Kopais 2: 247ff about Homeric *Hyle* (Homer, *Il.* II 500 and V 707-710), which he attempts to identify with the archaeological remains on the slope of Mount Skopià within the bay of Akraiphia. According to Homer, *Hyle* was '*sloping towards the Kephisian lake*' (*Il.* V 709, cf. Pausanias IX 38.7). Knauss 1987 notes as the strongest support for the identification, "*the discovery of some remarkable remains of a small but typically Minyan fortification, with walls in Cyclopean masonry. The prehistoric fortification is within the lower area of ancient Akraiphia and is surrounded by the polygonal walls of the archaic city*" (see map in Kopais 2 fig. 8.2). At the foot of Mount Skopià, in the deposit of a small torrent that descends from Mt. Ptoion, numerous surface finds confirm occupation in antiquity. Among them a number of gray and yellow Minyan sherds would constitute the evidence for a settlement in the MH period, bordering the Copais (Knauss 1987 and Kopais 2: 247ff). Knauss (Kopais 2: 250 and figs. 8.2 and 8.5) reports on the possible existence of a lower settlement in the Bronze Age, at the foot of the acropolis hill towards the bay.

3 to 10. Akraiphnion burial areas: *Components AK_54 to AK_59* (3); *AK_7 and AK_8* (4); *AK_11* (5); *AK_12 to AK_17* (6); *AK_13 and AK_14* (7); *AK_48 to AK_52* (8); *AK_18* (9); *AK_9 and AK_10* (10)

ID	Period	Type	Toponym	Discovery	Position
AK_7	G	burial place	Karditsa /Akraiphnion	accidental	certain
AK_8	A	burial place	Karditsa /Akraiphnion	accidental	certain
AK_9	A	burial place	Akraiphnion Gouni	accidental	approximate
AK_10	C	burial place	Akraiphnion Gouni	accidental	approximate
AK_11	R	burial place	Akraiphnion SE	accidental	certain
AK_12	A	? burial place	Akraiphnion W foot	accidental	certain
AK_13	H	burial place	Akraiphnion W	accidental	certain
AK_14	R	burial place	Akraiphnion W	accidental	certain
AK_15	A	burial place	Akraiphnion SW	accidental	certain
AK_16	H	burial place	Akraiphnion SW	accidental	certain
AK_17	R	burial place	Akraiphnion SW foot	accidental	approximate
AK_18	R	burial place	Akraiphnion S foot	accidental	certain
AK_48	PG	burial place	Akraiphnion Grava	Rescue excavation	approximate
AK_49	G	burial place	Akraiphnion Grava	Rescue excavation	approximate
AK_50	A	burial place	Akraiphnion Grava	Rescue excavation	approximate
AK_51	C	burial place	Akraiphnion Grava	Rescue excavation	approximate
AK_52	H	burial place	Akraiphnion Grava	Rescue excavation	approximate
AK_54	H	burial place	Karditsa /Akraiphnion	accidental	approximate
AK_59	R	burial place	Karditsa /Akraiphnion	Rescue excavation	certain

Around the ancient polis of Akraiphia several cemetery areas have been found, ranging from LG to R date.
In the area of the village lies at least one of the *polis* cemeteries (Fossey 1988: 268 and no.1 in fig.35 – *components AK_7* and *AK_8* and probably also *AK_54 and AK_59*, discovered in 1994 - AD 1994: 278-281).
A burial area has been located also at the W foot of the Akraiphia acropolis hill (Fossey 1988: 268 and no. 8 in fig.35 – *component AK_12*), another one at its SE foot (Fossey 1988: 268 and no. 6 in fig.35 – *component AK_11*), another at its S foot (Fossey 1988: 268 and no. 5[5] in fig.35 – *component AK_18*), and another one at its SW foot (Fossey 1988: 268 – *component AK_17*). Other burials are known to the SW of ancient Akraiphia (Fossey 1988: 268 and no. 4 in fig.35 – *components AK_13 and AK_14*[6]), and to the SE foot of the Pholies hill (Fossey 1988: 268 and no. 3 in fig.35 – *components AK_15 and AK_16*[7]).
Another burial area would be at the E end of the Copais Karditsa bay, in an area called *Gouni* (Fossey 1988: 268 and no. 2 in fig.35 – *components AK_9 and AK_10*). It could be linked to the *polis* but, located further than 1.5km from the city site, could also refer to the rural occupation in the *chora*.
Lauffer RE 1959 s.v. *Ptoion* summarises all information concerning cemeteries at his time, except the more recent (mainly from Andreiomenou,[8] listed by Fossey, and some other discoveries, more recent than Fossey's book, always mostly from Andreiomenou and from the most recent excavations by the *ephoreia*). To the ones listed above, therefore, one can add the more recent discoveries (not in Fossey's book) mainly to the S of the city in the Karditsa bay.
At the Grava locality[9] (Andreiomenou Ergon 1991: 37-45; Ergon 1992: 53-59; AD 1993: 176-180; AD 1994: 278-281 - *components AK_48 to AK_52*), a section of a very large cemetery site was uncovered (more than 2000 tombs excavated up to 1994), other sections of which are probably referred to in some of the aforementioned components.
As outlined by Aravantinos (2004a: 82), since 1994, work for the enlargement of the National Road Athens-Lamia in correspondence to Akraiphnio and in particular the Grava

[5] It probably corresponds to the Andreiomenou's '*Gypedo*' excavation (Praktika 1989: 125-145).
[6] They probably belong to the Grava cemetery area (see below in the text). See also *component AK_57* (Akraiphnio-Nat.Road junction).
[7] They probably belong to the Grava cemetery area (see below in the text).
[8] AAA 1974: 95-112; 1977: 273-286 and summary in her book 'To Κεραμεικόν Εργαστήριον της Ακραιφίας' 1980
[9] The exact location is not specified, but it is probably located below the foothills S of Akraiphnio, in the Copais Karditsa bay.

locality allowed for the realization of an extensive excavation (four years, 1994-1998). In that locality, previous research (rescue excavations mainly, or accidental discovery, since 1974: to the S of the road and along the Kephisos river-bed, during its enlargement, as well as to the N of it) had already shown the presence of archaeological remains –see above in text– belonging to the large A, C and H cemetery of ancient Akraiphia, located at the edge of the Copais lake.

As Aravantinos points out, the cemetery is not characterized by organization of the area in tombs enclosed in circuits (*perivoloi*) with burials positioned along lines as in other Boeotian areas where large cemeteries are known (Haliartos, Thebes, Tanagra). The graves were located in groups in the open area along the edge of the lake, below the steep slope of the acropolis. Graves dated from the late G to H period, from 6th C to 3rd-2ndC BC mainly.[10]

An inscribed funerary relief (end of 6thC BC) was ploughed up SE of the acropolis of ancient Akraiphia (Moraites field) in 1993. The discovery was followed by a rescue excavation which brought to light other graves belonging to the extensive cemetery area of ancient Akraiphia (AD 1993: 176-180). The funerary monument probably stood on the road from Akraiphia to Thebes, which was lined with extensive cemeteries (from Early G to the 3rd C BC). The actual stone-paved road (3.95m) was located in the NW part of the field [AE1722]. Apparently, in later periods (Late Hellenistic and Roman times), the dead were buried on the lower slope of the ancient acropolis (AD 1993: 176-180) and in its N foothill, in the area of the modern village (*components AK_54 and AK_59*).

11. Akraiphnion Grava: *Component AK_53*

ID	Period	Type	Toponym	Discovery	Position
AK_53	LR	rural site/activity	Akraiphnion Grava	Rescue excavation	approximate

In the locality of Grava (the same of the cemetery site represented by the *components AK_48 to AK_52*) the remains of a *'villa rustica'* (eight rooms) with an *opus spicatum* floor and an olive press were found during an extensive rescue excavation associated with works for the '*Agogos Physikou Aeriou* DEPA' (gas pipe line), in 1993 (AD 1994: 278-281). The villa was abandoned by the end of the 4thC AD. Aravantinos (2004a) also reports of other building remains of uncertain attribution, datable also to the LR period.[11]

12. Akraiphnion-Nat.Road junction: *Components AK_55, AK_57 and AK_58*

ID	Period	Type	Toponym	Discovery	Position
AK_55	G	activity focus	Akraiphnion-Nat.Road junction	Rescue excavation	approximate
AK_57	R	burial place	Akraiphnion-Nat.Road junction	Rescue excavation	certain
AK_58	R	special activity	Akraiphnion-Nat.Road junction	Rescue excavation	certain

Component AK_55: During rescue excavations for the construction of the new bridge, at the Akraiphnion National Road junction, remains of buildings and sherds in excavated layers, representing a G activity focus, were found (AD 1996: 275-278). The evidence has been interpreted as a cult place destroyed by fire[12] (Aravantinos 2004a: 86).

Component AK_57: by the Akraiphnion junction of the National Road Athens-Lamia, during rescue excavations a burial building of R date was found with multiple burials, probably to be identified as *heroon*, as epigraphical evidence seems to attest (Aravantinos 2004a: 88). It could possibly be a cult place associated with a special cult for a dead person, as Aravantinos (2004a: 88) would suggest, in association with the cemetery area around it (see *component AK_14*).

Component AK_58: a pottery kiln installation of R date has been found by the burials (Aravantinos 2004a: 88) and probably consequently, I would presume, by the road running through the Grava area [AE1722].

13. Akraiphnion-Nat.Road S: *Component AK_56*

ID	Period	Type	Toponym	Discovery	Position
AK_56	N	burial place	Akraiphnion-Nat.Road S	Rescue excavation	approximate

Below ancient Akraiphia, on the left side of the National Road Athens-Lamia, at the deepest strata of the gr-rom cemetery excavation, Andreiomenou (Praktika 1989: 125-145) found two built graves with two pots from the Neolithic (probably Late) period.[13]

14. Phtelio-Vristika: *Component AK_21*

ID	Period	Type	Toponym	Discovery	Position
AK_21	gr-rom	fort	Phtelio-Vristika	Travellers	approximate

On the SW slopes of Phtelio, in the Vristika locality, in the vicinity of a deserted village (*Skoino Megalo*) lie the remains of an ancient fort, in a quite low location.

Lauffer (Kopais I: 259-261) points out how the fortification here is similar to the Kiapha fortification (between Pavlo and Stroviki) - see appendix I.5 *component HO_17*. C sherds [AE1349] have been found in association with the fortification remains [AE1348], in polygonal walling. The fortification belongs probably to the Boeotian system of 4thC forts, related to the Theban egemony, as well as the one at Kiapha pass probably. In this case the fortification was overlooking a passage from Akraiphia to the basin and to ancient Copai (Lauffer Kopais I: 261, sketch 253), described already by Noack (1894: 406f, fig.17). Lauffer (Kopais I: 262, sketch 254) noticed traces of an ancient wall [AE1350], probably marking an old path, by the S corner of the fortification. He suggests that the fortification probably belonged to the territory of Akraiphia, and I would agree, also taking into account that the border was most probably at Phtelio, as the *oros* (boundary stone) would lead us to believe.

---

[10] Several are the reports on the most recent excavations on the cemetery area: AD 1996: 272-275 (amongst the graves were found shallow circular wells, probably associated with the drainage of the rain water [AE1821]) and 275-278; AD 1995: 301-304; AD 1994: 278-281; AD 1993: 176-80.

[11] Probably some of them are workshops (loomweights and bone needles were found) (Aravantinos 2004).

[12] In the same area ritual fire places (*pires*) –A to C period– full of figurines (continue cultic function) and buildings (not yet studied) from C period (probably cult place with associated workshops).

[13] The excavator hypothesizes the existence of a Neolithic settlement nearby, pointing out that up to then Neolithic graves in Boeotia had been found only at the Magoula Balomenou site (*component CH_21*). Sampson (2000) notes the similarity of the two pots with the ones of the Final Neolithic periods found fragmented in the cave *Spilia tou Sarakinou* (*component AK_25*).

15/16/17. Mytikas: *Component AK_22* (15); Yliki NE - Vathy Spithari: *Component AK_23* (16); Yliki NE: *Component AK_24* (17)

ID	Period	Type	Toponym	Discovery	Position
AK_22	gr-rom	? Fort	Mytikas	Ext. topographical survey	approximate
AK_23	gr-rom	fort	Yliki NE - Vathy Spithari	Ext. topographical survey	approximate
AK_24	gr-rom	fort	Yliki NE	Ext. topographical survey	approximate

Three gr-rom (probably 4thC) forts are shown by Fossey 1988 in fig. 35, though without mention in the text (probably from Guillon's map in Guillon 1943, vol. II, pl. V). One is on the Mytikas promontory, jutting into the Copais at the N side of the entrance to the Karditsa bay (in correspondence to and above the N end of the Akraiphia ancient dam).[14] The other two are in the area between Akraiphia and the NE bay of Yliki lake (one on top of Vathy Spithari – *component AK_23*, the other one on a hilltop between Pontiki and Kori mountain – *component AK_24*).

18. Spilia tou Sarakinou: *Components AK_25 to AK_27*

ID	Period	Type	Toponym	Discovery	Position
AK_25	N	settlement	Spilia tou Sarakinou	Personal or group interest	certain
AK_26	EH	activity focus	Spilia tou Sarakinou	Personal or group interest	certain
AK_27	MH	settlement	Spilia tou Sarakinou	Personal or group interest	certain

Spilia tou Sarakinou is a large cave site on the SW side of Pholies hill on the Mytikas ridge, about 500m W of the point where the side road to Karditsa turns off the New National Road (Fossey 1988: 273), about 100m above the level of Copais basin, along the 200m contour line, in a quite high up location, marked on the 1:50,000 GYS map. From the cave site one can enjoy a beautiful view of the Copais basin.
Excavation at the cave started with Th.Spyropoulos (AAA 1973: 201ff), and continued, after illegal digging activities at the site, with more systematic excavations carried out by A. Sampson (Speleological Service) since 1994, with modern methodologies and techniques.
The occupation at the site goes back to the Paleolithic period. The site is also important in the archaeological panorama of Boeotian prehistory, as it presents Neolithic pottery *in stratu*, unlike the majority of sites in Boeotia (with the exclusion of Eutresis, and few other cases). The cave of Sarakinos presents a quite wide opening (that allows light to come in) facing W. The inside is deep and flattish, quite high and with almost no stalactitic ornaments. Pottery, from the Neolithic and Bronze Ages (EH, MH), was also found deeply inside, while in tiny spaces below fallen rocks in the form of Neolithic idols have also been found (Sampson 2000). The cave was abandoned during MH (probably after the drainage of the Copais, as suggested by Sampson[15]). No traces of occupation in the historical period have been noticed, unlike other Boeotian caves (see Agia Triada cave, for instance – appendix I.1).

Neolithic period (Late N I and II, and deeper Middle N): long use (various heart places and several floors), Neolithic figurines. Bone tools and jewels. Arrow heads made of obsidian and *pirolithos* show hunting activities, that are also testified by wild goat's teeth and deer bones. A tool made of deer bone was also found.
EH period (EH II): nice Boeotian ware of this period (especially small *phiales* were found), not very long occupied.
MH period: the presence of grinding stones and spindle whores shows habitation activities in the cave. Bones of goats and sheep, but also shells and fishes of the lake were found.

19. Megalovouno S slope: *Component AK_19*

ID	Period	Type	Toponym	Discovery	Position
AK_19	R	burial place	Megalovouno S slope	accidental	certain

On the S slope of the Megalovouno ridge R burial areas have been noticed and are reported in Fossey (1988: 268 and no. 7 in fig.35). They are probably not related to the city site, since they are quite further away (more than 1.5km). They could be related to villa sites or hamlets.

20. Megalovouno E slope: *Component AK_20*

ID	Period	Type	Toponym	Discovery	Position
AK_20	gr-rom	fort	Megalovouno E slope	Travellers	certain

A fortification site was noticed by Noack (1894: 452ff), and then examined by Lauffer (Kopais I: 264ff , sketch 258) on the E slope of Megalovouno, probably on the Anemomilos height (or nearby), to the SE side of the hilltop (above the road to Kokkino). C sherds lie in the area. As Lauffer (Kopais I: 267) suggests, the fortification probably belonged to Thebes or to the Akraiphia *polis* soon absorbed by Thebes. Lauffer (Kopais I: 267) lists what one would be able to see from the fort: until the city of Thebes, the W part of Yliki lake, towards Larymna and the sea. The Megalovouno fort would guard especially the passage towards the NE bay of Copais.

PTOION Sanctuary
Two topographical locations are mainly linked to the Ptoion sanctuary of Apollo: Perdikovrisi and Kastraki.
The French School at Athens first began excavations in 1884 at Perdikovrisi and in 1903 at Kastraki.

21. Ptoion Kastraki: *Components AK_29 to AK_31*

ID	Period	Type	Toponym	Discovery	Position
AK_29	A	cult place	Ptoion Kastraki	Personal or group interest	approximate
AK_30	C	cult place	Ptoion Kastraki	Personal or group interest	approximate
AK_31	H	cult place	Ptoion Kastraki	Personal or group interest	approximate

This is the site of the cult place dedicated to the hero Ptoios. It is located on terraces on the N slopes of the Kastraki hill, on the S side of the valley which connects ancient Akraiphia with the sanctuary at Perdikovrisi (Fossey 1988: 273).
Excavations at the site were carried out by Guillon (1943) and then by Llinas (BCH 1964: 856-9; 1965: 914; 1966: 941-3). See also Papachatzis (1981:152-5) for useful plates and reproductions of plans (for this site as well as for the Ptoion Perdikovrisi site). As summed up by Fossey (1988: 273), the history of this religious site appears to begin early in the 6thC BC and run through to the early Hellenistic period.

[14] Strangely Lauffer does not seem to mention it.
[15] Sampson BCH 2000: 869-870 and Sampson 2000. He seems to accept the theory that drainage work in the Copais started in MH period.

22. Ptoion Perdikovrysi: *Components AK_32 to AK_37*

ID	Period	Type	Toponym	Discovery	Position
AK_32	G	cult place	Ptoion Perdikovrysi	Personal or group interest	approximate
AK_33	A	cult place	Ptoion Perdikovrysi	Personal or group interest	approximate
AK_34	C	cult place	Ptoion Perdikovrysi	Personal or group interest	approximate
AK_35	H	cult place	Ptoion Perdikovrysi	Personal or group interest	approximate
AK_36	R	cult place	Ptoion Perdikovrysi	Personal or group interest	approximate
AK_37	LR	? cult place	Ptoion Perdikovrysi	Personal or group interest	approximate

The main Apollo sanctuary lies on the SE slope of Ptoion ridge, by the Perdikovrysi spring, an hour's walk above Akraiphnio. It lies in a favourable dominant position at the cross-point of roads linking upland plateaus (Lauffer Kopais I: sketch 264 showing the cross road point) and probably on the upland pass linking E Boeotia (Thebes) with W Boeotia (Copais area) – see fig.1. In the vicinity is a perennial spring that comes out at two points, SE and W of the sanctuary site. The sanctuary is built up on terraces that continue until the chapel of Ag. Paraskevi nearby (Lauffer Kopais I: 268).

Excavations and research were carried out at the site since Orlandos (AD 1915: 94ff). His work was followed by a long tradition of French researches on the site (as well as at the Ptoion Kastraki site – see above), since 1923 (BCH 1923: 521), and even earlier with M. Holloeaux who in 1885 originally begun the work on the site.

An archaeological map was sketched by Guillon in 1943 (see '*Le trepied du Ptoon*'), before the Ducat book ('*Les Kouroi du Ptoion*') appeared in 1971, giving a long list of the various excavation reports and a good summary of the site history. The French School is still working at the site and reports come out every year, especially in the BCH journal.

The history of the sanctuary begins in the Late Geometric period and runs well into the Imperial period, if not later, since the cult centre seems to have moved to the Christian monastery of Ag. Paraskevi nearby (Ducat 1971). Cult activities at the Perdikovrysi site seem to start a bit earlier than that at the Ptoion hero shrine at Kastraki, where excavated layers dated back to the 6thC BC (A period).

------

23. Ptoion Perdikovrysi - W slope: *Components AK_38 to AK_40*

ID	Period	Type	Toponym	Discovery	Position
AK_38	N	? activity focus	Ptoion Perdikovrysi - W slope	Int. topographical surveys	approximate
AK_39	EH	settlement	Ptoion Perdikovrysi - W slope	Int. topographical surveys	approximate
AK_40	MH	? human presence	Ptoion Perdikovrysi - W slope	Int. topographical surveys	approximate

To the W of the sanctuary, on the terraced slope early research found wall and foundation remains, close to the W spring. Close to them obsidian flints and EH pithos fragments were found on the surface (probably noticed by Lauffer during a personal visit), as well as only one Neolithic potsherd (which could be put in connection with the obsidian flints found at the site – for which no date nor detailed study is available - Lauffer Kopais I: 268 and RE 1959: 1520-21). In RE 1959: 1520-21, Lauffer mentioned MH sherds from the area, while in Kopais I: 268-9 he did not mention MH.

According to the information, the place was therefore in use also in prehistoric times. According to Lauffer, the evidence can be interpreted as referring to a quite large primeval mountain settlement. The site here is close to a perennial fresh water spring and the area is characterised by fertile terraces. In considering the upland position, the evidence of Neolithic and especially EH occupation could be linked to the occupation of upland areas in the transition between Final Neolithic and EH, due to the increase and specialisation of pastoral activities (*Secundary Product Revolution* – as elsewhere in Boeotia, see chapter II.3.1).[16]

24/25. Perdykovrisi valley – plateau: *Components AK_41 to AK_43* (24) and Perdykovrisi valley: *Component AK_44* (25)

ID	Period	Type	Toponym	Discovery	Position
AK_41	C	fort	Perdykovrisi valley - plateau	Travellers	approximate
AK_42	gr-rom	fort	Perdykovrisi valley - plateau	Travellers	approximate
AK_43	Preh	human presence	Perdykovrisi valley - plateau	Personal or group interest	approximate
AK_44	LH	activity focus	Perdykovrisi valley	Personal or group interest	imprecise

On the flattish plateau above the Perdikovrysi stream, facing to the S the Ptoion sanctuary site, a fort was first noticed by Noack (1894: 454f).[17] The place probably overlooked the pass between E (Thebes) and W Boeotia (Copais area) – see fig.1, as noticed by Lauffer (Kopais I: 274).

Component AK_41 and AK_42: At the site by the fortification a large amount of C sherds [AE548] and later material [AE1359] can be noticed. Pot-findings help in the understanding of the use of the site in historical times as a gr-rom fort (probably 4thC BC – *component AK_41*) connected to the Theban hegemony. The fort was also in use in later times – gr-rom *component AK_42* (Lauffer Kopais I: 272f). Lauffer describes the view from the fort location, which I could not map though with precision. He suggests (Lauffer Kopais I: 285) that it may represent an 'acropolis' rather than a watchtower like the other four forts known in the area (*components AK_21 to AK_24*), since it lies in a quite low location and apparently not in a strategic position. Fossey does not map it in his map with forts (1988: fig. 35).

Component AK_43: De la Coste and Seyrig (BCH 1923: 521) found isolated obsidian flints [AE547] at the site by the fortification, that lead them (and later on Lauffer Kopais I: 285) to identify the site as a prehistoric fort reused in historical periods. Lauffer would suggest that the fort would belong to the prehistoric settlement site found to the W of the Apollo sanctuary (see above), as already suggested by de la Coste and Seyrig (BCH 1923: 521). Noack (1894) noticed a wall going down the slopes towards the location of the prehistoric settlement (100m ca below). Fossey (1988: 272) points out how several attempts have been made to associate the prehistoric

[16] Fossey (1988: 273), on the contrary, is tempted to associate the Neolithic an Early-Middle Helladic settlement at Perdikovrisi with the prehistoric settlement on the NW bay of Lake Yliki at Sengaina; the distance, he says, between them is too short for us to regard them easily as separate settlements - similar to the obviously small prehistoric settlement at the W end of Klimataria facing Sengaina across a small stretch of Yliki only – appendix I.12, *components T_113 and T_114*.

[17] See Lauffer 1940: 187 and Kopais I: 269ff.

material found in the area with various elements of the fortifications on the surrounding mountains (Lauffer RE 1959: 1520-21; Hope Simpson 1965: 121 no.415; Syriopoulos 1968: 118 YE63 refers briefly to this 'Mycenaean' citadel but contributes nothing new), though in the majority of them the datable material which may normally be observed consists of tile pieces of the Classical period. This site could be an exception, as Lauffer in other cases interprets the fort correctly as historical, while in this case he believes in the possibility that the fort could also be prehistoric (Lauffer Kopais I: 272), reused later in historical times. Lauffer points out, to support his thesis, that he noticed a fixing in a wall, witness of a later (historical) use, which could also testify additional work on the fort in a later period. This would not lead us to suppose automatically that the fort would have been built in a prehistoric period, but simply that it was built in a previous period. Furthermore, at the prehistoric site to the W of the sanctuary, Lauffer does not seem to list LH material, but MH at the most. On the other hand, Fossey (1988: 272) reports the information that Mycenaean sherds [AE550] come from this area, probably referring to Hope Simpson 1965 (no.415), but the connection with this fort is unclear (see *component AK_44*). The *component AK_43*, recorded as 'Prehistoric human presence', represents simply the obsidian flints found at the site.

Component AK_44: In the area of Perdikovrysi stream, to the S of the Ptoion sanctuary site, LH sherds have been reported (Hope Simpson 1965: no.415) as coming from one of the forts on Mt.Ptoion. It is probably the semicircular wall on the Ptoion highest peak described by Kenny 1935: 203ff, a little to the S of the 'round tower of polygonal masonry' probably Hellenistic (Philippson-Kirsten 1951 no.176). The evidence AE550 (LH sherds) could be also associated to the possible prehistoric fort (see above).

Two other forts were noticed in the area, one on each side of the valley above the temple of Apollo Ptoios (Kenny 1935: 203ff) – *components AK_45 and AK_47.* The forts in the area would guard the approaches from Anthedon (via Paralimni lake), from Thebes and from Larymna (see fig.1).

26. Ptoion Pelagia: *Component AK_45*

ID	Period	Type	Toponym	Discovery	Position
AK_45	gr-rom	fort	Ptoion Pelagia	Travellers	certain

On a rocky point to the S of the Ag. Pelagia monastery, to the E of the Apollo Ptoios sanctuary, Noack (1894: 454) noticed traces of a fortification [AE1363 and AE1364 - Lauffer Kopais I: 283, sketch 276], with surface tile fragments everywhere on the hilltop [AE1365 - Lauffer Kopais I: 284]. The walling is similar to the one of the Megalovouno fort (*component AK_20*). This fortification, as well as others in the area listed in this catalogue, is to be dated to the gr-rom period, probably to the 4thC BC. It overlooked the passage towards the S and the Yliki lake and the Theban plain, as the Megalovouno fort guarded the passage towards the NE bay of Copais. The view from the fort is described by Lauffer (Kopais I: 286). From the location one could see until Haliartos.

27. Ptoion Pelagia Monastery: *Component AK_46*

ID	Period	Type	Toponym	Discovery	Position
AK_46	unknown	activity focus	Ptoion Pelagia Monastery	Int. topographical surveys	certain

To the W of the monastery and to the NW of a water source nearby, surface walls were noticed by Lauffer (Kopais I: 282-3, sketch 275).[18]

28. Tsekoureli: *Component AK_47*

ID	Period	Type	Toponym	Discovery	Position
AK_47	gr-rom	fort	Tsekoureli	Travellers	certain

Tsekoureli is the steep sided hill to the N of the monastery of Ag. Pelagia, to the E of the Apollo Ptoios sanctuary. The Northern slope of the Tsekoureli leads into the NE Copais bay. A round shaped tower (watch tower) was noticed in the N part of its flat top (200m NS and 50m EW) by Noack (1894: 452f), overlooking the NE Copais bay (Lauffer Kopais I: sketch 271).[19] The view shed from the fort is described by Lauffer (Kopais I: 281): the three peninsulas by the Spitia Katavothra (Chantza, Agios Ioannis), Topolia/Kastro, Gla, Tourlogianni... Lauffer notes how the polygonal walling technique here seems older than the one employed in the round shape tower on Megalovouno (*component AK_20*), while it looks more recent than that employed at Perdikovrysi (*component AK_42*). The fort could therefore be dated to the gr-rom period, and should probably again be linked with the Theban hegemony (4thC BC).

29. Ag. Nikolaos: *Components AK_61 to AK_65*

ID	Period	Type	Toponym	Discovery	Position
AK_61	gr-rom	? activity focus	Ag.Nikolaos	Travellers	approximate
AK_62	MH	activity focus	Ag.Nikolaos	Personal or group interest	approximate
AK_63	C	? burial place	Ag.Nikolaos	Ext. topographical surveys	approximate
AK_64	EH	activity focus	Ag.Nikolaos	Personal or group interest	approximate
AK_65	MH	activity focus	Ag.Nikolaos	Personal or group interest	approximate

At the very edge of the Karditsa bay, when it juts into the larger basin, along the rocky line by the *katavothra* named Ag. Nikolaos, is a destroyed chapel of Ag. Nikolaos. Knauss (Kopais 2: 267ff and fig. 8.1) reports on two column bases, few cut blocks and two inscribed funerary stelai, already noticed by Lolling during his visit in the summer 1876, as well as terrace walls and sherds lying in the fields below. Among the sherds are byzantine glazed wares and few gray and yellow Minyan sherds. Lolling said that the chapel, still standing at his time, was built on ancient foundations. In the vineyards by the terracing walls have been noticed C sherds that the discovers linked to the presence of a cemetery, as well as the funerary stelai already noticed by Lolling (Knauss Kopais 2: 268). Three components have been created: a only possible activity focus in the gr-rom period, since the architectural fragments probably refer to a later (Byzantine?) use of the place, as the presence of the destroyed chapel would attest, in association with the surface sherds (*component AK_61*); a MH activity focus, testified by the presence of Minyan wares (*component AK_62*); a C? burial area revealed by the association of the C sherds with funerary stelai with names inscribed (*component AK_63*).

[18] I visited the place, which is full of limestone rocks (from bedrock). It is difficult to see whether some of them are walls as well, and whether they are ancient.

[19] Lauffer (Kopais I: 281) notes how from nowhere else the NE bay of Copais is so visible as from Tsekoureli hilltop.

Component AK_64 and AK_65: To the S, at the foothills of Phoinikon mountain, in a valley to the S of the ruined chapel of Ag. Nikolaos, between the chapel and the *katavothra* known as Palaiomylos, building foundations were found (one of them with an absidal shape), along with EH and MH pottery (Knauss Kopais 2: 269 and fig.8.1).

*Sites on the N edge of Yliki lake, though probably belonging to the territory of ancient Akraiphia, are listed and discussed under the *chora* of Thebes, where the Yliki lake and the surrounding settlement are discussed as a topographical unit.

Appendix I.8
The Copais area: Haliartia

THE ARCHAEOLOGICAL RECORD

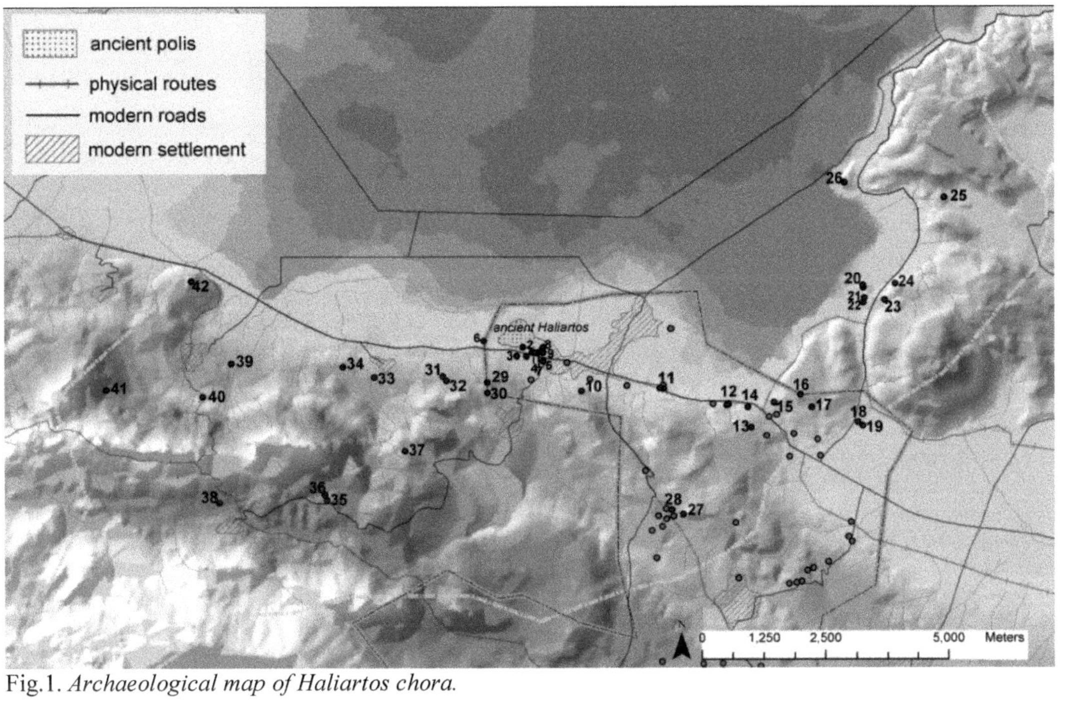

Fig.1. *Archaeological map of Haliartos chora.*

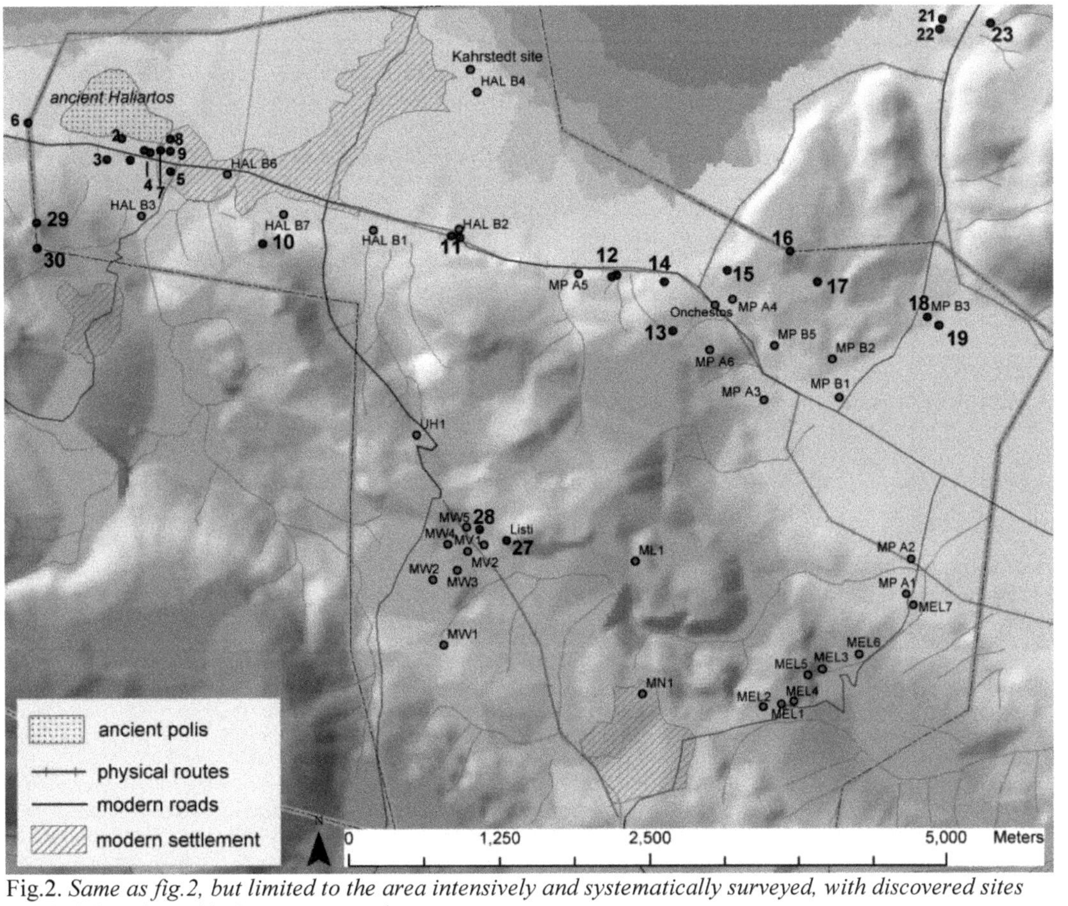

Fig.2. *Same as fig.2, but limited to the area intensively and systematically surveyed, with discovered sites marked (listed in table 1 - SURVEY SITES) .*

1. HALIARTOS: *Components HA_1 to HA_9*

ID	Period	Type	Toponym	Discovery	Position
HA_1	N	activity focus	Haliartos Kastron	Personal or group interest	approximate
HA_2	EH	activity focus	Haliartos Kastron	Personal or group interest	approximate
HA_3	MH	settlement	Haliartos Kastron	Personal or group interest	approximate
HA_4	LH	settlement	Haliartos Kastron	Ext. topographical surveys	approximate
HA_5	G	settlement	Haliartos Kastron	Ext. topographical surveys	approximate
HA_6	A	settlement	Haliartos Kastron	Travellers	certain
HA_7	C	settlement	Haliartos Kastron	Travellers	certain
HA_8	H	settlement	Haliartos Kastron	Travellers	certain
HA_9	R	settlement	Haliartos Kastron	Ext. topographical surveys	certain

The ancient *polis* of Haliartos lies on an abrupt limestone ridge which forms a terrace above the Copais basin, very close to the lake edge (currently the channelled Kephisos). From it the spring Kisoussa springs out (Philippson 1951: 474).[1] The city controlled the pass between the acropolis spur and the lake that even in summer periods, as Philippson (1951: 474) remarks, was probably occupied by marshes. The area is in fact below the 120m contour line, which would correspond to the marshy lane, according to the lake fluctuation digital model – see chapter II.3.1.
The terrace-promontory, on which the acropolis of the ancient *polis* used to be, lies to the right side of the road Thebes-Levadeia, immediately after the modern village.[2] It is well physically fortified at its N and E sides overlooking the lake. The ancient city used to lie on the natural slopes on the S (until the modern road) and SE, marked by a large quantity of visible surface wall alignment (similar to the situation of ancient Tanagra, as they are both buried ancient cities and protected sites). The lower town is surrounded by a circuit wall (for fortification sketch see Fossey 1988: 303, fig. 41). A smaller circuit also marks the small proper acropolis on top of the promontory, naturally fortified, as said earlier, at its N side.

The historical site was known already to travellers (Frazer 1913: 164-6, which summarizes previous accounts, especially Leake), while Fimmen (1912) had noticed the presence of pre-Hellenic material at Haliartos.
The city was excavated by Austin at the beginning of the 20th century.[3] Another small excavation at the site was carried out by Andreiomenou (AE 1976: 17-20). Tomlinson reports on the prospection carried out in 1986 on the Haliartos city site (BCH 1987: 541-542). Afterwards, work on the city site has been carried out by Bintliff and Snodgrass in the framework of the Boeotia Survey project (Bintliff-Snodgrass 1985a and 1985b; Bintliff 1994a).

A brief summary of the site history can be found in Austin (BSA 1927: 128-140) and later in Roesch 1976: 374-5. Generally speaking, we only have scattered references for Haliartos in official history, as it probably never played a leading part in Boeotian affairs (Austin BSA 1927: 129).[4]
Haliartos' earliest defence system, and the best surviving one, is the great 'cyclopean wall' around the acropolis. Typically Mycenaean, it would argue a major settlement at the site, together with the pot sherds [AE662] found on the acropolis (in the area of a later Archaic sanctuary, where MH and LH layers were excavated [AE658 and AE1443] – Austin BSA 1927: 129. As for prehistory, N and EH is known from the site, from the acropolis. Apart from this area very few prehistoric sherds are known, such as a few isolated MH from various parts of the site. Although G seemed to be poorly represented at Haliartos with only a few sherds of uncertain date and origin in the sanctuary and sanctuary area, intensive survey at the city confirmed the occupation from Proto Geometric to Late Geometric times on the acropolis of the hill, together with some very small rural sites in its immediate vicinity (survey sites HAL B4, HAL B5, HAL B6, HAL B7) – Bintliff 1994b: 213. Archaic material has also been found, especially in the sanctuary area on the acropolis.[5] In the C period, Haliartos was one of the *poleis* taking part in the Boeotian confederation (Kirsten 1956: 670). In the Boeotian league were (1) Levadeia, (2) Koroneia, (3) Haliartos (with the same rights in the same position); then the smaller places, as (4) Alalkomenai (small settlement) and (5) Okalia (with Tilphossion) and also Medeion/Phoinikis dependent from Haliartos, and also (6) Onchestos, probably also independent.[6]
Abundance of 4thC BC material at Haliartos shows the city flourishing at this time (Austin 1927: 210). To the H period dates the majority of graves discovered by the Haliartos city site. While the destruction of Thebes by Alexander in 335 deprived the confederate cities of political importance, it did not imperil their existence.
As is widely known, Haliartos was destroyed in 171 BC by the Roman praetor C. Lucretius, because the Haliartians had joined the cause of Perseus against the Romans.[7] In the R period, the evidence known is mainly to be associated with a rural occupation of the landscape, maybe hamlet-like (*components HA_25, HA_81, HA_82*), and of this kind was probably the Haliartos reoccupation after its destruction.[8]

[1] A small rivulet, which gushes forth from the N side of the rock, runs into the swamps of the Copais (Schliemann 1881: 127).
[2] Today the modern village has been renamed Haliartos. In the past it was called Gribas and was joined with the older village of Moulki (Philippson 1951: 474).
[3] Austin BSA 1925/6: 81-91; 1926/7: 128-140 (summary in JHS 1926: 234-5) and BSA 1931-2: 180-212 (summary in JHS 1931: 189-190).

[4] The first mention is in Homer –Catalogue of Ships *Il.* II 511ff- where Haliartos comes between Koroneia and Plataea, as 'grassy' Haliartos. Nonnus (*D.* XIII.53) calls it 'watery/moist Haliartos'.
[5] By the Archaic period (Austin 1927: 206) the late prehistoric Copais drainage was in disuse and Copais again impassable. In winter it was much more a lake, in summer rather a marsh.
[6] At the beginning of the 4thC BC is reported the attack of Lysander the Spartan on Haliartos and his defeat and death before its walls. In the Plutarch version (*Lysander* 28) some Thebans attacked via the gates, others came round by the spring of Kissousa, keeping the city on their left, and attacked the outermost portion of the enemy forces. As Austin (1927: 207-9) points out, the latter account would prove that there was a path between the edge of the lake and the cliffs that form the N extremities of the acropolis (see fluctuation model of the lake in fig.1 and described in chapter II.3.1 – PERILACUSTRINE MARSHY LANDSCAPE).
[7] In the first half of the 2nd C BC Rome was in conflict with Macedonia. In 171 Boeotians had to decide whose part they would take. Livy (XIV 46) tells of Macedonians winning back Boeotian cities - Thisbe, Koroneia and Haliartos. The latter two were won over and asked for garrisons. Haliartos people and people of Koroneia defied Romans and were besieged and were sacked awfully (Austin 1927: 211).
[8] Two years after the sack Athens asked for Haliartos land, that was given - several boundary stones were found. No new city was built, but it seems that a community of settlers tilled soil for benefit of Athenians (Austin 1927: 211-212). Also a hunting club is reported, which suits the Copais area (see chapter II.3.8 - RESOURCES).

2 to 6. Haliartos Burial Places*: Components HA_10, HA_11, HA_12, HA_13* (2)*; HA_23* (3)*, HA_24 and HA_80* (4)*; HA_26* (5)*; HA_27* (6)

ID	Period	Type	Toponym	Discovery	Position
HA_10	C	burial place	Haliartos Kastron	Rescue excavation	certain
HA_11	H	burial place	Haliartos Kastron	Rescue excavation	certain
HA_12	C	burial place	Haliartos Kastron	accidental	imprecise
HA_13	H	burial place	Haliartos Kastron - South	Personal or group interest	certain
HA_23	gr-rom	burial place	Haliartos Kastron - South	accidental	certain
HA_24	H	burial place	Haliartos Kastron - SE	Personal or group interest	certain
HA_26	H	burial place	Haliartos Kastron - SE road	accidental	certain
HA_27	gr-rom	burial place	Haliartos Kastron - West	Travellers	approximate
HA_80	C	burial place	Haliartos - East	accidental	certain

Ancient burial areas have been found (mainly accidentally or by means of rescue excavations) in the vicinity of the ancient *polis*, in the majority of cases to its S, along the old national road (Thebes-Lebadeia), which probably marks the path of the ancient road.[9] The graves have been found usually clustered into funeral *periboloi* (as in the case of other *poleis* excavated cemetery sites, for instance Tanagra's ones – see appendix I.14), some of which monumental, with burials dated until the H period,[10] not later than 171 BC, when the town was destroyed by the Roman praetor C. Lucretius.

7. Haliartos Kastron – SE: *Component HA_25*

ID	Period	Type	Toponym	Discovery	Position
HA_25	gr-rom	activity focus	Haliartos Kastron - SE	Personal or group interest	certain

Immediately to the N of the road Thebes-Levadeia, and SE of the ancient site, in 1966 building activities (construction of a store-room) lead to a rescue excavation. Walls belonging to a building with three phases of occupation (R?) were found (at the same time with graves – *component HA_24*). They could be interpreted as a R farm living after the destruction of Haliartos city in 171 BC (AE 1967: 20-8; AD 1967: 242).

8/9. Haliartos Kastron – East: *Components HA_81* (8) *and HA_82* (9)

ID	Period	Type	Toponym	Discovery	Position
HA_81	R	activity focus	Haliartos Kastron - East	Rescue excavation	approximate
HA_82	R	activity focus	Haliartos Kastron - East	Rescue excavation	approximate

[9] Fossey (1988: 305) reports on a very early 'excavation' of a grave near Haliartos: the Spartans occupying the Theban Kadmeia, 383-379 BC, excavated the 'Tomb of Alkmena' according to Plutarch (*de genio Socratis* 5-7; cf. Austin's discussion, BSA 1931/2: 209f). The tomb was found to contain a bronze bracelet, two earthenware jars, and a bronze tablet inscribed with 'Egyptian' script.

[10] To the H period dates the majority of graves discovered by the Haliartos city site.

Component HA_81: At the E edge of the ancient site parts of four rooms associated with R material were excavated (AD 1980: 213-5).

Component HA_82: To the SE of the ancient site, on the N side of the Thebes-Levadeia road, remains were found of three building phases: 5thC BC [AE1729], H [AE1730] and R (*component HA_82*). The construction of the R phase includes the use of *spolia* and part of a hypocaustal room (AD 1980: 213-5). The first and the second phase have been included in the *polis* components. The third phase confirms the reoccupation of the Haliartos city area after the destruction of the *polis* in 171 BC, probably with a small size focus rural in character (see *component HA_25*).

10. Mazi Pyrgakos: *Components HA_14 and HA_15*

ID	Period	Type	Toponym	Discovery	Position
HA_14	Preh	human presence	Mazi Pyrgakos	Int. topographical surveys	certain
HA_15	C	activity focus	Mazi Pyrgakos	Int. topographical surveys	certain

To the S of modern Haliartos and to the NE of Mazi village, on a hill with a flat top 20x50m large, the remains of a medieval round shaped tower are visible.

Component HA_14: On the E slope of the hill scattered obsidian flints were noticed by Lauffer (Kopais I: 48, sketch 49). Lauffer (Kopais I: 48) suggests an interpretation of the evidence as an early prehistoric site, maybe Neolithic. A few obsidian flints cannot be an evidence representing a settlement site, as Lauffer seems to suggest. An intensive and systematic revisit of the site would clarify the character of the evidence (and the presence of eventually datable pottery), which for the time has been left as human presence generally dated to the prehistoric period.

Component HA_15: On the hilltop itself and on the slopes C sherds and tile fragments, as well as Byzantine sherds, are to be noticed on the surface (Lauffer Kopais I: 48, sketch 49). The place could have been occupied by a farmstead in gr-rom (C only?) times.

11. Pyrgos Haliartos – cave: *Component HA_16* and Pyrgos Haliartos*: Component HA_17*

ID	Period	Type	Toponym	Discovery	Position
HA_16	N	activity focus	Pyrgos Haliartos - cave	Ext. topographical surveys	certain
HA_17	gr-rom	activity focus	Pyrgos Haliartos	Int. topographical surveys	certain

A 15m high rocky eminence can be easily distinguished above the road of Thebes-Levadeia to its left, before approaching the modern village of Haliartos, because of the presence of a quite well preserved Frankish tower on its top, after which the place is named. Underneath are some caves, 1km ca to the W of Seidi cave site. Among them, a cave large and high, used until recently as a sheep-fold and now by a new café by the road, marks the site of a prehistoric (Neolithic) activity focus. A seasonal water spring is nearby, 20m to the E (see Lauffer Kopais I: 42, sketch 40). Paleolithic material was also found at the site. An intensive artefact surface survey carried out at the place (Boeotia Survey project) showed the presence of a EH activity focus, and a G-A-C cemetery or shrine (see below, table 1 – *Survey sites*: HAL B2).

Component HA_16: Obsidian flints, and probably also sherds (Philippson 1951[11]), testify the occupation of the site in a prehistoric period (Neolithic). It was probably a cave site with fresh water nearby, and the marshes and the lake very close. Intensive artefact surface survey carried out at the place showed the presence of an EH activity focus (see below, table 1 – *Survey sites*: HAL B2). Fimmen (1912: 522) had noticed the site already and mentioned it as the more ancient site in Boeotia.[12] Fossey (1988: 319) notes how early excavations at the site have never been fully published, though it seems that material found was predominantly pre-Neolithic, as at the Seidi cave (see below no.12). Lauffer (Kopais I: 42-3, sketch 41) describes the section of an excavation trench (November 1937) in front of the cave, to the E slopes of the Pyrgos hill. In the upper layer (*humus*) C material was found (no Neolithic), but it probably would come from above (see results of intensive surface survey in table 1 – *Survey sites*: HAL B2). Underneath it is a 50cm thick layer of pebbles, and then the layer with Neolithic and earlier material. The site is therefore a quite important site in the Neolithic period of Copais (until Late N), together with Orchomenos and the Magoula Pyrgos in the NW bay. The abandonment after the Neolithic period can be noticed also at the Megali Katavothra site and at Magoula Balomenou, as Lauffer notices (Kopais I: 45-6) pointing out that in the three cases the abandonment would have been probably caused by the weak defence from water and marshes that such locations could offer.

Component HA_17: In historical periods, Lauffer suggests that probably a small hamlet or a fort was to be found at the site, on the hilltop. The top of the rocky hill is flat, extending 100m towards the S and 20m wide. On it can be noticed a concentration of artefacts, probably gr-rom (Lauffer Kopais I: 42). On the other hand, Lauffer notes, this material might also have been Frankish only, associated to the settlement around the tower. C material found in the superficial level of the trench excavated below, in front of the cave, according to Lauffer (Kopais I: 42-3) had probably come from the hilltop above. An intensive artefact surface survey carried out at the place (Boeotia Survey project) showed the presence of a G-A-C cemetery or shrine (see below, table 1 – *Survey sites*: HAL B2), confirming therefore Lauffer's hypothesis and adding to it more details. The reused gr-rom material in the Frankish tower could be in association with the material found at the Pyrgos site, but could also come from either Haliartos or Onchestos nearby.

12. Seidi: *Components HA_18 to HA_22*

ID	Period	Type	Toponym	Discovery	Position
HA_18	N	? activity focus	Seidi	Ext. topographical surveys	certain
HA_19	C	rural site/activity	Seidi	Ext. topographical surveys	certain
HA_20	G	rural site/activity	Seidi	Ext. topographical surveys	certain
HA_21	A	rural site/activity	Seidi	Ext. topographical surveys	certain
HA_22	H	rural site/activity	Seidi	Ext. topographical surveys	certain

500m ca from the Steni pass/Onchestos is a rocky emergence called, as well as the area immediately in front of it, Seidi. In the rock is a cave, where an epipaleolithic hunting *abri* was dug by E. Schmid in the 1950s and 1960s.

Component HA_18: Signs of a Neolithic occupation are reported by Fossey (1988: 319) and have been noticed later by intensive artefact surface surveys carried out at the place (Boeotia Survey project – see below, table 1 – *Survey sites*: HAL B1). Certainly Paleolithic material was also found at the site. Both Seidi and Pyrgos Haliartos cave sites are listed (separately) in Philippson 1951: 475 notes 2 and 3.

Components HA_19 to HA_22: In front of the cave, on the slopes as well as on top of the rock, lie C sherds (glaze), fragments from Megaric cups, as well as tile fragments. Most of the potsherds lie in front of the rocky hill (Lauffer Kopais I: 41 – who believes they might belong to an ancient settlement). According to Lauffer, material is only from C period, and there are no surface sherds from earlier or later periods, while, according to Fossey (1988: 319), who recognises and distinguishes better gr-rom wares, materials go from Geometric to Hellenistic times. The intensive artefact surface survey carried out at the place (Boeotia Survey project) confirmed the presence of a G-A-C-Early Hellenistic site, which can be assigned a rural character (see below, table 1 – *Survey sites*: HAL B1).[13]

13. Steni South: *Components HA_28 to HA_32* (ONCHESTOS sanctuary)

ID	Period	Type	Toponym	Discovery	Position
HA_28	A	cult place	Steni South	Other	certain
HA_29	C	cult place	Steni South	Travellers	certain
HA_30	H	cult place	Steni South	Travellers	certain
HA_31	R	cult place	Steni South	Travellers	certain
HA_32	LR	rural site/activity	Steni South	Rescue excavation	certain

The Onchestos sanctuary site lies at the 91st km to the left side of the road Thebes-Haliartos-Levadeia, on top of the ridge immediately to the S of the road.

Components HA_28 to HA_31: Though architectural fragments and walling from the area [AE688] were known already to travellers (for a summary of earlier accounts see Frazer 1913: 139-140), the Greek local *ephoreia* discovered the site in 1964 (AD 1964: 200-201 and AD 1967: 242) and in 1971 an excavation was carried out (AAA 1971: 379-381 and 1973: 379-381; AD 1973: 269-271).
The sanctuary life starts at the end of the 6thC BC (Spyropoulos AAA 1973: 379-381; Wallace 1979: 134-6). Excavated structures and surface sherds attest the use of the cult place until the R period. The sanctuary was in a strategic position and played an important active and symbolic role in Boeotian

[11] It is not clear whether they noticed N sherds on top or below the Pyrgos hill [AE1390].

[12] Lauffer (Kopais I: 46) notices that we know only few sites where Paleolithic and Early Neolithic are attested, as this one, and we know those mainly by accidental reasons only.

[13] A well carved block built in the church of Ag. Varvara chapel, 100m W of Seidi, is related by Lauffer (Kopais I: 41) to this gr-rom site. In my opinion, it can be easier related to the Onchestos site, and therefore in the database this Archaeological Evidence [AE1313] does not appear in association with any Seidi component.

history (see among others Kirsten RE 1939 s.v. *Onchestos* 412-7; Buck 1979: 10; Schachter 1986 s.v. *Poseidon (Onchestos)*).[14] A common Boeotian festival was celebrated at the Onchestos sanctuary, and its governing amphictiony is thought to have met there at the shrine of Poseidon (Strabo IX 2.33).

Component HA_32: In the LR period the area of the sanctuary was turned into a villa establishment (excavation report by Spyropoulos AAA 1973: 379-381; Fossey 1988: 309).

Surface walls [AE689] to the S of the modern road Thebes-Levadeia, built in polygonal walling (interpreted as a possible circuit wall) may belong to the town of Onchestos rather than to the sanctuary site. Also architectural fragments visible on the surface and already noticed by travellers [AE688; AE1447], could belong to the settlement site rather than to the proper sanctuary site. In any case, the town that developed here, to the S and the N of the Steni passage, was linked to the sanctuary and developed in connection with it.

14. Steni SW: *Component HA_35*

ID	Period	Type	Toponym	Discovery	Position
HA_35	gr-rom	burial place	Steni SW	Rescue excavation	approximate

To the S of the modern road Thebes-Levadeia, on the sanctuary ridge's W slope, some works apparently revealed a cemetery area, but they are unpublished and already Fossey was not able to acquire any further detail (Fossey 1988: 309).

15. Steni North – Tsumbitses West: *Components HA_33 to HA_34 and HA_85*

ID	Period	Type	Toponym	Discovery	Position
HA_33	C	settlement	Steni North – Tsumbitses West	Rescue excavation	certain
HA_34	H	settlement	Steni North – Tsumbitses West	Rescue excavation	certain
HA_85	R	settlement	Steni North – Tsumbitses West	Int. topographical surveys	certain

The remains of the ancient town of Onchestos lie on the SW terracing slopes of the Tsumbitses hill, at the Steni pass between the Copais and Teneric plains. The Tsumbitses flat top is ca 100m long, 50m wide and on it can be seen today the remains of buildings from the Ottoman village settled in this location (Kazarma locality – see below no.16).
Ploughed-up stones [AE687] showing ancient occupation were noticed on the SW slopes by travellers (Ross 1851 vol. I: 25), and gr-rom sherds [AE1373] were noticed by Lauffer on the surface, and dated as C by him (Lauffer Kopais I: 19, from his early visits). A white marble block built in a building of the deserted village has also been noticed (Lauffer Kopais I: 19), as well as the ruins of a large oblong building with its long axis orientated E-W, jointly with tile fragments (indication of a roofed building) [AE695] and a Doric capital of the late 5th or early 4thC BC and two inscriptions from the same area (AD 1964: 200-1, pl.237; BCH 1966: 856-7, fig.7; JHS 1965-6: 13).

Rescue excavation by the local Ephorate at the site (AD 1973-4: 442-443 and later on AD 1991: 333) discovered a complex dated to the 4thC BC, enclosed by porches with shops and interpreted as the ancient *agora* [AE448; AE1450].[15] The area belonged to the town of Onchestos and would be connected with the sanctuary of Poseidon (600m to the SE). According to Christopoulou 1995, it was probably a public open space (*agora*) for the private, civilian and cultural needs of the Boeotian *koinon*. Meetings and religious celebrations took place in the sanctuary, and the *agora* served all these. Christopoulou says that the expenses for the construction of the *agora* buildings were not covered by the town of Onchestos only, but by the confederation (*koinon*).[16]
An earlier phase (C) is also represented in the excavation layers [AE1449]. Christopoulou (1995: 429-445) would suggest that they are earlier constructions with the same function as the later ones. This seems to be just an interpretation though without strong evidence.[17]

The town of Onchestos is mentioned in Pausanias IX 26.5 as well as in Strabo IX 410-412. In the C period it was probably only a habitation area associated to the sanctuary, which later on became a small town site, mentioned in the R texts (Diodorus XVII 10.4) and replacing C-H Haliartos destroyed by the Romans (see above - HALIARTOS).

16. Tsumbitses-Kazarma: *Component HA_36*

ID	Period	Type	Toponym	Discovery	Position
HA_36	gr-rom	? Fort	Tsumbitses-Kazarma	Travellers	certain

On the Tsumbitses hilltop, by Kazarma deserted village, remains of an ancient tower were noticed by travellers: Ross 1851 Vol. I: 25; Vischer 1875: 561; Leake 1835: 215, and reported by Lauffer (Kopais I: 19).

17. Steni North - Tsumbitses East: *Components HA_37 and HA_38*

ID	Period	Type	Toponym	Discovery	Position
HA_37	EH	settlement	Steni North - Tsumbitses East	Int. topographical surveys	approximate
HA_38	MH	settlement	Steni North - Tsumbitses East	Int. topographical surveys	approximate

On the SE slopes of Tsumbitses (at Steni pass), Lauffer (Kopais I: 20) saw traces of prehistoric occupation (EH and MH) as well as obsidian flints [AE1374]. Lauffer could recognize MH material and only one EH sherd with certainty, though this is probably due to the fact that MH potsherds (especially the well

[14] On the sanctuary as a meeting point of Boeotian amphyctiony and a place for equestrian races (*agones*) see ancient sources: Homer *Il.* II 506; Homer *h. Ap.* 230; Pindarus *I.* I 33 with Schol.; Apollonius Rodius III 1242; Pausanias I 39; Stephanus Byzantius s.v. *Onchestos*.

[15] A system of porches with a Greek Π form [AE1450] was brought to light, and interpreted as *agora* surrounded by *stoai* with shops (on both sides of the road, at the 92nd km of the road Thebes-Levadeia, dated to the end of the 4th and middle 3rdC BC. A nice picture of one of the excavated rooms opening from the porch is shown in Christopoulou (1995: 422).

[16] The chronology of the later phase goes also in accordance with the end of the Theban leadership on the *koinon* and the need of a new meeting place for it that had to be chosen in a neutral place. Onchestos' sanctuary of Poseidon, with its central position, would be suitable for the task (Christopoulou 1995).

[17] In the excavated layers of the C period [AE1449] architectural fragments and remains of structures were found, as well as movable finds of two phases and small finds (two copper voting tokens from the C period; C black glazed tiles).

known Minyan and brown glaze ware) are more easily recognizable than the EH ones. The site is reported in Hope Simpson 1965 as Kazarma.
The place corresponds to the Onchestos prehistoric site surveyed during intensive and systematic artefact surface surveys (Boeotia Survey Project) and listed as 'Onchestos' in table 1 -*Survey sites*- below. The site showed not only presence of EH and MH material, but also Neolithic and less strong LH. The EH and MH activity focuses became 'settlement/habitation' in character, as recognized as such during intensive survey.

18/19. Tsumbitses Ag.Panagia: *Components HA_39, HA_40 and HA_76* (18); *HA_41* (19)

ID	Period	Type	Toponym	Discovery	Position
HA_39	MH	activity focus	Tsumbitses Ag.Panagia	Personal or group interest	certain
HA_40	gr-rom	? activity focus	Tsumbitses Ag.Panagia	Int. topographical surveys	certain
HA_41	gr-rom	? burial place	Tsumbitses Ag.Panagia	Int. topographical surveys	certain
HA_76	LH	activity focus	Tsumbitses Ag.Panagia	Personal or group interest	certain

On the SE spur of Tsumbitses/Vounalakia hill, a 50m large small plateau ca 15m above the plain juts S from low hills which form a saddle between Copais and the Theban plain, ca. 1km N of Thebes-Levadeia road (HS&D 1979: G31). On it lies the Ag. Panagia chapel, by the ruins of an older one. As Lauffer (Kopais I: 17) notes, Tsumbitses/Vounalakia is a calcareous hill, apparently with no water.[18] The small plateau overlooks the Teneric plain, which lies 8m under the level of the Copais basin, and easily turns into a swamp (see appendix II and Philippson 1951). Therefore, the saddle by the plateau, in use today by the railway and called Northern Pass (through Kiapha, see Philippson 1951: 17f), was preferred in antiquity to the Southern Pass by Steni. According to Lauffer (Kopais I: 17), the path was running nearby the Ag. Panagia location. The site corresponds to the MP B3 survey site (see below, table 1 – *Survey sites*), where an EH activity focus, as well as G-A traces of activity and a C to LR rural site have been recognized by means of intensive systematic surveys (Boeotia Survey Project).

Components HA_39 and HA_76: Prehistoric gazetteers reports MH sherds from the site (French 1972: fig. 19; Syriopoulos 1968; Hope Simpson and Lazenby 1970: 30-1; HS&D 1979: G31). Fossey (1988: 310) reports only MH material, but in French and HS&D 1979 LH is also listed. EH has been recognized by means of intensive systematic survey (see below, table 1 – *Survey sites*: MP B3).

Components HA_40 and HA_41: The chapel marks the site of a medieval settlement and probably of an older hamlet site. Lauffer (Kopais I: 17) reports ancient sherds on the surface, noticed also later during the intensive artefact surface survey (Boeotia Survey project) that recognized G-A traces of activity and a C to LR rural site (see below, table 1 – *Survey sites*: MP B3).[19] Lauffer also reports remains of tile graves at the foot of the spur (*component HA_41*).

20. Megalo Kastraki: *Components HA_44 to HA_49* and *HA_83*

ID	Period	Type	Toponym	Discovery	Position
HA_44	EH	activity focus	Megalo Kastraki	Int. topographical surveys	precise
HA_45	MH	settlement	Megalo Kastraki	Int. topographical surveys	precise
HA_46	LH	? activity focus	Megalo Kastraki	Int. topographical surveys	precise
HA_47	A	activity focus	Megalo Kastraki	Ext. topographical surveys	certain
HA_48	C	activity focus	Megalo Kastraki	Other	certain
HA_49	H	activity focus	Megalo Kastraki	Other	certain
HA_83	MH	burial place	Megalo Kastraki	Personal or group interest	precise

To the NW of the Sphingion mountain, between the railway and the higher Prophitis Ilias, lies a promontory with two hilltops overlooking the basin and linked to the mountain through a small pass. The higher one is named Megalo Kastraki after the remains lying on its top, while S of it is the lower one, called Mikro Kastraki (Lauffer Kopais I: 21ff).
Megalo Kastraki's hilltop is 10x30m large with a NW terrace which is also quite wide and fortified (Lauffer Kopais I: 23, sketch 12). Several surface house walls and terrace walls (apparently not noticed earlier by any traveller), as well as remains of fortification and surface sherds (from prehistoric to C period), were known (on the hilltop, on the NW terrace and on the S gentle slope[20]) before the systematic excavation carried out by the local *ephoreia* on the hilltop in 1995-97 (AD 1995: 299-301; AD 1996: 270-2; Sarri 2000). Lauffer (AM 1938/9: 177-185, pll. 71-2) gave the first very good and detailed account on the site. A detailed account is also from the same author in Kopais I: 21ff. While prehistoric materials lie especially on the W slopes, the later historical material is to be found on the SE slopes mainly (Lauffer Kopais I: 31; Sarri 2000).
The site is known also as *Medeon*, mentioned in Homeric Ilias, although the identification is just a possibility, as Fossey (1988: 314) and Lauffer (Kopais I: 34) remark. The place is also known as *Davlosis*, after the name of the area below, with a small hamlet.
The place in the MH period was a fortified hilltop settlement, probably with a low settlement area on the S slopes (*component HA_45*). Three MH cist tombs (child graves – *component HA_83*) have been excavated within the settlement site, as it is often the case in MH habitation sites (AD 1995: 299-301). The site was already occupied in the EH period (*component HA_44*), probably as a quite important settlement at that time already, though EH layers were not excavated during the recent excavations by the local *ephoreia* (1995-1997). A wall around the hilltop is visible but uncertain is the date. It seems not to be Mycenaean, though accounts on the site (earlier than the systematic excavations) talk about a fortified hilltop of the MH and LH period as well (Lauffer AM 1938/9; Lauffer Kopais I: 21ff), and LH sherds are reported from the site in various prehistoric gazetteers (HS&D 1979: G14; Hope Simpson 1965). A component for the LH period has been recorded and left as possible activity focus, since we cannot be sure on the location and quantity of the surface sherds reported.

[18] This is what Lauffer states, though two water springs are marked at the Steni pass on the IGME geological map, but they could be not permanent.
[19] Lauffer (Kopais I: 17) reports also of an ancient large block rebuilt in the chapel (which could be from Onchestos nearby).
[20] Respectively the acropolis and the lower settlement according to Lauffer (Kopais I: 31).

As mentioned earlier, material from historical times has been noticed in a lower area, on the SE hill slope. It would probably represent a site with rural character, according to Lauffer nothing more than a small hamlet, since there is no fresh water very close (the first one being at ca 1km to the NE), as he notes.[21] We would suggest though a possible interpretation as a sheep-fold, as there is one even today there, and the hill slopes as well as the surrounding area are suitable for pasturage. A systematic collection of material at the site would probably clarify its function.

21. Kastraki area: *Component HA_50*

ID	Period	Type	Toponym	Discovery	Position
HA_50	unknown	activity focus	Kastraki area	Int. topographical surveys	approximate

Between Megalo and Mikro Kastraki, in a lower area, traces of a single building foundation were noticed on the surface (Lauffer Kopais I: 22, who does not mention any associated material nor chronology). Associated material might be R in date (personal visit and pers.comm. from the excavator at the site for the local *ephoreia*, K. Sarri).

22. Mikro Kastraki: *Component HA_51*

ID	Period	Type	Toponym	Discovery	Position
HA_51	Preh	? activity focus	Mikro Kastraki	Int. topographical surveys	certain

Mikro Kastraki is a lower hilltop, to the NW of the Sphiggion mountain (see above -Megalo Kastraki- for location details). Lauffer (AM 1938/9: 177-185, pll. 71-2 and Kopais I: 22) reports a small fortified site (a small prehistoric fort?) which could be part of the main fortification on Megalo Kastraki, on the basis of house foundations on the surface on top of the hill [AE1380] and scattered sherds on the ground [AE1381], both of which only he noticed. On the other hand, my personal visit at the site, recently, as well as personal communication from the excavator at the site for the local *ephoreia*, K. Sarri, did not report anything.

23/24. Sphingion W: *Component HA_52* (23) and Davlosis: *Component HA_53* (24)

ID	Period	Type	Toponym	Discovery	Position
HA_52	LH	burial place	Sphingion W	Personal or group interest	approximate
HA_53	LH	burial place	Davlosis	Personal or group interest	approximate

Prehistoric gazetteers mention a LH cemetery on the W slope of Sphiggion as well as by Davlosis,[22] where locals report idols and rings found above the hamlet (see also Lauffer Kopais I: 35). At the Kalimbaki locality, at the W foot of the Sphiggion mountain, two LH chamber tombs [AE1736] were excavated by the local *ephoreia* in 1995 (AD 1996: 270-2); the evidence has been associated to *component HA_52*.

As Lauffer points out (Kopais I: 35), this reminds one of the typical situation which occurs in the Copais area, of prehistoric cemeteries in relation to the settlements they were attached to. Burial areas were usually at the foothills or on the slopes of the 'mainland' opposite the fortified settlement by the lake (on a spur, promontory, or 'islet'). In this case, the settlement site would be Megalo Kastraki. On the other hand, considering the apparent absence of LH layers on Megalo Kastraki, archaeologists from the local *ephoreia*, before the excavation of the chamber tombs, had hypothesized a location for the Mycenaean acropolis on the Kalimbaki hill (AD 1995: 299-301). This is on the basis of a strong retaining wall that was found facing the stream at the foot of Mt. Sphiggion [AE1734 - infrustructure] and possibly connected with prehistoric drainage works in the gulf of Davlosis (AD 1996: 270-2).

25. Kokkoretsa South: *Components HA_54 to HA_59*

ID	Period	Type	Toponym	Discovery	Position
HA_54	gr-rom	? Fort	Kokkoretsa South	Personal or group interest	approximate
HA_55	LH	? Fort	Kokkoretsa South	Personal or group interest	approximate
HA_56	Preh	activity focus	Kokkoretsa South	Ext. topographical surveys	approximate
HA_57	EH	? activity focus	Kokkoretsa South	Ext. topographical surveys	approximate
HA_58	C	activity focus	Kokkoretsa South	Ext. topographical surveys	approximate
HA_59	LR	human presence	Kokkoretsa South	Ext. topographical surveys	approximate

At the NE corner of the Davlosis bay, S of the Kokkoretsa/Samaroi promontory, lies a 15m high hill, 50m wide, which is probably the site of a fort overlooking the passage from Copais to Lake Yliki area. The place was visited by Lauffer (Kopais I: 36) who dates the remains to the LH period (*components HA_55*), while HS&D (1979: G15) would prefer to date the fort to the H period[23] (*component HA_54*), though they report some probable LH sherds from the site.[24] Fossey (1988: 313) describes it as a square tower of rough polygonal masonry, around which are remains of a few other buildings. The date is still uncertain. Evidence AE672, AE684 and AE1383 would refer to the same remains of fortification.

Additionally, sherds have been noticed at the site including EH (Fossey 1988: 313 only, apparently), obsidian and red flints, as well as C (Fossey 1988: 313 only seems to report C black glaze, which could be associated with the fort, if gr-rom) and LR material (Fossey 1988: 313 only. Material could also be off-site). Only obsidian and "early sherds" are recorded in the field report, as pointed out by HS&D (1979: G15).

26. Kokkoretsa - Proph.Ilias: *Component HA_60*

ID	Period	Type	Toponym	Discovery	Position
HA_60	unknown	activity focus	Kokkoretsa - Proph.Ilias	Int. topographical surveys	certain

On a hill W of Kokkoretsa, by the Proph.Ilias chapel, traces of a tower [AE1384 - AE1385] are reported by Lauffer (Kopais I: 37), though they constitute slighter signs compared to

[21] Lauffer compares this site with the Onchestos one, which was a much more important site. A small town/large village grew up next to the Poseidon sanctuary (see above in the text no.15).

[22] In the small hamlet at Davlosis, at the foothills of the Sphingion mountain, is a perennial fresh water spring.

[23] They say: "*polygonal walls appear to indicate a H fort*". Being the fort close to Medeon-Kastraki site (*component HA_49*; HS&D 1979: G14), it is possibly subsidiary to it (HS&D 1979: G15).

[24] HS&D 1979: G15 have LH? with question mark. LH is claimed in AM 1938-9: 183 and by Ålin 1962.

Kokkoretsa South (see above no.25). I preferred not to mark the site as a certain fort, and I left the component as a generic activity focus.
Kokkoretsa Southern and Western forts would mark the N border of the SW Copais bay, which belonged to Haliartos.

27/28. Listi: *Components HA_77 and HA_78* (27) and Listi nearby: *Component HA_79* (28)

ID	Period	Type	Toponym	Discovery	Position
HA_77	EH	activity focus	Listi	Ext. topographical surveys	certain
HA_78	LH	? activity focus	Listi	Ext. topographical surveys	certain
HA_79	N	? activity focus	Listi nearby	Ext. topographical surveys	approximate

Listi is named a hill 1.5km ca N of Mavromati village. At the W foot of the hill EH material is reported; LH material is also reported,[25] though the accounts on the site are very vague (Fossey 1988: 146). Buck (1972) is apparently the only account of Listi before an intensive survey, as he mentions the site in an attempt to identify it with ancient *Keressos* (Plutarch *Cam.* 19; Pausanias IX 1-4), though probably Listi was not even in Thespian lands (according to the borders of the *chorai* as we have defined them, it lays in Haliartos *chora*). Buck also mentions a Neolithic site nearby (*component HA_79*).
The Boeotia Survey project's preliminary results attest for the Listi site a gr-rom presence (A – C – early Hellenistic undetermined activity focus – see below, table 1 – *Survey Sites*: 'Listi').

29/30. Kariopoula: *Components HA_61* (29) *and HA_62* (30)

ID	Period	Type	Toponym	Discovery	Position
HA_61	gr-rom	fort	Kariopoula	Travellers	certain
HA_62	gr-rom	burial place	Kariopoula	Int. topographical surveys	certain

Component HA_61: On the slopes NW of Mazi village, by the road Mazi-Petra, remains of a large gr-rom fortification are visible, as well as surface walls probably belonging to the fortification on a quite vast area, at least 100m wide (Lauffer Kopais I: 50ff, sketches 52-53-54). The fort overlooks the way out from the upland valley.
Component HA_62: Tiles and bones were found there by the small hilltop in the N area of the large fortification (see Lauffer Kopais I: sketch 54), but sherds are also all over the slopes and graves were found on the E and W slopes.
The site had already been noticed by Vischer 1875: 558, who identified it with Lisandros' military camps before the attack to Haliartos, mentioned by Plutarch (*Lys.* 29.11) as on a hill named *Orchalides*. If one follows this interpretation, the fortification would not belong to the Haliartos *polis*, but would have been built for an 'independent' military meaning.

[25] Few sherds now at Thebes museum (Buck 1972).

31/32. Stokthi Keramidissa: *Components HA_63* (31) *and HA_64* (32)

ID	Period	Type	Toponym	Discovery	Position
HA_63	gr-rom	activity focus	Stokthi Keramidissa	Int. topographical surveys	certain
HA_64	N	? activity focus	Stokthi Keramidissa	Int. topographical surveys	certain

1.5km ca E of Petra are three terraces, with a water spring (Keramidissa/Keramida).

Component HA_63: On the ancient blocks probably *in situ* have been noticed, unless they have been regularly arranged in more recent times (Lauffer Kopais I: 53ff, sketches 58 and 56), as well as surface walls (very long, probably terracing walls) and a large quantity of ancient sherds. According to Lauffer, on the upper terrace was probably a large building, and building material and sherds in lower terraces come probably from the upper one. Among the architectural fragments is also a pillar, probably inscribed once. According to Lauffer (Kopais I: 54) the ensemble of remains may lead to the interpretation of the site as a cult place (sanctuary?). He suggests an identification with the ancient sanctuary of Kekrops of Haliartos (whose distance from the site is 1.5km ca), mentioned by Pausanias IX 33.1. I would not go further into identification problems, because this is not the purpose of the present research, and furthermore we cannot even be sure as to the interpretation of the site as a cult place. Evidence, as they are reported in the bibliography, could also be associated to a rural site. A visit to the place and appropriate intensive and systematic artefact surface surveys might help in the interpretation of site typology and function.

Component HA_64: A stone tool (axe) dated to the Neolithic period has been found at the site 50m S of the water source. Once again, states Lauffer (Kopais I: 55), we could be facing an open and not naturally defended site occupied in earlier periods. Could a single stone tool found be indication of a settlement at the site? In my opinion, a discovery like this, in the location of a later period activity focus, could mean: first, that the discovery occurred because of the interest in other (later) evidences at the site (more visible, if not more interesting for the archaeologist); second, that some kind of activities was going on at a time in the Neolithic period in that landscape, comprehensive of this place, but we should take into account that it is a movable object and hence is not a sure indication of activity; third, that the axe tool could be used for different purposes, and does not immediately mean a habitation site; fourth, slope erosion may have brought to the surface a single sign of a much intensive prehistoric (Neolithic?) activity which still lies buried.

33. Petra: *Component HA_65*

ID	Period	Type	Toponym	Discovery	Position
HA_65	gr-rom	? activity focus	Petra	Int. topographical surveys	imprecise

Petra is the name of the village mountain (Petra/Siacho) to the W of Haliartos, just above the road Thebes-Levadeia.
Ancient blocks and architectural fragments are built in two chapels immediately to the SW of Petra village (Ag. Taxiarchoi and Ag. Panagia). In front of the Ag. Panagia chapel lie marble blocks, probably dug up during renovation work of the chapel in 1926. According to Lauffer (Kopais I: 56), who reports of the remains, they could mark the site of a gr-rom cult place site or a

village site. Lauffer seems to prefer the latter hypothesis and attempts to identify the village with ancient *Okalea*, mentioned by Strabo IX 410.[26]
In the vicinity of the Ag. Panagia chapel a *miliarium* (BCH 1905: 98ff [AE1410]) was also found, dated in Hadrian times, marking a mile distance from the town of Alalkomenai (text reported by Lauffer Kopais I: 56). It could mark the passage of a road, but the evidence is too movable to be reliable and, moreover, the distance of the finding spot from the Alalkomenai site is longer than a mile.

34. Vigla: *Component HA_66*

ID	Period	Type	Toponym	Discovery	Position
HA_66	gr-rom	fort	Vigla	Int. topographical surveys	approximate

1km W from Petra, on a small terrace, remains of a fortification in the form of a tower and remains of a round wall are visible around the hilltop. Tile fragments have also been noticed at the site, as well as mortar remains, which probably indicate a reuse of the fortification in Roman or later periods (Lauffer Kopais I: 62, sketch 62).

35/36. Linopyrgo: *Components HA_67* (35) *and HA_68* (36)

ID	Period	Type	Toponym	Discovery	Position
HA_67	gr-rom	fort	Linopyrgo	Historical geography	approximate
HA_68	gr-rom	activity focus	Linopyrgo	Int. topographical surveys	approximate

NW of Goulas mountain, and SE of Kouphospithari, NE of Evangelistria, on the E side of a small hollow, facing the Evangelistria monastery to its E side (700m far away on a straight-line distance), and immediately off the road (to the N of it) from Haliartos to the village Zagora/Evangelistria, were found remains of a square shaped tower (*component HA_67*, known also as an Evangelistria tower), dated to the 4thC by Lauffer (Kopais I: 60) because of the walling technique (squared blocks). Few surface walls have been noticed in the vicinity (Fossey 1988: plan 43). No pottery has yet come to light at this site, and only one inscription, IG VII 2851 [AE712], in honour of the emperor Hadrian, comes from the area (Fossey 1988: 314). Lauffer (Kopais I: 60) notes that tile fragments can be seen on some lower terraces at quite a distance (*component HA_68*).
The fort is in a very good position, by the communication routes from the upland to Petra and Haliartos and to Evangelistria and Koroneia (fig.1). It also controls the ancient path which led from Thespiae -through Askra and the valley of the Muses- to Evangelistria and then Koroneia. Attempts have been made to identify the site with ancient *Okalea* (see chapter II.3.8 - VILLAGE LEVEL).

37. Lykophio: *Component HA_84*

ID	Period	Type	Toponym	Discovery	Position
HA_84	C	fort	Lykophio	Personal or group interest	approximate

On the Lykophio hill, on a small plateau at the SW corner, is a small fort of the C or H period. From it is visible the location of the Linopyrgo/Evangelistria tower (*component HA_67*). Built probably during the Sacred War (353-346), it would belong to the system of forts controlled by Thebes. The small fort linked the tower and the city of Haliartos (BCH 1991: 883-884 - see Kallet-Marx 1989).

38. Mavri: *Components HA_69 and HA_70*

ID	Period	Type	Toponym	Discovery	Position
HA_69	gr-rom	burial place	Mavri	Int. topographical surveys	certain
HA_70	gr-rom	? activity focus	Mavri	Int. topographical surveys	certain

On the North side of the Kalamaki/Ksirorrema river, to the NW of the Evangelistria village, on a river terrace, there is a place that was noticed by Lauffer (Kopais I: 61, sketch 69). Remains from tile graves and bones, along with ancient and Byzantine sherds and a ring with '*gemma*' [AE1414], as well as a squared block [AE1415] have been found.
Certainly there is evidence for a burial place (*component HA_69*). In addition, one could hypothesize some other kind of activity focus in connection with it (*component HA_70*).
Lauffer (Kopais I: 61) suggests an identification of the place with ancient *Hippotai* (see appendix I.1 no.31), on the basis of Plutarch (*Amat. Narr.* 4). I would rather suggest an interpretation of the place as a burial site, which could even be associated to a villa site on the river terrace, since it seems rich, if one considers the ring that was found (for which though no date is given, and we can only suppose a Roman origin). The place seems ideal for an upland site with an agro-pastoral vocation, though we cannot totally exclude the presence of a cult place (with associated burials).

39/40. Vrastamites/Ypsilantis: *Component HA_71* (39) and Ypsilantis Pyrgos: *Component HA_72* (40)

ID	Period	Type	Toponym	Discovery	Position
HA_71	gr-rom	? activity focus	Vrastamites/ Ypsilantis	Int. topographica l surveys	certain
HA_72	gr-rom	activity focus	Ypsilantis Pyrgos	Travellers	certain

On the E slope of the Palaiothiva-Petra/Tilphousion ridge, the modern village of Vrastamites/Ypsilantis lies at the bottom of a small valley running down the Palaiothiva ridge from the church of Ag. Nikolaos. A medieval tower very well preserved lies mid-slope in the valley, ca 1km S of the village.
Component HA_71: Ancient blocks rebuilt in the Ag. Panagia chapel E of the village [AE1416] and lying below the tower to its NE [AE1417], as well as surface walls and ancient foundations [AE1418] noticed under the chapel itself (Lauffer Kopais I: 63), seem to testify the presence of an ancient site by the modern village, though its date cannot be clarified through the available information.[27]
Component HA_72: By the tower Lauffer (Kopais I: 64) reports C sherds and tile fragments as well as a surface wall visible on the surface. Ross (1851: 31) had already noticed surface walls (probably disappeared today), as well as Stephani (1843: 73), who describes here on the road large stone foundations and surface walls. Personal visits of the Boeotia Survey team (J.L. Bintliff – pers. comm.) revealed below the tower the presence of

[26] See chapter II.3.8 - VILLAGE LEVEL, for the various attempts of identification of ancient *Okalea*.

[27] The structures could even relate to the large Frankish feudal village that goes on and flourishes in the early Ottoman period, whose remains have been noticed above the modern village (J.L. Bintliff –pers. comm.- and Bintliff-Kiel in preparation).

potsherd concentrations, which could imply remains of a possible gr-rom village.

Lauffer (Kopais I: 64) suggests an identification with ancient *Okalea*. Among the many possibilities of identification advanced (see chapter II.3.8 - VILLAGE LEVEL), he sees as the most probable one the Vrastamites-Ypsilantis site, since it is 6km distance to Haliartos and Alalkomenai, in line with Strabo IX 240. The identification had already been suggested by Ross (1851: 31). This interpretation is accepted also by Philippson (1951: 474).

41. Ypsilantis Ag. Nikolaos: *Component HA_73*

ID	Period	Type	Toponym	Discovery	Position
HA_73	gr-rom	activity focus	Ypsilantis Ag. Nikolaos	Int. topographical surveys	approximate

On a 200m wide terrace 400m below the chapel of Ag. Nikolaos, on the E slope of Palaiothiva ridge, ca 1.8km SW of Vrastamites-Ypsilantis. The place is marked, according to Lauffer (Kopais I: 65), by a *platanus* tree and a fresh water spring.
A large quantity of glaze sherds [AE1423] and remains of terracing wall [AE1424] mark the presence of a gr-rom activity focus. Lauffer (Kopais I: 66) notes how the area available for a hamlet here is probably too small, therefore he interprets the evidence as representing a fort or a cult place site (for the latter, he suggests the identification with the sanctuary of Praxidike, after Pausanias IX 33.4 'πρός the Tilphosion mountain'). Probably here it used to run, as it still does today (see fig.1), the passage from the small valley of Vrastamites/Ypsilantis (and Haliartos *chora*) to Solinari (and Koroneia *chora*),[28] and from the terrace can be seen not only the whole Vrastamites-Ypsilantis valley, but also the whole Copais until its N border (from Pyrgos to Sphiggion Mt.). It could easily be a cult place site on an open position in an upland terrace, as can usually be found in the Copais area, as often remarked by Lauffer. On the other hand, the place could also have a different character, and only a closer look to the material found at the site could probably give a more reliable answer.

42. Petra/Tilphousion: *Components HA_74 and HA_75*

ID	Period	Type	Toponym	Discovery	Position
HA_74	gr-rom	fort	Petra/Tilphousion	Travellers	certain
HA_75	LH	? Fort	Petra/Tilphousion	Int. topographical surveys	certain

The Petra mountain, the rocky ridge between Vrastamites-Ypsilantis and Solinari village, that juts into the Copais basin, has been identified with the ancient *Tilphousion*,[29] mentioned by Pausanias (IX 33.2) as lying ca 50 stadia from Haliartos.

On top of the Petra-Tilphousion spur, on a plateau ca. 80m EW and 100m NS, Ross (1851: 31) had already noticed a section of polygonal walling. Lauffer describes the site to be interpreted as a large fort (Lauffer Kopais I: 68ff, sketch 76). As he notices (Kopais I: 73) there is nowhere in Copais a place where one can have a better look over the entire basin.

Component HA_74: Glaze sherds and tile fragments [AE1428], as well as walling technique, lead to the interpretation of the site as a gr-rom fort probably of the Late Classical period (4thC BC?). Strabo (IX 413) refers to the *Tilphousion* as a fort pure and simple, without an associated settlement.
Component HA_75: Traces of an older, probably Mycenaean fortification [AE1429], have also been noticed by Lauffer (Kopais I: 71, sketch 80) at some places, especially to the E side. Lauffer suggests a probable Mycenaean fortified site, as many others in the Copais. I would rather locate here a simple fort (without associated settlement). See Fossey 1972, 1990d and 1992: 112-114.

Fortification walls are visible on the E side of the top, while the W side is naturally fortified. Three phases of walling can probably be distinguished: Mycenaean (LH), gr-rom, modern.[30] It is an important fort which played a role in Boeotian history and legend (Strabo IX 2.36; Leake 1835: 205, with references). The spur of Petra played an important role during the so-called 'last' battle of the independence war from the Turks in September 1829 (Papachatzis 1981: 209). Between the rocky hill and the lake, to the E, there was a tiny passage that in 1829 was controlled by the Greek forces and stopped Turkish troops leading to Lamia. In ancient times, the same passage would have been used as the passage for the road Haliartos-Koroneia, followed by Pausanias as well.
There is discussion on which of the two forts (the higher Palaiothiva,[31] or the lower Petra, described here) was actually named *Tilphousion*, since ancient texts could refer to both (Lauffer Kopais I: 86; Fossey 1972 and 1992). Only the name of the source *Tilphousia*[32] at the foot of Petra proffers something, but nothing else can provides further help.

A cult place is also known at the foot of the ridge (see chapter II.3.1 for details).

What follows is a preliminary version of the list of sites discovered and recorded during the Systematic Intensive Surface Survey carried out in the area (area surveyed mapped in fig.2) in the 1980s by J.L. Bintliff and A.M. Snodgrass and their équipe (Bintliff-Snodgrass AR 1986-87: 23ff.).

SurveySITE code	CHRONOLOGY	CHARACTER
HAL 0	A	?activity focus
HAL 0	G	?activity focus
HAL 0	Preh	activity focus
HAL B1	A	rural site
HAL B1	C	rural site
HAL B1	C-Ehl	rural site
HAL B1	EH	settlement
HAL B1	G	rural site
HAL B1	N	settlement
HAL B2	A	activity focus
HAL B2	C	activity focus
HAL B2	EH	?settlement
HAL B2	G	activity focus
HAL B3	A	rural site
HAL B3	EH	?settlement

[28] See Lauffer Kopais I: sketch 73.
[29] There are several variant spellings for the name. For reference on ancient sources see Fossey 1972 and Hansen 1996: 76. In particular, Demostenes XIX 141-148. See also in this work Solinari Agios Ioannis (*Tilphousion*) in Koroneia area (appendix I.1), especially for the Tilphousia spring.
[30] see Fossey 1992: 112-114; Papachatzis 1981: pl. 252.
[31] See *component KO_80* in Koroneia *chora* (appendix I.1).
[32] On the water source see chapter II.3.1.

SurveySITE code	CHRONOLOGY	CHARACTER
HAL B3	G	rural site
HAL B4	A	activity focus
HAL B4	C	activity focus
HAL B4	C-Ehl	activity focus
HAL B4	G	activity focus
HAL B4	LH	?activity focus
HAL B4	Preh	activity focus
Kahrstedt site	LR	rural site
HAL B5	A	?activity focus
HAL B5	C-Ehl	rural site
HAL B5	G	?activity focus
HAL B6	G	?activity focus
HAL B7	A	rural site
HAL B7	C	activity focus
HAL B7	C-Ehl	?activity focus
HAL B7	EH	settlement
HAL B7	G	rural site
HAL B7	LH	?activity focus
HAL B7	N	settlement
Listi	A	activity focus
Listi	C	activity focus
Listi	C-Ehl	activity focus
ME1	Lhl-ER	?activity focus
MN1	C-Ehl	rural site
MN1	Lhl-ER	rural site
MN1	LR	rural site
MP A3	LR	rural site
MP A4	A	rural site
MP A4	C	rural site
MP A5	C-Ehl	activity focus
MP A5	EH	activity focus
MP A5	LH	activity focus
MP A5	Lhl-ER	activity focus
MP A5	LR	activity focus
MP A5	MH	activity focus
MP A6	C-Ehl	rural site
MP A6	Lhl-ER	activity focus
MP A6	Preh	?activity focus
MP B1	C-Ehl	rural site
MP B2	A	?activity focus
MP B2	C	?activity focus
MP B2	C-Ehl	rural site
MP B3	A	activity focus
MP B3	C-Ehl	rural site
MP B3	EH	settlement
MP B3	G	activity focus
MP B3	LH	settlement
MP B3	Lhl-ER	rural site
MP B3	LR	rural site
MP B3	MH	settlement
MP B4	EH	settlement
MP B4	LH	settlement
MP B4	MH	?settlement
MP B5	A	?activity focus
MP B5	G	?activity focus
MP B5	LH	activity focus
MW1	LR	rural site
MW2	C	activity focus
MW3	C-Ehl	?activity focus
MW4	C	?rural site
MW4	C-Ehl	rural site
MW5	C	?activity focus
MW5	H	?activity focus
MW5	R	?activity focus
Onchestos	EH	settlement
Onchestos	LH	activity focus
Onchestos	MH	settlement
Onchestos	N	Settlement
Palaiomazi	A	?activity focus
Palaiomazi	C-Ehl	activity focus
Palaiomazi	H	activity focus
Palaiomazi	R	activity focus
Palaiomazi	LR	activity focus

Table 1. *List of intensive survey sites discovered during intensive and systematic survey work* (Bintliff-Snodgrass AR 1986-87: 23ff) – *mapped in fig.2.*

Appendix I.9
Eastern Helikon: Thespike

THE ARCHAEOLOGICAL RECORD

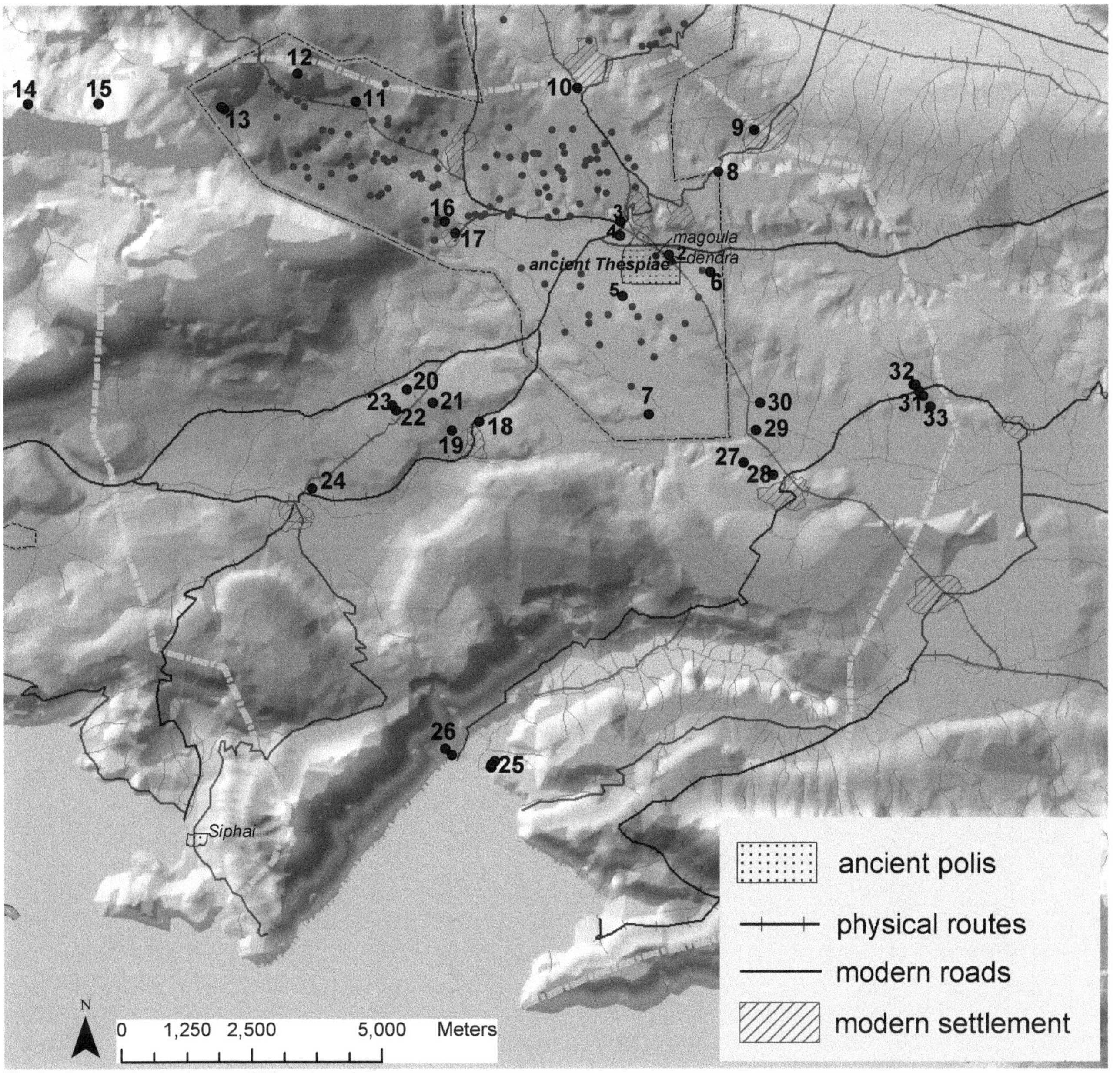

Fig.1. *Archaeological map of Thespiae chora.*

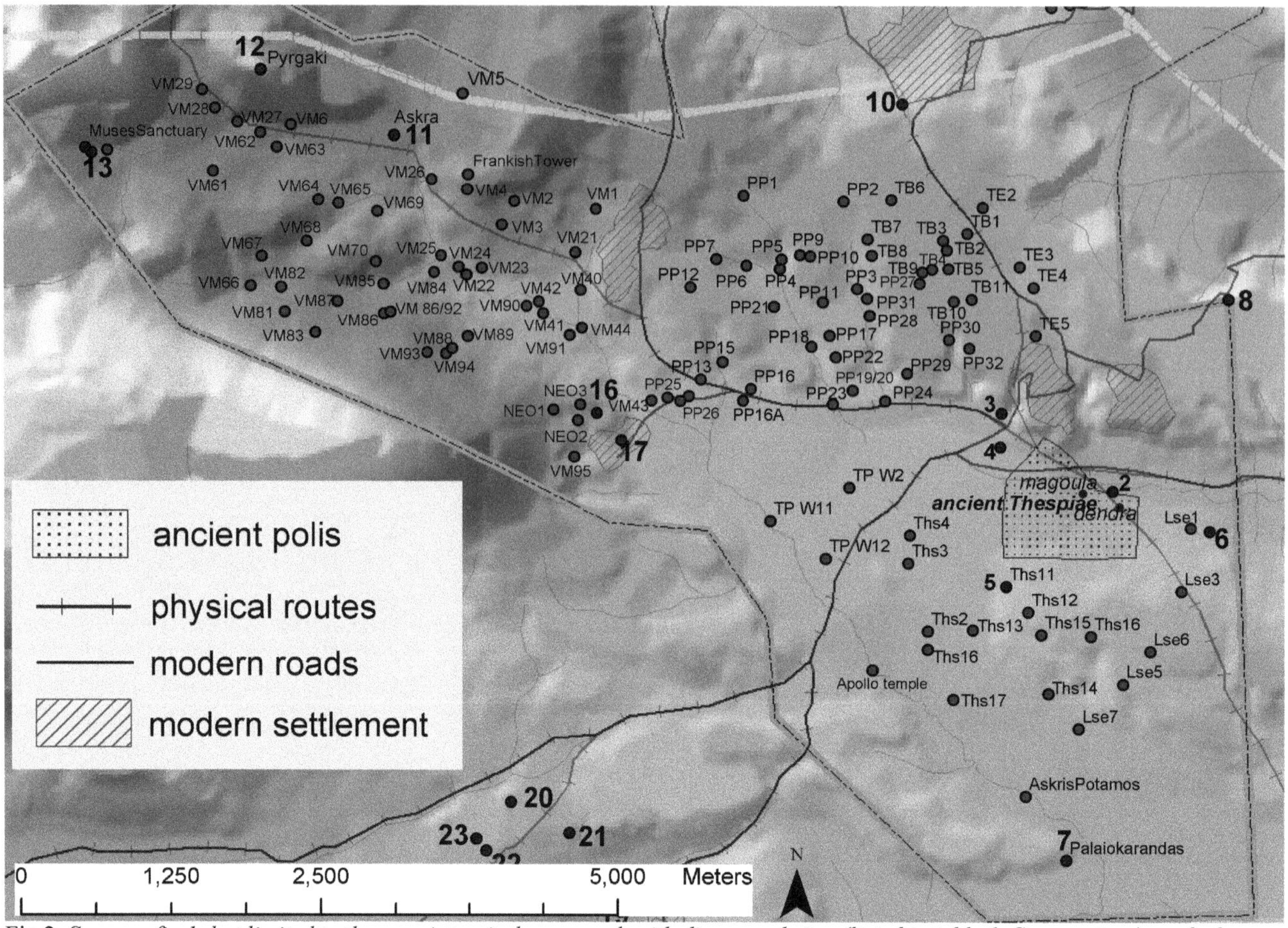

Fig.2. *Same as fig.1, but limited to the area intensively surveyed, with discovered sites (listed in table 1 -Survey sites-) marked.*

1. THESPIAE: *Components TH_1 to TH_12*

ID	Period	Type	Toponym	Discovery	Position
TH_1	G	settlement	Thespiae	Personal or group interest	certain
TH_2	A	settlement	Thespiae	Travellers	certain
TH_3	C	settlement	Thespiae	Travellers	certain
TH_4	H	settlement	Thespiae	Travellers	certain
TH_5	R	settlement	Thespiae	Travellers	certain
TH_7	LR	settlement	Thespiae	Int. and sys. artefact surface survey	certain
TH_8	LR	fort	Thespiae	Travellers	certain
TH_9	N	settlement	Thespiae magoula	Personal or group interest	certain
TH_10	EH	settlement	Thespiae magoula	Personal or group interest	certain
TH_11	MH	activity focus	Thespiae magoula	Personal or group interest	certain
TH_12	LH	? activity focus	Thespiae magoula	Personal or group interest	certain
TH_16	Preh	human presence	Thespiae Dendra	Ext. topographical surveys	approximate

To the S of modern Thespies and Leondari villages, across the Kanavaris river, on a low ridge, is still visible today, to an expert eye, the site of the ancient *polis* of Thespiae, on a heavily-cultivated area.

The area once occupied by the ancient city is marked by a dense scatter of potsherds and architectural fragments as well as building material [AE1044] scattered all over the area.[1] Heavy agricultural cultivation activities, in a known but not strictly protected site, have slowly destroyed the archaeological deposits. The city limit, which at some point exceeds the hypothesized city wall line, has been traced and drawn for the different periods of occupation by J.L.Bintliff and A.M.Snodgrass as a result of their intensive survey at the site (see Bintliff-Howard-Snodgrass 2007).

The area was already known as the site of ancient Thespiae by travellers (Lolling 1889, Frazer 1913 who quotes earlier travellers). Early excavations were also carried out at the site in a very small area by de Ridder (AD 1967: 241 and BCH 1953: 219).

A *protopolis* in G periods, the site was a *polis* from A to LR period.[2] Traces of the LR fort (*component TH_8*, dismantled by Jamot in 1888-1891 'excavations' [AE1547] see Fossey 1988: 136) are to be noticed in the middle of the ceramic city (*component TH_7*), on the highest part of the ridge, and its extension has been mapped by the Boeotia Survey project through intensive survey work in the 1980s. For more recent reports see AD 1981: 186-188, AD 1996: 286 and AReports 1986-87: 23-24 (Boeotia Survey Project report). A full account of the intensive survey work is in preparation.

[1] Sculptured pieces ranging from Archaic to Roman times are very frequent (detailed bibliography in Fossey 1988: 138 - note 7; see also recent work on architectural pieces from the site conducted by M.Bonanno and a team from Tor Vergata University - Rome).

[2] A lot of epigraphical evidence is known from and for the site and from the Thespiae area. See Fossey 1988: 138 and more recent reports, for instance: AD 1995: 310; AReports 1988-89: 45-46; BCH 1990: 763; BCH 1988: 642.

Prehistoric is also known from the site.
Components TH_9 to TH_12: The so-called *magoula* site lies at the edge of the gr-rom *polis*, as a low mound above and on the S bank of the Kanavaris stream, which creates a swamp in the lower area at that point. The site, already known to Blegen (1920), was investigated by the Americans (Caskey 1951: 289-290 and 1960: 159-160) and by the Boeotia Survey Project in 1985 and 1986.
N to EH occupation is attested, as well as MH and LH. As for the Neolithic, the major period of occupation at the site, plenty of MN and LN, and several E-MN, including figurines have been found (Boeotia survey project). Caskey (Hesperia 1960: 159-160) compared pottery to that found at Eutresis site, and suggested that probably at Thespiae we have got Middle and Late Neolithic. As for EH, Hope Simpson 1965 and French 1972 have EH I?, EH II-II. The Boeotia survey project attests EH but sub-periods remain unclear (I and II?). As for MH occupation, Hope Simpson 1965 and French report MH, including Grey and Yellow Minyan and Matt-painted. Hope Simpson 1965 and French report LH III A-B from the site, while the Boeotia survey project has few sherds dated to the LH period on the *magoula* site.
Another Prehistoric focus was found by the Boeotia Survey Project higher up to the S, in the middle of the later historical city area.
Component TH_16: at Dendra locality, another location in the Thespiae city area, located approximately to the E of the site, stone artefacts were found (BCH 1963: 219). As they were found isolated, they may represent only a certain human presence in the area in the Prehistoric period.

2. Thespiae NE: *Component TH_6*

ID	Period	Type	Toponym	Discovery	Position
TH_6	R	burial place	Thespiae	Rescue excavation	approximate

Component TH_6: by the magoula along the Kanavaris river (opposite Thespies and Leondari modern villages) the evidence AE1045 (excavated *peribolos* and a statue - BCH 1953: 219) could refer either to the R city settlement site (*component TH_5*) or, most probably, to a burial area (a probable urban high class cemetery) nearby, with statue and funeral *peribolos*.

3. Erimokastro W: *Component TH_13*

ID	Period	Type	Toponym	Discovery	Position
TH_13	LH	burial place	Erimokastro W	Ext. topographical surveys	approximate

In the rock at the foot of the hill W of Erimokastro/Thespies modern village, an opened chamber tomb is visible (Fossey 1988: 137).

4. Thespiae vicinity: *Component TH_15*

ID	Period	Type	Toponym	Discovery	Position
TH_15	LR	burial place	Thespiae vicinity	accidental	imprecise

In an unspecified location in the vicinity of ancient Thespiae, probably approximately to the NW of the city site (G.Selianiti plot), a LR (5th-6th C AD) tomb was found during ploughing (AD 1970: 228).

5. Thespiae S: *Components TH_17 to TH_19*

ID	Period	Type	Toponym	Discovery	Position
TH_17	H	burial place	Thespiae S	Personal or group interest	approximate
TH_18	A	burial place	Thespiae S	Personal or group interest	approximate
TH_19	C	burial place	Thespiae S	Personal or group interest	approximate

1600m to the S of the modern Thespies village, Andreiomenou (AD 1981: 186-7) excavated a number of H tombs, from the end of the 4th C BC. Among H graves an A and a C tomb were excavated. The area could also have been a cemetery in an earlier period, but regardless, it has always certainly been linked to Thespiae city. It corresponds to the intensive survey site THS11 (see table 1 –*Survey sites*– below).

6. Thespiae E (*poliandrion*): *Component TH_14*

ID	Period	Type	Toponym	Discovery	Position
TH_14	C	burial place	Thespiae E	Travellers	approximate

E of ancient Thespiae, a small rectangular base on which stood the statue of a lion marks the site of a *poliandrion* (probably for the Thespians killed in battle of Delion – Fossey 1988: 137), a C monumental burial place. Excavations were carried out (Stamatakis, Praktika 1882: 67-72; Keramopoulos, Praktika 1911: 153-163).

7. Palaiokarandas: *Components TH_ 29 to TH_32*

ID	Period	Type	Toponym	Discovery	Position
TH_29	H	? cult place	Palaiokarandas	Int. and sys. artefact surface survey	certain
TH_30	C	burial place	Palaiokarandas	accidental	certain
TH_31	H	burial place	Palaiokarandas	accidental	certain
TH_32	R	? burial place	Palaiokarandas	accidental	certain

The Palaiokarandas ridge is 4km S of modern Thespies village. Around it Askris river makes a turn before leading SW towards the Domvraina *polje*.
As stated by Fossey (1988: 147), the inscriptions and sculpted blocks found there [AE1027; AE1536] could be from ancient Thespiae, but he suggested that the site could also be a candidate for a missing village such as *Leontarne* or *Ellopia*. On the other hand, a cemetery could certainly fit some of the inscriptions listed in Fossey (1988: 147), such as the tombstones AE1028-AE1029-AE1030.
Only epigraphical evidence, and some sculpted blocks found on the surface, marked this area as a possible site. Fossey (1988) does not report any pottery evidence. In the 1980s, the Boeotia Survey Project surveyed the site intensively and recognised it as a Classical-Early Hellenistic burial place (cemetery) with probably a shrine or cult activities associated (see *Palaeokarandas* site in table 1 –*Survey sites*– below). As for the R period, only epigraphical evidence would attest the continuity of use of the burial area, with apparently no surface pottery associated. We know though that R cemeteries are much less recognisable on the surface than earlier (C-H) ones.

Despite the fact that at Palaiokarandas site Philippson (1951: 459 and 504) reports "*der* (mittelalterlichen?) *Ruine Palaiokarantas*" and the deserted village of Pyrgos on top of the ridge, intensive survey research at the site (Bintliff-Howard-

Snodgrass 2007 – Palaiokarandas) did not discover traces of any medieval settlement, whilst recording one of the major Prehistoric settlements of the area (*Palaeokarandas* site in table 1 –*Survey sites*- below).

8. Leondari NE: *Component TH_20*

ID	Period	Type	Toponym	Discovery	Position
TH_20	gr-rom	burial place	Leondari NE	accidental	approximate

Along the road running NE from Kaskaveli-Leondari to Vagia, tombs lining the road are reported in Fossey 1988: 137, without exact mention of his source of information.

9. Vagia: *Components TH_21 and TH_22*

ID	Period	Type	Toponym	Discovery	Position
TH_21	H	burial place	Vagia	accidental	imprecise
TH_22	R	burial place	Vagia	accidental	imprecise

In the area of modern Vagia village, epigraphical evidence[3] seems to attest the presence of a village and a cemetery area, although, as known, inscriptions are a very movable trace. The fact that there is only inscriptional evidence available makes uncertain the existence of a village here. Both Bintliff (pers.comm.) and Fossey (1988: 145) seem to hypothesize the existence of a very small (insignificant, as defined by Fossey) village here. As for the cemetery, we could be more certain that there were burials somewhere in the area. The Boeotia Survey project did not cover the area intensively.

10. Mavromati: *Components TH_26 to TH_28*

ID	Period	Type	Toponym	Discovery	Position
TH_26	C	? burial place	Mavromati	accidental	imprecise
TH_27	H	? burial place	Mavromati	accidental	imprecise
TH_28	R	? burial place	Mavromati	accidental	imprecise

The same as for the Vagia case (above) can be considered for the Mavromati case, where a 4th C sculpture of a woman's head and gr-rom inscriptions were found. More than in the case of Vagia, the materials here seem to come from ancient Thespiae. On the other hand, since the inscriptional evidence is constituted by tombstones mainly (and a 4th C woman's head – AD 1964: 201), a cemetery could have also been in this area, probably along a road from Thespiae city. Inscriptions are listed in Fossey 1988: 146.
The Boeotia Survey project covered the area intensively, and found a site at the immediate N outskirts of Mavromati modern village. On the surface, material was found from C to Medieval periods, with picks in C and LR period, as well as an early R quite strong occupation (see site MN1 in table 1 - *Survey sites*- below).

11. Askra Episkopi (ancient ASKRA): *Components TH_ 33 to TH_39*

ID	Period	Type	Toponym	Discovery	Position
TH_33	PG	settlement	Askra Episkopi	Int. and sys. artefact surface survey	certain
TH_34	G	settlement	Askra Episkopi	Int. and sys. artefact surface survey	certain
TH_35	A	settlement	Askra Episkopi	Int. and sys. artefact surface survey	certain
TH_36	C	settlement	Askra Episkopi	Int. and sys. artefact surface survey	certain
TH_37	H	settlement	Askra Episkopi	Int. and sys. artefact surface survey	certain
TH_38	R	settlement	Askra Episkopi	Int. and sys. artefact surface survey	certain
TH_39	LR	settlement	Askra Episkopi	Int. and sys. artefact surface survey	certain

At the SE and E feet of the Pyrgaki hill, by the river.[4]
A thick scatter of surface potsherds mainly (with some upright stone features [AE926]) testifies the presence of an extensive settlement, a large village site in this location, with evidence from PG to LR period. As for LR period, remains of a LR/Early Christian church, marking the settlement site, are to be seen at the site (Bintliff - Snodgrass 1985).
The final identification of the site with the ancient *kome* of Askra has been made by the intensive and systematic artefact surface survey work carried out in the framework of the Boeotia Survey Project (Bintliff 1996b).

12. Askra Pyrgaki: *Components TH_ 40 to TH_44*

ID	Period	Type	Toponym	Discovery	Position
TH_40	LH	activity focus	Askra Pyrgaki	Ext. topographical surveys	certain
TH_41	G	fort	Askra Pyrgaki	Ext. topographical surveys	certain
TH_42	A	fort	Askra Pyrgaki	Ext. topographical surveys	certain
TH_43	C	fort	Askra Pyrgaki	Ext. topographical surveys	certain
TH_44	H	fort	Askra Pyrgaki	Travellers	certain

To the W of the site of ancient Askra rises a high peak marking the landscape of the valley. The hilltop is called Pyrgaki because of the presence of an ancient fortification.
A watchtower of the typical S Boeotian style (mid-late 4th C BC – Fossey 1988: 142-3) and a surrounding circuit probably earlier in date (possibly in existence since G period) are to be noticed on the top.
LH material (and earlier Bronze Age probably – see table 1 *Survey sites*) has also been found on the hill (Fossey 1988: 143 from earlier accounts, supported by Boeotia Survey project data – Bintliff 1996b). The hilltop could have been in use as a fort (or fortified settlement) since the LH period, though the position is quite elevated and steep compared to the usual location of known LH forts in the region.
Pyrgaki site could be identified with ancient Keressos, mentioned in Plutarch (*Cam.* 19) and Pausanias IX 1-4.

13. Askra - Sanctuary of the Muses: *Components TH_ 45 to TH_49 and TH_23*

[3] Inscriptions are listed in Fossey 1988: 146.

[4] A description of the site (SE and E feet of the Pyrgaki hill) is in Bintliff and Snodgrass' publications (Bintliff - Snodgrass 1985b: 49-70; Snodgrass 1985: 87-95).

ID	Period	Type	Toponym	Discovery	Position
TH_23	LR	activity focus	Askra - Sanctuary of the Muses	Int. and sys. artefact surface survey	certain
TH_45	A	cult place	Askra - Sanctuary of the Muses	Personal or group interest	certain
TH_46	C	cult place	Askra - Sanctuary of the Muses	Personal or group interest	certain
TH_47	H	cult place	Askra - Sanctuary of the Muses	Personal or group interest	certain
TH_48	R	cult place	Askra - Sanctuary of the Muses	Personal or group interest	certain
TH_49	EH	? activity focus	Askra - Sanctuary of the Muses	Other	approximate

On the slopes of Helicon, high up and in an inner position in the Valley of the Muses, lay the so-called Sanctuary of the Muses. Activities are attested since A period. A probable cult place seems to be represented by votive material from 6th C BC (BCH 1922: 288-9 nos. 145-149), though no A layers are mentioned in the later excavation reports from Roux, who reported on Jamot's earlier excavations. In C, H and R period is certain the use of the place as a sanctuary and the main known structures dated to these periods. See Roux (BCH 1954: 22-45) for a detailed description of the site. The Boeotia Survey Project visited and surveyed the site (*MusesSanctuary* site in table 1 – *Survey Sites*- below; Bintliff 1996b). LR material is also present.

Component TH_49: Earlier use of the place seems to be attested by surface sherds of EH found in the area by Fossey together with some obsidian flakes (Fossey 1988: 141). There is no mention of EH in the preliminary data available from the Boeotia Survey Project.

14. Askra Helicon: *Custers TH_50 and TH_51*

ID	Period	Type	Toponym	Discovery	Position
TH_50	A	activity focus	Askra Helicon	Travellers	approximate
TH_51	C	activity focus	Askra Helicon	Travellers	approximate

N of the topographical point (1530m a.s.l.) N of Hippocrene (summit of Diakopi or Motsara or Zagara - Kryo Pigadi locality). Today the site is marked by the chapel of Proph. Ilias. The place was already known to travellers (Frazer, Vischer, Ulrichs, Bursian, Hitzig-Blumner and Welcke), that were looking for the altar of Zeus Helikonios mentioned by Hesiod, and noticed ancient structures underneath the chapel (Wallace 1974). They thought it could be the Altar of Zeus Helikonios mentioned by Hesiod, but it is most probably a watch tower controlling the Gulf of Corinth. More recent topographical work carried out by the local *ephoreia* showed that the component in the A period could be interpreted as cult building -pottery bearing graffiti dedications to Zeus-, then reused as a watch-tower in the C period (AD 1994: 263-4). On the other hand, Aravantinos (1996) reports it as a fort built sometime in the A period (graffiti pots and Lesbian masonry dated it to the A period) and then in use also in C times (A and C tiles found at the site).

15. Askra Hippokrene: *Component TH_52*

ID	Period	Type	Toponym	Discovery	Position
TH_52	gr-rom	activity focus	Askra Hippokrene	Personal or group interest	certain

At the spot where the water spring, identified with the Hippokrene mentioned by Hesiod, springs out, recent work of the local *ephoreia* (Aravantinos 1996) noticed that the stone blocks that today form the mouth of the well are ancient and the tracing of the wall is of elaborate masonry of polygonal or Lesbian style. He suggests a date for the A period, because of a similarity with the masonry style of the building at the summit (*component TH_50*). Aravantinos also suggests a cultual use for the spring, since its monumentality could not be justified by the use of water by soldiers and sheperds. Today the spring and a ruined chapel mark the site on the ridge of Diakopi (or Motsara or Zagara).

16/17. Neochori: *Components TH_96* (16) *and TH_97* (17)

ID	Period	Type	Toponym	Discovery	Position
TH_96	gr-rom	? activity focus	Neochori	Travellers	certain
TH_97	C	? burial place	Neochori	accidental	certain

At the SE end of the Valley of the Muses lays the modern village of Neochori.
Component TH_96: Wheler (1862: 471) found remains (surface walls? [AE1776] and pieces reused [AE1531]) above Neochori village including the inscription IG VII 1862 [AE1777].
Component TH_97: in *Antike Plastik* 12: 18-22 a 5th C stele from Neochori is reported. The very movable and single evidence has been associated with a possible burial place for the period.
Neochori area has been surveyed intensively and systematically, and Prehistoric, gr-rom and medieval settlement and occupation have been located in the area of modern Neochori village (see table 1 – *Survey sites* –).

ID	Period	Type	Toponym	Discovery	Position
TH_94	C	? burial place	Ellopia - Koraki	accidental	imprecise
TH_95	C	activity focus	Ellopia – Agia Kyriaki	accidental	imprecise

18/19. Ellopia – Koraki: *Component TH_94* (18) and Ellopia – Agia Kyriaki: *Component TH_95* (19)

To the N bank of Askris river, after its turning by Palaiokarandas, lays the modern village of Ellopia. In the area traces of ancient activities have been found: inscriptional evidence (Classical stelai) from Koraki (near Ellopia) - AD 1964: 201- and Classical sherds[5] from Agia Kyriaki (near Ellopia) - AD 1972: 323 - are reported.

20 - 24. Tatiza: *Components TH_53* (20), *TH_54* (21), TH_55 to TH_57 (22), *TH_58* (23) and Xironomi: *Components TH_59 and TH_60* (24)

ID	Period	Type	Toponym	Discovery	Position
TH_53	EH	activity focus	Tatiza	Personal or group interest	approximate
TH_54	LH	burial place	Tatiza	accidental	imprecise
TH_55	C	? burial place	Tatiza	accidental	approximate

[5] Black glazed kantharoi sherds and floral-ware kylix (2nd half 4th C BC) – AD 1972: 323.

ID	Period	Type	Toponym	Discovery	Position
TH_56	H	? burial place	Tatiza	accidental	approximate
TH_57	R	burial place	Tatiza	Rescue excavation	approximate
TH_58	LR	rural site/activity	Tatiza	Ext. topographical surveys	approximate
TH_59	H	? burial place	Xironomi	accidental	imprecise
TH_60	R	? burial place	Xironomi	accidental	imprecise

The deserted village of Tatiza lies NW of Karandas and NE of Xironomi. House walls and the church of Agia Triada are the visible remains of the village. A number of water sources spring out to the S of the site.

Components TH_55 to TH_60: Artefact concentrations are visible on the surface, to the E of the village, and to the S of the river (Fossey 1988: 147), confirmed by intensive survey work (Boeotia Survey Project). Surface walls and building material, as well as well known architectural pieces (including a Doric frieze) were noticed in the area.
The gr-rom site was probably on a lower terrace of a river (tributary of the Askris). A LR artefact concentration was found, probably marking a large rural establishment or a hamlet, occupied until Turkish and Early Modern period (Vroom 2003 no.21; Boeotia Survey project - preliminary results[6]). The low-lying setting of the LR site (*component TH_58*) is a quite ordinary location in the Boeotian landscape (see the results of the Tanagra survey project in appendix I.14).
Probably to the N of the old road, excavations in the 1960s and 1970s (Spyropoulos AD 1971: 221), albeit of limited extent, did not bring to the light any structural remains, but many human bones and Roman pottery - sure traces of a cemetery area here, at least in the R period. As for the C and H period, mainly epigraphical evidence (very movable) is known, while the cemetery in the R period is better attested. It could also have been a burial place in earlier periods.
Epigraphical evidence (leading to burial areas) comes from the area, both from Tatiza and Xironomi. Actually the hypothesis (suggested also by Fossey 1988: 148) that Xironomi ancient blocks were merely taken from Tatiza by villagers, that move from one village to the other when Tatiza got deserted (and the latter area is farmed by inhabitants of Xironomi), is less likely (Bintliff pers.comm.) as now funerary inscriptions (4th C, H and R mainly, from both Tatiza and Xironomi) and relief tomb stelai (from Xironomi) are reported. Probably there is enough evidence for two cemetery areas, one by Tatiza and the other one closer to Xironomi. A revisit of the actual findspots would be needed.
According to the evidence, one or two settlement sites could be hypothesized for this area for H and R period, albeit the slight evidences (see chapter II.3.9). Identifications have been suggested (*Donakon*, *Leontarne*) and they are reported in Fossey (1988: 149). In AD 1966: 202 a small village within the area of Thespiae is hypothesised for Tatiza area, as a sort of satellite settlement of Thespiae, like Askra.

Components TH_53 and TH_54: On the lower ridge to the N (of Tatiza), traces of an EH activity focus have been found (Fossey 1988: 148) as well as rich Mycenaean tombs found looted nearby Tatiza (Threpsiadis, AE 1973: 82-3, who locates them between Koryni, Domvraina and Ksironomi).[7]

[6] Intensive survey work was aimed to the Medieval and Early Modern hamlet of Tatiza, and discovered presence of earlier periods as well.
[7] A Mycenaean bronze spear point and a dagger were found in the tombs (given to the local ephorate by the plot owner). From the field above few Mycenaean sherds [AE1770] were collected, probably coming from the destroyed tombs (Threpsiadis AE 1973: 82-3).

25/26. Livadostro E – Kastro: *Components TH_61 to TH_69 and TH_24* (25) and Livadostro W: *Components TH_70 to TH_74* (26)

ID	Period	Type	Toponym	Discovery	Position
TH_24	C	? burial place	Livadostro E – Kastro	Ext. topographical surveys	approximate
TH_61	EH	activity focus	Livadostro E - Kastro	Personal or group interest	approximate
TH_62	MH	activity focus	Livadostro E - Kastro	Personal or group interest	approximate
TH_63	LH	activity focus	Livadostro E - Kastro	Personal or group interest	approximate
TH_64	Preh	burial place	Livadostro E - Kastro	Ext. topographical surveys	approximate
TH_65	A	activity focus	Livadostro E - Kastro	Ext. topographical surveys	approximate
TH_66	C	activity focus	Livadostro E - Kastro	Ext. topographical surveys	approximate
TH_67	H	activity focus	Livadostro E - Kastro	Ext. topographical surveys	approximate
TH_68	R	activity focus	Livadostro E - Kastro	Ext. topographical surveys	approximate
TH_69	unknown	burial place	Livadostro E	Ext. topographical surveys	approximate
TH_70	A	? Fort	Livadostro W	Personal or group interest	approximate
TH_71	C	fort	Livadostro W	Travellers	approximate
TH_72	H	fort	Livadostro W	Travellers	approximate
TH_73	LR	? Fort	Livadostro W	Personal or group interest	approximate
TH_74	gr-rom	special activity	Livadostro W	Personal or group interest	certain

The Livadostro bay is the Easternmost bay on Boeotia's southern coast. Livadostro "*is also the name of the torrent which flows from the Thespian plains, down a narrow gorge between ridges of Mount Korombili to the West and Mount Kithairon to the East; these same mountains' masses provide the steep slopes enclosing the plain and the triangular bay*" (Fossey 1988: 157).
As Fossey (1988: 157) notes, no real settlement is in the bay today, while summer houses have been built all over the area.
Ancient activity focuses are located at the two ends of the shore (Fossey 1988: 157). At the W end is a large ancient fortress, while at the E end a site which shows occupation from EH to R is marked by the location of a Frankish tower.[8]
A 'mole' from antiquity is visible (Philippson 1951: 505), testifying the use of the bay as a port of a considerable importance (mainly linked to the history of Thespiae city).

Livadostro East site (by Frankish tower).
On a low hill by the Frankish tower (named Kastro), to the E side of Livadostro bay, traces of activities from different periods have been found.

Components TH_61 to TH_64: In prehistoric time the site at the E end of the bay was in a strategic position, probably controlling the access from the sea to inland. It is not surprising,

[8] A sketch plan of plain and sites is in Roesch 1965: 218, map 4. In the Frankish middle ages there was here a port named Rivadostia, which served the communication between Athens and the West. In the 14th century it was a base of Catalan pirates (Philippson 1951: 505).

therefore, the finding of prehistoric material in this high up position (in the area of the Frankish tower). See also the traces of the LH road noticed by Heurtley 1923-25. EH, MH and LH activity focuses found at the site are mentioned in prehistoric gazetteers and were visited during extensive surveying (AD 1969: 185-6; Fossey 1988: 161). Two cist graves on the hilltop (*component TH_64*), probably from MH, would attest the existence of a MH habitation site here.

Components TH_65 to TH_68: A material from the site is reported only in Bintliff-Snodgrass (pers.comm.), while C material is well reported (AD 1972: 324; Fossey 1988: 161). Some of the material found (AD 1972: 324; Fossey 1988: 161) could refer to C burials (*component TH_24* – see also *component TH_69*). H and R traces of occupation have also been found (Fossey 1988: 161 from Spyropoulos AD 1969: 185-6). Apart from the Medieval tower on the summit, possible traces of an ancient circuit wall are reported on the E and the W side (Fossey 1988: 158).

Component TH_69: To the N of the hill with the Frankish tower (Livadostro E – Kastro site), along the feet of the Kithairon slopes, Fossey (1988: 158) reports many (undated) tombs robbed.

Livadostro West site (ancient fortress).
In the Livadostro bay, at the W end of the shore is a large ancient fortress.
Components TH_70 to TH_74: A, C, and H material have been found on the site (Fossey 1988: 161 from his own detailed survey - see Fossey-Gauvin 1985: 77-85). A plan of the fortress is in Fossey 1988: 158 fig.19. The fortress was probably in use also in the LR period (fortified settlement?).

Ancient sources lead to the identification of Livadostro with ancient *Kreusis*, and all agree Kreusis belonged to Thespiae and was its harbour (according to Pausanias IX 32.1). There is no mention of it in the Homeric Catalogue except as a variant of *Isos*. The site was known to ancient travellers (Leake 1835[9] among others – see Fossey 1988: 158), and is mentioned in Heurtley's work on Boeotian harbours (BSA 1923/5: 38ff).[10]
The area played an important role in the preliminaries to the battle of Leuktra in 371 BC (Xenophon *Hell.* V 4.16f).

27 - 30. Parapoungia/Leuktra: *Components TH_75* (27), *TH_76* (28), *TH_77* (29) *and TH_25* (30)

ID	Period	Type	Toponym	Discovery	Position
TH_25	C	special activity	Parapoungia /Leuktra	Travellers	approximate
TH_75	R	activity focus	Parapoungia /Leuktra - Grammatikes	accidental	imprecise
TH_76	H	activity focus	Parapoungia /Leuktra - Tzachanani	accidental	imprecise
TH_77	gr-rom	? activity focus	Parapoungia /Leuktra	accidental	imprecise

The modern Parapoungia/Leuktra village lies, as Fossey (1988: 145) notes, at a "curious watershed zone" between Askris[11] and the small river Oeroe which rises by Plataea and runs to Livadostro. Further North the two water systems are separated by rolling country overlooked by the village of Parapoungia/Leuktra.
As Fossey reports, some scholars have sought to place in the area the ancient settlement of Leuktra (see Pritchett 1965: 49-52, with previous references), if it ever existed. Pritchett suggests that the ancient settlement could be either under the modern village or covered by recent alluvium, and this would be the reason why the remains of it cannot be traced today.
Although Pausanias presents it as an inhabited town in 371 BC, and Strabo seems to regard it as a *polis*, it is highly doubtful if it is a real place rather than only a battlefield, or perhaps the name of a plain. Some 3rd century finds and R figurines in the area (localities Grammatikes, Tzachanani, Megali Goritza /Antonakia), however, as well as the famous *Tropaion – component TH_25*[12] could revive the belief in a site here (AD 1972: 323).
The question of identification is reported in Fossey 1988: 155, as well as the list of inscriptions from the villages of the area (Parapoungia/Leuktra, Baltza, Caparelli).[13] Fossey (1988: 155) suggests that the fact that many of the inscriptions can be found in Parapoungia village may result simply from its greater proximity to Eutresis, only 2.5km to the NW. Also Goldman (1931: 5) attests Eutresis site as a quarry of building stones for villages in the area, described as a pool of stones. The epigraphical evidences AE1534-AE1535-AE1537, therefore, have not been included in any component, because of the uncertainty of their provenance.

Fossey (1988: 154) reports several scatters of potsherds (4th C – R period) from AD 1972: 323. Only special finds or fine/decorated wares are reported from H and R period. We can only hypothesize the existence of larger artefact concentrations (including coarse wares also) in the area, usually not reported by extensive survey or accidental findings.

31/32/33. Arkopodi – Eutresis: *Components TH_ 78 to TH_86, TH_84 and TH_89* (31)*; TH_87, TH_88 TH_90 to TH_92* (32)*; TH_93* (33)

ID	Period	Type	Toponym	Discovery	Position
TH_78	N	settlement	Arkopodi - Eutresis	Personal or group interest	certain
TH_79	EH	settlement	Arkopodi - Eutresis	Personal or group interest	certain
TH_80	MH	settlement	Arkopodi - Eutresis	Personal or group interest	certain
TH_81	LH	settlement	Arkopodi - Eutresis	Personal or group interest	certain
TH_82	A	cult place	Arkopodi - Eutresis	Other	certain
TH_83	A	activity focus	Arkopodi - Eutresis	Ext. topographical surveys	approximate
TH_84	C	cult place	Arkopodi - Eutresis	Other	certain
TH_85	C	? Settlement	Arkopodi - Eutresis	Ext. topographical surveys	approximate

[9] Leake (1835) described it as 'a bad harbour in a large bay'. The British Admiralty Handbook 1944/45 has it as a port too exposed. Is a coastal change possible, which would have changed probably the character of the harbour?
[10] Heurtley notes remains of a road running inland from the bay, which he dates to the prehistoric period. The date is though questioned by Pritchett 1965: 52-6.
[11] The Askris river, which probably constituted once the upper course of Asopos (see chapter II.3.9 – TOPOGRAPHICAL SETTING).
[12] The monument has been excavated and restored (Praktika 1924: 38-40; 1958: 43-4; AD 1970: 229).
[13] To the list should be added inscriptions of unknown provenence within Leuktra area reported and brought to the Thebes museum (AD 1982: 173).

ID	Period	Type	Toponym	Discovery	Position
TH_86	G	human presence	Arkopodi - Eutresis	Other	approximate
TH_87	H	rural site/activity	Arkopodi - Eutresis	Personal or group interest	certain
TH_88	R	rural site/activity	Arkopodi - Eutresis	Personal or group interest	certain
TH_89	H	? cult place	Arkopodi - Eutresis	Personal or group interest	certain
TH_90	C	burial place	Arkopodi - Eutresis NW	accidental	certain
TH_91	H	? burial place	Arkopodi - Eutresis	accidental	approximate
TH_92	R	? burial place	Arkopodi - Eutresis	accidental	approximate
TH_93	C	? burial place	Arkopodi - Eutresis SE	accidental	certain

The ridge of Arkopodi (named after the spring of Arkopodi - marked on 1:50,000 IGME geological maps) is the site of ancient Eutresis, which extends also to the N of the road from Leuktra to Melissochori, running initially N-E and then bending to the E by the spring (Goldman 1931 fig.1, reproduced by Papachatzis 1981: 96).

The site was extensively excavated at the beginning of the 20th century (Goldman 1931). Excavations concerning prehistoric levels especially were carried out also by Caskey (Caskey AJA 1959: 187; Caskey, Hesperia 1960: 126-167).

The site is well known as a large Prehistoric settlement, with traces of occupation since Neolithic period, and reaching its occupation peak probably in EH, MH periods (MH levels were more extensively excavated). In LH period (LH III), the site was surrounded by a fortification wall enclosing a larger area than that actually occupied (Goldman 1931; Caskey, Hesperia 1960: 126-167).[14]

Fossey (1988: 152) notes how occupation at the site after the prehistoric period (known mainly because of the interest in the main prehistoric settlement) was never extensive, or at least little remains of it. However, we must remember that parts of the hill have been considerably denuded by erosion. We cannot be sure, therefore, that Eutresis was a settlement site in historic periods.
Inscriptions from the site are also known (listed in Fossey 1988: 153): a H boundary stone, a H dedication (to Apollo), tombstones (4th C, H and R[15]), which represent though, as always, a very movable evidence. A cemetery area (that the tombstones would represent - *components TH_90 to TH_93*) could have been nearby, considering also the presence of cult buildings on the hill.
The A[16] and C[17] cult places (*components TH_82* –on the hill summit- *and TH_84* –on the SE slope-) could be linked to the settlement (*components TH_83 and TH_85*). Especially as for *component TH_82*, its position on the hill summit can lead to the conclusion that it was a temple within the probable village site.
Cult activities are attested for the A and C period (*components TH_82 and TH_84*) and were probably present also in H period. A H dedication to Apollo was found at the site, as well as H ceramic material, though position -or association with the cult place excavated structures- is not specified in Fossey (1988: 151). Fossey (1988: 153) suggests an identification of the site in the gr-rom period with a sanctuary of Apollo (because of the dedication from the H period found). If one considers its position on the route Thespiae-Plataea, says Fossey, an Apollo Sanctuary could fit.
Generally speaking, habitation traces are evident, though they could be interpreted also as associated with the sanctuary site only. On the other hand, they may represent a real village site, filling in the gap between Thespiae and Plataea village (see chapter II.3.9 for a discussion). For H and R period, a courtyard building has been excavated, leading to the assumption either of a wealthy rural site here or a settlement. Since chronology does not seem to fit the dating of the cult place, *components TH_87 and TH_88* are listed as rural sites.

Component TH_86: Traces of G material have been found [AE959], attesting at least a human presence in that period. The place could have been somehow in use since that period, but two G sherds cannot constitute as evidence which one can rely on. We cannot say whether there was a gap of occupation in the Dark Age at the site.

As Fossey reports (1988: 152), traces of walling were found [AE954] to the NE of the hill. The first excavators thought they could be part of a circuit walling, in considering its similarity to circuit walls in East Locris, and they assumed that the settlement in C times lay in this NE area and was surrounded by fortification (having as a *terminus post quem* the early 5th C BC from pottery found in association with the wall). However, Fossey notes how the costruction of the wall found may rather suggest a mid 4th C BC date, and the fortification perhaps could be another part of the ashlar defense system of the South coast – see Fossey 1988: appendix 2 and 1992.

What follows is a preliminary version[18] of the list of sites discovered and recorded during the Systematic Intensive Surface Survey carried out in the area (area surveyed mapped in fig.2) in the 1980s by J.L.Bintliff and A.M.Snodgrass and their équipe (Bintliff-Snodgrass 1985a and Bintliff 1996b; Bintliff-Howard-Snodgrass 2007).

[14] For a summary see Fossey 1988: 151.
[15] See also AD 1994: 291.
[16] No structures remained, but offerings were found in a pit dug down into Mycenaean levels.
[17] Terrace walls and foundation remains, with pot material associated.
[18] Final are the information concerning the sites discovered in the S sector of the survey, at the immediate S outskirt of the city site (published in Bintliff-Howard-Snodgrass 2007).

Table 1. *List of intensive survey sites discovered during intensive and systematic survey work (Bintliff-Snodgrass 1985a and Bintliff 1996b; Bintliff-Howard-Snodgrass 2007) – mapped in fig.2.*

SurveySITE code	CHRONOLOGY	CHARACTER
ApolloTemple	C-Ehl	cult place
AskrisPotamos	A	settlement
AskrisPotamos	C-Ehl	settlement
AskrisPotamos	G	activity focus
AskrisPotamos	H	rural site
AskrisPotamos	Lhl-ER	rural site
AskrisPotamos	LR	settlement
AskrisPotamos	MR	settlement
Magoula	N	settlement
Magoula	Preh	activity focus
Lse1	A	rural site
Lse1	C-Ehl	rural site
Lse1	ER	rural site
Lse1	G	activity focus
Lse1	H	rural site
Lse1	Lhl-ER	rural site
Lse1	LR	rural site
Lse1	MR	rural site
Lse1	Preh	settlement
Lse3	A	?rural site
Lse3	C-Ehl	hamlet
Lse3	G	activity focus
Lse3	H	rural site
Lse3	Preh	settlement
Lse4	C-Ehl	burial place
Lse5	A	rural site
Lse5	C-Ehl	hamlet
Lse5	G	activity focus
Lse5	H	hamlet
Lse5	Preh	activity focus
Lse6	A	?rural site
Lse6	C-Ehl	settlement
Lse6	ER	rural site
Lse6	G	activity focus
Lse6	H	settlement
Lse6	Lhl-ER	rural site
Lse6	LR	activity focus
Lse6	MR	rural site
Lse6	Preh	settlement
Lse7	C-Ehl	rural site
Lse7	ER	rural site
Lse7	LR	rural site/villa-hamlet
Lse7	MR	rural site/villa
Lse7	Preh	activity focus
MEL2	C-Ehl	rural site
MEL4	C-Ehl	rural site
MN1	C-Ehl	rural site
MN1	Lhl-ER	rural site
MN1	LR	rural site
Palaeokarandas	C-Ehl	burial place
Palaeokarandas	Preh	settlement
PP1	C	activity focus
PP2	gr-rom	activity focus
PP3	gr-rom	activity focus
PP4	gr-rom	activity focus
PP5	gr-rom	activity focus
PP6	LR	activity focus
PP7	gr-rom	?activity focus
PP9	C	rural site
PP10	A	activity focus
PP10	C	activity focus
PP11	A	burial place
PP11	C	burial place
PP12	C-Ehl	?activity focus
PP12	Lhl-ER	activity focus
PP13	C-Ehl	activity focus
PP13	Lhl-ER	?activity focus
PP15	C-Ehl	rural site
PP15	LR	rural site
PP16	LR	?activity focus
PP17	H	rural site
PP17	Lhl-ER	rural site
PP18	C	?activity focus
PP18	C-Ehl	rural site
PP18	H	rural site
PP18	Lhl-ER	rural site
PP18	LR	rural site
PP19/20	EH	activity focus
PP19/20	MH	activity focus
PP19/20	N	activity focus
PP21	C-Ehl	rural site
PP21	Lhl-ER	activity focus
PP21	LR	?activity focus
PP22	H	?rural site
PP22	Lhl-ER	rural site
PP22	LR	rural site
PP23	C-Ehl	rural site
PP23	Lhl-ER	activity focus
PP23	LR	activity focus
PP23	Preh	?human presence
PP24	C-Ehl	?rural site
PP24	Lhl-ER	rural site
PP24	LR	?rural site
PP25	C-Ehl	rural site
PP25	EH	settlement
PP25	Lhl-ER	activity focus
PP25	MH	settlement
PP26	C-Ehl	rural site
PP26	Lhl-ER	?activity focus
PP26	LR	activity focus
PP27	C-Ehl	?activity focus
PP27	Lhl-ER	?activity focus
PP27	LR	rural site
PP28	C-Ehl	rural site
PP28	Lhl-ER	rural site
PP28	LR	rural site
PP29	C-Ehl	activity focus
PP29	Lhl-ER	activity focus
PP29	LR	activity focus
PP30	C-Ehl	burial place
PP31	C-Ehl	activity focus
PP31	Lhl-ER	activity focus
PP31	LR	activity focus
PP32	C-Ehl	rural site
PP32	Lhl-ER	?activity focus
Ths2	ER	rural site
Ths2	LR	rural site/villa
Ths2	MR	rural site
Ths3	C-Ehl	?burial place
Ths3	ER	activity focus
Ths3	LR	hamlet
Ths3	MR	activity focus
Ths3	Preh	activity focus
Ths3	R	settlement
Ths4	ER	rural site
Ths4	Lhl-ER	rural site

SurveySITE code	CHRONOLOGY	CHARACTER
Ths4	LR	hamlet
Ths4	MR	rural site
Ths4	Preh	activity focus
Ths11	A	?burial place
Ths11	C-Ehl	burial place
Ths11	G	activity focus
Ths11	Preh	activity focus
Ths12	A	rural site
Ths12	C-Ehl	rural site
Ths12	ER	rural site
Ths12	G	activity focus
Ths12	H	rural site
Ths12	Lhl-ER	rural site
Ths12	LR	rural site/villa
Ths12	MR	rural site/villa
Ths12	Preh	activity focus
Ths13	ER	rural site
Ths13	LR	rural site/villa
Ths13	MR	rural site
Ths13	Preh	settlement
Ths14	LR	rural site/villa-hamlet
Ths14	Preh	activity focus
Ths15	A	?burial place
Ths15	C-Ehl	burial place
Ths15	G	?burial place
Ths16	A	?rural site
Ths16	C-Ehl	hamlet
Ths16	ER	settlement
Ths16	G	activity focus
Ths16	H	hamlet
Ths16	Lhl-ER	rural site
Ths16	MR	settlement
Ths16	Preh	activity focus
Ths17	A	activity focus
Ths17	C-Ehl	rural site
Ths17	G	activity focus
Ths17	H	activity focus
T B1	C-Ehl	rural site
T B1	Lhl-ER	rural site
T B1	LR	?rural site
T B2	C-Ehl	rural site
T B2	Lhl-ER	?rural site
T B2	LR	rural site
T B3	C-Ehl	?rural site
T B3	Lhl-ER	rural site
T B3	LR	rural site
T B4	A	?activity focus
T B4	C-Ehl	rural site
T B4	G	?activity focus
T B4	Lhl-ER	?rural site
T B5	C-Ehl	rural site
T B6	C-Ehl	rural site
T B6	Lhl-ER	?activity focus
T B6	LR	?activity focus
T B7	A	?activity focus
T B7	C-Ehl	rural site
T B7	G	?activity focus
T B7	Lhl-ER	?activity focus
T B7	LR	rural site
T B8	C-Ehl	rural site
T B8	LR	rural site
T B9	C-Ehl	rural site
T B10	C-Ehl	rural site
T B11	C-Ehl	activity focus
T B11	LR	activity focus
T E2	C-Ehl	activity focus
T E2	LR	activity focus
T E3	C	burial place
T E4	A	rural site
T E4	C	rural site
T E5	C	activity focus
T E5	C-Ehl	activity focus
T E5	Lhl-ER	activity focus
T E5	LR	activity focus
TP W2	LR	activity focus
TP W2	R	activity focus
TP W11	R	activity focus
TP W12	A	?activity focus
TP W12	C	?activity focus
VM 86/92	A	?activity focus
VM 86/92	C-Ehl	rural site
VM 86/92	Lhl-ER	?rural site
VM 86/92	LR	rural site
VM1	A	?activity focus
VM1	C-Ehl	rural site
VM1	Lhl-ER	rural site
VM1	LR	rural site
VM1	Preh	?activity focus
VM2	A	rural site
VM2	C-Ehl	rural site
VM2	LH	activity focus
VM2	Lhl-ER	?rural site
VM2	LR	?rural site
VM21	A	rural site
VM21	C-Ehl	rural site
VM21	LH	?settlement
VM21	Lhl-ER	rural site
VM21	LR	rural site
VM21	MH	?settlement
VM22	A	activity focus
VM22	C-Ehl	activity focus
VM22	G	activity focus
VM22	LH	activity focus
VM22	Lhl-ER	activity focus
VM22	LR	activity focus
VM23	A	?activity focus
VM23	C-Ehl	?activity focus
VM23	G	?activity focus
VM23	Lhl-ER	?activity focus
VM23	LR	rural site
VM24	C-Ehl	?activity focus
VM24	Lhl-ER	?activity focus
VM24	LR	?activity focus
VM25	C-Ehl	?activity focus
VM26	C-Ehl	?activity focus
VM26	Lhl-ER	?activity focus
VM27	C-Ehl	rural site
VM27	LR	rural site
VM28	A	?burial place
VM28	C-Ehl	?burial place
VM29	C-Ehl	activity focus
VM29	Lhl-ER	rural site
VM29	LR	rural site
VM3	C-Ehl	rural site
VM3	Lhl-ER	rural site
VM3	LR	rural site
VM4	A	?activity focus
VM4	C-Ehl	?activity focus
VM4	EH	activity focus
VM4	G	activity focus
VM4	LH	activity focus
VM4	Lhl-ER	?activity focus

SurveySITE code	CHRONOLOGY	CHARACTER
VM4	LR	?activity focus
VM4	MH	activity focus
VM4	N	activity focus
VM40	C-Ehl	?activity focus
VM40	Preh	activity focus
VM41	C-Ehl	rural site
VM42	Lhl-ER	?activity focus
VM42	LR	activity focus
VM43	LR	activity focus
VM44	LR	rural site
VM5	A?	?burial place
VM5	C	?burial place
VM6	A	?activity focus
VM6	C-Ehl	rural site
VM6	Lhl-ER	?rural site
VM6	LR	rural site
VM6	Preh	activity focus
VM61	A	activity focus
VM61	C-Ehl	activity focus
VM61	G	?activity focus
VM61	Lhl-ER	rural site
VM61	LR	rural site
VM63	C-Ehl	?activity focus
VM63	Lhl-ER	rural site
VM63	LR	rural site
VM64	LR	rural site
VM64	R	rural site
VM65	A	rural site
VM65	C-Ehl	?activity focus
VM65	EH	activity focus
VM65	G	rural site
VM65	LH	activity focus
VM65	Lhl-ER	?activity focus
VM65	LR	rural site
VM65	MH	activity focus
VM66	C-Ehl	?rural site
VM66	Lhl-ER	?rural site
VM66	LR	rural site
VM67	unknown	
VM68	A	?activity focus
VM68	C-Ehl	rural site
VM68	LR	activity focus
VM69	C-Ehl	?activity focus
VM69	Lhl-ER	?activity focus
VM69	LR	rural site
VM70	A	?activity focus
VM70	C	rural site
VM70	G	?activity focus
VM70	Lhl-ER	?activity focus
VM70	LR	?activity focus
VM81	C-Ehl	rural site
VM81	LR	?activity focus
VM82	C-Ehl	rural site
VM82	LR	rural site
VM83	C-Ehl	rural site

SurveySITE code	CHRONOLOGY	CHARACTER
VM84	C-Ehl	?rural site
VM84	Lhl-ER	?activity focus
VM85	C-Ehl	activity focus
VM85	Lhl-ER	activity focus
VM85	LR	rural site
VM86	C-Ehl	?rural site
VM86	Lhl-ER	?activity focus
VM86	LR	rural site
VM87	C-Ehl	activity focus
VM87	LR	activity focus
VM88	Lhl-ER	?rural site
VM88	LR	?activity focus
VM89	C-Ehl	rural site
VM89	Lhl-ER	rural site
VM89	LR	rural site
VM90	unknown	
VM91	A	activity focus
VM91	C-Ehl	?activity focus
VM91	G	activity focus
VM91	Lhl-ER	?activity focus
VM91	LR	activity focus
VM93	C-Ehl	rural site
VM94	C-Ehl	rural site
VM94	Lhl-ER	rural site
VM94	LR	rural site
VM95	C-Ehl	rural site
VM96	C-Ehl	rural site
MusesSanctuary	A	?cult place
MusesSanctuary	C-Ehl	cult place
MusesSanctuary	Lhl-ER	cult place
MusesSanctuary	LR	activity focus
Pyrgaki	EH	?activity focus
Pyrgaki	MH	?activity focus
Pyrgaki	LH	activity focus
MW1	LR	rural site
MW2	C	activity focus
MW3	C-Ehl	?activity focus
MW4	C	?rural site
MW4	C-Ehl	rural site
MW5	C	?activity focus
MW5	H	?activity focus
MW5	R	?activity focus
Neo1	A	?activity focus
Neo1	C-Ehl	activity focus
Neo1	LR	activity focus
Neo1	MH	settlement
Neo1	Preh	settlement
Neo2	C-Ehl	rural site
Neo2	LR	?activity focus
Neo3	C-Ehl	rural site
Neo3	EH	settlement
Neo3	LR	?activity focus
Neo3	MH	?settlement

Appendix I.10
Three small *chorai* to the Gulf of Corinth: Siphai, Thisbe, Chorseiai

THE ARCHAEOLOGICAL RECORD

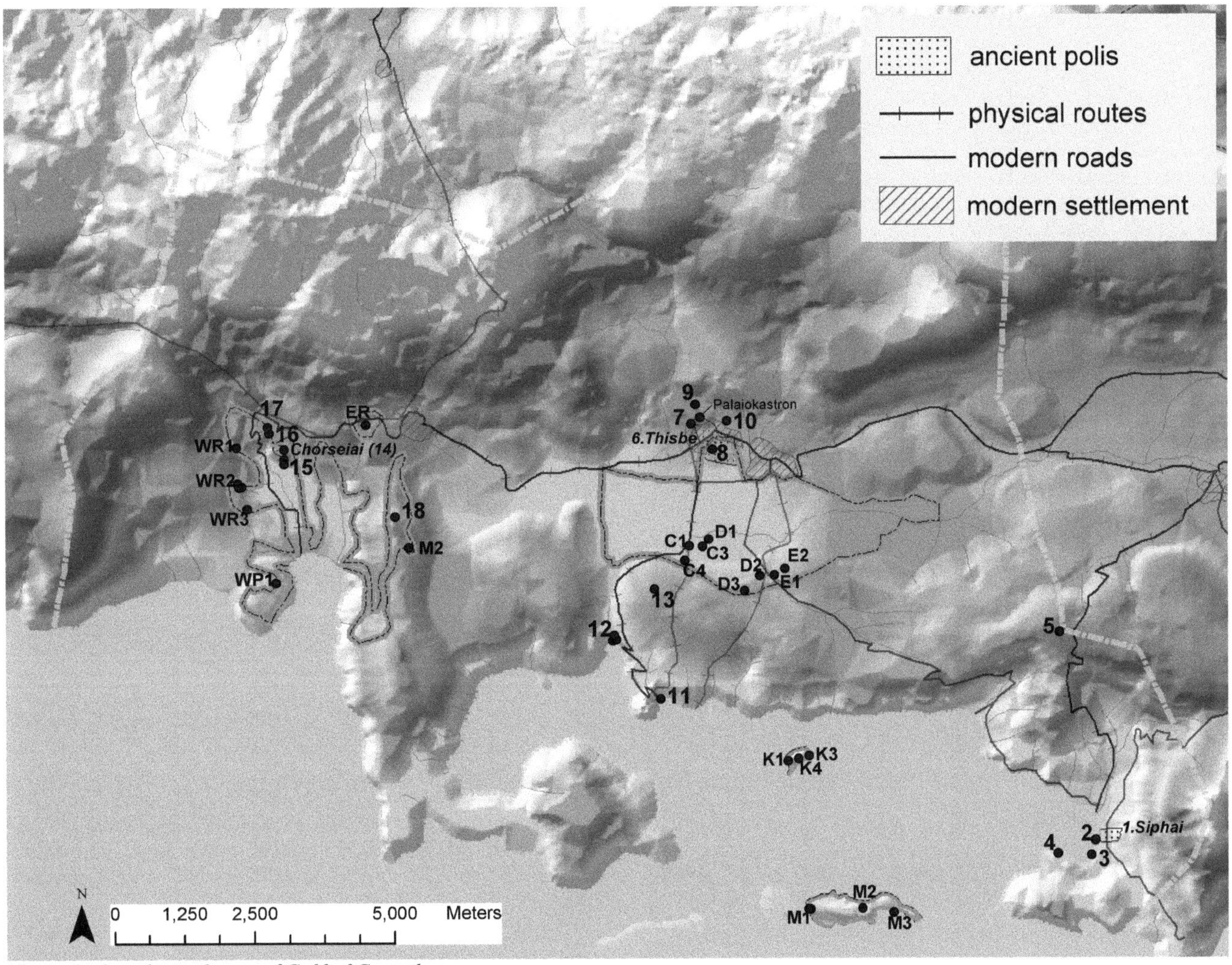

Fig.1. *Archaeological map of Gulf of Corinth area*

1. Aliki (ancient SIPHAI): *Components GC_1 to GC_8*

ID	Period	Type	Toponym	Discovery	Position
GC_1	EH	activity focus	Aliki	Other	certain
GC_2	MH	activity focus	Aliki	Other	certain
GC_3	LH	activity focus	Aliki	Travellers	certain
GC_4	A	settlement	Aliki	Personal or group interest	certain
GC_5	C	settlement	Aliki	Travellers	certain
GC_6	H	settlement	Aliki	Personal or group interest	certain
GC_7	R	settlement	Aliki	Personal or group interest	certain
GC_8	LR	settlement	Aliki	Personal or group interest	certain

The small bay of Aliki and the surrounding mountains has been identified with the settlement chamber of historical **Siphai**,[1] whose acropolis lay on top of the W peak of a rocky hill that reaches the very sea in the middle of Aliki bay, just above the modern settlement, and cuts the sandy beach into two sections, the larger being to the N (Fossey 1988: 168).

The site was already known to travellers (Leake 1835: 503; Frazer 1913: 164; Forchhammer 1857: 17-20, 30-32; Lebègue 1875: 89-97). Schwandner AA 1977: 513-551 constitutes a very detailed and useful study of the visible remains, quoted by Fossey 1988: 168. Fossey also visited the site,[2] and surface

[1] Siphai (or Tipha). *Siphai* is attested in Thucidides IV 76.2 and IG VII 207, *Tipha* in Pausanias (IX 32.4). The acropolis was walled in 424 (Thucidides IV 76.3).

[2] Fossey gives a short mention of the site in Tomlinson-Fossey 1970 and an account in Fossey-Giroux, *Teiresias Archaeologica* 1979: 3.

sherds information comes mainly from his visit (Fossey 1988: 171).
The site shows continuity of occupation, from the EH to the LR period.[3] On the acropolis site also Prehistoric material is known: EH, MH, LH.[4] Archaeological remains, also from the lower site [AE828], give the impression of a flourishing city in the Roman Imperial period, as already stated by Tomlinson and Fossey (BSA 1970: 243-263). Fossey (1988: 170) reports, for instance, that local inhabitants have spoken of the finding of building remains with tessellated floors (dated to the Imperial period) in the plain N of the acropolis. Also a small temple dating to the Imperial period has been noticed at the S end of Aliki beach (see *component GC_11* below).

Component GC_8: Evidence for the LR period does not seem to be so clear from the available information. Fossey (1988: 171) mentions Late Roman ripple ware seen during his own visit to the site. LR evidence from the area by the harbour, already in use in earlier periods, suggests a development of activities there (see below – *components GC_9 and GC_10*).

Fossey (1988: 171) notes how no tombs are reported as having been found in the area of Aliki, although several tombstones seem to come from the area (listed by Fossey 1988: 171 – C/H/R).

2. Aliki SW: *Components GC_9 and GC_10*

ID	Period	Type	Toponym	Discovery	Position
GC_9	LR	infrastructure	Aliki SW	Ext. topographical surveys	certain
GC_10	gr-rom	infrastructure	Aliki SW	Ext. topographical surveys	certain

LR evidence from the area by the harbour (*component GC_9*), already in use in earlier periods (*component GC_10*),[5] suggest a development of activities (probably also habitation) there.[6]

3. Aliki S: *Component GC_11*

ID	Period	Type	Toponym	Discovery	Position
GC_11	R	cult place	Aliki S end	Ext. topographical surveys	certain

Badly preserved remains of a small temple in the Corinthian order (extramural) and probably dating to the Imperial period (opinion of R.A. Tomlinson - from Fossey 1988: 171) have been noticed at the S end of Aliki beach.

4. Alonnisi Alikis: *Component GC_12*

ID	Period	Type	Toponym	Discovery	Position
GC_12	EH	settlement	Alonnisi Alikis	Personal or group interest	approximate

On the small promontory which dominates the entrance of a small sea lake (*limnothalassa*) on the Aliki shore on the Corinthian gulf, Sampson (AD 1981: 194) reports on the identification of an EH II settlement (on the basis of an EH artefact concentration [AE1479]) on the NE side of the small promontory.

5. Mavrovouni Kastron: *Components GC_13 to GC_18 and GC_79*

ID	Period	Type	Toponym	Discovery	Position
GC_13	G	cult place	Mavrovouni Kastron	Ext. topographical surveys	certain
GC_14	A	cult place	Mavrovouni Kastron	Ext. topographical surveys	certain
GC_15	C	cult place	Mavrovouni Kastron	Ext. topographical surveys	certain
GC_16	H	cult place	Mavrovouni Kastron	Ext. topographical surveys	certain
GC_17	C	fort	Mavrovouni Kastron	Travellers	approximate
GC_18	LR	fort	Mavrovouni Kastron	Ext. topographical surveys	approximate
GC_79	H	fort	Mavrovouni Kastron	Ext. topographical surveys	certain

On a hill on the W slopes of Mavrovouni ridge (named *Kastron* on 1:50,000 GYS map), NW of Aliki, on the left side of the road Aliki-Ksironomi.
The site was known already to the travellers (Lolling). Tomlinson and Fossey (BSA 1970: 243-263) visited the site during their topographical research in the area. Schwandner (AA 1977: 516-9) gives a preliminary account of the excavation of two buildings at the site, known also as *Ano Siphai*.
As noted by Fossey (1988: 173), we are dealing with a place with two different characters during its history of occupation: a cult place from G to H times, and a fort in C and Late C periods, and probably again in LR times. Schematically:
- G to H small sanctuary (*components GC_13 to GC_16*)
- C spartan military base (*component GC_17*)
- 4th C BC watch tower (*component GC_79*)
- LR small fortlet with tiny Christian chapel (*component GC_18*)

Fossey (1988:173) also notes that unlike in other sanctuary sites (Onchestos -Haliartos *chora*- or Ptoion - Akraiphiai *chora*- for instance), there is no indication that the Mavrovouni site was ever a settlement (cf. Lebègue 1875: 88), and he remarks how its history was purely religious with military intervals. Fossey (1988:167) includes, rather arbitrarily as he admits, the Mavrovouni sanctuary in the area of ancient Siphai, but it would lie actually on the border between Thespiae and Siphai and we do not know which city it belonged to.

[3] Geyer (RE 1929: 262-3 s.v. *Siphai*) collects ancient sources on Siphai.
[4] Fossey (1988: 171) reports LH II, IIIA-B; HS&D 1979: G37 report LH I/II? – LH IIIA-B. MH and LH sherds were found on the S slopes. Heurtley noticed remains of a wall in large blocks just below the summit; he suggested that it might be Mycenaean (Heurtley BSA 1923/4, 1924/5), but Fossey has not been able to find these remains.
[5] Harbour structures noticed underwater, remains of a small harbour in the gr-rom period, as well as a lot of debris and surface material of Roman date and mostly LR – 5th C AD or later, constitute the archaeological evidence available (Fossey 1988: 171).
[6] For the harbours in the Beotian area facing the gulf of Corinth (as well as for communication routes) see Heurtley's work in BSA 1923/4 - 1924/5: 40-1 and 44.

6. Kakosi-Thisvi (ancient THISBE): *Components GC_19 to GC_25, GC_77 and GC_78*

ID	Period	Type	Toponym	Discovery	Position
GC_19	EH	? activity focus	Kakosi/Thisvi Palaiokastron	Personal or group interest	certain
GC_20	MH	activity focus	Kakosi/Thisvi Palaiokastron	Personal or group interest	certain
GC_21	LH	settlement	Kakosi/Thisvi Palaiokastron	Personal or group interest	certain
GC_22	A	settlement	Kakosi/Thisvi	Personal or group interest	certain
GC_23	C	settlement	Kakosi/Thisvi	Travellers	certain
GC_24	H	settlement	Kakosi/Thisvi	Personal or group interest	certain
GC_25	R	settlement	Kakosi/Thisvi	Personal or group interest	certain
GC_29	R	burial place	Kakosi/Thisvi area	accidental	imprecise
GC_77	G	? Settlement	Kakosi/Thisvi	Int. and sys. artefact surface survey	certain
GC_78	LR	? Settlement	Kakosi/Thisvi	Int. and sys. artefact surface survey	certain

The modern village of Kakosi/Thisvi, W of Domvraina village, lies in a hollow between a flat plateau to the S and a roughly conical projection of Helicon to the NW. The inland basin, occupied in modern times by the two villages of Domvraina and Kakosi/Thisvi, has been identified with the settlement chamber of historical **Thisbe**.

The site was known to travellers (Dodwell 1819: 257-259; Leake 1835: 501-513; Frazer 1913: 160-164). It was known, apart from the fortification remains, also because of a series of rock-cut tombs and inscriptions (Gregory 1992 and Fossey 1988: 179).

The small *polis* of Thisbe was independent at least from the Hellenistic period, at the beginning of the Third Boeotian Federation (IG VII 2724b.6, ca 280-270 BC – Hansen 1996), while earlier was probably a dependent *polis* dominated by Thespiae.[7]

Both heights to the S (known as Neokastro or Pyrgos) and the NW (called Palaiokastro) of the modern village of Kakosi/Thisbe are fortified, but the proper acropolis was, according to Fossey (1988: 179), on the hill to the NW, not very high but quite steep-sided, except to the NW side.[8] From there one can have a good view over the whole inland plain (Fossey 1988: 179). The city wall was probably built in the Hellenistic period (Gregory 1992: 19 [AE1516]). On the other hand, Fossey gives also an A, Late C (4th C) and R date to the city walling [AE853, AE851, AE852].
The lower area of the city was lying in the saddle between the two fortified heights. It extended though also outside the city wall, apparently, until 160m E from the NE corner of the fortification, and to the S heavy artefact density continued to the edge of the survey area (ca. 1500m from the S tower of the city circuit), though the actual limit of the city has not been defined (Gregory 1992: fig. 3.3). These assumptions are based on the artefact density measured by Gregory (1992: 22) during his intensive systematic survey in the area (in 1979 and following years), though we know that intensive extra-mural, periurban area characterized several Mediterranean *poleis*, and Boeotian city sites as well (see Tanagra, for instance – appendix I.14 – THE TANAGRA SURVEY). Gregory (1992: 25) himself remarks that "*outside the fortification it seems possible to suggest an inverse relationship between the distance from the fortifications and artefact density*". Ancient artefacts had been reported before Gregory's survey from the far extremities of both Kakosi/Thisvi and Domvraina village. Sherds from the site[9] (only 2% of collected material could be dated with confidence, says Gregory 1992: 26) show occupation from the Mycenaean period onwards, with no significant historical period unrepresented. Mycenaean to Archaic surface finds were found especially on the Paliokastro hill.[10] EH (only Fossey 1988: 180 EH I-II as well as obsidian [AE1483]) and MH (Heurtley 1923/25: 41; Syriopoulos 1968; HS&D 1979: G35, from the acropolis) are also reported from the site, though not by the recent intensive work of Gregory.
A Frankish tower was built at the SE corner of the gr-rom fortification. On the plateau surface walls built with blocks reused from the fortification may date the buildings to a later period of occupation at the site, after the fortification had gone out of use (Gregory 1992: 26).

7/8/9. Kakosi-Thisvi: *Components GC_26* (7), *GC_27* (8) *and GC_28* (9)

ID	Period	Type	Toponym	Discovery	Position
GC_26	LH	burial place	Kakosi/Thisvi Palaiokastro W	Personal or group interest	certain
GC_27	LH	burial place	Kakosi/Thisvi Plateau	accidental	certain
GC_28	LH	burial place	Kakosi/Thisvi NW	Personal or group interest	approximate

Components GC_26 and GC_27: At the W foot of Kakosi-Thisvi Palaiokastro hill, and on the S and W side of the Pyrgos plateau S of Kakosi-Thisvi, rock-cut tombs cemeteries were already known to local inhabitants and to Frazer (1913: 162) – *component GC_27*, before archaeological researches were carried out on the site (Heurtley 1923/25: 41; Hope-Simpson 1965; AD 1970: 233). At least one of them is known as having produced Mycenaean material (Heurtley 1923/5: 41 - *component GC_26*).

Component GC_28: about half a mile NW of Kakosi-Thisvi village, near Palaiokastro hill, other tombs (apart from the rock-cut ones reported by Heurtley 1923/25 and others) have been noticed (HS&D 1979). Perhaps from here came the genuine elements of the so-called 'Thisbe Treasure' (Evans JHS 1925: 1-42).[11]

10. Kakosi-Thisvi area: *Component GC_29*

ID	Period	Type	Toponym	Discovery	Position
GC_29	R	burial place	Kakosi/Thisvi area	accidental	imprecise

[7] See Fiehn, RE 1936: 287-291 s.v. *Thisbe* for reference to Thisbe in ancient texts. For a preliminary investigation in the city area see Rolfe-Tarbell 1890: 112.

[8] See plan in Fossey 1988: fig.23, based on Maier, AM 1958 pl.19.1, and the plan in Gregory 1992: 20 fig. 3.2.

[9] From the acropolis and from a plateau to the S (see HS&D 1979: G35 *Paleokastro* and *Neokastro*).

[10] No explicit mention in previous accounts nor in Fossey 1988 nor in Gregory's survey account (1992) of G or LR surface finds at the city site (*components GC_77 and GC_78*).

[11] The so-called 'Thisbe Treasure' is a royal treasure of gold rings and bead seals in the same metal found in 1915 in a Mycenaean rock tomb by the site of Thisbe, by the village of Domvraina, when war conditions diverted the course of discovery from official channels (Evans JHS 1925: 1-42).

Accidental findings probably belonging to R burials (BCH 1920: 387-8), as well as Roman tombs have been found around the village (Fossey 1988: 181, though their exact location is not specified - BCH 1955: 257).

11. Ormos Ag. Ioannis: *Component GC_30*

ID	Period	Type	Toponym	Discovery	Position
GC_30	gr-rom	infrastructure	Ormos Ag. Ioannis	Ext. topographical surveys	certain

The small harbour of Agios Ioannis, in Domvraina bay, to the SE of Vathy harbour, must have been the secondary harbour of ancient Thisbe. The main one was Vathy (Heurtley 1923/25; Gregory 1992). Surface walls, evidence of ancient use, have been found (Gregory 1992: 26), as well as traces of a road running S of the modern village of Kakosi/Thisvi to the small harbour of Agios Ioannis (Fossey 1988: 180).

12. Vathy (gulf of Domvraina): *Components GC_31 to GC_36*

ID	Period	Type	Toponym	Discovery	Position
GC_31	gr-rom	infrastructure	Vathy	Ext. topographical surveys	certain
GC_32	C	infrastructure	Vathy	Personal or group interest	certain
GC_33	LR	infrastructure	Vathy	Personal or group interest	certain
GC_34	LR	? Settlement	Vathy	Personal or group interest	certain
GC_35	gr-rom	fort	Vathy	Personal or group interest	approximate
GC_36	gr-rom	fort	Vathy	Personal or group interest	approximate

Vathy location consists in a tiny but safe and drinking-water-provided beach, as described by Heurtley in his work on the Boeotian ports on the Gulf of Corinth (1923/25: 41). Vathy has been identified with the main harbour of ancient Thisbe (see above).
The plain is ca 72m N-S and 55 E-W; on three sides it is surrounded by steep cliffs and on the S it is open to the sea (Gregory 1992: 28). According to Gregory most traces of ancient occupation have been obscured by heavy alluviation. A powerful fresh-water spring has been noticed by Gregory (1992: 29) on the E side of the harbour.
An ancient road has been noticed [AE1488], leading from ancient Thisbe to the port of Vathy, through a saddle at the SW corner of the basin (Gregory 1992: 28). The road was following a narrow stream-bed cut (see fig.1).
A fortified gate[12] to control the access between the sea at Vathy and the Thisbe basin above is reported (Gregory 1992: 28, and already Heurtley 1923/25: 41) at the N end of the Vathy beach (*component GC_36*).

Component GC_35: Above Vathy harbour on the perpendicular cliff to the E, a gr-rom tower, described by Heurtley (1923/25: 41), commanded a view of both harbours (Vathy and also Agios Ioannis to the E) and the whole bay of Domvraina, including its entrance from the Gulf of Corinth (when preserved to its original height, as Gregory remarks). The tower is described also in Gregory (1992: 28), who notes how probably it was in visual communication with Thisbe through the gr-rom tower seen by some travellers on Mt. Kivadio (see *component GC_37*).

Some structures can be linked to the gr-rom occupation and use of the harbour site, in addition to the fortifications (mainly from Gregory 1992 and Heurtley 1923/25): a cistern [AE875] to the SW corner of the beach; along the waterfront a series of walls with associated LR material [AE1492 + AE1493]; building foundations along both sides of the harbour [AE1494], which constitute *component GC_31*.
C sherds testify the presence of harbour installation here in C period (Gregory 1992) - *component GC_32*. Along the waterfront a series of walls with associated LR material [AE1492 + AE1493] - *component GC_33* (Gregory 1992), as well as a LR artefact concentration, could testify the presence of a habitation site by the harbour installation (*component GC_34*) – see above the development in the LR period at Aliki harbour site.

13. Kivadio: *Component GC_37*

ID	Period	Type	Toponym	Discovery	Position
GC_37	gr-rom	fort	Kivadio	Travellers	approximate

Gr-rom tower seen by the travellers (Leake 1835: 507, who reported it as 'Hellenic') on Mt. Kivadio, but apparently no more visible today (reported in Gregory 1992: 28).[13] Its position could not be located with precision, but probably it was on the ridge looking down upon the port of Thisbe (Vathy), ridge which separates the plain of Thisbe from the coast (marked as Kimadio in the map reported in Fossey and Gauvin 1985, fig.4 and as Megali Rachi in the 1:50,000 GYS map). The tower was overlooking the road from Thisbe to the port, similar in function to that upon the ridge above Aliki (see above).

THISBE BASIN survey sites: *Components GC_38 to GC_54*
What follows is a list of sites located by the intensive and systematic survey work carried out by T.E. Gregory in the area of ancient Thisbe and its harbour(s), in 1979 and following years, and published in 1992. Several discrete artefact concentrations were noticed during surveying in the plain (Gregory 1992: 32-3, map in fig. 3.5).

ID	Period	Type	Toponym	Discovery	Position
GC_38	C	rural site/activity	Thisbe basin D-3	Int. and sys. artefact surface survey	certain
GC_39	C	settlement	Thisbe basin C-1	Int. and sys. artefact surface survey	certain
GC_40	H	settlement	Thisbe basin C-1	Int. and sys. artefact surface survey	certain
GC_41	LR	settlement	Thisbe basin C-1	Int. and sys. artefact surface survey	certain
GC_42	C	rural site/activity	Thisbe basin C-3	Int. and sys. artefact surface	certain

[12] A gr-rom wall noticed just before the roadway reaches the Vathy plain (Gregory 1992: 28, and already Heurtley 1923/25: 41).

[13] Leake does not mention the name of the ridge, which has been identified though with Mt. Kivadio.

ID	Period	Type	Toponym	Discovery	Position
				survey	
GC_43	H	rural site/activity	Thisbe basin C-3	Int. and sys. artefact surface survey	certain
GC_44	gr-rom	activity focus	Thisbe basin C-4	Int. and sys. artefact surface survey	certain
GC_45	C	rural site/activity	Thisbe basin D-1	Int. and sys. artefact surface survey	certain
GC_46	H	rural site/activity	Thisbe basin D-1	Int. and sys. artefact surface survey	certain
GC_47	R	rural site/activity	Thisbe basin D-1	Int. and sys. artefact surface survey	certain
GC_48	LR	rural site/activity	Thisbe basin D-1	Int. and sys. artefact surface survey	certain
GC_49	C	rural site/activity	Thisbe basin D-2	Int. and sys. artefact surface survey	certain
GC_50	H	rural site/activity	Thisbe basin D-2	Int. and sys. artefact surface survey	certain
GC_51	LR	rural site/activity	Thisbe basin E-2	Int. and sys. artefact surface survey	certain
GC_52	C	rural site/activity	Thisbe basin E-1	Int. and sys. artefact surface survey	certain
GC_53	H	rural site/activity	Thisbe basin E-1	Int. and sys. artefact surface survey	certain
GC_54	LR	rural site/activity	Thisbe basin E-1	Int. and sys. artefact surface survey	certain

Thisbe basin D-3: *component GC_38*
In the SE corner of the plain, on a pocket of good land extending into a cleft in the S ridge at an elevation of up to 200m, a concentration of C potsherds [AE1495] was found. In the same area remains of one or two structures [AE1496] that are probably to be dated to the C period (Gregory 1992: 33).

Thisbe basin C-1: *components GC_39 to GC_41*
A large rural agglomeration, ca 3ha in extension, which could have been a hamlet or a small village of recognized C, H and LR date (Gregory 1992: 33).

Thisbe basin C-3, D-1, D-2, E-2, E-1: *components GC_42 and GC_43, GC_45 to GC_54*
Rural sites/activities have been recognized during survey work (Gregory 1992: 33). Site E-2 (*component GC_51*) shows occupation only in LR times (836sqm in extension and with a quite high average artifact density – 6.97 sherds per sqm).

Thisbe basin C-4: *component GC_44*
Only gr-rom architectural fragments were found on the surface. They can represent an activity focus by themselves or they would probably be in connection with C-1, according to the director of the survey project (Gregory 1992: 33).

Except one site (*component GC_51*: LR site), the totality of rural sites found in the survey show C (namely Late C - Early H) occupation, confirming the pattern of the Boeotian rural landscape given by the results of the Stanford-Cambridge and Leiden-Ljubljana Boeotia survey project, which sees a progressive infill of the landscape in C-Early Hellenistic period in other areas of Boeotia. The establishment of *component GC_51*, jointly with continuity of occupation of earlier sites in the LR period, give also a picture of flourishing in late antiquity, which has to do with the usual settlement pattern recurrent in the Boeotian countryside, but also with the increased importance of the port of Thisbe at Vathy influencing all the area of the ancient city.

Kouveli island: *Components GC_55, GC_56 and GC_86*

ID	Period	Type	Toponym	Discovery	Position
GC_55	LR	rural site/activity	Kouveli island K-1	Int. and sys. topographical surveys	certain
GC_56	LR	settlement	Kouveli island K-3	Int. and sys. topographical surveys	certain
GC_86	LR	? Settlement	Kouveli island K-4	Int. and sys. topographical surveys	certain

Kouveli is a small island which lies in the Domvraina bay, 2.5km ca to the SE of the small harbour of Ag. Ioannis. It is roughly triangular in shape, less than 1km in length, low-lying in its S extremity, but rising to a height of 60m near the centre of the island. It is absolutely barren and desolate, with steep rocky sides rising sharply from the sea and only scrub vegetation (Gregory 1986: 18).
A preliminary note (Gregory 1980: 39) mentions surface pottery showing continuous occupation from the 3rd to the 12th C AD, as well as extensive architectural remains. Activities in the area are therefore attested from LR period onwards. Fossey (1988: 184) suggests it may have served as a refuge for people from Thisbe and nearby sites. On the other hand, in 1981 the Ohio State University and Kenyon College, led by T.E. Gregory, carried out intensive and systematic fieldwork, during which all the island, apart from the steepest slopes, was walked by individuals spaced at five-meter intervals (Gregory 1986). The project discovered four sizable 'sites', which turned out to be something different than simple refuge sites (in the period of Slavic invasion), as attested also by the settlement presence on the Makronisos island (see below) and by their early occupation in LR period. All sites show occupation between the later years of the Roman Empire and the early years of Byzantium, roughly from the 3rd to the 7th C AD.
The larger site is K-3 (*component GC_56*), where 21 buildings or part of buildings were recognized, some large of common use, some others smaller, presumably houses (Gregory 1986: 19).

Makronisos: *Components GC_87 to GC_90*

ID	Period	Type	Toponym	Discovery	Position
GC_87	LR	settlement	Makronisos M-1 (Diporto)	Int. and sys. topographical surveys	certain
GC_88	LR	? Settlement	Makronisos M-3	Int. and sys. topographical surveys	certain

ID	Period	Type	Toponym	Discovery	Position
GC_89	LR	rural site/activity	Makronisos M-2	Int. and sys. topographical surveys	precise
GC_90	LR	infrastructure	Makronisos M-1 (Diporto)	Int. and sys. topographical surveys	certain

Makronisos (Long Island) is the easternmost of the three islands which block the entrance to the Gulf of Domvraina. It is about 1km long and half a kilometre wide, dominated by a ridge as much as 120m high. Its landscape is similar to that of Kouveli island, though pine trees are considerably present (Gregory 1986: 18).

In 1982 and 1984 the Ohio State University and Kenyon College (led by T.E. Gregory) carried out intensive and systematic fieldwork, in which all the island, apart from the steepest slopes, was walked by individuals spaced at five-meter intervals (Gregory 1986: 20-21). The survey work discovered a considerable settlement, showing urban planning and economic differentiation, by a landing place (at Diporto, site M-1), as well as ten other sites, among which a small harbour settlement (M-3), a hamlet or villa site – M-2[14] – and a number of isolated farms for agricultural and mainly pasturage and bee-keeping activities, most of them dating from Late Antiquity and the Early Bizantine period (4th through 7th C AD).[15]

The most impressive site is M-1 (Diporto – *component GC_90*): on the W end of Makronisos, at a spot called Diporto, with high ridges shutting it on three sides and a harbour on the W. Surface walls, pottery, glass, and coins are reported. 57 buildings or part of buildings, most of them apparently houses, but also dock facilities and a church (4th to 7th C AD) - Gregory 1986: 20.

------

14. Prodromos Kastron (ancient CHORSEIAI): *Components GC_57 to GC_66*

ID	Period	Type	Toponym	Discovery	Position
GC_57	N	? activity focus	Prodromos Kastron	Personal or group interest	approximate
GC_58	EH	human presence	Prodromos Kastron	Personal or group interest	approximate
GC_59	MH	activity focus	Prodromos Kastron	Personal or group interest	approximate
GC_60	LH	activity focus	Prodromos Kastron	Personal or group interest	approximate
GC_61	G	settlement	Prodromos Kastron	Personal or group interest	approximate
GC_62	A	settlement	Prodromos Kastron	Personal or group interest	approximate
GC_63	C	settlement	Prodromos Kastron	Travellers	approximate
GC_64	H	settlement	Prodromos Kastron	Personal or group interest	approximate
GC_65	R	settlement	Prodromos Kastron	Personal or group interest	approximate
GC_66	LR	settlement	Prodromos Kastron	Personal or group interest	approximate

According to Fossey's description (1988: 187), 2.4km from the sea terminates a rocky spur of Helicon, between two valleys (the larger ones to the W) descending into the Sarandi bay. The spur is very steep on all sides except the N where it connects with the main massif. The area has been identified with the settlement chamber of ancient **Chorseiai**, and it is occupied by the modern village of Prodromos/Chostia, to the E of the ancient site.

The site was known to travellers (Leake 1835: 514-6; Forchhammer 1857: 24-27; Lebègue 1875: 107-112). Heurtley (1923/25) gives an account on the site in his work on Boeotian ports.

Fossey carried out at the site intensive fieldwork and excavation, published in Fossey Khostia I. Information recorded in the database is mainly from his work at the site, which added new information and confirmed older visits (for prehistory Heurtley 1923/25: 42; for gr-rom period Austin in February 1926 and de Ridder). A good plan of the site is available in Fossey 1988: 189, as well as a good description of the walling phases (190-1).

Occupation in prehistoric periods is attested at the site, although the entity of the remains (surface scattered sherds mainly) do not allow us to give a character of the occupation. Heurtley had apparently picked up only a few Mycenaean sherds at the site, as well as did Burn (1949).[16] We do not know more, though, not even through the excavations carried out at the site focusing on historical periods.

The gr-rom fortress of Chorseiai was guarding the frontier between Boeotia and Phokis (Heurtley 1923/25: 42). In 347/7 Chorseiai got captured by the Phokians (Diodoros XVI 58 and 60). Life at the city site though continued for the whole H period. Excavation work reports (Fossey 1988: 191) the abandonment of the site (after a destruction) in the 2nd or 1st C BC and a reoccupation from the 2nd to at least the 5th C AD. Construction of the imperial (2nd to 5th C AD) period seems to have been removed all but traces of earlier buildings. The most noticeable feature of this Roman work is the series of buildings backed onto the E and W circuit walls (Fossey 1988: 192).

15. Prodromos Kastron S: *Components GC_67 and GC_68*

ID	Period	Type	Toponym	Discovery	Position
GC_67	H	special activity	Prodromos Kastron - South	Personal or group interest	approximate
GC_68	R	burial place	Prodromos Kastron - South	Personal or group interest	approximate

Traces of an ancient aqueduct [AE1521] have been noticed, carrying water probably from the water source at Ag.Taxiarchoi nearby to the N, already noticed by Leake (1835: 514-6), as well as extramural industrial activity (*component GC_67*), on the W side of an ancient road descending from the S gate of Prodromos Kastron fortification (Fossey 1988: 192).

[14] M-2: villa site, with four phases (two of which dated to late antiquity). In latest periods used for seasonal herding, while in the earlier periods (LR-Early Byzantine) houses and outbuildings testify the presence of a settlement contemporary to that at Diporto (4th to 7th C AD), devoted at least partly to farming (Gregory 1986: 20-21).

[15] Some of them are occupied in later periods only (more Byzantine than LR), such as site M-4 (mapped on map in Gregory 1986: 18), an Early Byzantine cave site occupied in the period of Slavic invasions or Arab incursions to Greece during the 7th to 8th C AD.

[16] Burn 1949 mentions 'Chostia', but probably meaning the ancient historical site (*Chorseiai*).

At various points near the ancient roadway descending from the S gate are at least three rock-cut sarcophagi dated after the 2nd C AD (Fossey 1988: 192 – *component GC_68*).

16. Prodromos Kastron N: *Component GC_69*

ID	Period	Type	Toponym	Discovery	Position
GC_69	gr-rom	cult place	Prodromos Kastron - West	Personal or group interest	approximate

Remains of an elongated, aperipteral temple, partly concealed by a modern threshing floor, have been noticed on the W side of the ridge on a flat, exposed rock face (Fossey 1988: 192 and plan fig. 24). To its NE side runs the aqueduct [AE1521]. A mention of the temple is also in AA 1972: 85.[17]

17. Prodromos Kastron NW: *Component GC_70*

ID	Period	Type	Toponym	Discovery	Position
GC_70	gr-rom	special activity	Prodromos Kastron - NW	Int. topographical surveys	approximate

Along the E side of the Prodromos Kastron ridge, to the side of the aqueduct [AE1521], is a large rock outcrop on which can be seen extensive traces of ancient quarrying (noticed by Fossey 1988: 192).

18. Mali: *Components GC_71 to GC_74*

ID	Period	Type	Toponym	Discovery	Position
GC_71	LH	activity focus	Mali	Int. topographical surveys	approximate
GC_72	PG	settlement	Mali	Int. topographical surveys	approximate
GC_73	G	activity focus	Mali	Int. topographical surveys	approximate
GC_74	A	activity focus	Mali	Int. topographical surveys	approximate

To the S of the modern road between Chostia and Sarandi valley rises the ridge of Malia Senga, the first peak of which is known as Mali (Fossey 1988: 194). In 1980 the Canadian Chostia excavation team first noticed here extensive remains of an ancient fortified site, during intensive survey in the area of ancient Chorseiai (Fossey-Gauvin 1985a: 71-76). Apart from surface walls constituting buildings [AE899], small quantities of surface sherds have been recorded as belonging to PG, G and A periods. Also LH material does not seem to be quantitatively considerable.

Fossey (1988: 194-5) notes how "*it is tempting to suppose that it was the settlement for the area at precisely a period when the* Kastron *does not appear to have been inhabited, namely the 'Dark Ages'* ". Since the site shows traces of fortification [AE898], it might be a suitable explanation, though there are not enough elements to attest it.

Chorseiai Countryside survey: *Components GC_80 to GC_85, GC_75 and GC_76*

ID	Period	Type	Toponym	Discovery	Position
GC_80	C	rural site/activity	Chostia WR-1	Int. and sys. topographical surveys	certain
GC_81	LR	rural site/activity	Chostia WR-2	Int. and sys. topographical surveys	certain
GC_82	C	rural site/activity	Chostia WR-2A	Int. and sys. topographical surveys	certain
GC_83	C	rural site/activity	Chostia WR-2C	Int. and sys. topographical surveys	certain
GC_84	C	rural site/activity	Chostia WR-3	Int. and sys. topographical surveys	certain
GC_85	C	rural site/activity	Chostia M-2	Int. and sys. topographical surveys	certain
GC_75	C	rural site/activity	Chostia WP-1 (Sarandi Peninsula)	Int. topographical surveys	approximate
GC_76	R	? rural site/activity	Chostia ER (Eastern Ridge)	Int. and sys. topographical surveys	certain

Various remains (structural remains mainly) of rural sites/activities, mainly of C/H date, were noticed during survey work by the Canadian team in 1983, over an extended area on a ridge to the W of the main *Kastron* site and facing the latter across the Phikeza valley (Gauvin 1986: 51-81; Fossey 1988: 195 – WR sites[18]), as well as on the so-called Western peninsula (overlooking Sarandi bay from W – WP site) and in the area of Mali site (M site) and on the so-called Eastern ridge. In Gauvin (1986) descriptions are given for each single site.

Chostia WR-1: *Component GC_80*
On the W ridge in ancient Chorseiai area. 800m W/NW of Kastron site. In a narrow valley of a stream, marked by a modern route, below an escarp (260m ca asl), surface walls and sherds (coarse ware + glaze tiles) were noticed (Gauvin 1986: site WR-1; plan fig. 5.2).

Chostia WR-2: *Components GC_81, GC_82 and GC_83*
On the W ridge in ancient Chorseiai area. 300m S of the stream running by site WR-1 (*component GC_80*), at the forming of another small stream (200m ca asl). Building foundations, terracing walls and potsherds ('Late') were noticed (Gauvin 1986: site WR-2) – *component GC_81.*
At the forming of the small stream (200m ca asl), where a steep slope starts, the structure identified as site WR-2A was found, with surface walls marking an enclosure and few glaze tiles and few coarse ware sherds (Gauvin 1986: site WR-2A; plan fig. 5.3) – *component GC_82.*
On the opposite bank of the stream compared to WR-2A (ca 100m W-NW of it), the small building identified as site WR-2C was found, along with few glaze tiles and few coarse ware sherds (Gauvin 1986: site WR-2C; plan fig. 5.4) – *component GC_83.*

Chostia WR-3: *Component GC_84*
On the W ridge in ancient Chorseiai area. On the slopes above the last hill before the descent towards the Sarandi bay, to the W

[17] At Paralia Sarandi, below ancient Chorseiai, an inscription dated to the 2nd C BC, mentioning the construction of a temple, is reported in AD35 1980: 224-5 [AE1826].

[18] In Fossey 1988: 195 the sites are not listed separately but in a single 'site' named 'The Western ridge'.

of the small plateau which constitutes the summit of the Western ridge, surface walls marking buildings (with two round shape towers), terrace walls along with glaze tiles and few coarse ware sherds were found (Gauvin 1986: site WR-3; plan fig. 5.5).

Chostia M-2: *Component GC_85*
750m SE of Mali, on the highest slopes, in an area marked by small terraces (marked by small walls), overlooking the Thisbe basin, only a wall clearly visible on the surface and a large quantity of glaze tiles were noticed (Gauvin 1986: site M-2; plan fig. 5.7).

Chostia WP-1 (Sarandi Peninsula): *Component GC_75*
As Fossey (1988: 195) describes, the W side of the Sarandi bay is enclosed by a small, essentially oval-shaped, peninsula and hill whose summit (Rachi Tsourgias on the 1:50,000 GYS map) rises a little over 100m above sea level.[19]
The Canadian team in 1983, during survey work in the area, noted on the peninsula ancient remains (Gauvin 1986: 70-71 – site WP-1). They represent a large farm site, marked by surface Late C- Early H tiles and remains of an extensive enclosure.[20]

Chostia Eastern Ridge: *Component GC_76*
In the Northern part of the so-called Eastern ridge, above the modern village of Chostia/Prodromos, a small Roman site has been discovered during survey work by the Canadian team in 1980-81 (Fossey and Gauvin 1983: 237, map fig.1).

As seen above, LR scattered sherds have been noticed at the site named Chostia WR-2 (*component GC_81*). Gauvin (1986: 81), in concordance with Fossey (1988: 195), notes how "*surface sherds are rare but some are reminiscent of those found in the latest occupation deposits at the* Kastron *and suggest the possibility that after the destruction there in Late Roman times the population, or its remnants, may have moved to this Western Ridge*" (Fossey 1988: 195). This interpretation should be verified, since the picture given by the results of the survey seem very much close to the usual pattern in Boeotian countryside given by intensive and systematic surface survey in the framework of the Boeotia Survey project as well as in the Thisbe basin survey project presented earlier (C-H rural infill, then again a strong occupation of the countryside in LR times), which usually has little to do with the abandonment of the main places in the area.

[19] W of this lies another, much more open bay approached from Sarandi by a track over the saddle which separates the peninsula hill from the rest of the ridge to the N (Fossey 1988: 195).
[20] For a possible parallel, see farm sites found by Lohmann in Attica – Athene survey (Lohmann 1985).

Appendix I.11
The upper Asopos basin: Parasopia and Plataea

THE ARCHAEOLOGICAL RECORD

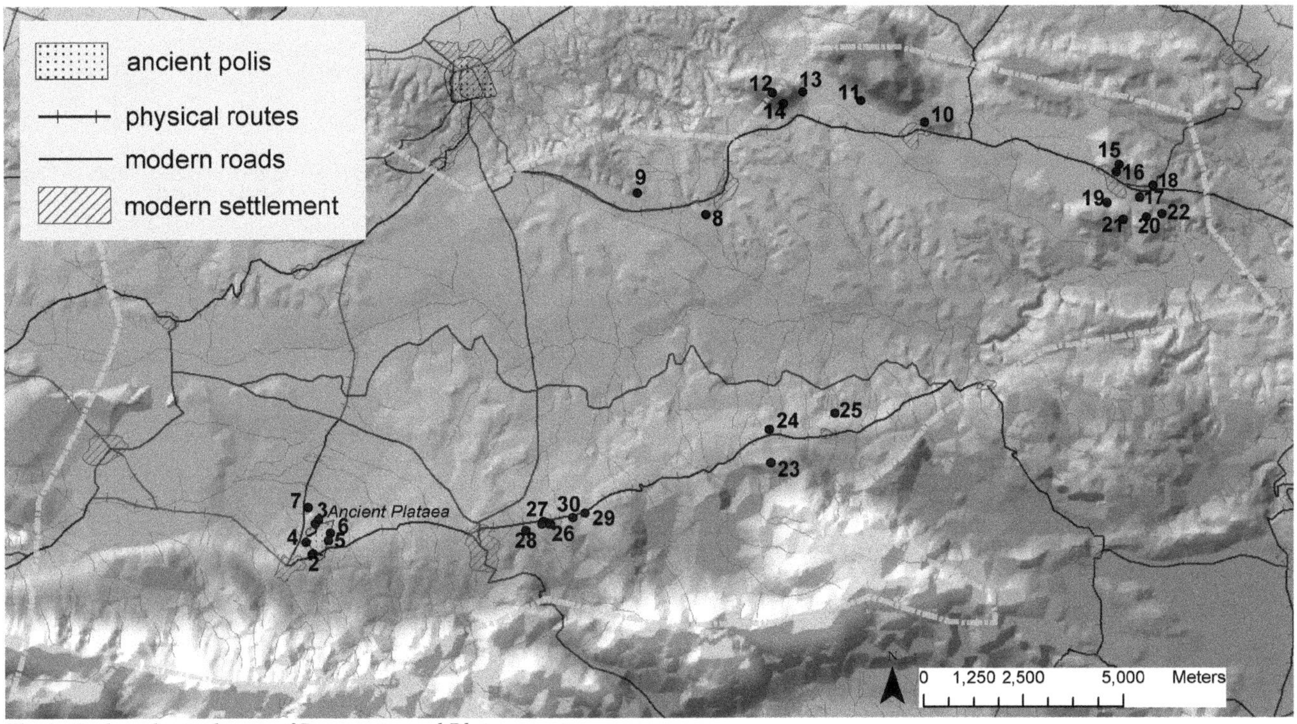

Fig.1. *Archaeological map of Parasopia and Plataea region.*

1. PLATAEA: *Components PP_1 to PP_10; PP_76 and PP_77*

ID	Period	Type	Toponym	Discovery	Position
PP_1	N	human presence	Plataea	Personal or group interest	certain
PP_2	EH	? activity focus	Plataea	Personal or group interest	certain
PP_3	MH	? activity focus	Plataea	Ext. topographical surveys	certain
PP_4	LH	activity focus	Plataea	Ext. topographical surveys	certain
PP_5	G	settlement	Plataea	Ext. topographical surveys	certain
PP_6	A	settlement	Plataea	Ext. topographical surveys	certain
PP_7	C	settlement	Plataea	Travellers	certain
PP_8	H	settlement	Plataea	Int. and sys. artefact surface survey	approximate
PP_9	R	settlement	Plataea	Ext. topographical surveys	certain
PP_10	LR	settlement	Plataea	Ext. topographical surveys	certain
PP_11	LR	burial place	Plataea	Personal or group interest	certain
PP_76	LH	activity focus	Plataea	Personal or group interest	certain
PP_77	MH	? activity focus	Plataea	Personal or group interest	certain

The ancient city of Plataea lay by the modern village of Koukkla/Plataea, to the NE of it, on a large plateau which projects Northward from the steep slopes of Kithairon (Fossey 1988: 102). The modern village lies on the W flank of this plateau. Below, at the SW fringes, a large permanent spring blows out, and several other water sources are present in the area (Philippson 1951: 504 and IGME geological map).

The localization of the ancient Plataea site, already known to travellers (Leake and others, listed in Fossey 1988: 102), was based on Strabo (IX 412), with the epigraphic testimonial as decisive evidence (Fossey 1988: 112).

Afterwards, the site was of interest to different extensive research and excavations, which are reported in the archaeological reports (a summary in Fossey 1988: 109). In the last decade on the site have been carried out intensive surveys and excavations[1] by an Austrian team (Konečny – Austrian Archaeological Institute) in collaboration with the local *ephoreia* (Aravantinos).[2]

The so-called 'acropolis' in the NW part of the site was the area of the earlier settlement at Plataea, since G period (Konečny

[1] Excavations have been carried out at the W gate and at the SW frindges of acropolis, by the great spring.

[2] In addition to the AD and AReports on the project (from 1997 onwards), see reports on Austrian journals: *Jahreshefte des Österreichischen archäologischen Instituts in Wien* (*OJH*) 68 and 69. See also Konečny et al. 2008.

AReports 1997-98: 60).[3] Different phases of walling have been recognized in the city wall surrounding the city in different periods (Fossey 1988: 102-108 and fig.14). The first phase of fortification has been recognized as Archaic (Konečny AReports 1988-1999: 56-7), surrounding the NW area, while the largest circuit wall dates to C period.[4]
A Classical temple [AE1553], with an earlier (A) temple underneath [AE1552], was excavated by the Americans in 1891 (AJA 1891: 390-399). The *katagogion* for the temple, partially excavated, became later the Roman *agora* (Praktika 1889: 42-56) [AE217].
The settlement probably reduced in size and life after the destruction of Plataea by the Thebans in 426 BC[5] (Late Classical – Hellenistic period), and flourished again in R times.[6]
A LR settlement is also known at the site (Kirsten 1956: 2316; AJA 1890: 459; Fossey 1988: 103-4 and note n.11), probably concentrated in the NW area.[7] Within the city area, outside the NW 'acropolis', LR tombs are reported (*component PP_11* - Spyropoulos AAA 1973: 375-9; Fossey 1988: 104), testifying of the probable shrinking of the city at that time.

Prehistoric occupation is also present at the site. In the NW part of the city area MH and LH[8] sherds were noticed earlier (HS&D 1979: G39; Fossey 1988: 103 – *components PP_3 and PP_4*), while the new Greek-Austrian intensive research excavated prehistoric deposits at the SW fringes of acropolis – *components PP_1, PP_2, PP_76 and PP_77* (listed below no.2).

2. Plataea SW frindges: *Components PP_1, PP_2, PP_76 and PP_77*

ID	Period	Type	Toponym	Discovery	Position
PP_1	N	human presence	Plataea	Personal or group interest	certain
PP_2	EH	? activity focus	Plataea	Personal or group interest	certain
PP_76	LH	activity focus	Plataea	Personal or group interest	certain
PP_77	MH	? activity focus	Plataea	Personal or group interest	certain

The new Greek-Austrian intensive research excavated LH layers at the SW fringes of acropolis, just above the great spring (*Megali Vrisi*), with MH pottery, EH pottery, and few Neolithic sherds found in the excavation though not within stratigraphic layers (AD 1995: 310).[9]
A Neolithic or EH site is reported on a low hill in the plain to the N of the ancient city (see below, Plataea N - Kotronaki).

3 - 6. Plataea burial places: *Components PP_12 and PP_13, PP_14* (3); *PP_17* (4), *PP_18* (5) *and PP_19* (6)

ID	Period	Type	Toponym	Discovery	Position
PP_12	H	burial place	Plataea	accidental	imprecise
PP_13	R	burial place	Plataea	accidental	imprecise
PP_14	gr-rom	burial place	Plataea N	Personal or group interest	approximate
PP_17	gr-rom	burial place	Plataea W	Travellers	certain
PP_18	gr-rom	burial place	Plataea E	accidental	approximate
PP_19	LR	burial place	Plataea E	Personal or group interest	approximate

Fossey (1988: 109) reports cemeteries to N, E and W of Plataea city walls (*components PP_14; PP_17 to PP_19*), mainly located by the Americans (AJA 1891: 390-1).[10] Later (LR) tombs are known within the wall circuit, especially to the E (*component PP_11* – see above), as well as Early Christian tombs in the E extramural area with buildings of the same date (*component PP_19* - AJA 1889: 442; 1890: 451; Spyropoulos AAA 1973: 375-9; Fossey 1988: 104 and plan fig. 13).
Epigraphical evidence also attests the presence of cemetery areas (*components PP_12 and PP_13*).

7. Plataea N – Kotronaki: *Components PP_15 and PP_16*

ID	Period	Type	Toponym	Discovery	Position
PP_15	N	activity focus	Plataea N - Kotronaki	Ext. topographical surveys	approximate
PP_16	EH	activity focus	Plataea N - Kotronaki	Ext. topographical surveys	approximate

On a low hill, steep on three sides, in the plain to the N of the ancient city, a Prehistoric site was reported, likely Neolithic or EH (AD 1969: 186; HS&D 1979: G39). A more recent report in AD 1981: 194 lists Late Neo and EH I/II in locality Kotronaki and refers to the same location.

NEOCHORAKI - KALLITHEA - SOROS

8/9. Neochoraki: *Components PP_20 to PP_22* (8); *PP_33 and PP_75* (9)

ID	Period	Type	Toponym	Discovery	Position
PP_20	G	burial place	Neochoraki	Historical geography	approximate
PP_21	A	burial place	Neochoraki	Historical geography	approximate
PP_22	C	burial place	Neochoraki	Historical geography	approximate
PP_33	A	burial place	Neochoraki W	Rescue excavation	approximate
PP_75	C	burial place	Neochoraki W	Rescue excavation	approximate

Pritchett (1965: 107-108) mentions a G-A-C cemetery in locality Neochoraki. He remarks on the depth at which the C sherds occurred in the cemetery area (from Fossey 1988: 122

---

[3] In contrast with earlier excavators who argued that the oldest area was the S part of site while the N part was a later expansion. On the other hand, Pritchett (1965: 119) already showed that the N area has focus for all periods, from Bronze Age to Byzantine.
[4] Leake (1835) already noted the contrast between the vast walled area and Thucydides (II 2.1-2) description of it as 'a small city'.
[5] Sketch map of the battle of Plataea (Skolos site, etc.) in Pritchett 1980: 293.
[6] The circuit wall is smaller in R times but occupation was found also outside the city wall.
[7] Nevertheless, buildings and Byzantine churches are spread all over the site (see AJA 1890 plan XXIII and 1889: 439-442; Fossey 1988: 103).
[8] LH layers were also excavated there.
[9] By the spring were found also H-R-LR, as well as Byzantine water conducts and few Byzantine walls (Konečny pers.comm.) [AE1549, AE1550, AE1551]. Architectural remains of the spring itself were also found during excavation. Yet by *Megali Vrisi*, to the W of the 'rampart', fragments "*d'entablements et de stèles d'époque romaine*" were accidentally discovered (AD 1994: 290) [AE1788].

[10] To the W of the ancient city were found stone carved sarcophagi still seen by Fossey (1988: 109) - photographs in Papachatzis 1981: 11.

note 84). In 1994, two A-C tombs[11] were found 2km W of the modern village (AD 1994: 284 – *component PP_33*). A pythos-burial of A period and a cremation burial of C period (*components PP_33 and PP_75*) were found 2km W of the modern village (AD 1994: 284; AD 1996: 284) during rescue excavation (DEPA gas pipeline work for pipeline Asopos-Oinophyta).

Fossey argues (1988: 122) that a settlement[12] should lie near Neochoraki in addition to the Soros site, considering the different tomb areas reported in the area, from Neochoraki and in general from Soros. The area is though interesting due to heavy alluvium deposits, and Fossey would argue that probably settlement traces disappeared. It seems that Fossey deliberately confuses here finds from an area too vast (modern Neochoraki and Kallithea and Soros areas) to be a single site. One should probably suppose the existence of two villages in the area, and more intensive and systematic research could probably be able to locate it. As for the present knowledge, the likely local material would probably be C, H, R, LR (Soros – see below) and G, A, C (as for Neochoraki burial area), but probably the corresponding settlement is not as close to the village as argued by Fossey (see chapter II.3.11 – LONG TERM SETTLEMENT TRENDS – for a discussion).

10. Moustaphades/Kallithea - Pyrgaki: *Components PP_23 and PP_ 24*

ID	Period	Type	Toponym	Discovery	Position
PP_23	LH	burial place	Moustaphades/Kallithea - Pyrgaki	Rescue excavation	approximate
PP_24	LR	? burial place	Moustaphades/Kallithea - Pyrgaki	Other	approximate

The village of Moustaphades/Kallithea (Thivon) is isolated in a great swathe of country, with some good land. Kallithea's physical landscape is constituted by mostly excellent land in Neogen, and the village has abundant freshwater too (see water sources mapped in the 1:50,000 IGME geological map – Thiva sheet).

Three LH tombs (*component PP_23*) cut in the rock and excavated[13] in 1963, 1966 and 1970 are reported by Syriopoulos 1968 and then by HS&D 1979 at the S foothill of the ridge Pyrgaki (locality Koros) to the N of modern Moustaphades/Kallithea village, but could not be found by Snodgrass (pers. comm.). In 1983, three new tombs with *dromos* were found (AD 1983: 134) in locality Rizo, 500m ca W of the water-reservoir of Kallithea village where the previous tombs where found. In the Early Byzantine period the tombs were reused (*component PP_24*), but it is not clear whether for the same use (burial). LH finds in Thebes museum are reported from the area, without clear provenience though [AE420]. They have been included as likely in the LH burial component (*component PP_23*), since they could be from these tombs. On the other hand, Fossey, as noted before, confuses Kallithea with Neochoraki area, and therefore the provenience remains unclear, and the LH finds could also be from Soros tombs (*component PP_74*).

[11] Between the two tombs sporadic offerings of pots and figurines were found (AD 1994: 284).

[12] In Fossey 1988: 123-126 is a long discussion about identification issues, with an attempt to identify the supposed settlement with *Therapnai/Skolos*.

[13] The tombs were found during the excavation of a cistern for the village (AD 1964: 199).

As Fossey (1988: 119) notes, rough and dominant mountains dominated the area, and he would recognize the hilltop 548m asl (mapped as Soros in the 1:50,000 GYS map) as a clearly important acropolis. In considering the LH tombs on the S slopes of Pyrgaki hill, and the name itself of the ridge, one could look for the LH settlement on that hilltop or in the vicinity (see below - Soros). On the other hand, signs of a LH site on the Soros top have not been found yet, and also no later (C) site is yet known in the area of Kallithea, though the settlement site in later prehistory and historical period could be nearby. On the other hand, the later occupation, the historical settlement of the area, could also have been somewhere else within the wider area, and the fortification on the Soros Easterly higher peak (*components PP_26 to PP_28*) is probably to be considered only a fort (and not the acropolis of a settlement) in historical periods (for a discussion on villages in the area see chapter II.3.11– LONG TERM SETTLEMENT TRENDS).

11. Kallithea Metochi: *Components PP_71 to PP_73*

ID	Period	Type	Toponym	Discovery	Position
PP_71	H	rural site/activity	Kallithea Metochi	accidental	approximate
PP_72	R	rural site/activity	Kallithea Metochi	accidental	approximate
PP_73	LR	rural site/activity	Kallithea Metochi	accidental	approximate

Nearby Kallithea village, in locality Metochi, HL/R surface walls are reported (AD 1981: 194) over a large area. The remains include Early Christian / Byzantine structures as well as H/R structures (including a well [AE1781]). The site, is said, may have a connection with drainage channels on the neighbouring hill.

12. Soros: *Components PP_25 to PP_28*

ID	Period	Type	Toponym	Discovery	Position
PP_25	Preh	? Fort	Soros	Ext. topographical surveys	certain
PP_26	A	fort	Soros	Ext. topographical surveys	certain
PP_27	C	fort	Soros	Ext. topographical surveys	certain
PP_28	H	fort	Soros	Ext. topographical surveys	certain

On Soros Eastern higher peak -geodetic point 548m asl-, to the NW of modern Kallithea village.

On the hilltop traces of at least three wall circuits have been noticed (Fossey 1988: 20f – plan fig.16),[14] with associated material (A to H). On the S slopes, below the fortification, large quantities of C and H potsherds [AE423 and AE424] have been noticed (Fossey 1988: 122). They could have easily come from the fortified top (the evidence has been associated to the components with a question mark).

Components have been created as listed above. Probably the fortification, with associated material, has to be considered a fort only as for historical periods (A to H). We could not completely exclude, though, the existence here of a sort of refuge settlement in some of the periods. There is no apparent need though, to consider the site as a fortified acropolis of a settlement nearby, as Fossey seems to suggest.

[14] Fossey 1988: 20f dates the three circuits as: Phase I Cyclopaean prehistoric walling; Phase II polygonal A - early C; Phase III polygonal H date (see also Fossey BICS 1971: 106-9).

In prehistoric times, Fossey (1988: 120f) would locate the LH fortified settlement site here, though he himself notes that 'little or no' prehistoric was found, and he seems to use walling style only to date the structural remains and to assign a prehistoric use to the site.
The fortified place probably also marks the border between the *chorai* of Plataea and Thebes, and could have belonged to the Plataeans before the destruction of the city by the Thebans (426 BC), and afterwards could have been taken by the Thebans and used as a watch place.

13. Soros: *Components PP_32 and PP_74*

ID	Period	Type	Toponym	Discovery	Position
PP_32	H	burial place	Soros	accidental	imprecise
PP_74	LH	burial place	Soros	accidental	approximate

On the Soros hill excavation of destroyed LH Mycenaean tombs and eight robbed H tombs was carried out in 1965 (AD 1965: 242).

14. Soros SE: *Components PP_29 to PP_31*

ID	Period	Type	Toponym	Discovery	Position
PP_29	gr-rom	? cult place	Soros SE	Travellers	approximate
PP_30	H	activity focus	Soros SE	Ext. topographical surveys	approximate
PP_31	LR	activity focus	Soros SE	Ext. topographical surveys	approximate

At the SE foot of East Soros peak, below the fortification (*components PP_26 to PP_28*), H [AE425] and LR [AE426] potsherds are reported in connection with the large building [AE190[15]] found here (Fossey 1988: 122). A R dedication [AE428] has also been found in the vicinity and could be connected with the abovementioned building [AE190] interpreted by scholars as a sanctuary (of the *Mater Theon*). The structures had already been seen by the travellers, in association with sculptures from C to R era (listed in Fossey 1988: 123).
I created a general gr-rom component as probable cult place (including the R dedication [AE428]), and two period-specific (H and LR) components since, even if the interpretation as cult place is not correct, we do have here a certain activity focus in those two periods, due to the presence of surface material [AE425 and AE426], which I left also in apparent association with the large building remains.

------

15 – 19. Asopia: *Components PP_56 and PP_57* (15); *PP_58 to PP_64* (16); *PP_65 and PP_66* (17); *PP_67 to PP_69* (18); *PP_70* (19)

ID	Period	Type	Toponym	Discovery	Position
PP_56	gr-rom	fort	Asopia - Prophitis Ilias	Travellers	certain
PP_57	C	fort	Asopia - Prophitis Ilias	Ext. topographical surveys	certain
PP_58	A	activity focus	Asopia	Ext. topographical surveys	certain
PP_59	C	activity focus	Asopia	Ext. topographical surveys	certain
PP_60	H	activity focus	Asopia	Ext. topographical surveys	certain
PP_61	EH	? activity focus	Asopia	Ext. topographical surveys	approximate
PP_62	MH	? activity focus	Asopia	Ext. topographical surveys	approximate
PP_63	LH	? activity focus	Asopia	Ext. topographical surveys	approximate
PP_64	G	? activity focus	Asopia	Ext. topographical surveys	certain
PP_65	gr-rom	burial place	Asopia vicinity	accidental	imprecise
PP_66	C	burial place	Asopia vicinity	accidental	imprecise
PP_67	gr-rom	burial place	Asopia E	accidental	approximate
PP_68	C	burial place	Asopia E	accidental	approximate
PP_69	H	burial place	Asopia E	accidental	approximate
PP_70	gr-rom	burial place	Asopia S	accidental	approximate

The modern village of Klembotsari/Asopia lays at the E end of the Asopos middle valley, and it is rather enclosed by hills and mountains which surround the Parasopia and face W into it. The village lies very close to the Asopos ravine through which the river enters the East Tanagra Plain. A road to Bratsi/modern Tanagra runs from it, but otherwise, as Fossey (1988: 127) notes, the area is cut off in most directions by the mountains which surround the Parasopia. Asopia's communications are, thus, mostly with the rest of the Asopos' upper basin to the W. Above the village to the N is the hill of Prophitis Ilias (422m asl), where traces of presence in antiquity have been noticed.
The place was visited by travellers (quoted in Fossey 1988: 127), who noticed the fortified hilltop of Prophitis Ilias (a small flat top, 120m E-W and 50m N-S –plan in Fossey 1988 fig. 17). Fossey (1988: 127) defines it as the 'acropolis' of an ancient settlement, and he suggests that the settlement would probably lay in the area of the modern village - on the S slopes of Prophitis Ilias (Fossey 1988: 128). On the other hand, as a lot of tile fragments but no potsherds were found, probably Prophitis Ilias hilltop was just a fort, either isolated or serving a settlement below, whenever it actually existed.
Many burial areas are reported (especially by locals) in the village or in the vicinity (*components PP_65 to PP_70; PP_79 and PP_80*), which would support the hypothesis of a settlement site nearby. The ancient tombs noticed by Lolling at the S foot of the hill could be visible cisterns.

Attempts of identification have been made (see Fossey 1988: 130), in particular with *Eilesion*,[16] while Wallace (1979) would suggest *Eteonos*.

Components PP_61 to PP_63: Also prehistoric occupation is attested, since prehistoric potsherds were found on the slopes.

[15] The building is 24x18m and is made of large worked limestone blocks.

[16] In fact Philippson-Kirsten 1951 would place *Eilesion* at Bratsi/modern Tanagra. Eilesion is mentioned in the Homeric *Catalogue of Ships* with Harma and Herythrai, while later sources are only etymological. Proof of attribution would be the continuity of the name, as fields at the E edge of village are given the name Eilesion in modern times.

They could have rolled down from the top, though a revisit would be needed to clarify it. The site apparently is not listed in prehistoric gazetteers and therefore the note for Prehistoric pottery is probably only from Fossey (1988: 128). No trace of LH fortification on the top.

20/21/22. Asopia S: *Components PP_78* (20); *PP_79* (21); *PP_80* (22)

ID	Period	Type	Toponym	Discovery	Position
PP_78	gr-rom	rural site/activity	Asopia S	Rescue excavation	approximate
PP_79	C	burial place	Asopia S	Rescue excavation	approximate
PP_80	C	burial place	Asopia S	Rescue excavation	approximate

Few discoveries have been made S of the village due to rescue excavations carried out when the gas pipe-line work was carried out.

Component PP_78: S of Klembotsari/Asopia village, to the S of the road Asopia-Kleidi, along the gas pipe-line, by the church of Ag. Stilianos, stone buildings associated with coarse ware sherds have been found (AD 1977: 376 – *thesi* B). They have been interpreted as sheds for pastoral or agricultural activities, also because of their position close to fresh water springs.

Component PP_79 and 80: S of Klembotsari/Asopia village, to the S of the road Asopia-Kleidi, to the left and the right of *component PP_78* along the pipeline, C graves have been found (AD 1977: 376 - *thesi* A and *thesi* Γ).

23. Daphni Kastron: *Components PP_47 to PP_ 49*

ID	Period	Type	Toponym	Discovery	Position
PP_47	gr-rom	fort	Daphni Kastron	Ext. topographical surveys	certain
PP_48	C	? human presence	Daphni Kastron	Ext. topographical surveys	certain
PP_49	LR	? human presence	Daphni Kastron	Ext. topographical surveys	certain

To the SW of the modern village of Daphni is the Kastron hilltop (656m asl), reachable from a path that rises from the village to the S. On it remains of a circuit wall [AE408] have been noticed (Fossey 1988: 116 - plan in fig.15).

The site could be interpreted as a simple military fort, in considering also the almost total lack of surface material - probably sign of a period of occupation reduced in time. Only two datable sherds (one C [AE1569] and one LR [AE1570]) have been recovered (Fossey 1988: 118). If they cannot date with certainty the fortification, they would attest a probable human presence in that period.

On the other hand, Fossey (1988: 116) would suggest that the site was the acropolis of an ancient settlement which would have lain on the plateau at the mountain's N foot (see below no.24 – Daphni plateau).

24. Daphni plateau (ancient ERYTHRAI?): *Components PP_50 to PP_54*

ID	Period	Type	Toponym	Discovery	Position
PP_50	LH	activity focus	Daphni plateau	Ext. topographical surveys	approximate
PP_51	gr-rom	? burial place	Daphni plateau	accidental	approximate
PP_52	C	activity focus	Daphni plateau	Ext. topographical surveys	approximate
PP_53	H	activity focus	Daphni plateau	Ext. topographical surveys	approximate
PP_54	LR	? activity focus	Daphni plateau	Ext. topographical surveys	approximate

On the plateau at the N foot of Kastron hill (see above no.23), ca 800m to the W of Daphni village, used to lie the *metochion* of the monastery of Hosios Meletios (to the S of the Pastra ridge). All that remains of it today are the walls of the church of Agios Athanasios. The road from Kriekouki-Erythrai to Darimari-Daphni crosses the S side of the plateau.

An ancient site on the plateau was known to the travellers already (Leake 1835: 330 and 369, and Ulrichs 1863: 74 quoted in Fossey 1988: 116). They noticed building blocks [AE413] which could be later also and traces of a fortification around the plateau (Leake 1835: 369) which could not be traced by Fossey during his visit (apart from the ruined church of Agios Athanasios).[17] Fossey noticed the presence of tiles everywhere and no frequent potsherds. All these factors suggested a post-Medieval scenario, very similar to the case of site CN4 for instance (Hyettos survey in the framework of the Boeotia Survey project - Bintliff pers. comm.). The evidences [AE413 – AE1571 – AE1572] have therefore been recorded as gr-rom (with a question mark as they could be later, and even in possible connection with the *metochion*). I have associated them to C, H and LR activity focuses (*components PP_52, PP_53, PP_54*) in a possible association only, as they could even not correspond, as discussed, to anything from gr-rom period.

The plateau could have been occupied by a cemetery site in a certain period of history (unknown), since local people talk about tombs with pots and bones found on the plateau and especially at its edge (Fossey 1988: 118). I therefore created a component gr-rom 'cemetery?' (with a question mark which relates to chronology, as surely existed a cemetery there but it could also be in association with the later *metochion*), and other components as C and H 'activity focus', in case there was another kind of activity going on there not linked to the burials and testified by the presence of surface potsherds, albeit few (probably fine wares, since easily recognized, but not enough to represent a settlement site). A component LR 'activity focus?' has also been created, mainly to keep track of the info which reports LR potsherds (Fossey 1988: 118), though it is most probable that the remains of mortared building should be linked to the later occupation at the site.[18]

A Diocletian's Price Tariff (AE 1931: 163-4 no.5) was found at the site, but it could also be possible that it moved here later from the site of ancient Plataea (Fossey 1988: 118).

Component PP_50: Few LH (III?) sherds are also reported from the site (HS&D 1979; Fossey 1988: 118).

An identification with ancient *Erythrai* has been suggested (see Fossey 1988: 119, who agrees with Pritchett's identification –

[17] On the plateau below the Kastron, where the *metochion* was, are traces of a large subterranean building near the road from Kriekouki-Erythrai to Darimari-Daphni, which crosses the S side of the plateau (maybe earlier than the monastic buiding - but not earlier than Roman because of the mortar) – Fossey 1988: 118.

[18] It is not clear either whether the LR pottery listed in Fossey 1988: 118, and certainly corresponding at least to the single sherd found on the Kastron hilltop [AE1570 - *component PP_49*], can be connected to the plateau also.

AJA 1957: 15). The series of settlements, reported in ancient texts, which position *Erythrai* after *Hysiai* from Plataea would seem to confirm it. Wallace tries to identify this site with *Skolos*. On the other hand, we should keep in mind that we cannot even be sure that the settlement was here, rather than somewhere nearby.

25. Daphni: *Component PP_ 55*

ID	Period	Type	Toponym	Discovery	Position
PP_55	gr-rom	burial place	Daphni	Ext. topographical surveys	approximate

NE of the modern village of Daphni a tomb area is reported by Kirsten (RE 1950 s.v. *Plataiai*). Terracotta figurines are reported from the tombs (Fossey 1988: 118).

26. Erythrai - Pantanassa: *Components PP_34 to PP_38*

ID	Period	Type	Toponym	Discovery	Position
PP_34	LH	activity focus	Erythrai - Pantanassa	Ext. topographical surveys	certain
PP_35	A	activity focus	Erythrai - Pantanassa	Ext. topographical surveys	certain
PP_36	C	activity focus	Erythrai - Pantanassa	Travellers	certain
PP_37	H	activity focus	Erythrai - Pantanassa	Ext. topographical surveys	certain
PP_38	R	? activity focus	Erythrai - Pantanassa	Ext. topographical surveys	certain

On a ridge jutting out into the Asopos valley to the E of modern Kriekouki/Erythrai village, in the area around the Pantanassa chapel.
Early travellers visited the site and noted remains of ancient structures and traces of a circuit of walling (no more visible). The main bibliographical reference for the site is Pritchett's article on AJA 1957: 12-15 that presents the result of his own fieldwork and reports on previous accounts and travellers.
For the site has been suggested an identification with ancient *Erythrai*, but more probably it has to be identified with ancient *Hysiai*[19] (see Fossey 1988: 114-115 for a discussion on identification).

LH potsherds are reported from the site (Pritchett AJA 1957: 12-15; Syriopoulos 1968; HS&D 1979: G40).
In historical periods, evidence is too limited (surface sherds and no more visible traces of fortification) to create 'settlement' components, and created components are of 'activity focus' type (from A to R period). It looks like a site which could never have been very defensible or impressive. If this was its actual location, *Hysiai* would have benefited from the fertility of Parasopia. This site would seem 'poor' (or the bit we know), but we only see the possible 'acropolis' (see *component PP_39* also).
As for R period, R material from the site seems to be mentioned only in Fossey (1988: 114) from Pritchett, while not in other gazetteers (HS&D 1979: G40 for instance). The material could also come from the Erythrai-Pyrgos site (see *component PP_43*), since usually in his work Fossey lists all the material together from the wider site area.
No remains have been noted at Kriekouki/Erythrai village itself (Pritchett AJA 1957: 15; Fossey 1988: 114).[20]

27. Erythrai - Pantanassa foothill: *Components PP_ 39 to PP_40*

ID	Period	Type	Toponym	Discovery	Position
PP_39	gr-rom	activity focus	Erythrai - Pantanassa foothill	Travellers	certain
PP_40	C	? cult place	Erythrai - Pantanassa foothill	Travellers	certain

At the foothill of Pantanassa ridge, E of modern Erythrai, surface walls along with architectural fragments mark certainly an activity focus in the gr-rom period. Probably some of the A to R potsherds mentioned in the bibliography as generally referring to Erythrai site [AE227 - AE228 - AE407 - AE229] are to be found in this foothill area (lower settlement area?), though we cannot be more precise since location of pottery is not specified in the available accounts.

Component PP_40: At the W foot of the ridge, near to a well containing ancient material noted by Leake (1835: 327), two buildings and associated potsherds and material suggest a sanctuary area of at least 5th C BC date (Pritchett AJA 1957: 12-15). The evidence could match the description given by Pausanias (IX 2. 1-3) of a half-finished Apollo temple, with the φρέαρ ιερόν which Pausanias recorded at *Hysiai* (from Fossey 1988: 115).

Component PP_39: The evidence available for the area are probably not only the remains of the building interpreted as a temple (recorded as *component PP_40*), or could not be interpreted as such, but could be linked to other activities. I therefore created also a generic gr-rom 'activity focus' component. There is no need to identify the structures and potsherds found with the remains of a cult place. They could even represent habitation structures if not a lower settlement area.

28. Erythrai - Pantanassa area: *Components PP_41 and PP_42*

ID	Period	Type	Toponym	Discovery	Position
PP_41	C	? burial place	Erythrai - Pantanassa area	accidental	imprecise
PP_42	H	? burial place	Erythrai - Pantanassa area	accidental	imprecise

Only two tombstones from the period could attest the presence of a C-H burial place in the area of Erythrai-Pantanassa, a Hellenistic one (AE 1936: 45 and AJA 1957: 15) and another one dated to the 4th C BC (AJA 1957: 15).

29. Erythrai – Pyrgos: *Component PP_43*

ID	Period	Type	Toponym	Discovery	Position
PP_43	R	? Fort	Erythrai - Pyrgos	Ext. topographical surveys	imprecise

---

[19] Modern Erythrai is a misnomer as the next settlement to Plataea, according to ancient sources, ought to be *Hysiai* - and therefore Pantanassa site likely. Hysiai was a dispute settlement between Thebes and Plataea. Ancient sources give three settlements on foot of Kithairon, from E to W Erythrai, Hysiai and Plataea.

[20] Apart from a note that Pritchett makes on 'possible prehistoric from town (?)'.

Pritchett (AJA 1957: 12-15) reports of remains of a tower (gr-rom or medieval?) 1km ca NE of the Pantanassa church on Pyrgos hill. Only few R sherds and a possible gr-rom tower (seen both by Pritchett AJA 1957: 11 and Fossey 1988: 113) mark this activity focus.

30. Erythrai - Pantanassa NE: *Components PP_44 to PP_46*

ID	Period	Type	Toponym	Discovery	Position
PP_44	C	rural site/activity	Erythrai - Pantanassa NE	Ext. topographical surveys	approximate
PP_45	H	rural site/activity	Erythrai Pantanassa NE	Ext. topographical surveys	approximate
PP_46	R	rural site/activity	Erythrai Pantanassa NE	Ext. topographical surveys	approximate

Half way between Erythrai-Pantanassa and Erythrai-Pyrgos, probably on the lower slopes, by the road, traces of activity focus (artefact concentration) have been noticed (Fossey 1988: 113). It could be interpreted as a probable rural site, living from the C to the R period (traces of walling with mortar were also found).

Appendix I.12
The Theban plain and the area of the lakes: Thebais

THE ARCHAEOLOGICAL RECORD

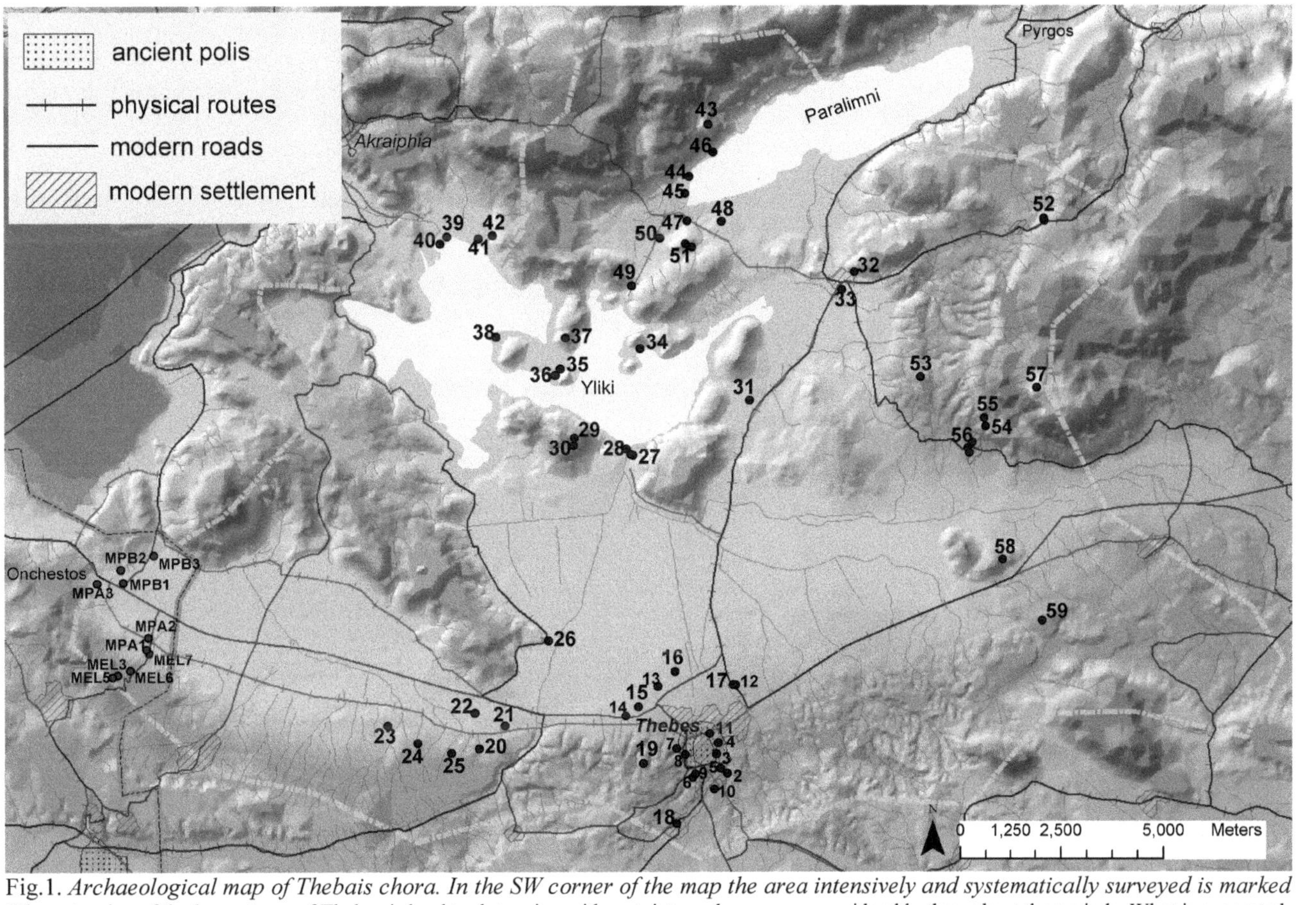

Fig.1. *Archaeological map of Thebais chora. In the SW corner of the map the area intensively and systematically surveyed is marked* (The extension of the lower town of Thebes is hard to determine with certainty and may vary considerably throughout the periods. What is suggested in the map is just a possibility to give an idea to the reader of the city's considerable extension. Within it, the Kadmeia is marked).

1. THEBES: *Components T_1 to T_14*

ID	Period	Type	Toponym	Discovery	Position
T_1	N	settlement	Thebes	Rescue excavation	certain
T_2	EH	settlement	Thebes	Rescue excavation	certain
T_3	MH	settlement	Thebes	Rescue excavation	certain
T_4	LH	settlement	Thebes	Personal or group interest	certain
T_5	SubM	settlement	Thebes	Rescue excavation	certain
T_6	PG	settlement	Thebes	Rescue excavation	certain
T_7	G	settlement	Thebes	Rescue excavation	certain
T_8	A	settlement	Thebes	Rescue excavation	certain
T_9	C	settlement	Thebes	Rescue excavation	certain
T_10	H	settlement	Thebes	Rescue excavation	certain
T_11	R	settlement	Thebes	Rescue excavation	certain
T_12	LR	settlement	Thebes	Rescue excavation	certain
T_13	EH	burial place	Thebes Ampheion	Personal or group interest	approximate
T_14	MH	burial place	Thebes	Rescue excavation	certain

The ancient *polis* of Thebes[1] is located to the S of its wide plain, on the slopes of the elevated areas of the so-called Tertiary *tafel* (see chapter II.3.12 – TOPOGRAPHICAL SETTING). The focus of the ancient settlement is on the Kadmeia hill. The toponym is

[1] The ancient name of *Thebai* has never been lost, being preserved in the demotic form Φήβα during Turkish times and probably also in a Frankish form 'Estives' (see Fossey 1988: 208).

probably to be included in the series of orographical names (mainly related to ridges and mountains whose slope or foothills host a settlement or city site) derived from the military equipment, for instance Aspis at Argo, Kerykion at Tanagra, Akontion at Orchomenos, yet in Boeotia.[2]

Continuity of settlement is known at Thebes since Neolithic times until the LR period (and onwards). N,[3] EH,[4] MH,[5] LH,[6] SubM, PG and G settlement is known on the Kadmeia, while A to LR on and around Kadmeia hill (*components T_1* to *T_12*).[7] In this work, the settlement of Thebes in different periods will not be examined in detail, since a separate work would be needed for it. For the purpose of our research, only the principal characters of the occupation have been listed in the database and considered in the analysis. The archaeological record from Thebes is rich and articulated and exploring in detail the particular urban site falls outside the purpose of the present research. In the database I collected the more meaningful information (in terms of time and space) in order to get an idea of the kind of occupation at the site in the different periods under study, remaining, that way, in a framework coherent with the rest of the work.

Unlike other sites known for Boeotia, in Thebes almost all the available archaeological information comes from excavations (mainly rescue excavations, like in the case of Levadeia for instance). The Kadmeia is today completely covered by the modern town, and therefore the archaeology of the ancient settlement can be considered in the category of 'urban archaeology' with all implications[8] and disadvantages (see chapter II.2; Aravantinos 1996 and 2000).

As for Prehistoric periods, archaeological reports as well as monographic works are available.[9] Generally speaking, Bronze Age periods are certainly the most examined and investigated at the site, due to the personal research interest of scholars and the availability of a rich archaeological record.

As for historical period, a good topographical work is Symeonoglou 1985, which gives an account of the monuments excavated within the city site but obviously does not include the more recent discoveries. A list of inscriptions found in the city up to the 1980s can be found in Fossey 1988: 206ff.

The city of Thebes flourished in C/H times, especially at the end of the 4th C BC, when its power reached its peak and the city controlled a wide area of the region. After the defeat by Macedonian hands is a period of decline, testified by archaeological remains (the city shrank on the Kadmeia) and also by historical sources. Strabo (IX 2.6) describes it as a mere village. In Pausanias' time its citadel alone was inhabited. Pausanias (IX 5.6) records a 'myth' according to which Kadmeia began as a settlement but later it became an acropolis with the growth of the lower town. He reports on a story about Sylla giving half Theban land 'to Gods' as looted treasures. The *Periegetis* describes the lower city of Thebes as deserted in his time (with the exception of the sanctuaries and the gymnasium) and the population living on the citadel, which they call Thebes and not Kadmeia.

2 – 7. LH burials: *Components T_15* (2), *T_16* (3), *T_17* (4), *T_35* (5), *T_45* (6), *T_46* (7)

ID	Period	Type	Toponym	Discovery	Position
T_15	LH	burial place	Thebes S and SE	Personal or group interest	approximate
T_16	LH	burial place	Thebes E	Personal or group interest	certain
T_17	LH	burial place	Thebes - Megalo Kastelli	Personal or group interest	certain
T_35	LH	burial place	Thebes Ismenion	Personal or group interest	certain
T_45	LH	burial place	Thebes Kolonaki	Personal or group interest	certain
T_46	LH	burial place	Thebes Kalogeros	Personal or group interest	certain

LH cemeteries (mainly Mycenaean chamber tombs) are known especially on the hills surrounding the Kadmeia (S, SE, at Megalo Kastelli, by the temple of Apollo Ismenios, in the Kolonaki suburb and in Kalogeros area).[10]

5. *Ismenion*: *Components T_30 to T_34 and T_36 and T_37*

ID	Period	Type	Toponym	Discovery	Position
T_30	G	cult place	Thebes Ismenion	Travellers	certain
T_31	A	cult place	Thebes Ismenion	Travellers	certain
T_32	C	cult place	Thebes Ismenion	Travellers	certain
T_33	H	cult place	Thebes Ismenion	Travellers	certain
T_34	R	? cult place	Thebes Ismenion	Ext. topographical surveys	certain
T_36	LR	burial place	Thebes Ismenion	Other	certain
T_37	LR	? Settlement	Thebes Ismenion	Other	certain

On the hill by the modern cemetery to the SE of the Kadmeia lie the remains of the extramural sanctuary of Apollo Ismenios, for which use is attested from Late Geometric to H or R times (with a possible break in the 7th C BC – Fossey 1988: 204).

Near the Temple of Apollo Ismenios, on the hill by the modern cemetery to the SE of the Kadmeia (AD 1969: 177), there are a probable habitation focus in LR times, as well as traces of tombs from the same period (*component T_36*).

2 Esichius, s.v. *Kadmos*, has 'spear, helmet, shield, in Cretan'. The name, according to Schachter 1985 would have been initially the name of the Theban citadel, and would have later become the name of its mythical founder (Kadmos).

3 N excavated layers.

4 EH excavated layers and buildings, especially to the S of the Kadmeia. Probably also a fortification [AE1582]. Recent excavations beyond the museum (for the construction of a new museum building) have thrown light to the EH settlement on the Kadmeia (Aravantinos 2004b). An EH burial (known as the tomb of Zethus and Amphion mentioned also by Pausanias IX 17.4) is a small mound of earth still recognizable today on the N slopes of the Kadmeia (*component T_13*).

5 MH: excavated layers and buildings (all over the Kadmeia). MH burial places within the settlement area (*component T_14*).

6 LH: excavated layers and buildings are all over the Kadmeia, where there are also palace(s) with linear B tablet archive, and workshops. The ruins of the so-called 'Palace of Kadmos' have been excavated below the modern town on the Kadmeia. Mycenaean fortification have also been excavated (recently behind the museum for the construction of the new museum building). To the LH III B2 / beginning of LH III C is dated the final destruction of the Kadmeia citadel (AD1983: 131).

7 The Kadmeia is actually formed by three hills. At times they would correspond to the whole site, other times to the acropolis of the larger town (Fossey 1988: 199).

8 An advantage could be the fact that through excavations we may know information about periods otherwise difficult to recognize (SubM or PG for instance).

9 Among others, a special reference can be made to the book of Konsola 1981, on pre-Mycenaean Thebes.

10 For bibliography see: AD 1967: 227; AD 1967: 240; AD 1968: 232; JHS Reports 1969: 17-19; BCH 1970: 1031-1037; AD 1970: 211-220; BCH 1972: 694-699. For a map see Symeonoglou 1985: fig.2.2.

8 - 14. Historical period burials: ***Components T_18*** **(8),** ***T_142*** **(9),** ***T_143*** **(10),** ***T_28 and T_29*** **(11),** ***T_36*** **(LR tombs at** ***Ismenion*** **- 5),** ***T_24 to T_27 and T_47*** **(12),** ***T_141*** **(13),** ***T_20 to T_23*** **at Pyri (14)**

ID	Period	Type	Toponym	Discovery	Position
T_18	C	burial place	Thebes	accidental	approximate
T_20	G	burial place	Thebes Pyri	Personal or group interest	approximate
T_21	A	burial place	Thebes Pyri	Personal or group interest	approximate
T_22	C	burial place	Thebes Pyri	Personal or group interest	approximate
T_23	H	? burial place	Thebes Pyri	Personal or group interest	approximate
T_24	G	burial place	Thebes Railway	Rescue excavation	certain
T_25	A	burial place	Thebes Railway	Rescue excavation	certain
T_26	C	burial place	Thebes Railway	Travellers	certain
T_27	H	burial place	Thebes Railway	Rescue excavation	certain
T_28	A	burial place	Thebes Ampheion E	Ext. topographical surveys	approximate
T_29	H	burial place	Thebes Ampheion E	Ext. topographical surveys	approximate
T_36	LR	burial place	Thebes Ismenion	Other	certain
T_47	R	burial place	Thebes Railway	Rescue excavation	certain
T_141	gr-rom	burial place	Thebes NW	Rescue excavation	certain
T_142	gr-rom	burial place	Thebes - Kolonaki	Rescue excavation	certain
T_143	gr-rom	burial place	Thebes - Kolonaki SE	Rescue excavation	certain

As far as the cemetery areas, information can be found also in more recent accounts on the recent rescue excavation work carried out by the local *ephoreia* in several areas around Thebes, but especially in the N area by the Railway station, where tombs from the 8th C BC until the 3rd C AD have been excavated in 2000-2001, in the occasion of the construction of the underground passage Thebes-Mouriki and mainly Thebes-National Road (Aravantinos 2004a – *components T_24 to T_27 and T_47*).[11] An Archaic tomb and more Hellenistic ones have been located E of the Ampheion hill (BCH 1953: 218; BCH 1954: 132; JHS 1954: 39, reported in Fossey 1988: 204 – *components T_28 and T_29*).

For the cemetery at Pyri locality (*components T_20 to T_23*), currently occupied by a suburb NW of Thebes,[12] see summary of excavation finds in Andreiomenou 1995. The majority of the tombs belong to the last quarter of the 7th C – first half of the 6th C BC (AD 1976: 119-20).

15. Pyri: ***Component T_19***

ID	Period	Type	Toponym	Discovery	Position
T_19	N	activity focus	Thebes Pyri	Other	approximate

A Neolithic activity focus has been noticed through surface sherds in the area of Pyri, to the NW of Kadmeia (known for the large gr-rom cemetery – *components T_20 to T_23*) - AD 1969: 176-7; JHS 1970-1: 15; BCH 1971: 921-8; Konsola 1981: 41.

16. Kapanitsa: ***Component T_139***

ID	Period	Type	Toponym	Discovery	Position
T_139	H	burial place	Kapanitsa	accidental	imprecise

A poorly built and without grave goods Macedonian-type tomb was excavated in the Kapanitsa area (Mazonaki-Faraklas AE 1967: 15-19 quoted in JHS 89 1969: 20; AD 1964: 200 and AD 1966: 145-6), between Pyri and *Dio Litharia* locality, to the N of Pyri and to the N of the National Road Athens-Lamia, along an agricultural track which starts at the E edge of Pyri suburb and leads straight into the plain. Because of the presence of the tomb the path of the agricultural road is believed to follow an ancient route.

17. Thebes Railway: ***Components T_153 to T_154***

ID	Period	Type	Toponym	Discovery	Position
T_153	N	settlement	Thebes Railway	Rescue excavation	certain
T_154	EH	settlement	Thebes Railway	Rescue excavation	certain

At the N outskirt of Thebes, underneath the cemetery of historical periods (*components T_24 to T_27 and T_47*) remains of Final Neolithic and the beginning of the Early Helladic settlement were excavated (3000-2000 BC), during the rescue excavations carried out at the occasion of the construction of the underground passage Thebes-Mouriki and mainly Thebes-National Road (Aravantinos 2004a – see also above no.8-14).

18. Tachi: ***Components T_38 to T_43; T_148 and T_149***

ID	Period	Type	Toponym	Discovery	Position
T_38	PG	? burial place	Tachi	Personal or group interest	approximate
T_39	G	burial place	Tachi	Rescue excavation	approximate
T_40	A	burial place	Tachi	Rescue excavation	approximate
T_41	C	activity focus	Tachi	Travellers	approximate
T_42	C	burial place	Tachi	Rescue excavation	approximate
T_43	R	? burial place	Tachi vicinity	accidental	imprecise

[11] Aravantinos 2004a refers to an extended cemetery area, probably the largest one belonging to ancient Thebes (p.89ff – see fig. p.235), known from locals since the end of the 19th and the beginning of the 20th century. Excavated in small areas during the 1960s, later it was interested by small rescue excavations in private plots, and in 1996-7 by digging activities for the sewage-pipe (*Apoxeteutikos Agogos*) of Thebes (AD 1996: 267-269 and AD 1997: 364-368) excavated in its Northern edge. The cemetery was crossed by a road S-N, running from Thebes to the Chalkis area (mentioned in Pausanias IX 18.1). The main road to Chalkis runs ca 200m to the E of the ancient one. From the ancient road would detach roads to Tanagra, Dilion, Oropos, Avlidas, passing by the centres of Mikalessos, Eleon, Arma. The road seems to have existed yet before the A graves. Graves are along the road and extended far beyond it (the extension of the cemetery is indicated in the figure on p.235 in Aravantinos 2004a). Graves from Late Geometric to Early Roman period are also reported (from the 8th C BC until the 3rd AD).

[12] Near the small church of Agia Eleousa - Pyri/Thebes, to the right of the Old National Road Athens-Lamia and ca 1.2km to the NW of the town centre. The location is marked in Aravantinos 2004a fig. p.235.

ID	Period	Type	Toponym	Discovery	Position
T_148	H	burial place	Tachi	Rescue excavation	approximate
T_149	LH	burial place	Tachi	Rescue excavation	approximate

In a S suburb of modern Thebes, called Tachi,[13] excavations have been carried out by the local *ephoreia* since 1973 (Andreiomenou AE 1976: 12-17). Beforehand, a number of ancient worked blocks were known from the area as mentioned by many travellers (among others Ulrichs 1863: 12-13 [AE1756]). Gr-rom potsherds also lie on the surface, mainly fine wares apparently (Fossey 1988: 209 from AE 1976: 12-17). An excavation in 1973 found C sherds in association with building remains badly preserved (*component T_41*). Andreiomenou, the excavator, had interpreted them initially as remains of a sanctuary, later on as a cemetery. Fossey (1988: 209) agrees with the latter interpretation because of other tombs found in the area [AE1589] which are probably along the ancient road from Thebes to Eutresis and the port of Livadostro.[14]

As Papadaki (2000) points out, excavations, since 1973, found (in Lalioti plot) a grave *peribolos* in use since the Early Geometric period (attested by an early Geometric amphora with burial function) up to the Early Hellenistic. The interpretation is confirmed by more recent rescue excavations, directed since 1993 by V. Aravantinos at the site (mainly on the site of new 'council flats' - *Organismos Ergatikis Katoikias Thiva II*), which found more graves from the LG to the A period (AD 1993: 167-173; AD 1994: 269-271; AD 1995: 295-296; AD 1997: 370 – *components T_38 to T_40; T_42 and T_148*).[15] The majority of graves date from the LG period up to the 7th C BC, while rich graves are dated to the 6th to the 4th C BC.

A R tombstone (IG vii 2693) is also reported from the area [AE1591]. According to Fossey (1988: 209) it could easily have travelled from a Theban cemetery. On the other hand, since we do have evidence from C-H and most probably also earlier cemeteries in the area, there could have been also a R burial area (*component T_43*).

A small group of Mycenaean graves (dated to the LH III A1 period) are also known from the area (*component T_149*) and have been excavated (AD 1993: 167-173; AD 1994: 269-271).

Part of an aqueduct [AE1759] was also found, as well as a well [AE1760 and AE1761] which contained sherds from LG and early A period (AD 1994: 269-271; AD 1995: 295-296). The aqueduct, excavated in the bedrock, cuts the necropolis in a N-S direction (a 100m long section has been uncovered) and carried water to the N (in use from C to post-Byzantine times). The well is dated to the LG and Early Archaic period on the base of the infill, and was probably connected with the burial rituals in the LG period.

Those evidences could lead to the assumption of the presence of a settlement nearby, probably flourishing in the 8th C BC, if one accepts the arguments and conclusions of Papadaki 2000. The presence of water installations may lead to the explanation of Pausanias (IX 8.1-2) who refers to the presence of cult in the Theban periphery (AD 1994: 269-271), if one accepts the original interpretation of the excavator as a cult place (sanctuary) site in the C period (AE 1976: 12-17; Fossey 1988: 209-210); a possible identification with ancient *Potniai* has been suggested (Papadaki 2000[16]).

19. Thebes W: *Component T_44*

ID	Period	Type	Toponym	Discovery	Position
T_44	C	rural site/activity	Thebes W	Ext. topographical surveys	certain

About 1.5Km W of Thebes, in the range of hills which separate the Parasopia from the Theban plain, on the upper North slopes of a low peak (height 266m - marked on the GYS 1:50,000 map), Fossey (1988: 211) observed, in 1965, a thick spread of sherds (Classical black glaze and probably also tile fragments), without structural remains.

We should consider the possibility that coarse ware (less easily datable) is also present at the site, which could be interpreted, as Fossey himself notes, as a rural site/activity, "*a small homestead of essentially agricultural nature*" in his words.

20. Kabeirion: *Components T_49 to T_54*

ID	Period	Type	Toponym	Discovery	Position
T_49	N	activity focus	Kabeirion	Other	certain
T_50	G	activity focus	Kabeirion	Personal or group interest	certain
T_51	A	cult place	Kabeirion	Travellers	certain
T_52	C	cult place	Kabeirion	Travellers	certain
T_53	H	cult place	Kabeirion	Travellers	certain
T_54	R	cult place	Kabeirion	Travellers	certain

The Kabeirion sanctuary lays W of Thebes, "*among the range of hills bordering the South side of the Theban or Teneric plain, in a valley just to the West of the valley of the Kanavaris, which flows from Thespies through these hills into the Theban plain*" (Fossey 1988: 211).

Excavations have been carried out at the site during the 20th century (Judeich 1988), while cleaning up operations have been carried out recently by the local *ephoreia* in the sanctuary area.[17]

Components T_50 to T_54: According to the structural remains, the life of the site as a cult place begins in the 6th C BC

[13] The location of the Tachi cemetery is marked in Aravantinos 2004a fig. p.235.

[14] It can be supposed that this area is crossed at least for the whole C period by the main way from the city to its *chora* towards S.

[15] 4th C BC tombs in the area of Tachi village are reported already in AD 1970: 233 [AE1589].

[16] Papadaki, in her paper presented at the 3th Boeotian Conference in 1996 (published in 2000), concludes that:

A. In Mycenaean times there was probably a settlement in the area called Potnia after the divinity, whose existence could be proved by the graves. Here was probably also the sanctuary of the Potnia goddess, present since Mycenaean times (see tablets from the palace on the Kadmeia, the latest published by Aravantinos-Godart-Sacconi in 2001).

B. The Early Geometric amphora with burial function may testify the hypothesis that in the Dark Ages population live outside the city wall of the Kadmeia not only to the N, as Simeonoglou suggests, but also to the S.

C. LG burials are numerous and quite rich and probably belonged to a prosperous settlement of the 8th C with commercial relationships with Corinth, Euboea, Attica and Boeotian pottery workshops. The city at that time extended probably until there, and included the surrounding rural communities. The site is between the city (again inhabited by that period) and its *chora*.

[17] For the German excavations, see Wolters - Bruns 1940 and other reports: AM 1888: 81-99, 412-28; AM 1890: 355-419; AA 1893: 63-4, 129-31; AD 1963: 115-21; AD 1966: 207-9; AD 1968: 224-5; AA 1964: 231-65; AA 1967: 228-63, 456.

and continues up to the Roman Imperial period.[18] On the other hand, also Middle Geometric pottery finds are attested at the site.

Component T_49: Neolithic material is also reported from the site. Looking at the landscape of the area, one could easily notice similarities with the Tanagra Asopos tributaries area (see appendix I.14 – THE TANAGRA SURVEY). The site is in a slightly inner position compared to the modern course of the Kanavaris river, by the confluence of a small stream with a tributary of Kanavaris. This area could constitute, therefore, a very probable ideal 'geomorphological window' for a Prehistoric landscape.

21. Kabeirion - Agios Nikolaos: *Component T_48*

ID	Period	Type	Toponym	Discovery	Position
T_48	gr-rom	? activity focus	Kabeirion - Agios Nikolaos	Personal or group interest	approximate

Judeich (1988: 84) locates the sanctuary of Demetra Kabeiria (positioned in Judeich map, p.83) mentioned by Pausanias on the basis of limestone worked blocks and gipsum fragments reused in the small chapel (*ikonostasio*) of Agios Nikolaos.
The location of the place is along the probable route which along the foothills crossed the Teneric plain to the E of Thebes towards Onchestos and the Copais.

22. Vagia-Provatari: *Components T_144 to T_147*

ID	Period	Type	Toponym	Discovery	Position
T_144	A	burial place	Vagia - Provatari	Rescue excavation	approximate
T_145	C	burial place	Vagia - Provatari	Rescue excavation	approximate
T_146	R	rural site/activity	Vagia - Provatari	Rescue excavation	approximate
T_147	R	rural site/activity	Vagia - Provatari	Rescue excavation	approximate

Components T_144 and T_145: In occasion of the gas-pipe installation (*Agogos Physikou Aeriou*, which runs from NW Boeotia to the area of Thebes), at the 68.4-68.8 km, in the area N of the Kabeirion valley, in Provatari locality, 8km W of the modern town of Thebes, part of a cemetery of the Late A and C periods was investigated, including a total of 50 tombs. They are probably to be related to a settlement site in the vicinity, among the hills between Helicon and Thebes (AD 1994: 284-286), considering the considerable distance from the ancient town of Thebes (see also BCH 1999: 734). From the same locality AD 1966: 202 reports architectural fragments (funerary *stelae*) delivered to the museum, found during ploughing [AE1753].

Components T_146 and T_147: Two *villae rusticae* were also partly uncovered by the cemetery site described above (AD 1994: 284-286). They reused building material from the cemetery (*components T_144 and T_145*), especially Late A and C inscribed funerary *stelae* (one of them reused as an oil-press [AE1751 and AE1752]).

23 - 25. Vagia-Mavrokampos, Vagia-Kritsola and Vagia-Darani: *Components T_55* (23), *T_56* (24), *T_57* (25)

ID	Period	Type	Toponym	Discovery	Position
T_55	C	activity focus	Vagia-Mavrokampos	Int. topographical surveys	certain
T_56	C	activity focus	Vagia-Kritsola	Int. topographical surveys	approximate
T_57	C	activity focus	Vagia-Darani	Int. topographical surveys	approximate

Nearby the modern village of Vagia, to its NE, are three ridges jutting out into the plain: from W to E, Mavrokampos (Mourokambos) – Kritsola – Darani. On all three Lauffer (Kopais I: 15-16) observed scattered C sherds (black glaze) and tile fragments (and also some Corinthian wares), especially on the N slopes of Mavrokampos hill, nearby which is also a deserted village/hamlet carrying the same name and was reported by travellers in the 19th century. Architectural fragments have also been noticed on Mavrokampos hill [AE1369].

As for *component T_55* (Vagia-Mavrokampos), Lauffer identifies it as the cult place dedicated to Herakles Hyppodetes, mentioned by Pausanias (IX 26.1) as a large sanctuary situated in the Teneric plain, right-hand of the Kabeirion.[19] The sanctuary would have marked the W border of Thebes area.
Vroom (2003) suggests that the name Mourokambos/Mavrokambos is probably related to *mouries* trees which were at the site until recently, according to locals. Intensive survey work in the area, in the 1980s, in the framework of the Boeotia Survey project, found C-H artefact concentration as well as LR (continuing into Mid-byzantine and probably until Turkish period).

Though the three locations are quite close to each other and the archaeological evidence noticed are reported together by Lauffer, we cannot be sure that the three activity focuses recognized are to be linked together. Probably the largest focus was at Mavrokampos (even if not accepted as a cult place), while the other two could also be recognised as rural sites (farmsteads?). Unfortunately, we cannot get any clue from Lauffer's text about the actual extension on space of the evidence. On site intensive and systematic survey would give information on the extension of the sites and possibly on their function. The three components are by the road that was probably running along the foothills.

26. Toumpi: *Component T_58*

ID	Period	Type	Toponym	Discovery	Position
T_58	N	settlement	Toumpi	Personal or group interest	approximate

Toumpi is a low hill at the SE edge of Sphiggion mountain jutting into the E part of the Teneric plain, W of the channelled Kanavaris brook, near a modern quarry. The mound (8km NW of Thebes) preserves several strata, and constitutes an important Neolithic site identified by D. French (1972: fig.19) and is reported in detail by Sampson (AD 1981: 194). Sampson reports pottery from LNeo I and MNeo. Obsidian flakes, stone tools and a head of Neo figurine (*eidolio*) were also found.[20]

[18] For a plan see Papachatzis (1981: 153-157). For a recent reworking of the site see Schachter 2003.

[19] See Pappadakis AD 1916: 217ff for a list of cults of Herakles in the Copais. For the myth of Herakles Hyppodetes and the destruction of the Orchomenian power by the Thebans lead by Herakles, see among others Schachter 1986.

[20] As Sampson notes, the site seems to be the Southern site in Greece known up to then which shows influence from Thessaly. The pottery

27/28. Lithares*: Components T_60 to T_68* (27), *and T_69* (28)

ID	Period	Type	Toponym	Discovery	Position
T_60	N	? activity focus	Lithares	Other	approximate
T_61	EH	settlement	Lithares	Travellers	approximate
T_62	MH	activity focus	Lithares	Personal or group interest	approximate
T_63	LH	activity focus	Lithares	Personal or group interest	approximate
T_64	H	activity focus	Lithares	Other	approximate
T_65	R	? activity focus	Lithares	Other	approximate
T_66	EH	burial place	Lithares N&W	Personal or group interest	approximate
T_67	MH	burial place	Lithares N&W	Personal or group interest	approximate
T_68	LH	? burial place	Lithares N&W	Personal or group interest	approximate
T_69	A	burial place	Lithares shore	Personal or group interest	approximate

Otherwise called Likeri (by Syriopoulos 1968: 70 ΠΕ28, for instance), the site of Lithares lay on a small hill projecting as peninsula into the Lake Yliki-Likeri from the S, E of the common outlet of the rivers Thespios and Ismenios into the lake.
Fossey (1988: 244) describes the landscape as characterized by a number of rivers and torrents that culminates in the Theban plain to flow together into the SE corner of Lake Yliki at a spot known as Anakoli. Those water courses are the Kanavaris (from the Erimokastro/Thespies area), the Kalamitis (from the feet of Mt. Sagmatas) and the two streams which flow to each side of Thebes (ancient Dirkis and Ismene-Knopos). "*The whole area around the mouth of this combined stream and along the lake's margin is liberally covered with large stones, whence the name Lithares. Many of these are building stones from the extensive ancient settlement which lay here*" (Fossey 1988: 244).

As Fossey notes (198: 245), Prehistoric surface finds were originally noticed (Papadimitriou, Praktika 1931: 274-6) on the small hill/peninsula to the E of the stream outlet. Known were also cist and chamber tombs at the N and W foot of this hill. Afterwards, the site was seen (Spyropoulos and Tzavella-Evjen's excavation reported in AAA 1973: 371-5 and summarised in Tzavella-Evjen's book *Lithares* 1985) as extending on both sides of the outlet, especially as far as the EH I and II (and also EH III, probably on a smaller area) periods are concerned (41,400 sqm the total extension of the site). Widely known from the site is 'The sanctuary of the Bull', an EH III large room containing terracotta figurines of bulls without parallels from the period. More than 37 rooms were dug with some 'houses' identified.
Hope Simpson and Dickinson (HS&D 1979: G29 and G30) report EH to LH from the site, known from earlier reports also (AD 1969: 28-46). Prehistoric burial places are also known (*component T_66*: EH cist and circular graves visible on the surface - Syriopoulos 1968 from PAE 1931: 274-6 one of which is MH; *component T_67*). LH tombs are reported by one of the earliest visitors of the site (Papadimitriou, Praktika 1931), but apparently without mention in later accounts.
Neolithic from the site is also briefly mentioned by Marinatos only (AAA 1968), as Neolithic and Chalcolithic material.

found was similar to the one found at Toumba Balomenou site (Chaironeiake *chora* – appendix I.3).

The main publication on the site, especially concerning the EH layers, is Tzavella-Evjen's book *Lithares* 1985. See also Tzavella-Evjen 1987.

Evidence from historical periods are also known at the site. During excavations of EH settlement, also H buildings were discovered. They could correspond to a rural site (*component T_64*). Not clear is the character of the R presence (*component T_65*). By the Yliki-Likeri lake shore to the W of the Lithares peninsula A burials were excavated partially by Spyropulos in 1968 (*component T_69*).
A G cemetery on the Stroulongo hill to the W of Lithares (see below no.30 - *component T_151*), as well as the A cemetery here (*component T_69*), could lead to assume the presence of a settlement (or sort of, such a large rural site) in the area. As Fossey (1988: 246) notes, though, the suggested identification of the site with ancient *Trapheia*[21] does not make much sense since there is no real evidence of a long-life settlement here in historical periods.

29/30. Lithares Stroulongo: *Components T_70 to T_72* (29) *and T_151* (30)

ID	Period	Type	Toponym	Discovery	Position
T_70	EH	activity focus	Lithares Stroulongo	Personal or group interest	approximate
T_71	LH	activity focus	Lithares Stroulongo	Personal or group interest	approximate
T_72	H	activity focus	Lithares Stroulongo	Personal or group interest	approximate
T_151	G	burial place	Lithares Stroulongo	Personal or group interest	approximate

On Stroulongo/Serlongos hill, W of Lithares peninsula site (Fossey 1988: 245) EH, LH and Hellenistic deposits were excavated (probably not published – Syriopoulos 1983-4: EH 226, MH 173, LH III 138 – *components T_70 to T_72*). Syriopoulos (1983-4) mentions also a Geometric cemetery (G160) at Stroulongo locality (*component T_151*).

31. Vouliagma*: Component T_59*

ID	Period	Type	Toponym	Discovery	Position
T_59	LH	burial place	Vouliagma	accidental	approximate

In locality Vouliagma, along the road towards Mouriki village from Thebes, and N of the National Road Athens-Lamia, half way between the road and the village, left to the incoming, LH chamber tombs - totally looted - and a possible *tholos* tomb are reported (AD 1967: 242; HS&D 1979). The tombs create a depression on the ground, after which the area was named.

32/33. Mouriki and Mouriki S*: Components T_75 to T_76* (32) *and T_78* (33)

ID	Period	Type	Toponym	Discovery	Position
T_75	A	settlement	Mouriki	Ext. topographical surveys	certain
T_76	C	settlement	Mouriki	Ext. topographical surveys	certain
T_77	H	settlement	Mouriki	Ext. topographical surveys	certain
T_78	gr-rom	burial place	Mouriki S	Ext. topographical surveys	certain

21 On the basis of a Nicander passage – *Ther.* 887-889.

On the N edge of the seasonally marshy Tymbanokambos plain, to the NE of the Yliki lake, lies the modern village of Mouriki, exploiting the land of the bay. Above the village to its NE rises the mountain of Sengouna. A lower peak, today marked by the presence of the Prophitis Ilias chapel, was probably the site of the ancient acropolis (Fossey 1988: 229).
Surface potsherds and walls alignment, as well as traces of a fortification (Fossey 1988: 229-232 with earlier accounts), mark the site of an ancient settlement, from the A to the H period, on the lower S slopes by the modern village (*components T_75 to T_77*). The same chronological range can probably be assigned to the gr-rom cemetery area linked to the settlement, to its S (*component T_78*). The extensive cemetery area was noticed by Symeonoglou in 1965. The settlement had a fortified 'acropolis' in the C period [AE156] and the fortification was probably still in use in the H period.
A possible identification has been suggested with ancient *Skoinos*, to which refer ancient texts listed in Fossey 1988: 232, among which Strabo IX 408 and Nonnus *D.* XIII 63.

34. Yliki E peninsula: *Components T_73 to T_74*

ID	Period	Type	Toponym	Discovery	Position
T_73	Preh	activity focus	Yliki E peninsula	Ext. topographical surveys	certain
T_74	H	rural site/activity	Yliki E peninsula	Ext. topographical surveys	certain

A peninsula juts into the Yliki lake at its N side, opposite Lithares, to the E of the larger peninsula of Klimataria. As Fossey (1988: 243) notes, the hook-like shape of the small peninsula forms a bay. On the shore are traces of ancient activities.

Component T_74: In 1966, a year of exceptional low water level since the drainage of the Copais, Fossey (1988: 243) saw the remains of a single L-shaped building with its longer axis running N/NE-S/SW [AE170], dated by Hellenistic potsherds found there [AE171]. Fossey interprets it as a small homestead, whose main occupation would have been in the Hellenistic period.[22]
Component T_73: Fossey (1988: 243) observed also obsidian flakes at the site, a sign of a Prehistoric activity there.

35/36/37. Klimataria: *Components T_ 109, T_110 and T_140* (35); Klimataria slopes: *Component T_111* (36) and Klimataria N: *Component T_112* (37)

ID	Period	Type	Toponym	Discovery	Position
T_109	H	activity focus	Klimataria	Ext. topographical surveys	certain
T_110	R	settlement	Klimataria	Travellers	certain
T_111	R	burial place	Klimataria slopes	Ext. topographical surveys	approximate
T_112	gr-rom	? Fort	Klimataria N	Travellers	certain
T_140	LR	activity focus	Klimataria	Int. and sys. artefact surface survey	approximate

[22] Fossey (1988: 243) notes how it may be part of the settlement organization in the area of the lakes which took place after the abandonment of the Oungra site.

Klimataria is named the L-shape W peninsula which juts into the Yliki lake from Mt. Ptoion.
Settlement remains at the angle formed by the L-shape peninsula were already known to travellers (Bolte RE 1916: 117-9 s.v. *Hyle* gives a list of previous accounts). A Frankish tower had been noted on Klimataria peninsula by Philippson, in the 19th century. Lock (1986: 113ff) remarks that the tower is under water, as the Yliki level rose by 25m after Copais drainage, but the top of the tower is visible in very dry summers just off the western tip of the Klimataria peninsula.
Fossey (1988: 239) mentions a fortification on the hill N of settlement site, which would correspond to the tower mentioned by Spyropoulos (AAA 1973: 206-9), and thought to be Medieval (Frankish?), as well as remains of various buildings and a cemetery area (Fossey 1988: 239 and fig.27).
In the framework of the Boeotia Survey project, and in a period when the lake water level had fallen down 10m (in 1989), an extensive R imperial settlement associated with the cemetery (*component T_111* - previously thought to be Medieval) was uncovered (AReports 1990-91: 34). Hellenistic and Roman (Imperial), and later on medieval (Frankish) occupation is known until now. LR occupation is mentioned by Vroom 2003 (from Boeotia survey project 1989 results).
The identification of the Klimataria site with ancient *Hyle* is discussed in Fossey (1988: 240-243).

38. Klimataria W: *Components T_ 113 and T_114*

ID	Period	Type	Toponym	Discovery	Position
T_113	EH	activity focus	Klimataria W	Ext. topographical surveys	certain
T_114	MH	? Settlement	Klimataria W	Ext. topographical surveys	certain

Fossey (1988: 247) describes the site as on a small flat area at the very W end of Klimataria peninsula, completely separated by a rocky mountain from the Klimataria site.
EH surface sherds have been noticed at the site, as well as MH material which, together with a cist grave, probably MH, may represent a MH settlement site (Fossey 1988: 247 and intensive survey work in the framework of the Boeotia Survey project).
Fossey (1988: 247) notes how the site had not been noticed previously, and how the single ruined house on the site was until recently a shepherd's shelter belonging to people of Akraiphnio. He comments how it is a very small site. He is tempted to consider it simply as an offshoot of the Yliki Sengaina site (in the EH period at least), and linked to it by the water of the lake. It could have been simply a small separated activity focus of some sort, even a small rural site exploiting that small flat area, but it could easily have been larger and gravitating towards the lake and lake-linked activities. The intensive work carried out in the framework of the Boeotia Survey project on the site showed a quite important Bronze Age phase.

39/40. Yliki Sengaina: *Components T_104 and 105* (39), *T_106* (40)

ID	Period	Type	Toponym	Discovery	Position
T_104	EH	activity focus	Yliki Sengaina	Ext. topographical surveys	approximate
T_105	C	rural site/activity	Yliki Sengaina	accidental	approximate
T_106	H	rural site/activity	Yliki Sengaina	accidental	approximate

In the NW bay of lake Yliki-Likeri, on the edge of the lake, Fossey (1988: 246) noticed a large number of EH sherds (EH I and II) on the surface, in Sengaina locality, which are traces of an activity focus (*component T_104*). He notes how the site had not apparently been noticed previously.
In AD 1981: 194 remains (probably surface potsherds) of an extensive C and H settlement are reported, again in locality Sengaina, "*until recently covered by water*" (*components T_105 and T_106*).
We cannot establish a spatial relationship between the two, the Prehistoric and the Historical site, since no exact location in space is given. On the other hand, we know that they both were on the edge of the lake and probably visible only when the water level is low. As noted in AD 1981: 194, the historical settlement (probably a hamlet) was agricultural in function, exploiting the valley which became a lake after the modern drainage of lake Copais (and it is now a reservoir for Athens) and the subsequent increase of the water level of Yliki lake. The same may be true for the EH activity focus as well.

41/42. Yliki Boubali*: Components T_ 107* (41) *and T_108* (42)

ID	Period	Type	Toponym	Discovery	Position
T_107	H	rural site/activity	Yliki Boubali	accidental	imprecise
T_108	R	burial place	Yliki Boubali	accidental	imprecise

On the N shore of Yliki-Likeri, in Boubali locality. The place is mentioned in AD 1981: 194, together with Yliki Sengaina (see above no.39-40).
In this locality R tombs by the shore and, on a nearby low hill, traces (remains of structures and associated pottery) of a Hellenistic settlement site, as well as an oil press, worned by the lake fluctuations, are reported as having appeared in 1981, until recently covered by the water. Location is not more precisely given, and therefore the position assigned on the map (fig.1) is only guessed. The component represents probably a hamlet with agricultural functions, as Yliki Sengaina.

Paralimni Oungra

A small plain at the SW extremity of lake Paralimni, formed by lake deposits and bordered by the same ridges that enclose the lake itself and by the lake, whose fluctuations the NE area of the plain is very much dependent on. In the middle of the plain used to lay the small family hamlet of Yliki (formerly Oungra). On the N side of the bay is the steep Chelonokastro hill, a spur of Mt. Ptoion fortified in some periods of history.

The site was already known to Noack, who reports it in his study on Prehistoric fortifications in the area (Noack 1894: 458-460 – see chapter II.3.7 and appendix I.7, Akraiphia *chora*). Remains of the settlement site by the lake shore were visible in 1966 after the first recorded drop of the level of the lake since the drainage of the Copais (AD 1966: 198-202). Surface explorations and excavations were carried out on the area then and afterwards (AD 1966: 198-202; AAA 1971: 319-328; AAA 1973: 207-8 figs. 11 and 13). HS&D (1979: G41 and G42) confuse material from the Paralimni - Pyrgos site at the NE end of Lake (included in Anthedon *chora* in the present work – *components AN_26 to AN_35*, appendix I.13) with material from the Paralimni - Oungra site at the SW end of Lake (included here in the present *chora*: Thebais).

43. Oungra Chelonokastro: *Components T_85 and T_86*

ID	Period	Type	Toponym	Discovery	Position
T_85	LH	fort	Oungra Chelonokastro	Travellers	certain
T_86	gr-rom	fort	Oungra Chelonokastro	Travellers	certain

Chelonokastro hill is a steep and rocky spur of Mt. Ptoion, on the N side of the Paralimni Oungra bay.
The site was already known to Noack, who reports it in his study on Prehistoric fortifications in the area (AM 1894: 458-460). Fossey (1988: 236) mentions LH as well as A sherds found in the wall fabric. Both could work as *terminus post quem* for different phases of walling.

44/45. Paralimni Oungra (lake shore site): *Components T_87 to T_91* (44)*; T_92 to T_95* (45)

ID	Period	Type	Toponym	Discovery	Position
T_87	N	? activity focus	Paralimni Oungra	Personal or group interest	approximate
T_88	EH	activity focus	Paralimni Oungra	Personal or group interest	approximate
T_89	MH	activity focus	Paralimni Oungra	Personal or group interest	approximate
T_90	LH	activity focus	Paralimni Oungra	Personal or group interest	approximate
T_91	SubM	activity focus	Paralimni Oungra	Personal or group interest	approximate
T_92	PG	activity focus	Paralimni Oungra	Personal or group interest	certain
T_93	G	settlement	Paralimni Oungra	Personal or group interest	certain
T_94	A	settlement	Paralimni Oungra	Personal or group interest	certain
T_95	C	settlement	Paralimni Oungra	Personal or group interest	certain

The large settlement site in the small plain at the SW extremity of lake Paralimni, by the shore, is mainly from PG/G to A/C period.[23] Fossey (1988: 239) notes how the settlement life lasted until the 4th C BC. The settlement seems to have been swallowed up by the lake at the end of the C period (on the basis of the absence of more recent material – AD 1967: 243), and according to Fossey it would have moved to Klimataria.
The bay was probably occupied also in Prehistoric period (since EH). Material is reported by Fossey (1988: 235-8) from Paralimni Oungra site, from surface explorations and excavations carried out in the area (AD 1966: 198-202; AD 1967: 243; AAA 1971: 319-328; AAA 1973: 207-8).
The area was occupied also in Byzantine period: a church lies few meters N of the stream that runs from Yliki to Paralimni lake.

46. Paralimni Oungra N: *Components T_96 to T_98*

ID	Period	Type	Toponym	Discovery	Position
T_96	MH	burial place	Paralimni Oungra N	Personal or group interest	certain
T_97	LH	burial place	Paralimni Oungra N	Personal or group interest	certain
T_98	PG	burial place	Paralimni Oungra N	Personal or group interest	certain

[23] Remains of a temple dated to the A period and part of the city wall were also found (AD 1967: 243).

Along the Paralimni lake shore, at the SW extremity of lake, below Chelonokastro hill, to the immediate N of Paralimni Oungra settlement site, a large cemetery area has been excavated (AD 1966: 198-202; AAA 1971: 319-328; AAA 1973: 207-8), mainly along a road running in a NE direction (AD 1967: 243), probably towards Mouriki area (Fossey 1988: 238). As mentioned in AD 1973: 265-6, grave enclosures are mainly in the elongated strip defined to the S by the coast of the lake. Below G graves, also some MH and Mycenaean cist graves have been found.

47. Paralimni Oungra area: *Component T_99*

ID	Period	Type	Toponym	Discovery	Position
T_99	C	burial place	Paralimni Oungra area	Other	approximate

Illegal clearing of Prehistoric and C tombs revealed by the receding water level is reported in the area of Oungra (*Epikairoteta* 31.7.90 from AReports 1990-91: 34). It is not clear whether they have to be linked to the Prehistoric and PG cemetery to the N of the settlement 'city wall' (*components T_96 to T_98*). Already in AD 1973: 265-6 there was a mention of Classical graves, spoiled by water or other causes, found at points between lakes Paralimni and Yliki (position no better indicated) – see *component T_116*.

48/49/50. Paralimni S hills: *Components T_100 to T_102* (48) and Paralimni Yliki: *Components T_116* (49) *and T_117* (50)

ID	Period	Type	Toponym	Discovery	Position
T_100	G	burial place	Paralimni S hills	Other	certain
T_101	A	burial place	Paralimni S hills	Other	certain
T_102	C	burial place	Paralimni S hills	Other	certain
T_116	C	burial place	Paralimni Yliki	accidental	approximate
T_117	R	? burial place	Paralimni Yliki	accidental	imprecise

Components T_100 to T_102: Cemetery area at the foot and on the slopes of the hills S of Paralimni lake, overlooking the Oungra settlement by the lake shore from the SE. Tombs were discovered through illicit plundering, after which the local *ephoreia* conducted excavation in 1965 (Fossey 1988: 236, fig.27).

Components T_116 and T_117: Classical graves have been found at points between lakes Paralimni and Yliki (AD 1973: 266) – see above *component T_99*, while a single R(?) tomb has been found near the modern hamlet of Yliki/Oungra (AD 1970: 233).

51. Mouriki – Kamelovrisi: *Components T_103 and T_152*

ID	Period	Type	Toponym	Discovery	Position
T_103	LH	burial place	Mouriki - Kamelovrisi	Personal or group interest	certain
T_152	MH	? activity focus	Mouriki - Kamelovrisi	Personal or group interest	certain

On a low hill on S shore of Lake Paralimni, ca. 5km NW of Mouriki, HS&D 1979 report on a possible MH settlement on the hill and LH chamber tombs on E slope.

52. Platanaki: *Components T_80 to T_84* and *T_79*

ID	Period	Type	Toponym	Discovery	Position
T_79	gr-rom	burial place	Platanaki road	Ext. topographical surveys	approximate
T_80	EH	activity focus	Platanaki	Ext. topographical surveys	approximate
T_81	MH	activity focus	Platanaki	Ext. topographical surveys	approximate
T_82	A	? activity focus	Platanaki	Ext. topographical surveys	approximate
T_83	gr-rom	activity focus	Platanaki	Ext. topographical surveys	approximate
T_84	R	? activity focus	Platanaki	Ext. topographical surveys	approximate

To Platanaki modern village (founded around 1922 by Turkish-Greek refugees[24]) is available a small plain trapped, as Fossey (1988: 233) describes it, between Mts. Kakousi and Kandili and the W slopes of Ktipas. A low hill ca 600m high marks the landscape NE of the modern village of Platanaki, NE of Mouriki, by the left side of the road to Loukisia.

The provenience of material within the area, either by the road below or on the hilltop or slopes, is very confused, therefore we cannot rely much on clustering. A list of archaeological evidence known from the site is much more meaningful and less biased. I created components for the gr-rom (probably C) cemetery area (*component T_79* – AD 1968: 223). Then, for each archaeological evidence (reported by Fossey 1988: 233), I created a corresponding component, as generic probable 'activity focus'. The evidence AE163 (LR or later surface walls) has not been included in any component because of uncertainty in chronology (it probably falls out of the chronological range under the present study).

As we see from the components listed above, also Prehistoric material (EH and MH) has been noticed on the surface by Fossey (1988: 233).

Less easily we can get a clue for the exact chronology and the character of occupation of the site in gr-rom periods. As noted, for historical periods the only sure evidence is the cemetery site (probably mainly Classical). Archaic surface sherds reported [AE168] may also come from the cemetery area, if in use since earlier periods.

We could infer the existence of a settlement site on the hilltop only by considering the cemetery area found below, by the road. Sherds (with no chronology strictly specified, except for some Archaic Corinthian wares and some generally R) have been found though on all sides of the hill, and lead to the possible existence of an activity focus of some sorts on the top. Chronology of potsherds found is not strictly specified, and it is not clear where the mentioned Archaic Corinthian wares and some generally R sherds have been found. We cannot either infer from Fossey's account whether sherds observed on the hill are fine wares or coarse wares. They could be coarse wares, since Fossey does not directly specify chronology for them and coarse wares are namely less easily datable.

Identification has been attempted with ancient *Peteon*, to which refers Strabo IX 410 (see Kirsten RE 1937: 1128 s.v. *Peteon* – from Fossey 1988: 233-4). There is no need though to hypothesize the existence here of a large settlement. It could

[24] In Platanaki village, Byzantine architectural fragments have been found reused in a church (AD 1967: 226-245).

have been also a large rural site with associated cemetery. The site would need a systematic revisit to clarify extension, chronology and function.

53. Sambali: *Component T_115*

ID	Period	Type	Toponym	Discovery	Position
T_115	gr-rom	fort	Sambali	Personal or group interest	approximate

On Mt. Sambali (S of Mouriki) Fossey (1988: 234 and *Teiresias* suppl. ii 1979: 9-13) mentions "*another rubble fortification of the type common in the environs of Skroponeri*", commonly dated to the 4th C BC.

54/55/56. Ypaton Tourleza: *Components T_118* (54) *and T_129* (55) and Ypaton: *Components T_119 to T_129* (56)

ID	Period	Type	Toponym	Discovery	Position
T_118	gr-rom	fort	Ypaton Tourleza	Travellers	certain
T_119	N	activity focus	Ypaton	Personal or group interest	approximate
T_120	EH	settlement	Ypaton	Personal or group interest	approximate
T_121	EH	burial place	Ypaton	accidental	approximate
T_122	MH	activity focus	Ypaton	Personal or group interest	approximate
T_123	LH	activity focus	Ypaton	Personal or group interest	approximate
T_124	G	burial place	Ypaton	Rescue excavation	approximate
T_125	C	? Settlement	Ypaton	Personal or group interest	approximate
T_126	gr-rom	? Settlement	Ypaton	Personal or group interest	approximate
T_127	C	burial place	Ypaton	Rescue excavation	approximate
T_128	R	? activity focus	Ypaton	Personal or group interest	approximate
T_129	C	burial place	Ypaton Tourleza N slope	accidental	approximate

Component T_118: At the SW foot of Mt. Sagmatas, above the modern village of Syrtzi/Ypaton, on the small, flat-topped hill called Tourleza (alb. Turle = tower, as Fossey notes), a gr-rom fortification is visible on the hilltop, datable to the C/H period from potsherds (a circuit wall of polygonal masonry - plan in Fossey 1988 fig.26). As Fossey notes, in 1838, Ulrichs (1863: 27) could still see remains of the wall surrounding the lower settlement, on the S and SW slopes; apparently this was built with large blocks in a rough polygonal style. For earlier accounts see Frazer 1913: 60-61.

Components T_124 and T_127 and T_129: Some hundreds of meters S of the village are reported G (Late Geometric) and C cemetery in the area (AD 1970: 224-6, as well as BCH 1972: 704; JHS 1971/2: 12). Furthermore, some tombs of the Classical or Archaic period had been plundered on the N slope of Tourleza in 1965 and 1966 (personal observation of Fossey 1988: 220).

Components T_125, T_126 and T_128: A possible lower settlement in association with the fortified hilltop (*component T_118*) and with the cemetery (*component T_127*) could be represented by surface potsherds. C and H dates are given by HS&D 1979 G28 from earlier accounts also (AD 1970: 224), but only for the hill summit. We do not have precise information on the material found in the lower area if not the one directly in association with the cemeteries. I created a C and a generic gr-rom component of activity focus with question mark, but also G and A might be present. LG material, though, seems to have come only from burial areas (Syriopoulos 1983: 84 G90), as well as Archaic (Fossey 1988: 220, without indication of provenance). An attempt at identification has been made with ancient *Glisas* (see Fossey 1988: 221-3, HS&D 1979: G28 from AD 1970: 224).
As far as the R period is concerned, R sherds are reported in the list of material from the site only by Fossey (1988: 220). Though, it could be that the settlement life stops in the C period and from Late C / H (4th C) onwards only a fort remains at this place. In R period one can suggest that not even the fort is in use, but another sort of activity would exist in the lower area. In order to be more precise, one should check where exactly Fossey noticed R potsherds, and in general a systematic revisit of the place would be needed, although nowadays the area is mostly occupied by modern buildings.

Components T_119 to T_123: Prehistoric material has been noticed on the SW slopes of Tourleza hill, in the area of the modern village, besides the new cemetery. Neolithic, EH and MH surface sherds as well as EH rock-cut tombs (in the area of the village school) are reported (AD 1970: 224). HS&D 1979 have a EH settlement here. Syriopoulos 1968 (reported by HS&D 1979) mentions also LH surface sherds and obsidian from the SW slopes.

57. Moni Sagmata: *Component T_130*

ID	Period	Type	Toponym	Discovery	Position
T_130	gr-rom	? cult place	Moni Sagmata	Ext. topographical surveys	certain

The large, flat top of Mt. Sagmatas is half rocky covered with holm-oak, and half (the SW) covered with a slight stratum of soil and covered with some deciduous trees. On the SW half is the large Byzantine monastery of Moni Sagmata reachable by a track from Ypaton village (Fossey 1988: 223).
No ancient sherds have been noticed at the site, but some architectural fragments and carved blocks lying in the monastery yard, as well as many gr-rom blocks reused and remains of an ancient wall visible on the surface, can be an indication of a H/R cult place here. Chronological indication is given only through the presence of a bench support in the yard, datable to the H/R period.
The site was already known to travellers (Frazer 1913: 61-2), mainly because of the possible identification (Fossey 1988: 224-5) with the sanctuary of Zeus Ypatos mentioned by Pausanias (IX 19.3) and probably linked with *Glisas* (see above – Ypaton Tourleza). The period of occupation seems though to differ, although no potsherds are available here.

58. Mesovouni: *Components T_131 to T_135*

ID	Period	Type	Toponym	Discovery	Position
T_131	EH	activity focus	Mesovouni	Other	certain
T_132	Preh	? activity focus	Mesovouni	Other	certain
T_133	C	settlement	Mesovouni	Travellers	certain
T_134	R	? Settlement	Mesovouni	Travellers	certain
T_135	LR	settlement	Mesovouni	Travellers	certain

Situated at the entrance to the Theban plain from the NE, i.e. from either Chalkis or the Tanagran plain, Mesovouni is a double hill in the middle of the small plain between two

mountain blocks: Mt. Soros to the S and Mt. Sagmatas to the N. The Western half is rocky and steep, and today mostly cut by rock mining, while the Eastern one has a flat top and gentle slopes on its E side (Fossey 1988: 212).
Ancient remains (building foundations and marble chips – AD 1976: 128) are to be seen especially on the E part. The site on the Easterly hill was already known to travellers (Ulrichs for instance), all listed in Frazer (1913: 60) apart from Gell 1819: 129 (quoted by Fossey 1988: 213), who probably saw more extensive remains also in the plain, today destroyed by intensive cultivation. An area near the S side of the hill is known as '*archaio patima*' possibly a reference to the presence in the past of ancient building remains here (Fossey 1988: 213) or rather to the passage of a road.[25] There is a water source nearby, to the N of the hill in the direction of the highway (marked on 1:50,000 IGME geological map).
The site has been identified, probably correctly, with ancient *Teumessos* (see Fossey 1988: 213-6; mentioned in Pausanias IX 19.1), associated with the cult of Athena Telchinia (AD 1976: 128).[26]

Component T_131: Beside the gr-rom occupation, also EH traces of activities have been noticed at the site (EH sherds on the flat top and on the E slopes and on the saddle between the two peaks – Fossey 1988: 213). The site could probably have been a focus of activity also in other less recognisable Prehistoric periods (flakes of obsidian and chert are reported in Fossey 1988: 213).

59. Stroma*: Components T_155 and T_136 to T_138*

ID	Period	Type	Toponym	Discovery	Position
T_155	N	activity focus	Stroma	Personal or group interest	certain
T_136	LH	activity focus	Stroma	Personal or group interest	certain
T_137	C	settlement	Stroma	Ext. topographical surveys	certain
T_138	H	settlement	Stroma	Ext. topographical surveys	certain

On a low hill at the S end of the valley formed by the Vlachorrema stream between Soros range and Mesovouni, on the S side of the Theban plain. It is probably the Stroma hill on the 1:50,000 GYS map, although Fossey (1988: 216) says that the valley is called Stroma and the hill Soules.
Neolithic and LH occupation is known at the site (HS&D 1979: G24), as well as C and H (HS&D 1979: G24 and Fossey 1988: 217).
This site has been identified with ancient *Teumessos* (Pappadakis BCH 1921: 523, for instance),[27] though others prefer to identify it with the Mesovouni site (see above no.58). Fossey (1988: 217), as well as Wallace (1979: 96-8), sees in the two sites complementary periods of occupation, and therefore suggests a shifting of settlement in time between the two locations.

What follows is a preliminary version of the list of sites discovered and recorded during the Systematic Intensive Surface Survey carried out in the 1980s by J.L. Bintliff and A.M. Snodgrass and their équipe (Bintliff-Snodgrass 1985a), albeit it covered, as for the Thebais *chora*, only the SW edge of the region (limits of the surveyed area mapped in fig.1).

SurveySITE code	CHRONOLOGY	CHARACTER
MP A1	C-Ehl	rural site
MP A1	Lhl-ER	rural site
MP A1	LR	rural site
MP A2	H	rural site
MP A3	LR	rural site
MP B1	C-Ehl	rural site
MP B2	A	?activity focus
MP B2	C	?activity focus
MP B2	C-Ehl	rural site
MP B3	A	activity focus
MP B3	C-Ehl	rural site
MP B3	EH	settlement
MP B3	G	activity focus
MP B3	LH	settlement
MP B3	Lhl-ER	rural site
MP B3	LR	rural site
MP B3	MH	settlement
MEL3	C-Ehl	rural site
MEL5	C-Ehl	rural site
MEL6	C-Ehl	rural site
MEL7	C-Ehl	?rural site
MEL7	LR	?rural site

Table 1. *List of intensive survey sites discovered during intensive and systematic survey work* (Bintliff-Snodgrass 1985a) *– mapped in fig.1.*

[25] The site is on or near the ancient road between Thebes and Chalkis (AD 1976: 128)
[26] For another hypothesis see also below (no.59 - Stroma) in the text.
[27] The site would seem compatible with Antimachos' description of Teumessos (quoted if not accepted by Strabo IX 2.24) as a 'windy little hill'. Teumessos is also mentioned with the epithet 'grassy' in the Homeric *Hymn to Apollo* 228, and Euripides *Ph.* 1106 seems to imply that it was in sight of Thebes. Pausanias (IX 19.1) tells a story on a fox about Teumessos, and all this leads some scholars to the identification of the place with this ancient settlement.

Appendix I.13
The Boeotian coastal area facing the Euboean sea N of Chalkis: ANTHEDONIA and SKROPONERI bay

THE ARCHAEOLOGICAL RECORD

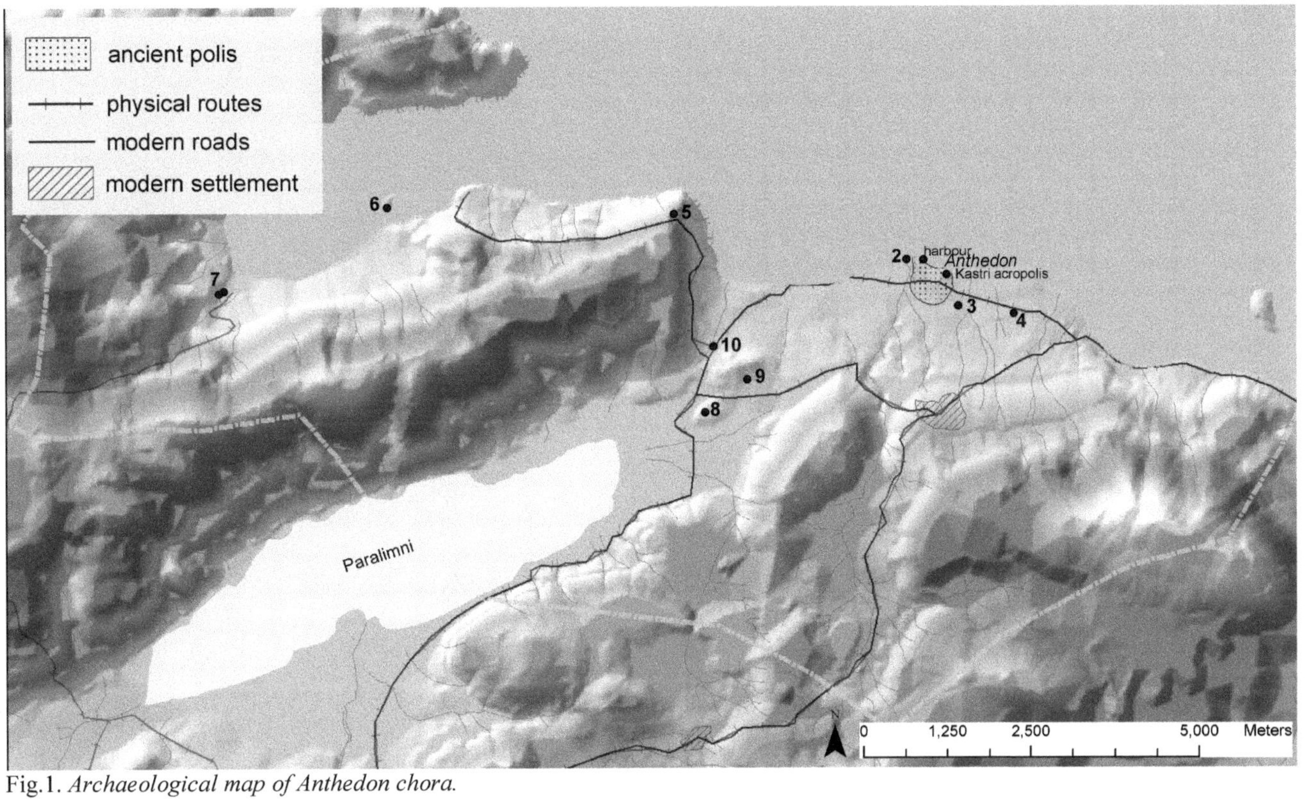

Fig.1. *Archaeological map of Anthedon chora.*

1. ANTHEDON Mandraki - Kastri: ***Components AN_1 to AN_11***

ID	Period	Type	Toponym	Discovery	Position
AN_1	EH	activity focus	Anthedon/Mandraki	Personal or group interest	certain
AN_2	MH	activity focus	Anthedon/Mandraki	Personal or group interest	certain
AN_3	LH	activity focus	Anthedon/Mandraki	Personal or group interest	certain
AN_4	G	settlement	Anthedon/Mandraki	Personal or group interest	certain
AN_5	A	settlement	Anthedon/Mandraki	Personal or group interest	certain
AN_6	C	settlement	Anthedon/Mandraki	Personal or group interest	certain
AN_7	H	settlement	Anthedon/Mandraki	Personal or group interest	certain
AN_8	R	settlement	Anthedon/Mandraki	Personal or group interest	certain
AN_9	LR	settlement	Anthedon/Mandraki	Personal or group interest	certain
AN_10	gr-rom	infrastructure	Mandraki harbour	Personal or group interest	certain
AN_11	LR	infrastructure	Mandraki harbour	Personal or group interest	certain
AN_12	gr-rom	cult place	Anthedon/Mandraki SE	Personal or group interest	approximate
AN_41	R	burial place	Anthedon - Kastri	accidental	certain

About 2 km below Loukisia/Anthedon village[1], "*at the coast, a low ridge running out from Ktipas ends in a small hill just over 20m high, right above the shore, recognized as the acropolis of ancient Anthedon*" (Fossey 1988: 252). The area is sometimes called Mandraki, properly the name of the harbour (Schlager et al. AA 1968: 21-102; plan 1 from Fossey 1988: 252), and the acropolis ridge is known as Kastri, to the S.

[1] Loukisia village overlooks the coastal plain from the NW slopes of Mount Ktipas (Fossey 1988: 252).

The wall circuit of town [AE559] encloses roughly a square area of ca. 36ha, and the acropolis is in the NE corner. For a plan of circuit wall see Fossey (1988: 253, fig. 32).
Fresh water springs out to the S of the site (mapped on 1:50,000 IGME geological map).

Anthedon showed Prehistoric (see for a summary HS&D 1979: G43 - EH to LH[2]) and historic (gr-rom) in great abundance. Traces of Neolithic occupation have also been found outside the circuit wall to the W/NW, on a small hill by the shore (*component AN_13* – Buck 1889).

It is included in the Homeric Catalogue of Ships and the *polis* is mentioned also in historical sources. There is as yet no indication of settlement in the early part of the DarkAge (HS&D 1979: G43). In historical times the *polis* settlement was associated with the harbour (*component AN_10*), and knew probably a great splendour as naval base for Thebes in 4th C BC (after its importance as a harbor in Homeric period, with a mention in the Homeric *Catalogue of Ships*). The artificial port was protected by two strong stone-dumps (Philippson 1951: 496). An extra-mural cult place has been found to the SE of the town (*component AN_12*).
A LR settlement (*component AN_9*) is known at the site (with a basilica [AE563]), associated with the harbour, whose remains, still visibile today, are said to be of LR date[3] (*component AN_11*).

The site has been extensively excavated by anglo-american-german équipes, since Rolfe (1890 and AJA 1889: 443-460). The small harbour was especially the subject of the Anglo-German investigations, particularly in 1966. A full report is published in Schlager et all. (AA 1968: 21-102). The excavation turned out to be disappointing for prehistoric periods, and BA is known only from surface sherds (AA 1968: 25 n.25), with the exception of a Mycenaean hoard of bronze tools and other objects, while historical periods are well represented.
Reused material in Loukisia village [AE582] probably comes from ancient Anthedon, despite the fact that it is 2km away. Leake though had found by Loukisia village some building blocks and inscriptions – some of which are of Late Hellenistic – Roman date (Fossey 1974) which, if they do not come from Anthedon site, could even represent a satellite settlement there (A.M. Snodgrass pers.comm.), whose 'ceramic' traces would have been cancelled by the modern settlement.

Component AN_41: A R grave was ploughed up in a private plot on Kastri area, within the city circuit (AD 1993: 197). It could be an indication of a shrinking of the town site in the R period.

2. Anthedon Mandraki W: *Component AN_13*

ID	Period	Type	Toponym	Discovery	Position
AN_13	N	settlement	Anthedon/Mandraki W	Ext. topographical surveys	approximate

Traces of Neolithic occupation have been found (Rolfe AJA 1889: 443-460; AD 1973-74: 447; Syriopoulos 1983: Neo246) outside the Anthedon circuit wall to the W/NW, on a small hill by the shore.

3. Anthedon Mandraki SE: *Component AN_12*

ID	Period	Type	Toponym	Discovery	Position
AN_12	gr-rom	cult place	Anthedon/Mandraki SE	Personal or group interest	approximate

Outside the Anthedon city circuit, to the SE, on a small knoll, an extramural sanctuary site was excavated, and interpreted as a sanctuary of Dionysos by the excavator (Rolfe AJA 1889: pl.XIV).

4. Anthedon Paliambela: *Components AN_40*

ID	Period	Type	Toponym	Discovery	Position
AN_40	R	? burial place	Anthedon - Paliambela	accidental	approximate

R tombstones plough up from a field in Anthedon - Paliabela locality. They constitute only an epigraphical evidence, very movable and not *in situ*. We can only suppose the existence of a cemetery in the area (AD 1988: 202-3).

5. Tourlo: *Components AN_14 to AN_16 and AN_39*

ID	Period	Type	Toponym	Discovery	Position
AN_14	LH	? activity focus	Tourlo	Personal or group interest	certain
AN_15	EH	? activity focus	Tourlo	Personal or group interest	certain
AN_16	C	? Fort	Tourlo	Ext. topographical surveys	certain
AN_39	N	activity focus	Tourlo	Personal or group interest	certain

A small conical hill to the N of the coastal road from Loukisia into Skroponeri bay, at the NE end of the bay, by the outlet of a small stream. This hill is separated from the Ptoion ridge by a small saddle on which runs the road.
On the S slopes of the hill towards the modern road are traces of ancient activities, as Fossey notes (1988: 261). Several traces of walling have lead Fossey to interprete the evidence as one of the 4th C BC watch posts protecting the Skroponeri naval base (Fossey 1988: 26). It has been argued that the bay could have been a base -perhaps the main one- for 4th C Boeotian fleets (see Skroponeri Kastro site – *component AN_24*).
Prehistoric (EH and LH) surface materials are reported by Fossey at the site, while Sampson (AD 1973: 447) seems to mention only obsidian and coarseware sherds (Neolithic).

6. Skroponeri Peninsula: *Component AN_17 to AN_23*

ID	Period	Type	Toponym	Discovery	Position
AN_17	N	activity focus	Skroponeri Peninsula	Ext. topographical surveys	approximate
AN_18	EH	activity focus	Skroponeri Peninsula	Ext. topographical surveys	approximate
AN_19	MH	activity focus	Skroponeri Peninsula	Ext. topographical surveys	approximate
AN_20	LH	activity focus	Skroponeri Peninsula	Ext. topographical surveys	approximate

[2] Philippson (1951: 496) mentions it as a port already in the Minyan period.
[3] Justinian period.

ID	Period	Type	Toponym	Discovery	Position
AN_21	C	activity focus	Skroponeri Peninsula	Ext. topographical surveys	approximate
AN_22	H	activity focus	Skroponeri Peninsula	Ext. topographical surveys	approximate
AN_23	LR	activity focus	Skroponeri Peninsula	Ext. topographical surveys	approximate

On a small peninsula (100m ca N-S) on the S side of the Skroponeri bay, in some parts quite rocky while in others with soil cover (Fossey 1988: 262), with a very narrow and low bit of land that connects it to the shore, only Fossey (1988: 262) seems to report archaeological evidence. Late Classical – Early Hellenistic and LR occupation at the site (Fossey 1988: 262), as well as prehistoric material (N, EH, MH, LH), have been noticed only by Fossey and not by Prehistoric gazetteers, which seem not to know the site.
It has been suggested (Fossey, *Teiresias* suppl. II 1979: 10) a possible identification of the site with ancient *Phokai* (*Ptolemais*), a gr-rom village/*kome*, arguing that the Skroponeri bay was the main base of the 4th C Boeotian fleet (see the Skroponeri Kastro site – *component AN_24*, that Fossey mentions together with this peninsula site).
Traces of a possible circuit wall on the E side [AE570] may be an indication of a fortified settlement, but masonry is not recognisable and therefore dating is impossible, as Fossey notes.

7. Skroponeri Kastro: *Component AN_24 to AN_25*

ID	Period	Type	Toponym	Discovery	Position
AN_24	gr-rom	fort	Skroponeri Kastro	Ext. topographical surveys	approximate
AN_25	LR	activity focus	Skroponeri Kastro	Ext. topographical surveys	approximate

On the hill of Kastro, at the head of the bay of Skroponeri and in the middle of the long beach, Fossey (1988: 262) only seems to report archaeological evidence. He reports remains of a fortification which encircles the summit itself, as well as extensive remains of surface walls [AE580], on the SW slopes of the hill, which according to Fossey might be dated through the artifact concentration (LR material – ripple ware [AE581]) found in connection with them. That the fortification which encircles the summit itself might be in association with the other surface walls remains just a hypothesis.

PARALIMNI area

Communication along the Paralimni lake was not easy. It is a lake of tectonic formation, and therefore with bordering slopes that are steep and rocky (see chapter II.1). There are openings only at the two ends, with very fertile zones formed by lacustrine deposits. The NE end can be interpreted as belonging to the area controlled by the *polis* of Anthedon (and this becomes valid especially when considering the difficulties in land communication between the two ends of the lake). On the other hand, the SW end has been included in the Thebais *chora* (chapter II.3.12), since it is much more easily approachable from there (and it was probably under the direct control of Thebes in the gr-rom period).

8. Paralimni Pyrgos: *Components AN_26 to AN_35*

ID	Period	Type	Toponym	Discovery	Position
AN_26	N	activity focus	Paralimni Pyrgos	Personal or group interest	certain
AN_27	EH	activity focus	Paralimni Pyrgos	Personal or group interest	certain
AN_28	MH	activity focus	Paralimni Pyrgos	Personal or group interest	certain
AN_29	LH	settlement	Paralimni Pyrgos	Travellers	certain
AN_30	G	activity focus	Paralimni Pyrgos	Ext. topographical surveys	certain
AN_31	A	activity focus	Paralimni Pyrgos	Personal or group interest	certain
AN_32	C	settlement	Paralimni Pyrgos	Personal or group interest	certain
AN_33	H	settlement	Paralimni Pyrgos	Personal or group interest	certain
AN_34	R	? activity focus	Paralimni Pyrgos	Ext. topographical surveys	certain
AN_35	LR	? activity focus	Paralimni Pyrgos	Personal or group interest	certain

Known also as *Palaiometochi*, the place is marked by a Frankish tower crowning a hill at the NE end of Paralimni lake, in a small plain formed by lacustrine deposits.
According to Fossey (1988: 257), the hill was probably the acropolis of an ancient settlement extending around the feet of the hill. The site was already known to travellers (Ross 1851: 132). It is marked by a continuity of occupation, from Final Neolithic to Medieval times, though it is quite sure a gap after the Hellenistic period, until LR/Early Christian times (see below).
Noack 1894: 457-8 (on acropolis), Lauffer RE 1959: 1575 s.v. *Ptoion* and Pharaklas AAA 1968: 139-140 (on the lower settlement) are the main accounts for the archaeological evidence at the site.
Prehistoric (EH and LH) surface materials are reported by Sampson (AD 1973-4: 448). His account is useful to check data against Hope-Simpson and Dickinson's gazetteer (HS&D 1979: G42) that confuses material from the Paralimni - Pyrgos site at the E end of Lake (Anthedon *chora* in the present work) with material from the Paralimni - Oungra site at the W end of Lake (Thebais *chora* in the present work). The lower part of both sites had disappeared after the drainage of the Copais lake, and the subsequent rise in level of Paralimni. As Fossey (1988: 257) says, they were revealed by the considerable drop of the lake in 1966.
At Paralimni Pyrgos, the lower settlement presents surface walls marking buildings to the SE and SW, the SW ones built on terraces, clearly visible because they were washed clean by the lake's waters (Fossey 1988: 258). They were investigated by the local ephorate (AAA 1968: 139-140), and in particular by Pharaklas, who notes how the lower area seems to have been inhabited from Mycenaean to Hellenistic times (and probably also in Early Christian period), withouth giving though indication on surface material.
According to that information, there might have been a gap in occupation between H and LR times. Anyhow, as we cannot be confident on this, I have created a component as possible R (*component AN_35*).

Pharaklas (AAA 1968: 139-140) identified the site with ancient *Isos* on the basis of Strabo IX 2.14, who says it was near Anthedon and abandoned (only traces of a *polis* had remained) in his times. Its name would have been an emendation for *Nisa* in the Catalogue of Ships. Fossey (1988: 257) agrees with the identification, while Snodgrass (pers. comm.) seems skeptical.

Wallace (1979: 62-64) needlessly sees continuity throughout the R era (not proved though!) as disproving an identification with ancient *Isos*.

9. Paralimni Pyrgos E: *Component AN_38*

ID	Period	Type	Toponym	Discovery	Position
AN_38	EH	burial place	Paralimni Pyrgos E	Personal or group interest	approximate

Half a kilometer to the E/NE of Paralimni Pyrgos site, against the foothills of Kandeli height, Early Bronze Age tombs of Cycladic type (known in Euboea and at Lithares in Boeotia) are reported in AD 1969: 174-5. They would have been connected to a settlement site nearby, but it is not certain that this was the EH site (*component AN_27*) at Paralimni Pyrgos (see tombs and settlement at Lithares – appendix I.12).

10. Paralimni Pyrgos NE: *Components AN_36 and AN_37*

ID	Period	Type	Toponym	Discovery	Position
AN_36	unknown	activity focus	Paralimni Pyrgos NE	Personal or group interest	approximate
AN_37	LH	burial place	Paralimni Pyrgos NE	Personal or group interest	approximate

NE of Paralimni Pyrgos site, at the foot of Mt. Ptoion, at the N foothills of promontory called Botsikoula, ca 500-700m NE of Paralimni Pyrgos E site, a large area with *poros* quarries [AE1464] has been noticed during research work by the local ephorate (Pharaklas AAA 1968: 139-140).
In the same area, *poros* foundations of buildings were noticed, as well as a LH chamber tombs cemetery area, with more than 30 chamber tombs with *dromos*. No information about surface material is given by Pharaklas. LH tombs were most probably in connection with *component AN_29* – LH settlement site.

Appendix I.14
The Tanagra plain and Eastern Boeotia

THE ARCHAEOLOGICAL RECORD

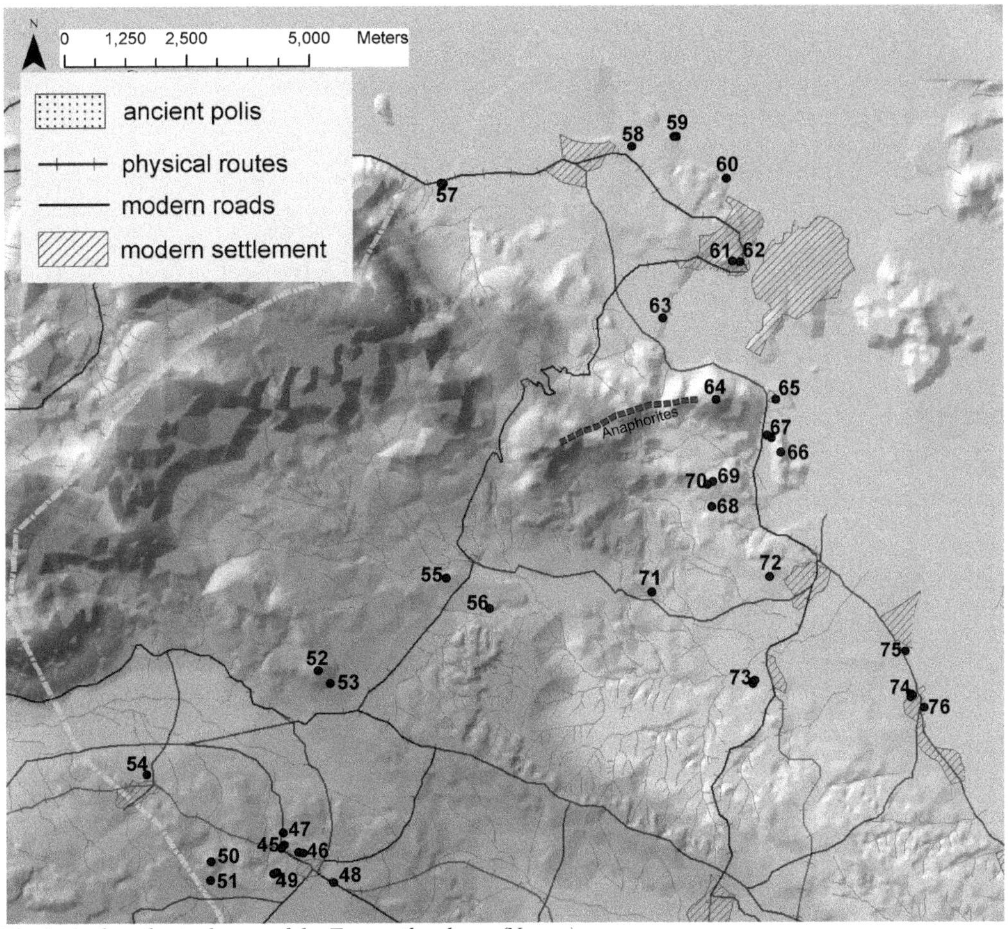

Fig.1. *Archaeological map of the Tanagrike chora (N part).*

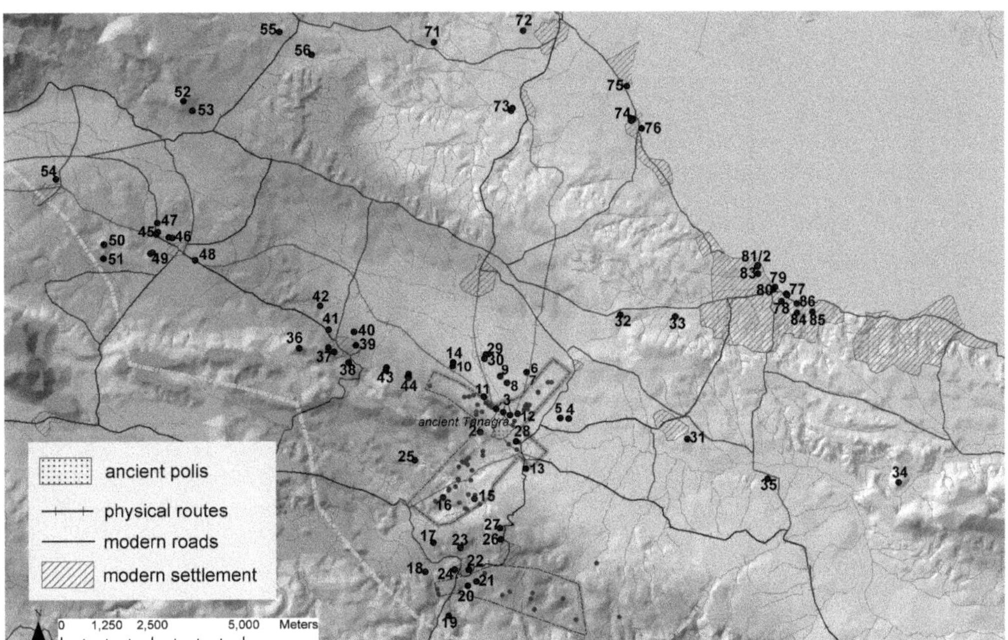

Fig.2. *Archaeological map of the Tanagrike chora (S part).*

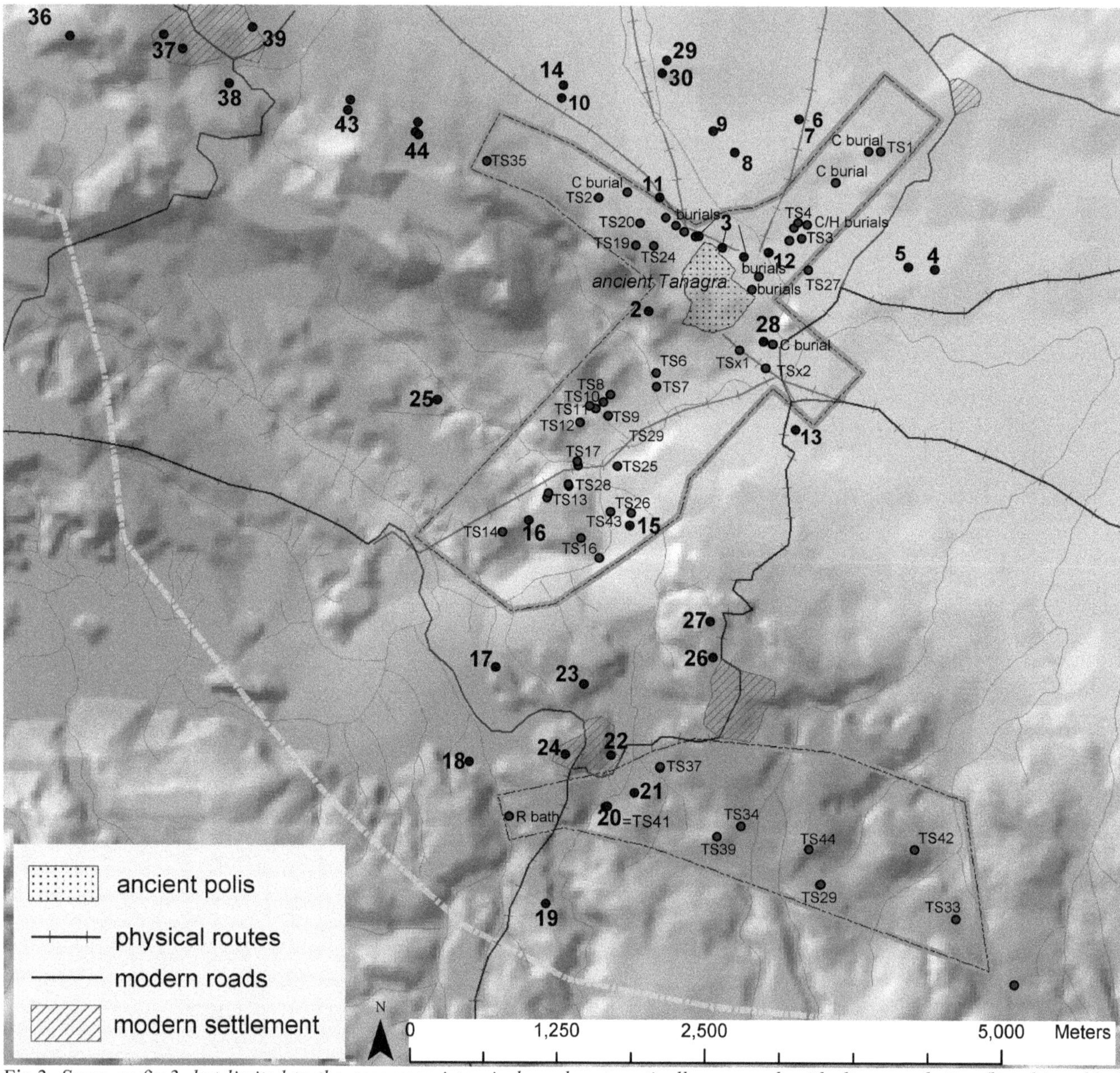

Fig.3. *Same as fig.3, but limited to the area area intensively and systematically surveyed, with discovered sites (listed in table 2) marked.*

1. TANAGRA Grimada: *Components TA_1 to TA_8*

ID	Period	Type	Toponym	Discovery	Position
TA_1	N	activity focus	Tanagra Grimada	Int. and sys. artefact surface survey	precise
TA_2	EH	activity focus	Tanagra Grimada	Int. and sys. artefact surface survey	precise
TA_3	MH	activity focus	Tanagra Grimada	Int. and sys. artefact surface survey	precise
TA_4	LH	human presence	Tanagra Grimada	Int. and sys. artefact surface survey	precise
TA_5	C	settlement	Tanagra Grimada	Travellers	precise
TA_6	H	settlement	Tanagra Grimada	Travellers	precise
TA_7	R	settlement	Tanagra Grimada	Travellers	precise
TA_8	LR	settlement	Tanagra Grimada	Travellers	precise

At the spot where the 'corridor' formation (see chapter II.3.14 – TOPOGRAPHICAL SETTING and Philippson 1951: 512) opens up into the lower valley of Asopos, in the corner between Asopos river and Lari, between two wide valleys (Philippson 1951: 514), is Tanagra ancient city site,[1] built on the very end slopes

[1] On the identification of Tanagra through ancient sources see Fossey 1988: 53-4.

of Soros,[2] in Grimada locality. The highest point of the city wall is 189m asl and then the area occupied by the city descends down the slope towards NE with terraces, until the edge of Lari banks. The urban area is crossed (NW-SE) by a ridge, which constitutes the acropolis of the ancient city site (see fig.5a).
The ancient city of Tanagra was therefore located at the intersection of two rivers, in a well defended spot and central for the E part of Boeotia of which it was the most important town. It controls two fertile valleys, those of Laris and Asopos, and also has available different topographic zones, plains, valleys and the hilly country of Tetartogene, mainly SW of the city.

The Homeric *Catalogue of Ships* omits Tanagra, although it does include other sites from the vicinity. It was not until late in the 6th century that the name of the *polis* was even alluded to (earliest inscription from Olympia – Schachter 2004: 63). Two stamped roof tiles have been dated as belonging to the first half of the 7th C BC (Schachter 2004: 63). Among the many graves both looted and excavated around the city, the earliest seem to be dated ca. 625-600 BC (Higgings 1986). After the Persian War Tanagra was under the control of Thebes. In 458, during the Peloponnesian war, took place the Tanagra battle (Thucidides I 108.1),[3] just before the Oinophyta battle, after which the walls of Tanagra which the Athenians tore down were later rebuilt. In the Boeotian confederacy of the time (end of 5th C – beginning of 4th C) Tanagra had the rights to provide one of the eleven boiotarchs. In later times (H and R) the city controlled all the coastal area and the area of the *Tetrakomia.*
Intensive and systematic artifact surface survey work was carried out in 2000-2003, in the framework of the Tanagra Survey project, on the city site. It monitored the actual presence of gr-rom material and the extension of the site in the different periods, and discovered a Prehistoric occupation on the later city site (for details see below - THE TANAGRA SURVEY). As for LH, only few sherds have been collected even during intensive and systematic artifact surface survey at the site, which cannot probably state more than a human presence or a small focus of activity.

2. Grimada: *Component TA_220*

ID	Period	Type	Toponym	Discovery	Position
TA_220	gr-rom	special activity	Tanagra Grimada	Ext. topographical surveys	approximate

A quarry of ancient Tanagra, with gray-blue limestone, was identified by Roller (AReports 1985/86: 39-41) SW of Tanagra ancient city.

3. Extramural buildings by ancient Tanagra: *Components TA_224 to TA_226*

ID	Period	Type	Toponym	Discovery	Position
TA_224	H	activity focus	Tanagra N	Rescue excavation	certain
TA_225	H	activity focus	Tanagra N	Rescue excavation	certain
TA_226	H	activity focus	Tanagra N	Rescue excavation	certain

Excavated structures and layers probably belonging to extramural buildings of ancient Tanagra have been excavated in the occasion of works for the gas pipe-line in 1997 (AD 1997: 371, map fig.8). Mainly storage vessels have been found during excavations, which could probably testify a storage use of those buildings.

4/5. Kokkali: *Components TA_9 to TA_11* (4); *TA_12, TA_13* and *TA_198* (5)

ID	Period	Type	Toponym	Discovery	Position
TA_9	N	? activity focus	Kokkali	Personal or group interest	certain
TA_10	EH	activity focus	Kokkali	Personal or group interest	certain
TA_11	LH	activity focus	Kokkali	Personal or group interest	certain
TA_12	A	burial place	Kokkali	accidental	approximate
TA_13	C	burial place	Kokkali	accidental	approximate
TA_198	H	burial place	Kokkali	accidental	certain

A low hill to the NE of ancient Tanagra site, to the opposite site of Lari river compared to Tanagra location, to the N side of the road from Oinoi to Agios Thomas and to the N of Asopos river (ca 750m N-NW of Ag.Thomas church –TS5 in table 2 *Survey sites* below- and 3.5km from Ag.Thomas village).
The hill is widely known though because of the presence of cemeteries from historical periods (A to H) excavated legally and illegally, where the famous Tanagrines (terracotta figures) were found (see mainly Higgings 1986).
Traces of Prehistoric activity focus were found on top of the low hill, though they are not visible anymore today, probably due to the intensive illegal robbing activity taking place on the hill because of the presence of the graves with the Tanagrines.
Neolithic occupation (reported by HS&D 1979) seems still in doubt, while more certain are traces of activity focus in EH and LH period (LH reported also in AD 1969: 187).

6 – 13. Other BURIAL PLACES: *Components TA_14 to TA_16* (6); *TA_221* (7); *TA_202* (8); *TA_199 to TA_201* (9); *TA_204 and TA_222* (10) ; *TA_203* (11) ; *TA_17 and TA_18* (12); *TA_205 to TA_207* (13)

ID	Period	Type	Toponym	Discovery	Position
TA_14	A	burial place	Tanagra N	Rescue excavation	approximate
TA_15	C	burial place	Tanagra N	Rescue excavation	approximate
TA_16	H	burial place	Tanagra N	Rescue excavation	
TA_17	C	burial place	Tanagra - Sewer Plantage	Rescue excavation	approximate
TA_18	H	burial place	Tanagra - Sewer Plantage	Rescue excavation	approximate
TA_199	A	burial place	Tanagra Airport	Rescue excavation	certain
TA_200	C	burial place	Tanagra Airport	Rescue excavation	certain
TA_201	H	burial place	Tanagra Airport	Rescue excavation	certain
TA_202	gr-rom	burial place	Tanagra area	accidental	certain
TA_203	gr-rom	burial place	Tanagra - Agios Polykarpos	accidental	certain
TA_204	gr-rom	burial place	Tanagra - Bali	accidental	certain
TA_205	A	burial place	Tanagra - Ghelezi	accidental	certain

[2] The limestone Soros mountain rises from the Tetartogene deposits E of Thebes and with an EW direction runs along 16km to the E where it meets the Tanagra basin. The ancient city is built in the last eastern foothills of the limestone mountain and before the limestone disappears below the deposits of the Tetartogene which occupy the wider Tanagra basin is built the ancient city (Philippson 1951: 512-3).
[3] For inscriptions dealing with the battle of Tanagra, see Schachter 2004: 69f.

ID	Period	Type	Toponym	Discovery	Position
TA_206	C	burial place	Tanagra - Ghelezi	accidental	certain
TA_207	H	burial place	Tanagra - Ghelezi	accidental	certain
TA_221	H	burial place	Tanagra N	Rescue excavation	certain
TA_222	A	burial place	Tanagra - Bali	Rescue excavation	certain

An excavated cemetery area is known to the N of the ancient city (BCH 1970: 699 - *components TA_14 to TA_16* and AD 1997: 371, map fig.8 – *component TA_221*), as well as by the National Aircraft Factory (AReports 1977-78: 36 and 1984-85: 30), where 170 tombs belonging to an A to H cemetery were excavated by Andreiomenou (AD 1976: 120-121 – *components TA_199 to TA_201*). A small cult place dedicated to Herakles is also mentioned by Andreiomenou within the burial area by the Military Airport (AReports 1985-86: 39-41). Another burial area (*components TA_17 and TA_18*), closer to the city site, has been recently excavated by the local ephorate (AD 1994: 286; AD 1995: 305-6; AD 1997: 377) 300m ca far from the city wall to the N, inside the sewer-plantage (*viologikos katharismos Schimatariou*).

In Praktika 1911: 132ff digs are mentioned around the ancient city in five places (burial areas – *components TA_202 to TA_207*). The majority of them are mapped in Higgings 1986 (Makrovouni,[4] Agios Polycarpos, Bali, Kokkali –see above– and Gkeliazi/Ghelezi).[5] In the last area (*components TA_205 to TA_207*), across the Asopos, tombs run from 6th C BC on (A-C-H). As this is further from the city compared to the other burial areas, it could be related to the rural occupation of the landscape (with burials along the main roads – see map in fig.2). Few burials of the A period found during works for the gas pipe-line (AD 1997: 371, map fig.9 – *component TA_222*) belong probably to the gr-rom necropolis already known (Bali - *component TA_204*) and positioned in Higgings 1986.

14. Tanagra N: *Component TA_227*

ID	Period	Type	Toponym	Discovery	Position
TA_227	LR	rural site/activity	Tanagra - Bali	Rescue excavation	approximate

A quite large rural site of LR times (with oil presses, storage rooms and clay aqueduct) has been excavated during works for the gas pipe-line in 1997 (AD 1997: 371, map fig.9) at Bali locality (NW of Tanagra), to the N of the known cemetery area (see above - *components TA_204 and TA_222*).

15. Agios Konstantinos: *Components TA_19 to TA_29*

ID	Period	Type	Toponym	Discovery	Position
TA_19	N	activity focus	Agios Konstantinos	Personal or group interest	certain
TA_20	EH	activity focus	Agios Konstantinos	Personal or group interest	certain
TA_21	MH	settlement	Agios Konstantinos	Personal or group interest	certain
TA_22	LH	settlement	Agios Konstantinos	Ext. topographical surveys	certain
TA_23	PG	activity focus	Agios Konstantinos	Ext. topographical surveys	certain
TA_24	G	activity focus	Agios Konstantinos	Ext. topographical surveys	certain
TA_25	A	activity focus	Agios Konstantinos	Other	certain
TA_26	C	activity focus	Agios Konstantinos	Other	certain
TA_27	H	activity focus	Agios Konstantinos	Other	certain
TA_28	R	activity focus	Agios Konstantinos	Ext. topographical surveys	certain
TA_29	LR	settlement	Agios Konstantinos	Ext. topographical surveys	certain

2 km. N of the villages of Kleidi and Ag.Thomas, at the N end of a low ridge projecting from S and only 2.5-3km from ancient Tanagra, is a low peak (with a small chapel of Prophitis Ilias) and an adjacent lower plateau with the church of Ag.Konstantinos. Near the church of Ag. Konstantinos a new monastery of Theotokos (*Iera Mither Theou*) and on the hilltop a small chapel of Prophitis Ilias have been recently built. The Asopos flows below.

On the site is available an account of Fossey (1988: 53) who reports C and H sherds but no structural remains on the terrace of the church, and an absence of sherds on the hill top, due to thick soil cover and heavy cultivation. The situation, though, has considerably changed with the construction of the new monastery, after which the Tanagra Survey project conducted a systematic survey on the site. According to Fossey's remark on the thick soil cover, finds on the top did not seem to have been known before the construction of the monastery, when the hilltop was cleaned up in order to build the chapel dedicated to Proph.Ilias. Potsherds from different periods are now visible in large quantity all over the hill top and the higher slopes. On the other hand, C and H sherds reported on the church plateau below (and apparently missing from Tanagra survey collection in summer 2003) could have disappeared during construction work and probable levelling activities.

Fortifications were not noticed by Fossey but were already reported in the visit of Bintliff and Snodgrass (Journal 1978: 64-5) and then seen during the Tanagra survey project in 2003 (Bintliff et al. 2004).

After the intensive research carried out by the Tanagra survey team, the site turned out to be, as preliminary accounts state, an extended prehistoric site covering the period from the Neolithic to the end of the Late Bronze Age, where among a large collection of fine wares typical boetian prehistoric wares were found, such as EH Urfirnis and Agia Marina ware, fine Minyan and matt-painted MH wares, fine plain and decorated LH pottery (Bintliff et al. 2004). As K. Sarri reports (pers. comm.), the general view of the prehistoric material from Agios Konstantinos indicates that on the top and higher terraces of the hill dominate the examples of the Bronze Age pottery, while on the lower terraces - close to the river - the same repertory coexists with an increasing number of Neolithic coarse wares, a phenomenon, which can be compared with a similar situation around the city region.

As for the graeco-roman period, PG and G material is reported (probably representing a cult place rather than a settlement, though still difficult to tell), as well as C (only probably) and H, which might represent a settlement site, as material does not seem to be distinctive enough for a sanctuary (V. Stissi, pers. comm.). Agios Konstantinos might therefore represent a hilltop refuge site rather then a rural village/*kome*.

[4] S of Mavrovouni (C necropolis between Schimatari and Kokkali) - AD 1969: 187.

[5] Some bibliographical reference for the cemeteries, until the 1980s, is also in Fossey 1988: 47 note 14.

Fossey (1988: 56) indicates the Agios Konstantinos site as a possible candidate for the ancient *kome* of *Schedia*.

16 – 26. Kleidi burials: *Components TA_30* (16); *TA_36* (17); *TA_34* (18); *TA_35* (19) ; *TA_32* (20); *TA_37 and TA_38* (21); *TA_31* (22); *TA_33* (23)

ID	Period	Type	Toponym	Discovery	Position
TA_30	LR	burial place	Agios Konstantinos	accidental	imprecise
TA_31	LH	burial place	Kleidi	accidental	approximate
TA_32	A	burial place	Kleidi	accidental	approximate
TA_33	C	burial place	Kleidi	accidental	approximate
TA_34	C	burial place	Kleidi	accidental	approximate
TA_35	C	burial place	Kleidi	accidental	approximate
TA_36	R	burial place	Kleidi	accidental	approximate
TA_37	LH	burial place	Kleidi	accidental	imprecise
TA_38	H	burial place	Kleidi	accidental	imprecise
TA_208	A	? cult place	Kleidi	accidental	approximate
TA_228	gr-rom	special activity	Asopia - Kleidi	Rescue excavation	approximate

Several finds are available to local people in the village coming from tomb robbing and illegal excavations mainly in the surrounding of Kleidi as well as Agios Thomas (formerly Liatani) village. Local information suggests several A-C-H burial focuses,[6] and probably R also as well as Mycenaean, in addition to the ones already located and excavated (AD 1964: 199-200). Some of them are mapped in Fossey (1988: 53 and map fig.6), who suggests they may be related to the Ag.Konstantinos site.[7] Some of them, though, are probably too far away. The burials seem to create a circle round Kleidi village.

24/25. Kleidi: *Components TA_208* (24) *and TA_228* (25)

ID	Period	Type	Toponym	Discovery	Position
TA_208	A	? cult place	Kleidi	accidental	approximate
TA_228	gr-rom	special activity	Asopia - Kleidi	Rescue excavation	approximate

Component TA_208: During intensive survey work in the area (Tanagra survey project), building remains (which could represent an Archaic (?) construction – a sanctuary?) were reported to the survey team by local people, who showed the team pictures of a structure found during the excavation for the construction of the water reservoir for the village (at the highest point of the village). A probable date to the Archaic period is based on the picture and some sherds shown and reported as coming from the site.
It could probably prove (along with the cemetery making a circle around the village) the existence of a settlement just below the modern Kleidi village. There was with more certainty the site of a Turkish/early modern settlement, judging from the material collected during survey at the W edge of the village.

Component TA_228: in the wider area of Kleidi, S of the Soros ridge, during works for the gas pipe-line ancient quarry installations have been found (AD 1997: 376).

26/27. Agios Thomas: *Components TA_40* (26); *TA_41 and TA_42* (27)

ID	Period	Type	Toponym	Discovery	Position
TA_40	LH	burial place	Agios Thomas	accidental	approximate
TA_41	EH	? activity focus	Agios Thomas	Personal or group interest	imprecise
TA_42	N	? activity focus	Agios Thomas	Personal or group interest	imprecise

Component TA_40: LH burials reported by locals (Fossey 1988: 57). They are probably to be linked to the other Mycenaean burials known from the Kleidi area (see above no.21-22).
Components TA_41 and TA_42: EH is reported from the area of Liatani/Agios Thomas village, where probably has to be located an EH focus, different from the Ag.Konstantinos site (Fossey 1988: 57 from HS&D 1979: F60; AD 1967: 242-3; French 1972: figs. 9-11). Also Neolithic is reported (only by HS&D 1979: F60).

28. Lari: *Component TA_39*

ID	Period	Type	Toponym	Discovery	Position
TA_39	gr-rom	? activity focus	Agios Thomas - Lari	accidental	approximate

In locality Lari Ag.Thomas, mechanical agricultural activities destroyed two sides of a large building made of large squared blocks [AE72]. Clay tubes (*fistulae*) were also found. C sherds were noticed around [AE74] (AD 1967: 226-245; Fossey 1988: 57). It could be a probable rural farmstead located by Lari stream, below ancient Tanagra, though further details on position are not available, and correspond to some of the Tanagra survey rural and/or burial sites.

29/30. Kesseri : *Components TA_43 and TA_44* (29) and *TA_223* (30)

ID	Period	Type	Toponym	Discovery	Position
TA_43	EH	burial place	Kesseri	accidental	imprecise
TA_44	C	activity focus	Kesseri	Ext. topographical surveys	imprecise
TA_223	gr-rom	? burial place	Tanagra N	Ext. topographical surveys	approximate

In Kesseri locality, probably in Schimatari area,[8] though position is not sure.
EH rock cut tombs,[9] found robbed, are reported from the site, from the S slope (Fossey 1988: 57; HS&D 1979: F60 from AD 1969: 187), as well as a small focus of A/C potsherds (noticed by Bintliff and Snodgrass in 1978, as a small and curiously-sited place, found while looking for Proph.Ilias site, near to the NW – pers. comm.). C-HL sherds S of a tomb area, reported in AD 1967: 226-245 along the road between Schimatari and Kokkali hill [AE1854] could probably be linked to this gr-rom activity focus(*component TA_223*), since location could correspond.

[6] The Archaic burial place near Kleidi, reported by Fossey, was found also during intensive survey work in 2004.
[7] Fossey (1988: 53) merges the discussion of A.Konstantinos site 2km to N with the cemeteries around these villages.

[8] In an area belonging to Schimatari and bordering the S Euboean gulf, in a thick pine-trees area (AD 1969: 187). Fossey (1988: 57) seems to have misplaced the site nearby Agios Thomas village, but he also mentiones it under Dilesi (62-6).
[9] Similar and probably almost contemporary to the ones found at Lithares (see appendix I.12).

31. Oinophyta: *Components TA_45 to TA_49*

ID	Period	Type	Toponym	Discovery	Position
TA_45	N	human presence	Oinophyta	Ext. topographical surveys	approximate
TA_46	EH	? human presence	Oinophyta	Ext. topographical surveys	approximate
TA_47	A	human presence	Oinophyta	Ext. topographical surveys	approximate
TA_48	C	human presence	Oinophyta	Ext. topographical surveys	approximate
TA_49	H	human presence	Oinophyta	Ext. topographical surveys	approximate

As Fossey (1088: 58) describes, the modern village of Oinophyta lies at the NW foot of the small conical hill of *Dichalotos Pyrgos*.[10] The Asopos river flows 0,5km to the S.
On the hill, a cistern has been excavated to provide most of the village's drinking water. There is no available information on the excavation, at which point antiquities might have been found, as Fossey comments. On the other hand, potsherds and material have been found on the hill's E slope in slit trenches (Fossey 1988: 59, probably from his own visit). In the church of Ag.Paraskevi (Oinophyta cemetery) a part of a frieze of R period was found reused (AE 1981: 195); the evidence [AE1803] has not been assigned to any component due to its mobility.
Oinophyta is known from ancient sources as a place where a battle took place in 457 BC, won by the Athenians (Thucydides I 108.2-3; Diodoros XI 81-83.3).[11] For a summary of the identification question see Fossey 1988: 59-60.
There is not, very probably, enough evidence for a settlement there.

32/33. Prophitis Ilias: *Components TA_50 to TA_54* (32) *and TA_55* (33)

ID	Period	Type	Toponym	Discovery	Position
TA_50	N	activity focus	Prophitis Ilias	Personal or group interest	approximate
TA_51	EH	activity focus	Prophitis Ilias	Personal or group interest	approximate
TA_52	MH	activity focus	Prophitis Ilias	Personal or group interest	approximate
TA_53	LH	settlement	Prophitis Ilias	Personal or group interest	approximate
TA_54	A	? burial place	Prophitis Ilias	Other	approximate
TA_55	MH	burial place	Prophitis Ilias E	accidental	approximate

The site of Prophitis Ilias has been given different locations in different gazetteers and reports on the site:
- a hill 2km to NE of Schimatari according to Prehistoric gazetteers.
- a hill SE of Schimatari (4km!) overlooking the valley leading to Dilesi by the chapel of Ag.Paraskevi (mentioned by Brown, the excavator) according to Fossey (1988: 57).
- a hill 2km from Ag.Dimitrios, at Schimatari station, to the right of the direct road to Dilesi, much nearer to the sea (to the NE of the shrine of Ag.Paraskevi on hill) according to Brown (Brown BSA 1905/6: 93ff).
The Brown location is probably the right one. He excavated a LH house (BSA 1905/6: 94-5). At the site material from other Prehistoric periods (Late Neolithic, EH, MH) has also been found (Prehistoric gazetteers: HS&D 1979; Syriopoulos 1968; French 1972).
Fossey (1988: 57) remarks how the Prehistoric site may have always been no more than a farmstead.
1.5km to the E of the Prophitis Ilias site tombs of MH date are reported by locals (*component TA_55*) – Fossey 1988: 58.

34. Bathiza: *Components TA_195 to TA_197*

ID	Period	Type	Toponym	Discovery	Position
TA_195	EH	activity focus	Bathiza	Personal or group interest	approximate
TA_196	MH	activity focus	Bathiza	Personal or group interest	approximate
TA_197	H	rural site/activity	Bathiza	Other	approximate

French (1972) describes the site as "*on high isolated rock (visible at least 10km away on West side); ca 3km N of the Ethniki Odos and ca 6km N of Kakosalesi the river Asopos runs around the base of the rock (sheer on S and E sides) on the S side; ridge on the West, leading to pine forests and low hills*".
Fossey (1988: 62) notes how the top of the hill is fairly flat, surrounded by steep slopes. The hill lies on the orographic left side of the Asopos river.
EH II and MH (Grey Minyan) potsherds [AE64 and AE65] and obsidian parallel-sided blade [AE1668] were found (HS&D 1979: F58; Fossey 1988: 62), marking a small Prehistoric site.
Fossey (1988: 62) reports also on H or R tiles at the site, probably to be associated to a rural activity.

35. Asopos S: *Component TA_219*

ID	Period	Type	Toponym	Discovery	Position
TA_219	H	burial place	Asopos S	accidental	imprecise

On a hill, S of Asopos river, to the right of the old National Road Lamia-Athens (200th km), a rescue excavation after robbery discovered eight chamber tombs open in the rock of H period (4th C BC) (AD 1980: 232).

36 - 42. Bratsi/modern Tanagra: *Components TA_56* (36); *TA_57, TA_68 to TA_70* (37), *TA_62* (38) ; *TA_65 to TA_67* (39)*; TA_64* (40)*; TA_63* (41); *TA_58 to TA_61* (42)

ID	Period	Type	Toponym	Discovery	Position
TA_56	gr-rom	fort	Bratsi/Modern Tanagra	Travellers	certain
TA_57	gr-rom	? activity focus	Bratsi/Modern Tanagra	Travellers	approximate
TA_58	EH	activity focus	Bratsi/Modern Tanagra	Ext. topographical surveys	certain
TA_59	A	activity focus	Bratsi/Modern Tanagra	Ext. topographical surveys	certain
TA_60	C	activity focus	Bratsi/Modern Tanagra	Ext. topographical surveys	certain

[10] A tower, apparently Medieval in date, once crowned the hill (locals say it got destroyed in World War II, 1944 by German troops) - Fossey 1988: 58. There are caves marked on the 1:50,000 GYS map.
[11] See Kirsten RE 1937: 2257-8 s.v. *Oinophyta*. The locality where the battle took place could have been around the village of Staniates/Oinophyta that, according to last decades' policy of 'hellenizing' place-names, was recently renamed Oinophyta. In the area around the village the local toponym Oineia (that could be survival of Oinophyta) seems quite frequently in use (as noted by travellers, as Leake and Ulrichs quoted by Fossey 1988: 59 note 58). The attribution is not based on any certain basis and could easily be applied to almost any place in Eastern Tanagrike (for a summary of the identification question see Fossey 1988: 59-60). For inscriptions dealing with the battle of Oinophyta, see Schachter 2004: 70f.

ID	Period	Type	Toponym	Discovery	Position
TA_61	LH	? burial place	Bratsi/Modern Tanagra	Ext. topographical surveys	certain
TA_62	LH	burial place	Bratsi/Modern Tanagra	accidental	approximate
TA_63	LH	burial place	Bratsi/Modern Tanagra - church	accidental	approximate
TA_64	LH	burial place	Bratsi/Modern Tanagra - Elies	accidental	approximate
TA_65	C	burial place	Bratsi/Modern Tanagra	accidental	approximate
TA_66	A	burial place	Bratsi/Modern Tanagra	accidental	approximate
TA_67	H	burial place	Bratsi/Modern Tanagra	accidental	approximate
TA_68	C	? activity focus	Bratsi/Modern Tanagra	Ext. topographical surveys	imprecise
TA_69	A	? activity focus	Bratsi/Modern Tanagra	Ext. topographical surveys	imprecise
TA_70	R	? activity focus	Bratsi/Modern Tanagra	Ext. topographical surveys	imprecise

To the W side of the modern Tanagra village (formerly Bratsi and recently renamed Tanagra), 4.5km SW of Schimatari, rises a peak, in the E section of the Soros range, marked as *Kastron* on the 1:50,000 GYS map. The modern village lies at the E foot of it.

Component TA_56: Fossey (1988: 50, plan fig.5) describes and shows a plan of the fortification on the hilltop [AE395]. He speaks about a refuge and not a real settlement. The area enclosed by the circuit is only 0.16ha. The total absence of surface sherds and of inner surface remains may lead to this conclusion. Masonry is rough polygonal, as described by Fossey, and a tower is visible by the entrance. It could have been simply a fort, as many others in Boeotia. The similarity with the long walls at Orchomenos, suggested by Fossey 1988: 50, could lead to assigning a H date.
An attempt has been made to identify Bratsi site with ancient *Stephon*=summit (discussion in Fossey 1988: 56, who admits this identification is just a guess but there is no other suitable site).

Component TA_57: At the foot of the hill near the modern village Ulrichs (1863: 76-7 – reported in Fossey 1988: 50) describes traces of buildings (not visible at all in 1966, as Fossey comments) and of a possible ancient road.

Components TA_58 to TA_60: To the NW of the modern village on a low hill are visible the remains of a Frankish tower (marked on 1:50,000 GYS map), ca 750m NW of end of village (by the Harma road fork). By the tower potsherds were noticed by J.L. Bintliff and A.M. Snodgrass during their visit (pers. comm.).

Components TA_61 to TA_64 (LH burial places): In the area of the modern Tanagra village have been found several focuses of LH tombs with LH III painted *larnakes* (most of them now in the Thebes museum) – some of them at the E foot of the Kastron hill (*component TA_62*). Spyropoulos, very much interested in the *larnakes*, have been investigating for years the area looking for other tombs and the Mycenaean settlement. Recent survey work (see below – THE TANAGRA SURVEY) in the area of ancient Tanagra cleared out that the main Mycenaean settlement, to which the rich tombs with the *larnakes* related to, was not at the site of the historical *polis* of Tanagra. Since the majority of LH *larnakes* cemetery areas seem to have been found in the surroundings of Bratsi / modern Tanagra, probably the main Mycenaean settlement in the area may have to be looked for in this area. A detailed and precise mapping of all the LH tombs excavations (legal and illegal) would help in the localising of the ancient settlement. Unfortunately illegal excavations or accidental findings detailing positions are not available, and furthermore, the legal excavations accounts on locations are not always very precise or easily locatable on maps from bibliographical records only. In Praktika 1971, Spyropoulos reports on cemetery excavations, with a list of the previous bibliography. To Spyropoulos' list may be added the various brief accounts: AD 1970: 221-2; BCH 1970: 1037-1040; 1971: 929-932; 1972: 699-704; JHS Arch 1969/70: 16-17; 1970/1: 15; 1971/2: 12; *Teiresias* 71.1.1; AD 1971: 214-5 (from Fossey 1988). The latest reports on cemetery excavations are in Spyropoulos, Ergon 1973: 11-14; 1975: 17-26; 1976: 8-14; 1977: 14-19; 1979: 5-6; 1980: 21-22.

Components TA_65 to TA_70 (gr-rom burial places and activities): Cemetery areas of historical period are also reported. Sherds (from A, C[12] and R) period listed in gazetteers and reports could be simply related to burial area (but they could also represent other kinds of activities in the landscape). Related units of Archaeological Evidence have been included in cemetery components (with question mark) and they also form 'activity focus?' components by themselves. They could also be in association with the surface walls and the traces of a road seen by Ulrichs by the village [AE364 – AE365].

43/44. Gephyra/Ledeza and Dendron/Grava: *Components TA_71 and TA_72* (43)*; TA_73, TA_74, TA_209 and TA_210* (44)

ID	Period	Type	Toponym	Discovery	Position
TA_71	LH	burial place	Gephyra/Ledeza	Personal or group interest	approximate
TA_72	LH	settlement	Gephyra/Ledeza	Personal or group interest	approximate
TA_73	LH	burial place	Dendron/Grava	Personal or group interest	approximate
TA_74	LH	settlement	Dendron/Grava	Personal or group interest	approximate
TA_209	A	activity focus	Dendron/Grava	Other	approximate
TA_210	R	rural site/activity	Dendron/Grava	Other	approximate

To the E/SE of Bratsi/modern Tanagra village, two large Mycenaean chamber tomb cemeteries, with graves cut into limestone, are reported: the first 400m ca. E of Bratsi (Gephyra/Ledeza), and the other one 700m SE of that cemetery (Dendron/Grava), both along with habitation focuses[13] that go with the chamber tombs (HS&D 1979: F63 that summarize previous accounts – mainly Praktika 1969: 6; Ergon 1974: 10-16; 1975: 17-26; 1976: 8-14; AD 1973: 266-7). Later excavations of graves also by Spyropoulos at Dendron in the large cemetery on the W side of the slope (AReports 1976-77;

[12] Remains from tombs (vases and inscriptions of 5th C BC) accidentally discovered by locals looking for tiles - from Higgings 1986: 140, as well as a well, probably ancient.

[13] One settlement is located 100-150m E of the Gephyra cemetery (Ergon 1975: 12-26). A small section of the settlement site was excavated by the cemetery (Ergon 1974: 10ff). The other settlement, by the Dendron cemetery, extends on a slope covered by thick olive wood of a low hill, rising along the S border of the fertile valley where now the military airport is. Two rooms were excavated, part of a building complex of Mycenaean period.

1978-79: 23; 1982-3: 31; 1983-4: 33-34; 1984-5: 30; Ergon 1984: 49-53).
Remains of Mycenaean occupation (including one megaron-like building) date from advanced Mycenaean times (LH III A2-B), while the first use of the cemeteries may be as early as LH II (Fossey 1988: 52).[14]

Components TA_209 and TA_210: Later habitation structures (A period, but mainly R period) at the site of the Mycenaean settlement at Dendron/Grava destroyed the earlier layer (Ergon 1974: 10ff).

45 - 48. Dritsa – Pyrgos (ancient ELEON): *Components TA_75 to TA_85 and TA_88* (45)*; TA_86 and TA_87* (46)*; T_89 and TA_90* (47) *; and TA_211* (48)

ID	Period	Type	Toponym	Discovery	Position
TA_75	N	? activity focus	Dritsa - Pyrgos	Other	certain
TA_76	EH	activity focus	Dritsa - Pyrgos	Personal or group interest	certain
TA_77	MH	settlement	Dritsa - Pyrgos	Personal or group interest	certain
TA_78	LH	settlement	Dritsa - Pyrgos	Personal or group interest	certain
TA_79	PG	? Settlement	Dritsa - Pyrgos	Ext. topographical surveys	certain
TA_80	G	settlement	Dritsa - Pyrgos	Ext. topographical surveys	certain
TA_81	A	settlement	Dritsa - Pyrgos	Travellers	certain
TA_82	C	settlement	Dritsa - Pyrgos	Travellers	certain
TA_83	H	settlement	Dritsa - Pyrgos	Travellers	certain
TA_84	R	settlement	Dritsa - Pyrgos	Personal or group interest	certain
TA_85	LR	settlement	Dritsa - Pyrgos	Personal or group interest	certain
TA_86	H	burial place	Dritsa village	accidental	approximate
TA_87	gr-rom	burial place	Dritsa village	accidental	approximate
TA_88	gr-rom	? activity focus	Dritsa - Pyrgos SW	Ext. topographical surveys	approximate
TA_89	gr-rom	burial place	Dritsa - Pyrgos N	Ext. topographical surveys	approximate
TA_90	A	burial place	Dritsa - Pyrgos N	Ext. topographical surveys	approximate
TA_211	gr-rom	burial place	Dritsa	accidental	approximate

W of the village of modern Harma (formerly Andritsa/Dritsa), on a small ridge (ca 200m length from NW to SE - 120m width), overlooking the E part of the Theban plain. The W part of the ridge is called Pyrgos after the medieval tower at its highest W point (Papachatzis 1981: 126, pl.144; Fossey 1988: 88 and fig.10).
Fresh water springs out to the SW of Pyrgos hill in front of the chapel [AE1653].

The site was known to travellers already (Frazer 1913: 63-6 - with earlier accounts: Leake 1835: 466-9; Ulrichs 1863: 78-9; Ross 1851: 107-9) and it has been identified with ancient *Eleon* (Strabo IX 404; Pausanias I 29.6 – see Fossey 1988: 94-5), a member of the τετρακωμία with Mykalessos, Harma (both known as having been deserted by Roman times) and Pharai. The name Eleon now, after the 'hellenization' of village names, is attached to a village to the NW, formerly called Sphaides, and this causes confusion between sites in some accounts (ex: HS&D 1979 and Syriopoulos 1968).
The etymology of Eleon is claimed 'marshes' by Strabo but apparently there were no swampy areas there in his time, as Fossey (1988: 97-9) notes. Eleon is mentioned as still inhabited in R times by Plutarch (*Moralia* II.2). The settlement has a natural fertile bloc available in the undulating plain to the S and SW.

The site shows surface walls and a rich variety of painted wares as well as a remarkable Lesbian wall, of debated date, probably symbolic rather than defensive *stricto sensu*. Prehistoric occupation is also known at the site, especially MH and LH (represented by abundant sherds of excellent quality - Syriopoulos 1968). It was an important LH settlement (I/IIIC), probably second only to Thebes in the Theban plain, as remarked by HS&D 1979: G25.[15]

Component TA_88 represents traces of a foundation of big, squared blocks, which could be ancient, below a small chapel, now rebuilt, at the SW foot of Pyrgos hill, noticed by Fossey (1988: 90) during a personal visit in 1996.

Components TA_86; TA_87; TA_89; TA_90; TA_211 (burials): several burial locations are known within and by the modern village, probably to be related to the Pyrgos site (as well as for the burials at Proph.Ilias site) – Fossey 1988: 92. By the modern village of Dritsa/Harma, along the railway, 20 minutes away from the ancient city wall seven tombs were excavated[16] (Higgings 1986: 140 - *component TA_211*).

49. Dritsa – Proph.Ilias: *Components TA_91 to TA_95*

ID	Period	Type	Toponym	Discovery	Position
TA_91	A	burial place	Dritsa – Proph.Ilias	accidental	certain
TA_92	C	burial place	Dritsa – Proph.Ilias	accidental	certain
TA_93	R	burial place	Dritsa – Proph.Ilias	accidental	certain
TA_94	LH	? activity focus	Dritsa – Proph.Ilias	accidental	approximate
TA_95	EH	? activity focus	Dritsa – Proph.Ilias	Other	approximate

On Prophitis Ilias hill (named after the church on its summit), to the S of the site of the ancient settlement on Pyrgos hill, and especially on its N and W slopes (Fossey 1988: 92, plan fig.10), are tombs cut into the rock and mostly robbed (A-C and R period), to be related with the settlement site at Dritsa-Pyrgos location (see above).
On the same hill are reported also scattered sherds from the EH and LH periods (Hope Simpson n.427 repeated by HS&D 1979: G25). The archaeological evidences for the EH period probably represent an activity focus in EH times (settlement/habitation?) and not a burial area (this is not clear in Fossey, and could not be checked out against HS&D 1979: G25, who do not mention

[14] Gephyra cemetery is slightly later than Dendron cemetery (Praktika 1971: 7ff; Ergon 1974: 10ff)

[15] The LH settlement occupied the flat top and the N slopes of the steep-sided hill to the NW of Dritsa/Harma village (HS&D 1979: G25).

[16] Vases, same as the more recent type of Ritsona, were found in the graves.

the site separately from Dritsa-Pyrgos). The LH evidence could represent halo activities of the settlement at Dritsa-Pyrgos site.

50/51. Meg. Kastron: *Components TA_96* (50); *TA_97 and TA_98* (51)

ID	Period	Type	Toponym	Discovery	Position
TA_96	EH	activity focus	Meg. Kastron	Ext. topographical surveys	certain
TA_97	gr-rom	burial place	Meg. Kastron S	accidental	approximate
TA_98	LR	activity focus	Meg. Kastron S	Other	approximate

Component TA_96: Half-way between modern Harma (formerly Dritsa) and Eleon (formerly Sphaides) villages is the hill Kastron (marked as Meg.Kastron on 1:50,000 GYS map).
Fossey (1988: 96) notes how to the N the site overlooks the watershed between the Theban and Tanagran plains. The site is naturally defended (very steep slopes from all sides). Few EH sherds are reported from the surface.

Components TA_97 and TA_98: In the fields to the S of the Meg.Kastron hill many stones and bones are said to be ploughed up by locals. Sherds in this low area are Late Roman and mainly Byzantine (Fossey 1988: 96). A gr-rom burial place component has been created (*component TA_97*, formed by AE47 + ?AE48), in addition to a LR activity focus (*component TA_98* from AE48), which could be either the cemetery recorded in *component TA_97* or some other kind of activity (even a rural site / villa to which the burial area could be associated).

52/53. Kastron Lykovouni (ancient HARMA?): *Components TA_ 99 to TA_104* (52); *TA_105 and TA_106* (53)

ID	Period	Type	Toponym	Discovery	Position
TA_99	EH	activity focus	Kastron Lykovouni	Personal or group interest	certain
TA_100	LH	settlement	Kastron Lykovouni	Personal or group interest	certain
TA_101	G	activity focus	Kastron Lykovouni	Ext. topographical surveys	certain
TA_102	A	? Fort	Kastron Lykovouni	Ext. topographical surveys	certain
TA_103	C	fort	Kastron Lykovouni	Travellers	certain
TA_104	H	? Fort	Kastron Lykovouni	Ext. topographical surveys	certain
TA_105	gr-rom	burial place	Kastron Lykovouni slopes	accidental	approximate
TA_106	gr-rom	activity focus	Kastron Lykovouni slopes	Travellers	approximate

Kastron is the SE spur (288m asl) of Lykovouni ridge (Lykovouni is called the main peak - 451m asl). Fossey (1988: 85) notes how the ridge provides the N boundary to the small plain which connects the main Theban plain to the E. The location can be described as a high barren scrub-covered hill to the N of Thebes-Chalkis road, where the road begins to ascend towards the Anaphorites pass (HS&D 1979: G27).

Remains of a fortification on the hilltop (Fossey 1988, plan fig.9) were known already to travellers (Ulrichs, Leake and Frazer – listed in Fossey 1988: 86). Less than 1ha is included in the enclosure. The extensive summit (ca 300m NS and 200m EW) is covered with wall foundations.
A lower settlement is reported on the S and E feet of the hill (Fossey 1988: 86-88). An identification has been suggested with ancient *Harma* (Fossey 1988: 89). Sources[17] mentioned Harma as in this area and abandoned in R times (and archaeological evidence seems to confirm it), while a possible inscription confirms the name attribution (IG VII 97). If the settlement site is really one of the *poleis* of the τετρακωμία (together with Mykalessos, Eleon and Pharai), it could have had its own resource zone in the flat plain below to the SE, S and SW, though it is constituted of poor soils, limestone and older fill clays.
The name *Harma* now, after the 'hellenization' of village names, is attached to the village formerly called Dritsa (see above), and this causes confusion between sites in some accounts (ex: HS&D 1979 and Syriopoulos 1968).

The fact that the fort on the hilltop is so small (less than 1ha enclose) can lead to the interpretation of the evidence as a military fort, rather than a settlement. In this case, the remains below (on S and W hill slopes – *components TA_105* and *TA_106*), interpreted as the lower settlement, could also belong to other kinds of activities in the landscape such as a cult place (see surface walls seen by Ulrichs [AE1658]) or a A-C rural site [potsherds AE307 and AE309] jointly with a burial areas (see tombstones [AE1659]). Only more intensive and systematic research, as well as functional analysis on surface remains, could probably give a more appropriate answer.
Snodgrass (pers. comm.) sees it as a poor place and better as a military place controlling the major road junctions of Chalkis and Oropos routes. On the other hand, the location of the settlement of Kapandriti, exploiting the same area (not very fertile, as seen above), should cause reflection.[18] Today the area is quite waterless, but formerly the situation could have been different, as Leake (1835) mentions a fountain at the foot of the hill.

Components TA_99 and TA_100: Prehistoric material is also reported from the site, mainly from the S end of the hill. Near the top some LH III sherds and obsidian chips were noticed [AE308]. Probably the LH site was small, at the higher S end of the hill (HS&D 1979: G27).

54. Sphaides/modern Eleon: *Component TA_212*

ID	Period	Type	Toponym	Discovery	Position
TA_212	H	burial place	Sphaides/modern Eleon	accidental	imprecise

By Sphaides/modern Eleon village H robbed tombs were indagated (AD 1965: 242). They can be linked to a rural settlement, probably lining a route leading from ancient Eleon (modern Dritsa/Harma) site towards the Theban plain.

[17] Strabo IX 404-5 mentions it as a *kome,* already *'erimos'*, which was 'Tanagraean', a deserted village near Mykalessos; Pausanias IX 19.4 mentions the ruins of the city of Harma (=chariot). Pausanias (I 34.2) notes how men of Tanagra use Harma territory (in addition to other places, i.e. Mykalessos basin) for grazing.
[18] In the Turkish era the village of Kapandriti used to be a short distance to the E and may have used the fountain mentioned by Leake at the fooot of the hill.

55. Ritsona: *Components TA_107 to TA_114 and TA_213*

ID	Period	Type	Toponym	Discovery	Position
TA_107	EH	activity focus	Ritsona	Personal or group interest	precise
TA_108	MH	activity focus	Ritsona	Personal or group interest	precise
TA_109	LH	activity focus	Ritsona	Personal or group interest	precise
TA_110	G	? Settlement	Ritsona	Personal or group interest	precise
TA_111	A	settlement	Ritsona	Travellers	precise
TA_112	C	settlement	Ritsona	Travellers	precise
TA_113	H	settlement	Ritsona	Travellers	precise
TA_114	LR	? Settlement	Ritsona	Ext. topographical surveys	precise
TA_213	R	? Settlement	Ritsona	Personal or group interest	precise

In the centre of a fertile basin, on the road from Thebes to Chalkis, at the point where the road to Aulis/Vathy forks off Eastwards, is the hamlet of Ritsona. Fossey (1988: 80) already noted how the hamlet is constituted of but a few houses, and currently the place is just a crossroad with a small simple taverna (named 'Ritsona' – the only way to find the place nowadays) and a couple of rural houses opposite. The deserted village lies on top of a small height to the NE of the crossroad, above the taverna.
The Ritsona basin (a quite easily recognisable settlement chamber – see chapter II.3.14) is surrounded by mountains and ridges (Fossey 1988: 80)[19]: to the W and NW Mount Ktipas, an outlyeer of which, the Lykovouni ridge, closes it on the SW; to the N Galatsideza ridge, one of the largest ridges radiating from Mt. Ktipas; to the S and E the lower hills of that ridge which runs eastwards separating the Tanagran plain from the coastal plain of Vathy to the N. The basin is constituted by the Attic poorer lowland limestone and conglomerates.

On a knoll (lower Eastern hill of Kazarma ridge) to the SW (400m ca) of the Ritsona hamlet, to the W side of the road from Thebes to Chalkis, archaeological traces of activities are visible and were already known to travellers (Ulrichs 1863: 30) - mainly surface sherds and traces of circuit walls and other ancient foundations, including C houses (AD 1965: 243). The evidence represent a settlement site from the A to the H period, with an apparent gap in R periods due to lack of surface material and a different character (probably as a rural nucleation) of occupation in the LR period. Probably, as in the case of other Boeotian *poleis*, a shrink of settlement occurred. As far as the G (Late G) period is concerned, a settlement at the site is in doubt. No G material is reported on the town ridge (Fossey 1988: 82), and probably Syriopoulos (1983-84: G85 (=Late G)) and HS&D (1979: G45) refers to the cemetery area. On the other hand though, they might not have noticed G (Late G)[20] or G (Late G) might not be visible on the town ridge. A settlement could have been there related to the rich cemetery found nearby (Ritsona Kamileza – see below no.56).

The excavations in the nearby G and later cemetery (BSA 14 1907-8: 216 – Ritsona Kamileza), as stated by HS&D (1979: G45), provide support for the identification as ancient *Mykalessos*. Ancient sources are listed and examined in Fossey (1988: 83-5). Mykalessos is mentioned as a member of the τετρακωμία (together with Harma, Eleon and Pharai). Despite the large basin available, poor limestone and conglomerates lowland which characterize it would never support anything larger than a *komopolis,* dependent from Tanagra since H times.[21]
Before the Persian War, Mykalessos was one of several Boeotian *poleis* to mint a full range of coins (Head 1911: 346),[22] as well as Tanagra in the same period (Head 1911: 348). This suggests, as stated by Schachter (2004: 52), that Tanagra and Mykalessos were more or less equal in status at that time.
The town flourished in mid 6th C BC (rich necropolis) and declined in 5th C BC. Destroyed during the Peloponnesian war, it was still in life in the 4th C BC, although under the control of Tanagra.

Prehistoric material is also known at the site (EH, MH and LH surface material are reported by Fossey 1988: 82 – personal visit in the 1960s). HS&D (1979: G45) refer to Fossey's visit account and suggest how the LH site would presumably have been confined to the knoll, whereas the historic site was certainly more extensive.

56. Ritsona Kamileza: *Components TA_115 to TA_118*

ID	Period	Type	Toponym	Discovery	Position
TA_115	G	burial place	Ritsona Kamileza	accidental	approximate
TA_116	A	burial place	Ritsona Kamileza	accidental	approximate
TA_117	C	burial place	Ritsona Kamileza	accidental	approximate
TA_118	H	burial place	Ritsona Kamileza	accidental	approximate

Along the S side of Kamileza ridge (a lower part of the Kazarma ridge to the E side of the road) lining the old path to Vathy remains of an extensive necropolis were found.
Sparkes (JHS 1967: 128-180) published a catalogue of ca 90 published graves from Rhitsona, from Late G to Hellenistic (late 3rd C BC), but especially late 6th and early 5th C BC, mainly excavated by Ure and his wife (BSA 14 1907-8: 216 and other accounts, listed in Fiehn, RE 1933: 1005-1015 s.v. *Mykalessos*; Ure, RE Suppl. VII 1939: 495-510 s.v. *Mykalessos*).
More recently rescue excavations were carried out on the C necropolis (AD 1992: 172-4).

57. Lithosoros: *Components TA_119 to TA_125*

ID	Period	Type	Toponym	Discovery	Position
TA_119	N	? activity focus	Lithosoros	Personal or group interest	certain
TA_120	EH	settlement	Lithosoros	Personal or group interest	certain
TA_121	MH	settlement	Lithosoros	Personal or group interest	certain
TA_122	LH	settlement	Lithosoros	Personal or group interest	certain
TA_123	LH	burial place	Lithosoros	Personal or group interest	approximate
TA_124	Preh	burial place	Lithosoros	Personal or group interest	approximate
TA_125	C	activity focus	Lithosoros	Other	certain

At the very NW corner of the coastal plain of Chaleia and, as Fossey (1988: 78) describes it, "*jammed between Mt.Ktipas and the sea*" is the Lithosoros natural hill, a landmark in the flat

[19] A view of the basin is in BSA 1907-8: 233, fig.4.
[20] G material at the site, less diagnostic than the one from the cemetery, could be less easily noticed than later wares.

[21] Pausanias (I 34.2) notes how men of Tanagra use the land of the Ritsona basin (in addition to other places, i.e. ancient Harma area) for grazing.
[22] Mykalessos minted coins from late in the 5th C until some time in the 4th C (Head 1911: 346-347), but this is not necessarily an argument for autonomy, as Hansen points out.

area, even if only 24m heigh. The flat top is ca 160x140m, and it is separated from the beach only by a road's width. A spring is on the shore nearby. The site is also known as 'tomb of Salganeus' (*tymvos tou Salganea*).
Earlier travellers had noticed remains of walling including a 'circuit' (see Bakhuizen 1970) but, as Fossey (1988: 79) notes, it appears rather to be the line to which stones have been moved during cultivation of the flat top.
The Prehistoric occupation on the site is well known, and excavation activities have been carried out by Pappadakis in 1912 (AA 1914: 122-123; AD 1915: n.55-6), and later by Spyropoulos (AD 1970: 222) and Touloupa (AD 1978: 117; AD 1984: 125-126)[23] – see also Sapouna-Sakellaraki (AAA 1988: 77-90[24]). Neolithic is mentioned apparently only by Pappadakis.
Pappadakis excavation accounts (AA 1914 and AD 1915) report also tombs from the site, probably contemporary to the site (possibly MH rather than LH, see also Touloupa's excavation in 1978) - *component TA_124.*
The Lithosoros site, together with Glypha (see below – no.63), constitutes one of the most important Prehistoric sites of the Boeotian coast (Bakhuizen 1970). It is a coastal site along the physical route ancient Anthedon-Chalkis, in a fertile plain closed up to the W by the Messapion oros (Ktypas).

Occupation in historical periods is also known, probably in C period on the flat top,[25] and could be characterized as a rural site (probably with associated burials which seem to be attested by some black glaze found at the site – AE17 – Fossey 1988: 80).
The mound has been identified (first by Leake 1835 - Wallace AJA 1969) with the town of *Salganeus*, mentioned by Strabo (I 1.17; IX 2.9) (Buck 1979: 21; Bakhuizen 1970: 6-12; Wallace 1979: 94-6; Fossey 1988: 99f).

58. Drosia area: *Components TA_126 and TA_127*

ID	Period	Type	Toponym	Discovery	Position
TA_126	H	? burial place	Drosia area	accidental	imprecise
TA_127	R	? burial place	Drosia area	accidental	imprecise

Burial places of gr-rom date are known from the area of modern Drosia (formerly Chalia) village. H and R tombstones are reported, though only attested by epigraphical evidence (very movable!).

59. Drosia – Ag.Minas: *Components TA_128 to TA_131*

ID	Period	Type	Toponym	Discovery	Position
TA_128	EH	settlement	Drosia – Ag.Minas	Personal or group interest	approximate
TA_129	G	activity focus	Drosia – Ag.Minas	Other	certain
TA_130	A	activity focus	Drosia – Ag.Minas	Other	certain
TA_131	C	activity focus	Drosia – Ag.Minas	Other	certain

Evidence from historical periods is reported by the headland of Ag.Minas, a rocky promontory that dominated from the N a small bay (NE of Drosia village). No characteristics of the pottery found there is unfortunately specified, therefore we cannot infer the character of occupation (the components have been left as 'activity focus' – *components TA_129 to TA_131*). The historical material from the site (to the W of the church) is reported only by the latest account (Sampson AE 1974: 14).[26]
Interest in the area is due mainly to the EH I settlement (*component TA_128*) reported at the site already by Theokares (*Archeion Eyboikon Meleton* 1959: 308-9), contemporary with Lithosoros site (see above no.57). The site is on the Boeotian coast, in front of peninsula Manika and Kaki Kephali (Euboea), and his location is quite suitable for a Prehistoric settlement, with a small harbour to the S side.

60. Rodia: *Components TA_132 and TA_133*

ID	Period	Type	Toponym	Discovery	Position
TA_132	Preh	? activity focus	Rodia	Personal or group interest	approximate
TA_133	gr-rom	? activity focus	Rodia	Other	approximate

Sampson (AE 1974: 14 n.18) reports to the SE of Drosia - Ag.Minas site along the coast, NW of Chalkis, at Rodia (marked on 1:50,000 GYS map), another possible Prehistoric site, as well as probable traces of occupation from later (gr-rom) periods. No diagnostic material was noticed, and this is mainly the reason of the uncertain date, but of a sure presence of some kind.

61/62. Karababa: *Components TA_142* (61) *and TA_143* (62)

ID	Period	Type	Toponym	Discovery	Position
TA_142	R	activity focus	Karababa	Historical geography	certain
TA_143	gr-rom	? burial place	Karababa	Historical geography	certain

Karababa is namely a hill occupied by a Venetian castle, on the Boeotian side, directly across the Euripos from Chalkis.
Karababa has R surface finds [AE124] and gr-rom reused [AE383], as well as probable gr-rom rock-cut tombs (Wallace AJA 1969: 246ff. and Wallace 1979: 41). An identification with historical *Salganeus* has been attempted (see also above no.57 – Lithosoros site).

63. Glypha: *Components TA_134 to TA_141*

ID	Period	Type	Toponym	Discovery	Position
TA_134	N	activity focus	Glypha	Personal or group interest	certain

[23] In 1970 was carried out a small stratigraphical excavation in the N slope, while in 1978 on the NE side were found by the local *ephoreia* remains of a MH absidal house, with Minyan gray and red as well as polychrome matt-painted pottery (AD 1978: 119). The 1978 test excavation (E.Touloupa) uncovered a building plot on the NE slope of the prehistoric mound known as the 'tomb of Salganeus', where parts of MH building [AE324] were found and a MH child's cist tomb [AE1820] without offerings (AD 1978: 117). Few meters higher on the W side of the hill, in 1983-84 were noted remains of a circuit wall of Mycenaean period and few earlier remains (AD 1984: 125-126), that enforced the idea that the hill and its surroundings were inhabited both in MH and Mycenaean periods. A destruction phase (fire) was noticed, which dates back to the LH IIIB, fact that constitutes a parallel with the same situation attested at Glypha site – see below no.63.

[24] In a small excavation on B.Kontogianni plot, two parallel walls were found. Mainly LH and few sherds of MH pottery were collected. Important EH, MH and LH remains on the hilltop as well as in the surroundings (AD 1988: 203).

[25] Sherds and tiles [AE17], building blocks and pieces of marble columns reused [AE14] in chapels nearby (noticed by travellers, reported by Bakhuizen 1970: 6-12). Those evidence might though correspond to different activities in the landscape (AE14 was attached to the C activity focus with a question mark).

[26] At cape Gaidaros are reported EH II potsherds, probably referring to a Prehistoric installation (AD 38 (1983): 154).

ID	Period	Type	Toponym	Discovery	Position
TA_135	EH	settlement	Glypha	Personal or group interest	certain
TA_136	MH	settlement	Glypha	Personal or group interest	certain
TA_137	LH	settlement	Glypha	Personal or group interest	certain
TA_138	G	? activity focus	Glypha	Other	certain
TA_139	C	activity focus	Glypha	Other	certain
TA_140	R	? activity focus	Glypha	Other	certain
TA_141	LR	rural site/activity	Glypha	Other	certain

On the low hill of Tseloneri, also known as Glypha or Vlicha (marked as Glypha on 1:50,000 GYS map), on the Boeotian side of the S harbour of Chalkis and about 2km S of the potteries, by the road Chalkis-Aulis (in the centre of the W side of Vourko Bay facing Chalkis). On the N and W slopes especially (Fossey 1988: 75) Prehistoric[27] and Historical evidence of occupation were noticed. Historical occupation is probably known because of the presence of the large Prehistoric focus at the site.
The first mention is from Theokares (*Archeion Eyboikon Meleton* 1959; 1966: 57) interested in Euboean Prehistoric settlements. Excavations were carried out in 1977 (AD 1977: 98) at the occasion of the installation of gas from the owner (rescue excavation).[28] In 1988 twentytwo trenches were excavated in different spots on the hill, and the continuation of the circuit wall previously discovered was found. The wall dates back to the Mycenaean period while were found also house walls of at least two phases of the same period (LH II-III A-B) - AD 1988: 203; Demakopoulou 1988. Important is the MH pottery that was found below the LH strata (Red Minyan and polychrome matt-painted), which constitutes a parallel with the one at Lithosoros and Dramesi along the Boeotian coast and with the one at Magoula and Amarinthos in Euboea (AD 1988: 203). Also four MH cist tombs, perhaps for child burials made within MH houses, were found (AD 1977: 98-100; AReports 1986-86: 39-41) [AE1819]. Few are sherds from the EH period, collected in 1987 (similar to pottery from Manika, Magoula and Lefkanti in Euboea (AAA 1987: 191-210).
Reference is made to a neighbouring LH cemetery, near the Artemis sanctuary (AReports 1988-89 from *Akropolis* 20.I.89 – see below no.65), which could be connected with this settlement site.

An identification has been suggested with ancient *Hyria* (Fossey 1988: 76).[29] I would prefer to see it as a rural site, and a rural site / activity component has been created for the LR period (probably a villa?), while components from other historical periods represented have been left as generic 'activity focus'. A systematic survey at the site would be preferable to check evidence.

64. Mikro Vathy - Meg. Vouno: *Components TA_144; TA_146 and TA_147*

ID	Period	Type	Toponym	Discovery	Position
TA_144	gr-rom	fort	Mikro Vathy - Meg. Vouno	Travellers	approximate
TA_146	H	activity focus	Mikro Vathy - Meg. Vouno	Ext. topographical surveys	approximate
TA_147	R	activity focus	Mikro Vathy - Meg. Vouno	Ext. topographical surveys	approximate

On the Meg.Vouno ridge, to the W of Mikro Vathy bay, remains of a gr-rom fortification are mapped in Fossey (1988 plan fig.7), who also reports on H and R potsherds on Meg. Vouno, and traces of activities seen by Blegen 1949 (*components TA_146 and TA_147*). If they are in connection with *component TA_144* (fort) they could attest a H and R use of the fort.

ANAPHORITES fortified line (fig.1): A component has not been created since the fortification is probably in connection with Meg.Vouno fort (*component TA_144*), and has been registered only as evidence [AE96]. The Anaphorites fortified line is located E of Aulis and above Chaleia, on the mountain ridge, and continues till it reaches the enclosure of Meg.Vouno (Scranton 1941: 146-7). The remains of the fortification wall (noticed by travellers – Frazer, Leake, Gell) oddly are not mentioned in Fossey 1988 though they were reported in Fossey 1970. As pointed out by Philippson (1951: 498) through the Anaphorites pass (400m asl) would have run the road from Chalkis to Thebes. The long wall could have constituted a historically defined border for Theban territory, but presumably this line of constructions marks the boundary between Boeotia and Chalkis when the latter had control of the plain of Chaleia below, and the Meg.Vouno would constitute the key fort.

AULIS sites

65. Mikro Vathy N: *Component TA_145*

ID	Period	Type	Toponym	Discovery	Position
TA_145	LH	burial place	Mikro Vathy N	Personal or group interest	certain

On the small peninsula N of Mikro Vathy bay, Mycenaean tombs were found (Fossey 1988: 70 and map fig.7 from Ergon 1956: 37).

66. Vathy Nisi: *Components TA_149 to TA_150*

ID	Period	Type	Toponym	Discovery	Position
TA_149	gr-rom	fort	Vathy Nisi	Travellers	certain
TA_150	LH	? Fort	Vathy Nisi	Personal or group interest	certain

On the rocky promontory of Vesalas, also known as Nisi (on this name see Fossey 1988: 68 n.92) or Geladovouni, between the main Aulis Bay (Megalo Vathy) to the S and a smaller bay (Mikro Vathy) to the N (plan in Fossey 1988: figs.7 and 8, from Bakhuizen 1970: 96-100), have been noticed the remains of an ancient fortress or acropolis. No sherds or small finds have been located on the 'acropolis'.[30]
Component TA_150: Low on the W slopes of Vesalas was found a wall of large stones, apparently Mycenaean in date

[27] Neolithic (reported in BSA 1966: 66 n.94) has been confirmed by Sampson's surface observation (AE 1974: 14 no.16), though not noticed by Fossey (1988: 75), French 1972 nor Vermeule (BSA 1966: 66 n.94).
[28] Excavations continued in 1986-87 with a small amount of money from the local *ephoreia* and mainly paid by the owner (AD 1988: 203).
[29] Chalia and Hyria are coupled by Theopompus (*FGrH* 115 F211, 212) as Boeotian towns close to one another.

[30] HS&D 1979: NISI (LH, G, A, C, H) seem to misplace the entries which are recorded here under Ag.Nikolaos, or what it was thought to be the site of the Aulis sanctuary.

(Praktika 1959: 32 and pl. 32), because of associated material (HS&D 1979; Fossey 1988: 70).

67. Mikro Vathy (Artemis sanctuary): *Components TA_151 to TA_155*

ID	Period	Type	Toponym	Discovery	Position
TA_151	C	cult place	Mikro Vathy	Personal or group interest	certain
TA_152	H	cult place	Mikro Vathy	Personal or group interest	certain
TA_153	R	cult place	Mikro Vathy	Personal or group interest	certain
TA_154	LR	special activity	Mikro Vathy	Personal or group interest	certain
TA_155	gr-rom	settlement	Mikro Vathy	Personal or group interest	approximate

In a small valley connecting the heads of the two Vathy bays, to the W of Nisi rocky promontory. Until recently the N part of this valley was in fact a southern arm of the bay of Mykro Vathy (Wallace 1979: 31).
Threpsiadis (Ergon 1960) excavated at this site, which turned out to be the site of the Aulis sanctuary of Artemis (Fossey 1988: 73-4), in use from C to R times. In LR times, excavated baths (*component TA_154*) turned out to have taken over the sanctuary, and were probably in association with a small settlement/villa site nearby. Settlement (*component TA_155*) and associated activities are probably to be considered also for the previous period, around the sanctuary. The H workshop [AE291] could be one of those, as well as the excavated R building [AE1 and AE2]. Activities of people living there were strictly associated with the sanctuary but also fishing activities and a small harbour seem to be attested (the surrounding area seems too poor to provide subsistence).
Remains of ancient habitation were excavated in the bottom of the valley (Fossey 1988: 72). Buildings have been partly uncovered (Threpsiadis, Ergon 1960) in an area between the two gulfs of Mikro and Megalo Vathy.

68/69/70. Vathy - Ag. Nikolaos: *Components TA_156* (68); TA_157 to TA_163 (69); *TA_164* (70)

ID	Period	Type	Toponym	Discovery	Position
TA_156	LR	rural site/activity	Vathy - Ag. Nikolaos	Ext. topographical surveys	certain
TA_157	SubM	activity focus	Vathy - Ag. Nikolaos	Ext. topographical surveys	approximate
TA_158	G	activity focus	Vathy - Ag. Nikolaos	Ext. topographical surveys	certain
TA_159	A	activity focus	Vathy - Ag. Nikolaos	Ext. topographical surveys	certain
TA_160	C	activity focus	Vathy - Ag. Nikolaos	Ext. topographical surveys	certain
TA_161	H	activity focus	Vathy - Ag. Nikolaos	Ext. topographical surveys	certain
TA_162	LR	? activity focus	Vathy - Ag. Nikolaos	Ext. topographical surveys	approximate
TA_163	LH	activity focus	Vathy - Ag. Nikolaos	Ext. topographical surveys	approximate
TA_164	gr-rom	burial place	Vathy - Ag. Nikolaos	Ext. topographical surveys	certain

A low ridge (named Vessalas – 100m asl) to the N of the convent of Ag.Nikolaos (now Ag.Stephanos), which reaches the Euboic sea[31] and separates the Megalo Vathy gulf area from the Mykro Vathy and the valley where the Aulis sanctuary is. On the ridge between Agios Nikolaos convent and Schoinousa, a valley to the S of Meg.Vouno, marked as Skina on 1:50,000 GYS map (cfr. Fossey 1988: map fig.7).
At Ag.Nikolaos was placed, before Threpsiadis excavation at Mikro Vathy, the sanctuary of Artemis (Ergon 1958/Ergon1959).
Potsherds from Ag.Nikolaos are mixed up in Fossey's account, listed all together under 'Agios Nikolaos' both from potsherd scatters and from the burial area mentioned in the text (1988: 73). I created a component 'activity focus' for each of the periods mentioned, joining the information with the account of what was thought to be the site of the Aulis sanctuary. As for the LR period, it cannot be sure either that the LR material listed refers not only to the LR *component TA_156* by the convent itself but also to another component at this location.

Component TA_164: On the same ridge, tombs from several periods are reported as cut by the road (mapped in Fossey 1988 fig.7).
------

71. Vathy - Proph.Ilias: *Component TA_148*

ID	Period	Type	Toponym	Discovery	Position
TA_148	EH	activity focus	Vathy - Proph.Ilias	Ext. topographical surveys	approximate

On a small rounded hill to the N of the remains of the chapel of Proph.Ilias, to the N side of the Ritsona-Vathy road and ca 3km from Vathy, Fossey (1988: 67-8 – the unique mention of the site) found a small number of Prehistoric (EH I/II) potsherds on the top, which he interpreted as a possible small farm.
The evidence does not seem to have been noticed before Fossey's visit in 1965.

72. Vathy - Koulooureza: *Component TA_214*

ID	Period	Type	Toponym	Discovery	Position
TA_214	H	burial place	Vathy - Koulooureza	accidental	imprecise

From locality Koulooureza (of Vathy Avlidos), one tomb is reported as robbed and dated to the H period or the beginning of Roman Imperial period, probably belonging to a necropolis (AD 1983: 153).

73. Agios Panteleimon (ancient PHARAI?): *Components TA_165 to TA_168*

ID	Period	Type	Toponym	Discovery	Position
TA_165	A	activity focus	Agios Panteleimon	Ext. topographical surveys	approximate
TA_166	C	activity focus	Agios Panteleimon	Ext. topographical surveys	approximate
TA_167	A	burial	Agios	accidental	approximate

[31] The coastal road at this point runs inner and not along the sea coast.

ID	Period	Type	Toponym	Discovery	Position
		place	Panteleimon		
TA_168	C	burial place	Agios Panteleimon	accidental	approximate

Ca 5km NE of Schimatari, on Ag.Panteleimon ridge, the hills bordering the S side of Vathy Plain are full of tombs, the majority of which were illegally dug.
The burial site is mentioned only by Fossey (1988: 96-7), which reports on a concentration of such tombs especially around the ridge at the N end of which is the chapel of Ag.Pandeleimon. The tombs have contained so far only Archaic and Classical (black glaze) material according to local information.

On the ridge of Agios Panteleimon itself traces of A-C occupation were revealed (a good spread of A-C surface material - Fossey 1988: 97), which, jointly with the recovery, through local tomb-robbing, of A-C tombs could lead to the interpetaton of the site as an habitation site of unspecified kind and size. The surrounding area could constitute a natural agricultural block for a village.

A possible interpretation of the site as a settlement and an identification with ancient *Pharai* is discussed in Fossey (1988: 97-8). Pharai was one of the members of the τετρακωμία (Strabo IX 405) along with Mykalessos, Harma and Eleon. This suggested identification would allow Pharai to explit the surrounding area but also the coastal Vathy plain, and would allow for a coherent grouping from a geographical point of view: ancient Tanagra, ancient Eleon at modern Harma to the W, then ancient Harma (at Lykovouni site?) to the NW, followed by ancient Mykalessos by modern Ritsona to the N-NW and then would come Pharai in this suggested location at Ag.Panteleimon.

74. Dramesi/Paralia Avlidos: *Components TA_169 to TA_176*

ID	Period	Type	Toponym	Discovery	Position
TA_169	N	activity focus	Dramesi/Paralia Avlidos	Personal or group interest	approximate
TA_170	EH	settlement	Dramesi/Paralia Avlidos	Personal or group interest	approximate
TA_171	MH	settlement	Dramesi/Paralia Avlidos	Personal or group interest	approximate
TA_172	LH	settlement	Dramesi/Paralia Avlidos	Personal or group interest	approximate
TA_173	LH	burial place	Dramesi/Paralia Avlidos	accidental	approximate
TA_174	MH	burial place	Dramesi/Paralia Avlidos	Personal or group interest	approximate
TA_175	LH	burial place	Dramesi/Paralia Avlidos	Personal or group interest	approximate
TA_176	C	activity focus	Dramesi/Paralia Avlidos	Other	approximate

On a hill NW of modern Paralia Avlidos (formerly Dramesi) village, ca 500m from the shore. The hill is called Magoula in Prehistoric gazetteers (a prominent 'high mound' site – HS&D 1979: F64), while Fossey notes how the hill is called Ag.Paraskevi after the chapel at its SE foot (Fossey 1970). The hill measures 300x90m ca, 15m heigh.
A small trial excavation on the hill of Agia Paraskevi showed a stratigraphic sequence which runs from Neolithic times right to the end of the Bronze Age (Pappadakis, Praktika 1911: 142 and AD 1915: 55, who reports a Mycenaean site, but badly destroyed). The top yields sherds of all periods from NEO to LH IIIC (AR 1939-45: 24).
Also crouched graves are reported (AA 1914), probably of MH date (see Lithosoros site – no.57), and evidence for LH burials (*components TA_173* and *TA_175*).
Traces of a probable fortification have been identified, and a four-sided stone stele, decorated with incised representations of ships, was recovered from illicit excavations; its significance is obscure (HS&D 1979: F64).

Component TA_176: Small quantities of black glaze were noticed on the hill (Fossey 1988: 67). The evidence could represent a rural cemetery with a farm associated.

The LH site has been identified with Homeric *Graia* (see Fossey, Euphrosyne 1970: 3-22 for a discussion). On the other hand, Blegen 1949 felt this was ancient *Hyria*, disagreeing with Frazer's (1913) attribution of this to Meg.Vouno area (Vathy - Agios Nikolaos site, see above no.69).

75. Pharos Avlidos: *Component TA_177 to TA_179*

ID	Period	Type	Toponym	Discovery	Position
TA_177	LH	burial place	Pharos Avlidos	accidental	approximate
TA_178	EH	activity focus	Pharos Avlidos	Personal or group interest	approximate
TA_179	LH	activity focus	Pharos Avlidos	Personal or group interest	approximate

On a small hill by the coast, ca 1km NW of Dramesi/Paralia Avlidos site, reported in the bibliography as Pharos Avlidos or Yerali as it was formerly called. LH burials were found, constituting a cemetery of chamber tombs, whose goods include much jewellery (Spyropoulos AD 1971: 218; HS&D 1979: F64A), as well as an EH and LH IIIB habitation focus (represented by surface sherds).[32]

76. Dramesi/Paralia Avlidos shore: *Component TA_180*

ID	Period	Type	Toponym	Discovery	Position
TA_180	C	burial place	Dramesi/Paralia Avlidos shore	Rescue excavation	approximate

On the coast near Paralia Avlidos modern village, a C isolated tomb (of probably 4th C BC date) was found (AD 1973-4: 446).

77 - 86. Dilesi: *Components TA_181 and TA_182 and TA_193* (77)*; TA_183 to TA_185* (78); *TA_186* (79); *TA_216* (80); *TA_194* (81); *TA_217 and TA_218* (82); *TA_187* (83); *TA_188* (84); *TA_189, TA_190 and TA_215 (85); TA_191 and TA_192 (86)*

ID	Period	Type	Toponym	Discovery	Position
TA_181	LR	infrastructure	Dilesi	Personal or group interest	approximate
TA_182	gr-rom	infrastructure	Dilesi	Personal or group interest	approximate
TA_183	gr-rom	activity focus	Dilesi Palaiochori	Personal or group interest	approximate

[32] The LH tombs are reported in Fossey (1988: 66-7 and Euphrosyne 1970: 3-22) but he does not mention EH, nor the LH occupation as an habitation site.

ID	Period	Type	Toponym	Discovery	Position
TA_184	C	activity focus	Dilesi Palaiochori	Personal or group interest	approximate
TA_185	R	burial place	Dilesi Palaiochori	Personal or group interest	approximate
TA_186	R	burial place	Dilesi E	accidental	approximate
TA_187	gr-rom	activity focus	Dilesi - Ag.Paraskevi	Personal or group interest	approximate
TA_188	gr-rom	activity focus	Dilesi - Palaio Pigadi	Personal or group interest	approximate
TA_189	C	burial place	Dilesi Agrilesa	Personal or group interest	approximate
TA_190	LR	burial place	Dilesi Agrilesa	Personal or group interest	approximate
TA_191	gr-rom	activity focus	Dilesi S	Personal or group interest	approximate
TA_192	LR	? activity focus	Dilesi S	Personal or group interest	approximate
TA_193	R	? activity focus	Dilesi	Rescue excavation	imprecise
TA_194	LH	? burial place	Plaka Dilesi	accidental	approximate
TA_215	gr-rom	burial place	Dilesi Agrilesa	Rescue excavation	approximate
TA_216	R	special activity	Dilesi	Rescue excavation	approximate
TA_217	R	settlement	Plaka Dilesi	accidental	approximate
TA_218	LR	settlement	Plaka Dilesi	accidental	approximate

Along the shore, where the modern village of Dilesi and the attached Plaka Dilesi are, at the end of a small valley separated by low hills to the S from the head of the Asopos and its coastal plain and to the N from the plain of Vathy.
As Fossey (1988: 62) notes, in previous centuries the village seems to have been larger and to have spread inland (cfr. Frazer 1913: 74; Pritchett 1969: 29). Nowadays, since all the small plain is built over, the old hamlet disappeared as well as the landmarks visible in the 1960s-1970s, which showed a small harbour swallowed up since then.
A-systematic excavations (trial soundings) were carried out in different locations around the modern village by Brown (BSA 1905/6: 93ff), while looking for a suitable site for the Apollo sanctuary of ancient Delion.[33] Mainly prominent locations for the sanctuary and shore locations for the harbour were examined (11 in total). A detailed map is published in Pritchett 1969: 27-30.
The site has been identified with ancient *Delion* (Pritchett 1969: 27-30), the locality after which the battle of 424 BC was named. Trial excavations at the site (reported in BSA 1905/6: 93ff) showed that the identification was not supported by archaeological evidence. Many architectural fragments have been noticed around Dilesi (Fossey 1988: 64) but they could have been transported locally or even from Eretria (Brown BSA 1905/6: 93ff).[34]
Strabo (IX 2.7) describes Delion as a 'small town', while Pausanias (IX 20.1) seems to imply it was just a sanctuary. The site was anyway a harbour, probably used by people from Tanagra. The main pick of Delion, like Anthedon, is attested to the H and R times (AD 1981: 193).
The archaeological record available allows for the 'construction' of different components, mainly representing the harbour installations as well as several activities focuses in different location of the area, found during the quest of the Apollo sanctuary.

On the shore of Dilesi (below Palaiochori ridge – below Zoodochos Pygi location) considerable traces of an ancient mole were still visible in Brown's time, and some of them still are (Brown BSA 1905/6: 97) – *components TA_181 and TA_182*. It appears that the coastline changed considerably and the sea level rose since ancient times, due to earthquakes doubtless, to which this region is particularly subject.
Rescue excavations discovered a Roman building (large) and amphorae sherds (3rd – 6th C AD) by the shore, by the terrace wall supporting the long sea road of village (AD 1983: 135-6). They could be related to the harbour activities (the harbour was probably to the S of it - AD 1961: 193) – *component TA_192 and TA_193*. Also in the centre of Dilesi, during the construction of a hotel, a sandy layer full of sherds, mainly from commercial vases (4th C BC until Early Byzantine period), as well as wood remains were found (AD 1981: 193) [AE1815].[35]
Again in Dilesi village, at the OTE plot, architectural remains of commercial and workshop character were excavated, dating from the R period (AD 1995: 305) – *component TA_216.*

At Paralia Dilesi, apart from the column known from BSA 1905-6: 96-97. architectural pieces were found [AE1812]), probably from a large '*naiko ktirio*' (probably connected with the sanctuary of Apollo Delios - AD 1982: 172 . Remains of a Hellenistic stoa were recently discovered by the shore, and identified as part of the Apollo sanctuary (Piteros 2000).

On Palaiochori ridge[36] / Zoodochos Pygi location (500m S of Dilesi), large architectural blocks are reported (Brown BSA 1905/6: 93ff), as well as traces of walling (terrace walling?) and some potsherds reported as C (Fossey 1988: 63). The site is candidate for the Apolline sanctuary (Pritchett 1969: 27-30) – *components TA_183 and TA_184.*

One R tomb was found S of the village by Brown (BSA 1905/6: 99; Fossey 1988: 64) at Palaiochori, 1.2km S of Dilesi – *component TA_185*, and two other R tombs were found to the E of the village during the construction of a road (AD 1972: 32) – *component TA_186*. Furthermore, on Agrilesa hill, 3-400m E of Dilesi, a C (?) tomb was found by Brown (BSA 1905/6: 93ff) – *component TA_189*, as well as LR lamps nearby - *component TA_190*. On the same hill, later ca 50 tombs or more were excavated by the local *ephoreia* (pers. comm.) – *component TA_215.*

On the hill above the church of Ag. Paraskevi, on a dominant position on the shore, few Hellenic tiles were found on the surface (Brown BSA 1905/6: 93ff) – *component TA_187.*

S of Dilesi and ca 1/2km from the sea (Palaio Pigadi location - where an ancient well is) sherds and tiles, as well as a surface wall, were noticed and interpreted as a pottery workshop, with the wall as part of a furnace (Brown BSA 1905/6: 93ff, mentioned in Fossey 1988: 64) – *component TA_188.*

100m S of village (precisely S of Dimitriou house), walls constructed of mortar, terracotta, Hellenic pot fragments, a vase probably post-Hellenic, lots of late tiles and pots, were noticed by Brown (BSA 1905/6: 93ff s.v. 'Middle field'). I created a gr-rom component and a LR one with a question mark (it could be

[33] Fossey (1988: 66 n.85) notes how the dedication to Apollo [AE1667] found in the church of Agios Dimitrios (near Schimatari Station), 3-5km W-SW of Dilesi (BSA 1901/5: 153-172) is interesting but proves nothing, though it has been used as the basis of the identification with temple of Apollo at Delium. The identification is based also on Herodotus (VI 118), who describes the Delion sanctuary as '*Chalkidos katantion*'.
[34] See AE1665; AE110; AE274; AE1663; AE275 (epigraphical evidence).

[35] Probably the sherds belong to a waste area (and not to a shipwreak, as initially thought).
[36] A ridge formed by a group of hills, appropriately called Palaiochori, dominates the coast at the point where are also preserved remains of an ancient quarry [AE125] – Fossey 1988: 63.

LR or Byzantine but we cannot guess from the available information) – *components TA_191 and TA_192*. At Paralia Dilesi, also remains of houses of the R and LR period (AD 1982: 172) are reported (*components TA_217 and TA_218*), indicating a settlement linked to the harbour installation.

At the sea end of the ridge 'Plaka Dilesi' W of the village (ca 720m W of Dilesi, 25m heigh), Fossey (1988: 64) reports on reliable sources that informed him of the finding of four Mycenaean tombs. This ridge had previously yelded nothing to Brown's investigations (BSA 1905/6: 93ff) – *component TA_194*.

The SKOURTA PLAIN survey

The Skourta plain, an upland valley within the mountain massif Kithairon-Parnes (see fig.1 in chapter II.1), and common pasture land shared by Boeotians, Athenians and even Corinthians (Munn & Zimmerman Munn 1990: 36-37), can be regarded as *methoria*, no man's land between the territories of Attica and Boeotia, as pointed out by Schachter 2004: 56 after the examination of historical sources. Today the area belongs administratively to Boeotia. Zangger notes the remarkable stability, since the last Ice Age, of much of the landscape of the Skourta plain (reported by Munn AD 1988: 192).
The shortest route between Thebes and Athens crosses this plain, and it was taken by many early travellers.

A survey project (namely the Stanford Skourta plain survey) has been conducted in the area from 1985 to 1989 (fieldwork in 1985 and 1987), and discovered a number of sites of different characters in and in the surroundings of the plain. The sites listed in table 1 and marked on the map in fig.4 come from the (preliminary) survey reports.[37]

Table 1. *Skourta plain survey sites.*

SurveySITE code	CHRONOLOGY	CHARACTER
A1	N	activity focus
A1	EH	activity focus
A1	MH	settl/habitation
A1	LH	settl/habitation
A1	PG	settl/habitation
A1	C	settlement
A1	C-Ehl	settlement
A10	EH?	settl/habitation?
A10	MH	settl/habitation
A10	N	settl/habitation
A11	EH?	settl/habitation?
A11	LR	isolated farm or hamlet
A11	N	settl/habitation
A14	C-Ehl	Isolated farm
A14	LR	isolated farm or hamlet
A15	C-Ehl	Isolated farm
A18	LR	isolated farm or hamlet
A2	PG	settl/habitation
A25	EH?	settl/habitation?
A25	N	settl/habitation
A28	C-Ehl	Isolated farm
A33	C-Ehl	Isolated farm
A36	LR	isolated farm or hamlet
A50	C	settl/habitation
A50	EH	settl/habitation
A50	LH	settl/habitation
A50	N	settl/habitation
A8	N	settl/habitation
B10	LR	hamlet
B10	N	settl/habitation
B10	PG	settl/habitation
B10	R	hamlet
B11	C-Ehl	Isolated farm
B11	LH	settl/habitation
B11	LR	isolated farm or hamlet
B14	N	settl/habitation
B18	C	settlement
B18	C-Ehl	settlement
B18	LH	settl/habitation
B18	LR	isolated farm or hamlet
B19	EH?	settl/habitation?
B19	N	settl/habitation
B2	C-Ehl	Isolated farm
B20	C-Ehl	Isolated farm
B20	EH?	settl/habitation?
B20	LR	hamlet
B20	N	settl/habitation
B20	R	hamlet
B21	LH	settl/habitation
B21	LR	isolated farm or hamlet
B22	C-Ehl	Isolated farm
B24	LR	cemetery
B25	EH?	settl/habitation?
B25	LR	hamlet
B25	N	settl/habitation
B25	R	hamlet
B26	C-Ehl	Isolated farm
B3	C-Ehl	Isolated farm
B30	LR	hamlet
B30	R	hamlet
B32	C-Ehl	Isolated farm
B33	EH?	settl/habitation?
B33	LH	settl/habitation
B33	N	settl/habitation
B33	PG	settl/habitation
B34	C-Ehl	Isolated farm
B36	C-Ehl	Isolated farm
B37	LR	isolated farm or hamlet
B38	C-Ehl	Isolated farm
B38	LR	hamlet
B38	R	hamlet
B39	C-Ehl	Isolated farm
B4	C	settl/habitation
B4	C-Ehl	Isolated farm
B4	LH	settl/habitation
B40	C-Ehl	Isolated farm
B41	C-Ehl	Isolated farm
B42	C-Ehl	Isolated farm
B43	C-Ehl	Isolated farm
B44	LR	isolated farm or hamlet

[37] Munn – Zimmerman Munn 1987-9 (AD reports); Munn – Zimmerman Munn 1989; Munn - Zimmerman Munn 1990. Intensive and systematic artefact surface survey was conducted in some areas (three areas intensively surveyed are marked on fig. 5.2 of Munn – Zimmerman Munn 1989 paper, though the map refers only to the 1985 season: an area around Panakton; an area NE of Stephani; an area to the NE of the latter, further away from Stephani). Otherwise, intensive and systematic topographical survey (see chapter II.2) was conducted, and in 1985 covered 25% of the whole area, with special attention to the areas of Pyli, Skourta and Stephani.
Prof. Munn kindly provided me of a final version of the map with sites discovered during survey (after which is based the map in fig.4). The chronology of sites is mainly taken from Munn-Zimmerman Munn's article in AD 1989: 154-6.

SurveySITE code	CHRONOLOGY	CHARACTER
B46	LR	isolated farm or hamlet
B47	EH?	settl/habitation?
B47	N	settl/habitation
B49	LR	hamlet
B49	R	hamlet
B50	EH?	settl/habitation?
B50	N	settl/habitation
B51	C-Ehl	Isolated farm
B53	C-Ehl	Isolated farm
B54	EH?	settl/habitation?
B54	N	settl/habitation
B57	LR	isolated farm or hamlet
B58	LR	Isolated farm
B6	EH?	settl/habitation?
B6	N	settl/habitation
B60	LR	hamlet
B60	N	settl/habitation
B60	R	hamlet
B62	LR	isolated farm or hamlet
B63	EH?	settl/habitation?
B63	N	settl/habitation
B7	C	settl/habitation
B7	LH	settl/habitation
B7	LR	isolated farm or hamlet
B9	C-Ehl	Isolated farm
C1	C-Ehl	Isolated farm
C1	LR	isolated farm or hamlet
C11	C-Ehl	Isolated farm
C11	N	settl/habitation
C14	C-Ehl	Isolated farm
C15	C-Ehl	Isolated farm
C17	C	settlement
C17	C-Ehl	settlement
C17	LH	settl/habitation
C17	LR	isolated farm or hamlet
C18	C-Ehl	Isolated farm
C25	C-Ehl	Isolated farm
C26	C-Ehl	Isolated farm
C3	A	settl/habitation
C3	C	settl/habitation
C3	G	settl/habitation
C3	LH	settl/habitation
C3	LR	isolated farm or hamlet
C7	C-Ehl	Isolated farm
J1	MH	fortif. Settlement
J2	LH	fortif. Settlement
J6	EH?	settl/habitation?
J6	N	settl/habitation
Megalovouno	C-Ehl	Fort

As Munn points out (Munn 1989: 231-244), the history of the area is very much related to and dependent on its bordering position between Boeotia and Attica.[38] In AD 1989: 154-6 as well as in Munn-Zimmerman Munn 1990 a history of the landscape is briefly traced, according to the survey results. I summarise here the main points.

NEOLITHIC period. In the Early Neolithic period dispersed settlements can be noted in all the areas above *magoulas* in the plain and on the hills around the plain (A10, B6, B10, B14, B33, B50, B60, B63, C11), while for the Middle Neolithic period only one site shows sure occupation (A8).[39] In the Late and Final Neolithic, several of the early Neolithic sites and yet the hilltops of the surrounding hills were settled (A10, A11, A25, A50, B6, B19, B20, B25, B33, B47, B50, B54, B63, J6).

EARLY HELLADIC period. EH I material was noticed in several of the aforementioned sites, while absence of EH II-III can be noted in all survey sites. Only at Daphnoula site (A50), at the bench of Asopos, far to the N and below the Skourta plain (see chapter II.3.11 of the present work), EH II has been found (saucepan and Urfirnis ware). This absence finds parallel in other periods of crucial development (in the nearby plain areas), such as the G and A periods. The absence has to do with political reasons rather than with environmental factors. In EH II period people were concentrated in political and economical centres, both in Boeotia and in Attica.

MIDDLE HELLADIC period. MH settlement sites have been found on two hilltops overlooking the plain (A1 – Panakton, and A10). A fortified MH settlement on a hilltop (J1 – Kastro-Malli Mesi) is also known at the back of the NE side of the plain (on an area difficult to access) between Mazareika and Voudima valleys.

LATE HELLADIC period. The biggest site of the area in LH period has to be located at Panakton (A1), while in Mazareika and Voudima valleys several sites have been discovered: B21 – Mazareika valley (LH II and IIIB), J2 – Agia Marina (LH IIIC acropolis),[40] B33 – Tsoukrati valley. The last three sites were abandoned either in LH IIIC or PG period. In other more reachable locations of the plain new sites were settled, the majority of which mark important locations also later on in C period (A1, B4, B7, B11 – Panagia, B18 – Agios Dimitrios, C17, A50?, C3[41]).

PROTO-GEOMETRIC period. Panakton (A1) seems to have been an important centre in that period, while occupation is also known at A2, B10 and B33 sites. The plain knew abandonment for four centuries. Only in the Kokkino valley, to the S, continous occupation is attested from G to A period (C3), while Boeotian *poleis* and Athens were forming. The absence seems once again to be related, as in EH II period, with political reasons rather than with environmental factors (Munn – Zimmerman Munn AD 1989: 155). In the area were common pasturage zones (shared between Boeotian and Attic communities), especially in the area around Panakton. In LH and in PG period Panakton plays the role of the central place. No sherds from the G or A period have been identified, and therefore one might suppose that Panakton lay inhabited for at least four centuries (Munn AReports 1991-92: 29-31).

A/Early C period. The area was inhabited again at the end of the 6th and the beginning of the 5th C BC. The area probably belonged to the Athenians in that period. The Panakton fortress was destroyed during the Peloponnesian war and then rebuilt as an outpost regularly used by military guards formed by young men before mid 4th C BC, as inscriptional evidence found at the

[38] The ancients likewise thought of this mountainous region as a barrier between the lands of the Athenians and the lands of the Boeotians: Herodotus VII 141.3; Xenophon *Mem.* III 5.25; Strabo IX 1.3 (Munn 1989: 231).

[39] This could be explained with the difficulty in dating the material of that period, and in general with the discovery of the hidden landscapes of some Prehistoric periods.

[40] The site is located nearby the MH fortified settlement (J1). The presence of a fortified settlement here in LH period could mean that they were economically independent probably from the Mycenaeans that controlled the plains (Munn – Zimmerman Munn AD 1989 (1995): 155).

[41] C3 seems to be the only site with continuity of occupation LH-G-A-C.

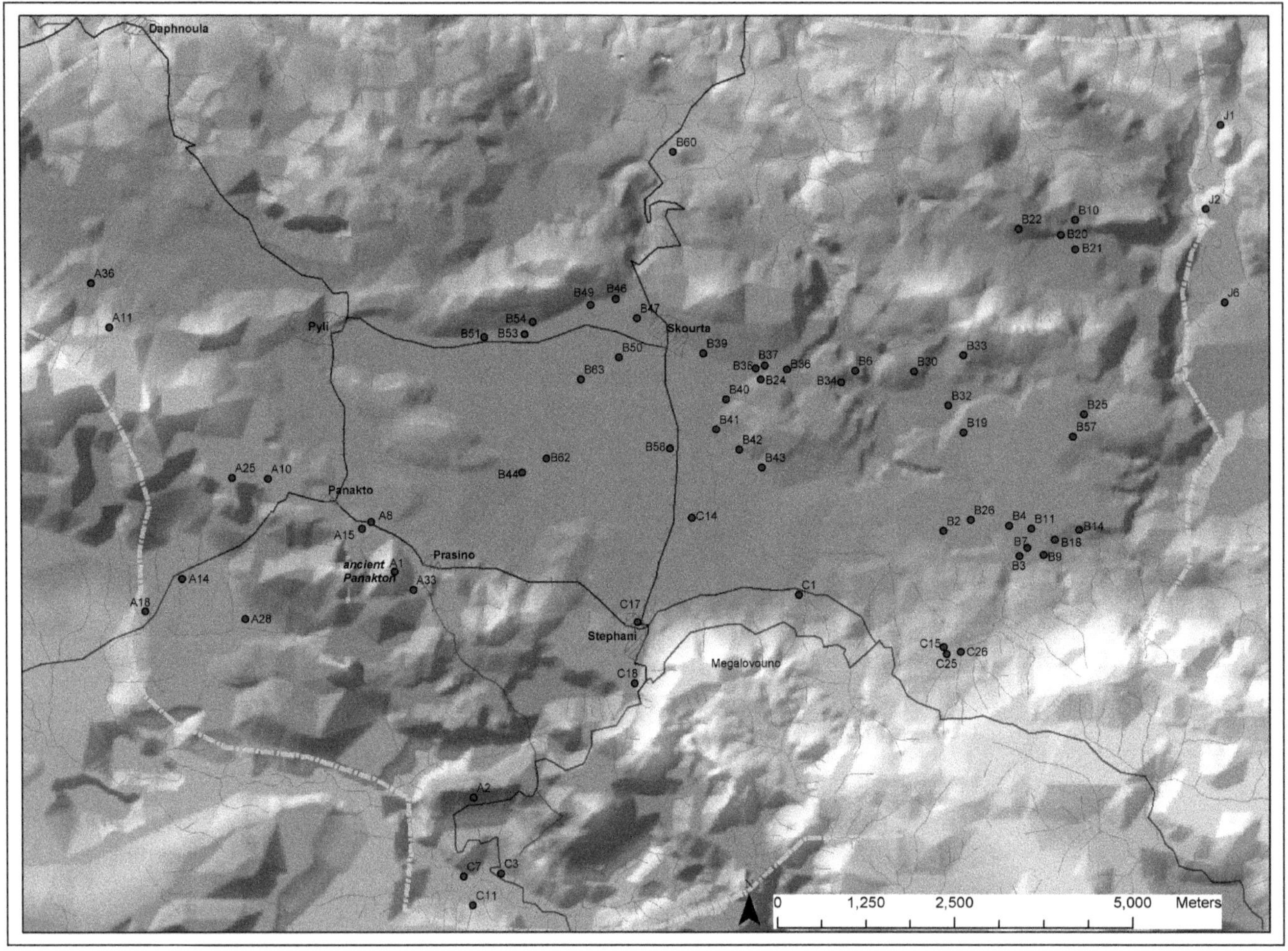

Fig.4. *Map with sites discovered in the framework of the Skourta plain survey.*

site attests. Sites which show occupation in this period are A1 – Panakton; C17 – Stefani; B18 – Agios Dimitrios.

LC/EHL period. Isolated farms around and outside the plain show up and the pick in the number of sites (A14, A15, A28 – Polonia, A33, B2, B3, B4, B9, B11, B20, B22, B26, B32, B34, B36, B38-43, B51, B53, C1, C7, C11, C14 – Ag. Giorgios, C15 – Pyrgari, C18, C25, C26) is registered by the end of the 4th and the beginning of the 3rd C BC, when the power of Thebes had receded and Athens was leaving its period of strength also on this inner land. Around 300 BC a fort (a mountain top rubble enclosure) was built on Megalovouno, in the SE part of the ridge.

HELLENISTIC period. We assist at changes in the landscape and settlement. The small rural sites mentioned above stopped existing during the 3rd C BC, and even Panakton was abandoned at the end of the 3rd C BC,[42] probably due to the decreasing of Athenian economic advantages from this area.

ROMAN period. On the contrary, from the 2nd C BC until the R imperial period, a number of small settlements along the N side of the plain can be noticed (B10, B20, B25, B30, B38, B49, B60). By this time, probably Tanagra was controlling this area after the Athenians, and the Skourta plain would administratively belong to Boeotia.

LATE ROMAN period. From 4th to 6th C AD new interest in the area would be reflected in the fact that settlements inhabited since Late Hellenistic period continue to be occupied, while new sites (isolated rural sites or hamlet/small villages) fill in the rest of the area (A11, A14, A18, A36, B7, B11, B18, B21, B24 – Koutsomiti Late Roman cemetery, B37, B44, B46, B57, B62, C1, C3, C17).

As for later periods, the site of Panakton is known as reoccupied in the Frankish period, as would attest a Medieval cemetery around the E end of the church (Munn AD 1991: 182-3).

Today, five modern settlements are in the area: Panakton, Prasino, Stephani, Pyli, and Skourta, belonging to three *koinotites*, namely Pyli, Skourta and Stephani, which all belong to the *nomos* (province) of Boeotia. While in the past the shortest and mostly used route joining Thebes and Athens crossed the area, today the basin constitutes a remote area with almost no modern development (see Philippson 1951: 522,531).

[42] At Panakton (A1), a fortified hilltop above Prasino village (formerly Kavasala), at the SW corner of the plain, excavations (1991 – extensive Greek-American excavation at the fortress) show extensive deposits preserved from the destruction and abandonment of Panakton in the spring of 420 BC (Thucydides V 39.3). Levels of the 4th and 3rd C BC are common. Amidst collapsed building debris that represent the abandonment of the site probably not far from 200 BC. The site was then reoccupied in the Frankish period, as would attest a Medieval cemetery around the E end of the church (Munn, AReports 1991-92: 29-31; AD 1991: 182-3; see also BCH 1992: 882-5 and a summary in AJA 1993: 364).

The TANAGRA survey project

Intensive and systematic artefact surface survey work was carried out in 2000-2006 on the ancient Tanagra city site and the surrounding areas (surveyed areas mapped in fig.3), in the framework of the Leiden-Ljubljana Boeotian cities Survey project, directed by J.L.Bintliff (and codirected by B.Slapšac – Ljubljana University as for research on the city site).[43]

The Tanagra intensive window (see chapter II.2) is marked by a high degree of intensity at two levels of research.

The first one concerns the intra-mural area of the ancient city, where intensive fieldwalking was carried out on the base of a regular grid.[44] Fig.5 shows the GIS plot of the overall density of surface ceramic material corrected by visibility. A careful digital recording of all the architectural features visible on the surface was undertaken, as was an intensive mapping of the morphology of the city surface (fig.5a – Farinetti-Sigalos 2002). Furthermore, geophysical prospections were carried out (jointly with a team from the University of Ljubljana, led by B. Slapšak and including the gephysicist B. Musič), deploying a range of geophysical devices to capture information about subsurface architectural and sedimentary structures across the intramural area as well as the city extention in different periods of history (fig.5b).

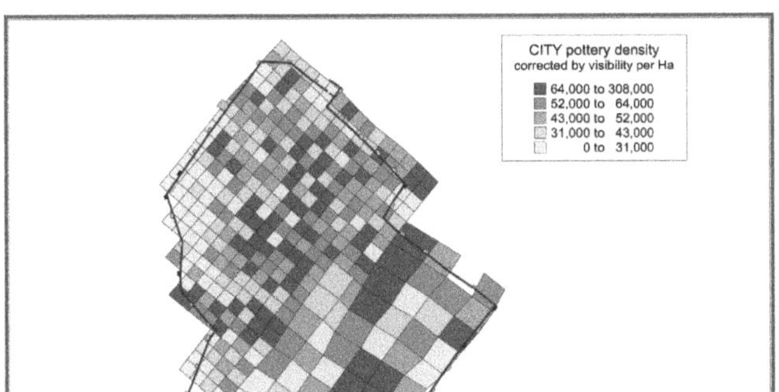

Fig.5. *A plot of the overall density of surface ceramic material corrected by visibility.*

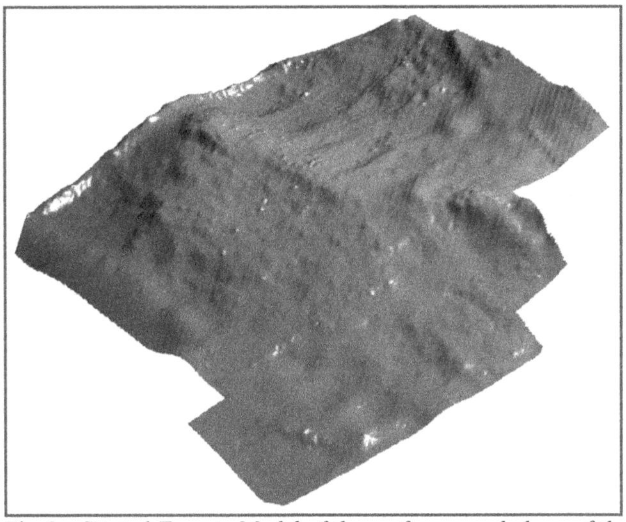

Fig.5a. *Digital Terrain Model of the surface morphology of the ancient city.*

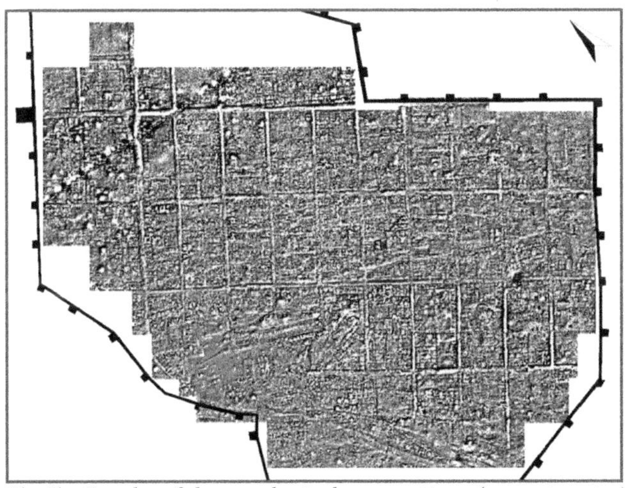

Fig.5b. *Results of the geophysical prospections (magnetometry) within the walled area of ancient Tanagra.*

The second level of research concerned the immediate surroundings of the walled city and the farming interland of the ancient city. Between 2000 and 2003 a series of long and narrow strips of land or transects were chosen, running in all major directions out of the city, for distances of one upto 3km, while in 2004 more distant transects were walked to investigate land use and rural settlement in the outer lands of the city's territory.[45] On fig.6 is visible the GIS based plot of the overall density of ceramic material on the surface, visibility corrected, and the location of the sites detected. Next, at a deeper level of investigation then, each site was gridded and fieldwalked, and sample material was collected and studied.

---

[43] Bintliff et al.2000; Bintliff et al.2001; Bintliff et al.2004; Bintliff 2006; Bintliff – Slapšak 2007; Bintliff et al. 2008. The writer is personally involved in the project as responsible for the GIS and database management.

[44] Fieldwork was directed by J.L.Bintliff and K.Sbonias.

[45] The area surveyed in 2004 were N of the modern village of Agios Thomas, S of the same village, and a basin between that village and the next modern settlement to the E at Kleidi.

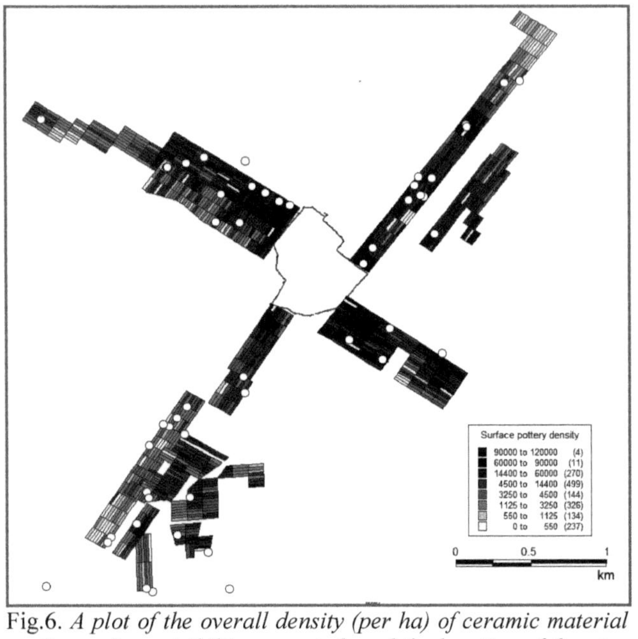

Fig.6. *A plot of the overall density (per ha) of ceramic material on the surface, visibility corrected, and the location of the sites detected.*

The ancient city of Tanagra, as seen earlier (no.1 - TANAGRA site), is strategically located at the intersection of two rivers, Laris and Asopos. Laris flows from W to E in a wide valley N of the Soros mountain mass and flows into the Asopos river in the vicinity of the ancient city. Asopos goes through the high Tetartogene plateau between Soros and the northern slopes of Parnis and Kithairon through a series of valleys, which are narrow and abrupt in the upper course (mainly in the Parasopia area where dominate the abrupt limestone hills), but then become wide with gentle slopes SW of Tanagra, where the river flows in the Tetartogene deposits of the E Tanagra basin, before the river's outlet into the Euboean sea.

The city controlled therefore two fertile valleys and also had available different topographic zones, plains, valleys and the hilly country of Tetartogene, mainly SW of the city, where in heights up to 200m are situated the present day villages of Ag.Thomas and Kleidi.

Those different landscape zones were covered in the survey in four transects which start from the city walls and direct E, W, N and S up to 3km from the city walls. The intensive landscape survey gathered a lot of data for the understanding of these micro-environments.

The transects N and E of the city cross the lower zone of the Tanagra basin and the low hills of the Tetartogene, while the W transect crosses the southern foothills of Soros and the upper part of Laris valley. The S transect covers part of the lower course of Asopos, before it begins to flow in the wide Tanagra basin.

The main archaeological characteristics within these transects is the wide distribution of material, which is dense in a distance of some hundred meters from the city walls and is related to the extra-urban activities outside the walls (extension of habitation in some cases, in the N transect extension of the classical city wall to the N, cemeteries and agricultural exploitation, confirmed also by rescue excavations – see no.8 and no.12 on archaeological map fig.2 and 3).

Afterwards and further from the city wall the density decreases and becomes a uniform distribution of 1-2 sherds per sqm, related with the rural sites, the rural exploitation of the wider area and in certain periods the transport from the city or the rurals sites of waste containing broken sherds for the fertilization of the soil (Bintliff et al. 2002). The results up to now point to the Classical and Early Hellenistic period as the main formation period of this uniform distribution and secondly to the Late Antiquity period.

With this background the location of sites and activity areas on the landscape is done through a combination of quantitative and qualitative criteria. The recording of the density and the dating and study of the pottery types permit the dating and identification of the type of sites (settlements and hamlets, cemeteries, isolated clusters of graves, rural sites of seasonal or permanent character, larger rural estates or villas, special purpose sites etc.).

What follows is a preliminary version of the list of sites discovered and recorded during the survey:

SurveySITE code	CHRONOLOGY	CHARACTER
TS1	C	rural site
TS1	C-Ehl	rural site
TS10	C	burial place
TS10	C-Ehl	burial place
TS11	A	burial place
TS11	C	burial place
TS11	H	burial place
TS11	N	?activity focus
TS11	R	burial place
TS12	R-LR	?rural site
TS13	A	burial place
TS13	C	burial place
TS13	H	burial place
TS13	Preh	?activity focus
TS14	EH	?activity focus
TS14	N	activity focus
TS16	LR	activity focus
TS17	A	activity focus
TS17	C-Ehl	rural site
TS17	EH	?activity focus
TS17	G	activity focus
TS17	N	settlement
TS19	R	?special activity
TS2	LR	rural site
TS2	R	?rural site
TS20	C	burial place
TS20	C-Ehl	burial place
TS21	C	burial place
TS21	Preh	activity focus
TS22	C	rural site
TS22	C-Ehl	rural site
TS23	C-Ehl	rural site
TS24	C-Ehl	activity focus
TS25	C	rural site
TS25	C-Ehl	rural site
TS26	LR	settlement
TS27	C	burial place
TS27	C-Ehl	burial place
TS28	C	burial place
TS28	C-Ehl	?rural site
TS28	LR	rural site
TS28	Preh	?settlement
TS29	C-Ehl	rural site
TS29	LR	rural site
TS3	LR	rural site
TS3	R	?rural site
TS33	C	rural site
TS34	C	rural site
TS35	C	rural site
TS37	C	rural site
TS37	R	rural site

SurveySITE code	CHRONOLOGY	CHARACTER
TS38	R	activity focus
TS39	C	?rural site
TS39	LR	rural site
TS39	Preh	?activity focus
TS4	LR	rural site
TS4	R	?rural site
TS41	A	burial place
TS43	Preh	activity focus
TS6	C	burial place
TS6	C-Ehl	burial place
TS7	C	rural site
TS7	C-Ehl	rural site
TS8	C	burial place
TS8	C-Ehl	burial place
TS9	LR	rural site
TS9	R	?rural site
TSx1	gr-rom	activity focus
TSx2	R	activity focus
TS44	C	rural site
R bath	R	rural site

Table 2. *Sites discovered in the framework of the Tanagra Survey project.*

PREHISTORY

On a radius of 3 kilometers around the city two sites constitute important prehistoric centres. First of all on the hill of the ancient city of Tanagra concentrations of Neolithic, EH II and III as well of MH material were noticed on the top and the slopes of the central ridge that runs E-W through the city (see above no.1 – TANAGRA site). They indicate the existence of a habitation site on the top, which is eroding on the slopes of the ridge. In contrast, the absence of extensive LH material does not support the presence of a Mycenaean settlement in this location. The Neolithic material was noticed in higher percentages than that of later prehistoric periods, indicating the presence of a main Neolithic settlement on the hill with coarse material of Middle and Late Neolithic date, similar to that from other Boeotian sites such as Chaironeia and Orchomenos.[46] The second large focus of prehistoric habitation in the wider area is the hill of Agios Konstantinos, whose defensible character is more marked. In particular, on the highest rocky top where the chapel of Profitis Ilias is, extensive prehistoric material was collected, dating from the Neolithic to the end of the Late Bronze Age (see above AGIOS KONSTANTINOS site, n.15 in the archaeological map in fig.1). Yet the bulk of the prehistoric material from the top and the higher terraces of the hill are dominated by the Middle and Late Bronze Age pottery, while on the lower terraces the Neolithic coarse wares coexist with later BA material.

Of particular interest is the area SW of Tanagra city. A predominant characteristic of the landscape S of Tanagra city is the Asopos valley with its system of streams and tributaries, which flow inside valleys ending in the Asopos river in this part of the plain. Almost next to every little stream that was crossed during fieldwork, the presence of prehistoric material was noticed, mainly coarse pottery and lithics, and several smaller or larger focuses of prehistoric activity were located. Therefore, although in the higher areas the prehistoric sherds were literally hidden, in the lower plain, pottery and lithics were abundant and easily recognized, providing thus a reliable picture of prehistoric settlement. These focuses of prehistoric habitation define a wider area of prehistoric activity of about 30 hectares around Asopos river and its tributaries.

The main characteristics of the landscape, in which the above mentioned concentrations are situated, are the lowland character of the area and the proximity to water resources, the Asopos river and its tributaries. This abundance of water is furthermore strengthened by the closed character of the Asopos valley at this part of the river course, which allows the flow of more water into the river. The focuses of prehistoric activity are closely linked to these water resources, but they spread on river terraces protected from stream flooding.

The geomorphological study of the area helped to understand the geomorphological history of landscape evolution, to identify geomorphological windows that enhance the visibility in small segments of the prehistoric landscape and to evaluate better the strategies of habitation, exploitation and use of the rural landscape. Fig.7 shows the results of the geomorphological work.[47] We can recognise three fluvial terraces and possibly a fourth very recent tiny terrace, which is linked to the acceleration of the presently active erosion cycle of the stream.

The Middle Terrace, visible today at several points along the Asopos river and its tributaries, is the one which marked the prehistoric landscape. This terrace was stretching up to the riverbank, which seems to have been in the past less steep and with a higher water table. Here we can observe signs of an older phase of incision of the valley, of ancient and middle Holocene date (up to 6-5.000 years BP). During this period the incision was moderate with scarce or no erosion of the slopes and accompanied by the formation of soils. The morphology was probably characterised by not very steep slopes and well-drained soils (because of the presence of flysch), good humidity year round, appropriate for a non-irrigated and non-intensive agriculture based on low technology. The riverbed of the stream was higher compared to the present one and the valley quite open. The aquifer linked to the stream was also higher in respect to the present one since the pumping of water for agricultural activities is very significant today. With these soils is related the distribution of the prehistoric material, which comes onto the surface as the present cultivation erodes and spreads the superficial levels of these strata.

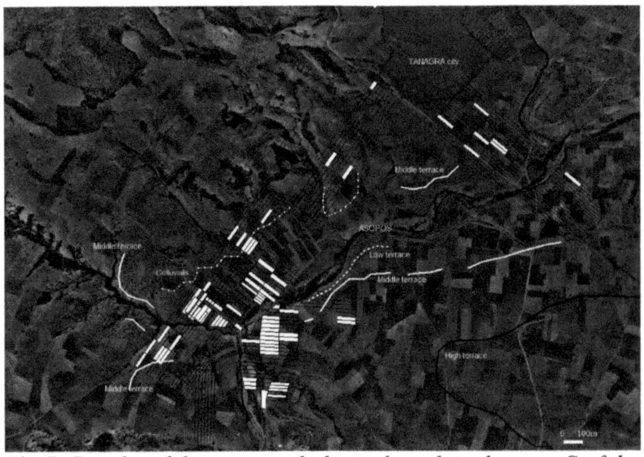

Fig.7. *Results of the geomorphological work in the area S of the ancient city of Tanagra and distribution of prehistoric sherds (in white).*

[46] See Bintliff et al. 2000; Bintliff et al. 2001; Sarri in Bintliff et al. 2008; Farinetti-Sarri-Sbonias 2006.

[47] The results presented here derive from a joint work carried out by the writer together with K. Sbonias, assistant director of the Tanagra Survey project, and the geoarchaeologist R.Sebastiani. The help of the archaeologist K.Sarri, expert in Prehistoric material, and the suggestions of the director of the project, J.L.Bintliff, were precious (Bintliff et al. 2006; Farinetti-Sbonias in preparation).

In the N transect, N of Tanagra city, the Lari valley is more open and dry and the river does not have the same quantity of water as the perennial streams of the Asopos valley S of the city. The fossil streams create quite open valleys, which do not collect the water. The dry soils here are not so poor and dry to limit cereal cultivation but they are not appropriate for wet agriculture. Therefore, this area is better for plough agriculture. The prehistoric material found here comes in accordance with this picture as the early material is scarce in comparison to the S transect (fig.7) and is attributed mainly to the Early and Middle Bronze Age.

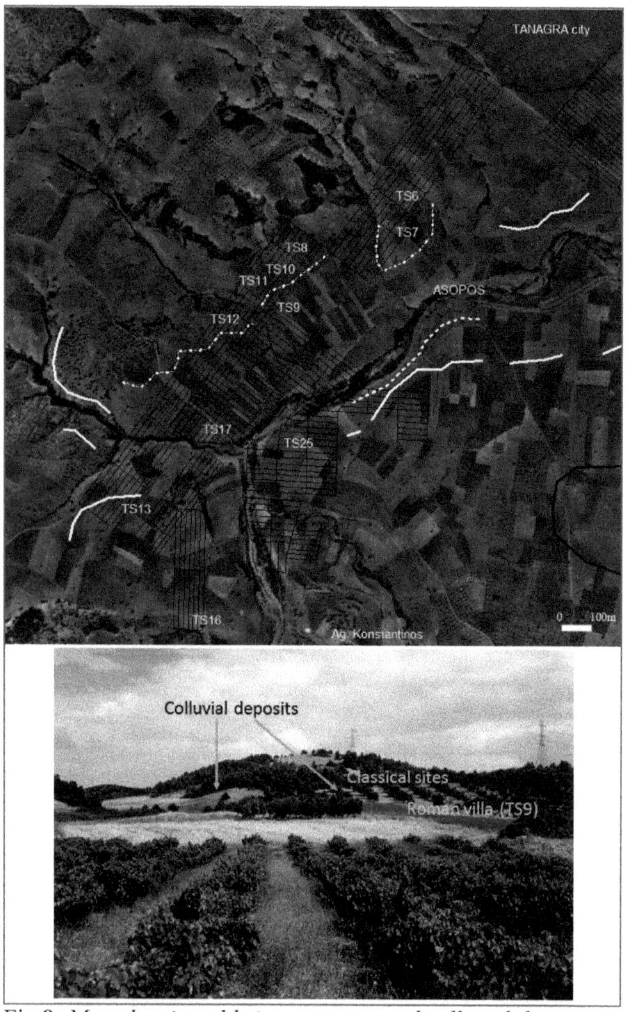

Fig.8. *Map showing old river terraces and colluvial deposits as well as the sites from the Greco-Roman period detected in the area S of the ancient city of Tanagra.*

On the contrary, the S transect is suitable for Neolithic agriculture which used the natural resources, the well drained light soils and the little perennial rivers which guaranteed the water supply. In the lower Asopos valley, the absence in the past of the present deeply incised riverbed as well as the more open character of the valley created a well-watered alluvial terrace with cultivable soils along the watersources. As we saw above, most of the prehistoric material here is associated with these well drained soils with good humidity which are suitable for non-intensive, non-irrigated agriculture based on low technology. The distribution of the prehistoric material along the larger tributaries and beside the main river is continuous and wide, corresponding to an extended dispersed flat settlement. Within it, smaller focuses can be recognised, reflecting a settlement pattern of small communities, which exploited the river valley. This kind of habitation pattern is known from Neolithic sites in Greece, as for example the Neolithic settlement of Nea Makri in Attica situated on a flat geological terrace and the Neolithic settlement of ancient Corinth, on a probable higher flat terrace by the sea, as well as from Late Neolithic large flat sites in Macedonia extending discontinuously over up to 45ha (Theocharis 1981: 55; Kotsakis 1999). The prehistoric activity spreads in a big area of 20-30ha because of either (or both) horizontal movement of the habitation in time or/and the discontinuous less packed habitation with open areas and farmland between the houses, resulting in a lack of separation between habitation and agricultural land (Kotsakis 1999: 71-73). For the time being the pottery has not been dated more accurately than that of generic Neolithic or Prehistoric date. The Late to Final Neolithic character of the material seems quite probable.

In terms of land use the pattern conforms better to a wet system of agriculture by the Asopos river and its tributaries related rather to the Neolithic habitation or to a continuation of former practices of land use, than to the settlement expansion of the Final Neolithic and Early Bronze Age. This could be indicated by the preference of wet areas, favoured during the earlier phases of Neolithic for an arable agriculture with simple tools (e.g. hoes), which took the character of fixed plot horticulture in comparison to the Final Neolithic/Early Bronze Age extended plough agriculture (Sherratt 1980: 313-330; Johnson 1996b: 267-295). The alternative would be to correlate this dispersed habitation with the changes that are noticed in the Final Neolithic and at the beginning of the Early Bronze Age. These changes are related to the introduction of the plough and the adoption of an agricultural system based on seasonal rains leading to the expansion of small short-lived communities, farmsteads or hamlets in the inland valleys, close to the main villages which were inhabited for longer periods (Halstead 1994: 198-200; Halstead 2000, 117; Wells 1994: 70). With this pattern we can connect two other kind of sites that were located: i) the small Bronze Age sites of the open landscape north of Tanagra city as well as some of the sites in the interfluvial area and on the stony slopes in the upper Asopos valley S of Tanagra city; ii) the main nucleations of the area, Tanagra city and the hilltop site of Agios Konstantinos, which yielded material of the Neolithic and especially the Early and Middle Bronze Age. This seems though less possible for the remnants of the alluvial terrace with Prehistoric material along Asopos and its tributaries, which must be part of, or continue, the earlier wetland farming.

GRAECO-ROMAN ANTIQUITY

To examine the landscape of Graeco-Roman antiquity as it comes out from the survey results, I will focus again on the river terraces which the geomorphological study identified and their correlation with sites dated to the Classical and Roman period.

From the Classical period were identified numerous cemeteries in the area outside the city walls and along the ancient roads that led from the city to the countryside (they start in the Archaic period and continue in the Classical and Hellenistic). This material come to light with the ploughing or is part of remains of older illegal excavations in the area.

Concentrations of cemeteries and of small burial areas of the Classical period were noticed outside the city walls and in the inland rural area, mainly across the hills of the S transect (Bintliff et al. 2004-5: 545-6). The cemeteries and burials, around 20, are situated a short distance from the city. The cemeteries appear in the first 500 meters from the city walls and continue up to 1600 meters away from these. None of the burial areas which were identified in the S transect are on the natural roads from the city but they are rather located higher on the hill slopes. Their position reflects more the position of family plots of people either resident in the city, who were going from it to the fields, or of people living in the countryside in nearby farms, of which some were identified by the survey (see Stissi in Bintliff et al. 2004-5: 556-560).

The small farms of Classical and Early Hellenistic date are all at a distance from the city (from 1.8km and further away). Site TS1, to the N of the city (at a distance of 1900m from the city wall) has the typical size of a small family farm of a peasant of the Classical-Early Hellenistic period (fig.9). It is about 2000 sq.m. large and contains all the finds which point towards permanent habitation and rural exploitation of the land. The surface finds studied by V.Stissi contain pieces of tiles, pithoi and amphorae, with the characteristic soft fabric of the period, as well as loom-weights, cooking pots, lamps and fine black glazed pottery. Few of this material can be accurately dated, but its date is between 350-250 BC or maybe a little bit later. From the functional point of view it belongs to a typical farmstead with quite a few storage pottery but pottery of daily use as well. Finds as the loom-weights, lamps and cooking pots indicate more or less permanent habitation. It could be identified as the farm of a citizen (see Stissi in Bintliff et al. 2008). Site TS25 can also be considered a residential farm in considering the resulting pot assemblage.
On the other hand, among the farms there are also sites without a variety of ceramic assemblage, where the finds contain mainly tiles and could be interpreted as seasonal sites.

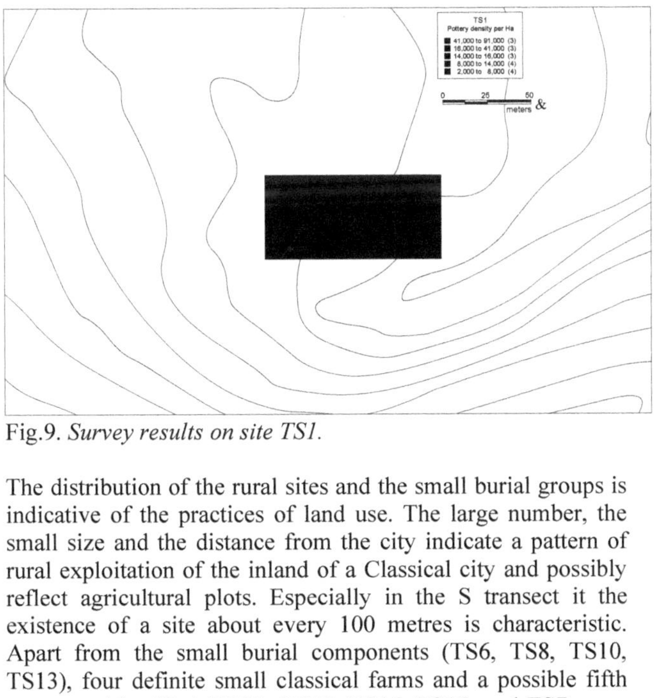

Fig.9. *Survey results on site TS1.*

The distribution of the rural sites and the small burial groups is indicative of the practices of land use. The large number, the small size and the distance from the city indicate a pattern of rural exploitation of the inland of a Classical city and possibly reflect agricultural plots. Especially in the S transect it the existence of a site about every 100 metres is characteristic. Apart from the small burial components (TS6, TS8, TS10, TS13), four definite small classical farms and a possible fifth one were identified (TS11, TS12, TS16, TS17, and TS7 – see figs. 3 and 8). Considering the topography the small farms and groups of burials, which as suggested could be related to family properties/plots, are located on the higher slopes of the Asopos valley, within the fertile hilly landscape of marls (fig.8). Such a choice of family farms typical of the C period for hill-slopes that naturally constitute suitable zones for olive trees and cereal cultivation has been noticed also by other research projects within Boeotia and elsewhere in Greece[48] and is also testified by the Tanagra project results. The relatively short distance of those activity focuses from the city wall testifies of the direct exploitation of the immediate outskirts by *polis* citizens. We should not forget that the population living in the rural sites in C period represent only a small percentage (estimated around 20-25%) of the whole *polis* (*asty* + *chora*) population.
The great 4th C BC plan and rewalling at Tanagra do attest to a large and wealthy city in Classical Greek times, and it is perhaps only with the arrival of Roman power in Late Hellenistic times (2nd to 1st centuries BC), when we might expect some radical change in its prosperity (in considering the shrinkage of urban areas at cities such as Thespiae and Hyettos and the temporary or permanent cessation of occupation at the town of Haliartos and the village-town of Askra, plus drastic depopulation at rural sites).

In the Roman period, we assist at the changes in the exploitation of the rural landscape, and in particular in the Late Roman period, when traces of these socio-economic changes become particularly visible archaeologically. Those changes concern site size, distance to the town, as well as topographical location, and testify to different socio-economic structures and transformations concerning land property which characterise the period. One very large villa site (TS9) occupied the area of the river floodplain and exploited the left side of the valley, replacing many smaller greek farms (Bintliff et al. 204: 40; Bintliff et al. 2004/5: 546-7 and figs.4-5).
In considering the topography, the first change one can observe is the preference of lower locations within the valley, in comparison with the higher level on the slopes which constitutes the preferred location during C times. In modern times this lower area along the axis of the valley as well as at its edge is used for the cultivation of cereals, grapes and vegetables. The location of the discovered sites (farms or villas) of R/LR period attests for the preference of these zones, in the S transect as well as in similar topographical settings/locations to the N and W of the city.

An impressive installation 900m from the city wall in the W transect is characterised by an extension of 0.35ha and included several architectural fragments, mainly column drams in piles (site TS2). To the N of the city two villa sites 600-700m from the city wall very much close one to the other (TS3 and TS4) gave a rich assemblage, including mosaic floors, plaster fragments and reused earlier (C) material. They could probably constitute a joint habitation group and would correspond to a larger estate. These impressive villa sites to the N and W of the city (Bintliff et al. 2002: 46 and figs.6-7), their proximity to the city site, as well as the clear difference in relation with the poor rural sites of C period, constitute a proof for the different socio-economic structures and the changes in land-property. The main period of use of such large rural sites is the Late Roman period (4th to 6th C AD), although some of them go back to the Early Imperial period (1st C BC - 1st C AD). They testify to socio-economic changes such as the abandonment of the small property system and gravitating around small rural sites typical of earlier periods and the reorganisation of the land use and property system from the local leading families.[49] One should point out, though, how apart from rich finds, fine architectural fragments and fluted columns, window glass and mosaic floors, those sites are marked by the absence of fine wares. Pot assemblages are mainly constituted by tile and large vase fragments,[50] which could be interpreted as sign of a main storage character of large parts of the site inhabited reasonably

[48] See for instance Bintliff-Howard-Snodgrass 2007 for the Thespiae area; Jameson– Runnels – van Andel 1994 for the Argolid; Cavanagh et al. 1996 for Laconia.

[49] It is customary to interpret this contrast as a shift by Roman times to a form of landownership favouring wealthier farmers over a previously dominant class of petty farmers.

[50] Mainly large amphorae, locally made food and drink containers, normally reserved for long-distance trade.

by workers, supervisors and coordinators. The main period of occupation of those sites in LR times goes in parallel with the pick of occupation of the LR *polis* of Tanagra, in which one could note high activity from the 4th to the 6th C AD. Within the city wall about ¾ of the ceramic material collected belongs to that period, as a result of the pick of Tanagra as the local center of a flourishing rural region within the Eastern Roman Empire. Generally speaking, the mid and large size Roman rural sites surrounding of the city reveal the flourishing of the countryside, while the noted preference of the axis of the valley (lower location within valleys and basins) reflects the character of specialised and intensive cultivations which characterise the period, most probably grapes, as the large quantity of amphora sherds which were found at the sites as well as in the city seem to attest.

As for the city site, the examination of ceramic material collected seems to indicate a fairly strong Flavian to Severan urban occupation. Taking the over-representation of the later Roman material into account, and the general undiagnostic features of contemporary cooking pots and transport wares, could give Strabo's (IX 2.5) consideration of Tanagra as a relatively well-to-do community somewhat more credibility. In the Late Roman period (ca. 400-600 AD) there seems to have been certainly a large and busy city, as the vast bulk of all surface finds from Tanagra belong to Late Roman forms (see Poblome et al. in Bintliff et al. 2008).

BIBLIOGRAPHY

References in the text usually refer to the last name of the author(s) or editor(s) and the year of publication; more publications by the same author in the same year are distinguished with letters after the year of publication. Some major works are referred to with an abbreviated title (e.g. Lauffer Kopais I). Generally, abbreviations for journals follow those indicated by the *American Journal of Archaeology*.

Adams, R.M. 1958. 'Survey of ancient water courses and settlements in central Iraq'. *Sumer* 14 (1-2): 101-103.

Agnew, J. – Livingstone, D.N. – Rogers, A. (eds.) 1996. *Human Geography. An essential Anthology.*

Amit, M. 1971. 'The Boeotian Confederation during the Pentakontaetia'. *Rivista storica dell'antichità* I: 49-64.

Alcock, S. E. 1993. *Graecia Capta. The landscape of Roman Greece.* Cambridge University Press, Cambridge.

Alcock, S. E. 1997. 'Changes on the ground in Early Imperial Boeotia', in J. L. Bintliff (ed.) *Recent Developments in the History and Archaeology of Central Greece*, BAR Int. Ser. 666. BAR Publishing, Oxford: 287-304.

Alcock, S.E. 2007. 'The essential countryside. The Greek world', in S. Alcock - R. Osborne (eds.) *Classical Archaeology.* Blackwell: 120-138.

Alcock, S.E. 1999. 'The Roman territory of Greek cities', in M. Brunet (ed.) *Territoire des Cités Grecques.* École française d'Athènes, BCH suppl. 34: 167-173.

Alcock, S.E. - Cherry, J.F. (eds.) 2004. *Side-by-Side Survey. Comparative Regional Studies in the Mediterranean World.* Oxbow Books, Oxford.

Alcock, S.E. - Osborne, R. (eds.) 1994 *Placing the Gods: sanctuaries and sacred space in ancient Greece.* Clarendon Press, Oxford.

Alcock, S.E. (ed.) 2002. *Archaeologies of the Greek past: landscapes, monuments, and memories.* Cambridge University Press, Cambridge.

Alcock, S. E. - Cherry, J. F. et al. 1994. 'Intensive survey, agricultural practice and the Classical landscape of Greece', in I. Morris (ed.) *Classical Greece. Ancient histories and modern archaeologies.* Cambridge University Press, Cambridge: 137-170.

Aldenderfer, M. – Maschner, H.D.G. (eds.) 1996. *Anthropology, Space, and Geographic Information Systems.* Spatial Information Systems, Oxford University Press.

Ålin, P. 1962. *Das Ende der mykenischen Fundstätten auf dem griechischen Festland.* Lund.

Allen, J.R.L. 1965. 'A review of the origin and characteristics of recent alluvial sediments'. *Sedimentology* 5: 89-191.

Allen, H. 1997. 'The environmental conditions of the Kopais basin, Boeotia during the Post Glacial with special reference to the Mycenaean period', in J. L. Bintliff (ed.) *Recent Developments in the History and Archaeology of Central Greece.* BAR Int. Ser. 666. BAR Publishing, Oxford: 39-58.

Allen, H. 1990. 'A Postglacial record from the Kopais Basin, Greece', in S. Bottema - G. Entjes-Nieborg - W. van Zeist (eds.) *Man's Role in the Shaping of Eastern Mediterranean Landscape.* Balkema, Rotterdam: 173-182.

Allen, K.M.S. – Green, S.W. – Zubrow, E.B.W. (eds.) 1990. *Interpreting Space: GIS and archaeology.* Taylor & Francis, London.

Andreiomenou, A.K. 1985. 'La nécropole classique de Tanagra', in Argoud - Roesch (eds.) 1985: 109-130

Andreiomenou, A.K. 1995. '*Ena archaiko nekrotapheio tis evriteras periochis ton Thivon. Agia Eleousa Piriou, Meros Γ*', in *B' diethnes synedro Boiotikon meleton*: 139-244.

Andreiomenou, A.K. 2007. *Tanagra. I anaskaphi tou nekrotafeiou (1976-1977, 1989).* Athens.

Andreou, S. - Kotsakis, K. (1994). 'Prehistoric rural communities in perspective: the Langadas survey project', in Doukellis - Mendoni (eds.) 1994: 17-21.

Antrop, M.- Vermeulen, F. et al. 2001. 'GIS and Spatial Analysis', in B. Slapšak (ed.) *Coast Action G2.* Ljubljana: 61-79.

Aravantinos, V. 1996. 'Tebe micenea. Recenti scoperte epigrafiche ed archeologiche (1993-1995)'. *SMEA* 38: 179-190.

Aravantinos, V. 2000. 'Le scoperte archeologiche ed epigrafiche micenee a Tebe: un bilancio riassuntivo di un quinquennio (1993-97) di scavi', in P.A. Bernardini (ed.) *Presenza e funzione della città di Tebe nella cultura greca: Atti del Convegno Internazionale (Urbino 7-9 luglio 1997).* Instituti Editoriali e Poligrafici Internazionali, Pisa: 27-59.

Aravantinos, V. 2004a. *Dio staseis gia anaskaphes sti Boiotia Akraiphia (ΠΑΘΕ) kai Thiva (ΕΡΓΟΣΕ). Archaiologikes Erevnes kai Megala Dimosia Erga* – 18-20 Semptember 2003 – Proceedings, Thessaloniki: 82-94.

Aravantinos, V. 2004b. 'New Evidence about the EH II Period in Thebes: A New Architectural Complex and a Group Burial within the Kadmeia', in E. Alram-Stern (ed.) *Die Ägäische Frühzeit 2. Serie: Forschungsbericht 1975-2002.* Verlag der Österreichischen Akademie der Wissenschaften, Wien: 1255-1259.

Aravantinos, V. – Godart, L. – Sacconi, A. 2001. *Thèbes. Fouilles de la Cadmée* I. Biblioteca di Pasiphae, Instituti Editoriali e Poligrafici Internazionali, Pisa / Rome.

Argoud, G. – Roesch, P. (ed.) 1985. *La Beotie Antique.* Colloques internationaux du Centre National de la Recherche Scientifique. Paris.

Arnoldus-Huyzendveld A. 2003. *I suoli di Roma.*Comune di Roma, Roma.

Arnoldus-Huyzendveld A. 2008. 'L'ambiente naturale', in A. Akerraz – E. Papi (eds.), *Sidi Ali ben Ahmed - Thamusida, 1. Les contextes-I contesti*, Quasar, Roma: 1-24.

Arroyo Bishop, D. - Lantada Zarzosa, M.T. 1995. 'To be or not to be: will an object-space-time GIS/AIS become a scientific reality or end up an archaeological entity?', in Lock-Stančič (eds.) 1995: 43-54.

Aston, M. – Gerrard, Ch. 1999. 'Unique, traditional and charming. The Shapwick Project, Somerset'. *The Antiquaries Journal*, vol. 79: 1-58.

Atherden, M. 2000. 'Human Impact on the vegetation of Southern Greece and problems of palynological interpretation: A case study from Crete. Landscape and Land Use in Postglacial Greece', in Halstead-Frederick (eds.) 2000: 62-78.

Attema, P.A.J. 1992. 'Measuring landscape perception in archaeology, A Model for the Pontine Region (Southern Lazio)', in *Caeculus I, Papers on Mediterranean Archaeology.* Archaeological Centre, Groningen University: 3-10

Attema, P.A.J. 1993. *An archaeological survey in the Pontine region. A contribution to the early settlement history of south Lazio 900-100 BC.* Groningen 1993

Attema, P.A.J. 2002. 'Two Challenges for Landscape Archaeology' in P.A.J. Attema – G.J. Burgers – E. van Joolen – M. van Leusen – B. Mater (eds.) *New Developments in Italian Landscape Archaeology*, BAR Int.Ser. 1091: 18-27

Attema et al. 2002. Attema, P.A.J. – Burgers, G.J. – van Joolen, E. – van Leusen, M. – Mater, B. (eds.) *New Developments in Italian Landscape Archaeology*, BAR Int.Ser. 1091

Attema, P.- van Leusen, M. 2004. 'Intra-regional and Inter-regional Comparison of Occupation Histories in Three Italian Regions: The RPC Project', in S.E. Alcock - J.F. Cherry (eds.) *Side-by-Side Survey. Comparative Regional Studies in the Mediterranean World*: 86-100.

Attolini, I. – Cambi, F. et al. 1991. 'Political geography and productive geography between the valleys of the Albegna and the Fiora in Northern Etruria', in G. Barker – J. Lloyd (eds.) *Roman Landscapes. Archaeological Survey in the Mediterranean Region.* British School at Rome Archaeological Monograph - London: 142-152

Babelon, E. 1907. *Traité des monnaies grecques et romaines.* Paris.

Bakhuizen, S.C. 1970. *Salganeus and the Fortification on its Mountains*, Groningen.

Backuizen, S.C. 1976. *Chalcis in Euboea. Iron and Chalcidians abroad.* Chalcidian Studies 3. Leiden.

Bakhuizen, S.C. 1979. 'On Boeotian iron'. *Teiresias* suppl. II: 19-20

Baladié, R. 1980. *Le Péloponnèse de Strabon. Étude de géographie historique.* Paris.

Bar Yosef, O.and Kra, R.S. (eds.) 1994 *Late Quaternary Chronology and Paleoclimate of Eastern Mediterranean.* The University of Arizona-Radiocarbon, Tucson.

Barceló, J.A. - M Pallarés 1998. 'Beyond GIS: The archaeology of social spaces'. *Archeologia e Calcolatori* 9: 47-80.

Barker, G. 1995. *A Mediterranean Valley. Landscape Archaeology and Annales History in the Biferno Valley.* Leicester University Press, Leicester.

Barker, G. - Lloyd J. (eds.) 1991. *Roman Landscapes. Archaeological Survey in the Mediterranean Region.* British School at Rome Archaeological Monograph 2, London.

Barker, G – Mattingly, D. 1999. 'General Editors'Introduction: The POPULUS Project', in Gillings et al. (eds.) *Geographical Information Systems and Landscape Archaeology.* The Archaeology of Mediterranean Landscape n.3. Oxbow books, Oxford: iii-ix.

Barker, G. – Symonds, J. 1984. 'The Montarrenti Survey, 1982-83'. *Archeologia Medievale* 11: 278-289.

Barrett, L. 1994. 'Not now, dear, I'm busy'. *New Scientist* (9 April): 30-34.

Baxter, M.J. 1994. *Exploratory Multivariate Analysis in Archaeology.* Edinburgh UP, Edinburgh.

Beex, W. 2004. 'Use and abuse of digital terrain/elevation models', in K.F. Ausserer – W. Börner et al. (eds.) *Enter the Past, CAA 2003* – Vienna. BAR Int.Ser. 1227: 240-242

Bell, C. 1992. *Ritual Theory, Ritual Practice.* Oxford University Press. Oxford.

Bell, M. 1982. 'The effects of land use and climate on valley sedimentation', in A. Harding (ed.) *Climate change in Later Prehistory*, Edinburgh University Press.

Bell, S. 1999. *Landscape pattern, perception and process.* E&FN Spon, London.

Bellwood, P. 2005. *First Farmers. The origin of agricultural societies.* Blackwell, Oxford.

Bender et al. 1997. Bender, B. – Hamilton, S.C. – Tilley, C. 'Leskernick: stone worlds; alternative narratives; ovested landscapes'. *PPS* 63: 147-78

Bertolani, M. - Marchetti, D. 1972. 'Prime notizie sulle ricerche flogistico-climatiche nell'area del Lago Copaide (Beozia-Grecia) sulla base di ricerche palinologiche', in *Atti XII Congresso Nazionale di Speleologia, Rassegna Speleologica Italiana*, Mem. XII: 207-210.

Bertolani, M. - Rossi, A. 1983. 'La zona speleologica del Lago Copaide (Beozia – Grecia)', in *Atti del XIV*

Congresso Nazionale di Speleologia, Le Grotte d'Italia, s.4 vol. XI: 205-248

Betancourt, P.P. - Hope-Simpson, R. 1992. 'The Agricultural System of Bronze Age Pseira'. *Cretan Studies* 3: 47-54

Bevan, A. 2002. 'The Rural Landscape of Neopalatial Kythera: A GIS Perspective'. *JMA* 15.2: 217-256

Bevan, A. – Conolly, J. 2002-2004. 'GIS, Archaeological Survey, and Landscape Archaeology on the Island of Kythera, Greece'. *JFA* 29.1-2: 123-138

Biasini A. 1978. 'Determinazione e classificazione della pendenza dei versanti tramite 'isoipsa allargata' con metodo fotografico'. *Bollettino Società Geologica Italiana*, vol. 97, issue 5-6.

Binford, L. R. 1964. 'A Consideration of Archaeological Research Design'. *American Antiquity* 29: 425-441

Bintliff, J.L. 1977. *Natural Environment and Human Settlement in Prehistoric Greece*, BAR 28. Oxford.

Bintliff, J.L. 1985. 'Greece: The Boeotia survey', in S. Macready – F.H. Thompson (eds.) *Archaeological Field Survey in Britain and Abroad.* Society of Antiquaries, London: 196 - 216

Bintliff, J.L. 1987. 'The Younger Fill: a timely review'. Paper delivered a the XII *Proceedings of the XII Congress of the International Quaternary Association.* Ottawa.

Bintliff, J.L. 1991a. 'The contribution of an Annaliste/Structural History approach to archaeology', in J.L.Bintliff (ed.) *The Annales School and Archaeology*: 1-33. Leicester University Press, Leicester.

Bintliff, J.L. 1991b. 'The roman countryside in Central Greece: observation and theories from the Boeotia survey (1978-1987)', in J. Barker – J. Lloyd (eds.) *Roman Landscapes. Archaeological Survey in the Mediterranean Region.* British School at Rome, London: 122-132

Bintliff, J.L. 1991c. 'Die Polis-Landschaften Griechenlands: Probleme und Aussichten der Bevolkerungsgeschichte', in E. Olshausen – H. Sonnabend (eds.) *Stuttgarter Kolloquium zur Historischen Geographie des Altertums 2, 1984 und 3, 1987. Geographica Historica* 5: 149 – 202.

Bintliff, J.L. 1992a. 'Appearance and reality: understanding the buried landscape trough new techniques in field survey', in M. Bernardi (ed.) *Archeologia del Paesaggio.* Edizioni all'Insegna del Giglio, Firenze: 89-140

Bintliff, J.L. 1992b. 'Erosion in the Mediterranean lands: a reconsideration of pattern, process and methodology', in J.Boardman and M.Bell (eds.) *Past and Present Soil Erosion.* Oxford Bools, Oxford: 125-131

Bintliff, J. 1992c. 'The history of the Greek countryside: As the wave breaks, prospects for future research', in P. N. Doukellis - L. G. Mendoni (eds.) *Structures Rurales et Societes Antiques*. Université de Besançon: 7-15

Bintliff, J.L. 1992d. 'The Boeotia Project 1991: survey at the city of Hyettos'. *University of Durham and University of Newcastle upon Tyne Archaeological Reports*, University of Durham, Durham: 23-28

Bintliff, J.L. 1994a. 'Reflections on nine years with the Bradford-Cambridge Boeotia Project', in J.M. Fossey (ed.) *Boeotia Antiqua I: Papers on Recent Work in Booetian Archaeology and History*, Amsterdam: 13-21

Bintliff, J.L. 1994b. 'Territorial Behaviour and the Natural History of the Greek Polis'. *Geographica Historica* 7: 207-249

Bintliff, J. L. 1995. '"Whither Archaeology?" revisited', in M. Kuna - N. Venclova (eds.) *Whither Archaeology? Papers in Honour of Evzen Neustupny*. Institute of Archaeology, Praha: 24-35

Bintliff, J.L.1996a. 'Catastrophe, Chaos and Complexity: the Death, Decay and Rebirth of Towns from Antiquity to Today', in E. Olshausen - H. Sonnabend (eds.) *Stuttgarter Kolloquium zur Historischen geographie des Altertums* 6, *Geographica Historica* 10: 417-438 (published also in *Journal of European Archaeology* 1997, vol.5: 67-90).

Bintliff, J.L. 1996b. 'The archaeological survey of the Valley of the Muses and its significance for Boeotian History', in A. Hurst and A. Schachter (eds.) *La montagne des Muses.* Faculté des lettres de Genève 7: 193-210

Bintliff, J.L. (ed.) 1997a. *Recent Developments in the History and Archaeology of Central Greece. Proceedings of the 6th International Boeotian Conference*, BAR Int. Ser. 666. BAR Publishing, Oxford.

Bintliff, J. L. 1997b. 'Further considerations on the population of ancient Boeotia', in J.L.Bintliff (ed.) *Recent Developments in the History and Archaeology of Central Greece.* BAR Int.Ser. 666: 231-252

Bintliff, J. L. 1997c 'Regional survey, demography, and the rise of complex societies in the ancient Aegean. Core-periphery, neo-Malthusian, and other interpretive models'. *Journal of Field Archaeology* 24: 1-38

Bintliff, J.L. 1999a. 'The hidden landscape of prehistoric Greece'. *JMA* 12: 139-68

Bintliff, J.L. 1999b. 'The concepts of 'site' and 'offsite' archaeology in surface artefact survey', in M.Pasquinucci - F.Tremens (eds.) *Non destructive Techniques Applied to Landscape Archaeology.* The Archaeology of Mediterranean Landscape n.3. Oxbow Books, Oxford: 200-215

Bintliff, J.L. 1999c. 'Chapter 13: Settlement and Territory', in G.Barker (ed.) *The Routledge Companion Encyclopedia of Archaeology.* Routledge, London: 505-545.

Bintliff, J.L. 1999d. 'Pattern and process in the city landscapes of Boeotia, from Geometric to Late Roman times', in M. Brunet (ed.) *Territoire des Cités Grecques.* École française d'Athènes, BCH suppl. 34: 15-33

Bintliff, J. L. 2000a. 'Deconstructing 'the sense of place'? Settlement systems, field survey, and the historic record: a case-study from Central Greece'. *PPS* 66: 123-149

Bintliff, J.L. 2000b 'Beyond dots on the map: future directions for surface artefact survey in Greece' in J. Bintliff – M.Kuna - N.Venclova (eds.) *The Future of Surface Artefact Survey in Europe*. Sheffield Academic Press, Sheffield: 3-20

Bintliff, J.L. 2000c. 'Settlement and Territory: a socio-ecological approach to the evolution of settlement systems', in G. Bailey - R. Charles - N. Winder (eds.) *Human Ecodynamics*, Proceedings of the Association for Environmental Archaeology Conference 1998 held at the University of Newcastle upon Tyne: 21-30

Bintliff, J.L. 2000d. 'Landscape change in Classical Greece: a review', in F. Vermeulen - M. D. Dapper (eds.) *Geoarchaeology of the Landscapes of Classical Antiquity*. Peeters Presse, Leuven: 49-70

Bintliff, J.L. 2000e. 'Reconstructing the Byzantine countryside. New approaches from landscape archaeology', in *Byzanz als Raum*. Osterreichische Akademie der Wissenschaften, Wien: 37-55.

Bintliff, J.L. 2002. 'Time, process and catastrophism in the study of Mediterranean alluvial history: a review'. *WorldArch* 33: 417-435

Bintliff, J.L. 2006. 'The Leiden University Ancient Cities of Boeotia Project: 2005 season at Tanagra'. *Pharos* 13: 29-38

Bintliff, J.L. - Howard, Ph. 1999. 'Studying needles in haystacks – Surface survey and the rural landscape of Central Greece in Roman times'. *Pharos* 7: 51-91

Bintliff, J.L. – Howard, Ph. – Snodgrass, A.M. 1999. 'The hidden landscape of prehistoric Greece'. *JMA* 12.2: 139-168

Bintliff, J.L. - Howard, Ph. – Snodgrass, A. 2007. *Testing the Hinterland: the work of the Boeotia Survey (1989-1991) in the southern approaches to the city of Thespiai.* McDonald Institute Monographs.

Bintliff, J.L. – Kiel, M. in preparation. *Gazetteer of Ottoman sites in Boeotia.*

Bintliff, J.L. – Kuna, M. et al. 2000. Bintliff, J.L. - Kuna, M. - Venclovà, N. (eds.), *The Future of Surface Artefact Survey in Europe*. Sheffield Academic Press.

Bintliff, J.L. et al. in preparation. *Artefact surface survey at the ancient city of Thespiae*. McDonald Institute Monographs.

Bintliff, J.L. – Snodgrass, A.M. 1985a. 'The Cambridge/Bradford Boeotian Expedition: the First Four Years'. *JFA* 12: 123-160

Bintliff, J.L. – Snodgrass, A.M. 1985b. 'The development of settlement in South-West Boeotia', in Argoud - Roesch (eds.) 1985: 49-70

Bintliff J.L - Snodgrass, A.M. 1988. 'Off-Site pottery distributions: a regional and interregional perspective'. *Current Anthropology* 29: 506-513

Bintliff, J.L. - Snodgrass, A.M. 1989. 'From Polis to Chorion in South-West Boeotia', in H.Beister and J.Bucklet (eds.) *Boiotika*. Editio Maris, Munich: 285-299

Bintliff, J.L. - Slapšak, B. 2007. The Leiden-Ljubljana Ancient Cities of Boeotia Project: Season 2006. *Pharos* 14: 15-27

Bintliff, J.L. et al. 2000: Bintliff, J.L. - Farinetti, E. - Mušič, B. - Sbonias, K. - Sigalos, L. - Slapšak, B. 'The Tanagra Survey. Report on the 2000 season'. *Pharos* 8: 95-129

Bintliff, J.L. et al. 2001. Bintliff, J.L. - Farinetti, E. - Sbonias, K. - Sigalos, L. - Slapšak, B. 'The Leiden Ancient Cities of Boeotia Project: preliminary report on the 2001 season'. *Pharos* 9: 33-74

Bintliff, J.L. et al. 2002. Bintliff, J.L. - Farinetti E. - Howard P. - Sarri, K. - Sbonias K. 'Classical farms, hidden prehistoric landscapes and Greek rural survey: a response and an update'. *JMA* 15.2: 259-265

Bintliff, J.L. et al. 2004. 'The Leiden-Ljubljana Tanagra Project: the 2003 season'. *Pharos* 11: 35-43

Bintliff, J.L. et al. 2006. Bintliff, J.L. – Farinetti, E. - Sarri, K. - Sebastiani, R. 'Landscape and Early Farming Settlement Dynamics in Central Greece'. *Geoarchaeology* 21.7: 665-674

Bintliff, J.L. et al. 2008 (2004-2005). 'The Tanagra project: investigations at an ancient city and its countryside (2000-2002)'. *BCH* 128-129: 541-606

Blackman, D. - Branigan, K. 1977. 'An archaeological survey of the Ayiofarango valley'. *BSA* 72: 13-84

Blegen, C.W. 1920. 'Corinth in Pre-historic Times' in *AJA* 24: 1-13

Blegen, C.W. 1949. 'Hyria'. *Hesperia* Suppl. 8: 39-42

Bloch, M. 1974. *Lavoro e tecnica nel medioevo*. Laterza, Bari.

Bommeljé et al. 1987. Bommeljé, S. – Doorn, P. et al. *Aetolia and the Aetolians. Towards the interdisciplinary study of a Greek region*, Studia Aetolica I. Parnassus Press, Utrecht.

Bordieu, P. 1977. *Outline of a theory of practice*. Cambridge University Press, Cambridge.

Borza, E.N. 1979. 'Some observations on malaria and the ecology of Central Macedonia in antiquity'. *AJAH* 4: 102-124

Borza, E. N. 1989. 'Some toponym problems in Eastern Macedonia'. *Ancient History Bulletin* 3: 60-67

Bottema, S.- Entjes-Nieborg, G. - van Zeist, W. (eds.) 1990. *Man's Role in the Shaping of Eastern Mediterranean Landscape*, Rotterdam, Balkema.

Bottema, S. 1994. The Prehistoric environment of Greece: a review of the palynological record, in P.N. Kardulias (ed.) *Beyond the Site. Regional Studies in the Aegean Area*, University Press of America, Lanham: 45-68

Bradley, R. 1994. 'Symbols and signposts - understanding the prehistoric petroglyphs of the British Isles', in C. Renfrew - E. B. W. Zubrow (eds.) *The ancient mind. Elements of cognitive archaeology*: 95-106

Braudel, F. 1972. *The Mediterranean and the Mediterranean World in the Age of Philip II*. Fontana/Collins, London.

Braudel, F. 1980. *On History*. The University of Chicago Press. Chicago.

British Admiralty 1944/45. *The Naval Intelligence Handbooks: Greece*, vols. I-III, London.

Brochier, J. E. 1992. 'Shepherds and sediments: Geo-ethnoarchaeology of pastoral sites'. *JAnthArch* 11: 47-102

Broodbank, C. 1999. 'Kythera Survey: Preliminary Report on the 1998 season'. *BSA* 94: 191-214

Brown, A.G. 1997. *Alluvial Geoarchaeology*. Cambridge Manuals in Archaeology, Cambridge University Press Cambridge.

Bruck, J. - Goodman, M. 1999a. *Making places in the prehistoric world: themes in settlement archaeology*. UCL Press (Routledge), London.

Bruck, J. and Goodman, M. 1999b. 'Introduction: themes for a critical archaeology of prehistoric settlement', in M. Goodman – J. Bruck (eds.) *Making places in the prehistoric world: themes in settlement archaeology*, UCL Press (Routledge) London: 1-19

Bruckner, H. 1986. 'Man's impact on the evolution of the physical environment in the mediterranean region in historical times'. *GeoJournal* 13: 7-17

Buck, C.D. 1889. 'Discoveries at Anthedon in 1889'. *AJA* 5: 443-460

Buck R.J. 1970. 'The Athenian Domination of Boeotia'. *CP* 65: 217-27

Buck, R.J. 1972. 'The Formation of the Boeotian League'. *CP* 68: 94-101

Buck, R.J. 1979. *A History of Boeotia*. The University of Alberta Press, Edmonton.

Buckler, J. 1980. *The Theban Hegemony, 371-362 BC*. Harvard Historical Studies vol. 98. Harvard University Press.

Budja M. and Mlekuz D. 2001. 'GIS support in reconstructing Neolithic flood-plain dynamics - the Ljubljana Moor case', in B. Slapšak (ed.) *Coast Action G2*. Ljubljana: 117-126

Bulle, H. 1908. *Orchomenos I*, Munich.

Burn, A.R. 1949. 'Helikon in History'. *BSA* 44 (1949): 313-324

Bursian, C. 1862. *Geographie von Griechenland* I. Leipzig

Butzer K.W. 1974. 'Accelerated soil erosion: a problem of man-land relationship', in J.R. Mannery - M.W. Mikesell (eds.) *Perspectives and Environment*. Association of American Geographers, Washington: 57-78

Butzer, K. W. 1982. *Archaeology as Human Ecology*. Cambridge.

C.E.C. 1985. *Soil map of the European Communities 1:1,000,000*. Commission of the European Communities: Luxembourg.

C.N.R. 1978. *Metodi di cartografia geo-ambientale e di cartografia della vegetazione*, Proceedings of the conference. CNR, Rome.

CAA 1995. Kamermans, H. – Fennema, K. (eds.) 1996. *Interfacing the Past. Computer Applications and Quantitative Methods in Archaeology, CAA95* – Leiden. Analecta Praehistorica Leidensia 28. University of Leiden, Leiden: 275-292

CAA 2000. Stančič, Z. - Veljanovski, T. (eds.) *Computing Archaeology for understanding the past, CAA2000* - Ljubljana. BAR Int.Ser. 931.

CAA 2001. Burenhult, G. (ed.) 2002. *Archaeological Informatics: Pushing the Envelope, CAA 2001* – Gotland. BAR Int.Ser. 1016.

CAA 2002. Doerr, M. - Sarris, A. (eds.) 2003. *The Digital heritage of Archaeology, CAA 2002* – Heraklion-Crete. Archive of Monuments and Publications, Hellenic Ministry of Culture.

CAA 2003. Ausserer, K.F. - Börner, W. - Goriany, M. - Karlhuber-Vöckl, L. (eds.) 2004. *Enter the Past, CAA 2003* – Vienna. BAR Int.Ser. 1227.

CAA 2004 in press. *Beyond the artefact CAA 2004* – Prato.

CAA 2005. Figueiredo, A. - Leite Velho, G. (eds.) 2007. *The world is in your eyes, CAA 2005* – CAA Portugal, Tomar.

CAA 2006. Clark, J. T. - Hagemeister, E.M. (eds.) 2007. *Digital Discovery. Exploring New Frontiers in Human Heritage, CAA2006* – Budapest. Archaeolingua, Budapest.

Campana, S. 2003. 'Ikonos-2 Multispectral Satellite Imagery to the Study of Archaeological Landscapes: an integrated Multi-sensor Approach in combination with 'Traditional' Methods', in *CAA2002*: 219-225

Camp II, J. McK. 1991. 'Notes on the towers and borders of Classical Boiotia'. *AJA* 95: 193-202

Camp et al. 1992. Camp, J. - Ierardi, M. - McInerney, J. – Morgan, K. - Umholtz, G. 'A trophy from the battle of Chaironeia of 86 B.C'. *AJA* 96: 443-455

Canarache, A. - Vintila, I. - Munteanu, I. (eds.) 2006. *Elsevier's Dictionary of Soil Science*. Elsevier. Boston, San Diego, London.

Casevitz, M. 'Pausanias en Beotie: a propos de Lebadeia', in A.Ch. Christodoulou (ed.) *B' diethnes sinedrio Boiotikon Meleton*, vol. 2: 727-733

Caskey, J.L. 1951. 'Neolithic sherds from Thespiae'. *Hesperia* 20: 289-290

Caskey, J.L. – Caskey, E.G. 1960. 'The earliest settlements at Eutresis. supplementary excavations 1958'. *Hesperia* 29: 126-167

Castay, G. 1985. *Idrologia. Principi e metodi*. Flaccovio, Palermo.

Castiglioni, G.B. 1991. *Geomorfologia*. UTET, Torino.

Cavanagh, W. et al. 1996. Cavanagh, W. – Crouwel, J. – Shipley, G. *Continuity and Change in a Greek Rural Landscape: The Laconia Survey* vol.II. BSA suppl. vol. 27.

Cavanagh et al. 2002. Cavanagh, W. - Crouwel, J. – Catling, R.W.V. - Shipley, G. *Continuity and Change in a Greek Rural landscape: The Laconia Survey. Vol. I: Methodology and Interpretation:* vol.I. BSA suppl. vol. 26.

Cavanagh, W. – Mee, Ch. – James, P. 2005. *The Laconia rural sites projects*. BSA suppl. vol. 36.

C.E.C. 1985 *Soil map of the European Communities 1:1,000,000.* Commission of the European Communities, Luxembourg.

Ceccarelli, L. - Niccolucci, F. 2003. 'Modelling time through GIS technology', in M. Doerr – A. Sarris (eds.) *The Digital heritage of Archaeology, CAA2002*: 133-138

Chadwick, A.J. 1978. 'A computer simulation of Mycenaean Settlement', in I. Hodder (ed.) *Simulation studies in archaeology*. Cambridge: 47-57

Chang, C. 1992. 'Archaeological landscapes: the ethnoarchaeology of pastoral land use in the Grevena province of Northern Greece', in J.Rossignol - L.A.Wandsnider (eds.) *Place, Time and Archaeological Landscapes*. Plenum Press, New York: 65-89

Chang, C. - Tourtellotte, P.A. 1993. 'Ethnoarchaeological survey of pastoral transhumance sites in the Grevena region. Greece'. *JFA* 20: 29-2

Chapman, H. 2000. 'Understanding wetland archaeological landscapes: GIS, environmental analysis and landscape reconstruction; pathways and narratives', in Lock (ed.) 2000: 49-59

Chapman, H. 2006. *Landscape archaeology and GIS.* Tempus, Stroud.

Chatterton, L. – Chatterton, B. 1996. *Sustainable Dryland Farming: Combining Farmer Innovation and Medic Pasture in a Mediterranean Climate.* Cambridge University Press, London.

Cherry, J. F. 1983. 'Frogs round the pond: perspectives on current archaeological survey projects in the Mediterranean region', in D. R. Keller - D. W. Rupp (eds.) *Archaeological Survey in the Mediterranean Area.* Oxford, BAR Int.Ser. 155: 375-416.

Cherry, J.F. 1990. 'The first colonization of the Mediterranean islands: a review of recent research'. *JMA* 3.2: 145-221

Cherry, J.F. 1994. 'Regional surveys in the Aegean: the 'New Wave' (and After)', in N. Kardulias (ed.) *Beyond the Site: Regional Studies in the Aegean Area.* University Press of America, Lanham: 91-112

Cherry, J.F. 2003. 'Archaeology beyond the site: regional survey and its future', in J.K. Papadopoulos - R.M. Leventhal (eds.) *Theory and Practice in mediterranean Archaeology: Old World and New World Perspectives.* Cotsen Institute of Archaeology, University of California, Los Angeles: 137-159

Cherry, J. F. - Davis, J. L. et al. 1988. 'Archaeological survey in an artifact-rich landscape: A Middle Neolithic example from Nemea, Greece' *AJA* 92: 159-176

Cherry et al. 1991. Cherry, J. F. - Davis, J. L. - Mantzourani, E. (eds.) *Landscape Archaeology as Long-Term History: Northern Keos in the Cycladic Islands.* Institute of Archaeology, University of California, Los Angeles.

Christaller, W. 1933. *Die zentralen Orte in Süddeutschland.* Gustav Fischer, Jena.

Christidou, A. 1997. *Neolithic Boeotia.* Dissertation DAI 57 (1997) 4422-A/4423-A

Christopoulou, A. 1995. 'Idisis apo ti Steni Mavromatiou' in *B' diethes synedrio Boiotikon Meleton* 2.A: 429-445

Clarke, D.L. (ed.) 1977. *Spatial Archaeology.* Methuen. London.

Coccia C., Mattingly D. J. et al. 1993. 'Settlement History, Environment, and Human Exploitation of an intermontana Basin in the Central Appennines. The Rieti Survey 1988-1991, Part I'. *BSR* 60: 213-89

Coldstream, J.N. 1968. *Greek Geometric Pottery*, London.

Comel, A. 1972. *Il terreno: manuale di pedologia per agricoltori.* Edagricole, Bologna.

Conolly, J. 2002-4. 'GIS, Archaeological Survey, and Landscape Archaeology on the Island of Kythera, Greece'. *JFA* 29: 123-138

Conolly, J. - Lake, M. 2006. *Geographical Information Systems in Archaeology.* Cambridge Manuals in Archaeology. Cambridge: Cambridge University Press.

Corsin Jiménez, A. 2003. 'On space as a capacity'. *Journal of the Royal Anthropological Institute*: 137-153

Cosmopoulos, M.B. 2001. *The rural history of Ancient Greek city states. The Oropos survey project.* BAR Int. Ser. 1001. BAR Publishing, Oxford.

Cox, K. R. 1969. 'On the utility and definition of regions in comparative political sociology'. *Comparative Political Studies* (April): 68-98

Cox, K.R. - Mair, A. 1991. 'From localised social structures to localities as agents'. *Environment and Planning* 1991 vol.23: 197-213

Crescioli, M. - D'Andrea, A. - Niccolucci, F. 2000. 'A GIS-based analysis of the Etruscan cemetery of Pontecagnano using fuzzy logic', in Lock (ed.) 2000: 157-79

Curtius, E. 1892. *Die Deichbauten der Minyer.* Sitzungsbericht der Berliner Akademie der Wissenschaften, vol. 55. Berlin.

Daly, P.T. - Lock, G. 1999. 'Timing is everything: Commentary on managing Temporal Variables in Geographical Information Systems', in J.A. Barcelo - I. Britz - A. Vila (eds.) *New Techniques for Old Times, CAA 1998.* BAR Int.Ser. 757: 287-293

Dasios, Ph. 1995a. 'Peri tis Viotiko-Phokikis methorion', in A.Ch. Christodoulou (ed.) *B' diethnes synedrio Boiotikon Meleton*: 245-255

Dasios, Ph. 1995b. 'Diodoros Sikeliotes (XVI 58.4) and the fortress at Mavrokastro', in Fossey, J.M. – Smith, P.J. (eds.) *Boeotia Antiqua V. Studies on Boiotian Topography, Cults and Terracottas.* Amsterdam: 35-41

Daskalaki, P. - Voudouris, K. 2008. 'Groundwater quality of porous aquifers in Greece: a synoptic review'. *Environmental Geology* 54.3: 505-513

Daverio Rocchi, G. 1988. *Frontiera e confini nella Grecia Antica*. Roma.

Davidson, D.A. et al. 1994. Davidson D.A. - Theocaropoulos S.P. - Bloksma R.J., 'A land evaluation project in Greece using GIS and based on Boolean and fuzzy set methodologies'. *International Journal Geographical Information Systems* 8.4: 369-384

Davidson D.A. - Theocaropoulos, S.P. 1992. 'A Survey of soil erosion in Viotia, Greece', in M. Bell - J. Boardman (eds.) *Past and Present Soil Erosion*. Oxford:149-154

Davies, O. 1935. *Roman Mines in Europe*. Oxford.

Davis et al. 1997: Davis, J. L. - Alcock, S.E. - Bennet, J. - Lolos, Y.G. - Shelmerdine, C.W. 'The Pylos Regional Archaeological Project, part I: overview and the archaeological survey'. *Hesperia* 66: 391-494 and plates 85-92

Davis, J.L. (ed) 1998. *Sandy Pylos. An Archaeological History from Nestor to Navarino*. The University of Texas Press, Austin.

De Angelis, F. 1992. 'A forgotten inscription from Khlembotsari (Asopia), Boiotia' in j.M. Fossey – J. Morin (eds.) *Boeotia Antiqua II. Recent Papers in Boiotian Archaeology and Epigraphy*. Amsterdam: 53-56.

De Ridder, A. 1895. *Fouilles d'Orchomène*. *BCH* 19: 137-224

Decourt, J. C. 1992. 'Étude d'archéologie spatiale: essai d'application à la géographie historique en Béotie', in L.D.I. Blum – J.-Cl. Decourt – B. Helly – G. Lucas (eds.) *Topographie antique et géographie historique en pays grec* VII, Paris: 15-48

Demakopoulou, K. 1988. Glypha (or Vlicha), Chalkis: A Major Mycenaean Settlement on the Boeotian Shore near Aulis. *Teiresias 18*: 1-18

Dercourt, J. (ed.) 1977. 'Réunion extraordinaire de la Société Gèologique de France en Grèce, co-organisée avec la Société Gèologique de Grèce, 9-20/9/1976'. *Bulletin de la Société Gèologique de France*, 7e série, vol. XIX, n. 1: 1-164

De Ridder, A. 1895. 'Fouilles d'Orchomène', *BCH* 19: 150-155.

Derrau, M. 1974. *Précis de Géomorphologie*. Masson, Paris.

Desborough, V. R.d' A. 1952. *Protogeometric Pottery*. Oxford.

Desborough, V. R.d' A. 1972. *The Greek Dark Ages*. London.

De Silva, M. – Pizziolo, G. 2001. 'Setting up a 'human calibrated' anisotropic cost surface for archaeological landscape investigation', in Stančič, Z. - Veljanovski, T. (eds.) *Computing Archaeology for understanding the past, CAA2000*. BAR Int.Ser. 931: 279-286

Dewar, R. E. 1986. 'Discovering settlement systems of the past in New England site distributions'. *Man in the Norteast* 31: 77-88

Dewar, R. E. - McBride, K. A. 1992. 'Remnant Settlement Patterns', in J. Rossignol - L. A. Wandsnider (ed.) *Space, Time and Archaeological Landscapes*. New York: 227-255

Dickinson, O. 1994. *The Aegean Bronze Age*, Cambridge.

Dobres M. - Robb J.E. (eds) 2000. *Agency in Archaeology*. Routledge, London.

Dobrzanska H. - Jerem E. - Kalicki T. (eds.) 2004. *The geoarchaeology of river valleys*. Archaeolingua series minor 18. Budapest.

Dodwell, E. 1819. *A Classical and Topographical Tour through Greece during the Years 1801, 1805, 1806*. vol.1. London.

Doorn, P. 1993. 'Geographical Location and Interaction Models and the Reconstruction of Historical Settlement and Communication: The Example of Aetolia, Central Greece'. *Historical Social Research* 18.3: 35-71

Dreslerová, D. 1995. 'A settlement-economic model for a prehistoric microregion: settlement activities in the Vinoř-stream basin during the Hallstatt period' in M. Kuna - N. Venclová (eds.), *Wither Archaeology?*, Praha 1995: 145-60

Ducat, J. 1973. 'La Conféderation béotienne et l'expansion thébaine à l'époque archaïque'. *BCH* 97: 59-73

Duchaufour, P. 1970. *Précis dePédologie*. Masson, Paris.

Duke, Ch. 2003. 'Quantifying palaeolithic landscapes: computer approaches to terrain analysis and visualisation', in M. Doerr – A. Sarris (eds.) *The Digital heritage of Archaeology, CAA 2002*: 139-146

Doukellis, P.N. - Mendoni, L.G. (eds.) 1994. *Structures Rurales et Sociétés Antiques*, Actes du colloque de Corfou (14-16 mai 1992), Centre de Recherches d'Histoire Ancienne 126, Paris.

Doukellis, P.N. – Mendoni, L.G. (eds.) 2004. *Perception and Evaluation of Cultural Landscapes*. Athens.

Dunnel, R.C. – Dancey, W.S. 1983. 'The siteless survey : a regional scale data collection strategy', in M.B.Schiffer (ed.) *Advances in Archaeological method and Theory* 6: 267-287

Dunnell, R. C. 1992. 'The Notion Site', in J. Rossignol - L. A. Wandsnider (ed). *Place, Time and Archaeological Landscapes*. Plenum Press, New York: 21-37.

Dyson-Hudson, R. – Smith, E. A. 1978. 'Human territoriality: An ecological reassessment'. *American Anthropologist* 80: 21-41.

Embleton, C. (ed.) 1984. *Geomorphology of Europe*. Macmillan, London.

Étienne, R. - Knoepfler, D. 1976. *Hyettos de Beotie*. BCH Suppl. III.

Entrikin, N.J. 1994. 'Place and region'. *Progress in Human Geography* 18: 227-33.

Evans, J.G. 1999. *Environmental Archaeology: Principles and Methods*. Sutton.

Evans, J.G. 2003. *Environmental Archaeology and the Social Order*. Routledge, London/New York.

Fantasia, U. 1999. 'Aree marginali nella Grecia antica: paludi e bonifiche', in D. Vera (ed.) *Demografia, sistemi agrari, regimi alimentari nel mondo antico*: 65-116

FAO 1966. *Soil Map of Europe.* Rome.

FAO 1967. *Soil survey interpretation and its use.* Soil Bulletin 8, Rome.

FAO 1968. *Directives pour la descriptions des sols.* Rome.

FAO 1976. *A Framework for Land Evaluation.* Soil Bulletin 55, Rome.

FAO 1977. *Guidelines for soil profile description.* Rome.

FAO 1998. *World Reference Base for Soil Resources.* World Soil Resources Report 84, Rome.

FAO-UNESCO 1981. *Carte Mondiale des sols 1:5,000,000. Vol. V Europe*. Paris, UNESCO.

FAO-UNESCO 1994. *Soil Map of the World. Revised Legend with Corrections.* International Soil Reference and Information Center, Wageningen.

Farinetti, E. 2003. 'Boeotian Orchomenos: a progressive creation of a *polis* identity', in H. Hokwerda (ed.) *Constructions of Greek Past*, Groningen: 1-10

Farinetti, E. 2008. 'Fluctuating landscapes: the case of the Copais basin in ancient Boeotia', *ASAtene* LXXXVI: 115-138

Farinetti, E. - Sarri, K. - Sbonias, K. 2006. 'Detecting the prehistoric landscape in the area of a Greek-Roman *Polis*. The case of ancient Tanagra, Boeotia' (in greek) – *1st archaeological meeting of Thessaly and Central greece 2003* - Volos (Thessaly): 579-592

Farinetti, E. - Sbonias, K. 2004, 'Fields of Wheat back to the Land. A GIS environment for the study of medieval village history in Central Greece', in *CAA 2003*: 271-275

Farinetti, E. – Sbonias, K. in preparation. 'The Prehistoric landscape and the geomorphological work South of Tanagra city'.

Farinetti, E. – Sigalos, L. 2002. 'Detailed Topography and Surface Survey. What is the point? Tanagra City Survey 2000', in *CAA 2001*: 385-391

Farinetti, E. et al. in press. Farinetti, S. – Hermon, S. – Niccolucci, F. 'Fuzzy logic application to survey data', in F. Niccolucci (ed.), *Beyond the artefact - Proceedings of CAA2004 - Prato 13-17 April 2004.*

Fimmen, D. 1912. 'Die Besiedlung Böotiens bis in frühgriechische Zeit'. *NJbb* 29: 521-541

Flannery, K. V. (ed.) 1976 *The Early Mesoamerican Village.* Academic Press, New York.

Fleming, A. 1990. 'Landscape archaeology, prehistory and rural studies'. *Rural History* I.1: 5-15

Foley, R. 1981. 'Off-site archaeology: an alternative approach for the shortsided', in I.Hodder - G.Iaasc - N. Hammond (eds.) *Patterns of the Past.* Cambridge: 157-183

Forbes, H. 1989. 'Of grandfathers and grand theories: the hierarchised ordering of responses to hazard in a Greek rural community', in O.Halstead - J.O'Shea (eds.) *Bad Year Economics. Cultural Responses to Risk and Uncertainty*, Cambridge University Press, Cambridge: 87-97

Forbes, H. 1995. 'The identification of pastoralist sites within the context of estate-based agriculture in ancient Greece: beyond the 'transhumance versus agro-pastoralist' debate'. *BSA* 90: 325-338

Forbes H. 1996. "The Uses of the uncultivated Landscape in Modern Greece: a Pointer to the Value of the Wilderness in Antiquity?", in G. Shipley – J. Salmon (eds.) *Human Landscapes in Classical Antiquity. Environment and Culture.* London-New York: 68-97

Forbes, C. - Mee, H. (eds.) 1997. *A Rough and Rocky Place. The Landscape and Settlement History of the Methana Peninsula, Greece.* Liverpool university Press, Liverpool.

Forchhammer, P. W. 1837. *Hellenika. Griechenland im neuen das Alte.* Berlin.

Forchhammer, P.W. 1857. *Halkyonia, Wanderungen an den Ufern des Halkyonischen Meeres: Sendschreiben an Prof. Böckh.* Berlin.

Forni G. 1989a. 'Questioni di storia agraria pre-romana: le quattro fasi dell'agricoltura etrusca'. *II Congresso Internazionale Etrusco.* Roma: 1501-1515

Forni G. 1989b. 'Le più antiche evidenze dell'introduzione di vomeri d'aratro e di altri strumenti agricoli come documento dell'evoluzione della metallurgia in Italia' in N. Cuomo di Caprio – C. Simoni (eds.) *La siderurgia nell'antichità*, atti del Simposio Val Camonica 1988. Varese, Musei Civici di Villa Mirabello 20: 359-380

Forni 1990. *Gli albori dell'agricoltura.* Roma.

Forte M. - Kay S. - Perlingieri C. - Perlingieri R. 2003. 'Remote Sensing Technologies and Virtual Reconstruction of Archaeological Landscapes: New Developments of the Aksum Project', in *CAA2002*: 49-55

Fossey, J.M. 1970. 'The identification of Graia', *Euphrosyne* N.S. IV: 3-22

Fossey, J.M. 1972. 'Tilphossaion?' in J.M. Fossey (ed.) *Proceedings of the first International Conference on Boiotian Antiquities. Teiresias* Suppl. I: 1-16

Fossey, J.M. 1973. 'The Copaic Basin in the 2nd Century A.D.', in *Acts of the 6th International Congress og Greek and Latin Epigraphy*, München. Vestigia 17: 451-5

Fossey, J.M. 1974. 'Some unpublished Boiotian tombstones'. *BSA* 69: 121-128

Fossey J.M. 1978. 'The "Two Boeotias" in the Geometric-Archaic period: evidence of the ceramic imports', in *Acta of the XI International Congress of Classical Archaeology*, London: 188

Fossey, J.M. 1979. 'The city of the Kopais in the Roman Period'. *Aufstieg und Niedergang der Römischen Welt* II.7.1, Berlin: 549-591

Fossey, J.M. 1979b. 'Une base navale d' Épaminondas'. *Teiresias* Suppl. II: 9-13

Fossey, J.M. 1980. 'Mycenaean fortification of the Northeast Copais'. *OpAth* 13 (1980): 155-162 = in

Fossey J.M. 1990 *Papers in Boeotian Topography and History* : 72-89

Fossey, J.M. 1981-3. 'The city archive at Koroneia Boiotia', *Euphrosyne* XI (1981-3): 45-60

Fossey, J.M. 1983. 'The Topography of ancient Boiotia: the Extensive Survey', in D.R. Keller – W. Rupp (eds.) *Archaeological Survey in the Mediterranean area*, BAR Int.Series 155: 233-4

Fossey, J.M. 1986. *The ancient topography of Eastern Phokis*. Amsterdam.

Fossey KHOSTIA I. Fossey, J.M. (ed.) 1986. *Khostia. Resultats des explorations et fouilles canadiennes à Khostia en Béotie, Grèce Centrale*: I. Amsterdam.

Fossey, J.M. 1988. *Topography and Population of Ancient Boiotia*. Chicago.

Fossey, J.M. 1990a. *The ancient Topography of Opuntian Lokris*. Amsterdam.

Fossey, J.M. 1990b. 'Le fortifications de l'acropole de Chéronée', in J.M. Fossey *Papers in Boeotian Topography and History* : 100-124

Fossey J.M. 1990c. 'The cities of the Copais in the Roman period', in J.M. Fossey *Papers in Boeotian Topography and History* : 220-266

Fossey, J.M. 1990d 'The end of the bronze age in SW Copais', in J.M. Fossey *Papers in Boeotian Topography and History* : 53-71 (published earlier in *Euphrosyne* VI 1973-4: 7-21)

Fossey, J.M. 1990e 'Two mountain sites around Kopais', in J.M. Fossey *Papers in Boiotian Topography and History*: 205-211

Fossey, J.M. 1990f 'Tilphosaion?', in J.M. Fossey *Papers in Boiotian Topography and History*: 169-184

Fossey, J.M. 1991. *Epigraphica Boeotica 1: Studies in Boiotian Inscriptions*. J. C. Gieben, Amsterdam.

Fossey, J.M. 1992. 'The development of some defensive networks in eastern central Greece during the Classical period', in S. van de Maele – J.M. Fossey (eds.) *Fortificationes Antiquae*. Amsterdam: 109-32.

Fossey J.M. - Gauvin G. 1983. 'The topography of ancient Boeotia: intensive survey and excavation', in D.R. Keller – W. Rupp (eds.) *Archaeological Survey in the Mediterranean area*, BAR Int.Ser. 155: 237-9

Fossey, J.M. – Gauvin, G. 1985a, 'Mali en Béotie du Sud: un site des "Dark Ages"' in Argoud - Roesch (eds.) 1985: 71-76

Fossey, J.M. – Gauvin, G. 1985b. 'Livadhostro : un relevé topographique des fortifications de l'ancienne Kreusis', in Argoud - Roesch (eds.) 1985: 77-85

Fossey, J.M. - Morin, J. 1986. *KHOSTIA 1983. Rapport preliminaire sur la seconde campagne de fouilles canadienne a Khostia en Beotie, Grece Centrale*. Amsterdam.

Fossey, J.M. - Morin, J. 1987: 'The *Khostia* project: excavation and survey', in J.M. Fossey (ed.) *Boeotia Antiqua I: Papers on Recent Work in Booetian Archaeology and History*, Amsterdam: 165-174

Foth, H.D. 1990. *Fundamentals of soil science*. Wiley, New York.

Fox, C. 1923. *The Archaeology of the Cambridge Region*. Cambridge University Press, Cambridge.

Foxhall, L. 2004. 'Small, rural farmstead sites in Ancient Greece: a material cultural analysis', in F. Kolb (ed.), *Chora und Polis: Methoden und Ergebnisse der historische Landeskunde*. Proceedings of the Munich Greek Countryside Conference. Oldenbourg, 2004: 249-270

Frazer, G. 1913. *Pausania's Description of Greece*, V. London.

Freitag, K. 2000. *Der Gulf von Korinth. Historisch-Topographische Untersuchungen von der Archaik bis in das 1. Jh.V.Chr.* Munich.

French, D.H. 1972. *Notes on Prehistoric Pottery Groups from Central Greece*. Athens.

Fronza, V. et al. 2001. Fronza, V. - Nardini, A. - Salzotti, F. - Valenti, M. 'A GIS solution for Excavations: Experience of the Siena University LIAAM', in *CAA 2000*: 173-177

Fronza, V. et al. 2003. Fronza, V. - Nardini, A. - Valenti, M. 'An integrated information System for Archaeological Data management: latest Developments', in *CAA 2002*: 147-154

Gaffney, V. – Stančič, Z. 1991. *GIS approaches to regional analysis. A case study of the island of Hvar*. Znanstveni Institut Filozofske Facultete, Ljubljana.

Gaffney, V. - Stančič, Z. 1992. 'Diodorus Siculus and the island of Hvar, Dalmatia: testing the text with GIS', in G. Lock - J. Moffett (eds.) *Computer Applications and Quantitative Methods in Archaeology – CAA1991*. BAR Int. Ser. 577, BAR Publishing, Oxford: 113-125

Gaffney, V. - Tingle, M. 1984. 'The tiranny of the site: method and theory in field survey', in *Scottish Archaeological Review* 3: 134-140

Gaffney, V. - van Leusen, M. 1995. 'GIS and environmental determinism', in Lock-Stančič (eds.) 1995: 367-382

Gallant, T.W. 1986. 'Background noise and site definition: a contribution to site methodology'. *Journal of Field Archaeology* 13: 403-418

Gams, I. ed. 1975. *Karst processes and relevant landforms*. Commission on Karst Denudation at the International Speleological Union, Ljubljana.

Garlan, Y. 1974. *Recherches de poliorcetique greque*. Paris.

Gary, M. et al. 1973. Gary, M. – McAfee, Jr. R. – Wolf, C. L. (eds.) *Glossary of Geology*. 2nd edition. American Geological Institute, Washington D.C.

Gauvin, J. 1986. 'Quelque fermes en peripherie de la vallée' in J.M. Fossey (ed.) *KHOSTIA I*: 51-81

Gell, W. 1819. *The Itinerary of Greece*. London.

Gerrard, Ch. - Aston, M. 2007. *Shapwick Project Sommerset: a Rural Landscape Explored.* Society for Medieval Archaeology, Leeds.

Gilchrist, R. 2000. Archaeological Biographies: realising human lifecycles, -courses and –histories. *World Archaeology* 31 (3): 325-328

Gillings, M. 1995. 'Flood dynamics and settlement in the Tisza valley of North-East Hungary: GIS and the Upper Tisza Project', in Lock-Stančič (eds.) 1995: 67-84

Gillings, M. 2000. 'The Utility of the GIS Approach in the Collection, Management, Storage and Analysis of Surface Survey Data', in J.L. Bintliff – M. Kuna – N. Venclovà (eds.) *The Future of Surface Artefact Survey in Europe*: 105-120

Gillings, M. - Goodrik, G.T. 1996. 'Sensous and reflexive GIS: exploring visualisation and VRML'. *Internet Archaeology* 1 (http://intarch.ac.uk/)

Gillings, M. et al. 1999: Gillings M., Mattingly D., Dalen van J., (eds.), *Geographical Information Systems and Landscape Archaeology.* The Archaeology of Mediterranean Landscape n.3. Oxbow Books, Oxford.

Gillings M. - Wise A. (eds.) 1999. *GIS. Guide to good practice.* CARLSON General Coll. 1999 http://ads.ahds.ac.uk/project/goodguides/gis/

Gillings, M. - Wheatley, D. 2002. *Spatial Tecnology and Archaeology. The Archaeological Applications of GIS.* Taylor & Francis, London and New York.

Girard, P. 1881. *De Locris Opuntiis* 37. Apud E. Thorin.

Given, M. - Knapp, A. et al. 1999. 'The Sydney Cyprus Survey Project: An Interdisciplinary Investigation of Long-Term Change in the North Central Troodos, Cyprus'. *JFA* 26.1: 19-40

Given, M. - Knapp, A.B. 2003. *The Sydney Cyprus Survey Project. Social Appproaches to Regional Archaeological Survey.* Monumenta Archaeologica 21. The Cotsen Institute of Archaeology, University of California, Los Angeles.

Godard A. - Rapp A. (eds.) 1987 *Processus et mesure de l'érosion.* 25e Congrés International de Géographie, Paris 1984, Paris éd du CNRS.

Gojda, M. (ed.) 2004. *Ancient Landscape, Settlement Dynamics and Non-Destructive Archaeology A Czech Research Project: 1997-2002.* Academia, Prague.

Goldman, H. 1931. *Excavations at Eutresis in Boeotia.* Cambridge, Mass.

Gomme A.W. 1911/12. 'The Topography of Boeotia and the Theories of M.Bérard'. *BSA* 18: 189-210

Goodchild, M.F. 1995. 'Geographic Information Systems and geographic research', in J. Pickels (ed.) *Ground truth.* The Guilford Press, New York: 31-50

Goodwin, B.C. 1988. 'Organisms and minds: the dialectics of the animal-human interface in biology', in T. Ingold (ed.) *What is an animal?*. Unwin Hyman, London.

Gorenflo, L.J. - Gale, N. 1990. 'Mapping regional settlement in information space'. *Journal of Anthropological Archaeology* 9: 240-274.

Gosden, C. 1994. *Social Being and Time.* Blackwell, Oxford.

Gosden, C. 1999. *Anthropology and Archaeology. A changing relationship.* Routledge, London and New York.

Grant, E. 1986. 'Hillforts, central places and territories', in E. Grant (ed.) *Central Places, Archaeology and History.* Sheffield.

Gregory, T.E. 1980. 'Ohio Boeotia Expedition: Field Seasons 1979-1980'. *Teiresias Archaeologica* 1980: 31-41

Gregory, T.E. 1992. 'Archaeological Explorations in the Thisbe Basin', in J.M. Fossey – J. Morin (eds.) *Boeotia Antiqua II: Recent Papers in Boiotian Archaeology and Epigraphy.* Amsterdam: 17-34

Gregory, T.E 1986. 'A Desert Island Survey in the Gulf of Corinth'. *Archaeology* 39.3: 16-21 (and 78)

Greig, J.R.A - Turner, J. 1974. 'Some pollen diagrams from greece and their archaeological significance'. *JAS* 1: 177-94

Grenville, J. (ed.) 1999. *Managing the Historic Rural Landscape.* Routledge, London/New York.

Guillon, P. 1943. *Les Trépieds du Ptoion.* Paris.

Guillon, P. (ed.) 1948. *La Béotie antique.* Paris.

Guillot, D. - Leroy, G. 1995. 'The use of GIS for archaeological resource management in France: the SCALA Project', in Lock-Stančič (eds.) 1995: 15-26

Gustavsson, R. 1999. 'Local distinctiveness in Landscape Character', in M.B. Usher (ed.) *Landscape Character: Perspectives on Management and Change.* The Stationery Office, Edinburgh: 41-54

Hailer, U. – Zimmermann, M. 1998. 'Lykien. Stadt und Land in der Antike'. *Archaologie im Deutchland* 2.

Halstead, P. 1987. 'Traditional and ancient rural economy in Mediterranean Europe: plus ça change?'. *JHS* 107: 77-87

Halstead, P. 1994. 'The North-South divide: regional pathways to complexity in prehistoric Greece', in C. Mathers - S. Stoddart (eds.) *Development and Decline in the Mediterranean Bronze Age.* J.R.Collis Publications, Sheffield: 195-219

Halstead, P. 1995. 'Plough and Power: The Economic and Social Significance of Cultivation with the Ox-Drawn Ard in the Mediterranean'. *Bulletin of Sumerian Agriculture* 8: 11-22

Halstead, P. (ed.) 1999. *Neolithic Society in Greece.* Sheffield Academic Press, Sheffield.

Halstead, P. - Frederick, Ch. (eds.) 2000. *Landscape and Land Use in Postglacial Greece.* Sheffield Studies in Aegean Archaeology 3. Sheffield Academic press, Sheffield.

Hansen, J.M. 1991. *The Palaeoethnobotany of Franchti Cave.* Excavations at Franchti Cave, Greece, fasc.7. Indiana University Press, Bloomington and Indianapolis.

Hansen, M.H. 1995. 'KOME. A study in how the Greeks designated and classified settlements which were not POLEIS', in M.H. Hansen - K. Raaflaub (eds.) *Studies in the Ancient Greek Polis*. Historia Einzelschriften 95: 45-81

Hansen, M.H. 1996. 'An inventory of Boiotian POLEIS in the Archaic and Classical periods', in M.H. Hansen (ed.), *Introduction To An Inventory Of Poleis*. Acts of the Copenhagen Polis Centre 3: 73-116

Hansen, M.H. 1996a. 'Were the Boiotian POLEIS deprived of their AUTONOMIA during the first and second Boiotian federations? A reply', in M.H. Hansen - K. Raaflaub (eds.) *More Studies In The Ancient Greek Polis*. Historia Einzelschriften 108: 127-136

Hanson, C.B. 1980. 'Fluvial taphonomic processes: models and experiments', in A.K. Behrensmeyer - A.P. Hill (eds.) *Fossils in the Making*. University of Chicago Press, Chicago: 156-181

Harris, T.M. - Lock, G. 1992. 'Toward a Regional GIS Site Information Retrieval System: the Oxfordshire Site and Monuments Record (SMR) Prototype', in C.U. Larsen (ed.) *Sites and Monuments. National Archaeological Records*. Copenhagen. National Museum of Denmark:185-199

Hassan, F.A. 2004 'Ecology in Archaeology: From Cognition to Action', in J.L. Bintliff (ed.) *A Companion to Archaeology*. Blackwell Publishing, Oxford: 311-333

Hatzinikolaou, E. – Hatzichristos, Th. - Siolas, A. – Mantzourani, E. 2003. 'Predicting Archaeological Site Locations using GIS and Fuzzy Logic', in *CAA2002*: 169-177

Hatzinikolaou, E.G. 2006. 'Quantitative methods in archaeological prediction: from binary to fuzzy logic', in N.W. Mehrer - K.L. Wescott (eds.) *Gis and Archaeological Site Location Modeling*: 437-446

Heidegger, M. 1971. 'Building Dwelling Thinking', in *Poetry, language, thought*. Trans. A.Hofstadter. Harper and Row, New York: 145-61

Heidinga, H. A. 1987. *Medieval Settlement and Economy North of the Lower Rhine*. Van Gorcum.

Hellenic National Statistical Office 1982. *Πραγαμτικός πληθυσμός της Ελλάδος, κατά την απογραφή της 5 απριλίου 1981 / Population de fait de la Grèce, au recensement du 5 avril 1981*. Athens.

Helly, B. 1984. 'Le territoire de Larisa: ses limites, son extension, son organisation'. *Ktèma* 9 : 213-234

Hennig D. 1974. *Orchomenos*. RE Suppl. XIV: 333-55

Hermon et al. 2004. Hermon, S. – Niccolucci, F. – Alhaique, F. – Leonini, V. 'Archaeological typologies - an archaeological fuzzy reality', in K.F. Ausserer - W. Börner et al. (eds.) *Enter the Past, CAA 2003*: 30-34

Herz, N. - Garrison, G. 1998. *Geological Methods for Archaeology*. Oxford, Oxford University Press.

Heurtley, W.A. 1923/25. 'Notes on the Harbours of South Boeotia, and Sea-trade between Boeotia and Corinth in Prehistoric Times'. *BSA* 26: 38-45

Higgins, R. *Tanagra and the figurines*. Princeton 1986.

Higgins, M.D. - Higgins, R. 1996. *A Geological Companion to Greece and the Aegean*. Duckworth, London.

Hill, G.F. 1906. *Historical Greek Coins*. London.

Hodder, I. - Orton, C. 1976. *Spatial Analysis in Archaeology*. Cambridge University Press, Cambridge.

Hodkinson, S. 1990. 'Politics as a determinant of pastoralism: the case of Southern Greece, ca.800- 300 BC'. *RStLig* vol. 56: 139-163

Holstrom, S. - Voorrips, A. - Kamermans, H. (eds.) 2004a. *The Agro Pontino Archaeological Survey*. Archaeological Studies Leiden University (ASLU) 11. Leiden (CD-ROM).

Holstrom, S. - Voorrips, A. - Kamermans, H. (eds.) 2004b. 'The Agro Pontino Archaeological Survey. An annotated and illustrated catalogue', in S. Holstrom - A.Voorrips - H. Kamermans (ed.), *The Agro Pontino Archaeological Survey*. Archaeological Studies Leiden University (ASLU) 11. Leiden (CD-ROM).

Holt-Jensen, A. 1999. *Geography. Its History and Concepts*. SAGE Publications.

Houby-Nielsen, S. 1988. *Siedlungsgeographie des Kopaisgebietes*. AA 102 (dissertation)

Hope Simpson, R. 1965. *A Gazeteer and Atlas of Mycenaean Sites*. London.

Hope Simpson R. – Lazenby, J.F. 1970. *The Catalogue of Ships in Homer's Iliad*, Oxford.

Howard, Ph. 2007. 'Spatial Analysis of Boeotia Field-walking Data', in Bintliff - Howard –Snodgrass 2007: 111-127

HS&D 1979. Hope-Simpson, R. – Dickinson, O.T.P.K. *A Gazetteer on the Aegean Civilization. The Mainland and Islands*. Göteborg.

Huggett, J. 2000. 'Looking at intra-site GIS', in V.M. Birliba (ed.) *CAA96. Computer applications and quantitative methods in archaeology*. BAR Publishing, Oxford BAR Int.Ser. 945: 117-122

Hughes, R. - Buchan, N. 1999. 'The landscape character assessment of Scotland', in M.B. Usher (ed.) *Landscape Character: Perspectives on Management and Change*. The Stationery Office, Edinburgh: 1-11

Huntley, B. - Prentice, I.C. 1988. 'July temperatures in Europe from pollen data, 6000 years Before Present'. *Science* 241: 687-90

Iakovidis, S.E. 2001. *Gla and the Kopais in the 13th century B.C.* The Archaeological Society at Athens, Athens.

Inglis, F. 1977. 'Nation and community: a landscape and its morality'. *Sociological Review* 25: 489-514

Ingold, T. 1993. 'The temporality of the landscape'. *WorldArch* 25.2: 152-74

Ingold, T. 2000. *The perception of the environment. Essays in livelihood, dwelling and skill*. Routledge, London/New York.

Jablonka, P. 2004: 'Reconstructing Sites and Archives: Information and Presentation Systems at Troy', in *CAA 2003*: 281-285

Jacobshagen, V. 1986. *Geologie von Griechenland.* Gebrüde Borntraeger, Berlin, Stuttgart.

Jahnkuhn, H. 1955. 'Methoden und Probleme siedlungsarchaeologischer Forschung'. *Archaelogica geographica* 4: 73-84

Jameson, M.H. – Runnels, C.N. – van Andel, T.H. (eds.) 1994. *A Greek countryside. The southern Argolid from Prehistory to the present day.* Stanford.

Jameson, M.H. 1978. 'Agriculture and Slavery in Classical Athens'. *CJ* 73 (1977-1978): 122-145

Janda, K. – Gillies, R. 1983. 'How well does 'region' explain political party characteristics?'. *Political Geography Quarterly* Vol. 2, No. 3, July 1983: 179-203

Jeffery, L.H. 1961. *The Local Scripts of Archaic Greece.* Oxford.

Johnson, M. 1996a. 'The Berbati-Limnes archaeological survey. The Neolithic period', in Wells (ed.) 1996: 37-73

Johnson, M. 1996b. 'Water, Animals and Agricultural Technology: Study of Settlement Patterns and Economic Change in Neolithic Southern Greece'. *OJA* 15: 267-295.

Johnson, M. 1999. *Archaeological Theory: an introduction.* Oxford Blackwell Publishers.

Johnson, I. - North, M. (eds.) 1997. *Archaeological Applications of GIS. Proceedings of Colloquium II, UISPP XIIIth Congress*, Forlì-Italy, September 1996, Sidney (CD-ROM)

Johnson, I. – Wilson, A. 1998. 'The TimeMap Project: Developing Time-Based GIS Display for Cultural Data', at: http://www.timemap.net/indepth/documentation/esri_book_2001/esri_timemap_article_2000_images_linked.htm (last accessed in September 2008) 2nd

Johnston, R.J. 1991. *Geography and Geographers. Anglo-American Human Geography since 1945.*

Johnston, R. 1998. 'Approaches to the perception of landscape: philosophy, theory, methodology'. *Archaeological Dialogues* 5: 54-68

Judeich, W. 1988. 'Das Kabirenheiligtum bei Theben'. *AM* 13: 81-87

Kahrstedt, U. 1937. 'Der Kopaissee im Altertum und die Minyschen Kanäle'. *AA* 52: 1-20

Kahrstedt, U. 1954. *Das Wirtschaftliche Gesicht Griechenlands in der Kaiserzeit.* Bern.

Kalcyk, H. - Heinrich, B. 1986. 'Hochwasserschutzbauten in Arkadien'. *AntW* 2nd Sondernummer: 3-14.

Kalcyk, H. - Heinrich, B. 1989. 'The Munich Kopais-Project', in Fossey J.M. (ed.) *Boeotia Antiqua I: Papers on Recent Work in Booetian Archaeology and History*, Amsterdam: 55-71

Kallet-Marx, R. 1989. 'The Evangelistria watchtower at the defense of the Zagara Pass', in H.Beister and J.Bucklet (eds.) *Boiotika.* Editio Maris, Munich 1989: 301-311

Kambanis, M.L. 1892. 'Le Dessèchement du Lac Copais par les Anciens' (I). *BCH* 16: 121-137

Kambanis, M.L. 1893. 'Le Dessèchement du Lac Copais par les Anciens' (II). *BCH* 17: 322-342

Kamermans, H. 1993. *Archeologie en landevaluatie in de Agro Pontino (Lazio, Italië).* Academisch proefschrift, Amsterdam.

Kamermans, H. 1996. 'The Agro Pontino Survey project: a summary of the application of land evaluation', in S. Milliken - C. Peretto (eds.) *Archaeology, methodology and the organisation of research*, A.B.A.C.O. s.r.l., Forlì: 51-60

Kamermans, H. 2000. 'Land evaluation as predictive modelling: a deductive approach', in Lock (ed.) 2000: 124-146

Kamermans, H. 2001. *Strategic research into, and development of best practise for, predictive Cultural Resource Management.* Unpublished research proposal.

Kamermans, H. 2003. 'Predictive Maps and Land Quality Mapping', in J. Kunow - J. Müller (eds.), *Landschaftsarchäologie und geographische Informationssysteme: Prognosekarten, Besiedlungsdynamik und prähistorische Raumordnungen. The Archaeology of Landscapes and Geographic Information Systems: Predictive Maps, Settlement Dynamics and Space and Territory in Prehistory* (Forschungen zur Archäologie im Land Brandenburg 8. Archäoprognose Brandenburg I). Brandenburgisches Landesamt für Denkmalpflege und Archäologisches Landesmuseum. Wünsdorf: 151-160

Kamermans, H., 2004. 'Archaeology and land evaluation in the Agro Pontino (Lazio, Italy)', in S. Holstrom - A.Voorrips - H. Kamermans (eds.), *The Agro Pontino Archaeological Survey.* Archaeological Studies Leiden University (ASLU) 11. Leiden (CD-ROM).

Kamermans, H. in press: 'The Application of Predictive Modelling in Archaeology: Problems and Possibilities', in F. Niccolucci (ed.), *Beyond the artefact – Proceedings of CAA2004 – Prato 13-17 April 2004.*

Kardulias, P. N. 1992. 'Estimating population at ancient military sites: the use of historical and contemporary analogy'. *AmerAnt* 57: 276-287.

Kardulias, P. N. – Shutes, M. T. (eds.) 1997. *Aegean Strategies: Studies of Culture and Environment on the European Fringe.* Rowman Littlefield Publishers, Lanham.

Kenny, E.J.A. 1935. 'The Ancient Drainage of the Copais'. *AAA* 22: 189-206

Kirsten, E. 1951 'Beiträge zur historischen Landeskunde des östlichen Mittelgriechenlands und Euboeas', in A. Philippson, *Die griechische Landschaften, I: Der Nordosten der griechischen Halbinsel.* Frankfurt am Main: 645-729

Kirsten, E. 1956. *Die Griechische Polis als historisch-geographisches Problem des Mittelmeerraumes.* Bonn.

Klápště, J. 1991. 'Studies of structural change in medieval settlement in Bohemia'. *Antiquity* 65: 396-405

Knapp, A. B. (ed.) 1992. *Archaeology, Annales, and Ethnohistory*. CUP, Cambridge.

Knapp, A.B. - Ashmore, W. 1999. 'Archaeological Landscapes: Constructed, Conceptualized, Ideational', in A. B. Knapp - W. Ashmore (eds.) *Archaeologies of Landscape. Contemporary Perspectives*. Blackwell: 1-30

Knauss, J. 1987. 'Munich Kopais expedition, topographical surveys of October 1986 and May 1987: progress report'. *Teiresias* 17: 1-5

Knauss, J. 1988. 'Der Damm im Takka-See beim alten Tegea (Arkadien, Peloponnes)'. *AM* 103: 25-36

Knauss, J. 1992. 'Purpose and function of the ancient hydraulic structures at Thisbe', in J.M. Fossey – J. Morin (eds.) *Boeotia Antiqua II: Recent Papers in Boiotian Archaeology and Epigraphy*, Amsterdam: 35-45

Knauss, J. 1995. 'NE Kopais: Technical and Historical Aspects of the unfinished ancient drainage Tunnel at the outmost Northeast Corner of the Kopais Basin', in A.Ch. Christodoulou (ed.) *B' diethnes synedrio Boiotikon Meleton*: 83-95

Knauss, J. 2000. 'Praehistorische Landgewinnung beim boeotischen Medeon am Fuss des Sphinxberges', in *G' diethnes synedrio Boiotikon Meleton*: 242-252 (English version in *ArchDelt* 50 1995 (2000) B1: 242-252 'Prehistoric drainage of Medeon polder').

Knauss, J. 2001. *Späthelladische Wasserbauten: Erkundungen zu wasserwirtschaftlichen Infrastrukturen der mykenischen Welt: Zusammenfassung aller bisherigen Untersuchungsergebnisse*. Lehrstuhl für Wasserbau und Wasserwirtschaft, Institut für Wasserwesen, München.

Knauss, J. et al. Kopais 1 1984. *Die Wasserbauten der Mynier in der Kopais – die alteste Flussregulierung Europas*. Munchen-Obernach.

Knauss, J. Kopais 2 1987. *Die Melioration des Kopaisbeckens durch die Minver im 2. Jt.v.Chr.: Kopais 2 : Wasserbau und Siedlungsbedingungen im Altertum*. Technische Universitat Munchen, München.

Knauss, J. Kopais 3 1990. *Wasserbau und Geschichte: minische Epoche, bayerische Zeit* (vier Jahrhunderte, ein Jahrzehnt). Technische Universitat Munchen, München.

Knoepfler, D. 1992. '7 annes de researches sur l'epigraphie de la Beotie'. *Chiron* 22: 411-503

Koder, J. – Hild, F. 1976. *Tabula Imperii Byzantini, I. Hellas und Thessalia*. Wien.

Kolendo J. 1980. *L'agricoltura nell'Italia romana*. Editori Riuniti, Roma.

Konečny et al. 2008. Konečny, A.L. – Boyd, M.J. – Marchese, R.T. – Aravantinos, V. 'Plataiai in Boiotia: A Preliminary Report on Geophysical and Field Surveys Conducted in 2002–2005'. *Hesperia* 77.1 : 43-71

Konsola, D. 1981. *Promykenaiki Thiva*. Athens.

Kopanias, K. 2008. 'Das Strassenetz der Ostphokis und die Frage nach der Lokalisierung von Abai'. *AM* 123 (in press).

Koster, H.A. – Bouza Koster, J. 1976. 'Competition Or Symbiosis?: Pastoral Adaptive Strategies In The Southern Argolid, Greece', in *Competition or Symbiosis?* Annals of the New York Academy of Sciences 268.1: 275–285.

Kotsakis, K. 1999. 'What Tells can Tell: Social Space and Settlement in the Greek Neolithic', in P. Halstead (ed.) *Neolithic Society in Greece*. Sheffield Academic Press, Sheffield: 66-76.

Kretschmer, P. 1914. 'Literaturbericht fur das Jahr 1911'. *Glotta* 5: 265

Kuna M. - Venclova N. (eds.) 2000. *The Future of Archaeological Field Survey in Europe.*

Kuna, M. – Zvelebil, M. – Foster, P.J. – Dreslerova, D. 1993. 'Field survey and landscape archaeology research design methodology of a regional field survey in Bohemia'. *Pamatky Archaeologicke* 1993, n.2: 110-130

Kuna, M. 1991. 'The structuring of prehistoric landscape'. *Antiquity* 65: 332-47

Kuna, M. 2000. 'Surface artefact studies in the Czech Republic', in J. Bintliff – M. Kuna - N. Venclova (eds.) *The Future of Surface Artefact Survey in Europe*: 29-44

Kunze, E. 1931. *Orchomenos II*. München.

Kunze, E. 1934. *Orchomenos III*. München.

Kvamme, K. L. 1983. 'Computer Processing Techniques for Regional Modelling of Archaeological Site Locations'. *Advances in Computer Archaeology* I: 26-52

Kvamme, K.L. 1985. 'Determining empirical relationships between the natural environment and prehistoric site location: a huntergatherer example', in C. Carr (ed.) *For Concordance in Archaeological Analysis. Bridging Data Structure, Quantitative Technique, and Theory*. Westport Publishers, Kansas City: 208-238

Kvamme, K. L. 1989. 'Geographic Information Systems in Regional Archaeological Research and Data Management', in M.B. Schiffer (ed.) *Archaeological method and Theory* I. University of Arizona Press, Tucson: 139-203

Kvamme, K. L. 1990. 'The fundamental principles and practice of predictive archaeological modelling', in A. Voorrips (ed.) *Mathematics and Information Science in Archaeology, Volume 3*. Holos Verlag, Bonn: 257-295

Kvamme, K.L. 1999. 'Recent Directions and Developments in Geographical Information Systems'. *Journal of Archaeological Research* 7(2): 153-201

Lambeck, K. 1996. 'Sea-level change and shore-line evolution in Aegean Greece since Upper Palaeolithic time'. *Antiquity* 70: 588-611

Lambrakis, N. - Kallergis, G. 2001. 'Reaction of subsurface coastal aquifers to climate and land use changes in Greece. Modelling of groundwater refreshening patterns under natural recharge conditions'. *Journal of Hydrology* 245: 19-31

Lang, N. 2000. 'Beyond the map: harmonising research and Cultural Research Management', in Lock (ed.) 2000: 214-228

Lang, N. - Stead, S. 1992. 'Sites and Monuments Records in England - theory and Practice', in G. Lock, and J. Moffett (eds.) *Computer Applications and Quantitative Methods in Archaeology 1991*. BAR Int. Ser. 577: 69-76

Langdon, M.K. 1976. *A Sanctuary of Zeus on Mount Hymettos*. Hesperia Suppl. 16, Princeton.

Langran, G. 1992. *Time in geographical information systems*. Taylor&Francis, London.

Larsen, J.A.O. 1955. 'The Boiotian Confederacy and Fifth Century Oligarchic Theory'. *TAPA* 86: 40-50

Larsen, J.A.O. 1960. *Orchomenos and the Formation of the Boiotian Confederacy in 447 b.C.*. *CP* 55 IV: 9-18

Larsen, J.A.O. 1068. *Greek Federal States*, Oxford.

Larsen, C.U. (ed.) 1992. *Sites and Monuments. National Archaeological Records*. The National Museum of Denmark, DKC, Copenhagen.

Lauffer, S. 1940. 'Kopais-Gebiet, Neue Funde'. *JdI* 1940: 187-194

Lauffer, S. 1959. 'Ptoion'. *RE* XXIII 2: 1506-1537

Lauffer, S. 1974. 'Orchomenos'. *RE* Suppl. XIV: 295-331

Lauffer, S. 1976. 'Inschriften aus Bootien', *Chiron* 6: 11-51

Lauffer Kopais I. Lauffer, S. 1986. *Kopais I. Untersuchungen zur historisschen Landeskunde Mittelgriechenlands*. P. Lang, Frankfurt.

Leake, W.M. 1835. *Travels in Northern Greece*. Vol.II. London.

Lebègue, J.A. 1875. *De oppidis et portibus Megaridis ac Boeotiae in Corinthiaci sinus littore sitis*. Paris.

Lehmann, H. 1939. 'Die Siedlungsräume Ostkretas im Wandel der Zeiten'. *Geographische Zeitschrift* 45.6: 212-228

Léveau, Ph. 1993. 'Mentalité economique et grand travaux: le drainage du lac Fucin. Aux origines d'un modèle'. *Annales: Économies Sociétés Civilizations* 48.1 : 3-16

Levin, M. H. 1981. 'The roads to Gla'. *The Pennsylvania Gazette* 80.2: 25-29

Lewin, R. 1993. *Complexity. Life at the Edge of Chaos*. J.M.Dent, London.

Limp, W. F. 1987. 'The identification of archaeological site patterning through integration of remote sensing, geographic information systems and exploratory data analysis'. *Proceedings of US Army Corps of Engineers Sixth Remote Sensing Symposium*. Galveston: 232-262

Llobera, M. 1996. 'Exploring the topography of mind: GIS, social space and archaeology'. *Antiquity* 70 no.269: 612-622

Llobera, M. 2000. Understanding movement: a pilot model towards the sociology of movement, in Lock (ed.) 2000: 65-48

Loader, C. 1998. *Building in cyclopaean masonry. With special reference to the Mycenaean fortifications on mainland Greece*. SMA. Paul Åmströms Förlag.

Lock, G. 1998. 'The past, present and future of GIS-based cultural landscape research, in J. Peterson – B. Slapšak – Z. Stančič (eds.) *The use of Geographic Information Systems in the study of ancient landscapes and features related to ancient land use*, Workshop proceedings, Ljubljana - 27 April 1996, COST Action G2 (Paysages antiques et structures rurales), European Community: 31-35

Lock, G. (ed.) 2000. *Beyond the map. Archaeology and Spatial Technologies*. NATO Science Series A: Life Sciences 321. IOS Press, Amsterdam.

Lock, G. - Brown, K. (eds.) 2000. *On the Theory and Practice of Archaeological Computing*. Oxford University Committee for Archaeology Monograph No.51, Oxford.

Lock, G. - Harris, T. 2000, 'Analysing Change Through Time Within a Cultural Landscape: Conceptual and Functional limitations of a GIS Approach', at: http://www.arkeologi.uu.se/afr/projects/BOOK/lockframe.htm (last accessed in September 2008)

Lock, G. - Stančič, Z. (eds.) 1995. *Archaeology and GIS. A European Perspective*. Taylor & Francis, London.

Lock, G.R. - Harris, T.M. 1996. 'Danebury revisited: an English Iron Age hillfort in a digital landscape', in M. Aldenderfer – H.D.G. Maschner (eds.) *Anthropology, Space, and Geographic Information Systems*. Oxford University Press, New York: 214-240

Lock, P. 1986. 'The Frankish Towers of Central Greece'. *BSA* 81: 101-123

Lock, P. 1995. *The Franks in the Aegean 1204-1500*. Longman, New York.

Lohmann, H. 1985 'Landleben im klassischen Attika', Jahrbuch Ruhr-Universität Bochum: 71-96

Lohmann, H. 1992. 'Agriculture and Country Life in Classical Attica', in Wells (ed.) 1992: 29-60

Lohmann, H. 1993. *Atene. Forschungen zu Siedlungs- und Wirtschaftsstruktur des klassischen Attika*. Böhlau, Köln.

Lohmann, H. 1995. 'Die Chora Athens im 4.Jarhundert v Chr: Festungswesen, Bergbau und Siedlungen', in W. Eder (ed.) *Die atenische Demokratie im 4.Jarhundert v. Chr.*: 515-548

Lolling, H. 1887. 'Hellenische Landeskunde und Topographie', in I. Miller *Handbuch der Klassischen Altertumswissenschaft* III.

Lolling, H. 1889. *Reisenotizen aus Griechenland 1876 und 1877*. Berlin.

Martin, D. 1996. *Geographic Information Systems. Socioeconomic applications* (2nd edition). Routledge, London.

Massagrande, F.A. 1995. 'A GIS approach to the study of non-sistematically collected data: a case study from the Mediterranean', in J. Hugget and N. Ryan (eds.) *Computer Applications and Quantitative Methods in Archaeology, CAA 1994*. BAR Int.Ser.: 147-156

Mazonaki, E. – Faraklas, N. 1967. 'Neos "Makedonikos" taphos en Thibais'. *AE* 124: 15-19

McDonald, W. A. - G. R. Rapp (eds.) 1972. *The Minnesota Messenia Expedition. Reconstructing a Bronze Age regional environment* I. Minneapolis, University of Minnesota.

McGlade, J. 1995. 'Archaeology and the ecodynamics of human-modified landscapes'. *Antiquity* 69: 113-132

McInerney, J. 1999. *The folds of Parnassos. Land and ethnicity in ancient Phokis*. University of Texas Press, Austin.

Meduna, P. - Černá, E. 1991. 'Settlement structure of the early Middle Ages in Northwest Bohemia: investigations of the Pětipsy basin area'. *Antiquity* 65: 388-95

Mee, C.B. - Cavanagh, W.G. 1990. 'The spatial distribution of Mycenaean tombs'. *BSA* 85: 225-243

Megaw, J.S.V. (ed.) 1977. *Gatherers and first farmers beyond Europe.* Leicester University Press, Leicester.

Mehemet, O. 2002. *Modern and traditional irrigation technologies in the Eastern Mediterranean*, National Library of Canada, Ottawa.

Mehrer, N.W. - Wescott, K.L. (eds.) 2006. *Gis and archaeological site location modeling*. Tylor&Francis.

Merlo, S. 2004. 'The "Contemporary Mind". 3D GIS as a Challenge in Excavation Practice', in K.F. Ausserer – W. Börner et al. (eds.) *Enter the Past, CAA 2003* – Vienna. BAR Int.Ser. 1227: 276-279

Mesogaia 2001. Tsouni, K. (co-ordinating editor) *Mesogaia*. Athens International Airport S.A. 'Eleftherios Venizelos'. Athens.

Migeotte, L. 1994. 'Ressources financières des cités Béotiennes', in J.M. Fossey (ed.) *Boeotia Antiqua* IV. *Boiotian (and other) epigraphy*, Amsterdam: 3-15

Miller, W. 1921. *The Turkish restoration in Greece, 1718-1797*. New York.

Millet, M. 1991. 'Pottery: population or supply patterns? The *Ager Tarraconensis* approach', in G. Barker – J. Lloyd (eds.) *Roman Landscapes: Archaeological Survey in the Mediterranean region.* Archaeological Monographs of the British School at Rome 2, London: 18-26

Mitropoulou, E. 1995. 'I latreia ton Nimphon kai tou Panos stin Boiotia', in *B' diethnes sinedrio Boiotikon Meleton*: 331-387

ML 1897-1902. Roscher, W.H. *Ausführliches Lexicon der griechischen und römischen Mythologie*. Leipzig.

Morin, J. 1992/93. *La prehistoire de Khostia. DAI* 53 (1992-1993) 4434-A (dissertation).

Morris, I.M. 1987. *Burial and Ancient Society: the Rise of the Greek City-State.* Cambridge University Press.

Morris, I. (ed.) 1994. *Classical Greece. Ancient histories and modern archaeologies*. Cambridge University Press, Cambridge.

Morrison, I.A. 1968. 'Appendix I. Relative sea-level change in the Saliagos area since Neolithic times', in J.D. Evans - C. Renfrew, *Excavations at Saliagos near Antiparos*, British School at Athens, suppl.vol. 5, Thames and Hudson, London: 92-98

Müller, C. 1995. 'Le Ptoion et Akraiphia (Beotie)'. *BCH* 119: 655-660

Müller, C. 1996. 'Les travaux de l'Ecole francaise d'Athenes au sanctuaire du Val des Muses: le passé et l'avenir', in A. Hurst - A. Schachter (eds.) *La montagne des Muses*: 171-183

Munn, M. 1989. 'New Light on Panakton and the Attic-Boiotian Frontier', in H.Beister and J.Bucklet (eds.) *Boiotika*. Editio Maris, Munich: 231-244

Munn, M. 1996, 'The first excavations at Panakton on the Attic-Boeotian frontier', in J.M. Fossey – P.J. Smith (eds.) *Boeotia Antiqua* VI, Amsterdam: 47-58

Munn, M. – Zimmerman-Munn, M.H. 1987-9. 'Skourta Plain'. *ArchDelt 1987 B1*, 198-9; *ArchDelt 1988 B1*, 191-2; *ArchDelt 1989 B1*: 154-6.

Munn, M. – Zimmerman-Munn, M.H. 1989: 'Studies on the Attic-Boiotian frontier: the Stanford-Skourta Plain project, 1985', in J.M. Fossey (ed.) *Boeotia Antiqua I: Papers on Recent Work in Booetian Archaeology and History*, Amsterdam: 73-127

Munn, M. – Zimmerman-Munn, M.H. 1990, 'On the frontiers of Attica and Boiotia: the results of the Stanford Skourta plain project'. *Teiresias* Suppl. 3, Montreal: 33-39

Munn, N. 1990. 'Constructing regional worlds in experience: kula exchange, witchcraft and Gawan local events'. *Man* 25: 1-17

Murray, D.M. 1995. 'The management of archaeological information - a strategy', in J. Wilcock - K. Lockyear (eds.) *Computer Applications and Quantitative Methods in Archaeology 1993*. BAR Int.Ser. 598: 83-87.

Musič, B. 2001. 'An evaluation of the potential of geophysical prospection in difficult environments. The silent presence of GIS', in B. Slapšak (ed.) *Coast Action G2*. Ljubljana: 127-144

Musič, B. – Slapšak, B. - Farinetti, E. 2004. 'Ancient Tanagra (Grimadha): geophysics, DEM, modelling, statistics and classification for understanding of urban plan and on-site activity areas', in *CAA 2003*: 317-320

Nackaerts, K, - Govers, G. - Loots, L. 1999. 'The Use of Monte Carlo Techniques for the Estimation of Visibility', in L. Dingwall et al. (eds.) *Archaeology in the Age of the Internet. Computer Applications and Quantitative Methods in Archaeology*. BAR Int.Ser. 750: 63-5

Nerantzis, I. 2003. *I chora ton Aitolon*. PhD thesis, University of Crete-Rethymnon. Agrinio (abstract published in Pharaklas N. *I geopolitiki Organosi tis Peloponnisiakis Achaias*, Rethymno 2001: 204-5).

Neuenschwander, L.F. – Peters, W.J. 1988. *Slash and Burn: Farming in the Third World Forest.* Idaho Research Foundation.

Neustupny, E. 1991. 'Community areas of prehistoric farmers in Bohemia'. *Antiquity* 65: 326-31

Neustupny, E. 1994. 'The settlement area theory in Bohemian Archaeology'. *Památky Archeologické*, Suppl. I. Praha: 248-258

Neustupny, E. (ed.) 1998. *Space in Prehistoric Bohemia*. Praha.

Nicod, J. 1972. *Pays et paysage du calcaire*. Paris.

Nixon, L. - Price, S. 2001 'The diachronic analysis of pastoralism through comparative variables'. *BSA* 96: 395-424

Nixon et al. 1990. Nixon, L. - Moody, J. et al. (1990) 'Archaeological survey in Sphakia, Crete'. *Echos du Monde Classique/Classical Views* 34.9: 213-20

Noack, F. 1894. 'Arne'. *AM* 19 : 405-485

Nokas, N. 2000. 'Prehistoric Eutresis. Historical and Archaeological guide' (in greek), in *G' diethnes synedrio Boiotikon Meleton*: 219-223

Novakovič, P. 2001. 'Detecting Territoriality and Social Structure in the Bronze Ages: GIS and the Hillforts in the Kras Region', in B. Slapšak (ed.) *Coast Action G2*. Ljubljana: 101-115

Odum, E.P. 1959. *Fundamentals of Ecology*, 2nd ed. Saunders, Philadelphia.

Oikonomos, G.P. 'Anathimatikon kranos ton Orchomenion'. *AE* 1925/26: 87-94.

Oldfather, W.A. 1916. 'Studies in the History and Topography of Locris'. *AJA* 20: 33-61

Osborne, R. 1985a. 'The land-leases from Hellenistic Thespiai', in Argoud - Roesch (eds.) 1985: 317-323

Osborne, R. 1985b. 'Building and residence on the land in Classical and Hellenistic Greece: the contribution of epigraphy'. *BSA* 80: 119-128.

Osborne, R. 1987. *Classical landscape with figures. The ancient Greek city and its countryside*. London.

Osborne, R. 1992. 'Is it a farm? The definition of agricultural sites and settlements in ancient Greece', in Wells (ed.) 1992: 21-28.

Osborne, R. 1996. *Greece in the making 1200-479 BC*. London.

Osborne, R. 2001. 'Counting the cost: comments on David K. Pettegrew, 'Chasing the Classical farmstead'. *JMA* 14.2: 212-22.

Osborne, R. 2004. 'Greek archaeology: a survey of recent work'. *AJA* 108: 87-102.

Paasi, A. 1991. 'Deconstructing regions: notes on the scales of spatial life'. *Environment and Planning* A 1991 vol.23: 239-256.

Palm, C. - Vosti S.A. – Sanchez, P.A. – Ericksen, P.J. 2005. *Slash-and-burn agriculture : the search for alternatives*. Columbia University Press.

Papachatzis, N. 1981. *Pausaniou Ellados Periigisis*, vol.5 *Viotika kai Phokika*. Ekdotiki Athinon, Athens

Papadaki, A. 2000. 'On the Topography and Archaeology of Ancient Potniai (Tachi, Thebes)' (in greek), in *G' Diethnes Sinedrio Boiotikon meleton*: 357-369

Papadopoulou, K. A. - Gournellos, T. 1996. 'A study of the subsurface karst of Kopais (Central Greece)'. *Theoretical and Applied Karstology* 9: 69–76

Partsch, J. 1887. *Die Insel Corfu. Eine Geographische Monographie*. Gotha.

Partsch, J. 1889. *Die Insel Leukas. Eine Geographische Monographie*. Gotha.

Partsch, J. 1890. *Kephallenia und Ithaka. Eine Geographische Monographie*. Gotha.

Pasquinucci, M. – Tremens, F. (eds.) 1999. *Non destructive Techniques Applied to Landscape Archaeology*. The Archaeology of Mediterranean Landscape n.3. Oxbow Books, Oxford.

Patterson, H. - Millet, M. 1998. 'The Tiber Valley Project'. *BSR* 66: 1-20

Perlès, C. 2001. *The Early Neolithic in Greece*. Cambridge World Archaeology, Cambridge.

Petraschek, W.E. 1954. *Die Eisenerz- und Nickelerzlagerstätten von Lokris in Griechenland*. Athens.

Petropoulos, M. – Pontrandolfo, A. – Rizakis, A.D. 2004. 'Terza campagna di ricognizioni archeologiche in Egialea (ottobre 2004)'. *ASAtene* 2004 II: 783-805.

Pettegrew, D.K. 2001 'Chasing the Classical framstead: assessing the formation and signature of rural settlement in Greek landscape archaeology'. *Journal of Mediterranean Archaeology* 14: 189-209

Petts, G.E. – Moller, H. – Roux, A.L. (eds.) 1989. *Historical changes of large alluvial rivers: Western Europe*. Wiley, Chichester.

Philippson, A. 1894. *Der Kopais-See in Griechenland und seine Umgebung*, Zeitschrift der Gesellschaft fur Erdkunde zu Berlin 29.

Philippson, A. 1951. 'Das östliche Mittelgriechenland und die Insel Euboea', in Philippson 1950/9 Band I, Teil II.

Philippson, A. 1950/9. *Die Griechischen Landschaften. Eine Landeskunde*. Vittorio Klostermann, Frankfurt am Main.

Piérart, M. 1974. *Platon et la cité grecque*. Bruxelles.

Piteros, Ch. 1995. 'Anaskaphi oikopedon aphon Tolia sti Livadia', in *B' diethnes sinedrio Boiotikon Meleton*: 481-507

Piteros, Ch. 2000. 'Anaskaphi stoikou oikodomimatos sto Dilesi kai to iero tou Diliou Apollona', in Aravantinos V. (ed.) *Γ' diethnes sinedrio Boiotikon Meleton*: 592-623

Pizziolo, G. - De Silva, M. 2004. 'Gis Analysis of Historical Cadastral Maps as a Contribution in Landscape Archaeology', in K.F. Ausserer - W. Börner - M. Goriany - L. Karlhuber-Vöckl (eds.) *Enter the Past, CAA 2003*. BAR Int. Ser. 1227: 294-298

Podobnikar et al. 2004. Podobnikar, T. - Sneza, T.H. – Dular, J. 'Iterative approach to ancient paths modelling in the iron age study of the Dolenjska region (Slovenia)', in K.F. Ausserer - W. Börner - M. Goriany - L. Karlhuber-Vöckl (eds.) *Enter the Past, CAA 2003*. BAR Int. Ser. 1227: 254-257

Pope, K.O. - Van Andel, T.H. 1984. 'Late Quaternary alluviation and soil formation in the Southern Argolid: its history, causes and archaeological implications'. *JAS* 11: 281-306

Potter, T. 1979. *The changing landscape of South Etruria.* New York.

Pritchett W.K. 1958. 'Observations on Chaironeia'. *AJA* 62: 307-311

Pritchett, W.K. 1965. *Studies in Ancient Greek Topography part I.* Berkeley University of California Press, Berkeley.

Pritchett, W.K. 1969. *Studies in Ancient Greek Topography part II, Battlefields.* Berkeley University of California Press, Berkeley.

Pritchett, W.K. 1980. *Studies in Ancient Greek Topography part III, Roads.* Berkeley University of California Press, Berkeley.

Pritchett, W.K. 1982. *Studies in Ancient Greek Topography part IV, Passes.* Berkeley University of California Press, Berkeley.

Pritchett, W.K. 1985. *Studies in Ancient Greek Topography part V.* Berkeley University of California Press, Berkeley.

Pullen, D.J. 1992. 'Ox and plow in the Early Bronze Age Aegean'. *AJA* 96: 45-54

Rackham, O. 1983. 'Observations on the historical ecology of Boeotia'. *BSA* 78: 291-351

Rackham, O. 1990. 'Ancient Landscapes', in O. Murray - O. Price (eds.) *The Greek city from Homer to Alexander.* Oxford: 85-111

Rajala, U.M. - Harrison, A.R. - Stoddart, S.K.F. 1999. 'The enhancement of the South Etruria Survey: GIS in Faliscan Area', in L. Dingwall et al. (eds.) *Archaeology in the Age of the Internet, CAA 1997.* BAR Int.Ser. 750 (CD)

Rapp, G. Jr. – Hill, C.L. 1998. *Geoarchaeology.* Yale Univerity Press, London.

Reed, M. - Harvey, D.L. 1992. 'The new science and the old complexity and realism in the social sciences'. *Journal for the Theory of Social Research* 22: 353-80

Renfrew, A. C. 1972. *The Emergence of Civilisation. The Cyclades and the Aegean in the Third Millennium B.C.* Methuen, London.

Renfrew, A.C. - Level, E.V. 1979. 'Exploring dominance: predicting polities from centres', in C. Renfrew – K.L. Cooke (eds.) *Transformations: Mathematical Approaches to Culture Changes.* Academic Press, New York: 145-167

Renfrew, C. - Wagstaff, M. (eds.) 1982. *An Island Polity. The Archaeology of Exploitation in Melos.* Cambridge University Press, Cambridge.

Renfrew, C. - Zubrow, E. 1994. *The ancient mind: elements of cognitive archaeology (New directions in archaeology).* Cambridge University Press, Cambridge.

Rice, R.J. 1977. *Fundamentals of Geomorphology.* Longman, London.

Richards, J.D. 2000. Book review: *On the Theory and Practice of Archaeological Computing.* Oxford University Committee for Archaeology Monograph 51, Oxford - http://intarch.ac.uk/journal/issue9/reviews/richards.html (last accessed in September 2008)

Risotti, G. 1983. *Geologia e pedologia nell'assetto del territorio.* Edagricole, Bologna.

Rizakis, A. D. (ed.) 1992. *Paysage d'Achaie I. Le bassin du Peiros et la Plaine Occidentale.* Meletimata 15.

Rizakis et al. 2001. 'Exploring the distances', in B. Slapšak (ed.) *Coast Action G2.* Ljubljana: 57-60

Robert, L. 1985.'Decrets d'Akraiphia'. *BCH* 59: 438

Roberts, B.K. 1987. 'Landscape archaeology', in J.M. Wagstaff (ed.) *Landscape and Culture. Geographical and Archaeological Perspectives.* Blackwell, Oxford: 77-95

Roberts, B. K. 1996. *Landscapes of Settlement. Prehistory to the Present.* Routledge, London and New York.

Roberts, N. 1998. *The Holocene. An Environmental History.* 2nd ed. Blackwell, Oxford.

Robinson, H. 1993. 'The archaeological implications of a computerised integrated National Heritage Information System', in J. Andresen - T. Madsen - I. Scollar (eds.) *Computing the Past. Computer Applications and Quantitative Methods in Archaeology, CAA92.* Aarhus University Press, Aarhus: 139-150

Roesch, P. 1965. *Thespies et la confederation Beotienne.* Paris.

Roesch, P. 1965b. 'Notes d'épigraphie béotienne'. *RPhil* 39 (1965): 256-261

Roesch, P. 1976. 'Haliartos', in R. Stillwell (ed.) *The Princeton Encyclopedia of Classical Sites.* NJ Princeton University Press, Princeton.

Roesch, P. 1982. *Etudes béotiennes.* Paris.

Rolfe, J.C. 1890. 'Discoveries at Anthedon in 1889. Inscriptions from Anthedon'. *AJA* 6: 96-107

Rolfe, J.C. – Tarbell, E. 1890. 'Discoveries at Thisbe in 1889'. AJA 6: 112-120

Roller, D.W. 1986. 'Tanagra Survey project 1985: Preliminary Report'. *Classical Views* 5: 160-172

Roller, D.W. 1987. 'Tanagra Survey Project 1985. The site of Grimadha'. *BSA* 82: 129-163

Roller, D.W. 1988. *Early Travellers In Eastern Boiotia.* Amsterdam.

Roller, D.W. 1988a. 'Recent investigations at the Site of Grimadha (Tanagra)', in Classical Association of the Middle west and South, Eighty-fourth Annual Meeting, New Orleans, LA, April 1988. *CJ* 83: 363

Roller, D.W. 1989. *TANAGRAN STUDIES I: Sources and documents on Tanagra in Boiotia.* Amsterdam.

Romano, D.G. – Tolba, O. 1996. 'Remote Sensing and GIS in the Study of Roman Centuriation in the Corinthia, Greece', in H. Kammermans – K. Fennema (eds.) *Interfacing the Past, CAA 1995*: 457-463

Roorda, I.M. - Wiemer, R. 1992. 'Towards a new archaeological information system in the Netherlands', in G. Lock - J. Moffett (eds.) *CAA 1991.* BAR Int.Ser. 577: 85-88

Ross, L. 1851. *Wanderungen in Griechenland in Gefolge des Königs Otto und der Königinn Amalie, mit besonderer Rundsicht auf Topographie und Geschichte.* Halle.

Rossignol, J. - Wandsnider, L.A. (eds.) 1992. *Space, Time and Archaeological Landscapes.* New York.

Rougemont, G. 1991. 'Complémentarité entre les differentes parties du territoire dans les cités grecques de l'antiquité classique', in M.C. Cauvin (ed.) *Rythes et rythmes agraires*: 127-134

Roux, G. 1954. 'Hesiod et la vallée des Muses'. *BCH* 78: 22-48

Russett, B. M. 1967. *International Regions and the International System: A Study in Political Ecology.* Rand-McNally, Chicago.

Rutkowski, B. 1986. *The cult places of the Aegean.* London.

Sacket, L.H. et al. 1966. 'Prehistoric Euboea: contributions towards a Survey'. *BSA* 61: 33-112

Sallares, R. 1991. *The ecology of the ancient Greek world.* Ithaca.

Salmon, P. 1956. 'Les districts béotiens'. *REA* 1: 51-70

Sampson, A. 2000. 'Cave dwelling in the Kopais region. The cave of Sarakenos', in *G' diethnes sinedrio Boiotikon Meleton*: 133-155

Sanders, W. - Jeffrey, T. - Parsons, R. - Santley, R. 1979. *The Basin of Mexico.* Academic Press, New York.

Sapouna-Sakellaraki, E. 1987. 'Glypha i Vlycha Boiotias. I Mykinaiki Avlida?'. *AAA* 20: 191-210

Sapouna-Sakellaraki, E. 1993. 'O Proistorikos tumbos 'tou Salganea' (Lithosoros) stis aktes tis Boiotias'. *AAA* 21 (1988): 77-90

Sarandopoulou, N. 2000. 'Late Geometric and Early Archaic Pottery from a cemetery at Akontion, Boiotia', in *G' diethnes sinedrio Boiotikon Meleton*: 459-493

Sarri, K. 2000. 'Medeon. Excavations on Megalo Kastraki hill (1995)', in *G' diethnes sinedrio Boiotikon Meleton*: 224-241

Sarri, K. 2010. *Orchomenos IV: Orchomenos in der mittleren Bronzezeit.* AbhMünch 135. München.

Sauer, C. 1963. 'The morphology of landscape', in J. Leighly (ed.) *Land and Life: A Selection of the Writings of Carl Sauer.* University of California Press, Berkeley: 315-50

Schachter, A. 1966. 'Costituzione e sviluppo dell'ethnos beotico'. *Quaderni Urbinati di Cultura Classica* 52: 5-29

Schachter, A. 1985. 'Kadmos and the implications of the tradition for Boiotian history', in Argoud - Roesch (eds.) 1985: 145-153

Schachter, A. 1996. 'Reconstructing Thespiai', in A. Hurst and A. Schachter (eds.) *La montagne des Muses*: 99-126

Schachter, A. 1981-1986-1994. *Cults of Boiotia* Vols. 1 to 4. BICS. University of London, Institute of Classical Studies, Bulletin Supplement Number 38.1/38.2/38.3/38.4. London.

1981: Vol.1 (Acheloos to Hera)

1986: Vol.2 (Herakles to Poseidon)

1994: Vol.3 (Potnia to Zeus)

1981a: Vol.4 (Index of inscriptions)

Schachter, A. 2003. 'Evolution of a Mistery Cult: The Theban Kabiroi', in M.B. Cosmopoulos (ed.) *Greek Mysteries: The Archaeology and Ritual of Ancient Greek Secret Cults*: 112-142

Schachter, A. 2004. 'Tanagra: the Geographical and Historical context: part one'. *Pharos* XI: 45-74

Schlager, H. – Blackman, D.J. - Schafer, J. 1968. 'Der Hafen von Anthedon mit Beitragen zur Topographie und Geschichte der Stadt'. *AA* 1: 21-102

Schliemann H. 1881. *Orchomenos, Bericht uber meine Ausgrabungen,* Leipzig (English version in *JHS* 2 (1881): 122-163)

Schumm, S.A. 1972. 'Fluvial palaeochannels', in J.K. Rigby – W.K. Hamblin (eds.) *Recognition of Ancient Sedimentary Environments.* S.E.P.M. (Society of Economic Paleontologists and Mineralogists) 16: 98-107

Scottish Executive 2000. *The Rural Stewardship. Stationery Office.* Edinburgh.

Scranton, R.L. 1941. *Greek walls.* Mass, Cambridge.

Selkirk, A. 1997. 'Shapwick'. *Current Archaeology* 13(7): 244-254

Shanks, M. - Tilley, C. 1987. *Re-Constructing Archaeology.* Cambridge University Press, Cambridge.

Sherratt, A. 1980. 'Water, soil and seasonality in early cereal cultivation'. *WorldArch* 11: 313-330

Sherratt, A. 1981. 'Plough and Pastoralism: Aspects of the Secondary Products Revolution'. I. Hodder – I. Glynn - N. Hammond (eds.), *Pattern of the Past: Studies in Honour of David Clarke.* Cambridge University Press, Cambridge: 261-305

Shiel, R.S. 2000. 'Refuting the land degradation myth for Boeotia', in G. Bailey - R. Charles -N. Winder (eds.) *Human Ecodynamics.* Oxbow Books, Oxford: 55-62

Shiel, R.S. – Chapman, H. 1988. 'The extent of change in the agricultural landscape of Dalmatia, Yugoslavia, as a result of 8,000 years of land management', in J.C. Chapman - J. Bintliff - V. Gaffney - B. Slapšak (eds.) *Recent Developments in Yugoslav archaeology.* BAR Int.Ser. 431: 31-44

Shiel, R. - Stewart, A. 2007. 'Chapter 7: The soils and agricultural potential of the Thespiai area', in Bintliff - Howard –Snodgrass 2007: 95-109

Schoder, R. 1974. *Ancient Greece from the Air.* London.

Simoni, E. - Papagiannopoulos, K. 1998. 'Project for the topography of ancient Achaia, Greece: quantitative analysis and visualisation of the results of the intensive surface survey at Kamenitsa', in J.Peterson (ed.) *The use of Geographic Information Systems in the study of ancient landscapes and features related to ancient land use* COST ACTION G2 European commission: 43-55

Slapšak, B. (ed.) 2001. *On the good use of Geographic Information Systems in Archaeological Landscape Studies*. Workshop proceedings, Ljubljana - 27 April 1996, COST Action G2 (Paysages antiques et structures rurales), European Community.

Slapšak et al. 2001: Slapšak, B. – Erič, M. – Musič, B. – Plevnik, D. 'Landscape structures survey in the chora of Pharos: GIS support, visualisation and landscape micro-analysis', in B. Slapšak (ed.) *Cost Action G2*: 81-93

Snodgrass, A.M. 1980. *Archaic Greece. The age of experiment*. University of California Press.

Snodgrass, A.M. 1985. 'The Site of Askra', in Argoud - Roesch (eds.) 1985: 87-95

Snodgrass, A.M. 1987. *An archaeology of Greece: the present state and future scope of a discipline*. University of California, Berkeley.

Snodgrass, A.M. 1987/9. 'The rural landscape and its political significance', *Rivista Internazionale per la storia economica e sociale dell'antichità* vols. 6-8: 53-65

Snodgrass, A.M. 1990. 'Survey archaeology and the rural landscape of the greek city', in O. Murray and S.Price (eds.) *The Greek city from Homer to Alexander*. Oxford University press, Oxford: 113-136

Soil Survey Staff 1975. *Soil Taxonomy. A Basic System of Soil Classification for Making and Interpreting Soil Survey*. Agricultural Handbook 436. United States Department of Agriculture, Washington DC.

Soil Survey Staff 1993. *Soil Survey Manual of the United States*. United States Department of Agriculture, Washington DC.

Soil Survey Staff 1998. *Keys to Soil Taxonomy*. U.S.A.D.-Natural Resources Conservation Service, Washington DC.

Sordi, M. 1966. 'Mitologia e propaganda nella Beozia Arcaica'. *Atene e Roma* 11: 16-24

Sordi, M. 1968. '*Aspetti* del federalismo greco arcaico: autonomia e egemonia nel koinon beotico'. *Atene e Roma* 13: 66-75

Sordi, M. 1993. 'La battaglia di Ceresso e la secessione di Tespie', in Fossey, J.M. (ed.) *Boeotia Antiqua III. Papers in Boiotian History, Institutions and Epigraphy in Memory of Paul Roesch:* 25-32

Spawforth, A. 1995 'Boeotia and the wealth of Herodes Atticus : a fishy business ?', in *B' diethnes synedrio Boiotikon Meleton*: 469-476

Spencer-Wood, S.M. 2000. 'Strange Attractors. Feminist theory, nonlinear systems theory, and their implications for archaeological theory', in M.B.Schiffer (ed.) *Social Theory and Archaeology*. The University of Utah Press, Salt Lake City: 112-125

Spencer-Wood, S.M. 1996 'Feminist historical archaeology and the transformation of american culture by domestic reform movements, 1840-1925'. *Historical Archaeology and the Study of American Culture*: 397-445

Stančič, Z. - Kvamme, K.L. 1999. 'Settlement Pattern Modelling through Boolean Overlays of Social and Environmental Variables', in J. Barceló – A. Briz – A. Vila (eds.) *New Techniques for Old Times - CAA98. Computer Applications and Quantitative Methods in Archaeology*. BAR Int.Ser. 757: 231-237

Stančič, Z. - Veljanovski, T. 2000. 'Understanding Roman settlement patterns through multivariate statistics and predictive modelling', in Lock (ed.) 2000: 147-156

Stančič et al. 1997. Stančič, Z. - Gaffney, V. - Ostir-Sedej, K. - Podobnikar, T. 'GIS Analysis of Land-use, Settlement Patterns and Territories on the Island of Brac, in Johnson – North (eds.) 1997 (CDrom).

Stephani, L. 1843. *Reise durch einige Gegenden des nördlichen Griechenland*. Leipzig.

Stewart, I. 1989. *Does God Play Dice? The New Mathematics of Chaos*. Penguin Books, London.

Stewart, P. J. - Strathern, A. (eds.) 2003. *Landscape, Memory, and History: Anthropological Perspectives*. Anthropology, Culture, and Society Series. Pluto Press, London/New York.

Stiros, S. - Papageorgiou, S. 1994. 'Post Mesolithic evolution of the Thessalian landscape', in *La Thessalie. Quinze années de researches archeologiques 1975-1990. Bilans et perspectives*. Ministère de la Culture, Editions Kapon, Athens vol.1: 29-36

Stoddart S. – Belcher, M. - Harrison A. 1997. 'L'applicazione del GIS all'Etruria Meridionale. The South Etruria Survey. Why apply GIS?', in A. Bietti – A. Cazzella – I. Johnson – A. Voorrips (eds.) *Theoretical and Methodological Problems. Colloquium II. The present State of GIS Applications and analogous Systems in Prehistoric Archaeology. Preprints of the XIII Congress of the UISPP*. Forli, Abaco: 185-192

Stoddart S. - Whitehead, N. 1991. 'Cleaning the Iguvine stables: site and off-site analysis from a central Mediterranean perspective', in A.J. Schofield (ed.) *Interpreting Artefact Scatters. Contributions to Ploughzone Archaeology*. Oxford, Oxbow Books, Oxbow Monograph 4: 141-148

Stoddart, S. (ed.) 2000. *Landscapes from Antiquity*. Antiquities Publication.

Strathern, M. 1995. 'The nice thing about culture is that everyone has it', in M.Strathern (ed.) *Shifting Contexts: Transformations in Anthropological Knowledge*. Routledge, London: 153-76

Sweeting, M.M. 1972. *Karst landforms*. Macmillan, London.

Symeonoglou, S. 1985. *The topography of Thebes from the Bronze Age to Modern times*. Princeton.

Syriopoulos, K.Th. 1968. *Η Προϊστορία της Στερεάς Ελλάδος*. Athens.

Syriopoulos, K.Th. 1983-84. *Οι Μεταωατικοί Χρόνοι από της Μυκηναικής εις την Αρχαικήν Περίοδον 1200-700 π.Χ.*. Athens.

Tassinari, G. 1976. *Manuale dell'Agronomo*. Ed. Reda, Roma.

Tataris, A.A. 1970. *Vagia: Geological map of Greece*. 1:50 000 scale. IGSR, Athens.

Taylor, P.J. - Johnston, R.J. 1995. 'GIS and Geography', in J. Pickles (ed.) *Ground Truth. The social implications of Geographic Information Systems*. The Guilford Press, London: 51-67

Terkenli, T. 1997. *The cultural landscape: geographical perspective*. Papazissis Publications (in greek).

Terrenato, N. 2000. 'The visibility of sites and the interpretation of field survey results: towards an analysis of incomplete distributions', in Francovich, R. – Patterson, H. (eds.) *Extracting Meaning from Ploughsoil Assemblages*. Oxbow Books: 60-71

Terrenato, N. 2004. 'Sample Size Matters! The Paradox of Global Trends and Local Surveys', in Alcock, S.E. - Cherry, J.F. (eds.) *Side-by-Side Survey. Comparative Regional Studies in the Mediterranean World*. Oxbow Books, Oxford: 36-48

Terrenato, N. – Ammerman, A.J. 1996. 'Visibility and site recovery in the Cecina Valley Survey, Italy'. *JFA* 23 (1): 91-109

Theocharis, D. 1981. *Neolithikos Politismos*. MIET, Athens.

Thomas, K. (ed.) 1990. 'Soils and Early Agriculture'. *WorldArch* 22 (1).

Thomas, J. 1996. *Time, Culture and Identity*. Routledge, London.

Thomas, K. 1984. *Man and the Natural World: Changing Attitudes in England 1500-1800*. Penguin, Harmondsworth.

Thornthwaite, W. 1948. 'An approach toward a rational classification of climate'. *Geographical Review* 38(1): 55–94.

Tilley, C. 1994. *A Phenomenology of Landscape: Places, Paths and Monuments*. Berg, Oxford.

Tomkins, P. – Kokkinaki, L. – Soetens, S. – Sarris, A. in press. 'Settlement patterns and socio-economic differentiation in East Crete in the Final Neolithic', in *Proceedings of the CAA 2004 conference*.

Tomlinson, R.A. – Fossey, J.M. 1970. 'Ancient Remains on Mount Avrovouni, South Boeotia'. BSA 65: 243-63

Torello, C. 1586. *Ricordo di agricoltura*. Mantova.

Traina, G. 1988. *Paludi e bonifiche del mondo antico. Saggio di archeologia geografica*. Roma.

Tsigakou, F.M. 1981. *The rediscovery of Greece*. EkdothikiAthinos, Athens.

Tuan, Y.F. 1977. *Place and Space. The Perspective of Experience*. Edward Arnold Publishers Ltd., London.

Turner, B.L. – Brush, S.B. (eds.) 1987. *Comparative Farming Systems*. The Guilford Press, New York.

Turner, J. 1978. 'The vegetation of Greece during prehistoric times: the palynological evidence', in *Thera and the Aegean World*, vol. 1. Thera Foundation, London: 765-73

Turner, J. - Greig, J.R.A. 1975. 'Some Holocene pollen diagrams from Greece'. *Review of Palaeobotany and Palynology* 20: 171-204

Turner, L.A. 1992. 'The temple and sanctuary of Zeus Basileus at Livadeia'. *AJA* 96: 365

Turner, L.A. 1993. 'The Topography and Monuments of Livadia: intra and extra-urban landscapes'. *AJA* 97: 345

Turner, L.A. 1995. Turner's dissertation (Pennsylvania), *The history, monuments and topography of ancient Lebadeia in Boeotia, Greece*. DAI 55 (1994/1995) 1356-A (American Institute)

Turner, V. 1977. *The Ritual Process: Structure and Anti-Structure*. Cornell University Press. Ithaca, NY.

Tzanimis, A. 1995. 'To chroniko ton anaskaphon sti koilada ton Mouson', in *B' diethnes synedrio Boiotikon Meleton*: 407-427

Tzavella-Evjen, H. 1985. *Lithares: an Early Bronze Age settlement in Boeotia*, University of California, Los Angeles, Institute of Archaeology Occasional Paper 15. Los Angeles.

Tzavella-Evjen, H. 1987. 'Lithares revisited', in Fossey J.M. (ed.) *Boeotia Antiqua I: Papers on Recent Work in Booetian Archaeology and History*, Amsterdam: 5-12

Tzavella-Evjen, H. 1988. 'Dierevnisi tis Neolithikis toumbas Mbalomenou tis Chaironeias', in *A' diethes synedrio Boiotikon Meleton*: 37-41

Ucko P.J., Dimbleby G.W. (eds.) 1969. *Domestication and Explotation of Plants and Animals*. Duckworth, London.

Uleberg, E. 2004. 'Strange attractors in the High Mountains', in K.F. Ausserer - W. Börner et al. *Enter the Past, CAA 2003* – Vienna. BAR Int. Ser. 1227: 444-447

Ulrichs, H.N. 1840. *Reisen und Forschungen in Griechenland. Erster Teil – Reise über Delphi durch Phocis und Boeotien bis Theben*. Bremen.

Ulrichs, H.N. 1863. *Reisen und Forschungen in Griechenland. Zweiter Teil – Topographische und Archäologische Abhandlungen hierausgegeben von A.Passow*. Berlin.

UNESCO-FAO 1963. *Bioclimatic Map of Mediterranean Zone. Explanatory Notes. Arid Zone Research* XXI. Paris.

Vallas, St. 2000. *To Asklipieio tis Leivadias*, Morphotikos Syllogos Levadias, Athina.

Vallega, A. 1999. *Introduzione alla geografia umana*. Mursia, Roma.

Van Andel, T H. 1987. 'Part I, The landscape', in T. H. Van Andel - S.B. Sutton (eds.) *Landscape and People of the Franchti Region*, Excavations at Franchti Cave, Greece, 2, Indiana University Press, Bloomington and Indianapolis: 3-62

Van Andel, T. H. - Runnels, C.N. 1987. *Beyond the Acropolis: A Rural Greek Past*. Stanford University Press, Stanford.

Van Andel, T. H. - Runnels, C.N. 1995. 'The earliest farmers in Europe'. *Antiquity* 69: 481-500

Van Andel, T. H. - Zangger, E. et al. 1990a. 'Land use and soil erosion in prehistoric and historical Greece'. *JFA* 17: 379-396

Van Andel, T. H. - Zangger, E. et al. 1990b. 'Landscape Stability and Destabilization in the Prehistory of Greece', in S. Bottema - G. Entjes-Nieborg - W. Zeist van (eds.) *Man's Role in the Shaping of Eastern Mediterranean Landscape*. Balkema, Rotterdam: 139-57

Van der Leeuw, S.E. - McGlade, J. (eds.) 1997. *Time, Process and Structured Transformation in Archaeology*. Routledge, London.

Van Effenterre, H. 1989. *Les Béotiens*. Editions Errance, Paris.

Van Effenterre, H. 1997. 'Aspects of Archaic Boeotia', in J.L. Bintliff (ed.) *Recent Developments in the History and Archaeology of Central Greece*. Proceedings of the 6th International Boeotian Conference, BAR Int.Ser. 666: 135-137

Vanhaverbeke, H. - Waelkens, M. 2003. *The Chora of Sagalassos. The Evolution of the Settlement Pattern from Prehistoric until Recent Times*. Studies in Eastern Mediterranean Archaeology V. Brepols Publishers.

Van Hove, D. 'Agency and GIS: the Neolithic Land Use Hypothesis within Southern Italy', in *CAA2002*: 201-207

Van Joolen, E. 2003 (published on internet) *Reconstruction of ancient landscapes, agriculture and vegetation history in Central and South Italy during the first millennium BC*. Ph.D. thesis, University of Groningen.

Van Leusen, P. M. 1993. 'Cartographic Modelling in a cell-based GIS', in M. T. Andersen - J. Scollar - I. Aarhus (eds.) *CAA 1992*. Aarhus University Press: 105-124

Van Leusen, M. 1995. 'GIS and archaeological resource management: a European agenda', in Lock-Stančič (eds.) 1995: 27-42

Van Leusen, P.M. 1996. 'GIS and locational modeling in Dutch archaeology: a review of current approaches', in H.D.G. Maschner (ed.) *New Methods, Old Problems. Geographic Information Systems in modern archaeological research*. Carbondale: Southern Illinois University Center for Archaeological Investigations. Occasional Paper No. 23: 177-197

Van Leusen, M. 1999: Van Leusen, P. M., 'Viewshed and Cost Surface Analysis using GIS', in J. Barcelo - Vila A. (eds.) *New Techniques for Old Times, CAA 1998*. BAR Int.Ser. 757: 215-223

Van Leusen, M. 2002. *Pattern to Process. Methodological investigations into the formation and interpretation of spatial patterns in archaeological landscapes*. PhD thesis, University of Groningen (http://www.ub.rug.nl/eldoc/dis/arts/p.m.van.leusen).

Van Leusen, M. 2004. 'Visibility and the landscape: an exploration of GIS modelling techniques?', in K.F. Ausserer - W. Börner - M. Goriany - L. Karlhuber-Vöckl (eds.) *Enter the Past, CAA 2003*. BAR Int. Ser. 1227: 352

Van Leusen, M. - Kamermans, H. (eds.) 2005. *Predictive Modelling for Archaeological Heritage Management: a research agenda. National Service for Archaeological Heritage*, Amersfoort.

Vance, R. W. - Henderson, G. M. 1968. 'Culture area'. *International Encyclopedia of the Social Sciences,* vol. 3. Macmillan/The Free Press, New York: 563-568

Vatin, C. 1971. 'Le tarif des poissons d'Akraiphia', in F. Salviat - C. Vatin, *Inscriptions de Grèce centrale*. Paris: 95-109

Veljanovski, T. - Stančič, Z. 2006. 'Predictive modeling in archaeological location analysis and archaeological resource management: principles and applications', in N.W. Mehrer - K.L. Wescott (eds.) *Gis and Archaeological Site Location Modeling*: 394-41

Vencl, S. 1991. 'On the importance of spatio-temporal differences in the intensity of Palaeolithic and Mesolithic settlement in Central Europe'. *Antiquity* 65 (1991): 308-17

Venclová, N. 1995. 'Settlement area, production area and industrial zone' in M. Kuna – N. Venclová (eds.) *Wither Archaeology?* Praha: 161-169

Verhagen, P. 2006. 'Quantifying the qualified: the use of multicriteria methods and bayesian statistics for the development of archaeological predictive models' in N.W. Mehrer – K.L. Wescott (eds.) *Gis and Archaeological Site Location Modeling*: 192-214

Verhagen, P. – Wansleeben, M. - van Leusen, M. 2000. 'Predictive Modelling in the Netherlands. The prediction of archaeological values in Cultural Resource Management and Academic Research', in O. Harl (ed.) *Archäologie und Computer 1999. Forschungsgeselschaft Wiener Stadtarchäeologie* 4: 66-82

Verhagen, P. - Kamermans, H. - van Leusen, M. - Deeben, J. - Hallewas, D. - Zoetbrood, P. in press. 'First thoughts on the incorporation of cultural variables into predictive modelling', in F. Niccolucci (ed.), *Beyond the artefact - Proceedings of CAA2004 - Prato 13-17 April 2004.*

Verhagen, P. - Gazenbeek, M. 2006. 'The use of predictive modeling for guiding the archaeological survey of roman pottery kilns in the Argonne region (Northeast France)', in N.W. Mehrer – K.L. Wescott (eds.) *Gis and Archaeological Site Location Modeling*: 447-460

Vermeulen, F. 2001. 'The potential of GIS in Landscape Archaeology', in B. Slapšak (ed.) *Coast Action G2*. Ljubljana: 9-16

Vermeulen, F. 2002. 'The Potenza Valley Survey (Marche)', in Attema et al. (eds.) *New Developments in Italian Landscape Archaeology*, BAR Int. Ser. 109: 104-106

Vermeulen, F. 2003. 'The Potenza Valley Survey: preliminary report on field campaign 2002'. *BABesch* 78: 71-106

Vermeulen, F. 2006. 'Understanding lines in the roman landscape: a study of ancient roads and field systems based on GIS technology', in N.W. Mehrer - K.L. Wescott (eds.) *GIS and Archaeological Site Location Modeling*: 291-316

Vermeulen et al. 2001. Vermeulen, F. - Antrop, M. - Hageman, B. - Wiedemann, T. 'Ancient Roads and Fields in Northwestern Gaul – A GIS Based Analysis', in *CAA2000*: 187-196

Vermeulen, F. - Antrop, M. et al. (eds.) 2001. *Ancient lines in the landscape*. Leuven.

Vermeulen, F. - De Dapper, M. (eds.) 2000. *Geoarchaeology of the Landscapes of Classical* Antiquity. Bulletin van de Antieke Beschaving Suppl. 5. Stichting Babesch, Leiden.

Verstappen, H. Th. (ed.) 1971. *Manual of detailed geomorphological mapping*. International Geographical Union. Czechoslovak Academy, Brno.

Vionis, A.K. 2006. 'The Archaeology of Ottoman Villages in Central Greece: ceramics, housing and everyday life in Post-Medieval Boeotia', in A. Erkanal-Oktu - E. Ozgen - S. Gunel et al. (eds.), *Studies in Honor of Hayat Erkanal: Cultural Reflections*: 784-800

Vischer, W. 1875. *Erinnerungen und Eindruck aus Griechenland.*

Visher G.S. 1965. 'Fluvial processes as interpreted from ancient and recent fluvial deposits', in G.V. Middleton (ed.) *Primary Sedimentary Structures and their Hydrodynamic Interpretation.* S.E.P.M. (Society of Economic Paleontologists and Mineralogists)12: 116-132

Vita-Finzi, C. - Higgs, E.S. 1970. 'Prehistoric economy in the Mt.Carmel area of Palesatine: site catchment analysis'. *PPS* 36: 1-37

Vitkovsky, O.V. - Kolossov, V.A. 1980. 'Regional politico-geographical studies: some theoretical and practical considerations'. *International Political Science Review* 1: 531-539

Vlad Borrelli, L. 'Orchomenos'. *EAA* 5: 711-13

Von Thunen, J.H. 1826. *The Isolated State.* Loyola University Press, Chicago.

Vroom, J. 2003. *After antiquity. Ceramics and Society in the Aegean from the 7th to the 20th century A.C. A Case Study from Boeotia, Central Greece.* Archaeological Studies Leiden University, Leiden.

Vullo, N. - Barker, G. 1997. 'Regional Sampling and GIS: The Tuscanica Survey Project', in Johnson – North (eds.) 1997 (CDrom).

Wace, A.J.B. – Thompson, M.S. 1912. *Prehistoric Thessaly.*

Wagley, C. ed. 1974. *Man in the Amazon.* University Presses of Florida, Gainesville.

Wagner, R. 1991. 'The fractal person', in M. Strathern - M. Godelier (eds.) *Big Men and Great Men: The Personifications of power in Melanesia*: 159-73. Cambridge University Press, Cambridge.

Wagstaff, J. M. 1982. *The development of rural settlements. A study of the Helos Plain in Southern Greece.* Avebury.

Wagstaff, M. 1992. 'Agricultural terraces: the Vasilikos Valley, Cyprus', in M.G. Bell - J. Boardman (eds.) *Past and Present Soil Erosion. Archaeological and Geographical Perspectives.* Oxbow, Oxford: 155-162

Wallace, P.W. 1969. 'A Reidentification of the Boiotian City Salganeus and a Suggested Location for the Homeric Nisa'. *AJA* 73: 246

Wallace, P.W. 1974. 'Hesiod and the Valley of the Muses'. *GRBS* 15: 5-24

Wallace, P.W. 1979. *Strabo's Description of* Boiotia. Heidelberg.

Wallace, S. 2001. 'Case studies of settlement change in Early Iron Age Crete (c 1200-700 BC): economic models of cause and effect reassessed'. *Aegean Archaeology* 4 (1997-2000). Warsaw: 61-99

Wansleeben, M. 1988, 'Applications of geographical information systems in archaeological research', in S.P. Rahtz (ed.) *CAA 1988*, BAR Int.Ser. 446 (ii): 435-451.

Watrous, L.V. 1982. *Lasithi, a history of settlement on a highland plain in Crete.* American School of Classical Studies at Athens, Princeton, N.J.

Watrous, L.V. et al. 2004. Watrous, L.V. – Hadzi-Vallianou, D. – Blitzer, H. *The Plain of Phaistos. Cycles of Social Complexity in the Mesara Region of Crete.* Monumenta Archaeologica 23. Cotsen Institute of Archaeology at UCLA.

Weber, A. 1909. *Über den Standort der Industrie* (Theory of the Location of Industries).

Wells, B. (ed.) 1992. *Agriculture in Ancient Greece. Proceedings of the Seventh International Symposium at the Swedish Institute at Athens, 16-17 May 1990*, Paul Astroms Forlag.

Wells, B. 1994. 'A Prehistoric Environmental Catastrophe. The Case of Berbati and Limnes', in K.A. Sheedy (ed.) *Archaeology in the Peloponnese. New Excavations and Research.* Oxbow Monograph 48. Oxford: 65-76

Wells, B. (ed.) 1996. *The Berbati-Limnes Archaeological Survey 1988-1990.* Acta Instituti Atheniensis Regni Sueciae, Series IN 4, XLIV. Stockholm.

Wells, L.E. 2001. 'A geomorphological approach to reconstructing archaeological patterns based on superficial artefact distribution', in P. Goldberg – V. Holliday – C.R. Ferring (eds.) *Earth Sciences and Archaeology.* Kluwer Academic, New York: 107-41

West, M.L. 1997. *The East face of Helicon: West Asiatic elements in Greek poetry and myth.* Clarendon Press, Oxford.

Westcott, K.L. – Brandon, R.J. (eds.) 1999. *Practical applications of GIS in archaelogy*: a predictive modeling toolkit with CDrom. Taylor & Francis.

Wheatley, D.W. 1995. 'Cumulative viewsheds analysis: a GIS-based method for investigating intervisibility, and its archaeological application', in Lock-Stančič (eds.) 1995: 171-186

Wheatley, D.W. 2000. 'Spatial Technology and archaeological theory revisited', in K. Lockyear - T.J.T. Sly - V.Mihailescu-Birliba (eds.) *CAA96.* BAR Int.Ser. 845: 123-131

Wheatley, D.W. 1996. 'Between the lines: the role of GIS-based predictive modelling in the interpretation of extensive survey data', in H. Kamermans – K. Fennema (eds). *Interfacing the Past. Computer applications and quantitative methods in Archaeology CAA95* – Leiden. Analecta Praehistorica Leidensia 28: 275-292

Wheatley, D.W. 1993. 'Going over old ground: GIS, archaeological theory and the act of perception', in J. Andresen - T. Madsen - I. Scollar (eds) *Computing the Past. Computer Applications and Quantitative Methods in Archaeology, CAA92*. Aarhus: Aarhus University Press: 133-138

Wheatley, D.W. 1995. 'Cumulative Viewshed Analysis: a GIS-based method for investigating intervisibility, and its archaeological application', in Lock-Stančič (eds.) 1995: 171-186

Wheatley, D.W. 1996. 'Between the lines: the role of GIS-based predictive modelling in the interpretation of extensive survey data', in H. Kamermans - K. Fennema (eds.) *Interfacing the Past. Computer Applications and Quantitative Methods in Archaeology CAA95*. Analecta Praehistorica Leidensia 28: 275-292

Wheatley, D.W. 2000. 'Spatial technology and archaeological theory revisited', in K.Lockyear - T.J.T. Sly - V. Mihailescu-Birliba (eds.) *CAA 96. Computer applications and quantitative methods in archaeology*, BAR Int. Ser. 845. Oxford: 123-31

Wheler, G. 1862. *A journey into Greece.*

White, J.P. – Modjeska, N. 1978. 'Acquirers, users, finders, losers: the use of axe blades among the Duna'. *Mankind* 11: 276-287

White, L. jr. 1976. *Tecnica e società nel medioevo.* Il Saggiatore, Milano.

Whitelaw, T.M. 2000. 'Settlement Instability and Landscape Degradation in the Southern Aegean in the Third Millennium BC', in Halstead-Frederick (eds.) 2000: 135-161

Whitley, T.G. 2000. '*Dynamical Systems Modelling in Archaeology: A GIS Approach to Site Selection Processes in the Greater Yellowstone Region*'. Unpublished Dissertation, Department of Anthropology, University of Pittsburgh, PA.

Whittaker, C.R. (ed.) 1988. *Pastoral economies in classical antiquity*. Cambridge, Cambridge Philological Society.

Wilamowitz - Moellendorff, U. von 1922. *Pindaros*. Weidmann, Berlin.

Wilkinson, T. J. 2004. 'The Archaeology of Landscape' in Bintliff, J. (ed.) *A companion to Archaeology* Blackwell: 334-356

Willey, G.R. 1953. *Prehistoric Settlement Patterns in the Viru Valley, Peru*. Bureau of American Ethnology Bulletin 155. Washington, D.C.

Winter, F.E. 1997 'The use of artillery in fourth-century and Hellenistic towers'. *Classical Views* 41: 247-292

Wischmeier W.H. - Smith D.D. 1965. *Predicting rainfall – erosion losses from cropland east of Rocky Mountains – Guide for selection of practices for soil and water conservation*. Agricultural Handbook 282, US Dept. of Agriculture (U.S.D.A.)

Wise, A.L. - Miller, P. 1997. 'Why metadata matters in Archaeology', in *Internet Archaeology* 2. Available from http://intarch.ac.uk/journal/issue2/wise_index.html.

Wolters, P. - Bruns, G. 1940. *Das Kabirenheiligtum bei Theben*, vol.I. Berlin.

Wright et al. 1990. Wright, J. C. – Cherry, J.F. et al. 'The Nemea Valley Archaeological Project - A Preliminary Report'. *Hesperia* 59: 579-659

Young, J.H. 1956. 'Studies in South Attica, Country Estate at Sounion'. *Hesperia* 25: 122-146

Zangger, E. 1992. 'Neolithic to present soil erosion in Greece', in M. Bell -J. Boardman (eds.), *Past and Present Soil Erosion: Archaeological and Geographical Perspectives*. Oxbow Books: 133-48.

Ziehen, L. 1949. 'Pamboiotia'. *RE* XVIII 3, col.288.

Zubrow, E.B.W. 1984. 'Cultural Causality: A Multivariate Analysis of Prehistoric Western New York from Small-Scale Surveys'. *American Archeology* 4(1): 29-35

Zubrow, E.B.W. 1994. 'Knowledge representation and archaeology: a cognitive example using GIS', in C. Renfrew - E.B.W. Zubrow (eds.) *The ancient mind. Elements of cognitive archaeology*: 107-118

www.ingramcontent.com/pod-product-compliance
Lightning Source LLC
LaVergne TN
LVHW072246300926
849160LV00018B/1607